Preface

This book is a continuation of a previous one by the same author (*Differential Forms in Electromagnetics*, [1]) on the application of multivectors, multiforms, and dyadics to electromagnetic problems. Main attention is focused on applying the formalism to the analysis of electromagnetic media, as inspired by the ongoing engineering interest in constructing novel metamaterials and metaboundaries. In this respect the present exposition can also be seen as an enlargement of a chapter in a recent book on metamaterials [2] by including substance from more recent studies by this author and collaborators. The present four-dimensional (4D) formalism has proved of advantage in simplifying expressions in the analysis of general media in comparison to the classical three-dimensional (3D) Gibbsian formalism. However, the step from electromagnetic media, defined by medium parameters, to actual metamaterials and metaboundaries, defined by physical structures, is beyond the scope of this book.

The first four chapters are devoted to the algebra of multiforms and dyadics in order to introduce the formalism and useful analytic tools. Similar material presented in [1] has been extended. Chapter 5 summarizes basic electromagnetic concepts in the light of the present formalism. Chapter 6 discusses transformations useful for simplifying problems. In the final Chapters 7–10 different classes of electromagnetic media are defined on the basis of their various properties. Because the most general linear electromagnetic medium requires 36 parameters for its definition, it is not easy to understand the effect of all these parameters. This is why it becomes necessary to define medium classes with reduced numbers of parameters. In Chapter 7 the classes are defined in terms of a natural decomposition of the medium bidyadic in three components, independent of any basis representation. Chapter 8 considers media whose medium bidyadic can be expressed in terms of quadratic functions of dyadics defined by 16 parameters. In Chapter 9 medium classes are defined by the degree of the algebraic equation satisfied by the medium bidyadic. Finally, in Chapter 10 media are defined by certain restrictions imposed on plane waves propagating in the media.

Main emphasis lies on the application of the present formalism in the definition and analysis of media. It turns out that certain concepts cannot be easily defined through the 3D Gibbsian vector and dyadic representation. For example, the perfect electromagnetic conductor (PEMC) medium generalizing both perfect electric conductor (PEC) and perfect magnetic conductor (PMC) media appears as the simplest possible medium in the present formalism while in terms of conventional engineering representation with Gibbsian medium dyadics it requires parameters of infinite magnitude. As another example, decomposable bi-anisotropic media, defined to generalize uniaxially anisotropic media in which fields can be decomposed in transverse electric (TE) and transverse magnetic (TM) components, can be represented in a compact 4D form while the original analysis applying 3D Gibbsian formulation produced extensive expressions. In addition to the economy in expression, the present analysis is able to reveal novel additional solutions. A number of details in the analysis has been skipped in the text and left as problems for the reader. Solutions to the problems can be found at the end of the book, which allows the book to be used for self-study.

Because of the background of the author, the book is mainly directed to electrical engineers, although physicists and applied mathematicians may find the contents of interest as well. It has been attempted to make the transition from 3D Gibbsian vector and dyadic formalism, familiar to most electrical engineers, to the 4D exterior calculus involving multivectors, multiforms, and dyadics, as small as possible by showing connections to the corresponding Gibbsian quantities in an appendix. The main idea for adopting the 4D formalism is not to emphasize time-domain analysis of electromagnetic fields but to obtain compactness in expression and analysis. In fact, harmonic time dependence $\exp(j\omega t)$ is often tacitly assumed by allowing complex magnitudes for the medium parameters.

Compared to the previous book [1], the present approach shows some changes in the terminology followed by an effort to make the presentation more accessible. For example, to emphasize the most important dyadics defining electromagnetic media, they have been called bidyadics because they represent mappings between two-forms and/or bivectors.

The author thanks students of the postgraduate courses based on the material of this book for their comments and responses. Special thanks are due to professors Ari Sihvola and Friedrich Hehl and for doctors Alberto Favaro and Luzi Bergamin for their long-lasting interest and help in treating questions during the years behind this book.

Ismo V. Lindell

CHAPTER 1

Multivectors and Multiforms

1.1 VECTORS AND ONE-FORMS

Let us consider two four-dimensional (4D) linear spaces, that of vectors, $\mathbb{E}_1$ and that of one-forms $\mathbb{F}_1$. The elements of $\mathbb{E}_1$ are most generally denoted by boldface lowercase Latin letters,

$$\mathbf{a}, \mathbf{b}, \mathbf{c}, \ldots \in \mathbb{E}_1, \tag{1.1}$$

while the elements of $\mathbb{F}_1$ are most generally denoted by boldface lowercase Greek letters

$$\boldsymbol{\alpha}, \boldsymbol{\beta}, \boldsymbol{\gamma}, \ldots \in \mathbb{F}_1. \tag{1.2}$$

The space of scalars is denoted by $\mathbb{E}_0$ or $\mathbb{F}_0$ and its elements are in general represented by nonboldface Latin or Greek letters $a, b, c, \ldots$, $\alpha, \beta, \gamma, \ldots$.

Exceptions are made for quantities with established conventional notation. For example, the electric and magnetic fields are one-forms which are respectively denoted by the boldface uppercase Latin letters **E** and **H**.

1.1.1 Bar Product |

The product of a vector **a** and a one-form $\boldsymbol{\alpha}$ yielding a scalar is denoted by the "bar" sign | as $\mathbf{a}|\boldsymbol{\alpha} \in \mathbb{E}_0$. The product is assumed symmetric,

$$\mathbf{a}|\boldsymbol{\alpha} = \boldsymbol{\alpha}|\mathbf{a}. \tag{1.3}$$

Multiforms, Dyadics, and Electromagnetic Media, First Edition. Ismo V. Lindell.

Because of the sign, it will be called as the bar product. In the past it has also been known as the duality product or the inner product. The bar product should not be confused with the dot product. The dot product can be defined for two vectors as $\mathbf{a} \cdot \mathbf{b}$ or two one-forms as $\boldsymbol{\alpha} \cdot \boldsymbol{\beta}$ and it depends on a particular metric dyadic as will be discussed later.

1.1.2 Basis Expansions

A set of four vectors, $\mathbf{e}_1, \mathbf{e}_2, \mathbf{e}_3, \mathbf{e}_4$, is called a basis if any vector $\mathbf{a}$ can be expressed as

$$\mathbf{a} = a_1\mathbf{e}_1 + a_2\mathbf{e}_2 + a_3\mathbf{e}_3 + a_4\mathbf{e}_4, \tag{1.4}$$

in terms of some scalars a_i. Similarly, any one-form can be expanded in a basis of one-forms, $\boldsymbol{\varepsilon}_1, \boldsymbol{\varepsilon}_2, \boldsymbol{\varepsilon}_3, \boldsymbol{\varepsilon}_4$ as

$$\boldsymbol{\alpha} = \alpha_1\boldsymbol{\varepsilon}_1 + \alpha_2\boldsymbol{\varepsilon}_2 + \alpha_3\boldsymbol{\varepsilon}_3 + \alpha_4\boldsymbol{\varepsilon}_4. \tag{1.5}$$

The expansion of the bar product yields

$$\mathbf{a}|\boldsymbol{\alpha} = \sum_{i=1}^{4}\sum_{j=1}^{4} a_i\alpha_j\mathbf{e}_i|\boldsymbol{\varepsilon}_j. \tag{1.6}$$

The vector and one-form bases are called reciprocal to one another if they satisfy

$$\mathbf{e}_i|\boldsymbol{\varepsilon}_j = \delta_{i,j}, \tag{1.7}$$

with

$$\delta_{i,i} = 1, \qquad \delta_{i,j} = 0 \;\; i \neq j. \tag{1.8}$$

In this case the scalar coeffients in (1.4) and (1.5) satisfy

$$a_i = \boldsymbol{\varepsilon}_i|\mathbf{a}, \qquad \alpha_i = \mathbf{e}_i|\boldsymbol{\alpha}, \tag{1.9}$$

and the bar product can be expanded as

$$\mathbf{a}|\boldsymbol{\alpha} = \sum_{i=1}^{4} a_i\alpha_i = a_1\alpha_1 + a_2\alpha_2 + a_3\alpha_3 + a_4\alpha_4. \tag{1.10}$$

From here onwards we always assume that when the two bases are denoted by $\mathbf{e}_i$ and $\boldsymbol{\varepsilon}_j$, they are reciprocal.

Vectors can be visualized as yardsticks in the 4D spacetime, and they can be used for measuring one-forms. For example, measuring the electric field one-form $\mathbf{E} \in \mathbb{F}_1$ by a vector $\mathbf{a}$ yields the voltage U between the endpoints of the vector

$$\mathbf{a}|\mathbf{E} = U, \tag{1.11}$$

provided $\mathbf{E}$ is constant in space or $\mathbf{a}$ is small in terms of wavelength.

The bar product $\mathbf{a}|\boldsymbol{\alpha}$ is a bilinear function of $\mathbf{a}$ and $\boldsymbol{\alpha}$. Thus, $\mathbf{a}|\boldsymbol{\alpha}$ can be conceived as a linear scalar-valued function of $\boldsymbol{\alpha}$ for a given vector $\mathbf{a}$. Conversely, any linear scalar-valued function $f(\boldsymbol{\alpha})$ can be expressed as a bar product $\mathbf{a}|\boldsymbol{\alpha}$ in terms of some vector $\mathbf{a}$. To prove this, we express $\boldsymbol{\alpha}$ in a basis $\{\boldsymbol{\varepsilon}_i\}$ and apply linearity, whence we have

$$\mathbf{a}|\boldsymbol{\alpha} = f(\boldsymbol{\alpha}) = f\left(\sum \alpha_i \boldsymbol{\varepsilon}_i\right) = \sum \alpha_i f(\boldsymbol{\varepsilon}_i) = \sum f(\boldsymbol{\varepsilon}_i)\mathbf{e}_i|\boldsymbol{\alpha}, \tag{1.12}$$

in terms of the reciprocal vector basis $\{\mathbf{e}_i\}$. Thus, the vector $\mathbf{a}$ can be defined as

$$\mathbf{a} = \sum f(\boldsymbol{\varepsilon}_i)\mathbf{e}_i. \tag{1.13}$$

1.2 BIVECTORS AND TWO-FORMS

1.2.1 Wedge Product $\wedge$

The antisymmetric wedge product $\wedge$ between two vectors $\mathbf{a}$ and $\mathbf{b}$ yields a bivector, an element of the space $\mathbb{E}_2$ of bivectors,

$$\mathbf{a} \wedge \mathbf{b} = -\mathbf{b} \wedge \mathbf{a}. \tag{1.14}$$

This implies

$$\mathbf{a} \wedge \mathbf{a} = 0, \tag{1.15}$$

for any vector $\mathbf{a}$. In general, bivectors are denoted by boldface uppercase Latin letters,

$$\mathbf{A}, \mathbf{B}, \mathbf{C}, \ldots \in \mathbb{E}_2, \tag{1.16}$$

and they can be represented by a sum of wedge products of vectors,

$$\mathbf{A} = \mathbf{a} \wedge \mathbf{b} + \mathbf{c} \wedge \mathbf{d} + \cdots. \tag{1.17}$$

Similarly, the wedge product of two one-forms $\boldsymbol{\alpha}$ and $\boldsymbol{\beta}$ produces a two-form

$$\boldsymbol{\alpha} \wedge \boldsymbol{\beta} = -\boldsymbol{\beta} \wedge \boldsymbol{\alpha}. \tag{1.18}$$

Two-forms are denoted by boldface uppercase Greek letters whenever it appears possible,

$$\mathbf{\Gamma}, \mathbf{\Phi}, \mathbf{\Psi}, \ldots \in \mathbb{F}_2, \tag{1.19}$$

and they are linear combinations of wedge products of one-forms,

$$\mathbf{\Gamma} = \boldsymbol{\alpha} \wedge \boldsymbol{\beta} + \boldsymbol{\gamma} \wedge \boldsymbol{\delta} + \cdots . \tag{1.20}$$

A bivector which can be expressed as a wedge product of two vectors, in the form

$$\mathbf{A} = \mathbf{a} \wedge \mathbf{b}, \tag{1.21}$$

is called a simple bivector. Similarly, two-forms of the special form

$$\mathbf{\Gamma} = \boldsymbol{\alpha} \wedge \boldsymbol{\beta}, \tag{1.22}$$

are called simple two-forms.

For the 4D vector space as considered here, the bivectors form a space of six dimensions as will be seen below. It is not possible to express the general bivector in the form of a simple bivector.

1.2.2 Basis Expansions

Expanding vectors in a vector basis $\{\mathbf{e}_i\}$ induces a basis expansion of bivectors where the basis bivectors can be denoted by $\mathbf{e}_{ij} = \mathbf{e}_i \wedge \mathbf{e}_j$. Because $\mathbf{e}_{ii} = 0$ and six of the remaining twelve bivectors are linearly dependent of the other six,

$$\mathbf{e}_{12} = \mathbf{e}_1 \wedge \mathbf{e}_2 = -\mathbf{e}_{21}, \qquad \mathbf{e}_{23} = \mathbf{e}_2 \wedge \mathbf{e}_3 = -\mathbf{e}_{32}, \quad \text{etc.}, \tag{1.23}$$

the space of bivectors is six dimensional. Actually, the bivector basis need not be based on any vector basis. Any set of six linearly independent bivectors could do.

A bivector can be expanded in the bivector basis as

$$\begin{aligned} \mathbf{A} &= \sum_J A_J \mathbf{e}_J \\ &= A_{12}\mathbf{e}_{12} + A_{23}\mathbf{e}_{23} + A_{31}\mathbf{e}_{31} + A_{14}\mathbf{e}_{14} + A_{24}\mathbf{e}_{24} + A_{34}\mathbf{e}_{34}. \end{aligned} \tag{1.24}$$

Here, $J = ij$ is a bi-index containing two indices i, j taken in a suitable order. In the following we will apply the order

$$J = 12,\ 23,\ 31,\ 14,\ 24,\ 34. \tag{1.25}$$

Similarly, a basis of two-forms can be built upon the basis of one-forms as $\boldsymbol{\varepsilon}_J = \boldsymbol{\varepsilon}_{ij} = \boldsymbol{\varepsilon}_i \wedge \boldsymbol{\varepsilon}_j$.

It helps in memorizing if we assume that the index 4 corresponds to the temporal basis element and 1, 2, 3 to the three spatial elements. In this case the spatial indices appear in cyclical order $1 \rightarrow 2 \rightarrow 3 \rightarrow 1$ in J while the index 4 occupies the last position.

It is useful to define an operation $K(J)$ yielding the complementary bi-index of a given bi-index J as

$$K(12) = 34, \qquad K(23) = 14, \qquad K(31) = 24, \tag{1.26}$$

$$K(14) = 23, \qquad K(24) = 31, \qquad K(34) = 12. \tag{1.27}$$

Obviously, the complementary index operation satisfies

$$K(K(J)) = J. \tag{1.28}$$

The basis expansion (1.24) can be used to show that any bivector can be expressed as a sum of two simple bivectors, in the form

$$\mathbf{A} = \mathbf{a} \wedge \mathbf{b} + \mathbf{c} \wedge \mathbf{d}. \tag{1.29}$$

Such a representation is not unique. As an example, assuming $A_{23} \neq 0$ in (1.24), we can write

$$\mathbf{A} = \frac{1}{A_{23}}(A_{31}\mathbf{e}_1 - A_{23}\mathbf{e}_2) \wedge (A_{12}\mathbf{e}_1 - A_{23}\mathbf{e}_3) + \left(\sum_{i=1}^{3} A_{i4}\mathbf{e}_i\right) \wedge \mathbf{e}_4. \tag{1.30}$$

Thus, any bivector can be expressed in the form

$$\mathbf{A} = \mathbf{a}_1 \wedge \mathbf{a}_2 + \mathbf{a}_3 \wedge \mathbf{e}_4, \tag{1.31}$$

where the vectors $\mathbf{a}_i$ are spatial, that is, they satisfy $\mathbf{a}_i|\boldsymbol{\varepsilon}_4 = 0$. $\mathbf{a}_1 \wedge \mathbf{a}_2$ is called the spatial part of $\mathbf{A}$ and $\mathbf{a}_3 \wedge \mathbf{e}_4$ its temporal part. Similar rules are valid for two-forms. In particular, any two-form can be expanded in terms of spatial and temporal one-forms as

$$\boldsymbol{\Gamma} = \boldsymbol{\alpha}_1 \wedge \boldsymbol{\alpha}_2 + \boldsymbol{\alpha}_3 \wedge \boldsymbol{\varepsilon}_4, \quad \mathbf{e}_4|\boldsymbol{\alpha}_i = 0. \tag{1.32}$$

1.2.3 Bar Product

We can extend the definition of the bar product of a vector and a one-form to that of a bivector and a two-form, $\mathbf{A}|\boldsymbol{\Phi} = \boldsymbol{\Phi}|\mathbf{A}$. Starting from a simple bivector $\mathbf{a} \wedge \mathbf{b}$ and a simple one-form $\boldsymbol{\alpha} \wedge \boldsymbol{\beta}$ the bar product is a quadrilinear scalar function of the two vectors and two one-forms and it can be expressed in terms of the four possible bar products of vectors and one-forms as

$$(\mathbf{a}\wedge\mathbf{b})|(\boldsymbol{\alpha}\wedge\boldsymbol{\beta}) = (\mathbf{a}|\boldsymbol{\alpha})(\mathbf{b}|\boldsymbol{\beta}) - (\mathbf{a}|\boldsymbol{\beta})(\mathbf{b}|\boldsymbol{\alpha}) = \det\begin{pmatrix}\mathbf{a}|\boldsymbol{\alpha} & \mathbf{a}|\boldsymbol{\beta}\\ \mathbf{b}|\boldsymbol{\alpha} & \mathbf{b}|\boldsymbol{\beta}\end{pmatrix}. \tag{1.33}$$

Such an expansion follows directly from the antisymmetry of the wedge product and assuming orthogonality of the basis bivectors and two-forms as

$$\mathbf{e}_{ij}|\boldsymbol{\varepsilon}_{k\ell} = \delta_{i,k}\delta_{j,\ell}. \tag{1.34}$$

by assuming ordered indices. Equation (1.33) can be memorized from the corresponding rule for three-dimensional (3D) Gibbsian vectors denoted by $\mathbf{a}_g, \mathbf{b}_g, \mathbf{c}_g, \mathbf{d}_g \in \mathbb{E}_1$,

$$(\mathbf{a}_g\times\mathbf{b}_g)\cdot(\mathbf{c}_g\times\mathbf{d}_g) = (\mathbf{a}_g\cdot\mathbf{c}_g)(\mathbf{b}_g\cdot\mathbf{d}_g) - (\mathbf{a}_g\cdot\mathbf{d}_g)(\mathbf{b}_g\cdot\mathbf{c}_g). \tag{1.35}$$

Relations of multivectors and multiforms to Gibbsian vectors are summarized in Appendix B.

As examples of spatial two-forms we may consider the electric and magnetic flux densities, for which we use the established symbols **D** and **B**. Bivectors can be visualized as surface regions with orientation (sense of rotation). They can be used to measure the flux of a two-form through the surface region. For example, the magnetic flux Φ (a scalar) of the magnetic spatial two-form **B** through the bivector $\mathbf{a}\wedge\mathbf{b}$ is obtained as

$$\Phi = (\mathbf{a}\wedge\mathbf{b})|\mathbf{B}. \tag{1.36}$$

For more details on geometric interpretation of multiforms see, for example, [3, 4].

1.2.4 Contraction Products ⌋ and ⌊

Considering a bivector $\mathbf{a}\wedge\mathbf{b}$ and a two-form $\boldsymbol{\Phi}$, the bar product $(\mathbf{a}\wedge\mathbf{b})|\boldsymbol{\Phi}$ can be conceived as a linear scalar-valued function of the vector **a**. Thus, there must exist a one-form $\boldsymbol{\alpha}$ in terms of which we can express

$$\mathbf{a}|\boldsymbol{\alpha} = (\mathbf{a}\wedge\mathbf{b})|\boldsymbol{\Phi} = \boldsymbol{\Phi}|(\mathbf{a}\wedge\mathbf{b}) = -\boldsymbol{\Phi}|(\mathbf{b}\wedge\mathbf{a}) = \boldsymbol{\alpha}|\mathbf{a}. \tag{1.37}$$

Obviously, the one-form $\boldsymbol{\alpha}$ is a linear function of both $\mathbf{b}$ and $\boldsymbol{\Phi}$ so that we can express it as a product of the vector $\mathbf{b}$ and the two-form $\boldsymbol{\Phi}$ and denote it either

$$\alpha = \mathbf{b}\rfloor\boldsymbol{\Phi}, \tag{1.38}$$

or

$$\alpha = -\boldsymbol{\Phi}\lfloor\mathbf{b}. \tag{1.39}$$

The operation denoted by the multiplication sign $\rfloor$ or $\lfloor$ will be called contraction, because the two-form $\boldsymbol{\Phi}$ is contracted ("shortened") by the vector $\mathbf{b}$ from the left or from the right to yield a one-form. Thus, the contraction product obeys the simple rules

$$\mathbf{a}|(\mathbf{b}\rfloor\boldsymbol{\Phi}) = (\mathbf{a}\wedge\mathbf{b})|\boldsymbol{\Phi}, \tag{1.40}$$

$$(\boldsymbol{\Phi}\lfloor\mathbf{b})|\mathbf{a} = \boldsymbol{\Phi}|(\mathbf{b}\wedge\mathbf{a}). \tag{1.41}$$

Contraction of a bivector $\mathbf{A}$ by a one-form $\boldsymbol{\alpha}$ can be defined similarly. Applied to (1.33), with slightly changed symbols, yields

$$(\mathbf{d}\wedge\mathbf{a})|(\beta\wedge\gamma) - ((\mathbf{d}|\beta)(\mathbf{a}|\gamma) - (\mathbf{d}|\gamma)(\mathbf{a}|\beta)) =$$
$$\mathbf{d}|[\mathbf{a}\rfloor(\beta\wedge\gamma)) - ((\beta(\mathbf{a}|\gamma) - \gamma(\mathbf{a}|\beta)))] = 0, \tag{1.42}$$

which is valid for any vector $\mathbf{d}$. Choosing $\mathbf{d} = \mathbf{e}_i$ for $i = 1, \ldots, 4$, all components of the one-form expression in square brackets vanish. Thus, we immediately obtain the "bac-cab rule" valid for any vector $\mathbf{a}$ and one-forms $\boldsymbol{\beta}, \boldsymbol{\gamma}$,

$$\mathbf{a}\rfloor(\beta\wedge\gamma) = \beta(\mathbf{a}|\gamma) - \gamma(\mathbf{a}|\beta) = (\gamma\wedge\beta)\lfloor\mathbf{a}. \tag{1.43}$$

Equation (1.43) corresponds to the well-known bac-cab rule of 3D Gibbsian vectors, Appendix B,

$$\mathbf{a}_g\times(\mathbf{b}_g\times\mathbf{c}_g) = \mathbf{b}_g(\mathbf{a}_g\cdot\mathbf{c}_g) - \mathbf{c}_g(\mathbf{a}_g\cdot\mathbf{b}_g) = (\mathbf{c}_g\times\mathbf{b}_g)\times\mathbf{a}_g, \tag{1.44}$$

which helps in memorizing the 4D rule (1.43).

Useful contraction rules for basis vectors and one-forms can be obtained as special cases of (1.43) as

$$\mathbf{e}_j\rfloor(\varepsilon_i\wedge\varepsilon_j) = \mathbf{e}_j\rfloor\varepsilon_{ij} = \varepsilon_i, \tag{1.45}$$

$$(\mathbf{e}_i\wedge\mathbf{e}_j)\lfloor\varepsilon_i = \mathbf{e}_{ij}\lfloor\varepsilon_i = \mathbf{e}_j. \tag{1.46}$$

They can be easily memorized as a way of canceling basis vectors and one-forms with the same index from the contraction operation.

1.2.5 Decomposition of Vectors and One-Forms

Two vectors $\mathbf{a}, \mathbf{b} \neq 0$ are called parallel if they satisfy the relation

$$\mathbf{a} \wedge \mathbf{b} = 0. \tag{1.47}$$

Applying the bac-cab rule (1.43) for parallel vectors $\mathbf{a}, \mathbf{b}$,

$$\boldsymbol{\alpha}\rfloor(\mathbf{a} \wedge \mathbf{b}) = \mathbf{a}(\boldsymbol{\alpha}|\mathbf{b}) - \mathbf{b}(\boldsymbol{\alpha}|\mathbf{a}) = 0, \tag{1.48}$$

implies that parallel vectors are linearly dependent, that is, one is a multiple of the other one. Assuming $\mathbf{a} \wedge \mathbf{b} \neq 0$ and $\boldsymbol{\alpha}|\mathbf{a} \neq 0$, we can write the following decomposition for a given vector $\mathbf{b}$:

$$\mathbf{b} = \mathbf{b}_{\|} + \mathbf{b}_{\perp}, \tag{1.49}$$

$$\mathbf{b}_{\|} = \frac{\boldsymbol{\alpha}|\mathbf{b}}{\boldsymbol{\alpha}|\mathbf{a}}\mathbf{a}, \qquad \mathbf{b}_{\perp} = \frac{\boldsymbol{\alpha}\rfloor(\mathbf{b} \wedge \mathbf{a})}{\boldsymbol{\alpha}|\mathbf{a}}. \tag{1.50}$$

Here, $\mathbf{b}_{\|}$ can be interpreted as the component parallel to a given vector $\mathbf{a}$, while $\mathbf{b}_{\perp}$ can be called as the component perpendicular to a given one-form $\boldsymbol{\alpha}$, because it satisfies

$$\boldsymbol{\alpha}|\mathbf{b}_{\perp} = \boldsymbol{\alpha}|(\boldsymbol{\alpha}\rfloor(\mathbf{a} \wedge \mathbf{b})) = (\boldsymbol{\alpha} \wedge \boldsymbol{\alpha})|(\mathbf{a} \wedge \mathbf{b}) = 0. \tag{1.51}$$

Similarly, we can decompose a one-form $\boldsymbol{\beta}$ as

$$\boldsymbol{\beta} = \boldsymbol{\beta}_{\|} + \boldsymbol{\beta}_{\perp}, \tag{1.52}$$

$$\boldsymbol{\beta}_{\|} = \frac{\mathbf{a}|\boldsymbol{\beta}}{\mathbf{a}|\boldsymbol{\alpha}}\boldsymbol{\alpha}, \qquad \boldsymbol{\beta}_{\perp} = \frac{\mathbf{a}\rfloor(\boldsymbol{\alpha} \wedge \boldsymbol{\beta})}{\mathbf{a}|\boldsymbol{\alpha}}, \tag{1.53}$$

in terms of a given one-form $\boldsymbol{\alpha}$ and a given vector $\mathbf{a}$ satisfying $\boldsymbol{\alpha}|\mathbf{a} \neq 0$.

1.3 MULTIVECTORS AND MULTIFORMS

Higher-order multivectors and multiforms are produced through wedge multiplication. The wedge product is associative so that we have

$$\mathbf{a} \wedge (\mathbf{b} \wedge \mathbf{c}) = (\mathbf{a} \wedge \mathbf{b}) \wedge \mathbf{c} = \mathbf{a} \wedge \mathbf{b} \wedge \mathbf{c}, \tag{1.54}$$

and the brackets can be omitted. Thus, trivectors and three-forms are obtained as

$$\mathbf{k} = \mathbf{a}_1 \wedge \mathbf{a}_2 \wedge \mathbf{a}_3 + \mathbf{b}_1 \wedge \mathbf{b}_2 \wedge \mathbf{b}_3 + \cdots \in \mathbb{E}_3, \tag{1.55}$$

$$\boldsymbol{\pi} = \boldsymbol{\alpha}_1 \wedge \boldsymbol{\alpha}_2 \wedge \boldsymbol{\alpha}_3 + \boldsymbol{\beta}_1 \wedge \boldsymbol{\beta}_2 \wedge \boldsymbol{\beta}_3 + \cdots \in \mathbb{F}_3. \tag{1.56}$$

They will be denoted by lowercase Latin and Greek characters taken, if possible, from the end of the alphabets. Quadrivectors and four-forms can be constructed as

$$\mathbf{q}_N = \mathbf{a}_1 \wedge \mathbf{a}_2 \wedge \mathbf{a}_3 \wedge \mathbf{a}_4 + \mathbf{b}_1 \wedge \mathbf{b}_2 \wedge \mathbf{b}_3 \wedge \mathbf{b}_4 + \cdots \in \mathbb{E}_4, \tag{1.57}$$

$$\boldsymbol{\kappa}_N = \boldsymbol{\alpha}_1 \wedge \boldsymbol{\alpha}_2 \wedge \boldsymbol{\alpha}_3 \wedge \boldsymbol{\alpha}_4 + \boldsymbol{\beta}_1 \wedge \boldsymbol{\beta}_2 \wedge \boldsymbol{\beta}_3 \wedge \boldsymbol{\beta}_4 + \cdots \in \mathbb{F}_4. \tag{1.58}$$

The subscript N denoting the quadri-index $N = 1234$ is used to mark a quadrivector or a four-form.

Because the space of vectors is 4D, there are no multivectors of higher order than four. In fact, because any vector $\mathbf{a}_5$ can be expressed as a linear combination of a basis $\mathbf{a}_1 \ldots \mathbf{a}_4$ satisfying $\mathbf{a}_1 \wedge \mathbf{a}_2 \wedge \mathbf{a}_3 \wedge \mathbf{a}_4 \neq 0$, as will be shown below, we have $\mathbf{a}_1 \wedge \mathbf{a}_2 \wedge \mathbf{a}_3 \wedge \mathbf{a}_4 \wedge \mathbf{a}_5 = 0$. The spaces of trivectors and three-forms are 4D and, those of quadrivectors and four-forms, one dimensional.

1.3.1 Basis of Multivectors

The vector basis $\{\mathbf{e}_i\}$ induces the trivector basis

$$\mathbf{e}_{123} = \mathbf{e}_1 \wedge \mathbf{e}_2 \wedge \mathbf{e}_3, \quad \mathbf{e}_{234} = \mathbf{e}_{23} \wedge \mathbf{e}_4 = \mathbf{e}_2 \wedge \mathbf{e}_3 \wedge \mathbf{e}_4,$$
$$\mathbf{e}_{314} = \mathbf{e}_{31} \wedge \mathbf{e}_4 = \mathbf{e}_3 \wedge \mathbf{e}_1 \wedge \mathbf{e}_4, \quad \mathbf{e}_{124} = \mathbf{e}_{12} \wedge \mathbf{e}_4 = \mathbf{e}_1 \wedge \mathbf{e}_2 \wedge \mathbf{e}_4, \tag{1.59}$$

whence the space of trivectors is 4D. There is only a single basis quadrivector denoted by

$$\mathbf{e}_N = \mathbf{e}_{1234} = \mathbf{e}_1 \wedge \mathbf{e}_2 \wedge \mathbf{e}_3 \wedge \mathbf{e}_4, \tag{1.60}$$

based upon the vector basis. Similar definitions apply for the basis three-forms $\boldsymbol{\varepsilon}_{ijk} = \boldsymbol{\varepsilon}_i \wedge \boldsymbol{\varepsilon}_j \wedge \boldsymbol{\varepsilon}_k$ and the basis four-form $\boldsymbol{\varepsilon}_N = \boldsymbol{\varepsilon}_{1234} = \boldsymbol{\varepsilon}_1 \wedge \boldsymbol{\varepsilon}_2 \wedge \boldsymbol{\varepsilon}_3 \wedge \boldsymbol{\varepsilon}_4$.

Recalling the definition of the complementary bi-index (1.26), (1.27), and applying the antisymmetry of the wedge product we obtain

$$\mathbf{e}_{23} \wedge \mathbf{e}_{K(23)} = \mathbf{e}_{23} \wedge \mathbf{e}_{14} = \mathbf{e}_2 \wedge \mathbf{e}_3 \wedge \mathbf{e}_1 \wedge \mathbf{e}_4 = \mathbf{e}_{1234} = \mathbf{e}_N. \tag{1.61}$$

More generally, we can write

$$\mathbf{e}_J \wedge \mathbf{e}_{K(J)} = \mathbf{e}_{K(J)} \wedge \mathbf{e}_J = \mathbf{e}_N. \tag{1.62}$$

Defining the bi-index Kronecker delta by

$$\delta_{I,J} = 0, \quad I \neq J, \qquad \delta_{I,J} = 1, \quad I = J, \tag{1.63}$$

we can write even more generally,

$$\mathbf{e}_I \wedge \mathbf{e}_J = \delta_{I,K(J)} \mathbf{e}_N. \tag{1.64}$$

This means that, unless I equals $K(J)$, that is, J equals $K(I)$, the wedge product yields zero.

1.3.2 Bar Product of Multivectors and Multiforms

Extending the bar product to multivectors and multiforms, we can define the orthogonality relations for the reciprocal basis multivector and multiforms as

$$\mathbf{e}_{ijk} | \boldsymbol{\varepsilon}_{rst} = \delta_{i,r}\delta_{j,s}\delta_{k,t}, \tag{1.65}$$

$$\mathbf{e}_{ijk\ell} | \boldsymbol{\varepsilon}_{rstu} = \delta_{i,r}\delta_{j,s}\delta_{k,t}\delta_{\ell,u}, \tag{1.66}$$

when the indices are ordered. From the antisymmetry of the wedge product we obtain the expansion rules

$$(\mathbf{a} \wedge \mathbf{b} \wedge \mathbf{c}) | (\boldsymbol{\alpha} \wedge \boldsymbol{\beta} \wedge \boldsymbol{\gamma}) = \det \begin{pmatrix} \mathbf{a}|\boldsymbol{\alpha} & \mathbf{a}|\boldsymbol{\beta} & \mathbf{a}|\boldsymbol{\gamma} \\ \mathbf{b}|\boldsymbol{\alpha} & \mathbf{b}|\boldsymbol{\beta} & \mathbf{b}|\boldsymbol{\gamma} \\ \mathbf{c}|\boldsymbol{\alpha} & \mathbf{c}|\boldsymbol{\beta} & \mathbf{c}|\boldsymbol{\gamma} \end{pmatrix}, \tag{1.67}$$

$$(\mathbf{a} \wedge \mathbf{b} \wedge \mathbf{c} \wedge \mathbf{d}) | (\boldsymbol{\alpha} \wedge \boldsymbol{\beta} \wedge \boldsymbol{\gamma} \wedge \boldsymbol{\delta}) = \det \begin{pmatrix} \mathbf{a}|\boldsymbol{\alpha} & \mathbf{a}|\boldsymbol{\beta} & \mathbf{a}|\boldsymbol{\gamma} & \mathbf{a}|\boldsymbol{\delta} \\ \mathbf{b}|\boldsymbol{\alpha} & \mathbf{b}|\boldsymbol{\beta} & \mathbf{b}|\boldsymbol{\gamma} & \mathbf{b}|\boldsymbol{\delta} \\ \mathbf{c}|\boldsymbol{\alpha} & \mathbf{c}|\boldsymbol{\beta} & \mathbf{c}|\boldsymbol{\gamma} & \mathbf{c}|\boldsymbol{\delta} \\ \mathbf{d}|\boldsymbol{\alpha} & \mathbf{d}|\boldsymbol{\beta} & \mathbf{d}|\boldsymbol{\gamma} & \mathbf{d}|\boldsymbol{\delta} \end{pmatrix}. \tag{1.68}$$

All quadrivectors are multiples of a given basis quadrivector $\mathbf{e}_N$ and all four-forms are multiples of a given basis four-form $\boldsymbol{\varepsilon}_N$:

$$\mathbf{q}_N = q\mathbf{e}_N, \qquad \boldsymbol{\kappa}_N = \kappa\boldsymbol{\varepsilon}_N, \tag{1.69}$$

with

$$q = \mathbf{q}_N | \boldsymbol{\varepsilon}_N, \qquad \kappa = \mathbf{e}_N | \boldsymbol{\kappa}_N. \tag{1.70}$$

Applying the expansion rule for the determinant in (1.67), we can expand the bar product

$$(\mathbf{a} \wedge \mathbf{b} \wedge \mathbf{c}) | (\boldsymbol{\alpha} \wedge \boldsymbol{\beta} \wedge \boldsymbol{\gamma}) = \mathbf{a}|\boldsymbol{\alpha} \det \begin{pmatrix} \mathbf{b}|\boldsymbol{\beta} & \mathbf{b}|\boldsymbol{\gamma} \\ \mathbf{c}|\boldsymbol{\beta} & \mathbf{c}|\boldsymbol{\gamma} \end{pmatrix} - \mathbf{a}|\boldsymbol{\beta} \det \begin{pmatrix} \mathbf{b}|\boldsymbol{\alpha} & \mathbf{b}|\boldsymbol{\gamma} \\ \mathbf{c}|\boldsymbol{\alpha} & \mathbf{c}|\boldsymbol{\gamma} \end{pmatrix}$$
$$+ \mathbf{a}|\boldsymbol{\gamma} \det \begin{pmatrix} \mathbf{b}|\boldsymbol{\alpha} & \mathbf{b}|\boldsymbol{\beta} \\ \mathbf{c}|\boldsymbol{\alpha} & \mathbf{c}|\boldsymbol{\beta} \end{pmatrix}, \tag{1.71}$$

whence from (1.33) we obtain the rule

$$(\mathbf{a}\wedge\mathbf{b}\wedge\mathbf{c})|(\alpha\wedge\beta\wedge\gamma) = (\mathbf{a}|\alpha)(\mathbf{b}\wedge\mathbf{c})|(\beta\wedge\gamma) + (\mathbf{a}|\beta)(\mathbf{b}\wedge\mathbf{c})|(\gamma\wedge\alpha) + (\mathbf{a}|\gamma)(\mathbf{b}\wedge\mathbf{c})|(\alpha\wedge\beta)]. \tag{1.72}$$

1.3.3 Contraction of Trivectors and Three-Forms

Defining the contraction of a three-form by a bivector as arising from

$$(\mathbf{a}\wedge\mathbf{b}\wedge\mathbf{c})|(\alpha\wedge\beta\wedge\gamma) = \mathbf{a}|((\mathbf{b}\wedge\mathbf{c})\rfloor(\alpha\wedge\beta\wedge\gamma)) = ((\alpha\wedge\beta\wedge\gamma)\lfloor(\mathbf{b}\wedge\mathbf{c}))|\mathbf{a}, \tag{1.73}$$

from (1.72) we obtain the expansion rule

$$(\mathbf{b}\wedge\mathbf{c})\rfloor(\alpha\wedge\beta\wedge\gamma) = (\alpha\wedge\beta\wedge\gamma)\lfloor(\mathbf{b}\wedge\mathbf{c}) = ((\mathbf{b}\wedge\mathbf{c})|(\beta\wedge\gamma))\alpha + ((\mathbf{b}\wedge\mathbf{c})|(\gamma\wedge\alpha))\beta + ((\mathbf{b}\wedge\mathbf{c})|(\alpha\wedge\beta))\gamma. \tag{1.74}$$

From this it follows that if the three one-forms satisfy $\alpha\wedge\beta\wedge\gamma = 0$, they must be linearly dependent.

Rewriting (1.72) in the form

$$(\mathbf{b}\wedge\mathbf{c})|(\mathbf{a}\rfloor(\alpha\wedge\beta\wedge\gamma)) = ((\alpha\wedge\beta\wedge\gamma)\lfloor\mathbf{a})|(\mathbf{b}\wedge\mathbf{c}) = (\mathbf{b}\wedge\mathbf{c})|[(\beta\wedge\gamma)\mathbf{a}|\alpha + (\gamma\wedge\alpha)\mathbf{a}|\beta + (\alpha\wedge\beta)\mathbf{a}|\gamma], \tag{1.75}$$

which remains valid when $\mathbf{b}\wedge\mathbf{c}$ is replaced by any bivector $\mathbf{A}$ because of linearity, we obtain another contraction rule for contracting a three-form by a vector,

$$\mathbf{a}\rfloor(\alpha\wedge\beta\wedge\gamma) = (\alpha\wedge\beta\wedge\gamma)\lfloor\mathbf{a} = (\mathbf{a}|\alpha)(\beta\wedge\gamma) + (\mathbf{a}|\beta)(\gamma\wedge\alpha) + (\mathbf{a}|\gamma)(\alpha\wedge\beta). \tag{1.76}$$

If $\alpha\wedge\beta\wedge\gamma = 0$, the three two-forms $\beta\wedge\gamma$, $\gamma\wedge\alpha$, and $\alpha\wedge\beta$ must be linearly dependent, which also follows from the linear dependence of the three one-forms.

The contraction rules (1.74) and (1.76) are similar to the bac-cab rule (1.43) and they can be easily memorized because of the cyclic symmetry. Other similar forms are obtained by replacing vectors by one-forms and one-forms by vectors in (1.74) and (1.76). Commutation rules for the contraction product can be summarized as

$$\alpha\rfloor\mathbf{k} = \mathbf{k}\lfloor\alpha, \qquad \boldsymbol{\Phi}\rfloor\mathbf{k} = \mathbf{k}\lfloor\boldsymbol{\Phi}, \tag{1.77}$$

$$\mathbf{a}\rfloor\pi = \pi\lfloor\mathbf{a}, \qquad \mathbf{A}\rfloor\pi = \pi\lfloor\mathbf{A}. \tag{1.78}$$

Here, $\mathbf{a}$ is a vector, $\mathbf{A}$ is a bivector and $\mathbf{k}$ is a trivector while $\boldsymbol{\alpha}$ is a one-form, $\boldsymbol{\Phi}$ is a two-form and $\boldsymbol{\pi}$ is a three-form. Useful rules for the contraction operations involving basis trivectors and three-forms can be formed as

$$\mathbf{e}_{jk}\rfloor\boldsymbol{\varepsilon}_{ijk} = \mathbf{e}_j\rfloor\boldsymbol{\varepsilon}_{ij} = \boldsymbol{\varepsilon}_i, \tag{1.79}$$

$$\mathbf{e}_k\rfloor\boldsymbol{\varepsilon}_{ijk} = \boldsymbol{\varepsilon}_{ij}, \tag{1.80}$$

showing how similar indices are canceled in contraction operations.

1.3.4 Contraction of Quadrivectors and Four-Forms

Following the same path of reasoning, starting from (1.68) we can expand the contraction of a four-form by a trivector

$$\begin{aligned}(\mathbf{b}\wedge\mathbf{c}\wedge\mathbf{d})\rfloor(\boldsymbol{\alpha}\wedge\boldsymbol{\beta}\wedge\boldsymbol{\gamma}\wedge\boldsymbol{\delta}) &= -(\boldsymbol{\alpha}\wedge\boldsymbol{\beta}\wedge\boldsymbol{\gamma}\wedge\boldsymbol{\delta})\lfloor(\mathbf{b}\wedge\mathbf{c}\wedge\mathbf{d})\\ &= -(\mathbf{b}\wedge\mathbf{c}\wedge\mathbf{d})|(\boldsymbol{\alpha}\wedge\boldsymbol{\beta}\wedge\boldsymbol{\gamma})\boldsymbol{\delta} + (\mathbf{b}\wedge\mathbf{c}\wedge\mathbf{d})|(\boldsymbol{\beta}\wedge\boldsymbol{\gamma}\wedge\boldsymbol{\delta})\boldsymbol{\alpha}\\ &\quad + (\mathbf{b}\wedge\mathbf{c}\wedge\mathbf{d})|(\boldsymbol{\gamma}\wedge\boldsymbol{\alpha}\wedge\boldsymbol{\delta})\boldsymbol{\beta} + (\mathbf{b}\wedge\mathbf{c}\wedge\mathbf{d})|(\boldsymbol{\alpha}\wedge\boldsymbol{\beta}\wedge\boldsymbol{\delta})\boldsymbol{\gamma},\end{aligned} \tag{1.81}$$

the contraction of a four-form by a bivector,

$$\begin{aligned}(\mathbf{c}\wedge\mathbf{d})\rfloor(\boldsymbol{\alpha}\wedge\boldsymbol{\beta}\wedge\boldsymbol{\gamma}\wedge\boldsymbol{\delta}) &= (\boldsymbol{\alpha}\wedge\boldsymbol{\beta}\wedge\boldsymbol{\gamma}\wedge\boldsymbol{\delta})\lfloor(\mathbf{c}\wedge\mathbf{d})\\ &= (\mathbf{c}\wedge\mathbf{d})|(\boldsymbol{\beta}\wedge\boldsymbol{\gamma})(\boldsymbol{\alpha}\wedge\boldsymbol{\delta}) + (\mathbf{c}\wedge\mathbf{d})|(\boldsymbol{\beta}\wedge\boldsymbol{\alpha})(\boldsymbol{\gamma}\wedge\boldsymbol{\delta})\\ &\quad + (\mathbf{c}\wedge\mathbf{d})|(\boldsymbol{\gamma}\wedge\boldsymbol{\alpha})(\boldsymbol{\beta}\wedge\boldsymbol{\delta}) + (\mathbf{c}\wedge\mathbf{d})|(\boldsymbol{\gamma}\wedge\boldsymbol{\delta})(\boldsymbol{\alpha}\wedge\boldsymbol{\beta})\\ &\quad + (\mathbf{c}\wedge\mathbf{d})|(\boldsymbol{\alpha}\wedge\boldsymbol{\delta})(\boldsymbol{\beta}\wedge\boldsymbol{\gamma}) + (\mathbf{c}\wedge\mathbf{d})|(\boldsymbol{\beta}\wedge\boldsymbol{\delta})(\boldsymbol{\gamma}\wedge\boldsymbol{\alpha}),\end{aligned} \tag{1.82}$$

and the contraction of a four-form by a vector,

$$\begin{aligned}\mathbf{d}\rfloor(\boldsymbol{\alpha}\wedge\boldsymbol{\beta}\wedge\boldsymbol{\gamma}\wedge\boldsymbol{\delta}) &= -(\boldsymbol{\alpha}\wedge\boldsymbol{\beta}\wedge\boldsymbol{\gamma}\wedge\boldsymbol{\delta})\lfloor\mathbf{d}\\ &= (\mathbf{d}|\boldsymbol{\alpha})(\boldsymbol{\beta}\wedge\boldsymbol{\gamma}\wedge\boldsymbol{\delta}) + (\mathbf{d}|\boldsymbol{\beta})(\boldsymbol{\gamma}\wedge\boldsymbol{\alpha}\wedge\boldsymbol{\delta})\\ &\quad + (\mathbf{d}|\boldsymbol{\gamma})(\boldsymbol{\alpha}\wedge\boldsymbol{\beta}\wedge\boldsymbol{\delta}) - (\mathbf{d}|\boldsymbol{\delta})(\boldsymbol{\alpha}\wedge\boldsymbol{\beta}\wedge\boldsymbol{\gamma}).\end{aligned} \tag{1.83}$$

The above expressions appear invariant to cyclic permutation of the one-forms $\boldsymbol{\alpha}, \boldsymbol{\beta}, \boldsymbol{\gamma}$, which may help in memorizing and checking the formulas.

If $\boldsymbol{\alpha}\wedge\boldsymbol{\beta}\wedge\boldsymbol{\gamma}\wedge\boldsymbol{\delta} = 0$, from (1.81) it follows that the four one-forms are linearly dependent, from (1.82) it further follows that also the six two-forms are linearly dependent and from (1.83) it follows that the four three-forms are linearly dependent. The contraction of a four-form $\boldsymbol{\kappa}_N$ or quadrivector $\mathbf{q}_N$ obeys the commutation rules

$$\mathbf{a}\rfloor\boldsymbol{\kappa}_N = -\boldsymbol{\kappa}_N\lfloor\mathbf{a}, \qquad \boldsymbol{\alpha}\rfloor\mathbf{q}_N = -\mathbf{q}_N\lfloor\boldsymbol{\alpha}, \tag{1.84}$$

$$\mathbf{A}\rfloor\boldsymbol{\kappa}_N = \boldsymbol{\kappa}_N\lfloor\mathbf{A}, \qquad \boldsymbol{\Phi}\rfloor\mathbf{q}_N = \mathbf{q}_N\lfloor\boldsymbol{\Phi}, \tag{1.85}$$

$$\mathbf{k}\rfloor\boldsymbol{\kappa}_N = -\boldsymbol{\kappa}_N\lfloor\mathbf{k}, \qquad \boldsymbol{\pi}\rfloor\mathbf{q}_N = -\mathbf{q}_N\lfloor\boldsymbol{\pi}, \tag{1.86}$$

Equations (1.81)–(1.83) imply the following contraction rules for the basis multivectors and multiforms:

$$\mathbf{e}_\ell\rfloor\boldsymbol{\varepsilon}_{ijk\ell} = \boldsymbol{\varepsilon}_{ijk}, \qquad \mathbf{e}_{ijk\ell}\lfloor\boldsymbol{\varepsilon}_i = \mathbf{e}_{jk\ell} \tag{1.87}$$

$$\mathbf{e}_{k\ell}\rfloor\boldsymbol{\varepsilon}_{ijk\ell} = \boldsymbol{\varepsilon}_{ij}, \qquad \mathbf{e}_{ijk\ell}\lfloor\boldsymbol{\varepsilon}_{ij} = \mathbf{e}_{k\ell} \tag{1.88}$$

$$\mathbf{e}_{jk\ell}\rfloor\boldsymbol{\varepsilon}_{ijk\ell} = \boldsymbol{\varepsilon}_i, \qquad \mathbf{e}_{ijk\ell}\lfloor\boldsymbol{\varepsilon}_{ijk} = \mathbf{e}_\ell, \tag{1.89}$$

which can be applied for canceling indices in expressions involving contraction of basis multivectors and multiforms. From (1.81) to (1.83) we can see that contraction of a four-form can be applied to transform vectors to three-forms, bivectors to two-forms and trivectors to one-forms and conversely. The converse cases can be obtained by applying the rules

$$\mathbf{e}_N\lfloor(\boldsymbol{\varepsilon}_N\lfloor\mathbf{a}) = (\mathbf{a}\rfloor\boldsymbol{\varepsilon}_N)\rfloor\mathbf{e}_N = -\mathbf{a}, \tag{1.90}$$

$$\mathbf{e}_N\lfloor(\boldsymbol{\varepsilon}_N\lfloor\mathbf{A}) = (\mathbf{A}\rfloor\boldsymbol{\varepsilon}_N)\rfloor\mathbf{e}_N = \mathbf{A}, \tag{1.91}$$

$$\mathbf{e}_N\lfloor(\boldsymbol{\varepsilon}_N\lfloor\mathbf{k}) = (\mathbf{k}\rfloor\boldsymbol{\varepsilon}_N)\rfloor\mathbf{e}_N = -\mathbf{k}. \tag{1.92}$$

1.3.5 Construction of Reciprocal Basis

Given a set of basis vectors $\mathbf{a}_i$, $i = 1, \ldots, 4$, and a four-form $\boldsymbol{\kappa}_N$, we can form the reciprocal one-form basis as

$$\boldsymbol{\alpha}_i = \frac{\mathbf{a}_{K(i)}\rfloor\boldsymbol{\kappa}_N}{\mathbf{a}_N|\boldsymbol{\kappa}_N}, \tag{1.93}$$

where the $\mathbf{a}_{K(i)}$ are four three-forms defined by

$$\mathbf{a}_{K(1)} = \mathbf{a}_{234}, \quad \mathbf{a}_{K(2)} = \mathbf{a}_{314}, \quad \mathbf{a}_{K(3)} = \mathbf{a}_{124}, \quad \mathbf{a}_{K(4)} = -\mathbf{a}_{123}. \tag{1.94}$$

satisfying

$$\mathbf{a}_j \wedge \mathbf{a}_{K(i)} = -\mathbf{a}_{K(i)} \wedge \mathbf{a}_j = \mathbf{a}_N\delta_{i,j}. \tag{1.95}$$

The rule (1.93) is easily checked:

$$\mathbf{a}_j|\boldsymbol{\alpha}_i = \frac{(\mathbf{a}_j \wedge \mathbf{a}_{K(i)})|\boldsymbol{\kappa}_N}{\mathbf{a}_N|\boldsymbol{\kappa}_N} = \delta_{i,j}. \tag{1.96}$$

When the indices are ordered in the above sense, we can define the two-form basis reciprocal to $\{\mathbf{a}_{ij}\}$ as

$$\boldsymbol{\alpha}_{ij} = \frac{\mathbf{a}_{K(ij)}\rfloor\boldsymbol{\kappa}_N}{\mathbf{a}_N|\boldsymbol{\kappa}_N}. \tag{1.97}$$

In fact, from (1.64) we have

$$\mathbf{a}_{k\ell}|\boldsymbol{\alpha}_{ij} = \frac{(\mathbf{a}_{k\ell} \wedge \mathbf{a}_{K(ij)})|\boldsymbol{\kappa}_N}{\mathbf{a}_N|\boldsymbol{\kappa}_N} = \delta_{i,k}\delta_{j,\ell}. \tag{1.98}$$

Because all four-forms are equal except for a scalar factor, the definitions are independent of the chosen four-form $\boldsymbol{\kappa}_N$.

1.3.6 Contraction of Quintivector

Because any quintivector $\mathbf{Q} = \mathbf{a} \wedge \mathbf{b} \wedge \mathbf{c} \wedge \mathbf{d} \wedge \mathbf{e}$ based on the 4D vector space vanishes, continuing the previous pattern by expanding the contraction $\boldsymbol{\kappa}_N\rfloor\mathbf{Q}$ we can obtain the following relation between the five vectors:

$$\begin{aligned}&\mathbf{a}(\mathbf{b} \wedge \mathbf{c} \wedge \mathbf{d} \wedge \mathbf{e})|\boldsymbol{\kappa}_N - \mathbf{b}(\mathbf{a} \wedge \mathbf{c} \wedge \mathbf{d} \wedge \mathbf{e})|\boldsymbol{\kappa}_N + \mathbf{c}(\mathbf{a} \wedge \mathbf{b} \wedge \mathbf{d} \wedge \mathbf{e})|\boldsymbol{\kappa}_N \\ &\quad - \mathbf{d}(\mathbf{a} \wedge \mathbf{b} \wedge \mathbf{c} \wedge \mathbf{e})|\boldsymbol{\kappa}_N + \mathbf{e}(\mathbf{a} \wedge \mathbf{b} \wedge \mathbf{c} \wedge \mathbf{d})|\boldsymbol{\kappa}_N = 0.\end{aligned} \tag{1.99}$$

Assuming that the four vectors $\mathbf{a} \cdots \mathbf{d}$ are linearly independent, we have $(\mathbf{a} \wedge \mathbf{b} \wedge \mathbf{c} \wedge \mathbf{d})|\boldsymbol{\kappa}_N \neq 0$, whence from (1.99) we obtain a rule how a given vector $\mathbf{e}$ can be expressed as a linear combination of the other four vectors.

1.3.7 Generalized Bac-Cab Rules

From the expansions of the previous sections we can derive useful operational rules similar to the bac-cab rule (1.43) involving more general bivectors or two-forms. Let us start from (1.74) which can be written as

$$(\mathbf{b} \wedge \mathbf{c})\rfloor(\boldsymbol{\alpha} \wedge \boldsymbol{\beta} \wedge \boldsymbol{\gamma}) = \boldsymbol{\beta}(\mathbf{b} \wedge \mathbf{c})|(\boldsymbol{\gamma} \wedge \boldsymbol{\alpha}) + \boldsymbol{\alpha}(\mathbf{f}|\boldsymbol{\gamma}) - \boldsymbol{\gamma}(\mathbf{f}|\boldsymbol{\alpha}), \tag{1.100}$$

where we denote for brevity

$$\mathbf{f} = (\mathbf{b} \wedge \mathbf{c})\lfloor\boldsymbol{\beta}. \tag{1.101}$$

Applying the bac-cab rule (1.43) as

$$\boldsymbol{\alpha}(\mathbf{f}|\boldsymbol{\gamma}) - \boldsymbol{\gamma}(\mathbf{f}|\boldsymbol{\alpha}) = \mathbf{f}\rfloor(\boldsymbol{\alpha} \wedge \boldsymbol{\gamma}), \tag{1.102}$$

(1.100) can be rewritten in the form

$$(\mathbf{b} \wedge \mathbf{c})\rfloor(\boldsymbol{\alpha} \wedge \boldsymbol{\beta} \wedge \boldsymbol{\gamma}) = \boldsymbol{\beta}(\mathbf{b} \wedge \mathbf{c})|(\boldsymbol{\gamma} \wedge \boldsymbol{\alpha}) + (\boldsymbol{\gamma} \wedge \boldsymbol{\alpha})\lfloor((\mathbf{b} \wedge \mathbf{c})\lfloor\boldsymbol{\beta}). \tag{1.103}$$

Since this identity, valid for any vectors $\mathbf{b}, \mathbf{c}$ and one-forms $\boldsymbol{\alpha}, \boldsymbol{\beta}, \boldsymbol{\gamma}$, is linear in the simple bivector $\mathbf{b} \wedge \mathbf{c}$ it remains valid if we replace $\mathbf{b} \wedge \mathbf{c}$ by a sum of bivector products $\sum_i \mathbf{b}_i \wedge \mathbf{c}_i$. Thus, it is actually valid if we replace $\mathbf{b} \wedge \mathbf{c}$ by the general bivector $\mathbf{A}$. Similarly, since the expression is linear in the two-form $\boldsymbol{\gamma} \wedge \boldsymbol{\alpha}$, it can be replaced by the general two-form $\boldsymbol{\Gamma}$. Thus, we have arrived at the more general operational rule

$$\mathbf{A}\rfloor(\boldsymbol{\beta} \wedge \boldsymbol{\Gamma}) = \boldsymbol{\beta}(\mathbf{A}|\boldsymbol{\Gamma}) + \boldsymbol{\Gamma}\lfloor(\mathbf{A}\lfloor\boldsymbol{\beta}), \tag{1.104}$$

which is another bac-cab rule valid for any bivector $\mathbf{A}$, one-form $\boldsymbol{\beta}$ and two-form $\boldsymbol{\Gamma}$. A sister formula is obtained immediately by swapping vectors and forms as

$$\boldsymbol{\Gamma}\rfloor(\mathbf{b} \wedge \mathbf{A}) = \mathbf{b}(\boldsymbol{\Gamma}|\mathbf{A}) + \mathbf{A}\lfloor(\boldsymbol{\Gamma}\lfloor\mathbf{b}). \tag{1.105}$$

Through similar considerations we can derive the following bac-cab rules valid for any bivectors $\mathbf{B}, \mathbf{C}$, vector $\mathbf{b}$ and one-form $\boldsymbol{\alpha}$:

$$\boldsymbol{\alpha}\rfloor(\mathbf{b} \wedge \mathbf{C}) = \mathbf{b} \wedge (\boldsymbol{\alpha}\rfloor\mathbf{C}) + \mathbf{C}(\boldsymbol{\alpha}|\mathbf{b}), \tag{1.106}$$

$$\boldsymbol{\alpha}\rfloor(\mathbf{B} \wedge \mathbf{C}) = \mathbf{B} \wedge (\boldsymbol{\alpha}\rfloor\mathbf{C}) + \mathbf{C} \wedge (\boldsymbol{\alpha}\rfloor\mathbf{B}). \tag{1.107}$$

Equation (1.106) can be used to define a decomposition for a given bivector $\mathbf{B}$ in two components defined by a given vector $\mathbf{a}$ and one-form $\boldsymbol{\alpha}$ satisfying $\mathbf{a}|\boldsymbol{\alpha} \neq 0$ as

$$\mathbf{B} = \mathbf{B}_{\|} + \mathbf{B}_{\perp}, \qquad \mathbf{a} \wedge \mathbf{B}_{\|} = 0, \qquad \boldsymbol{\alpha}\rfloor\mathbf{B}_{\perp} = 0. \tag{1.108}$$

The components can be identified from the rule (1.106) as

$$\mathbf{B}_{\|} = -\frac{\mathbf{a} \wedge (\boldsymbol{\alpha}\rfloor\mathbf{B})}{\mathbf{a}|\boldsymbol{\alpha}}, \qquad \mathbf{B}_{\perp} = \frac{\boldsymbol{\alpha}\rfloor(\mathbf{a} \wedge \mathbf{B})}{\mathbf{a}|\boldsymbol{\alpha}}. \tag{1.109}$$

1.4 SOME PROPERTIES OF BIVECTORS AND TWO-FORMS

Bivectors and two-forms play very central roles in electromagnetics. Let us consider some of their properties.

1.4.1 Bivector Invariant

Two bivectors commute in the wedge product,

$$\mathbf{A} \wedge \mathbf{B} = \mathbf{B} \wedge \mathbf{A}, \tag{1.110}$$

and the result is a quadrivector. Any bivector $\mathbf{A}$ has a quadrivector-valued quadratic invariant which can be defined as

$$\mathbf{A}^{\wedge} = \frac{1}{2}\mathbf{A} \wedge \mathbf{A}. \tag{1.111}$$

For a vector representation in terms of four vectors

$$\mathbf{A} = \mathbf{a} \wedge \mathbf{b} + \mathbf{c} \wedge \mathbf{d}, \tag{1.112}$$

the invariant has the form

$$\mathbf{A}^{\wedge} = \mathbf{a} \wedge \mathbf{b} \wedge \mathbf{c} \wedge \mathbf{d}. \tag{1.113}$$

Inserting an expansion in a vector basis $\{\mathbf{e}_i\}$ we obtain

$$\begin{aligned} \mathbf{A}^{\wedge} &= \frac{1}{2}(A_{12}\mathbf{e}_{12} + A_{23}\mathbf{e}_{23} + \cdots) \wedge (A_{12}\mathbf{e}_{12} + A_{23}\mathbf{e}_{23} + \cdots) \\ &= (A_{12}A_{34} + A_{23}A_{14} + A_{31}A_{24})\mathbf{e}_N. \end{aligned} \tag{1.114}$$

Alternatively, we can write

$$\mathbf{A}^{\wedge} = \frac{1}{2}\sum A_I\mathbf{e}_I \wedge \sum A_J\mathbf{e}_J = \frac{1}{2}\sum A_I A_{K(I)}\mathbf{e}_N, \tag{1.115}$$

where summing goes over the ordered bi-indices. Obviously, the quadrivector invariant of a simple bivector vanishes,

$$(\mathbf{a} \wedge \mathbf{b})^{\wedge} = 0. \tag{1.116}$$

The converse is also true: if a bivector satisfies $\mathbf{A}^{\wedge} = 0$, it must be simple. A corresponding invariant can be defined for two-forms with similar properties.

1.4.2 Natural Dot Product

A bivector $\mathbf{A}$ is mapped to a two-form by a four-form $\boldsymbol{\varepsilon}_N$ and a two-form $\boldsymbol{\Gamma}$ is mapped by a quadrivector $\mathbf{e}_N$ to a bivector as

$$\boldsymbol{\Gamma} = \boldsymbol{\varepsilon}_N \lfloor \mathbf{A}, \qquad \mathbf{A} = \mathbf{e}_N \lfloor \boldsymbol{\Gamma}. \tag{1.117}$$

A natural dot product for two bivectors $\mathbf{A}, \mathbf{B}$ can be defined by

$$\mathbf{A} \cdot \mathbf{B} = \mathbf{A}|(\boldsymbol{\varepsilon}_N \lfloor \mathbf{B}) = \boldsymbol{\varepsilon}_N|(\mathbf{A} \wedge \mathbf{B}) = \mathbf{B} \cdot \mathbf{A}, \tag{1.118}$$

which yields a scalar. Similarly, a dot product for two two-forms is defined by

$$\boldsymbol{\Gamma} \cdot \boldsymbol{\Lambda} = \boldsymbol{\Gamma}|(\mathbf{e}_N \lfloor \boldsymbol{\Lambda}) = \mathbf{e}_N|(\boldsymbol{\Gamma} \wedge \boldsymbol{\Lambda}) = \boldsymbol{\Lambda} \cdot \boldsymbol{\Gamma}. \tag{1.119}$$

Here one must note that the magnitude and sign of the scalars depend on the choice of the quadrivector $\mathbf{e}_N$ and the reciprocal four-form $\boldsymbol{\varepsilon}_N$. A bivector $\mathbf{A}$ is simple exactly when $\mathbf{A} \cdot \mathbf{A} = 0$ and a two-form $\boldsymbol{\Gamma}$ is simple when $\boldsymbol{\Gamma} \cdot \boldsymbol{\Gamma} = 0$.

1.4.3 Bivector as Mapping

A given bivector $\mathbf{G}$ maps one-forms $\boldsymbol{\alpha}$ to vectors $\mathbf{a}$ as

$$\mathbf{G} \lfloor \boldsymbol{\alpha} = \mathbf{a}. \tag{1.120}$$

The mapping may have an inverse of the form

$$\boldsymbol{\Gamma} \lfloor \mathbf{a} = \boldsymbol{\alpha}. \tag{1.121}$$

To find the two-form $\boldsymbol{\Gamma}$ for a given bivector $\mathbf{G}$ we can first apply the bac-cab rule (1.107) which for $\mathbf{C} = \mathbf{B} = \mathbf{G}$ can be rewritten as

$$\begin{aligned} \mathbf{G} \wedge (\boldsymbol{\alpha} \rfloor \mathbf{G}) &= \frac{1}{2} \boldsymbol{\alpha} \rfloor (\mathbf{G} \wedge \mathbf{G}) = \boldsymbol{\alpha} \rfloor \mathbf{G}^{\wedge} = \boldsymbol{\alpha} \rfloor \mathbf{e}_N(\boldsymbol{\varepsilon}_N|\mathbf{G}^{\wedge}) \\ &= -\mathbf{e}_N \lfloor \boldsymbol{\alpha}(\boldsymbol{\varepsilon}_N|\mathbf{G}^{\wedge}). \end{aligned} \tag{1.122}$$

On the other hand from (1.120) we have

$$\mathbf{G} \wedge (\boldsymbol{\alpha} \rfloor \mathbf{G}) = -\mathbf{G} \wedge (\mathbf{G} \lfloor \boldsymbol{\alpha}) = -\mathbf{G} \wedge \mathbf{a}, \tag{1.123}$$

whence the two mappings are related by

$$\mathbf{e}_N \lfloor \boldsymbol{\alpha}(\boldsymbol{\varepsilon}_N|\mathbf{G}^{\wedge}) = \mathbf{G} \wedge \mathbf{a}. \tag{1.124}$$

Operating this by $\varepsilon_N \lfloor$ and applying (1.90) we obtain

$$\alpha(\varepsilon_N|\mathbf{G}^\wedge) = -\varepsilon_N\lfloor(\mathbf{G}\wedge\mathbf{a}) = -(\varepsilon_N\lfloor\mathbf{G})\lfloor\mathbf{a}. \qquad (1.125)$$

Comparing with (1.121) we can now find the two-form $\boldsymbol{\Gamma}$ as

$$\boldsymbol{\Gamma} = -\frac{\varepsilon_N\lfloor\mathbf{G}}{\varepsilon_N|\mathbf{G}^\wedge}. \qquad (1.126)$$

Obviously, this is valid only when the bivector $\mathbf{G}$ is not simple, $\mathbf{G}\wedge\mathbf{G}\neq 0$.

PROBLEMS

1.1 Show that for $\alpha\beta\neq 0$ the linear combination $\alpha\mathbf{A}+\beta\mathbf{B}$ of two simple bivectors $\mathbf{A}$ and $\mathbf{B}$ is simple exactly when $\mathbf{A}\wedge\mathbf{B}=0$ is valid.

1.2 Assuming a nonsimple bivector $\mathbf{C}$, show that, for a given simple bivector $\mathbf{A}$ satisfying $\mathbf{A}\wedge\mathbf{C}\neq 0$, we can find another simple bivector $\mathbf{B}$ and a scalar α so that we can express $\mathbf{C}=\alpha\mathbf{A}+\mathbf{B}$. This proves that any bivector is either simple or it can be expressed as a sum of two simple bivectors.

1.3 Expressing a bivector in terms of a given vector basis $\mathbf{a}_i$ as $\mathbf{B} = \sum B_{ij}\mathbf{a}_i\wedge\mathbf{a}_j$, $i<j$, find the relation between the coefficients B_{ij} so that $\mathbf{B}$ is simple.

1.4 Assuming $\mathbf{A}$, a simple bivector, and $\mathbf{B}$ a nonsimple bivector, using basis expansions find the condition for $\mathbf{B}$ satisfying $\mathbf{A}\wedge\mathbf{B}=0$.

1.5 Show that for a bivector $\mathbf{C}$ and one-form $\boldsymbol{\beta}$ the condition $\boldsymbol{\beta}\rfloor\mathbf{C}=0$ implies that either $\boldsymbol{\beta}=0$ or $\mathbf{C}$ is simple.

1.6 Assuming two simple bivectors $\mathbf{A}$ and $\mathbf{B}$, show that if they satisfy $\mathbf{A}\wedge\mathbf{B}=0$, there exist vectors $\mathbf{a},\mathbf{b},\mathbf{c}$ so that we can write $\mathbf{A}=\mathbf{a}\wedge\mathbf{c}$ and $\mathbf{B}=\mathbf{b}\wedge\mathbf{c}$.

1.7 Show that a linear combination of two nonsimple bivectors can be so defined that it is a simple bivector.

1.8 Prove that if $\mathbf{A}$ is a simple bivector, $\boldsymbol{\Phi}=\varepsilon_N\lfloor\mathbf{A}$ is a simple two-form.

1.9 Find the one-form $\boldsymbol{\alpha}$ solution for the equation

$$\mathbf{A}\lfloor\boldsymbol{\alpha} = \mathbf{a},$$

when $\mathbf{A}$ is a given nonsimple bivector and $\mathbf{a}$ is a given vector. Hint: write $\mathbf{A} = \mathbf{e}_1 \wedge \mathbf{e}_2 + \mathbf{e}_3 \wedge \mathbf{e}_4$ where $\mathbf{e}_i$ form a basis and expand $\mathbf{a} = \sum a_i\mathbf{e}_i$, $\boldsymbol{\alpha} = \sum \alpha_j\boldsymbol{\varepsilon}_j$ when $\boldsymbol{\varepsilon}_j$ is the reciprocal basis.

1.10 Derive the identity

$$(\mathbf{A} \wedge \mathbf{A})\lfloor\overline{\overline{\mathsf{I}}}^T = 2\mathbf{A} \wedge (\mathbf{A}\lfloor\overline{\overline{\mathsf{I}}}^T)$$

for a nonsimple bivector $\mathbf{A}$ and apply it for the solution of the equation of the previous problem. Hint: define $\mathbf{A} = \mathbf{e}_1 \wedge \mathbf{e}_2 + \mathbf{e}_3 \wedge \mathbf{e}_4$ in terms of basis vectors $\mathbf{e}_i$.

1.11 Show that if a bivector $\mathbf{A}$ satisfies

$$\mathbf{A}\lfloor\boldsymbol{\alpha} = 0, \qquad \mathbf{A}\lfloor\boldsymbol{\beta} = 0$$

for two one-forms satisfying $\boldsymbol{\alpha} \wedge \boldsymbol{\beta} \neq 0$, it can be represented as

$$\mathbf{A} = \mathbf{k}_N\lfloor(\boldsymbol{\alpha} \wedge \boldsymbol{\beta})$$

in terms of some quadrivector $\mathbf{k}_N$.

1.12 Defining six nonsimple bivectors in terms of simple basis bivectors $\mathbf{e}_{ij}$ as

$$\begin{aligned} \mathbf{E}_1 &= \mathbf{e}_{12} + \mathbf{e}_{34}, & \mathbf{E}_2 &= \mathbf{e}_{23} + \mathbf{e}_{14}, & \mathbf{E}_3 &= \mathbf{e}_{31} + \mathbf{e}_{24}, \\ \mathbf{E}_4 &= \mathbf{e}_{12} - \mathbf{e}_{34}, & \mathbf{E}_5 &= \mathbf{e}_{23} - \mathbf{e}_{14}, & \mathbf{E}_6 &= \mathbf{e}_{31} - \mathbf{e}_{24}, \end{aligned}$$

show that they form a basis, that is, any bivector can be expanded as

$$\mathbf{A} = \sum A_i\mathbf{E}_i,$$

and that they are orthogonal in the dot product $\mathbf{E}_i \cdot \mathbf{E}_j = \boldsymbol{\varepsilon}_N|(\mathbf{A}_i \wedge \mathbf{A}_j)$.

1.13 Derive (1.106) and (1.107).

1.14 Find the most general bivector $\mathbf{B}$ satisfying

$$\mathbf{a} \wedge \mathbf{B} = 0$$

for a given vector $\mathbf{a}$.

1.15 Show that a bivector satisfying the equation

$$\boldsymbol{\alpha}\rfloor\mathbf{B} = 0$$

for a given one-form $\boldsymbol{\alpha}$ must be of the form

$$\mathbf{B} = \boldsymbol{\alpha}\rfloor\mathbf{k},$$

where $\mathbf{k}$ may be any trivector. Show also that $\mathbf{B}$ must be simple bivector.

1.16 Derive the identity

$$\boldsymbol{\alpha}\rfloor(\mathbf{k}\wedge\mathbf{a}) = \mathbf{a}\wedge(\boldsymbol{\alpha}\rfloor\mathbf{k}) + (\boldsymbol{\alpha}|\mathbf{a})\mathbf{k},$$

where $\boldsymbol{\alpha}$ is a one-form, $\mathbf{a}$ is a vector and $\mathbf{k}$ is a trivector. Show that it is possible to decompose any trivector as $\mathbf{k} = \mathbf{k}_\alpha + \mathbf{k}_a$ in terms of two trivectors satisfying $\boldsymbol{\alpha}\rfloor\mathbf{k}_\alpha = 0$ and $\mathbf{a}\wedge\mathbf{k}_a = 0$ when $\mathbf{a}$ and $\boldsymbol{\alpha}$ are two given quantities satisfying $\mathbf{a}|\boldsymbol{\alpha} \neq 0$.

1.17 Derive the identity

$$\mathbf{e}_N\lfloor(\boldsymbol{\alpha}\wedge\boldsymbol{\Psi}) = -\boldsymbol{\alpha}\rfloor(\mathbf{e}_N\lfloor\boldsymbol{\Psi})$$

valid for any one-form $\boldsymbol{\alpha}$ and two-form $\boldsymbol{\Psi}$.

1.18 Derive the identity

$$(\boldsymbol{\varepsilon}_N\lfloor\mathbf{k})\rfloor\mathbf{A} = (\boldsymbol{\varepsilon}_N\lfloor\mathbf{A})\rfloor\mathbf{k},$$

valid for any bivector $\mathbf{A} \in \mathbb{E}_2$ and trivector $\mathbf{k} \in \mathbb{E}_3$. Check the identity for $\mathbf{k} = \mathbf{e}_{234}$, $\mathbf{A} = \mathbf{e}_{12}$.

1.19 Find solutions $\mathbf{X}$ to the bivector equation $\mathbf{A}\wedge\mathbf{X} = 0$ when $\mathbf{A}$ satisfies $\mathbf{A}\wedge\mathbf{A} \neq 0$.

1.20 Show that any simple bivector $\mathbf{a}\wedge\mathbf{b}$ can be expressed in the form $\mathbf{c}_s\wedge\mathbf{d}$ where $\mathbf{c}$ is a spatial vector and $\mathbf{d}$ another vector. Assume $\boldsymbol{\varepsilon}_4|\mathbf{a} \neq 0$ and $\boldsymbol{\varepsilon}_4|\mathbf{b} \neq 0$.

CHAPTER 2

Dyadics

Dyadics are algebraic objects performing linear mappings between multivectors and/or multiforms, $\mathbb{E}_p \rightarrow \mathbb{E}_q$, $\mathbb{F}_p \rightarrow \mathbb{F}_q$, $\mathbb{E}_p \rightarrow \mathbb{F}_q$, $\mathbb{F}_p \rightarrow \mathbb{E}_q$. Mapping between spaces of the same dimension, $p = q$ or $p = 4 - q$, appear to be the most interesting ones because they may be inverted.

2.1 MAPPING VECTORS AND ONE-FORMS

2.1.1 Dyadics

Any linear mapping between two vectors $\mathbf{x} \rightarrow \mathbf{y}$ can be expressed in the form

$$\mathbf{y} = \sum_{i=1}^{4} \mathbf{a}_i(\boldsymbol{\alpha}_i|\mathbf{x}) = \left(\sum_{i=1}^{4} \mathbf{a}_i\boldsymbol{\alpha}_i\right)|\mathbf{x} = \overline{\overline{\mathsf{A}}}|\mathbf{x}, \tag{2.1}$$

in terms of suitably chosen vectors $\mathbf{a}_i$ and one-forms $\boldsymbol{\alpha}_i$. Here we denote

$$\overline{\overline{\mathsf{A}}} = \sum \mathbf{a}_i\boldsymbol{\alpha}_i \tag{2.2}$$

which is called a dyadic polynomial or dyadic for short. Here we apply the original Gibbsian notation [5] denoting the dyadic product by "no sign" instead of the sign ⊗ encountered in mathematical literature. The representation (2.2) is not unique. By choosing a vector basis $\mathbf{e}_i$ or a one-form basis $\boldsymbol{\varepsilon}_j$ the same dyadic can be expressed in various forms,

$$\overline{\overline{\mathsf{A}}} = \sum \mathbf{e}_i\boldsymbol{\alpha}_i = \sum \mathbf{a}_i\boldsymbol{\varepsilon}_i = \sum\sum A_{ij}\mathbf{e}_i\boldsymbol{\varepsilon}_j. \tag{2.3}$$

Multiforms, Dyadics, and Electromagnetic Media, First Edition. Ismo V. Lindell.

The number of free parameters is 16 which can be defined through four one-forms $\boldsymbol{\alpha}_i$, four vectors $\mathbf{a}_i$, or 16 scalars A_{ij}. Dyadics mapping vectors to vectors form a linear space which is denoted by $\mathbb{E}_1\mathbb{F}_1$.

Similarly, dyadics mapping one-forms to one-forms form a linear space which is denoted by $\mathbb{F}_1\mathbb{E}_1$. They have the general form

$$\overline{\overline{\mathsf{A}}} = \sum \boldsymbol{\alpha}_i \mathbf{a}_i. \tag{2.4}$$

As a special example, the dyadic

$$\overline{\overline{\mathsf{I}}} = \mathbf{e}_1\boldsymbol{\varepsilon}_1 + \mathbf{e}_2\boldsymbol{\varepsilon}_2 + \mathbf{e}_3\boldsymbol{\varepsilon}_3 + \mathbf{e}_4\boldsymbol{\varepsilon}_4, \tag{2.5}$$

called the unit dyadic, maps any vector to itself for any choice of reciprocal vector and one-form bases $\mathbf{e}_i$, $\boldsymbol{\varepsilon}_j$,

$$\overline{\overline{\mathsf{I}}}|\mathbf{a} = \sum \mathbf{e}_i(\boldsymbol{\varepsilon}_i|\mathbf{a}) = \sum a_i\mathbf{e}_i = \mathbf{a}. \tag{2.6}$$

The corresponding unit dyadic mapping any one-form to itself is denoted by

$$\overline{\overline{\mathsf{I}}}^T = \boldsymbol{\varepsilon}_1\mathbf{e}_1 + \boldsymbol{\varepsilon}_2\mathbf{e}_2 + \boldsymbol{\varepsilon}_3\mathbf{e}_3 + \boldsymbol{\varepsilon}_4\mathbf{e}_4, \tag{2.7}$$

where $()^T$ denotes the transpose operation,

$$(\mathbf{a}\boldsymbol{\alpha})^T = \boldsymbol{\alpha}\mathbf{a}. \tag{2.8}$$

For dyadics of the space $\mathbb{E}_1\mathbb{F}_1$ one can define the bar product of two dyadics which yields a dyadic of the same space,

$$\overline{\overline{\mathsf{A}}}|\overline{\overline{\mathsf{B}}} = \sum \mathbf{a}_i\boldsymbol{\alpha}_i \mid \sum b_j\boldsymbol{\beta}_j = \sum\sum(\boldsymbol{\alpha}_i|\mathbf{b}_j)\mathbf{a}_i\boldsymbol{\beta}_j. \tag{2.9}$$

Dyadic powers are defined by

$$\overline{\overline{\mathsf{A}}}^2 = \overline{\overline{\mathsf{A}}}|\overline{\overline{\mathsf{A}}}, \quad \overline{\overline{\mathsf{A}}}^3 = \overline{\overline{\mathsf{A}}}|\overline{\overline{\mathsf{A}}}|\overline{\overline{\mathsf{A}}}, \quad \cdots \tag{2.10}$$

The inverse dyadic $\overline{\overline{\mathsf{A}}}^{-1}$, when it exists, satisfies

$$\overline{\overline{\mathsf{A}}}|\overline{\overline{\mathsf{A}}}^{-1} = \overline{\overline{\mathsf{A}}}^{-1}|\overline{\overline{\mathsf{A}}} = \overline{\overline{\mathsf{I}}}, \tag{2.11}$$

and negative powers can be understood as

$$\overline{\overline{\mathsf{A}}}^{-p} = (\overline{\overline{\mathsf{A}}}^{-1})^p = (\overline{\overline{\mathsf{A}}}^p)^{-1}. \tag{2.12}$$

Similar properties are valid for dyadics belonging to the space $\mathbb{F}_1\mathbb{E}_1$.

2.1.2 Double-Bar Product ||

The double-bar product of a dyadic $\overline{\overline{\mathsf{A}}} = \sum \mathbf{a}_i \boldsymbol{\alpha}_i$ and a dyadic $\overline{\overline{\mathsf{B}}} = \sum \mathbf{b}_j \boldsymbol{\beta}_j$ transposed is defined by

$$\overline{\overline{\mathsf{A}}}||\overline{\overline{\mathsf{B}}}^T = \sum \mathbf{a}_i \boldsymbol{\alpha}_i || \sum \boldsymbol{\beta}_j \mathbf{b}_j = \sum \sum (\mathbf{a}_i|\boldsymbol{\beta}_j)(\boldsymbol{\alpha}_i|\mathbf{b}_j). \tag{2.13}$$

The result is a scalar which is a linear function of both dyadics. The double-bar product satisfies

$$\overline{\overline{\mathsf{A}}}||\overline{\overline{\mathsf{B}}}^T = \overline{\overline{\mathsf{B}}}^T||\overline{\overline{\mathsf{A}}} = \overline{\overline{\mathsf{B}}}||\overline{\overline{\mathsf{A}}}^T = \overline{\overline{\mathsf{A}}}^T||\overline{\overline{\mathsf{B}}}. \tag{2.14}$$

Dyadics of the spaces $\mathbb{E}_1\mathbb{F}_1$ and $\mathbb{F}_1\mathbb{E}_1$ have a scalar invariant called the trace. For $\overline{\overline{\mathsf{A}}} = \sum \mathbf{a}_i \boldsymbol{\alpha}_i$ it is defined by

$$\mathrm{tr}\overline{\overline{\mathsf{A}}} = \overline{\overline{\mathsf{A}}}||\overline{\overline{\mathsf{I}}}^T = \sum \mathbf{a}_i|\overline{\overline{\mathsf{I}}}^T|\boldsymbol{\alpha}_i = \sum \mathbf{a}_i|\boldsymbol{\alpha}_i = \mathrm{tr}\overline{\overline{\mathsf{A}}}^T, \tag{2.15}$$

or

$$\mathrm{tr}\overline{\overline{\mathsf{A}}} = \mathrm{tr} \sum \sum A_{i,j} \mathbf{e}_i \boldsymbol{\varepsilon}_j = \sum \sum A_{i,j} \mathbf{e}_i|\boldsymbol{\varepsilon}_j = \sum A_{i,i}. \tag{2.16}$$

Because $\overline{\overline{\mathsf{I}}}$ is independent of the choice of the reciprocal bases, so is the trace of the dyadic. In particular, we have

$$\mathrm{tr}\overline{\overline{\mathsf{I}}} = \sum \mathbf{e}_i|\boldsymbol{\varepsilon}_i = 4. \tag{2.17}$$

Trace of a product of two dyadics of the space $\mathbb{E}_1\mathbb{F}_1$ satisfies

$$\mathrm{tr}(\overline{\overline{\mathsf{A}}}|\overline{\overline{\mathsf{B}}}) = \overline{\overline{\mathsf{A}}}||\overline{\overline{\mathsf{B}}}^T. \tag{2.18}$$

The trace of a single dyadic can be expressed as

$$\mathrm{tr}\overline{\overline{\mathsf{A}}} = \mathrm{tr}(\overline{\overline{\mathsf{A}}}|\overline{\overline{\mathsf{I}}}) = \overline{\overline{\mathsf{A}}}||\overline{\overline{\mathsf{I}}}^T = \overline{\overline{\mathsf{A}}}^T||\overline{\overline{\mathsf{I}}} = \overline{\overline{\mathsf{I}}}||\overline{\overline{\mathsf{A}}}^T = \overline{\overline{\mathsf{I}}}^T||\overline{\overline{\mathsf{A}}}. \tag{2.19}$$

Any dyadic $\overline{\overline{\mathsf{A}}} \in \mathbb{E}_1\mathbb{F}_1$ can be uniquely decomposed as a sum of a multiple of the unit dyadic and a trace-free part as

$$\overline{\overline{\mathsf{A}}} = \frac{1}{4}\mathrm{tr}\overline{\overline{\mathsf{A}}}\,\overline{\overline{\mathsf{I}}} + \overline{\overline{\mathsf{A}}}_o, \quad \mathrm{tr}\overline{\overline{\mathsf{A}}}_o = 0. \tag{2.20}$$

Similar properties apply for dyadics of the space $\mathbb{F}_1\mathbb{E}_1$.

2.1.3 Metric Dyadics

Dyadics of the spaces $\mathbb{F}_1\mathbb{F}_1$ and $\mathbb{E}_1\mathbb{E}_1$ mapping respectively vectors to one-forms and one-forms to vectors are called metric dyadics. They can be expanded as

$$\overline{\overline{\Gamma}} = \sum \Gamma_{ij}\boldsymbol{\varepsilon}_i\boldsymbol{\varepsilon}_j \in \mathbb{F}_1\mathbb{F}_1 \tag{2.21}$$

and

$$\overline{\overline{\mathsf{G}}} = \sum G_{ij}\mathbf{e}_i\mathbf{e}_j \in \mathbb{E}_1\mathbb{E}_1. \tag{2.22}$$

Because the transpose operation of a metric dyadic yields a dyadic in the same space, the dyadic eigenproblems

$$\overline{\overline{\mathsf{G}}}^T = \lambda\overline{\overline{\mathsf{G}}}, \quad \overline{\overline{\Gamma}}^T = \lambda\overline{\overline{\Gamma}} \tag{2.23}$$

make sense and have two eigenvalues $\lambda = 1, -1$ corresponding to symmetric and antisymmetric dyadics, respectively.

Symmetric dyadics can be used to define a dot product for vectors and one-forms as

$$\mathbf{a}\cdot\mathbf{b} = \mathbf{b}\cdot\mathbf{a} = \mathbf{a}|\overline{\overline{\Gamma}}|\mathbf{b} = \sum \Gamma_{ii}(\mathbf{a}|\boldsymbol{\varepsilon}_i)(\mathbf{b}|\boldsymbol{\varepsilon}_i), \tag{2.24}$$

$$\boldsymbol{\alpha}\cdot\boldsymbol{\beta} = \boldsymbol{\beta}\cdot\boldsymbol{\alpha} = \boldsymbol{\alpha}|\overline{\overline{\mathsf{G}}}|\boldsymbol{\beta} = \sum G_{ii}(\boldsymbol{\alpha}|\mathbf{e}_i)(\boldsymbol{\beta}|\mathbf{b}_i). \tag{2.25}$$

The dot products for vectors and one-forms are compatible when the dot product for the mapped vectors $\overline{\overline{\Gamma}}|\mathbf{a} \in \mathbb{F}_1$, $\overline{\overline{\Gamma}}|\mathbf{b} \in \mathbb{F}_1$ equals the dot product of the vectors,

$$(\overline{\overline{\Gamma}}|\mathbf{a})\cdot(\overline{\overline{\Gamma}}|\mathbf{b}) = \mathbf{a}|\overline{\overline{\Gamma}}|\overline{\overline{\mathsf{G}}}|\overline{\overline{\Gamma}}|\mathbf{b} = \mathbf{a}|\overline{\overline{\Gamma}}|\mathbf{b} = \mathbf{a}\cdot\mathbf{b} \tag{2.26}$$

for all vectors $\mathbf{a}, \mathbf{b}$. This requires that $\overline{\overline{\mathsf{G}}}|\overline{\overline{\Gamma}} = \overline{\overline{\mathsf{I}}}$, that is, that the two symmetric dyadics are inverses of each other,

$$\overline{\overline{\Gamma}} = \overline{\overline{\mathsf{G}}}^{-1}, \quad \overline{\overline{\mathsf{G}}} = \overline{\overline{\Gamma}}^{-1}. \tag{2.27}$$

The double-bar product $\overline{\overline{\mathsf{A}}}||\overline{\overline{\mathsf{B}}}$ can be defined for two metric dyadics when the dyadics are taken from different spaces, $\overline{\overline{\mathsf{A}}} \in \mathbb{E}_1\mathbb{E}_1$, $\overline{\overline{\mathsf{B}}} \in \mathbb{F}_1\mathbb{F}_1$. Because of the property

$$\overline{\overline{\mathsf{A}}}||\overline{\overline{\mathsf{B}}} = \overline{\overline{\mathsf{A}}}^T||\overline{\overline{\mathsf{B}}}^T, \tag{2.28}$$

the double-bar product vanishes if one of the metric dyadics is symmetric and the other one is antisymmetric.

2.2 MAPPING MULTIVECTORS AND MULTIFORMS

The previous definitions of dyadics can be extended to mappings between multivectors and/or multiforms.

2.2.1 Bidyadics

A dyadic defined by

$$\overline{\overline{\mathsf{D}}} = \sum D_{I,J}\mathbf{e}_I\boldsymbol{\varepsilon}_J \in \mathbb{E}_2\mathbb{F}_2 \tag{2.29}$$

where the bi-indices I, J go from 12 to 34, maps bivectors to bivectors,

$$\mathbf{B} = \overline{\overline{\mathsf{D}}}|\mathbf{A} = \sum D_{I,J}\mathbf{e}_I\boldsymbol{\varepsilon}_J| \sum A_K\mathbf{e}_K = \sum D_{I,J}A_J\mathbf{e}_I \in \mathbb{E}_2, \tag{2.30}$$

and it can be called a bidyadic. Similarly, a dyadic $\overline{\overline{\mathsf{D}}} \in \mathbb{E}_2\mathbb{E}_2$ mapping two-forms to bivectors

$$\mathbf{B} = \overline{\overline{\mathsf{D}}}|\boldsymbol{\Phi} = \sum D_{I,J}\mathbf{e}_I\mathbf{e}_J| \sum_K \boldsymbol{\varepsilon}_K\Phi_K = \sum D_{I,J}\Phi_J\mathbf{e}_I \in \mathbb{E}_2 \tag{2.31}$$

can be called a metric bidyadic. Because the electromagnetic fields are represented by two-forms, bidyadics play a central role in electromagnetics. In fact, electromagnetic media can be defined in terms of bidyadics.

Bivectors are six-dimensional objects, whence a bidyadic has $6 \times 6 = 36$ parameters in the most general case. Another form for the bidyadic $\overline{\overline{\mathsf{D}}} \in \mathbb{E}_2\mathbb{F}_2$ is the expansion in bivectors and two-forms as

$$\overline{\overline{\mathsf{D}}} = \sum \mathbf{A}_I\boldsymbol{\Phi}_I. \tag{2.32}$$

Choosing any linearly independent set of six bivectors $\mathbf{A}_I$ for a basis of bivectors, the bidyadic $\overline{\overline{\mathsf{D}}}$ is uniquely determined by the set of six two-forms $\boldsymbol{\Phi}_I$ and vice versa. The trace of a bidyadic is defined by

$$\mathrm{tr}\overline{\overline{\mathsf{D}}} = \sum \mathbf{A}_I|\boldsymbol{\Phi}_I = \sum D_{I,I}. \tag{2.33}$$

2.2.2 Double-Wedge Product $^\wedge_\wedge$

Multivectors and multiforms can be constructed from vectors and one-forms by applying the wedge product. Similarly, multidyadics can be formed from dyadics by applying the double-wedge product.

The double-wedge product between two dyadic products of vectors and one-forms can be introduced by

$$(\mathbf{a}\boldsymbol{\alpha}){}^\wedge_\wedge(\mathbf{b}\boldsymbol{\beta}) = (\mathbf{a} \wedge \mathbf{b})(\boldsymbol{\alpha} \wedge \boldsymbol{\beta}), \tag{2.34}$$

and the result lies in the space of bidyadics denoted by $\mathbb{E}_2\mathbb{F}_2$. More generally, we can define the double-wedge product of two dyadics as

$$\overline{\overline{\mathsf{A}}}{}^\wedge_\wedge\overline{\overline{\mathsf{B}}} = \left(\sum \mathbf{a}_i\boldsymbol{\alpha}_i\right){}^\wedge_\wedge\left(\sum \mathbf{b}_j\boldsymbol{\beta}_j\right) = \sum\sum(\mathbf{a}_i \wedge \mathbf{b}_j)(\boldsymbol{\alpha}_i \wedge \boldsymbol{\beta}_j). \tag{2.35}$$

Since the wedge product is anticommutative and associative, the double-wedge product is commutative,

$$\overline{\overline{\mathsf{A}}}{}^\wedge_\wedge\overline{\overline{\mathsf{B}}} = \overline{\overline{\mathsf{B}}}{}^\wedge_\wedge\overline{\overline{\mathsf{A}}}, \tag{2.36}$$

and associative,

$$(\overline{\overline{\mathsf{A}}}{}^\wedge_\wedge\overline{\overline{\mathsf{B}}}){}^\wedge_\wedge\overline{\overline{\mathsf{C}}} = \overline{\overline{\mathsf{A}}}{}^\wedge_\wedge(\overline{\overline{\mathsf{B}}}{}^\wedge_\wedge\overline{\overline{\mathsf{C}}}). \tag{2.37}$$

Thus, dyadics in a chain of double-wedge products can be set in any order and there is no need for brackets to show the order of multiplication. Of course, a product chain of five or more dyadics yields zero.

The corresponding double-cross product ${}^\times_\times$ was introduced by Gibbs already in 1886 [6], Figure 2.1.

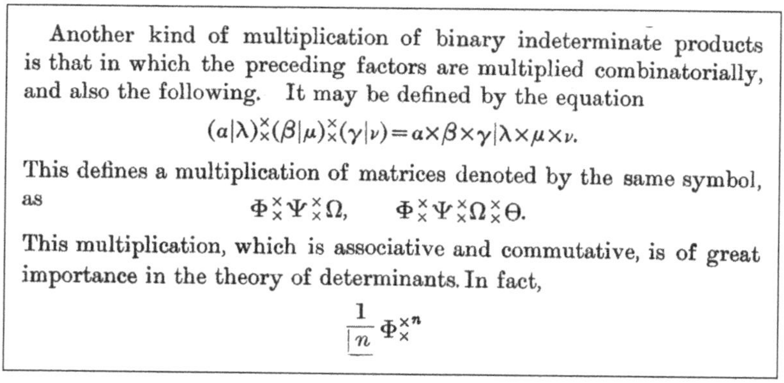

Another kind of multiplication of binary indeterminate products is that in which the preceding factors are multiplied combinatorially, and also the following. It may be defined by the equation

$$(\alpha|\lambda){}^\times_\times(\beta|\mu){}^\times_\times(\gamma|\nu) = \alpha\times\beta\times\gamma|\lambda\times\mu\times\nu.$$

This defines a multiplication of matrices denoted by the same symbol, as

$$\Phi{}^\times_\times\Psi{}^\times_\times\Omega, \qquad \Phi{}^\times_\times\Psi{}^\times_\times\Omega{}^\times_\times\Theta.$$

This multiplication, which is associative and commutative, is of great importance in the theory of determinants. In fact,

$$\frac{1}{\underline{|n}}\Phi^{\times n}_{\times}$$

Figure 2.1. The double-cross product ${}^\times_\times$ corresponding to the double-wedge product ${}^\wedge_\wedge$ was originally introduced by Gibbs in 1886 [6]. Here the bar corresponds to the dyadic product (later replaced by "no sign" by Gibbs [5]). The expression on the last line stands for the nth double-cross power of the dyadic $\boldsymbol{\Phi}$.

The basic rule for the double-wedge product of two dyadics can be formed by starting from the expansion

$$\begin{aligned}(\mathbf{a}\boldsymbol{\alpha}{}^{\wedge}_{\wedge}\mathbf{b}\boldsymbol{\beta})|(\mathbf{c}\wedge\mathbf{d}) &= ((\mathbf{a}\wedge\mathbf{b})(\boldsymbol{\alpha}\wedge\boldsymbol{\beta}))|(\mathbf{c}\wedge\mathbf{d})\\ &= (\mathbf{a}\wedge\mathbf{b})((\boldsymbol{\alpha}\wedge\boldsymbol{\beta})|(\mathbf{c}\wedge\mathbf{d}))\\ &= (\mathbf{a}\wedge\mathbf{b})((\boldsymbol{\alpha}|\mathbf{c})(\boldsymbol{\beta}|\mathbf{d})-(\boldsymbol{\alpha}|\mathbf{d})(\boldsymbol{\beta}|\mathbf{c}))\\ &= (\mathbf{a}\boldsymbol{\alpha}|\mathbf{c})\wedge(\mathbf{b}\boldsymbol{\beta}|\mathbf{d})-(\mathbf{a}\boldsymbol{\alpha}|\mathbf{d})\wedge(\mathbf{b}\boldsymbol{\beta}|\mathbf{c}). \end{aligned}\qquad(2.38)$$

Since the first and last expressions are linear in the dyadic products $\mathbf{a}\boldsymbol{\alpha}$ and $\mathbf{b}\boldsymbol{\beta}$, we can replace them by the respective general dyadics $\overline{\overline{\mathsf{A}}}\in\mathbb{E}_1\mathbb{F}_1$ and $\overline{\overline{\mathsf{B}}}\in\mathbb{E}_1\mathbb{F}_1$, whence we have arrived at the rule

$$(\overline{\overline{\mathsf{A}}}{}^{\wedge}_{\wedge}\overline{\overline{\mathsf{B}}})|(\mathbf{a}\wedge\mathbf{b}) = (\overline{\overline{\mathsf{A}}}|\mathbf{a})\wedge(\overline{\overline{\mathsf{B}}}|\mathbf{b})+(\overline{\overline{\mathsf{B}}}|\mathbf{a})\wedge(\overline{\overline{\mathsf{A}}}|\mathbf{b}).\qquad(2.39)$$

The right-hand side can be easily memorized by noticing that it is invariant in the interchange of $\overline{\overline{\mathsf{A}}}$ and $\overline{\overline{\mathsf{B}}}$ and changes sign in the interchange of **a** and **b**.

Multiplying (2.39) dyadically by $(\boldsymbol{\alpha}\wedge\boldsymbol{\beta})$ from the right we obtain

$$\begin{aligned}(\overline{\overline{\mathsf{A}}}{}^{\wedge}_{\wedge}\overline{\overline{\mathsf{B}}})|(\mathbf{a}\wedge\mathbf{b})(\boldsymbol{\alpha}\wedge\boldsymbol{\beta}) &= (\overline{\overline{\mathsf{A}}}{}^{\wedge}_{\wedge}\overline{\overline{\mathsf{B}}})|(\mathbf{a}\boldsymbol{\alpha}){}^{\wedge}_{\wedge}(\mathbf{b}\boldsymbol{\beta})\\ &= (\overline{\overline{\mathsf{A}}}|\mathbf{a}\boldsymbol{\alpha}){}^{\wedge}_{\wedge}(\overline{\overline{\mathsf{B}}}|\mathbf{b}\boldsymbol{\beta})+(\overline{\overline{\mathsf{B}}}|\mathbf{a}\boldsymbol{\alpha}){}^{\wedge}_{\wedge}(\overline{\overline{\mathsf{A}}}|\mathbf{b}\boldsymbol{\beta}).\end{aligned}\qquad(2.40)$$

Replacing again the dyadic products $(\mathbf{a}\boldsymbol{\alpha})$ and $(\mathbf{b}\boldsymbol{\beta})$ by the respective general dyadics $\overline{\overline{\mathsf{C}}}$ and $\overline{\overline{\mathsf{D}}}$ we obtain an identity valid for any four dyadics of the space $\mathbb{E}_1\mathbb{F}_1$,

$$(\overline{\overline{\mathsf{A}}}{}^{\wedge}_{\wedge}\overline{\overline{\mathsf{B}}})|(\overline{\overline{\mathsf{C}}}{}^{\wedge}_{\wedge}\overline{\overline{\mathsf{D}}}) = (\overline{\overline{\mathsf{A}}}|\overline{\overline{\mathsf{C}}}){}^{\wedge}_{\wedge}(\overline{\overline{\mathsf{B}}}|\overline{\overline{\mathsf{D}}})+(\overline{\overline{\mathsf{A}}}|\overline{\overline{\mathsf{D}}}){}^{\wedge}_{\wedge}(\overline{\overline{\mathsf{B}}}|\overline{\overline{\mathsf{C}}}).\qquad(2.41)$$

The form of (2.41) remains valid when the four dyadics are taken from the space $\mathbb{F}_1\mathbb{E}_1$. This can be shown by transposing both sides of the dyadic identity, applying

$$(\overline{\overline{\mathsf{A}}}{}^{\wedge}_{\wedge}\overline{\overline{\mathsf{B}}})^T = \overline{\overline{\mathsf{A}}}^T{}^{\wedge}_{\wedge}\overline{\overline{\mathsf{B}}}^T,\qquad(2.42)$$

and changing the symbols. Finally, (2.41) also remains to be valid for dyadics taken from metric spaces $\overline{\overline{\mathsf{A}}},\overline{\overline{\mathsf{B}}}\in\mathbb{E}_1\mathbb{E}_1$ and $\overline{\overline{\mathsf{C}}},\overline{\overline{\mathsf{D}}}\in\mathbb{F}_1\mathbb{F}_1$ or conversely.

Applying the double-wedge product one can produce multidyadics of higher order, $\overline{\overline{\mathsf{A}}}{}^{\wedge}_{\wedge}\overline{\overline{\mathsf{B}}}{}^{\wedge}_{\wedge}\overline{\overline{\mathsf{C}}}\in\mathbb{E}_3\mathbb{F}_3$, and $\overline{\overline{\mathsf{A}}}{}^{\wedge}_{\wedge}\overline{\overline{\mathsf{B}}}{}^{\wedge}_{\wedge}\overline{\overline{\mathsf{C}}}{}^{\wedge}_{\wedge}\overline{\overline{\mathsf{D}}}\in\mathbb{E}_4\mathbb{F}_4$. Going through algebraic steps similar to the previous ones, the corresponding basic multiplication rule for three dyadics and three vectors can be expressed as

$$\begin{aligned}(\overline{\overline{\mathsf{A}}}{}^\wedge_\wedge\overline{\overline{\mathsf{B}}}{}^\wedge_\wedge\overline{\overline{\mathsf{C}}})|(\mathbf{a}\wedge\mathbf{b}\wedge\mathbf{c}) = {} & (\overline{\overline{\mathsf{A}}}|\mathbf{a})\wedge(\overline{\overline{\mathsf{B}}}|\mathbf{b})\wedge(\overline{\overline{\mathsf{C}}}|\mathbf{c}) \\ & +(\overline{\overline{\mathsf{A}}}|\mathbf{b})\wedge(\overline{\overline{\mathsf{B}}}|\mathbf{c})\wedge(\overline{\overline{\mathsf{C}}}|\mathbf{a})+(\overline{\overline{\mathsf{A}}}|\mathbf{c})\wedge(\overline{\overline{\mathsf{B}}}|\mathbf{a})\wedge(\overline{\overline{\mathsf{C}}}|\mathbf{b}) \\ & -(\overline{\overline{\mathsf{A}}}|\mathbf{a})\wedge(\overline{\overline{\mathsf{B}}}|\mathbf{c})\wedge(\overline{\overline{\mathsf{C}}}|\mathbf{b})-(\overline{\overline{\mathsf{A}}}|\mathbf{c})\wedge(\overline{\overline{\mathsf{B}}}|\mathbf{b})\wedge(\overline{\overline{\mathsf{C}}}|\mathbf{a}) \\ & -(\overline{\overline{\mathsf{A}}}|\mathbf{b})\wedge(\overline{\overline{\mathsf{B}}}|\mathbf{a})\wedge(\overline{\overline{\mathsf{C}}}|\mathbf{c}),\end{aligned} \tag{2.43}$$

which gives rise to the general rule for two sets of three dyadics becomes

$$\begin{aligned}(\overline{\overline{\mathsf{A}}}{}^\wedge_\wedge\overline{\overline{\mathsf{B}}}{}^\wedge_\wedge\overline{\overline{\mathsf{C}}})|(\overline{\overline{\mathsf{D}}}{}^\wedge_\wedge\overline{\overline{\mathsf{E}}}{}^\wedge_\wedge\overline{\overline{\mathsf{F}}}) = {} & (\overline{\overline{\mathsf{A}}}|\overline{\overline{\mathsf{D}}}){}^\wedge_\wedge(\overline{\overline{\mathsf{B}}}|\overline{\overline{\mathsf{E}}}){}^\wedge_\wedge(\overline{\overline{\mathsf{C}}}|\overline{\overline{\mathsf{F}}}) \\ & +(\overline{\overline{\mathsf{A}}}|\overline{\overline{\mathsf{E}}}){}^\wedge_\wedge(\overline{\overline{\mathsf{B}}}|\overline{\overline{\mathsf{F}}}){}^\wedge_\wedge(\overline{\overline{\mathsf{C}}}|\overline{\overline{\mathsf{D}}})+(\overline{\overline{\mathsf{A}}}|\overline{\overline{\mathsf{F}}}){}^\wedge_\wedge(\overline{\overline{\mathsf{B}}}|\overline{\overline{\mathsf{D}}}){}^\wedge_\wedge(\overline{\overline{\mathsf{C}}}|\overline{\overline{\mathsf{E}}}) \\ & +(\overline{\overline{\mathsf{A}}}|\overline{\overline{\mathsf{D}}}){}^\wedge_\wedge(\overline{\overline{\mathsf{B}}}|\overline{\overline{\mathsf{F}}}){}^\wedge_\wedge(\overline{\overline{\mathsf{C}}}|\overline{\overline{\mathsf{E}}})+(\overline{\overline{\mathsf{A}}}|\overline{\overline{\mathsf{F}}}){}^\wedge_\wedge(\overline{\overline{\mathsf{B}}}|\overline{\overline{\mathsf{E}}}){}^\wedge_\wedge(\overline{\overline{\mathsf{C}}}|\overline{\overline{\mathsf{D}}}) \\ & +(\overline{\overline{\mathsf{A}}}|\overline{\overline{\mathsf{E}}}){}^\wedge_\wedge(\overline{\overline{\mathsf{B}}}|\overline{\overline{\mathsf{D}}}){}^\wedge_\wedge(\overline{\overline{\mathsf{C}}}|\overline{\overline{\mathsf{F}}}).\end{aligned} \tag{2.44}$$

The double-wedge product of four dyadics of the space $\mathbb{E}_1\mathbb{F}_1$ is always a multiple of $\mathbf{e}_N\boldsymbol{\varepsilon}_N$:

$$\overline{\overline{\mathsf{A}}}{}^\wedge_\wedge\overline{\overline{\mathsf{B}}}{}^\wedge_\wedge\overline{\overline{\mathsf{C}}}{}^\wedge_\wedge\overline{\overline{\mathsf{D}}} = (\boldsymbol{\varepsilon}_N|(\overline{\overline{\mathsf{A}}}{}^\wedge_\wedge\overline{\overline{\mathsf{B}}}{}^\wedge_\wedge\overline{\overline{\mathsf{C}}}{}^\wedge_\wedge\overline{\overline{\mathsf{D}}})|\mathbf{e}_N)\mathbf{e}_N\boldsymbol{\varepsilon}_N. \tag{2.45}$$

2.2.3 Double-Wedge Powers

The double-wedge square of a dyadic can be defined as

$$\overline{\overline{\mathsf{A}}}{}^{(2)} = \frac{1}{2}\overline{\overline{\mathsf{A}}}{}^\wedge_\wedge\overline{\overline{\mathsf{A}}}, \tag{2.46}$$

and, because of (2.39), it obeys the rule

$$\overline{\overline{\mathsf{A}}}{}^{(2)}|(\mathbf{a}\wedge\mathbf{b}) = (\overline{\overline{\mathsf{A}}}|\mathbf{a})\wedge(\overline{\overline{\mathsf{A}}}|\mathbf{b}). \tag{2.47}$$

The sign $()^{(p)}$ is due to Reichardt [7], see Figure 2.2. From

$$\overline{\overline{\mathsf{B}}}{}^{(2)}|\overline{\overline{\mathsf{A}}}{}^{(2)}|(\mathbf{a}\wedge\mathbf{b}) = \overline{\overline{\mathsf{B}}}{}^{(2)}|((\overline{\overline{\mathsf{A}}}|\mathbf{a})\wedge(\overline{\overline{\mathsf{A}}}|\mathbf{b})) = (\overline{\overline{\mathsf{B}}}|\overline{\overline{\mathsf{A}}}|\mathbf{a})\wedge(\overline{\overline{\mathsf{B}}}|\overline{\overline{\mathsf{A}}}|\mathbf{b}) \tag{2.48}$$

we come to the rule

$$(\overline{\overline{\mathsf{B}}}|\overline{\overline{\mathsf{A}}})^{(2)} = \overline{\overline{\mathsf{B}}}{}^{(2)}|\overline{\overline{\mathsf{A}}}{}^{(2)}. \tag{2.49}$$

This follows also from (2.41) when choosing $\overline{\overline{\mathsf{C}}} = \overline{\overline{\mathsf{A}}}$ and $\overline{\overline{\mathsf{D}}} = \overline{\overline{\mathsf{B}}}$.

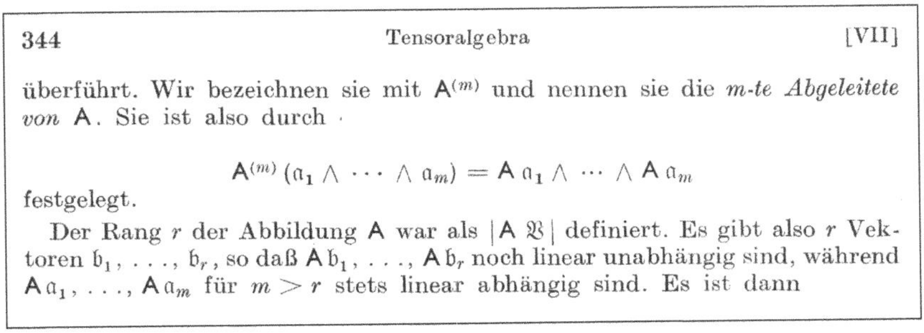

344 Tensoralgebra [VII]

überführt. Wir bezeichnen sie mit $\mathsf{A}^{(m)}$ und nennen sie die *m-te Abgeleitete von* A. Sie ist also durch

$$\mathsf{A}^{(m)}(\mathfrak{a}_1 \wedge \cdots \wedge \mathfrak{a}_m) = \mathsf{A}\,\mathfrak{a}_1 \wedge \cdots \wedge \mathsf{A}\,\mathfrak{a}_m$$

festgelegt.

Der Rang r der Abbildung A war als $|\mathsf{A}\,\mathfrak{B}|$ definiert. Es gibt also r Vektoren $\mathfrak{b}_1, \ldots, \mathfrak{b}_r$, so daß $\mathsf{A}\,\mathfrak{b}_1, \ldots, \mathsf{A}\,\mathfrak{b}_r$ noch linear unabhängig sind, während $\mathsf{A}\,\mathfrak{a}_1, \ldots, \mathsf{A}\,\mathfrak{a}_m$ für $m > r$ stets linear abhängig sind. Es ist dann

Figure 2.2. The notation (bracketed number in superscript) for the double-wedge power was introduced by Reichardt [7]. There is no sign denoting the bar product.

The double-wedge square of the unit dyadic $\overline{\overline{\mathsf{I}}}$

$$\overline{\overline{\mathsf{I}}}^{(2)} = \frac{1}{2}\sum\sum(\mathbf{e}_i \wedge \mathbf{e}_j)(\boldsymbol{\varepsilon}_i \wedge \boldsymbol{\varepsilon}_j) = \sum_{i<j} \mathbf{e}_{ij}\boldsymbol{\varepsilon}_{ij},$$

$$= \mathbf{e}_{12}\boldsymbol{\varepsilon}_{12} + \mathbf{e}_{23}\boldsymbol{\varepsilon}_{23} + \mathbf{e}_{31}\boldsymbol{\varepsilon}_{31} + \mathbf{e}_{14}\boldsymbol{\varepsilon}_{14} + \mathbf{e}_{24}\boldsymbol{\varepsilon}_{24} + \mathbf{e}_{34}\boldsymbol{\varepsilon}_{34} \qquad (2.50)$$

maps any bivector to itself, whence it serves as the unit bidyadic for bivectors. Similarly, its transpose, $\overline{\overline{\mathsf{I}}}^{(2)T}$ serves as the unit dyadic for two-forms. The trace has the value 6,

$$\mathrm{tr}\overline{\overline{\mathsf{I}}}^{(2)} = \mathrm{tr}\overline{\overline{\mathsf{I}}}^{(2)T} = \sum_{i<j} \mathbf{e}_{ij}|\boldsymbol{\varepsilon}_{ij} = 6. \qquad (2.51)$$

The trace of a bidyadic can be expressed as

$$\mathrm{tr}\overline{\overline{\mathsf{D}}} = \mathrm{tr}\sum \mathbf{A}_I\boldsymbol{\Phi}_I = \sum \mathbf{A}_I|\boldsymbol{\Phi}_I = \sum \mathbf{A}_I|\overline{\overline{\mathsf{I}}}^{(2)}|\boldsymbol{\Phi}_I = \overline{\overline{\mathsf{D}}}||\overline{\overline{\mathsf{I}}}^{(2)T}. \qquad (2.52)$$

Higher double-wedge powers of dyadics can be formed as

$$\overline{\overline{\mathsf{A}}}^{(3)} = \frac{1}{3!}\overline{\overline{\mathsf{A}}}{}^\wedge_\wedge\overline{\overline{\mathsf{A}}}{}^\wedge_\wedge\overline{\overline{\mathsf{A}}}, \qquad (2.53)$$

$$\overline{\overline{\mathsf{A}}}^{(4)} = \frac{1}{4!}\overline{\overline{\mathsf{A}}}{}^\wedge_\wedge\overline{\overline{\mathsf{A}}}{}^\wedge_\wedge\overline{\overline{\mathsf{A}}}{}^\wedge_\wedge\overline{\overline{\mathsf{A}}}, \qquad (2.54)$$

which belong to the respective higher dyadic spaces $\mathbb{E}_3\mathbb{F}_3$ and $\mathbb{E}_4\mathbb{F}_4$. The multiplication rule (2.49) can be generalized to

$$\overline{\overline{\mathsf{A}}}_1^{(p)}|\overline{\overline{\mathsf{A}}}_2^{(p)}|\ldots\overline{\overline{\mathsf{A}}}_q^{(p)} = (\overline{\overline{\mathsf{A}}}_1|\overline{\overline{\mathsf{A}}}_2|\ldots\overline{\overline{\mathsf{A}}}_q)^{(p)}. \qquad (2.55)$$

Obviously, we have $\overline{\overline{\mathsf{A}}}^{(p)} = 0$ for $p > 4$. From (2.55) we obtain

$$(\overline{\overline{\mathsf{A}}}^q)^{(p)} = (\overline{\overline{\mathsf{A}}}^{(p)})^q, \qquad (2.56)$$

which remains valid for negative q values if the inverse $\overline{\overline{\mathsf{A}}}^{-1}$ exists. Let us denote for simplicity

$$\overline{\overline{\mathsf{A}}}^{(-p)} = (\overline{\overline{\mathsf{A}}}^{-1})^{(p)} = (\overline{\overline{\mathsf{A}}}^{(p)})^{-1}. \tag{2.57}$$

The dyadics

$$\overline{\overline{\mathsf{I}}}^{(3)} = \sum_{i<j<k} \mathbf{e}_{ijk}\boldsymbol{\varepsilon}_{ijk}, \tag{2.58}$$

and

$$\overline{\overline{\mathsf{I}}}^{(4)} = \mathbf{e}_{1234}\boldsymbol{\varepsilon}_{1234} = \mathbf{e}_N\boldsymbol{\varepsilon}_N \tag{2.59}$$

serve as unit dyadics for trivectors and quadrivectors, respectively, and their transposed versions do the same for three-forms and four-forms. Trace of these dyadics yields

$$\mathrm{tr}\overline{\overline{\mathsf{I}}}^{(3)} = \sum_{i<j<k} \mathbf{e}_{ijk}|\boldsymbol{\varepsilon}_{ijk} = 4, \tag{2.60}$$

$$\mathrm{tr}\overline{\overline{\mathsf{I}}}^{(4)} = \mathbf{e}_N|\boldsymbol{\varepsilon}_N = 1. \tag{2.61}$$

The double-wedge fourth power of any dyadic of the space $\mathbb{E}_1\mathbb{F}_1$ is a multiple of $\overline{\overline{\mathsf{I}}}^{(4)}$:

$$\overline{\overline{\mathsf{A}}}^{(4)} = A\overline{\overline{\mathsf{I}}}^{(4)} = A\mathbf{e}_N\boldsymbol{\varepsilon}_N, \quad A = \mathrm{tr}\overline{\overline{\mathsf{A}}}^{(4)}. \tag{2.62}$$

We can alternatively write

$$\mathrm{tr}\overline{\overline{\mathsf{A}}}^{(4)} = \det\overline{\overline{\mathsf{A}}} \tag{2.63}$$

as a definition of the determinant of the dyadic $\overline{\overline{\mathsf{A}}}$. In fact, inserting the expansion

$$\mathrm{tr}\overline{\overline{\mathsf{A}}}^{(4)} = \mathrm{tr}\left(\sum A_{ij}\mathbf{e}_i\boldsymbol{\varepsilon}_j\right)^{(4)} = \det[A_{ij}], \tag{2.64}$$

the expression can be shown to equal the determinant of the coefficient matrix in any chosen basis.

2.2.4 Double Contractions $\lfloor\lfloor$ and $\rfloor\rfloor$

We can introduce double contractions $\lfloor\lfloor$ and $\rfloor\rfloor$ between a bidyadic $\mathbf{A}\boldsymbol{\Gamma}$ and a dyadic $\boldsymbol{\alpha}\mathbf{a}$ as

$$(\mathbf{A}\boldsymbol{\Gamma})\lfloor\lfloor(\boldsymbol{\alpha}\mathbf{a}) = (\mathbf{A}\lfloor\boldsymbol{\alpha})(\boldsymbol{\Gamma}\lfloor\mathbf{a}), \tag{2.65}$$

$$(\boldsymbol{\alpha}\mathbf{a})\rfloor\rfloor(\mathbf{A}\boldsymbol{\Gamma}) = (\boldsymbol{\alpha}\rfloor\mathbf{A})(\mathbf{a}\rfloor\boldsymbol{\Gamma}). \tag{2.66}$$

The result belongs to the dyadic space $\mathbb{E}_1\mathbb{F}_1$. More generally, double contractions between a bidyadic $\overline{\overline{\mathsf{A}}} \in \mathbb{E}_2\mathbb{F}_2$ and a dyadic $\overline{\overline{\mathsf{B}}} \in \mathbb{E}_1\mathbb{F}_1$ can be defined as

$$\overline{\overline{\mathsf{A}}} \lfloor\lfloor \overline{\overline{\mathsf{B}}}^T = \sum_{i<j} \mathbf{A}_{ij}\boldsymbol{\Gamma}_{ij} \lfloor\lfloor \sum_k \boldsymbol{\beta}_k \mathbf{b}_k = \sum_{i<j,k} (\mathbf{A}_{ij}\lfloor\boldsymbol{\beta}_k)(\boldsymbol{\Gamma}_{ij}\lfloor\mathbf{b}_k), \tag{2.67}$$

$$\overline{\overline{\mathsf{B}}}^T \rfloor\rfloor \overline{\overline{\mathsf{A}}} = \sum_k \boldsymbol{\beta}_k \mathbf{b}_k \rfloor\rfloor \sum_{i<j} \mathbf{A}_{ij}\boldsymbol{\Gamma}_{ij} = \sum_{i<j,k} (\boldsymbol{\beta}_k\rfloor\mathbf{A}_{ij})(\mathbf{b}_k\rfloor\boldsymbol{\Gamma}_{ij}). \tag{2.68}$$

Still more generally, double contractions can be applied to operate between multidyadics of different orders. For example, a bidyadic $\overline{\overline{\mathsf{D}}}_1 \in \mathbb{E}_2\mathbb{F}_2$ can be mapped to another bidyadic $\overline{\overline{\mathsf{D}}}_2 \in \mathbb{E}_2\mathbb{F}_2$ through double contraction of the quadrivector unit dyadic $\overline{\overline{\mathsf{I}}}^{(4)}$ as

$$\overline{\overline{\mathsf{D}}}_2 = \overline{\overline{\mathsf{I}}}^{(4)} \lfloor\lfloor \overline{\overline{\mathsf{D}}}_1^T = \mathbf{e}_N \boldsymbol{\varepsilon}_N \lfloor\lfloor \sum \sum D_{IJ} \boldsymbol{\varepsilon}_J \mathbf{e}_I = \sum \sum D_{IJ} \mathbf{e}_{K(J)} \boldsymbol{\varepsilon}_{K(I)}. \tag{2.69}$$

2.2.5 Natural Dot Product for Bidyadics

The natural dot product for two bivectors (1.118)

$$\mathbf{A} \cdot \mathbf{B} = \mathbf{A}|(\boldsymbol{\varepsilon}_N \lfloor \mathbf{B}) = \boldsymbol{\varepsilon}_N|(\mathbf{A} \wedge \mathbf{B}) \tag{2.70}$$

and two two-forms (1.119)

$$\boldsymbol{\Phi} \cdot \boldsymbol{\Psi} = \boldsymbol{\Phi}|(\mathbf{e}_N \lfloor \boldsymbol{\Psi}) = \mathbf{e}_N|(\boldsymbol{\Phi} \wedge \boldsymbol{\Psi}) \tag{2.71}$$

can be directly applied to define a dot product for two bidyadics. The dot products are defined by the two metric bidyadics $\overline{\overline{\mathsf{E}}}$ and $\overline{\overline{\mathsf{E}}}^{-1}$ defined by

$$\overline{\overline{\mathsf{E}}} = \mathbf{e}_N \lfloor \overline{\overline{\mathsf{I}}}^{(2)T} \in \mathbb{E}_2\mathbb{E}_2, \qquad \overline{\overline{\mathsf{E}}}^{-1} = \boldsymbol{\varepsilon}_N \lfloor \overline{\overline{\mathsf{I}}}^{(2)} \in \mathbb{F}_2\mathbb{F}_2, \tag{2.72}$$

which are symmetric,

$$\begin{aligned} \overline{\overline{\mathsf{E}}} &= \sum \mathbf{e}_N \lfloor \mathbf{e}_{ij} \boldsymbol{\varepsilon}_{ij} = \sum \boldsymbol{\varepsilon}_{K(ij)} \boldsymbol{\varepsilon}_{ij} \\ &= \mathbf{e}_{12}\mathbf{e}_{34} + \mathbf{e}_{23}\mathbf{e}_{14} + \mathbf{e}_{31}\mathbf{e}_{24} + \mathbf{e}_{14}\mathbf{e}_{23} + \mathbf{e}_{24}\mathbf{e}_{31} + \mathbf{e}_{34}\mathbf{e}_{12} \end{aligned} \tag{2.73}$$

$$\begin{aligned} \overline{\overline{\mathsf{E}}}^{-1} &= \sum \boldsymbol{\varepsilon}_N \lfloor \mathbf{e}_{ij} \boldsymbol{\varepsilon}_{ij} = \sum \boldsymbol{\varepsilon}_{K(ij)} \boldsymbol{\varepsilon}_{ij} \\ &= \boldsymbol{\varepsilon}_{12}\boldsymbol{\varepsilon}_{34} + \boldsymbol{\varepsilon}_{23}\boldsymbol{\varepsilon}_{14} + \boldsymbol{\varepsilon}_{31}\boldsymbol{\varepsilon}_{24} + \boldsymbol{\varepsilon}_{14}\boldsymbol{\varepsilon}_{23} + \boldsymbol{\varepsilon}_{24}\boldsymbol{\varepsilon}_{31} + \boldsymbol{\varepsilon}_{34}\boldsymbol{\varepsilon}_{12}. \end{aligned} \tag{2.74}$$

For two bidyadics $\overline{\overline{\mathsf{C}}}, \overline{\overline{\mathsf{D}}} \in \mathbb{E}_2\mathbb{F}_2$ there exist four possible dot products

$$\overline{\overline{\mathsf{C}}} \cdot \overline{\overline{\mathsf{D}}}^T = \overline{\overline{\mathsf{C}}}|(\mathbf{e}_N \lfloor \overline{\overline{\mathsf{D}}}^T) = \overline{\overline{\mathsf{C}}}|\overline{\overline{\mathsf{E}}}|\overline{\overline{\mathsf{D}}}^T \in \mathbb{E}_2\mathbb{E}_2, \tag{2.75}$$

$$\overline{\overline{\mathsf{D}}} \cdot \overline{\overline{\mathsf{C}}}^T = \overline{\overline{\mathsf{D}}}|(\mathbf{e}_N \lfloor \overline{\overline{\mathsf{C}}}^T) = \overline{\overline{\mathsf{D}}}|\overline{\overline{\mathsf{E}}}|\overline{\overline{\mathsf{C}}}^T \in \mathbb{E}_2\mathbb{E}_2, \tag{2.76}$$

$$\overline{\overline{\mathsf{C}}}^T \cdot \overline{\overline{\mathsf{D}}} = \overline{\overline{\mathsf{C}}}^T|(\boldsymbol{\varepsilon}_N\lfloor\overline{\overline{\mathsf{D}}}) = \overline{\overline{\mathsf{C}}}^T|\overline{\overline{\mathsf{E}}}^{-1}|\overline{\overline{\mathsf{D}}} \in \mathbb{F}_2\mathbb{F}_2 \tag{2.77}$$

$$\overline{\overline{\mathsf{D}}}^T \cdot \overline{\overline{\mathsf{C}}} = \overline{\overline{\mathsf{D}}}^T|(\boldsymbol{\varepsilon}_N\lfloor\overline{\overline{\mathsf{C}}}) = \overline{\overline{\mathsf{D}}}^T|\overline{\overline{\mathsf{E}}}^{-1}|\overline{\overline{\mathsf{C}}} \in \mathbb{F}_2\mathbb{F}_2. \tag{2.78}$$

The dot product of bidyadics has some obvious properties. The symmetric dyadics $\overline{\overline{\mathsf{E}}}$ and $\overline{\overline{\mathsf{E}}}^{-1}$ act as unit mappings in the dot product. For example, for the metric bidyadic $\overline{\overline{\mathsf{G}}} \in \mathbb{E}_2\mathbb{E}_2$ we have

$$\overline{\overline{\mathsf{E}}} \cdot \overline{\overline{\mathsf{G}}} = \overline{\overline{\mathsf{E}}}|\overline{\overline{\mathsf{E}}}^{-1}|\overline{\overline{\mathsf{G}}} = \overline{\overline{\mathsf{G}}}. \tag{2.79}$$

Also, we have for the metric bidyadics $\overline{\overline{\mathsf{G}}}_1, \overline{\overline{\mathsf{G}}}_2$

$$\overline{\overline{\mathsf{G}}}_1^{-1} \cdot \overline{\overline{\mathsf{G}}}_2^{-1} = \overline{\overline{\mathsf{G}}}_1^{-1}|\overline{\overline{\mathsf{E}}}|\overline{\overline{\mathsf{G}}}_2^{-1} = (\overline{\overline{\mathsf{G}}}_2|\overline{\overline{\mathsf{E}}}^{-1}|\overline{\overline{\mathsf{G}}}_1)^{-1} = (\overline{\overline{\mathsf{G}}}_2 \cdot \overline{\overline{\mathsf{G}}}_1)^{-1}. \tag{2.80}$$

Finally, we can easily prove

$$(\boldsymbol{\varepsilon}_N\boldsymbol{\varepsilon}_N\lfloor\lfloor\overline{\overline{\mathsf{G}}}_1) \cdot (\boldsymbol{\varepsilon}_N\boldsymbol{\varepsilon}_N\lfloor\lfloor\overline{\overline{\mathsf{G}}}_2) = \boldsymbol{\varepsilon}_N\boldsymbol{\varepsilon}_N\lfloor\lfloor(\overline{\overline{\mathsf{G}}}_1 \cdot \overline{\overline{\mathsf{G}}}_2). \tag{2.81}$$

Similar rules can be written for other bidyadics in slightly different form.

2.3 DYADIC IDENTITIES

Dyadic identities, equations valid for any dyadics, act as operational rules which can be applied in the analysis of equations encountered in electromagnetics. Use of identities replaces working with basis expansions of quantities and applying their relations. Let us now consider some dyadic identities which can be straightforwardly derived through basis expansions. Proofs of some of them are given here while the rest are left as exercises.

2.3.1 Contraction Identities

Three identities involving contractions of the double-wedge product of two, three and four dyadics form a basis from which special identities can be tailored. All dyadics $\overline{\overline{\mathsf{A}}}, \overline{\overline{\mathsf{B}}}, \overline{\overline{\mathsf{C}}}, \overline{\overline{\mathsf{D}}}$ belong to the space $\mathbb{E}_1\mathbb{F}_1$. The identities are also valid for dyadics taken from the space $\mathbb{F}_1\mathbb{E}_1$ but in this case the unit dyadic $\overline{\overline{\mathsf{I}}}$, wherever present, must be replaced by its transpose $\overline{\overline{\mathsf{I}}}^T$.

The identity

$$(\overline{\overline{\mathsf{A}}}{}^\wedge_\wedge\overline{\overline{\mathsf{B}}})\lfloor\lfloor\overline{\overline{\mathsf{C}}}^T = (\overline{\overline{\mathsf{A}}}||\overline{\overline{\mathsf{C}}}^T)\overline{\overline{\mathsf{B}}} + (\overline{\overline{\mathsf{B}}}||\overline{\overline{\mathsf{C}}}^T)\overline{\overline{\mathsf{A}}} - \overline{\overline{\mathsf{A}}}|\overline{\overline{\mathsf{C}}}|\overline{\overline{\mathsf{B}}} - \overline{\overline{\mathsf{B}}}|\overline{\overline{\mathsf{C}}}|\overline{\overline{\mathsf{A}}}, \tag{2.82}$$

can be derived by replacing the dyadics by three dyadic products of vectors and one-forms and applying the bac-cab rule twice,

$$\begin{aligned}
(\mathbf{a}\boldsymbol{\alpha}{}^\wedge_\wedge\mathbf{b}\boldsymbol{\beta})\lfloor\lfloor(\mathbf{c}\boldsymbol{\gamma})^T &= ((\mathbf{a}\wedge\mathbf{b})\lfloor\boldsymbol{\gamma})((\boldsymbol{\alpha}\wedge\boldsymbol{\beta})\lfloor\mathbf{c}),\\
&= (\mathbf{b}(\mathbf{a}|\boldsymbol{\gamma}) - \mathbf{a}(\mathbf{b}|\boldsymbol{\gamma}))(\boldsymbol{\beta}(\boldsymbol{\alpha}|\mathbf{c}) - \boldsymbol{\alpha}(\boldsymbol{\beta}|\mathbf{c})),\\
&= \mathbf{b}\boldsymbol{\beta}(\mathbf{a}\boldsymbol{\alpha}||\boldsymbol{\gamma}\mathbf{c}) + \mathbf{a}\boldsymbol{\alpha}(\mathbf{b}\boldsymbol{\beta}||\boldsymbol{\gamma}\mathbf{c})\\
&- \mathbf{a}\boldsymbol{\alpha}|\mathbf{c}\boldsymbol{\gamma}|\mathbf{b}\boldsymbol{\beta} - \mathbf{b}\boldsymbol{\beta}|\mathbf{c}\boldsymbol{\gamma}|\mathbf{a}\boldsymbol{\alpha}.
\end{aligned}\tag{2.83}$$

At the last stage the four terms were assembled to produce the original dyadic products. Since the resulting identity is linear in its three dyadic products of vectors and one-forms, replacing the products by general dyadics $\overline{\overline{\mathsf{A}}}, \overline{\overline{\mathsf{B}}}, \overline{\overline{\mathsf{C}}}$ yields the dyadic identity (2.82). The following two identities can be derived similarly with somewhat more effort.

$$\begin{aligned}
(\overline{\overline{\mathsf{A}}}{}^\wedge_\wedge\overline{\overline{\mathsf{B}}}{}^\wedge_\wedge\overline{\overline{\mathsf{C}}})\lfloor\lfloor\overline{\overline{\mathsf{D}}}^T &= (\overline{\overline{\mathsf{A}}}{}^\wedge_\wedge\overline{\overline{\mathsf{B}}})(\overline{\overline{\mathsf{C}}}||\overline{\overline{\mathsf{D}}}^T) + (\overline{\overline{\mathsf{B}}}{}^\wedge_\wedge\overline{\overline{\mathsf{C}}})(\overline{\overline{\mathsf{A}}}||\overline{\overline{\mathsf{D}}}^T) + (\overline{\overline{\mathsf{C}}}{}^\wedge_\wedge\overline{\overline{\mathsf{A}}})(\overline{\overline{\mathsf{B}}}||\overline{\overline{\mathsf{D}}}^T)\\
&- \overline{\overline{\mathsf{A}}}{}^\wedge_\wedge(\overline{\overline{\mathsf{B}}}|\overline{\overline{\mathsf{D}}}|\overline{\overline{\mathsf{C}}} + \overline{\overline{\mathsf{C}}}|\overline{\overline{\mathsf{D}}}|\overline{\overline{\mathsf{B}}}) - \overline{\overline{\mathsf{B}}}{}^\wedge_\wedge(\overline{\overline{\mathsf{C}}}|\overline{\overline{\mathsf{D}}}|\overline{\overline{\mathsf{A}}} + \overline{\overline{\mathsf{A}}}|\overline{\overline{\mathsf{D}}}|\overline{\overline{\mathsf{C}}})\\
&- \overline{\overline{\mathsf{C}}}{}^\wedge_\wedge(\overline{\overline{\mathsf{A}}}|\overline{\overline{\mathsf{D}}}|\overline{\overline{\mathsf{B}}} + \overline{\overline{\mathsf{B}}}|\overline{\overline{\mathsf{D}}}|\overline{\overline{\mathsf{A}}})
\end{aligned}\tag{2.84}$$

$$\begin{aligned}
(\overline{\overline{\mathsf{A}}}{}^\wedge_\wedge\overline{\overline{\mathsf{B}}}{}^\wedge_\wedge\overline{\overline{\mathsf{C}}}{}^\wedge_\wedge\overline{\overline{\mathsf{D}}})\lfloor\lfloor\overline{\overline{\mathsf{E}}}^T &= \mathrm{tr}(\overline{\overline{\mathsf{A}}}{}^\wedge_\wedge\overline{\overline{\mathsf{B}}}{}^\wedge_\wedge\overline{\overline{\mathsf{C}}}{}^\wedge_\wedge\overline{\overline{\mathsf{D}}})(\overline{\overline{\mathsf{I}}}^{(4)}\lfloor\lfloor\overline{\overline{\mathsf{E}}}^T)\\
&= (\overline{\overline{\mathsf{A}}}{}^\wedge_\wedge\overline{\overline{\mathsf{B}}}{}^\wedge_\wedge\overline{\overline{\mathsf{C}}})(\overline{\overline{\mathsf{D}}}||\overline{\overline{\mathsf{E}}}^T) + (\overline{\overline{\mathsf{A}}}{}^\wedge_\wedge\overline{\overline{\mathsf{B}}}{}^\wedge_\wedge\overline{\overline{\mathsf{D}}})(\overline{\overline{\mathsf{C}}}||\overline{\overline{\mathsf{E}}}^T) + (\overline{\overline{\mathsf{A}}}{}^\wedge_\wedge\overline{\overline{\mathsf{C}}}{}^\wedge_\wedge\overline{\overline{\mathsf{D}}})(\overline{\overline{\mathsf{B}}}||\overline{\overline{\mathsf{E}}}^T)\\
&+ (\overline{\overline{\mathsf{B}}}{}^\wedge_\wedge\overline{\overline{\mathsf{C}}}{}^\wedge_\wedge\overline{\overline{\mathsf{D}}})(\overline{\overline{\mathsf{A}}}||\overline{\overline{\mathsf{E}}}^T) - (\overline{\overline{\mathsf{A}}}{}^\wedge_\wedge\overline{\overline{\mathsf{B}}}){}^\wedge_\wedge(\overline{\overline{\mathsf{C}}}|\overline{\overline{\mathsf{E}}}|\overline{\overline{\mathsf{D}}} + \overline{\overline{\mathsf{D}}}|\overline{\overline{\mathsf{E}}}|\overline{\overline{\mathsf{C}}})\\
&- (\overline{\overline{\mathsf{A}}}{}^\wedge_\wedge\overline{\overline{\mathsf{C}}}){}^\wedge_\wedge(\overline{\overline{\mathsf{B}}}|\overline{\overline{\mathsf{E}}}|\overline{\overline{\mathsf{D}}} + \overline{\overline{\mathsf{D}}}|\overline{\overline{\mathsf{E}}}|\overline{\overline{\mathsf{B}}}) - (\overline{\overline{\mathsf{A}}}{}^\wedge_\wedge\overline{\overline{\mathsf{D}}}){}^\wedge_\wedge(\overline{\overline{\mathsf{B}}}|\overline{\overline{\mathsf{E}}}|\overline{\overline{\mathsf{C}}} + \overline{\overline{\mathsf{C}}}|\overline{\overline{\mathsf{E}}}|\overline{\overline{\mathsf{B}}})\\
&- (\overline{\overline{\mathsf{B}}}{}^\wedge_\wedge\overline{\overline{\mathsf{C}}}){}^\wedge_\wedge(\overline{\overline{\mathsf{A}}}|\overline{\overline{\mathsf{E}}}|\overline{\overline{\mathsf{D}}} + \overline{\overline{\mathsf{D}}}|\overline{\overline{\mathsf{E}}}|\overline{\overline{\mathsf{A}}}) - (\overline{\overline{\mathsf{B}}}{}^\wedge_\wedge\overline{\overline{\mathsf{D}}}){}^\wedge_\wedge(\overline{\overline{\mathsf{A}}}|\overline{\overline{\mathsf{E}}}|\overline{\overline{\mathsf{C}}} + \overline{\overline{\mathsf{C}}}|\overline{\overline{\mathsf{E}}}|\overline{\overline{\mathsf{A}}})\\
&- (\overline{\overline{\mathsf{C}}}{}^\wedge_\wedge\overline{\overline{\mathsf{D}}}){}^\wedge_\wedge(\overline{\overline{\mathsf{A}}}|\overline{\overline{\mathsf{E}}}|\overline{\overline{\mathsf{B}}} + \overline{\overline{\mathsf{B}}}|\overline{\overline{\mathsf{E}}}|\overline{\overline{\mathsf{A}}})
\end{aligned}\tag{2.85}$$

2.3.2 Special Cases

It is now easy to derive special cases of the above identities by replacing some of the dyadics by other dyadics. From (2.82) we obtain

$$\overline{\overline{\mathsf{A}}}^{(2)}\lfloor\lfloor\overline{\overline{\mathsf{B}}}^T = (\overline{\overline{\mathsf{A}}}||\overline{\overline{\mathsf{B}}}^T)\overline{\overline{\mathsf{A}}} - \overline{\overline{\mathsf{A}}}|\overline{\overline{\mathsf{B}}}|\overline{\overline{\mathsf{A}}}, \tag{2.86}$$

$$\overline{\overline{\mathsf{I}}}^{(2)}\lfloor\lfloor\overline{\overline{\mathsf{A}}}^T = (\mathrm{tr}\overline{\overline{\mathsf{A}}})\overline{\overline{\mathsf{I}}} - \overline{\overline{\mathsf{A}}}, \tag{2.87}$$

$$\overline{\overline{\mathsf{A}}}^{(2)}\lfloor\lfloor\overline{\overline{\mathsf{I}}}^T = (\mathrm{tr}\overline{\overline{\mathsf{A}}})\overline{\overline{\mathsf{A}}} - \overline{\overline{\mathsf{A}}}^2 = (\overline{\overline{\mathsf{I}}}^{(2)}\lfloor\lfloor\overline{\overline{\mathsf{A}}}^T)|\overline{\overline{\mathsf{A}}}, \tag{2.88}$$

Similarly, from (2.84) we obtain

$$\overline{\overline{\mathsf{A}}}^{(3)}\lfloor\lfloor\overline{\overline{\mathsf{B}}}^T = \mathrm{tr}(\overline{\overline{\mathsf{A}}}|\overline{\overline{\mathsf{B}}})\overline{\overline{\mathsf{A}}}^{(2)} - \overline{\overline{\mathsf{A}}}{}^\wedge_\wedge(\overline{\overline{\mathsf{A}}}|\overline{\overline{\mathsf{B}}}|\overline{\overline{\mathsf{A}}}), \tag{2.89}$$

$$\overline{\overline{\mathsf{I}}}^{(3)}\lfloor\lfloor\overline{\overline{\mathsf{A}}}^T = (\mathrm{tr}\overline{\overline{\mathsf{A}}})\overline{\overline{\mathsf{I}}}^{(2)} - \overline{\overline{\mathsf{I}}}{}^\wedge_\wedge\overline{\overline{\mathsf{A}}}, \tag{2.90}$$

$$\overline{\overline{\mathsf{A}}}^{(3)}\lfloor\lfloor\overline{\overline{\mathsf{I}}}^T = (\mathrm{tr}\overline{\overline{\mathsf{A}}})\overline{\overline{\mathsf{A}}}^{(2)} - \overline{\overline{\mathsf{A}}}{}^\wedge_\wedge\overline{\overline{\mathsf{A}}}^2 = (\overline{\overline{\mathsf{I}}}^{(3)}\lfloor\lfloor\overline{\overline{\mathsf{A}}}^T)|\overline{\overline{\mathsf{A}}}^{(2)}, \tag{2.91}$$

and, from (2.85),

$$\overline{\overline{\mathsf{A}}}^{(4)}\lfloor\lfloor\overline{\overline{\mathsf{B}}}^T = (\overline{\overline{\mathsf{A}}}||\overline{\overline{\mathsf{B}}}^T)\overline{\overline{\mathsf{A}}}^{(3)} - \overline{\overline{\mathsf{A}}}^{(2)}{}^\wedge_\wedge(\overline{\overline{\mathsf{A}}}|\overline{\overline{\mathsf{B}}}|\overline{\overline{\mathsf{A}}}), \tag{2.92}$$

$$\overline{\overline{\mathsf{I}}}^{(4)}\lfloor\lfloor\overline{\overline{\mathsf{A}}}^T = (\mathrm{tr}\overline{\overline{\mathsf{A}}})\overline{\overline{\mathsf{I}}}^{(3)} - \overline{\overline{\mathsf{I}}}^{(2)}{}^\wedge_\wedge\overline{\overline{\mathsf{A}}}, \tag{2.93}$$

$$\overline{\overline{\mathsf{A}}}^{(4)}\lfloor\lfloor\overline{\overline{\mathsf{I}}}^T = (\mathrm{tr}\overline{\overline{\mathsf{A}}})\overline{\overline{\mathsf{A}}}^{(3)} - \overline{\overline{\mathsf{A}}}^{(2)}{}^\wedge_\wedge\overline{\overline{\mathsf{A}}}^2 = (\overline{\overline{\mathsf{I}}}^{(4)}\lfloor\lfloor\overline{\overline{\mathsf{A}}}^T)|\overline{\overline{\mathsf{A}}}^{(3)}. \tag{2.94}$$

The unit dyadic satisfies the relations

$$\overline{\overline{\mathsf{I}}}^{(4)}\lfloor\lfloor\overline{\overline{\mathsf{I}}}^{(3)T} = \overline{\overline{\mathsf{I}}}, \quad \overline{\overline{\mathsf{I}}}^{(4)}\lfloor\lfloor\overline{\overline{\mathsf{I}}}^{(2)T} = \overline{\overline{\mathsf{I}}}^{(2)}, \quad \overline{\overline{\mathsf{I}}}^{(4)}\lfloor\lfloor\overline{\overline{\mathsf{I}}} = \overline{\overline{\mathsf{I}}}^{(3)}, \tag{2.95}$$

$$\overline{\overline{\mathsf{I}}}^{(3)}\lfloor\lfloor\overline{\overline{\mathsf{I}}}^{(2)T} = \overline{\overline{\mathsf{I}}}^{(2)}\lfloor\lfloor\overline{\overline{\mathsf{I}}}^T = 3\overline{\overline{\mathsf{I}}}. \tag{2.96}$$

Further identities are obtained by contracting previous identities by yet another dyadic and using other identities to expand the result.

$$\begin{aligned}
\mathrm{tr}(\overline{\overline{\mathsf{A}}}{}^\wedge_\wedge\overline{\overline{\mathsf{B}}}) &= \overline{\overline{\mathsf{I}}}^{(2)}||(\overline{\overline{\mathsf{A}}}{}^\wedge_\wedge\overline{\overline{\mathsf{B}}})^T = (\overline{\overline{\mathsf{I}}}^{(2)}\lfloor\lfloor\overline{\overline{\mathsf{A}}}^T)||\overline{\overline{\mathsf{B}}}^T, \\
&= (\mathrm{tr}\overline{\overline{\mathsf{A}}}\,\overline{\overline{\mathsf{I}}} - \overline{\overline{\mathsf{A}}})||\overline{\overline{\mathsf{B}}}^T = \mathrm{tr}\overline{\overline{\mathsf{A}}}\;\mathrm{tr}\overline{\overline{\mathsf{B}}} - \mathrm{tr}(\overline{\overline{\mathsf{A}}}|\overline{\overline{\mathsf{B}}}),
\end{aligned} \tag{2.97}$$

$$\begin{aligned}
\overline{\overline{\mathsf{I}}}^{(3)}\lfloor\lfloor(\overline{\overline{\mathsf{A}}}{}^\wedge_\wedge\overline{\overline{\mathsf{B}}})^T &= (\overline{\overline{\mathsf{I}}}^{(3)}\lfloor\lfloor\overline{\overline{\mathsf{A}}}^T)\lfloor\lfloor\overline{\overline{\mathsf{B}}}^T = (\mathrm{tr}\overline{\overline{\mathsf{A}}}\,\overline{\overline{\mathsf{I}}}^{(2)} - \overline{\overline{\mathsf{I}}}{}^\wedge_\wedge\overline{\overline{\mathsf{A}}})\lfloor\lfloor\overline{\overline{\mathsf{B}}}^T, \\
&= (\mathrm{tr}\overline{\overline{\mathsf{A}}}\;\mathrm{tr}\overline{\overline{\mathsf{B}}} - \mathrm{tr}(\overline{\overline{\mathsf{A}}}|\overline{\overline{\mathsf{B}}}))\overline{\overline{\mathsf{I}}} - (\mathrm{tr}\overline{\overline{\mathsf{A}}})\overline{\overline{\mathsf{B}}} \\
&\quad - (\mathrm{tr}\overline{\overline{\mathsf{B}}})\overline{\overline{\mathsf{A}}} + \overline{\overline{\mathsf{A}}}|\overline{\overline{\mathsf{B}}} + \overline{\overline{\mathsf{B}}}|\overline{\overline{\mathsf{A}}}, \\
&= \mathrm{tr}(\overline{\overline{\mathsf{A}}}{}^\wedge_\wedge\overline{\overline{\mathsf{B}}})\overline{\overline{\mathsf{I}}} - (\overline{\overline{\mathsf{A}}}{}^\wedge_\wedge\overline{\overline{\mathsf{B}}})\lfloor\lfloor\overline{\overline{\mathsf{I}}}^T.
\end{aligned} \tag{2.98}$$

$$\begin{aligned}
\overline{\overline{\mathsf{A}}}^{(3)}\lfloor\lfloor\overline{\overline{\mathsf{B}}}^{(2)T} &= \mathrm{tr}(\overline{\overline{\mathsf{A}}}^{(2)}|\overline{\overline{\mathsf{B}}}^{(2)})\overline{\overline{\mathsf{A}}} - \mathrm{tr}(\overline{\overline{\mathsf{A}}}|\overline{\overline{\mathsf{B}}})\overline{\overline{\mathsf{A}}}|\overline{\overline{\mathsf{B}}}|\overline{\overline{\mathsf{A}}} + \overline{\overline{\mathsf{A}}}|\overline{\overline{\mathsf{B}}}|\overline{\overline{\mathsf{A}}}|\overline{\overline{\mathsf{B}}}|\overline{\overline{\mathsf{A}}} \\
&= \mathrm{tr}(\overline{\overline{\mathsf{A}}}^{(2)}|\overline{\overline{\mathsf{B}}}^{(2)})\overline{\overline{\mathsf{A}}} - \overline{\overline{\mathsf{A}}}|(\overline{\overline{\mathsf{B}}}^{(2)}\lfloor\lfloor\overline{\overline{\mathsf{A}}}^T)|\overline{\overline{\mathsf{A}}}
\end{aligned} \tag{2.99}$$

$$\overline{\overline{\mathsf{A}}}^{(3)}\lfloor\lfloor\overline{\overline{\mathsf{I}}}^{(2)T} = \mathrm{tr}(\overline{\overline{\mathsf{A}}}^{(2)})\overline{\overline{\mathsf{A}}} - (\mathrm{tr}\overline{\overline{\mathsf{A}}})\overline{\overline{\mathsf{A}}}^2 + \overline{\overline{\mathsf{A}}}^3, \tag{2.100}$$

$$\overline{\overline{\mathsf{I}}}^{(3)}\lfloor\lfloor\overline{\overline{\mathsf{A}}}^{(2)T} = (\mathrm{tr}\overline{\overline{\mathsf{A}}}^{(2)})\overline{\overline{\mathsf{I}}} - (\mathrm{tr}\overline{\overline{\mathsf{A}}})\overline{\overline{\mathsf{A}}} + \overline{\overline{\mathsf{A}}}^2, \tag{2.101}$$

$$\overline{\overline{\mathsf{I}}}^{(4)}\lfloor\lfloor(\overline{\overline{\mathsf{A}}}{}^\wedge_\wedge\overline{\overline{\mathsf{B}}})^T = \mathrm{tr}(\overline{\overline{\mathsf{A}}}{}^\wedge_\wedge\overline{\overline{\mathsf{B}}})\overline{\overline{\mathsf{I}}}^{(2)} - ((\overline{\overline{\mathsf{A}}}{}^\wedge_\wedge\overline{\overline{\mathsf{B}}})\lfloor\lfloor\overline{\overline{\mathsf{I}}}^T){}^\wedge_\wedge\overline{\overline{\mathsf{I}}} + \overline{\overline{\mathsf{A}}}{}^\wedge_\wedge\overline{\overline{\mathsf{B}}}, \tag{2.102}$$

$$\overline{\overline{\mathsf{I}}}^{(4)}\lfloor\lfloor(\overline{\overline{\mathsf{A}}}{}^\wedge_\wedge\overline{\overline{\mathsf{B}}}{}^\wedge_\wedge\overline{\overline{\mathsf{C}}})^T = \mathrm{tr}(\overline{\overline{\mathsf{A}}}{}^\wedge_\wedge\overline{\overline{\mathsf{B}}}{}^\wedge_\wedge\overline{\overline{\mathsf{C}}})\overline{\overline{\mathsf{I}}} - (\overline{\overline{\mathsf{A}}}{}^\wedge_\wedge\overline{\overline{\mathsf{B}}}{}^\wedge_\wedge\overline{\overline{\mathsf{C}}})\lfloor\lfloor\overline{\overline{\mathsf{I}}}^{(2)T}. \tag{2.103}$$

These have the following simple interrelations

$$\overline{\overline{\mathsf{A}}}^{(4)}\lfloor\lfloor\overline{\overline{\mathsf{I}}}^{(3)T} = (\overline{\overline{\mathsf{I}}}^{(4)}\lfloor\lfloor\overline{\overline{\mathsf{A}}}^{(3)T})|\overline{\overline{\mathsf{A}}} = (\mathrm{tr}\overline{\overline{\mathsf{A}}}^{(4)})\overline{\overline{\mathsf{I}}}, \tag{2.104}$$

$$\overline{\overline{\mathsf{I}}}^{(4)}\lfloor\lfloor\overline{\overline{\mathsf{A}}}^{(3)T} = (\mathrm{tr}\overline{\overline{\mathsf{A}}}^{(3)})\overline{\overline{\mathsf{I}}} - \overline{\overline{\mathsf{A}}}^{(3)}\lfloor\lfloor\overline{\overline{\mathsf{I}}}^{(2)T}, \tag{2.105}$$

$$\overline{\overline{\mathsf{A}}}^{(3)}\lfloor\lfloor\overline{\overline{\mathsf{I}}}^{(2)T} = (\overline{\overline{\mathsf{I}}}^{(3)}\lfloor\lfloor\overline{\overline{\mathsf{A}}}^{(2)T})|\overline{\overline{\mathsf{A}}}, \tag{2.106}$$

$$\overline{\overline{\mathsf{I}}}^{(3)}\lfloor\lfloor\overline{\overline{\mathsf{A}}}^{(2)T} = (\mathrm{tr}\overline{\overline{\mathsf{A}}}^{(2)})\overline{\overline{\mathsf{I}}} - \overline{\overline{\mathsf{A}}}^{(2)}\lfloor\lfloor\overline{\overline{\mathsf{I}}}^T, \tag{2.107}$$

$$\overline{\overline{\mathsf{A}}}^{(2)}\lfloor\lfloor\overline{\overline{\mathsf{I}}}^T = (\overline{\overline{\mathsf{I}}}^{(2)}\lfloor\lfloor\overline{\overline{\mathsf{A}}}^T)|\overline{\overline{\mathsf{A}}}, \tag{2.108}$$

$$\overline{\overline{\mathsf{I}}}^{(2)}\lfloor\lfloor\overline{\overline{\mathsf{A}}}^T = (\mathrm{tr}\overline{\overline{\mathsf{A}}})\overline{\overline{\mathsf{I}}} - \overline{\overline{\mathsf{A}}}, \tag{2.109}$$

$$\overline{\overline{\mathsf{A}}}^{(4)}\lfloor\lfloor\overline{\overline{\mathsf{I}}}^{(2)T} = (\overline{\overline{\mathsf{I}}}^{(4)}\lfloor\lfloor\overline{\overline{\mathsf{A}}}^{(2)T})|\overline{\overline{\mathsf{A}}}^{(2)}. \tag{2.110}$$

Traces of different double-wedge powers of dyadics can be expanded as

$$\mathrm{tr}\overline{\overline{\mathsf{A}}}^{(2)} = \frac{1}{2}((\mathrm{tr}\overline{\overline{\mathsf{A}}})^2 - \mathrm{tr}\overline{\overline{\mathsf{A}}}^2), \tag{2.111}$$

$$\mathrm{tr}\overline{\overline{\mathsf{A}}}^{(3)} = \frac{1}{6}((\mathrm{tr}\overline{\overline{\mathsf{A}}})^3 - 3\mathrm{tr}\overline{\overline{\mathsf{A}}}\mathrm{tr}\overline{\overline{\mathsf{A}}}^2 + 2\mathrm{tr}\overline{\overline{\mathsf{A}}}^3), \tag{2.112}$$

and

$$\mathrm{tr}\overline{\overline{\mathsf{A}}}^{(4)} = \frac{1}{24}((\mathrm{tr}\overline{\overline{\mathsf{A}}})^4 - 6(\mathrm{tr}\overline{\overline{\mathsf{A}}})^2\mathrm{tr}\overline{\overline{\mathsf{A}}}^2 + 8\mathrm{tr}\overline{\overline{\mathsf{A}}}\mathrm{tr}\overline{\overline{\mathsf{A}}}^3 + 3(\mathrm{tr}\overline{\overline{\mathsf{A}}}^2)^2 - 6\mathrm{tr}\overline{\overline{\mathsf{A}}}^4). \tag{2.113}$$

2.3.3 More General Rules

The above identities are valid for dyadics in the space $\mathbb{E}_1, \mathbb{F}_1$. It is possible to generalize some of them to involve bidyadics. For example, since the identity (2.102) is linear in the bidyadic $\overline{\overline{\mathsf{A}}}{}^\wedge_\wedge\overline{\overline{\mathsf{B}}}$ we can replace it by an arbitrary bidyadic $\overline{\overline{\mathsf{C}}}$ and arrive at the more general rule

$$\overline{\overline{\mathsf{I}}}^{(4)}\lfloor\lfloor\overline{\overline{\mathsf{C}}}^T = \mathrm{tr}\overline{\overline{\mathsf{C}}}\,\overline{\overline{\mathsf{I}}}^{(2)} - (\overline{\overline{\mathsf{C}}}\lfloor\lfloor\overline{\overline{\mathsf{I}}}^T){}^\wedge_\wedge\overline{\overline{\mathsf{I}}} + \overline{\overline{\mathsf{C}}}, \quad \overline{\overline{\mathsf{C}}} \in \mathbb{E}_2\mathbb{F}_2 \tag{2.114}$$

Similarly, from (2.98) we obtain

$$\overline{\overline{\mathsf{I}}}^{(3)}\lfloor\lfloor\overline{\overline{\mathsf{C}}}^T = \mathrm{tr}\overline{\overline{\mathsf{C}}}\,\overline{\overline{\mathsf{I}}} - \overline{\overline{\mathsf{C}}}\lfloor\lfloor\overline{\overline{\mathsf{I}}}^T, \quad \overline{\overline{\mathsf{C}}} \in \mathbb{E}_2\mathbb{F}_2 \tag{2.115}$$

and (2.99) has the more general form

$$\overline{\overline{\mathsf{A}}}^{(3)}\lfloor\lfloor\overline{\overline{\mathsf{C}}}^T = \mathrm{tr}(\overline{\overline{\mathsf{A}}}^{(2)}|\overline{\overline{\mathsf{C}}})\overline{\overline{\mathsf{A}}} - \overline{\overline{\mathsf{A}}}|(\overline{\overline{\mathsf{C}}}\lfloor\lfloor\overline{\overline{\mathsf{A}}}^T)|\overline{\overline{\mathsf{A}}}, \quad \overline{\overline{\mathsf{C}}} \in \mathbb{E}_2\mathbb{F}_2. \tag{2.116}$$

This last result is based on the fact that an identity of the form $f(\overline{\overline{\mathsf{B}}}^{(2)}) = 0$ valid for any dyadic $\overline{\overline{\mathsf{B}}} \in \mathbb{E}_1\mathbb{F}_1$ where $f()$ is a linear function of a bidyadic, can be generalized to the form $f(\overline{\overline{\mathsf{C}}}) = 0$ for any bidyadic $\overline{\overline{\mathsf{C}}} \in \mathbb{E}_2\mathbb{F}_2$. The proof is left as an exercise.

Finally, we can show that the identity (2.103) can be generalized to any tridyadic $\overline{\overline{\mathsf{T}}}$ as

$$\overline{\overline{\mathsf{I}}}^{(4)}\lfloor\lfloor\overline{\overline{\mathsf{T}}}^T = \mathrm{tr}\overline{\overline{\mathsf{T}}}\,\overline{\overline{\mathsf{I}}} - \overline{\overline{\mathsf{T}}}\lfloor\lfloor\overline{\overline{\mathsf{I}}}^{(2)T}, \quad \overline{\overline{\mathsf{T}}} \in \mathbb{E}_3\mathbb{F}_3. \tag{2.117}$$

However, since tridyadics can be reduced to dyadics $\overline{\overline{\mathsf{A}}} \in \mathbb{E}_1\mathbb{F}_1$ as

$$\overline{\overline{\mathsf{T}}} = \overline{\overline{\mathsf{I}}}^{(4)}\lfloor\lfloor\overline{\overline{\mathsf{A}}}^T, \quad \overline{\overline{\mathsf{A}}} = \overline{\overline{\mathsf{I}}}^{(4)}\lfloor\lfloor\overline{\overline{\mathsf{T}}}^T, \tag{2.118}$$

(2.117) follows from (2.87).

2.3.4 Cayley–Hamilton Equation

Eliminating the contraction terms stepwise from (2.104)–(2.109) we obtain an important identity:

$$\begin{aligned}
(\mathrm{tr}\overline{\overline{\mathsf{A}}}^{(4)})\overline{\overline{\mathsf{I}}} &= (\overline{\overline{\mathsf{I}}}^{(4)}\lfloor\lfloor\overline{\overline{\mathsf{A}}}^{(3)T})\overline{\overline{\mathsf{A}}}\\
&= (\mathrm{tr}\overline{\overline{\mathsf{A}}}^{(3)})\overline{\overline{\mathsf{A}}} - (\overline{\overline{\mathsf{A}}}^{(3)}\lfloor\lfloor\overline{\overline{\mathsf{I}}}^{(2)T})|\overline{\overline{\mathsf{A}}}\\
&= (\mathrm{tr}\overline{\overline{\mathsf{A}}}^{(3)})\overline{\overline{\mathsf{A}}} - (\overline{\overline{\mathsf{I}}}^{(3)}\lfloor\lfloor\overline{\overline{\mathsf{A}}}^{(2)T})|\overline{\overline{\mathsf{A}}}^2\\
&= (\mathrm{tr}\overline{\overline{\mathsf{A}}}^{(3)})\overline{\overline{\mathsf{A}}} - (\mathrm{tr}\overline{\overline{\mathsf{A}}}^{(2)})\overline{\overline{\mathsf{A}}}^2 + (\overline{\overline{\mathsf{A}}}^{(2)}\lfloor\lfloor\overline{\overline{\mathsf{I}}}^{T})|\overline{\overline{\mathsf{A}}}^2\\
&= (\mathrm{tr}\overline{\overline{\mathsf{A}}}^{(3)})\overline{\overline{\mathsf{A}}} - (\mathrm{tr}\overline{\overline{\mathsf{A}}}^{(2)})\overline{\overline{\mathsf{A}}}^2 + (\overline{\overline{\mathsf{I}}}^{(2)}\lfloor\lfloor\overline{\overline{\mathsf{A}}}^{T})\overline{\overline{\mathsf{A}}}^3\\
&= (\mathrm{tr}\overline{\overline{\mathsf{A}}}^{(3)})\overline{\overline{\mathsf{A}}} - (\mathrm{tr}\overline{\overline{\mathsf{A}}}^{(2)})\overline{\overline{\mathsf{A}}}^2 + (\mathrm{tr}\overline{\overline{\mathsf{A}}})\overline{\overline{\mathsf{A}}}^3 - \overline{\overline{\mathsf{A}}}^4,
\end{aligned}$$

which can be written as

$$\overline{\overline{\mathsf{A}}}^4 - (\mathrm{tr}\overline{\overline{\mathsf{A}}})\overline{\overline{\mathsf{A}}}^3 + (\mathrm{tr}\overline{\overline{\mathsf{A}}}^{(2)})\overline{\overline{\mathsf{A}}}^2 - (\mathrm{tr}\overline{\overline{\mathsf{A}}}^{(3)})\overline{\overline{\mathsf{A}}} + (\mathrm{tr}\overline{\overline{\mathsf{A}}}^{(4)})\overline{\overline{\mathsf{I}}} = 0. \qquad (2.119)$$

This is called the Cayley–Hamilton equation and it is satisfied by any dyadic $\overline{\overline{\mathsf{A}}} \in \mathbb{E}_1\mathbb{F}_1$ operating in the 4D vector space.

The Cayley–Hamilton equation was shown to be valid for 2×2 matrices by Cayley in 1858 in his "Memoir on the theory of matrices." He also mentioned having proved it for 3×3 matrices and assumed that it would be valid for $n \times n$ matrices as well. Hamilton had given a version of the same equation for quaternions in 1853 in his "Lectures on quaternions." The proof for a more general form in terms of $n \times n$ matrices was given by Frobenius in 1878.

2.3.5 Inverse Dyadics

For dyadics $\overline{\overline{\mathsf{A}}} \in \mathbb{E}_1\mathbb{F}_1$ satisfying $\mathrm{tr}\overline{\overline{\mathsf{A}}}^{(4)} = \det\overline{\overline{\mathsf{A}}} \neq 0$, the following compact analytic rule for the inverse $\overline{\overline{\mathsf{A}}}^{-1} \in \mathbb{E}_1\mathbb{F}_1$ is obtained from the identity (2.104):

$$\overline{\overline{\mathsf{A}}}^{-1} = \frac{\overline{\overline{\mathsf{I}}}^{(4)}\lfloor\lfloor\overline{\overline{\mathsf{A}}}^{(3)T}}{\mathrm{tr}\overline{\overline{\mathsf{A}}}^{(4)}}. \qquad (2.120)$$

Similar expression is valid for dyadics of the space $\mathbb{F}_1\mathbb{E}_1$ when $\overline{\overline{\mathsf{I}}}^{(4)}$ is replaced by $\overline{\overline{\mathsf{I}}}^{(4)T}$. Multiplying (2.119) by $\overline{\overline{\mathsf{A}}}^{-1}|$ we obtain an expansion for the inverse dyadic (2.120),

$$\overline{\overline{\mathsf{A}}}^{-1} = \frac{-1}{\mathrm{tr}\overline{\overline{\mathsf{A}}}^{(4)}}[\overline{\overline{\mathsf{A}}}^3 - (\mathrm{tr}\overline{\overline{\mathsf{A}}})\overline{\overline{\mathsf{A}}}^2 + (\mathrm{tr}\overline{\overline{\mathsf{A}}}^{(2)})\overline{\overline{\mathsf{A}}} - (\mathrm{tr}\overline{\overline{\mathsf{A}}}^{(3)})\overline{\overline{\mathsf{I}}}]. \qquad (2.121)$$

Applying $\overline{\overline{\mathsf{A}}}^{(4)} = (\mathrm{tr}\overline{\overline{\mathsf{A}}}^{(4)})\overline{\overline{\mathsf{I}}}^{(4)}$, the identity (2.110) can be expressed as

$$(\mathrm{tr}\overline{\overline{\mathsf{A}}}^{(4)})\overline{\overline{\mathsf{I}}}^{(2)} = (\overline{\overline{\mathsf{I}}}^{(4)}\lfloor\lfloor\overline{\overline{\mathsf{A}}}^{(2)T})|\overline{\overline{\mathsf{A}}}^{(2)}, \tag{2.122}$$

which is valid for any dyadic $\overline{\overline{\mathsf{A}}} \in \mathbb{E}_1\mathbb{F}_1$. For $\mathrm{tr}\overline{\overline{\mathsf{A}}}^{(4)} \neq 0$, we obtain the second inverse formula

$$\overline{\overline{\mathsf{A}}}^{(-2)} = (\overline{\overline{\mathsf{A}}}^{(2)})^{-1} = (\overline{\overline{\mathsf{A}}}^{-1})^{(2)} = \frac{\overline{\overline{\mathsf{I}}}^{(4)}\lfloor\lfloor\overline{\overline{\mathsf{A}}}^{(2)T}}{\mathrm{tr}\overline{\overline{\mathsf{A}}}^{(4)}}, \tag{2.123}$$

where we have applied the convention (2.57). This can be inverted as

$$\overline{\overline{\mathsf{A}}}^{(2)} = \overline{\overline{\mathsf{I}}}^{(4)}\lfloor\lfloor(\overline{\overline{\mathsf{I}}}^{(4)T}\lfloor\lfloor\overline{\overline{\mathsf{A}}}^{(2)}) = (\mathrm{tr}\overline{\overline{\mathsf{A}}}^{(4)})\overline{\overline{\mathsf{I}}}^{(4)}\lfloor\lfloor\overline{\overline{\mathsf{A}}}^{(-2)T}, \tag{2.124}$$

which also follows from (2.123) when replacing $\overline{\overline{\mathsf{A}}}$ by $\overline{\overline{\mathsf{A}}}^{-1}$ and applying

$$\mathrm{tr}\overline{\overline{\mathsf{A}}}^{(-4)} = \frac{1}{\mathrm{tr}\overline{\overline{\mathsf{A}}}^{(4)}}. \tag{2.125}$$

Finally, (2.94) can be written as

$$(\mathrm{tr}\overline{\overline{\mathsf{A}}}^{(4)})\overline{\overline{\mathsf{I}}}^{(3)} = (\overline{\overline{\mathsf{I}}}^{(4)}\lfloor\lfloor\overline{\overline{\mathsf{A}}}^{T})|\overline{\overline{\mathsf{A}}}^{(3)}, \tag{2.126}$$

which yields the third inverse formula

$$\overline{\overline{\mathsf{A}}}^{(-3)} = \frac{\overline{\overline{\mathsf{I}}}^{(4)}\lfloor\lfloor\overline{\overline{\mathsf{A}}}^{T}}{\mathrm{tr}\overline{\overline{\mathsf{A}}}^{(4)}}. \tag{2.127}$$

This can also be obtained from (2.120) when replacing $\overline{\overline{\mathsf{A}}}$ by $\overline{\overline{\mathsf{A}}}^{-1}$.

The previous inverse rules can be given an easily memorizable form in terms of

$$\overline{\overline{\mathsf{A}}}^{(-4)} = \mathrm{tr}\overline{\overline{\mathsf{A}}}^{(-4)}\,\overline{\overline{\mathsf{I}}}^{(4)} = \frac{1}{\mathrm{tr}\overline{\overline{\mathsf{A}}}^{(4)}}\overline{\overline{\mathsf{I}}}^{(4)}, \tag{2.128}$$

as

$$\overline{\overline{\mathsf{A}}}^{(-p)} = \overline{\overline{\mathsf{A}}}^{(-4)}\lfloor\lfloor\overline{\overline{\mathsf{A}}}^{(4-p)T}, \tag{2.129}$$

valid for $0 < p < 4$. Similarly, we can write

$$\overline{\overline{\mathsf{A}}}^{(p)} = \overline{\overline{\mathsf{A}}}^{(4)}\lfloor\lfloor\overline{\overline{\mathsf{A}}}^{(p-4)T}. \tag{2.130}$$

Since a tridyadic $\overline{\overline{\overline{\mathsf{T}}}}$ can be expressed in terms of a dyadic $\overline{\overline{\mathsf{A}}}$ as in (2.118), one can show that the inverses satisfy the relation

$$\overline{\overline{\overline{\mathsf{T}}}}^{-1} = \overline{\overline{\mathsf{I}}}^{(4)}\lfloor\lfloor\overline{\overline{\mathsf{A}}}^{-1T}, \tag{2.131}$$

so that the inverse of the general tridyadic can be expressed in a compact analytic form as

$$\overline{\overline{\mathsf{T}}}^{-1} = \overline{\overline{\mathsf{I}}}^{(4)} \lfloor\lfloor \frac{\overline{\overline{\mathsf{I}}}^{(4)} \lfloor\lfloor \overline{\overline{\mathsf{A}}}^{(3)T}}{\mathrm{tr}\overline{\overline{\mathsf{A}}}^{(4)}} = \frac{(\overline{\overline{\mathsf{I}}}^{(4)} \lfloor\lfloor \overline{\overline{\mathsf{T}}}^{T})^{(3)}}{\mathrm{tr}(\overline{\overline{\mathsf{I}}}^{(4)} \lfloor\lfloor \overline{\overline{\mathsf{T}}}^{T})^{(4)}}. \tag{2.132}$$

There does not seem to exist a simple analytic inverse rule for the general bidyadic. Special cases will be discussed in Chapter 3.

Since any dyadic $\overline{\overline{\mathsf{A}}} \in \mathbb{E}_1\mathbb{F}_1$ can be expanded in terms of spatial quantities $\overline{\overline{\mathsf{A}}}_s, \mathbf{a}_s, \boldsymbol{\alpha}_s$ as

$$\overline{\overline{\mathsf{A}}} = \overline{\overline{\mathsf{A}}}_s + \mathbf{a}_s\boldsymbol{\varepsilon}_4 + \mathbf{e}_4\boldsymbol{\alpha}_s + a\mathbf{e}_4\boldsymbol{\varepsilon}_4, \tag{2.133}$$

the following expression for the inverse can be easily derived by solving a set of spatial linear equations,

$$\overline{\overline{\mathsf{A}}}^{-1} = \overline{\overline{\mathsf{A}}}_s^{-1} + \frac{(\overline{\overline{\mathsf{A}}}_s^{-1}|\mathbf{a}_s - \mathbf{e}_4)(\boldsymbol{\alpha}_s|\overline{\overline{\mathsf{A}}}_s^{-1} - \boldsymbol{\varepsilon}_4)}{a - \boldsymbol{\alpha}_s|\overline{\overline{\mathsf{A}}}_s^{-1}|\mathbf{a}_s}. \tag{2.134}$$

This expansion requires that the spatial inverse of $\overline{\overline{\mathsf{A}}}_s$ exists and that $a - \boldsymbol{\alpha}_s|\overline{\overline{\mathsf{A}}}_s^{-1}|\mathbf{a}_s$ does not vanish. The spatial inverse $\overline{\overline{\mathsf{A}}}_s^{-1}$ is defined as a dyadic satisfying

$$\overline{\overline{\mathsf{A}}}_s|\overline{\overline{\mathsf{A}}}_s^{-1} = \overline{\overline{\mathsf{A}}}_s^{-1}|\overline{\overline{\mathsf{A}}}_s = \overline{\overline{\mathsf{I}}}_s = \overline{\overline{\mathsf{I}}} - \mathbf{e}_4\boldsymbol{\varepsilon}_4. \tag{2.135}$$

One can show that the spatial inverse exists whenever $\mathrm{tr}\overline{\overline{\mathsf{A}}}_s^{(3)}$ does not vanish and it obeys the rule

$$\overline{\overline{\mathsf{A}}}_s^{-1} = \frac{1}{\mathrm{tr}\overline{\overline{\mathsf{A}}}_s^{(3)}}\overline{\overline{\mathsf{I}}}_s^{(3)} \lfloor\lfloor \overline{\overline{\mathsf{A}}}_s^{(2)T} = \overline{\overline{\mathsf{A}}}_s^{(-3)} \lfloor\lfloor \overline{\overline{\mathsf{A}}}_s^{(2)T}. \tag{2.136}$$

Similarly, the inverse of a spatial bidyadic $\overline{\overline{\mathsf{C}}}_s \in \mathbb{E}_2\mathbb{F}_2$ can be expressed in the form

$$\overline{\overline{\mathsf{C}}}_s^{-1} = \frac{1}{\Delta_{C_s}}(\overline{\overline{\mathsf{I}}}_s^{(3)} \lfloor\lfloor \overline{\overline{\mathsf{C}}}_s^{T})^{(2)}, \tag{2.137}$$

with

$$\Delta_{C_s} = \mathrm{tr}(\overline{\overline{\mathsf{I}}}_s^{(3)} \lfloor\lfloor \overline{\overline{\mathsf{C}}}_s^{T})^{(3)}. \tag{2.138}$$

Derivation of (2.136) and (2.137) are left as exercises.

2.4 RANK OF DYADICS

Any dyadic $\overline{\overline{\mathsf{A}}} \in \mathbb{E}_1\mathbb{F}_1$ operating in the 4D vector space can be represented in the form

$$\overline{\overline{\mathsf{A}}} = \mathbf{a}_1\boldsymbol{\alpha}_1 + \mathbf{a}_2\boldsymbol{\alpha}_2 + \mathbf{a}_3\boldsymbol{\alpha}_3 + \mathbf{a}_4\boldsymbol{\alpha}_4, \tag{2.139}$$

in terms of some vectors $\mathbf{a}_i$ and one-forms $\boldsymbol{\alpha}_i$. Choosing either one of the sets $\{\mathbf{a}_i\}$ or $\{\boldsymbol{\alpha}_i\}$ as a basis, the other set defines the dyadic uniquely. Dyadics can be classified in terms of their rank which equals the dimension of the space of the vectors $\mathbf{b}$ obtained through the mapping

$$\mathbf{b} = \overline{\overline{\mathsf{A}}}|\mathbf{a}, \tag{2.140}$$

when $\mathbf{a}$ goes through all possible vectors.

Dyadics of Rank 4

A dyadic $\overline{\overline{\mathsf{A}}}$ is of full rank 4 when it cannot be represented with less than four terms as (2.139). This requires that the four vectors and one-forms $\{\mathbf{a}_i\}$ and $\{\boldsymbol{\alpha}_i\}$ in (2.139) each form a basis, or, what is equivalent,

$$\overline{\overline{\mathsf{A}}}^{(4)} = (\mathbf{a}_1 \wedge \mathbf{a}_2 \wedge \mathbf{a}_3 \wedge \mathbf{a}_4)(\boldsymbol{\alpha}_1 \wedge \boldsymbol{\alpha}_2 \wedge \boldsymbol{\alpha}_3 \wedge \boldsymbol{\alpha}_4) \neq 0. \tag{2.141}$$

A dyadic of rank 4 has an inverse as expressed by (2.120). From

$$\mathrm{tr}\overline{\overline{\mathsf{A}}}^{(4)} = \boldsymbol{\varepsilon}_N|(\mathbf{a}_1 \wedge \mathbf{a}_2 \wedge \mathbf{a}_3 \wedge \mathbf{a}_4)\mathbf{e}_N|(\boldsymbol{\alpha}_1 \wedge \boldsymbol{\alpha}_2 \wedge \boldsymbol{\alpha}_3 \wedge \boldsymbol{\alpha}_4) \tag{2.142}$$

we see that the scalar condition $\mathrm{tr}\overline{\overline{\mathsf{A}}}^{(4)} \neq 0$ is sufficient for the dyadic $\overline{\overline{\mathsf{A}}}$ being of rank 4. The equations $\overline{\overline{\mathsf{A}}}|\mathbf{a} = 0$ and $\boldsymbol{\alpha}|\overline{\overline{\mathsf{A}}} = 0$ only have the solutions $\mathbf{a} = 0$, $\boldsymbol{\alpha} = 0$.

Dyadics of Rank 3

The rank of a dyadic $\overline{\overline{\mathsf{A}}}$ is 3 when it can be represented in the form

$$\overline{\overline{\mathsf{A}}} = \mathbf{a}_1\boldsymbol{\alpha}_1 + \mathbf{a}_2\boldsymbol{\alpha}_2 + \mathbf{a}_3\boldsymbol{\alpha}_3, \tag{2.143}$$

but not in less than three terms. Thus, the dyadic must satisfy

$$\overline{\overline{\mathsf{A}}}^{(4)} = 0, \qquad \overline{\overline{\mathsf{A}}}^{(3)} \neq 0, \tag{2.144}$$

since the latter requires that the vectors $\{\mathbf{a}_i\}$ and one-forms, $\{\boldsymbol{\alpha}_j\}$ be two linearly independent triplets. For a dyadic of rank 3 there exists a vector $\mathbf{a} \neq 0$ satisfying

$$\overline{\overline{\mathsf{A}}}|\mathbf{a} = 0. \tag{2.145}$$

In fact, in terms of a quadrivector $\mathbf{e}_N$ the vector $\mathbf{a}$ can be expressed as

$$\mathbf{a} = \mathbf{e}_N \lfloor (\boldsymbol{\alpha}_1 \wedge \boldsymbol{\alpha}_2 \wedge \boldsymbol{\alpha}_3) \neq 0, \tag{2.146}$$

because it satisfies

$$\begin{aligned}\overline{\overline{\mathsf{A}}}|\mathbf{a} &= \sum_{i=1}^{3} \mathbf{a}_i \boldsymbol{\alpha}_i|(\mathbf{e}_N \lfloor (\boldsymbol{\alpha}_1 \wedge \boldsymbol{\alpha}_2 \wedge \boldsymbol{\alpha}_3)) \\ &= \sum_{i=1}^{3} \mathbf{a}_i(\boldsymbol{\varepsilon}_N|(\boldsymbol{\alpha}_1 \wedge \boldsymbol{\alpha}_2 \wedge \boldsymbol{\alpha}_3 \wedge \boldsymbol{\alpha}_i)) = 0.\end{aligned} \tag{2.147}$$

Alternatively, choosing a vector $\mathbf{b}$ satisfying $\mathbf{a}_1 \wedge \mathbf{a}_2 \wedge \mathbf{a}_3 \wedge \mathbf{b} \neq 0$, we can write

$$\begin{aligned}\mathbf{a} &= (\overline{\overline{\mathsf{I}}}^{(4)} \lfloor \lfloor \overline{\overline{\mathsf{A}}}^{(3)T})|\mathbf{b} = \mathbf{e}_N \lfloor (\boldsymbol{\alpha}_1 \wedge \boldsymbol{\alpha}_2 \wedge \boldsymbol{\alpha}_3)(\boldsymbol{\varepsilon}_N \lfloor (\mathbf{a}_1 \wedge \mathbf{a}_2 \wedge \mathbf{a}_3))|\mathbf{b} \\ &= \mathbf{e}_N \lfloor (\boldsymbol{\alpha}_1 \wedge \boldsymbol{\alpha}_2 \wedge \boldsymbol{\alpha}_3)\boldsymbol{\varepsilon}_N|(\mathbf{a}_1 \wedge \mathbf{a}_2 \wedge \mathbf{a}_3 \wedge \mathbf{b}) \neq 0.\end{aligned} \tag{2.148}$$

Dyadics of Rank 2

The rank of a dyadic $\overline{\overline{\mathsf{A}}}$ is 2 when it can be expressed in the form

$$\overline{\overline{\mathsf{A}}} = \mathbf{a}_1\boldsymbol{\alpha}_1 + \mathbf{a}_2\boldsymbol{\alpha}_2, \tag{2.149}$$

but not as a single dyadic term. Thus, the dyadic must satisfy

$$\overline{\overline{\mathsf{A}}}^{(3)} = 0, \quad \overline{\overline{\mathsf{A}}}^{(2)} \neq 0. \tag{2.150}$$

Because in this case the vectors $\mathbf{a}_1, \mathbf{a}_2$ and the one-forms $\boldsymbol{\alpha}_1, \boldsymbol{\alpha}_2$ must be linearly independent, $(\mathbf{a}_1 \wedge \mathbf{a}_2)(\boldsymbol{\alpha}_1 \wedge \boldsymbol{\alpha}_2) \neq 0$, there must exist two linearly independent vectors $\mathbf{b}_1, \mathbf{b}_2$ so that

$$\overline{\overline{\mathsf{A}}}|\mathbf{b}_1 = 0, \quad \overline{\overline{\mathsf{A}}}|\mathbf{b}_2 = 0 \tag{2.151}$$

are satisfied. Another way to express the same is to require that

$$\overline{\overline{\mathsf{A}}}\rfloor\mathbf{B} = 0 \tag{2.152}$$

be valid for some simple bivector $\mathbf{B}$. In fact, writing $\mathbf{B} = \mathbf{b}_1 \wedge \mathbf{b}_2$ we can expand

$$\overline{\overline{\mathsf{A}}}\rfloor(\mathbf{b}_1 \wedge \mathbf{b}_2) = (\overline{\overline{\mathsf{A}}}|\mathbf{b}_2)\mathbf{b}_1 - (\overline{\overline{\mathsf{A}}}|\mathbf{b}_1)\mathbf{b}_2 = 0. \tag{2.153}$$

The two vectors can be defined as

$$\mathbf{b}_1 = \mathbf{e}_N \lfloor (\boldsymbol{\alpha}_1 \wedge \boldsymbol{\alpha}_2 \wedge \boldsymbol{\alpha}_3), \quad \mathbf{b}_2 = \mathbf{e}_N \lfloor (\boldsymbol{\alpha}_1 \wedge \boldsymbol{\alpha}_2 \wedge \boldsymbol{\alpha}_4), \tag{2.154}$$

in terms of two one-forms $\boldsymbol{\alpha}_3, \boldsymbol{\alpha}_4$ which added to $\boldsymbol{\alpha}_1, \boldsymbol{\alpha}_2$ complete the one-form basis.

Dyadics of Rank 1

The rank of $\overline{\overline{\mathsf{A}}}$ is 1 when it satisfies

$$\overline{\overline{\mathsf{A}}}^{(2)} = 0, \qquad \overline{\overline{\mathsf{A}}} \neq 0. \tag{2.155}$$

Such a dyadic can be presented as a single dyad,

$$\overline{\overline{\mathsf{A}}} = \mathbf{a}_1 \boldsymbol{\alpha}_1. \tag{2.156}$$

2.5 EIGENPROBLEMS

2.5.1 Eigenvectors and Eigen One-Forms

Any dyadic $\overline{\overline{\mathsf{A}}} \in \mathbb{E}_1\mathbb{F}_1$ defines an eigenvalue problem as

$$\overline{\overline{\mathsf{A}}}|\mathbf{x} = \lambda \mathbf{x}, \tag{2.157}$$

where λ is an eigenvalue and $\mathbf{x}$ is an eigenvector. Because the dyadic $\overline{\overline{\mathsf{A}}} - \lambda\overline{\overline{\mathsf{I}}}$ maps the corresponding eigenvector $\mathbf{x}$ to zero, its rank is less than 4 whence it satisfies

$$\mathrm{tr}(\overline{\overline{\mathsf{A}}} - \lambda\overline{\overline{\mathsf{I}}})^{(4)} = 0, \tag{2.158}$$

or

$$\lambda^4 - \lambda^3 \mathrm{tr}\overline{\overline{\mathsf{A}}} + \lambda^2 \mathrm{tr}\overline{\overline{\mathsf{A}}}^{(2)} - \lambda \mathrm{tr}\overline{\overline{\mathsf{A}}}^{(3)} + \mathrm{tr}\overline{\overline{\mathsf{A}}}^{(4)} = 0. \tag{2.159}$$

In the general case there are four eigenvalues λ_i satisfying

$$(\lambda - \lambda_1)(\lambda - \lambda_2)(\lambda - \lambda_3)(\lambda - \lambda_4) = 0. \tag{2.160}$$

Comparing this and (2.159) we obtain the relations

$$\mathrm{tr}\overline{\overline{\mathsf{A}}} = \lambda_1 + \lambda_2 + \lambda_3 + \lambda_4, \tag{2.161}$$

$$\mathrm{tr}\overline{\overline{\mathsf{A}}}^{(2)} = \lambda_1\lambda_2 + \lambda_2\lambda_3 + \lambda_3\lambda_1 + \lambda_1\lambda_4 + \lambda_2\lambda_4 + \lambda_3\lambda_4, \tag{2.162}$$

$$\mathrm{tr}\overline{\overline{\mathsf{A}}}^{(3)} = \lambda_1\lambda_2\lambda_3 + \lambda_2\lambda_3\lambda_4 + \lambda_3\lambda_1\lambda_4 + \lambda_1\lambda_2\lambda_4, \tag{2.163}$$

$$\mathrm{tr}\overline{\overline{\mathsf{A}}}^{(4)} = \lambda_1\lambda_2\lambda_3\lambda_4 = \det\overline{\overline{\mathsf{A}}}. \tag{2.164}$$

The corresponding eigenproblem for the transposed dyadic is

$$\overline{\overline{\mathsf{A}}}^T|\boldsymbol{\xi}_i = \boldsymbol{\xi}_i|\overline{\overline{\mathsf{A}}} = \mu\boldsymbol{\xi}_i, \tag{2.165}$$

where μ is the (left) eigenvalue and ξ the eigen one-form of the dyadic $\overline{\overline{\mathsf{A}}}$. Since the equation for the eigenvalue μ equals (2.159), the set of solutions is the same, $\mu_i = \lambda_i$. From

$$\xi_i|\overline{\overline{\mathsf{A}}}|\mathbf{x}_j = \lambda_i \xi_i|\mathbf{x}_j = \lambda_j \xi_i|\mathbf{x}_j \tag{2.166}$$

we obtain the orthogonality rule

$$\lambda_i \neq \lambda_j \;\Rightarrow\; \xi_i|\mathbf{x}_j = 0. \tag{2.167}$$

Solutions for the eigenproblem of the dyadic $\overline{\overline{\mathsf{A}}}^{(2)}$ can be found in terms of the eigensolutions of the dyadic $\overline{\overline{\mathsf{A}}}$ as

$$\overline{\overline{\mathsf{A}}}^{(2)}|(\mathbf{x}_i \wedge \mathbf{x}_j) = (\overline{\overline{\mathsf{A}}}|\mathbf{x}_i) \wedge (\overline{\overline{\mathsf{A}}}|\mathbf{x}_j) = \lambda_i\lambda_j \mathbf{x}_i \wedge \mathbf{x}_j, \tag{2.168}$$

$$(\xi_i \wedge \xi_j)|\overline{\overline{\mathsf{A}}}^{(2)} = \lambda_i\lambda_j \xi_i \wedge \xi_j. \tag{2.169}$$

2.5.2 Reduced Cayley–Hamilton Equations

For certain dyadics the Cayley–Hamilton equation (2.119) can be reduced to dyadic equations of lower order than four. Actually, there are 11 basic forms each defining a class of dyadics. Assuming that λ_i's are not equal, the different cases can be represented as

1. $(\overline{\overline{\mathsf{A}}} - \lambda_1\overline{\overline{\mathsf{I}}})|(\overline{\overline{\mathsf{A}}} - \lambda_2\overline{\overline{\mathsf{I}}})|(\overline{\overline{\mathsf{A}}} - \lambda_3\overline{\overline{\mathsf{I}}})|(\overline{\overline{\mathsf{A}}} - \lambda_4\overline{\overline{\mathsf{I}}}) = 0$
2. $(\overline{\overline{\mathsf{A}}} - \lambda_1\overline{\overline{\mathsf{I}}})|(\overline{\overline{\mathsf{A}}} - \lambda_2\overline{\overline{\mathsf{I}}})|(\overline{\overline{\mathsf{A}}} - \lambda_3\overline{\overline{\mathsf{I}}})^2 = 0$
3. $(\overline{\overline{\mathsf{A}}} - \lambda_1\overline{\overline{\mathsf{I}}})|(\overline{\overline{\mathsf{A}}} - \lambda_2\overline{\overline{\mathsf{I}}})|(\overline{\overline{\mathsf{A}}} - \lambda_3\overline{\overline{\mathsf{I}}}) = 0$
4. $(\overline{\overline{\mathsf{A}}} - \lambda_1\overline{\overline{\mathsf{I}}})|(\overline{\overline{\mathsf{A}}} - \lambda_2\overline{\overline{\mathsf{I}}})^3 = 0$
5. $(\overline{\overline{\mathsf{A}}} - \lambda_1\overline{\overline{\mathsf{I}}})^2|(\overline{\overline{\mathsf{A}}} - \lambda_2\overline{\overline{\mathsf{I}}})^2 = 0$
6. $(\overline{\overline{\mathsf{A}}} - \lambda_1\overline{\overline{\mathsf{I}}})|(\overline{\overline{\mathsf{A}}} - \lambda_2\overline{\overline{\mathsf{I}}})^2 = 0$
7. $(\overline{\overline{\mathsf{A}}} - \lambda_1\overline{\overline{\mathsf{I}}})|(\overline{\overline{\mathsf{A}}} - \lambda_2\overline{\overline{\mathsf{I}}}) = 0$
8. $(\overline{\overline{\mathsf{A}}} - \lambda_1\overline{\overline{\mathsf{I}}})^4 = 0$
9. $(\overline{\overline{\mathsf{A}}} - \lambda_1\overline{\overline{\mathsf{I}}})^3 = 0$
10. $(\overline{\overline{\mathsf{A}}} - \lambda_1\overline{\overline{\mathsf{I}}})^2 = 0$
11. $\overline{\overline{\mathsf{A}}} - \lambda_1\overline{\overline{\mathsf{I}}} = 0$

It is assumed that the equations for each case cannot be reduced to simpler equations on the lines below.

Let us just briefly consider some cases. The last class 11 defines dyadics of the form $\overline{\overline{\mathsf{A}}} = A\overline{\overline{\mathsf{I}}}$. Also, the classes 8–10 require that $\overline{\overline{\mathsf{A}}} = A\overline{\overline{\mathsf{I}}} + \overline{\overline{\mathsf{N}}}$, where $\overline{\overline{\mathsf{N}}}$ is any nilpotent dyadic of respective order $p = 4 \cdots 1$, satisfying $\overline{\overline{\mathsf{N}}}^p = 0$. For example, case 10 can be shown to correspond to dyadics of the general form

$$\overline{\overline{\mathsf{A}}} = A\overline{\overline{\mathsf{I}}} + \mathbf{a}_1\boldsymbol{\alpha}_1 + \mathbf{a}_2\boldsymbol{\alpha}_2, \quad \mathbf{a}_i|\boldsymbol{\alpha}_j = 0, \ \ i,j = 1,2. \tag{2.170}$$

The class 7 condition can be rewritten as

$$\left(\overline{\overline{\mathsf{A}}} - \frac{\lambda_1 + \lambda_2}{2}\overline{\overline{\mathsf{I}}}\right)^2 = \frac{(\lambda_1 - \lambda_2)^2}{4}\overline{\overline{\mathsf{I}}}, \tag{2.171}$$

whence the dyadic in brackets must be a multiple of a unipotent dyadic. Such dyadics will be considered in Chapter 4. Any class 1 dyadic with four distinct eigenvalues can be expanded in terms of its eigenvectors and eigen one-forms as

$$\overline{\overline{\mathsf{A}}} = \sum_i \lambda_i \mathbf{x}_i \boldsymbol{\xi}_i, \quad \mathbf{x}_i|\boldsymbol{\xi}_j = \delta_{i,j}. \tag{2.172}$$

2.5.3 Construction of Eigenvectors

The eigenvector $\mathbf{x}_i$ and the eigen one-form $\boldsymbol{\xi}_i$ corresponding to a given eigenvalue λ_i can be constructed in a straightforward manner as follows. Applying the inverse formula (2.120) in the form

$$\overline{\overline{\mathsf{A}}}|(\overline{\overline{\mathsf{I}}}^{(4)}\lfloor\lfloor\overline{\overline{\mathsf{A}}}^{(3)T}) = (\overline{\overline{\mathsf{I}}}^{(4)}\lfloor\lfloor\overline{\overline{\mathsf{A}}}^{(3)T})|\overline{\overline{\mathsf{A}}} = \mathrm{tr}\overline{\overline{\mathsf{A}}}^{(4)}\,\overline{\overline{\mathsf{I}}}, \tag{2.173}$$

and replacing $\overline{\overline{\mathsf{A}}}$ by the dyadic $\overline{\overline{\mathsf{A}}} - \lambda_i\overline{\overline{\mathsf{I}}}$ we have

$$\begin{aligned}(\overline{\overline{\mathsf{A}}} - \lambda_i\overline{\overline{\mathsf{I}}})|(\overline{\overline{\mathsf{I}}}^{(4)}\lfloor\lfloor(\overline{\overline{\mathsf{A}}} - \lambda_i\overline{\overline{\mathsf{I}}})^{(3)T}) &= (\overline{\overline{\mathsf{I}}}^{(4)}\lfloor\lfloor(\overline{\overline{\mathsf{A}}} - \lambda_i\overline{\overline{\mathsf{I}}})^{(3)T})|(\overline{\overline{\mathsf{A}}} - \lambda_i\overline{\overline{\mathsf{I}}}) \\ &= \mathrm{tr}(\overline{\overline{\mathsf{A}}} - \lambda_i\overline{\overline{\mathsf{I}}})^{(4)}\overline{\overline{\mathsf{I}}}.\end{aligned} \tag{2.174}$$

Because of (2.158) the last expression vanishes. Assuming

$$(\overline{\overline{\mathsf{A}}} - \lambda_i\overline{\overline{\mathsf{I}}})^{(3)} \neq 0, \tag{2.175}$$

the first expression yields

$$\overline{\overline{\mathsf{A}}}|(\overline{\overline{\mathsf{I}}}^{(4)}\lfloor\lfloor(\overline{\overline{\mathsf{A}}} - \lambda_i\overline{\overline{\mathsf{I}}})^{(3)T})|\mathbf{a} = \lambda_i(\overline{\overline{\mathsf{I}}}^{(4)}\lfloor\lfloor(\overline{\overline{\mathsf{A}}} - \lambda_i\overline{\overline{\mathsf{I}}})^{(3)T})|\mathbf{a}, \tag{2.176}$$

which must be valid for any vector $\mathbf{a}$. Thus, the eigenvector corresponding to the eigenvalue λ_i can be represented as

$$\mathbf{x}_i = (\overline{\overline{\mathsf{I}}}^{(4)}\lfloor\lfloor(\overline{\overline{\mathsf{A}}} - \lambda_i\overline{\overline{\mathsf{I}}})^{(3)T})|\mathbf{a}, \tag{2.177}$$

in terms of any vector **a** which yields a nonzero result. The left eigen one-form is obtained similarly as

$$\boldsymbol{\xi}_i = \boldsymbol{\alpha}|(\overline{\overline{\mathsf{I}}}^{(4)}\lfloor\lfloor(\overline{\overline{\mathsf{A}}} - \lambda_i\overline{\overline{\mathsf{I}}})^{(3)T}) \tag{2.178}$$

in terms of any one-form $\boldsymbol{\alpha}$ yielding a nonzero result. Actually, from (2.175) and (2.158) we conclude that the dyadic $\overline{\overline{\mathsf{A}}} - \lambda_i\overline{\overline{\mathsf{I}}}$ must be of rank 3, whence we can write

$$\overline{\overline{\mathsf{I}}}^{(4)}\lfloor\lfloor(\overline{\overline{\mathsf{A}}} - \lambda_i\overline{\overline{\mathsf{I}}})^{(3)T} = \mathbf{x}_i\boldsymbol{\xi}_i \tag{2.179}$$

not caring about magnitudes of the eigenvectors and eigen one-forms. Obviously, this procedure is only valid when the eigenvalue λ_i does not correspond to a multiple root of (2.159), because in the converse case we would have $(\overline{\overline{\mathsf{A}}} - \lambda_i\overline{\overline{\mathsf{I}}})^{(3)} = 0$.

If $\lambda_1 = \lambda_2$ is a double but not a triple eigenvalue, we have two linearly independent eigenvectors $\mathbf{x}_1, \mathbf{x}_2$ corresponding to the same eigenvalue and satisfying

$$(\overline{\overline{\mathsf{A}}} - \lambda_1\overline{\overline{\mathsf{I}}})\rfloor(\mathbf{x}_1 \wedge \mathbf{x}_2) = (\overline{\overline{\mathsf{A}}} - \lambda_1\overline{\overline{\mathsf{I}}})|(\mathbf{x}_2\mathbf{x}_1 - \mathbf{x}_1\mathbf{x}_2) = 0. \tag{2.180}$$

In this case, we have $(\overline{\overline{\mathsf{A}}} - \lambda_1\overline{\overline{\mathsf{I}}})^{(3)} = 0$ and $(\overline{\overline{\mathsf{A}}} - \lambda_1\overline{\overline{\mathsf{I}}})^{(2)} \neq 0$ whence $\overline{\overline{\mathsf{A}}} - \lambda_1\overline{\overline{\mathsf{I}}}$ must be of rank 2. It then follows that there must exist vectors **a**, **b** and one-forms $\boldsymbol{\alpha}, \boldsymbol{\beta}$ such that we can write

$$\overline{\overline{\mathsf{A}}} - \lambda_1\overline{\overline{\mathsf{I}}} = \mathbf{a}\boldsymbol{\alpha} + \mathbf{b}\boldsymbol{\beta}, \tag{2.181}$$

and

$$(\overline{\overline{\mathsf{A}}} - \lambda_1\overline{\overline{\mathsf{I}}})^{(2)} = (\mathbf{a} \wedge \mathbf{b})(\boldsymbol{\alpha} \wedge \boldsymbol{\beta}) \neq 0. \tag{2.182}$$

Comparing (2.180) to

$$\begin{aligned}(\overline{\overline{\mathsf{A}}} - \lambda_1\overline{\overline{\mathsf{I}}})\rfloor(\overline{\overline{\mathsf{I}}}^{(4)}\lfloor\lfloor(\overline{\overline{\mathsf{A}}} - \lambda_1\overline{\overline{\mathsf{I}}})^{(2)T}) &= (\mathbf{a}\boldsymbol{\alpha} + \mathbf{b}\boldsymbol{\beta})\rfloor(\mathbf{e}_N\lfloor(\boldsymbol{\alpha} \wedge \boldsymbol{\beta}))(\boldsymbol{\varepsilon}_N\lfloor(\mathbf{a} \wedge \mathbf{b}))\\ &= -(\mathbf{a}\mathbf{e}_N\lfloor(\boldsymbol{\alpha} \wedge \boldsymbol{\beta} \wedge \boldsymbol{\alpha}) + \mathbf{b}\mathbf{e}_N\lfloor(\boldsymbol{\alpha} \wedge \boldsymbol{\beta} \wedge \boldsymbol{\beta}))(\boldsymbol{\varepsilon}_N\lfloor(\mathbf{a} \wedge \mathbf{b})) = 0,\end{aligned} \tag{2.183}$$

shows us that $\mathbf{e}_N\lfloor(\boldsymbol{\alpha} \wedge \boldsymbol{\beta})$ must be a multiple of $\mathbf{x}_1 \wedge \mathbf{x}_2$. Similarly, $\boldsymbol{\varepsilon}_N\lfloor(\mathbf{a} \wedge \mathbf{b})$ must be a multiple of $\boldsymbol{\xi}_1 \wedge \boldsymbol{\xi}_2$, whence ignoring normalization we may write

$$\overline{\overline{\mathsf{I}}}^{(4)}\lfloor\lfloor(\overline{\overline{\mathsf{A}}} - \lambda_1\overline{\overline{\mathsf{I}}})^{(2)T} = (\mathbf{x}_1 \wedge \mathbf{x}_2)(\boldsymbol{\xi}_1 \wedge \boldsymbol{\xi}_2). \tag{2.184}$$

This defines the two-dimensional (2D) subspaces of eigenvectors and eigen one-forms corresponding to the double eigenvalue λ_1. Another way to express the relation is

$$(\overline{\overline{\mathsf{A}}} - \lambda_1\overline{\overline{\mathsf{I}}})^{(2)} = \overline{\overline{\mathsf{I}}}^{(4)}\lfloor\lfloor(\boldsymbol{\xi}_1 \wedge \boldsymbol{\xi}_2)(\mathbf{x}_1 \wedge \mathbf{x}_2) = \mathbf{e}_N\lfloor(\boldsymbol{\xi}_1 \wedge \boldsymbol{\xi}_2)\,\boldsymbol{\varepsilon}_N\lfloor(\mathbf{x}_1 \wedge \mathbf{x}_2). \tag{2.185}$$

This process can be continued so that, for $\lambda_1 = \lambda_2 = \lambda_3 \neq \lambda_4$ the three linearly independent eigenvectors $\mathbf{x}_1, \mathbf{x}_2, \mathbf{x}_3$ and eigen one-forms $\boldsymbol{\xi}_1, \boldsymbol{\xi}_2, \boldsymbol{\xi}_3$ can be obtained as

$$\overline{\overline{\mathsf{I}}}^{(4)}\lfloor\lfloor(\overline{\overline{\mathsf{A}}} - \lambda_1\overline{\overline{\mathsf{I}}})^T = (\mathbf{x}_1 \wedge \mathbf{x}_2 \wedge \mathbf{x}_3)(\boldsymbol{\xi}_1 \wedge \boldsymbol{\xi}_2 \wedge \boldsymbol{\xi}_3), \tag{2.186}$$

or

$$\begin{aligned}\overline{\overline{\mathsf{A}}} - \lambda_1\overline{\overline{\mathsf{I}}} &= \overline{\overline{\mathsf{I}}}^{(4)}\lfloor\lfloor(\boldsymbol{\xi}_1 \wedge \boldsymbol{\xi}_2 \wedge \boldsymbol{\xi}_3)(\mathbf{x}_1 \wedge \mathbf{x}_2 \wedge \mathbf{x}_3)\\ &= \mathbf{e}_N\lfloor(\boldsymbol{\xi}_1 \wedge \boldsymbol{\xi}_2 \wedge \boldsymbol{\xi}_3)\,\boldsymbol{\varepsilon}_N\lfloor(\mathbf{x}_1 \wedge \mathbf{x}_2 \wedge \mathbf{x}_3).\end{aligned} \tag{2.187}$$

The last dyadic term is actually a multiple of $\mathbf{x}_4\boldsymbol{\xi}_4$, the product of the eigenvector and eigen one-form corresponding to the eigenvalue λ_4.

2.6 METRIC DYADICS

Metric dyadics are dyadics mapping vectors to one-forms or one-forms to vectors, that is, members of the spaces $\mathbb{F}_1\mathbb{F}_1$ or $\mathbb{E}_1\mathbb{E}_1$. Because the transpose operation maps dyadics to the same space, symmetric and antisymmetric dyadics form subspaces in the general space of metric dyadics.

Any metric dyadic $\overline{\overline{\mathsf{B}}} \in \mathbb{E}_1\mathbb{E}_1$ has a bivector invariant which can be defined in terms of its vector expansion as

$$\begin{aligned}\overline{\overline{\mathsf{B}}}^{\wedge} &= \sum_i (\mathbf{a}_i\mathbf{b}_i)^{\wedge} = \sum_i \mathbf{a}_i \wedge \mathbf{b}_i \in \mathbb{E}_2\\ &= \mathbf{a}_1 \wedge \mathbf{b}_1 + \mathbf{a}_2 \wedge \mathbf{b}_2 + \mathbf{a}_3 \wedge \mathbf{b}_3 + \mathbf{a}_4 \wedge \mathbf{b}_4.\end{aligned} \tag{2.188}$$

Similarly, any metric dyadic $\overline{\overline{\Theta}} \in \mathbb{F}_1\mathbb{F}_1$ has the two-form invariant

$$\overline{\overline{\Theta}}^{\wedge} = \sum_i (\boldsymbol{\alpha}_i\boldsymbol{\beta}_i)^{\wedge} = \sum_i \boldsymbol{\alpha}_i \wedge \boldsymbol{\beta}_i \in \mathbb{F}_2. \tag{2.189}$$

A metric dyadic $\overline{\overline{\mathsf{B}}}$ is of full rank 4 if it satisfies $\overline{\overline{\mathsf{B}}}^{(4)} \neq 0$. In the converse case, the rank is 3 for $\overline{\overline{\mathsf{B}}}^{(3)} \neq 0$, 2 for $\overline{\overline{\mathsf{B}}}^{(3)} = 0$, $\overline{\overline{\mathsf{B}}}^{(2)} \neq 0$ and 1 for $\overline{\overline{\mathsf{B}}}^{(2)} = 0$, $\overline{\overline{\mathsf{B}}} \neq 0$.

The metric dyadic $\overline{\overline{\mathsf{B}}} \in \mathbb{E}_1\mathbb{E}_1$ of full rank has an inverse $\overline{\overline{\mathsf{B}}}^{-1}$ which is in the space $\mathbb{F}_1\mathbb{F}_1$. Transposing the equation

$$\overline{\overline{\mathsf{B}}}|\overline{\overline{\mathsf{B}}}^{-1} = \overline{\overline{\mathsf{I}}} \Rightarrow \overline{\overline{\mathsf{B}}}^{-1T}|\overline{\overline{\mathsf{B}}}^{T} = \overline{\overline{\mathsf{I}}}^{T}, \tag{2.190}$$

and assuming symmetric or antisymmetric dyadic, $\overline{\overline{\mathsf{B}}}^T = \pm\overline{\overline{\mathsf{B}}}$, we obtain

$$\pm\overline{\overline{\mathsf{B}}}^{-1T}|\overline{\overline{\mathsf{B}}} = \overline{\overline{\mathsf{I}}}^{T} \Rightarrow \overline{\overline{\mathsf{B}}}^{-1T} = \pm\overline{\overline{\mathsf{B}}}^{-1}, \tag{2.191}$$

whence also the respective inverse is symmetric or antisymmetric. The same is also valid for metric dyadics of the spaces $\mathbb{E}_p\mathbb{E}_p$, $\mathbb{F}_p\mathbb{F}_p$ for $p = 2, 3, 4$.

2.6.1 Symmetric Dyadics

Symmetric dyadics $\overline{\overline{\mathsf{S}}} \in \mathbb{E}_1\mathbb{E}_1$ can be expanded as

$$\overline{\overline{\mathsf{S}}} = \sum_i \mathbf{a}_i\mathbf{b}_i = \overline{\overline{\mathsf{S}}}^T = \sum_i \mathbf{b}_i\mathbf{a}_i = \frac{1}{2}\sum_i(\mathbf{a}_i\mathbf{b}_i + \mathbf{b}_i\mathbf{a}_i), \tag{2.192}$$

whence the bivector invariant of any symmetric dyadic vanishes,

$$\overline{\overline{\mathsf{S}}}^{\wedge} = \frac{1}{2}\sum_i(\mathbf{a}_i \wedge \mathbf{b}_i + \mathbf{b}_i \wedge \mathbf{a}_i) = 0. \tag{2.193}$$

Conversely, if the bivector invariant of a the metric dyadic $\overline{\overline{\mathsf{B}}}$ vanishes, $\overline{\overline{\mathsf{B}}}^{\wedge} = 0$, the dyadic is symmetric:

$$\begin{aligned}\overline{\overline{\mathsf{B}}} &= \sum_i(\mathbf{a}_i\mathbf{b}_i) = \frac{1}{2}\sum(\mathbf{a}_i\mathbf{b}_i + \mathbf{b}_i\mathbf{a}_i) + \frac{1}{2}\sum(\mathbf{a}_i\mathbf{b}_i - \mathbf{b}_i\mathbf{a}_i)\\ &= \frac{1}{2}\sum(\mathbf{a}_i\mathbf{b}_i + \mathbf{b}_i\mathbf{a}_i) - \frac{1}{2}\sum(\mathbf{a}_i \wedge \mathbf{b}_i)\lfloor\overline{\overline{\mathsf{I}}}^T = \frac{1}{2}\sum(\mathbf{a}_i\mathbf{b}_i + \mathbf{b}_i\mathbf{a}_i).\end{aligned} \tag{2.194}$$

Symmetric dyadics satisfying $\overline{\overline{\mathsf{S}}}^{(4)} \neq 0$ can be used to define the dot product for one-forms as

$$\boldsymbol{\alpha} \cdot \boldsymbol{\beta} = \boldsymbol{\alpha}|\overline{\overline{\mathsf{S}}}|\boldsymbol{\beta} = \boldsymbol{\beta} \cdot \boldsymbol{\alpha}. \tag{2.195}$$

Similarly, a dot product for vectors can be obtained in terms of a symmetric dyadic $\overline{\overline{\Sigma}} \in \mathbb{F}_1\mathbb{F}_1$ satisfying $\overline{\overline{\Sigma}}^{(4)} \neq 0$ as

$$\mathbf{a} \cdot \mathbf{b} = \mathbf{a}|\overline{\overline{\Sigma}}|\mathbf{b} = \mathbf{b} \cdot \mathbf{a}. \tag{2.196}$$

In Minkowskian metric, two complementary symmetric dyadics can be expressed in terms of reciprocal basis expansions as

$$\overline{\overline{\Gamma}} = \boldsymbol{\varepsilon}_1\boldsymbol{\varepsilon}_1 + \boldsymbol{\varepsilon}_2\boldsymbol{\varepsilon}_2 + \boldsymbol{\varepsilon}_3\boldsymbol{\varepsilon}_3 - \boldsymbol{\varepsilon}_4\boldsymbol{\varepsilon}_4, \quad \overline{\overline{\Gamma}}^{(4)} = -\boldsymbol{\varepsilon}_N\boldsymbol{\varepsilon}_N, \tag{2.197}$$

$$\overline{\overline{\mathsf{G}}} = \mathbf{e}_1\mathbf{e}_1 + \mathbf{e}_2\mathbf{e}_2 + \mathbf{e}_3\mathbf{e}_3 - \mathbf{e}_4\mathbf{e}_4, \quad \overline{\overline{\mathsf{G}}}^{(4)} = -\mathbf{e}_N\mathbf{e}_N. \tag{2.198}$$

They satisfy the relations

$$\overline{\overline{\mathsf{G}}}|\overline{\overline{\Gamma}} = \overline{\overline{\mathsf{I}}}, \quad \overline{\overline{\mathsf{G}}} = \overline{\overline{\Gamma}}^{-1}, \tag{2.199}$$

$$\overline{\overline{\Gamma}} = -\boldsymbol{\varepsilon}_N\boldsymbol{\varepsilon}_N\lfloor\lfloor\overline{\overline{\mathsf{G}}}^{(3)}, \quad \overline{\overline{\mathsf{G}}} = -\mathbf{e}_N\mathbf{e}_N\lfloor\lfloor\overline{\overline{\Gamma}}^{(3)}. \tag{2.200}$$

2.6.2 Antisymmetric Dyadics

Expanding the general antisymmetric dyadic $\overline{\overline{\mathsf{A}}} \in \mathbb{E}_1\mathbb{E}_1$ as

$$\begin{aligned}\overline{\overline{\mathsf{A}}} &= \sum_i \mathbf{a}_i\mathbf{b}_i = \frac{1}{2}\sum_i(\mathbf{a}_i\mathbf{b}_i - \mathbf{b}_i\mathbf{a}_i)\\ &= -\frac{1}{2}\sum_i(\mathbf{a}_i\wedge\mathbf{b}_i)\lfloor\overline{\overline{\mathsf{I}}}^T = -\frac{1}{2}(\overline{\overline{\mathsf{A}}}^\wedge)\lfloor\overline{\overline{\mathsf{I}}}^T,\end{aligned} \tag{2.201}$$

it is seen that an antisymmetric dyadic is determined by its bivector invariant (2.188). Thus, any dyadic of the form

$$\overline{\overline{\mathsf{A}}} = \mathbf{A}\lfloor\overline{\overline{\mathsf{I}}}^T, \tag{2.202}$$

is antisymmetric and, conversely, any antisymmetric dyadic $\overline{\overline{\mathsf{A}}} \in \mathbb{E}_1\mathbb{E}_1$ can be expressed in terms of a bivector,

$$\mathbf{A} = -\frac{1}{2}\overline{\overline{\mathsf{A}}}^\wedge. \tag{2.203}$$

Considering 3D spatial antisymmetric dyadics the bivector $\mathbf{A}$ is spatial and it can be expressed in terms of a spatial one-form $\boldsymbol{\alpha}$ as can be seen from the expansion

$$\begin{aligned}\overline{\overline{\mathsf{A}}} &= A_{12}(\mathbf{e}_1\mathbf{e}_2 - \mathbf{e}_2\mathbf{e}_1) + A_{23}(\mathbf{e}_2\mathbf{e}_3 - \mathbf{e}_3\mathbf{e}_2) + A_{31}(\mathbf{e}_3\mathbf{e}_1 - \mathbf{e}_1\mathbf{e}_3)\\ &= -(A_{12}\mathbf{e}_1\wedge\mathbf{e}_2 + A_{23}\mathbf{e}_2\wedge\mathbf{e}_3 + A_{31}\mathbf{e}_3\wedge\mathbf{e}_1)\lfloor\overline{\overline{\mathsf{I}}}_s^T\\ &= -(\mathbf{e}_{123}\lfloor(A_{12}\boldsymbol{\varepsilon}_3 + A_{23}\boldsymbol{\varepsilon}_1 + A_{31}\boldsymbol{\varepsilon}_2))\lfloor\overline{\overline{\mathsf{I}}}^T = (\mathbf{e}_{123}\lfloor\boldsymbol{\alpha})\lfloor\overline{\overline{\mathsf{I}}}_s^T.\end{aligned} \tag{2.204}$$

The most general 3D spatial antisymmetric bidyadic $\overline{\overline{\mathsf{A}}} \in \mathbb{E}_2\mathbb{E}_2$ can be expanded in terms of a spatial vector $\mathbf{a}$, as can be seen from the expansion

$$\overline{\overline{\mathsf{A}}} = A_1(\mathbf{e}_{31}\mathbf{e}_{12} - \mathbf{e}_{12}\mathbf{e}_{31}) + A_2(\mathbf{e}_{12}\mathbf{e}_{23} - \mathbf{e}_{23}\mathbf{e}_{12}) + A_3(\mathbf{e}_{23}\mathbf{e}_{31} - \mathbf{e}_{31}\mathbf{e}_{23})$$
$$= -\mathbf{e}_{123}\lfloor\overline{\overline{\mathsf{I}}}_s^T \wedge (A_1\mathbf{e}_1 + A_2\mathbf{e}_2 + A_3\mathbf{e}_3) = \mathbf{e}_{123}\lfloor\overline{\overline{\mathsf{I}}}^T \wedge \mathbf{a}. \quad (2.205)$$

2.6.3 Inverse Rules for Metric Dyadics

The expression (2.120) for the inverse of a dyadic $\overline{\overline{\mathsf{A}}} \in \mathbb{E}_1\mathbb{F}_1$ can be rewritten for the metric dyadic $\overline{\overline{\mathsf{B}}} \in \mathbb{E}_1\mathbb{E}_1$ in the form

$$\overline{\overline{\mathsf{B}}}^{-1} = \frac{\boldsymbol{\varepsilon}_N\boldsymbol{\varepsilon}_N\lfloor\lfloor\overline{\overline{\mathsf{B}}}^{(3)T}}{\boldsymbol{\varepsilon}_N\boldsymbol{\varepsilon}_N||\overline{\overline{\mathsf{B}}}^{(4)}}, \quad \boldsymbol{\varepsilon}_N\boldsymbol{\varepsilon}_N||\overline{\overline{\mathsf{B}}}^{(4)} \neq 0. \quad (2.206)$$

Other inverse rules based on the same metric dyadic are

$$\overline{\overline{\mathsf{B}}}^{(-2)} = \frac{\boldsymbol{\varepsilon}_N\boldsymbol{\varepsilon}_N\lfloor\lfloor\overline{\overline{\mathsf{B}}}^{(2)T}}{\boldsymbol{\varepsilon}_N\boldsymbol{\varepsilon}_N||\overline{\overline{\mathsf{B}}}^{(4)}}, \quad (2.207)$$

$$\overline{\overline{\mathsf{B}}}^{(-3)} = \frac{\boldsymbol{\varepsilon}_N\boldsymbol{\varepsilon}_N\lfloor\lfloor\overline{\overline{\mathsf{B}}}^{T}}{\boldsymbol{\varepsilon}_N\boldsymbol{\varepsilon}_N||\overline{\overline{\mathsf{B}}}^{(4)}}. \quad (2.208)$$

Since the inverse of a metric dyadic $\overline{\overline{\mathsf{B}}} \in \mathbb{E}_1\mathbb{E}_1$ is in the space $\mathbb{F}_1\mathbb{F}_1$ and since the inverse of an antisymmetric dyadic (2.202) is antisymmetric, there must exist a rule of the form

$$(\mathbf{A}\lfloor\overline{\overline{\mathsf{I}}}^T)^{-1} = \boldsymbol{\Phi}\lfloor\overline{\overline{\mathsf{I}}}, \quad (2.209)$$

where $\boldsymbol{\Phi}$ is a two-form depending on the bivector $\mathbf{A}$. Leaving the details as an exercise, the result can be expressed in the form

$$\boldsymbol{\Phi} = -\frac{\boldsymbol{\varepsilon}_N\lfloor\mathbf{A}}{\boldsymbol{\varepsilon}_N|\mathbf{A}^\wedge}, \quad (2.210)$$

which is valid if the bivector invariant $\mathbf{A} = -\overline{\overline{\mathsf{A}}}^{\wedge}/2$ of the antisymmetric dyadic is not simple, $\mathbf{A}^\wedge = \frac{1}{2}\mathbf{A}\wedge\mathbf{A} \neq 0$. Because every spatial bivector is simple, spatial antisymmetric dyadics have no inverse.

Spatial inverse for a spatial metric dyadic $\overline{\overline{\mathsf{B}}}_s \in \mathbb{E}_1\mathbb{E}_1$ can be expressed following the pattern of (2.136) as

$$\overline{\overline{\mathsf{B}}}_s^{-1} = \frac{1}{\boldsymbol{\varepsilon}_{123}\boldsymbol{\varepsilon}_{123}||\overline{\overline{\mathsf{B}}}_s^{(3)}}\boldsymbol{\varepsilon}_{123}\boldsymbol{\varepsilon}_{123}\lfloor\lfloor\overline{\overline{\mathsf{B}}}_s^{(2)T}, \quad (2.211)$$

and it is valid for dyadics $\overline{\overline{\mathsf{B}}}_s$ satisfying $\overline{\overline{\mathsf{B}}}_s^{(3)} \neq 0$. Because antisymmetric spatial dyadics satisfy $\overline{\overline{\mathsf{B}}}_s^{(3)} = 0$, they do not have a spatial inverse.

PROBLEMS

2.1 Prove the rule $\mathrm{tr}(\overline{\overline{\mathsf{I}}}{}^{\wedge}_{\wedge}\overline{\overline{\mathsf{A}}}) = 3\mathrm{tr}\overline{\overline{\mathsf{A}}}$, $\overline{\overline{\mathsf{A}}} \in \mathbb{E}_1\mathbb{F}_1$ (1) by starting from the expansions $\overline{\overline{\mathsf{I}}} = \sum \mathbf{e}_i\boldsymbol{\varepsilon}_i$ and $\overline{\overline{\mathsf{A}}} = \sum \mathbf{e}_j\boldsymbol{\alpha}_j$ and (2) by applying the rule $\overline{\overline{\mathsf{B}}}^{(2)}\lfloor\lfloor\overline{\overline{\mathsf{A}}}^T = (\overline{\overline{\mathsf{B}}}||\overline{\overline{\mathsf{A}}}^T)\overline{\overline{\mathsf{B}}} - \overline{\overline{\mathsf{B}}}|\overline{\overline{\mathsf{A}}}|\overline{\overline{\mathsf{B}}}$.

2.2 Assuming a given dyadic $\overline{\overline{\mathsf{A}}} \in \mathbb{E}_1\mathbb{F}_1$, solve the equation

$$\overline{\overline{\mathsf{I}}}^{(2)}\lfloor\lfloor\overline{\overline{\mathsf{B}}}^T = \overline{\overline{\mathsf{A}}}.$$

for the dyadic $\overline{\overline{\mathsf{B}}} \in \mathbb{E}_1\mathbb{F}_1$ by applying the rule of the previous problem.

2.3 Applying the multivector identity

$$(\mathbf{a}_1 \wedge \mathbf{a}_2 \wedge \mathbf{a}_3)\lfloor\boldsymbol{\delta} = ((\mathbf{a}_1 \wedge \mathbf{a}_2)\mathbf{a}_3 + (\mathbf{a}_2 \wedge \mathbf{a}_3)\mathbf{a}_1 + (\mathbf{a}_3 \wedge \mathbf{a}_1)\mathbf{a}_2)|\boldsymbol{\delta}$$

develop an expansion rule for $(\overline{\overline{\mathsf{A}}}_1{}^{\wedge}_{\wedge}\overline{\overline{\mathsf{A}}}_2{}^{\wedge}_{\wedge}\overline{\overline{\mathsf{A}}}_3)\lfloor\lfloor\overline{\overline{\mathsf{B}}}^T$ involving four dyadics $\overline{\overline{\mathsf{A}}}_1 \cdots \overline{\overline{\mathsf{B}}} \in \mathbb{E}_1\mathbb{F}_1$. Form also the special cases $\overline{\overline{\mathsf{A}}}_i = \overline{\overline{\mathsf{A}}}$ and $\overline{\overline{\mathsf{B}}} = \overline{\overline{\mathsf{A}}}^{-1}$. Check the last case by setting $\overline{\overline{\mathsf{A}}} = \overline{\overline{\mathsf{I}}}$ and taking the trace of both sides.

2.4 Applying the dyadic rule

$$(\overline{\overline{\mathsf{A}}}{}^{\wedge}_{\wedge}\overline{\overline{\mathsf{B}}})\lfloor\lfloor\overline{\overline{\mathsf{C}}}^T = (\overline{\overline{\mathsf{A}}}||\overline{\overline{\mathsf{C}}}^T)\overline{\overline{\mathsf{B}}} + (\overline{\overline{\mathsf{B}}}||\overline{\overline{\mathsf{C}}}^T)\overline{\overline{\mathsf{A}}} - \overline{\overline{\mathsf{A}}}|\overline{\overline{\mathsf{C}}}|\overline{\overline{\mathsf{B}}} - \overline{\overline{\mathsf{B}}}|\overline{\overline{\mathsf{C}}}|\overline{\overline{\mathsf{A}}}$$

valid for $\overline{\overline{\mathsf{A}}}, \overline{\overline{\mathsf{B}}}, \overline{\overline{\mathsf{C}}} \in \mathbb{E}_1\mathbb{F}_1$, find solution for $\overline{\overline{\mathsf{B}}}$ in the dyadic equation

$$\overline{\overline{\mathsf{A}}}{}^{\wedge}_{\wedge}\overline{\overline{\mathsf{B}}} = \overline{\overline{\mathsf{D}}} \in \mathbb{E}_2\mathbb{F}_2,$$

assuming that $\overline{\overline{\mathsf{A}}}^{-1}$ exists. Find a condition between $\overline{\overline{\mathsf{A}}}$ and $\overline{\overline{\mathsf{D}}}$ for the solution $\overline{\overline{\mathsf{B}}}$ to exist. Hint: try $\overline{\overline{\mathsf{C}}} = \overline{\overline{\mathsf{A}}}^{-1}$. Check the result by setting $\overline{\overline{\mathsf{D}}} = \overline{\overline{\mathsf{I}}}^{(2)}$ and $\overline{\overline{\mathsf{A}}} = \overline{\overline{\mathsf{I}}}$.

2.5 Prove the identity

$$\overline{\overline{\mathsf{A}}}^T|(\boldsymbol{\Phi}\lfloor\overline{\overline{\mathsf{A}}}) = (\boldsymbol{\Phi}|\overline{\overline{\mathsf{A}}}^{(2)})\lfloor\overline{\overline{\mathsf{I}}}^T,$$

where $\overline{\overline{\mathsf{A}}} \in \mathbb{E}_1\mathbb{E}_1$ is a metric dyadic and $\boldsymbol{\Phi} \in \mathbb{F}_2$ is a two-form, in two ways: (1) expanding $\overline{\overline{\mathsf{A}}} = \sum \mathbf{a}_i\mathbf{b}_i$ and (2) assuming that $\boldsymbol{\Phi}$ is a simple two-form.

2.6 Solve the eigendyadic problem

$$\overline{\overline{\mathsf{I}}}^{(2)}\lfloor\lfloor\overline{\overline{\mathsf{X}}}^T = \lambda\overline{\overline{\mathsf{X}}}.$$

for λ and $\overline{\overline{\mathsf{X}}} \in \mathbb{E}_1\mathbb{F}_1$.

2.7 Find the inverse of the general antisymmetric dyadic $\overline{\overline{\mathsf{A}}} \in \mathbb{E}_1\mathbb{E}_1$ which can be expressed in the form

$$\overline{\overline{\mathsf{A}}} = \mathbf{A}\lfloor\overline{\overline{\mathsf{I}}}^T = \overline{\overline{\mathsf{I}}}\rfloor\mathbf{A},$$

where $\mathbf{A} \in \mathbb{E}_2$ is a bivector. What is the condition for the existence of the inverse dyadic? Hint: apply the bac-cab rule

$$\boldsymbol{\alpha}\rfloor(\mathbf{B}\wedge\mathbf{C}) = \mathbf{B}\wedge(\boldsymbol{\alpha}\rfloor\mathbf{C}) + \mathbf{C}\wedge(\boldsymbol{\alpha}\rfloor\mathbf{B}).$$

2.8 Prove the identity satisfied by any bivector $\mathbf{A}$,

$$(\overline{\overline{\mathsf{I}}}\rfloor\mathbf{A})^{(2)} = \mathbf{A}\mathbf{A} - \frac{1}{2}(\mathbf{A}\wedge\mathbf{A})\lfloor\overline{\overline{\mathsf{I}}}^{(2)T}. \tag{2.212}$$

Hint: apply the expansion $\mathbf{A} = \mathbf{a}\wedge\mathbf{b} + \mathbf{c}\wedge\mathbf{d}$.

2.9 Expanding two dyadics $\overline{\overline{\mathsf{A}}}_{1,2} \in \mathbb{E}_1\mathbb{F}_1$ in 3D components as

$$\overline{\overline{\mathsf{A}}}_i = \overline{\overline{\mathsf{B}}}_i + \mathbf{b}_i\boldsymbol{\varepsilon}_4 + \mathbf{e}_4\boldsymbol{\beta}_i + B_i\mathbf{e}_i\boldsymbol{\varepsilon}_i, \quad i = 1, 2.$$

assuming $B_2 \neq 0$ and that the 3D inverse $\overline{\overline{\mathsf{B}}}_1^{-1}$ exists, show that $\overline{\overline{\mathsf{A}}}_1|\overline{\overline{\mathsf{A}}}_2 = 0$ implies the existence of $\mathbf{c} \in \mathbb{E}_1$ and $\boldsymbol{\gamma} \in \mathbb{F}_1$ such that $\overline{\overline{\mathsf{A}}}_2 = \mathbf{c}\boldsymbol{\gamma}$.

2.10 Prove that $\overline{\overline{\mathsf{D}}}\lfloor\lfloor\overline{\overline{\mathsf{I}}}^T = 0$ implies $\overline{\overline{\mathsf{D}}} = 0$ for a dyadic $\overline{\overline{\mathsf{D}}} \in \mathbb{E}_3\mathbb{F}_3$. Hint: any dyadic $\overline{\overline{\mathsf{D}}} \in \mathbb{E}_3\mathbb{F}_3$ can be expressed as $\overline{\overline{\mathsf{D}}} = \overline{\overline{\mathsf{I}}}^{(4)}\lfloor\lfloor\overline{\overline{\mathsf{A}}}^T$ in terms of some dyadic $\overline{\overline{\mathsf{A}}} \in \mathbb{E}_1\mathbb{F}_1$.

2.11 Derive the identities

$$\overline{\overline{\mathsf{I}}}^{(3)}\lfloor\lfloor\overline{\overline{\mathsf{A}}}^T = (\mathrm{tr}\overline{\overline{\mathsf{A}}})\overline{\overline{\mathsf{I}}}^{(2)} - \overline{\overline{\mathsf{I}}}{}^{\wedge}_{\wedge}\overline{\overline{\mathsf{A}}},$$

$$\overline{\overline{\mathsf{I}}}^{(4)}\lfloor\lfloor\overline{\overline{\mathsf{A}}}^T = (\mathrm{tr}\overline{\overline{\mathsf{A}}})\overline{\overline{\mathsf{I}}}^{(3)} - \overline{\overline{\mathsf{I}}}^{(2)}{}^{\wedge}_{\wedge}\overline{\overline{\mathsf{A}}},$$

valid for any dyadic $\overline{\overline{\mathsf{A}}} \in \mathbb{E}_1\mathbb{F}_1$, by applying the identity

$$\overline{\overline{\mathsf{A}}}^{(p)}\lfloor\mathbf{a} = (\overline{\overline{\mathsf{A}}}|\mathbf{a})\wedge\overline{\overline{\mathsf{A}}}^{(p-1)},$$

where $\mathbf{a}$ is a vector and p is 2, 3 or 4. Hint: start by setting $\overline{\overline{\mathsf{A}}} = \mathbf{a}\boldsymbol{\alpha}$.

2.12 Derive the identity

$$(\overline{\overline{\mathsf{I}}}^{(2)}{}^{\wedge}_{\wedge}\overline{\overline{\mathsf{A}}})\lfloor\lfloor\overline{\overline{\mathsf{B}}}^{T} = (\mathrm{tr}(\overline{\overline{\mathsf{A}}}|\overline{\overline{\mathsf{B}}}))\overline{\overline{\mathsf{I}}}^{(2)} + \overline{\overline{\mathsf{I}}}{}^{\wedge}_{\wedge}\overline{\overline{\mathsf{A}}}\ \mathrm{tr}\overline{\overline{\mathsf{B}}} - \overline{\overline{\mathsf{I}}}{}^{\wedge}_{\wedge}(\overline{\overline{\mathsf{A}}}|\overline{\overline{\mathsf{B}}} + \overline{\overline{\mathsf{B}}}|\overline{\overline{\mathsf{A}}}) - \overline{\overline{\mathsf{A}}}{}^{\wedge}_{\wedge}\overline{\overline{\mathsf{B}}},$$

valid for any dyadics $\overline{\overline{\mathsf{A}}}, \overline{\overline{\mathsf{B}}} \in \mathbb{E}_1\mathbb{F}_1$.

2.13 Derive the identity

$$\overline{\overline{\mathsf{I}}}^{(4)}\lfloor\lfloor\overline{\overline{\overline{\overline{\mathsf{C}}}}}^{T} = (\mathrm{tr}\overline{\overline{\overline{\overline{\mathsf{C}}}}})\overline{\overline{\mathsf{I}}}^{(2)} - (\overline{\overline{\overline{\overline{\mathsf{C}}}}}\lfloor\lfloor\overline{\overline{\mathsf{I}}}^{T}){}^{\wedge}_{\wedge}\overline{\overline{\mathsf{I}}} + \overline{\overline{\overline{\overline{\mathsf{C}}}}},$$

valid for any bidyadic $\overline{\overline{\overline{\overline{\mathsf{C}}}}} \in \mathbb{E}_2\mathbb{F}_2$. Hint: start by setting $\overline{\overline{\overline{\overline{\mathsf{C}}}}} = \overline{\overline{\mathsf{A}}}{}^{\wedge}_{\wedge}\overline{\overline{\mathsf{B}}}$ and use the identities of the two previous problems.

2.14 Show that the identity of the previous problem remains unchanged when both sides of the transposed identity are operated by $\overline{\overline{\mathsf{I}}}^{(4)}\lfloor\lfloor$ or when the bidyadic $\overline{\overline{\overline{\overline{\mathsf{C}}}}}$ in the original is replaced by $\overline{\overline{\mathsf{I}}}^{(4)}\lfloor\lfloor\overline{\overline{\overline{\overline{\mathsf{C}}}}}^{T}$.

2.15 Show that if an identity is of the form $f(\overline{\overline{\mathsf{A}}}^{(2)}) = 0,\ \overline{\overline{\mathsf{A}}} \in \mathbb{E}_1\mathbb{F}_1$ where f is a linear function, it is valid for any bidyadic $\overline{\overline{\overline{\overline{\mathsf{D}}}}} \in \mathbb{E}_2\mathbb{F}_2$ as $f(\overline{\overline{\overline{\overline{\mathsf{D}}}}}) = 0$.

2.16 A symmetric dyadic $\overline{\overline{\mathsf{S}}} \in \mathbb{E}_1\mathbb{E}_1$ and an antisymmetric dyadic $\overline{\overline{\mathsf{A}}} = \mathbf{A}\lfloor\overline{\overline{\mathsf{I}}}^{T} \in \mathbb{E}_1\mathbb{E}_1$ defined by a bivector $\mathbf{A}$ form a bidyadic $\overline{\overline{\mathsf{S}}}{}^{\wedge}_{\wedge}\overline{\overline{\mathsf{A}}} \in \mathbb{E}_2\mathbb{E}_2$ which is obviously antisymmetric. Find a trace-free dyadic $\overline{\overline{\mathsf{B}}}_o \in \mathbb{E}_1\mathbb{F}_1$ in terms of which we can express

$$\overline{\overline{\mathsf{S}}}{}^{\wedge}_{\wedge}(\mathbf{A}\lfloor\overline{\overline{\mathsf{I}}}^{T}) = (\overline{\overline{\mathsf{B}}}_o{}^{\wedge}_{\wedge}\overline{\overline{\mathsf{I}}})\rfloor\mathbf{e}_N.$$

Hint: because the expression is linear in both $\overline{\overline{\mathsf{S}}}$ and $\mathbf{A}$, so is $\overline{\overline{\mathsf{B}}}_o$. Replace $\overline{\overline{\mathsf{S}}} \rightarrow \mathbf{ss}$ and $\mathbf{A} \rightarrow \mathbf{e}_{12}$, find $\overline{\overline{\mathsf{B}}}_o(\mathbf{ss}, \mathbf{e}_{12})$ and express $\overline{\overline{\mathsf{B}}}_o(\overline{\overline{\mathsf{S}}}, \mathbf{A})$. Hint: the multivector rule

$$\mathbf{a}\rfloor(\boldsymbol{\alpha}\wedge\boldsymbol{\beta}\wedge\boldsymbol{\gamma}) = (\mathbf{a}|\boldsymbol{\alpha})(\boldsymbol{\beta}\wedge\boldsymbol{\gamma}) + \boldsymbol{\alpha}\wedge(\mathbf{a}\rfloor(\boldsymbol{\beta}\wedge\boldsymbol{\gamma}))$$

may be useful in the derivation.

2.17 Derive the expression (2.134) for the dyadic inverse.

2.18 Find an expression for the inverse of a metric dyadic of the form $\overline{\overline{\mathsf{A}}} + \mathbf{ab} + \mathbf{ba}$ where $\overline{\overline{\mathsf{A}}}$ is an antisymmetric dyadic possessing an inverse. Check the special case $\mathbf{b} = \mathbf{a}/2$.

2.19 Derive the spatial inverse rule (2.136). Hint: start from the Cayley–Hamilton equation for the spatial dyadic,

$$\overline{\overline{\mathsf{A}}}_s^3 - (\mathrm{tr}\overline{\overline{\mathsf{A}}}_s)\overline{\overline{\mathsf{A}}}_s^2 + (\mathrm{tr}\overline{\overline{\mathsf{A}}}_s^{(2)})\overline{\overline{\mathsf{A}}}_s - (\mathrm{tr}\overline{\overline{\mathsf{A}}}_s^{(3)})\overline{\overline{\mathsf{I}}}_s = 0.$$

2.20 Show that if a dyadic $\overline{\overline{\mathsf{A}}} \in \mathbb{E}_1\mathbb{F}_1$ satisfies (1) $\overline{\overline{\mathsf{A}}}|\mathbf{a} = 0$ for a vector $\mathbf{a} \neq 0$, it must satisfy $\overline{\overline{\mathsf{A}}}^{(4)} = 0$ and (2) if it satisfies $\overline{\overline{\mathsf{A}}}|\mathbf{a} = 0$ and $\overline{\overline{\mathsf{A}}}|\mathbf{b} = 0$ for two vectors with $\mathbf{a} \wedge \mathbf{b} \neq 0$, it must satisfy $\overline{\overline{\mathsf{A}}}^{(3)} = 0$.

2.21 Expressing the spatial bidyadic $\overline{\overline{\mathsf{C}}}_s \in \mathbb{E}_2\mathbb{F}_2$ in terms of a spatial dyadic $\overline{\overline{\mathsf{A}}}_s \in \mathbb{E}_1\mathbb{F}_1$ as $\overline{\overline{\mathsf{C}}}_s = \overline{\overline{\mathsf{I}}}_s^{(3)} \lfloor\lfloor \overline{\overline{\mathsf{A}}}_s^T$, derive the inverse rule (2.137) for $\overline{\overline{\mathsf{C}}}_s$ starting from that of $\overline{\overline{\mathsf{A}}}_s$ (2.136).

2.22 Prove the identity

$$\overline{\overline{\mathsf{A}}}{}^{\wedge}_{\wedge}\overline{\overline{\mathsf{S}}}^{(2)} = \mathbf{e}_N\mathbf{e}_N \lfloor\lfloor (\mathbf{\Phi} \lfloor \overline{\overline{\mathsf{I}}}),$$

where $\overline{\overline{\mathsf{A}}} = \mathbf{A}\lfloor\overline{\overline{\mathsf{I}}}^T \in \mathbb{E}_1\mathbb{E}_1$ is an antisymmetric dyadic and $\overline{\overline{\mathsf{S}}} \in \mathbb{E}_1\mathbb{E}_1$ is a symmetric dyadic. Show that the two-form $\mathbf{\Phi}$ depends on the bivector $\mathbf{A}$ as

$$\mathbf{\Phi} = -\mathbf{A}|(\boldsymbol{\varepsilon}_N\boldsymbol{\varepsilon}_N \lfloor\lfloor \overline{\overline{\mathsf{S}}}^{(2)}).$$

Hint: apply the result of Problem 2.5.

CHAPTER 3

Bidyadics

Bidyadics, elements of the spaces $\mathbb{E}_2\mathbb{F}_2$, $\mathbb{F}_2\mathbb{E}_2$, $\mathbb{E}_2\mathbb{E}_2$ or $\mathbb{F}_2\mathbb{F}_2$, perform mappings between bivectors and/or two-forms. Elements of the latter two spaces may be called metric bidyadics because they define scalar products between two bivectors or two two-forms like $\mathbf{\Phi}|\overline{\overline{\mathsf{C}}}_m|\mathbf{\Psi}$ for $\overline{\overline{\mathsf{C}}}_m \in \mathbb{E}_2\mathbb{E}_2$. There are simple transformations between bidyadics and metric bidyadics. In fact, for every bidyadic $\overline{\overline{\mathsf{C}}} \in \mathbb{E}_1\mathbb{F}_1$ there exist metric bidyadics $\overline{\overline{\mathsf{C}}}\rfloor\mathbf{e}_N \in \mathbb{E}_2\mathbb{E}_2$ and $\boldsymbol{\varepsilon}_N\lfloor\overline{\overline{\mathsf{C}}} \in \mathbb{F}_2\mathbb{F}_2$. Bidyadics play a central role in the analysis of electromagnetic media.

There is a natural dot product between metric bidyadics $\overline{\overline{\mathsf{C}}}_m, \overline{\overline{\mathsf{D}}}_m \in \mathbb{E}_2\mathbb{E}_2$ induced by the natural dot product of bivectors, defined by

$$\overline{\overline{\mathsf{C}}}_m \cdot \overline{\overline{\mathsf{D}}}_m = \overline{\overline{\mathsf{C}}}_m|\overline{\overline{\mathsf{E}}}^{-1}|\overline{\overline{\mathsf{D}}}_m. \tag{3.1}$$

The symmetric bidyadic $\overline{\overline{\mathsf{E}}} \in \mathbb{E}_2\mathbb{E}_2$ serves as the unit bidyadic in the dot product:

$$\overline{\overline{\mathsf{E}}} \cdot \overline{\overline{\mathsf{C}}}_m = \overline{\overline{\mathsf{C}}}_m \cdot \overline{\overline{\mathsf{E}}} = \overline{\overline{\mathsf{C}}}_m. \tag{3.2}$$

Powers of metric bidyadics must be understood in terms of the dot product as

$$\overline{\overline{\mathsf{C}}}{}^2_m = \overline{\overline{\mathsf{C}}}_m \cdot \overline{\overline{\mathsf{C}}}_m, \qquad \overline{\overline{\mathsf{C}}}{}^p_m \cdot \overline{\overline{\mathsf{C}}}{}^q_m = \overline{\overline{\mathsf{C}}}{}^{p+q}_m, \tag{3.3}$$

whence $\overline{\overline{\mathsf{C}}}{}^0_m = \overline{\overline{\mathsf{E}}}$. The inverse of the metric bidyadic is defined with respect to the bar product as

$$\overline{\overline{\mathsf{C}}}_m|\overline{\overline{\mathsf{C}}}{}^{-1}_m = \overline{\overline{\mathsf{I}}}{}^{(2)}, \qquad \overline{\overline{\mathsf{C}}}{}^{-1}_m|\overline{\overline{\mathsf{C}}}_m = \overline{\overline{\mathsf{I}}}{}^{(2)T}. \tag{3.4}$$

Multiforms, Dyadics, and Electromagnetic Media, First Edition. Ismo V. Lindell.

The inverse with respect to the dot product can be represented as

$$\overline{\overline{\mathsf{C}}}_m \cdot (\mathbf{e}_N \mathbf{e}_N \lfloor\lfloor \overline{\overline{\mathsf{C}}}_m^{-1}) = (\mathbf{e}_N \mathbf{e}_N \lfloor\lfloor \overline{\overline{\mathsf{C}}}_m^{-1}) \cdot \overline{\overline{\mathsf{C}}}_m = \overline{\overline{\mathsf{E}}}. \tag{3.5}$$

Like dyadics in the previous chapter, bidyadics can be classified in terms of their rank, which is defined as the dimension of the bivector or two-form subspace formed by the image of the bidyadic. A bidyadic has an inverse only when the rank is 6. Bidyadics form a most important topic because, in the 4D representation of electromagnetic fields, linear media are defined in terms of bidyadics.

3.1 CAYLEY–HAMILTON EQUATION

Any bidyadic $\overline{\overline{\mathsf{C}}} \in \mathbb{E}_2\mathbb{F}_2$ defines an eigenvalue problem

$$\overline{\overline{\mathsf{C}}}|\mathbf{A} = \lambda \mathbf{A}. \tag{3.6}$$

In the general case, there are six eigenvalues λ_i corresponding to six eigenbivectors $\mathbf{A}_i$. The trace of the bidyadic equals the sum of the six eigenvalues,

$$\mathrm{tr}\overline{\overline{\mathsf{C}}} = \sum_i \lambda_i, \tag{3.7}$$

and the determinant (to be defined below) equals the product of the eigenvalues,

$$\det\overline{\overline{\mathsf{C}}} = \lambda_1\lambda_2\lambda_3\lambda_4\lambda_5\lambda_6. \tag{3.8}$$

Similarly to the case handled in Chapter 2, the eigenvalues of the bidyadic satisfy a scalar characteristic equation of the sixth order, which can be written as

$$a_0\lambda^6 + a_1\lambda^5 + a_2\lambda^4 + a_3\lambda^3 + a_4\lambda^2 + a_5\lambda + a_6 = 0. \tag{3.9}$$

The corresponding sixth-order Cayley–Hamilton equation for the bidyadic $\overline{\overline{\mathsf{C}}} \in \mathbb{E}_2\mathbb{F}_2$ has the same form as the characteristic equation,

$$a_0\overline{\overline{\mathsf{C}}}^6 + a_1\overline{\overline{\mathsf{C}}}^5 + a_2\overline{\overline{\mathsf{C}}}^4 + a_3\overline{\overline{\mathsf{C}}}^3 + a_4\overline{\overline{\mathsf{C}}}^2 + a_5\overline{\overline{\mathsf{C}}} + a_6\overline{\overline{\mathsf{I}}}^{(2)} = 0. \tag{3.10}$$

For the metric bidyadic $\overline{\overline{\mathsf{C}}}_m = \overline{\overline{\mathsf{C}}}\rfloor\mathbf{e}_N \in \mathbb{E}_2\mathbb{E}_2$ (3.10) has a similar form in terms of powers defined by (3.3)

$$a_0\overline{\overline{\mathsf{C}}}_m^6 + a_1\overline{\overline{\mathsf{C}}}_m^5 + a_2\overline{\overline{\mathsf{C}}}_m^4 + a_3\overline{\overline{\mathsf{C}}}_m^3 + a_4\overline{\overline{\mathsf{C}}}_m^2 + a_5\overline{\overline{\mathsf{C}}}_m + a_6\overline{\overline{\mathsf{E}}} = 0. \tag{3.11}$$

The scalar equation (3.9) for $\lambda = \lambda_i$ is obtained when multiplying (3.10) by the eigenbivector as $()|\mathbf{A}_i$. When the bidyadic is transformed as

$$\overline{\overline{\mathsf{C}}} \rightarrow \overline{\overline{\mathsf{C}}}_1 = \overline{\overline{\mathsf{D}}}|\overline{\overline{\mathsf{C}}}|\overline{\overline{\mathsf{D}}}^{-1} \tag{3.12}$$

in terms of some bidyadic $\overline{\overline{\mathsf{D}}} \in \mathbb{E}_2\mathbb{F}_2$ possessing an inverse, the Cayley–Hamilton equation (3.10) and the eigenvalues λ are the same for the bidyadic $\overline{\overline{\mathsf{C}}}_1$. Equation (3.12) is called the similarity transformation. All bidyadics satisfying the same Cayley–Hamilton equation are obtained from one known solution through different similarity transformations [8].

3.1.1 Coefficient Functions

Derivation of the Cayley–Hamilton equation (3.10) is more complicated than that of the corresponding fourth-order equation (2.119). Let us just briefly summarize the result given in the literature. The coefficients a_i are functions of the bidyadic $\overline{\overline{\mathsf{C}}}$ and they can be found through the following recursion process [7]

$$a_0 = 1 \tag{3.13}$$

$$a_1 = -a_0\mathrm{tr}\overline{\overline{\mathsf{C}}} \tag{3.14}$$

$$2a_2 = -(a_1\mathrm{tr}\overline{\overline{\mathsf{C}}} + a_0\mathrm{tr}\overline{\overline{\mathsf{C}}}^2) \tag{3.15}$$

$$3a_3 = -(a_2\mathrm{tr}\overline{\overline{\mathsf{C}}} + a_1\mathrm{tr}\overline{\overline{\mathsf{C}}}^2 + a_0\mathrm{tr}\overline{\overline{\mathsf{C}}}^3) \tag{3.16}$$

$$4a_4 = -(a_3\mathrm{tr}\overline{\overline{\mathsf{C}}} + a_2\mathrm{tr}\overline{\overline{\mathsf{C}}}^2 + a_1\mathrm{tr}\overline{\overline{\mathsf{C}}}^3 + a_0\mathrm{tr}\overline{\overline{\mathsf{C}}}^4) \tag{3.17}$$

$$5a_5 = -(a_4\mathrm{tr}\overline{\overline{\mathsf{C}}} + a_3\mathrm{tr}\overline{\overline{\mathsf{C}}}^2 + a_2\mathrm{tr}\overline{\overline{\mathsf{C}}}^3 + a_1\mathrm{tr}\overline{\overline{\mathsf{C}}}^4 + a_0\mathrm{tr}\overline{\overline{\mathsf{C}}}^5) \tag{3.18}$$

$$6a_6 = -(a_5\mathrm{tr}\overline{\overline{\mathsf{C}}} + a_4\mathrm{tr}\overline{\overline{\mathsf{C}}}^2 + a_3\mathrm{tr}\overline{\overline{\mathsf{C}}}^3 + a_2\mathrm{tr}\overline{\overline{\mathsf{C}}}^4 + a_1\mathrm{tr}\overline{\overline{\mathsf{C}}}^5 + a_0\mathrm{tr}\overline{\overline{\mathsf{C}}}^6). \tag{3.19}$$

For example, in the special case $\overline{\overline{\mathsf{C}}} = \overline{\overline{\mathsf{I}}}^{(2)}$ these yield

$$a_0 = 1,\quad a_1 = -6,\quad a_2 = 15,\quad a_3 = -20,\quad a_4 = 15,\quad a_5 = -6,\ a_6 = 1, \tag{3.20}$$

whence (3.10) is reduced to

$$(\overline{\overline{\mathsf{C}}} - \overline{\overline{\mathsf{I}}}^{(2)})^6 = 0. \tag{3.21}$$

For another special case $\overline{\overline{\mathsf{C}}} = \overline{\overline{\mathsf{T}}}^{(2)}$, where $\overline{\overline{\mathsf{T}}}$ is the time reversion dyadic

$$\overline{\overline{\mathsf{T}}} = \overline{\overline{\mathsf{I}}}_s - \mathbf{e}_4\boldsymbol{\varepsilon}_4, \quad \overline{\overline{\mathsf{T}}}^{(2)} = \overline{\overline{\mathsf{I}}}_s^{(2)} - \overline{\overline{\mathsf{I}}}_s{}^\wedge_\wedge\mathbf{e}_4\boldsymbol{\varepsilon}_4, \tag{3.22}$$

we obtain

$$a_0 = 1,\ a_1 = 0,\ a_2 = -3,\ a_3 = 0,\ a_4 = 3,\ a_5 = 0,\ a_6 = -1, \tag{3.23}$$

whence (3.10) is reduced to

$$(\overline{\overline{\mathsf{C}}}^2 - \overline{\overline{\mathsf{I}}}^{(2)})^3 = (\overline{\overline{\mathsf{C}}} - \overline{\overline{\mathsf{I}}}^{(2)})^3|(\overline{\overline{\mathsf{C}}} + \overline{\overline{\mathsf{I}}}^{(2)})^3 = 0. \tag{3.24}$$

The recursion equations (3.13)–(3.19) can be solved for the coefficients as

$$a_1 = -\mathrm{tr}\overline{\overline{\mathsf{C}}} \tag{3.25}$$

$$2a_2 = (\mathrm{tr}\overline{\overline{\mathsf{C}}})^2 - \mathrm{tr}\overline{\overline{\mathsf{C}}}^2 \tag{3.26}$$

$$3!a_3 = -((\mathrm{tr}\overline{\overline{\mathsf{C}}})^3 - 3\mathrm{tr}\overline{\overline{\mathsf{C}}}\ \mathrm{tr}\overline{\overline{\mathsf{C}}}^2 + 2\mathrm{tr}\overline{\overline{\mathsf{C}}}^3) \tag{3.27}$$

$$4!a_4 = (\mathrm{tr}\overline{\overline{\mathsf{C}}})^4 - 6(\mathrm{tr}\overline{\overline{\mathsf{C}}})^2\mathrm{tr}\overline{\overline{\mathsf{C}}}^2 + 3(\mathrm{tr}\overline{\overline{\mathsf{C}}}^2)^2 + 8\mathrm{tr}\overline{\overline{\mathsf{C}}}\ \mathrm{tr}\overline{\overline{\mathsf{C}}}^3 - 6\mathrm{tr}\overline{\overline{\mathsf{C}}}^4 \tag{3.28}$$

$$\begin{aligned}5!a_5 = &-((\mathrm{tr}\overline{\overline{\mathsf{C}}})^5 - 10(\mathrm{tr}\overline{\overline{\mathsf{C}}})^3\mathrm{tr}\overline{\overline{\mathsf{C}}}^2 + 15(\mathrm{tr}\overline{\overline{\mathsf{C}}}^2)^2\mathrm{tr}\overline{\overline{\mathsf{C}}} + 20(\mathrm{tr}\overline{\overline{\mathsf{C}}})^2\ \mathrm{tr}\overline{\overline{\mathsf{C}}}^3 \\ &- 30\mathrm{tr}\overline{\overline{\mathsf{C}}}^4\ \mathrm{tr}\overline{\overline{\mathsf{C}}} - 20\mathrm{tr}\overline{\overline{\mathsf{C}}}^2\ \mathrm{tr}\overline{\overline{\mathsf{C}}}^3 + 24\mathrm{tr}\overline{\overline{\mathsf{C}}}^5)\end{aligned} \tag{3.29}$$

$$\begin{aligned}6!a_6 = &(\mathrm{tr}\overline{\overline{\mathsf{C}}})^6 - 15(\mathrm{tr}\overline{\overline{\mathsf{C}}})^4\mathrm{tr}\overline{\overline{\mathsf{C}}}^2 + 40(\mathrm{tr}\overline{\overline{\mathsf{C}}})^3\mathrm{tr}\overline{\overline{\mathsf{C}}}^3 + 45(\mathrm{tr}\overline{\overline{\mathsf{C}}})^2(\mathrm{tr}\overline{\overline{\mathsf{C}}}^2)^2 \\ &- 120\mathrm{tr}\overline{\overline{\mathsf{C}}}\ \mathrm{tr}\overline{\overline{\mathsf{C}}}^2\ \mathrm{tr}\overline{\overline{\mathsf{C}}}^3 - 90(\mathrm{tr}\overline{\overline{\mathsf{C}}})^2\mathrm{tr}\overline{\overline{\mathsf{C}}}^4 + 144\mathrm{tr}\overline{\overline{\mathsf{C}}}\ \mathrm{tr}\overline{\overline{\mathsf{C}}}^5 \\ &+ 40(\mathrm{tr}\overline{\overline{\mathsf{C}}}^3)^2 - 15(\mathrm{tr}\overline{\overline{\mathsf{C}}}^2)^3 + 90\mathrm{tr}\overline{\overline{\mathsf{C}}}^2\mathrm{tr}\overline{\overline{\mathsf{C}}}^4 - 120\mathrm{tr}\overline{\overline{\mathsf{C}}}^6.\end{aligned} \tag{3.30}$$

It is easy to verify that these expressions reduce to (3.20) and (3.23) for the respective two special bidyadics $\overline{\overline{\mathsf{C}}} = \overline{\overline{\mathsf{I}}}^{(2)}$ and $\overline{\overline{\mathsf{C}}} = \overline{\overline{\mathsf{T}}}^{(2)}$.

There is an interesting set of relations between the expressions of the coefficients a_i. If we consider $x = \mathrm{tr}\overline{\overline{\mathsf{C}}}$ to be independent of $\mathrm{tr}\overline{\overline{\mathsf{C}}}^p$ for $p > 1$, we can show that there exist the following relations

$$\partial_x a_i(x) = -a_{i-1}(x), \tag{3.31}$$

for all i from 1 to 6. Thus, starting from the expression of the coefficient $a_6(x)$, we can obtain the expressions of a_i, $i < 6$ through successive differentiations as

$$a_i(x) = (-\partial_x)^{6-i}a_6(x). \tag{3.32}$$

Thus, the Cayley–Hamilton equation can be expressed in terms of a bidyadic differential operator as

$$\overline{\overline{\mathsf{L}}}(\partial_x)a_6(x) = 0, \quad \overline{\overline{\mathsf{L}}}(\partial_x) = \sum_{i=0}^{6}\overline{\overline{\mathsf{C}}}^i(-\partial_x)^i. \tag{3.33}$$

Details of this are left as an exercise.

3.1.2 Determinant of a Bidyadic

The inverse of a bidyadic $\overline{\overline{\mathsf{C}}}$ can be given the following analytic form,

$$\overline{\overline{\mathsf{C}}}^{-1} = -\frac{1}{a_6}(a_0\overline{\overline{\mathsf{C}}}^5 + a_1\overline{\overline{\mathsf{C}}}^4 + a_2\overline{\overline{\mathsf{C}}}^3 + a_3\overline{\overline{\mathsf{C}}}^2 + a_4\overline{\overline{\mathsf{C}}} + a_5\overline{\overline{\mathsf{I}}}^{(2)}), \qquad (3.34)$$

which is valid for $a_6 \neq 0$. The determinant of a bidyadic $\overline{\overline{\mathsf{C}}}$ can now be defined as a multiple of a_6. The choice

$$\det\overline{\overline{\mathsf{C}}} = a_6, \qquad (3.35)$$

appears best because it yields $\det(\overline{\overline{\mathsf{I}}}_s \pm \mathbf{e}_4\boldsymbol{\varepsilon}_4)^{(2)} = \pm 1$ which coincides with the results for the corresponding 6×6 matrices consisting of three 1's and three ± 1's on its diagonal. The determinant function of a bidyadic $\overline{\overline{\mathsf{C}}}$ has quite an extensive analytic form,

$$\begin{aligned}\det\overline{\overline{\mathsf{C}}} = {} & \frac{1}{720}(\mathrm{tr}\overline{\overline{\mathsf{C}}})^6 - \frac{1}{48}((\mathrm{tr}\overline{\overline{\mathsf{C}}})^4 + (\mathrm{tr}\overline{\overline{\mathsf{C}}}^2)^2)\mathrm{tr}\overline{\overline{\mathsf{C}}}^2 + \frac{1}{16}(\mathrm{tr}\overline{\overline{\mathsf{C}}}\,\mathrm{tr}\overline{\overline{\mathsf{C}}}^2)^2 \\ & + \frac{1}{18}((\mathrm{tr}\overline{\overline{\mathsf{C}}})^3 + \mathrm{tr}\overline{\overline{\mathsf{C}}}^3)\mathrm{tr}\overline{\overline{\mathsf{C}}}^3 - \frac{1}{6}(\mathrm{tr}\overline{\overline{\mathsf{C}}}^6 + \mathrm{tr}\overline{\overline{\mathsf{C}}}\,\mathrm{tr}\overline{\overline{\mathsf{C}}}^2\,\mathrm{tr}\overline{\overline{\mathsf{C}}}^3) \\ & + \frac{1}{8}(\mathrm{tr}\overline{\overline{\mathsf{C}}}^2 - (\mathrm{tr}\overline{\overline{\mathsf{C}}})^2)\mathrm{tr}\overline{\overline{\mathsf{C}}}^4 + \frac{1}{5}\mathrm{tr}\overline{\overline{\mathsf{C}}}\,\mathrm{tr}\overline{\overline{\mathsf{C}}}^5. \end{aligned} \qquad (3.36)$$

For some special cases this expression can be simplified. For example, for a trace-free bidyadic $\overline{\overline{\mathsf{C}}}$ satisfying $\mathrm{tr}\overline{\overline{\mathsf{C}}} = 0$, (3.36) is reduced to

$$\det\overline{\overline{\mathsf{C}}} = -\frac{1}{48}(\mathrm{tr}\overline{\overline{\mathsf{C}}}^2)^3 + \frac{1}{18}(\mathrm{tr}\overline{\overline{\mathsf{C}}}^3)^2 + \frac{1}{8}\mathrm{tr}\overline{\overline{\mathsf{C}}}^2\mathrm{tr}\overline{\overline{\mathsf{C}}}^4 - \frac{1}{6}\mathrm{tr}\overline{\overline{\mathsf{C}}}^6. \qquad (3.37)$$

Let us check this for the time reversal bidyadic (3.22) which satisfies $\mathrm{tr}\overline{\overline{\mathsf{C}}} = 0$. Expanding

$$\begin{aligned}\overline{\overline{\mathsf{C}}}^i = \overline{\overline{\mathsf{I}}}^{(2)} \quad & \Rightarrow \quad \mathrm{tr}\overline{\overline{\mathsf{C}}}^i = 6, \quad i = 2, 4, 6, \\ \overline{\overline{\mathsf{C}}}^i = \overline{\overline{\mathsf{C}}} \quad & \Rightarrow \quad \mathrm{tr}\overline{\overline{\mathsf{C}}}^i = 0, \quad i = 1, 3, 5, \end{aligned} \qquad (3.38)$$

and substituting in (3.37) yields the known result $\det\overline{\overline{\mathsf{C}}} = -1$.

3.1.3 Antisymmetric Bidyadic

Considering an antisymmetric bidyadic $\overline{\overline{\mathsf{C}}}_m = \overline{\overline{\mathsf{C}}}\rfloor\mathbf{e}_N = -\overline{\overline{\mathsf{C}}}_m^T$, the bidyadic Cayley–Hamilton equation (3.11) can be split in its symmetric and antisymmetric parts as

$$a_0\overline{\overline{\mathsf{C}}}_m^6 + a_2\overline{\overline{\mathsf{C}}}_m^4 + a_4\overline{\overline{\mathsf{C}}}_m^2 + a_6\overline{\overline{\mathsf{E}}} = 0, \tag{3.39}$$

$$a_1\overline{\overline{\mathsf{C}}}_m^5 + a_3\overline{\overline{\mathsf{C}}}_m^3 + a_5\overline{\overline{\mathsf{C}}}_m = 0, \tag{3.40}$$

with $\overline{\overline{\mathsf{E}}} = \mathbf{e}_N \lfloor \overline{\overline{\mathsf{I}}}^{(2)T}$. Because we have $\mathrm{tr}(\overline{\overline{\mathsf{C}}}_m \rfloor \boldsymbol{\varepsilon}_N)^p = 0$ for odd values of p, the coefficients a_1, a_3 and a_5 actually vanish whence (3.40) is identically valid and the Cayley–Hamilton equation for an antisymmetric bidyadic $\overline{\overline{\mathsf{C}}}_m$ has the simpler form (3.39). Defining the bidyadic

$$\overline{\overline{\mathsf{D}}} = \overline{\overline{\mathsf{C}}}^2 = \overline{\overline{\mathsf{C}}}_m^2 \rfloor \boldsymbol{\varepsilon}_N = \overline{\overline{\mathsf{C}}}_m \cdot \overline{\overline{\mathsf{C}}}_m \rfloor \boldsymbol{\varepsilon}_N, \tag{3.41}$$

it satisfies

$$a_0\overline{\overline{\mathsf{D}}}^3 + a_2\overline{\overline{\mathsf{D}}}^2 + a_4\overline{\overline{\mathsf{D}}} + a_6\overline{\overline{\mathsf{I}}}^{(2)} = 0, \tag{3.42}$$

with the coefficients defined by

$$a_0 = 1 \tag{3.43}$$

$$a_2 = -\frac{1}{2}\mathrm{tr}\overline{\overline{\mathsf{D}}} \tag{3.44}$$

$$a_4 = \frac{1}{8}(\mathrm{tr}\overline{\overline{\mathsf{D}}})^2 - \frac{1}{4}\mathrm{tr}\overline{\overline{\mathsf{D}}}^2$$

$$a_6 = -\frac{1}{48}(\mathrm{tr}\overline{\overline{\mathsf{D}}})^3 + \frac{1}{8}\mathrm{tr}\overline{\overline{\mathsf{D}}}\,\mathrm{tr}\overline{\overline{\mathsf{D}}}^2 - \frac{1}{6}\mathrm{tr}\overline{\overline{\mathsf{D}}}^3. \tag{3.45}$$

Thus, any bidyadic $\overline{\overline{\mathsf{D}}}$ which can be expressed as a square of a bidyadic $\overline{\overline{\mathsf{C}}}$ of the form $\overline{\overline{\mathsf{C}}} = \overline{\overline{\mathsf{C}}}_m \rfloor \boldsymbol{\varepsilon}_N$, where $\overline{\overline{\mathsf{C}}}_m$ is an antisymmetric bidyadic, satisfies a bidyadic equation of the third order.

The form (3.42) of Cayley–Hamilton equation can also be expressed in operational form if we consider the coefficients as functions of $x = \mathrm{tr}\overline{\overline{\mathsf{D}}} = \mathrm{tr}\overline{\overline{\mathsf{C}}}^2$,

$$\overline{\overline{\mathsf{L}}}(\partial_x)\mathrm{det}\overline{\overline{\mathsf{C}}}(x) = 0, \quad \overline{\overline{\mathsf{L}}}(\partial_x) = \sum_{j=0}^{3} \overline{\overline{\mathsf{D}}}^j(-\partial_x/2)^j. \tag{3.46}$$

3.2 BIDYADIC EIGENPROBLEM

The double contraction $\overline{\overline{\mathsf{I}}}^{(4)} \lfloor\lfloor = \mathbf{e}_N \boldsymbol{\varepsilon}_N \lfloor\lfloor$ can be applied to map between bidyadics $\overline{\overline{\mathsf{C}}} \in \mathbb{E}_2\mathbb{F}_2$ and $\overline{\overline{\mathsf{D}}} \in \mathbb{E}_2\mathbb{F}_2$ as

$$\overline{\overline{\mathsf{D}}} = \overline{\overline{\mathsf{I}}}^{(4)} \lfloor\lfloor \overline{\overline{\mathsf{C}}}^T, \quad \overline{\overline{\mathsf{C}}} = \overline{\overline{\mathsf{I}}}^{(4)} \lfloor\lfloor \overline{\overline{\mathsf{D}}}^T. \tag{3.47}$$

Because of $\mathrm{tr}\overline{\overline{\mathsf{D}}}^p = \mathrm{tr}\overline{\overline{\mathsf{C}}}^p$, both $\overline{\overline{\mathsf{C}}}$ and $\overline{\overline{\mathsf{D}}}$ satisfy the same Cayley–Hamilton equation, an equation with the same coefficients a_i. Being independent of any basis, such a mapping creates a natural eigenproblem for bidyadics,

$$\overline{\overline{\mathrm{I}}}^{(4)}\lfloor\lfloor\overline{\overline{\mathsf{C}}}^T = \lambda\overline{\overline{\mathsf{C}}}. \tag{3.48}$$

From

$$\overline{\overline{\mathsf{C}}} = \overline{\overline{\mathrm{I}}}^{(4)}\lfloor\lfloor(\overline{\overline{\mathrm{I}}}^{(4)T}\lfloor\lfloor\overline{\overline{\mathsf{C}}}) = \lambda\overline{\overline{\mathrm{I}}}^{(4)}\lfloor\lfloor\overline{\overline{\mathsf{C}}}^T = \lambda^2\overline{\overline{\mathsf{C}}}, \tag{3.49}$$

valid for any eigenbidyadic $\overline{\overline{\mathsf{C}}}$, we arrive at two possible eigenvalues,

$$\lambda_\pm = \pm 1. \tag{3.50}$$

The trace of (3.48) yields

$$\mathrm{tr}(\overline{\overline{\mathrm{I}}}^{(4)}\lfloor\lfloor\overline{\overline{\mathsf{C}}}^T) = \mathrm{tr}\overline{\overline{\mathsf{C}}} = \lambda\mathrm{tr}\overline{\overline{\mathsf{C}}}, \tag{3.51}$$

whence the eigenbidyadics $\overline{\overline{\mathsf{C}}}_\pm$ corresponding to the eigenvalues ± 1 satisfy

$$\mathrm{tr}\overline{\overline{\mathsf{C}}}_\pm = \pm\mathrm{tr}\overline{\overline{\mathsf{C}}}_\pm. \tag{3.52}$$

The eigenbidyadic $\overline{\overline{\mathsf{C}}}_-$ is trace free:

$$\mathrm{tr}\overline{\overline{\mathsf{C}}}_- = 0. \tag{3.53}$$

Applying the bidyadic identity (2.114) as

$$\overline{\overline{\mathrm{I}}}^{(4)}\lfloor\lfloor\overline{\overline{\mathsf{C}}}^T = (\mathrm{tr}\overline{\overline{\mathsf{C}}})\overline{\overline{\mathrm{I}}}^{(2)} - (\overline{\overline{\mathsf{C}}}\lfloor\lfloor\overline{\overline{\mathrm{I}}}^T)^\wedge_\wedge\overline{\overline{\mathrm{I}}} + \overline{\overline{\mathsf{C}}} \tag{3.54}$$

to the two eigenbidyadics, we obtain the conditions

$$(\mathrm{tr}\overline{\overline{\mathsf{C}}}_\pm)\overline{\overline{\mathrm{I}}}^{(2)} - (\overline{\overline{\mathsf{C}}}_\pm\lfloor\lfloor\overline{\overline{\mathrm{I}}}^T)^\wedge_\wedge\overline{\overline{\mathrm{I}}} + \overline{\overline{\mathsf{C}}}_\pm = \pm\overline{\overline{\mathsf{C}}}_\pm. \tag{3.55}$$

Further, we have

$$\overline{\overline{\mathrm{I}}}^{(4)}\lfloor\lfloor\overline{\overline{\mathsf{C}}}_\pm^{Tp} = (\overline{\overline{\mathrm{I}}}^{(4)}\lfloor\lfloor\overline{\overline{\mathsf{C}}}_\pm^T)^p = (\pm 1)^p\overline{\overline{\mathsf{C}}}_\pm^p, \tag{3.56}$$

$$\mathrm{tr}\overline{\overline{\mathsf{C}}}_\pm^p = (\pm 1)^p\mathrm{tr}\overline{\overline{\mathsf{C}}}_\pm^p, \tag{3.57}$$

whence $\mathrm{tr}\overline{\overline{\mathsf{C}}}_-^p = 0$ for odd values of p. Let us consider the two eigendyadics separately.

3.2.1 Eigenbidyadic $\overline{\overline{\mathsf{C}}}_-$

Inserting (3.53) in (3.55) we obtain the following condition for the trace-free eigenbidyadic $\overline{\overline{\mathsf{C}}}_-$:

$$\overline{\overline{\mathsf{C}}}_- = \frac{1}{2}(\overline{\overline{\mathsf{C}}}_- \lfloor\lfloor \overline{\mathsf{I}}^T)^\wedge_\wedge \overline{\mathsf{I}}. \tag{3.58}$$

This means that there must exist a dyadic $\overline{\mathsf{B}}_o \in \mathbb{E}_1\mathbb{F}_1$ in terms of which the eigenbidyadic can be expressed in the form

$$\overline{\overline{\mathsf{C}}}_- = \overline{\mathsf{B}}_o{}^\wedge_\wedge \overline{\mathsf{I}}, \qquad \overline{\mathsf{B}}_o = \frac{1}{2}(\overline{\overline{\mathsf{C}}}_- \lfloor\lfloor \overline{\mathsf{I}}^T). \tag{3.59}$$

Applying again (3.53), we have

$$\mathrm{tr}\overline{\overline{\mathsf{C}}}_- = \mathrm{tr}(\overline{\mathsf{B}}_o{}^\wedge_\wedge \overline{\mathsf{I}}) = \overline{\mathsf{B}}_o||(\overline{\mathsf{I}}\rfloor\rfloor \overline{\mathsf{I}}^{(2)T}) = 3\mathrm{tr}\overline{\mathsf{B}}_o = 0, \tag{3.60}$$

whence the dyadic $\overline{\mathsf{B}}_o$ must be trace free.

The converse is also true: any bidyadic $\overline{\overline{\mathsf{C}}}_-$ expressed in the form (3.59) in terms of some trace-free dyadic $\overline{\mathsf{B}}_o \in \mathbb{E}_1\mathbb{F}_1$ is a solution of the eigenproblem (3.48) for $\lambda = -1$. In fact, inserting $\overline{\overline{\mathsf{C}}} = \overline{\mathsf{B}}_o{}^\wedge_\wedge \overline{\mathsf{I}}$ in (2.114) and applying the expansion

$$(\overline{\mathsf{B}}_o{}^\wedge_\wedge \overline{\mathsf{I}})\lfloor\lfloor \overline{\mathsf{I}}^T = (\mathrm{tr}\overline{\mathsf{B}}_o)\overline{\mathsf{I}} + (\mathrm{tr}\overline{\mathsf{I}})\overline{\mathsf{B}}_o - \overline{\mathsf{B}}_o|\overline{\mathsf{I}}|\overline{\mathsf{I}} - \overline{\mathsf{I}}|\overline{\mathsf{I}}|\overline{\mathsf{B}}_o = 2\overline{\mathsf{B}}_o, \tag{3.61}$$

we obtain

$$\overline{\mathsf{I}}^{(4)}\lfloor\lfloor(\overline{\mathsf{B}}_o{}^\wedge_\wedge \overline{\mathsf{I}})^T = -((\overline{\mathsf{B}}_o{}^\wedge_\wedge \overline{\mathsf{I}})\lfloor\lfloor \overline{\mathsf{I}}^T)^\wedge_\wedge \overline{\mathsf{I}} + \overline{\mathsf{B}}_o{}^\wedge_\wedge \overline{\mathsf{I}} = -\overline{\mathsf{B}}_o{}^\wedge_\wedge \overline{\mathsf{I}}, \tag{3.62}$$

whence $\overline{\overline{\mathsf{C}}}_- = \overline{\mathsf{B}}_o{}^\wedge_\wedge \overline{\mathsf{I}}$ is an eigenbidyadic for any trace-free dyadic $\overline{\mathsf{B}}_o$ corresponding to $\lambda = -1$.

3.2.2 Eigenbidyadic $\overline{\overline{\mathsf{C}}}_+$

For an eigenbidyadic $\overline{\overline{\mathsf{C}}}_+$ the identity (3.55) is reduced to

$$(\overline{\overline{\mathsf{C}}}_+ \lfloor\lfloor \overline{\mathsf{I}}^T)^\wedge_\wedge \overline{\mathsf{I}} = (\mathrm{tr}\overline{\overline{\mathsf{C}}}_+)\overline{\mathsf{I}}^{(2)}. \tag{3.63}$$

Writing

$$\overline{\mathsf{I}}^{(2)} = \frac{1}{2}\overline{\mathsf{I}}{}^\wedge_\wedge \overline{\mathsf{I}} = \frac{1}{6}(\overline{\mathsf{I}}^{(2)}\lfloor\lfloor \overline{\mathsf{I}}^T)^\wedge_\wedge \overline{\mathsf{I}}, \tag{3.64}$$

(3.63) can be expressed as

$$(\overline{\overline{\mathsf{C}}}_{+o}\lfloor\lfloor \overline{\mathsf{I}}^T)^\wedge_\wedge \overline{\mathsf{I}} = 0, \tag{3.65}$$

where

$$\overline{\overline{\mathsf{C}}}_{+o} = \overline{\overline{\mathsf{C}}}_{+} - \frac{1}{6}\mathrm{tr}\overline{\overline{\mathsf{C}}}_{+}\,\overline{\overline{\mathsf{I}}}^{(2)}, \tag{3.66}$$

is the trace-free part of $\overline{\overline{\mathsf{C}}}_{+}$. Thus, the eigenbidyadic $\overline{\overline{\mathsf{C}}}_{+}$ can be decomposed in two components as

$$\overline{\overline{\mathsf{C}}}_{+} = \overline{\overline{\mathsf{C}}}_{+o} + A\overline{\overline{\mathsf{I}}}^{(2)}, \quad A = \frac{1}{6}\mathrm{tr}\overline{\overline{\mathsf{C}}}_{+}, \quad \mathrm{tr}\overline{\overline{\mathsf{C}}}_{+o} = 0. \tag{3.67}$$

In fact, inserting (3.67) in (3.55) we obtain (3.65) as the condition for the trace-free part of the eigenbidyadic $\overline{\overline{\mathsf{C}}}_{+}$. Actually, (3.65) is equivalent with the condition

$$\overline{\overline{\mathsf{C}}}_{+o}\lfloor\lfloor\overline{\overline{\mathsf{I}}}^{T} = 0. \tag{3.68}$$

To see this, let us assume that a dyadic $\overline{\overline{\mathsf{B}}} \in \mathbb{E}_1\mathbb{F}_1$ satisfies $\overline{\overline{\mathsf{B}}}{}^{\wedge}_{\wedge}\overline{\overline{\mathsf{I}}} = 0$. From the rule (2.82) we have

$$(\overline{\overline{\mathsf{B}}}{}^{\wedge}_{\wedge}\overline{\overline{\mathsf{I}}})\lfloor\lfloor\overline{\overline{\mathsf{I}}}^{T} = (\mathrm{tr}\overline{\overline{\mathsf{B}}})\overline{\overline{\mathsf{I}}} + 2\overline{\overline{\mathsf{B}}} = 0. \tag{3.69}$$

The trace of this yields $6\mathrm{tr}\overline{\overline{\mathsf{B}}} = 0$, whence $\overline{\overline{\mathsf{B}}}{}^{\wedge}_{\wedge}\overline{\overline{\mathsf{I}}} = 0$ implies $\overline{\overline{\mathsf{B}}} = 0$.

To conclude, the eigenbidyadic $\overline{\overline{\mathsf{C}}}_{+}$ can be decomposed in two parts as (3.67) both of which satisfy the eigenequation (3.48) for $\lambda_+ = +1$. The bidyadic $\overline{\overline{\mathsf{C}}}_{+o}$ is trace free and satisfies the condition (3.68). Conversely, starting from a bidyadic $\overline{\overline{\mathsf{C}}}_{o}$ satisfying $\mathrm{tr}\overline{\overline{\mathsf{C}}}_{o} = 0$ and $\overline{\overline{\mathsf{C}}}_{o}\lfloor\lfloor\overline{\overline{\mathsf{I}}}^{T} = 0$, inserting $\overline{\overline{\mathsf{C}}} = \overline{\overline{\mathsf{C}}}_{o}$ in (3.54) we obtain

$$\overline{\overline{\mathsf{I}}}^{(4)}\lfloor\lfloor\overline{\overline{\mathsf{C}}}_{o}{}^{T} = \overline{\overline{\mathsf{C}}}_{o}, \tag{3.70}$$

whence $\overline{\overline{\mathsf{C}}}_{o}$ is a trace-free eigenbidyadic corresponding to $\lambda = +1$. Because of $\overline{\overline{\mathsf{I}}}^{(4)}\lfloor\lfloor\overline{\overline{\mathsf{I}}}^{(2)T} = \overline{\overline{\mathsf{I}}}^{(2)}$, $\overline{\overline{\mathsf{C}}}_{+} = \overline{\overline{\mathsf{C}}}_{o} + A\overline{\overline{\mathsf{I}}}^{(2)}$ is a more general form of an eigenbidyadic corresponding to $\lambda = +1$.

3.3 HEHL–OBUKHOV DECOMPOSITION

Because of its significance in representing electromagnetic media, let us consider the bidyadic $\overline{\overline{\mathsf{M}}} \in \mathbb{F}_2\mathbb{E}_2$ mapping two-forms to two-forms. The eigenproblem (3.48) takes now the form

$$\overline{\overline{\mathsf{I}}}^{(4)}\lfloor\lfloor\overline{\overline{\mathsf{M}}} = \lambda\overline{\overline{\mathsf{M}}}^{T}. \tag{3.71}$$

Any given bidyadic $\overline{\overline{\mathsf{M}}}$ can be uniquely expanded as a sum of eigenbidyadics as

$$\overline{\overline{\mathsf{M}}} = \overline{\overline{\mathsf{M}}}_+ + \overline{\overline{\mathsf{M}}}_-. \tag{3.72}$$

In fact, we can write

$$\overline{\overline{\mathsf{M}}}_\pm = \frac{1}{2}(\overline{\overline{\mathsf{M}}} \pm \overline{\overline{\mathsf{I}}}^{(4)T} \lfloor\lfloor \overline{\overline{\mathsf{M}}}^T). \tag{3.73}$$

Separating the trace-free and multiple of the unit bidyadic terms of $\overline{\overline{\mathsf{M}}}_+$, a more complete decomposition can be expressed as

$$\overline{\overline{\mathsf{M}}} = \overline{\overline{\mathsf{M}}}_1 + \overline{\overline{\mathsf{M}}}_2 + \overline{\overline{\mathsf{M}}}_3. \tag{3.74}$$

The terminology given here follows that introduced by Hehl and Obukhov in [9] and the decomposition will be called the Hehl–Obukhov decomposition for short.

Here we denote

$$\overline{\overline{\mathsf{M}}}_2 = \overline{\overline{\mathsf{M}}}_-, \tag{3.75}$$

which is called *the skewon part* of $\overline{\overline{\mathsf{M}}}$. Further,

$$\overline{\overline{\mathsf{M}}}_1 + \overline{\overline{\mathsf{M}}}_3 = \overline{\overline{\mathsf{M}}}_+, \tag{3.76}$$

in which $\overline{\overline{\mathsf{M}}}_3$ is called *the axion part* of $\overline{\overline{\mathsf{M}}}$, a multiple of the unit bidyadic,

$$\overline{\overline{\mathsf{M}}}_3 = \frac{1}{6}\mathrm{tr}\overline{\overline{\mathsf{M}}}\,\overline{\overline{\mathsf{I}}}^{(2)T}. \tag{3.77}$$

Finally, $\overline{\overline{\mathsf{M}}}_1$ is called *the principal part* of $\overline{\overline{\mathsf{M}}}$ and it equals the trace-free part of the eigenbidyadic $\overline{\overline{\mathsf{M}}}_+$. Because of the condition (3.68), the principal part satisfies

$$\overline{\overline{\mathsf{M}}}_1 \lfloor\lfloor \overline{\overline{\mathsf{I}}} = 0, \tag{3.78}$$

implying $\mathrm{tr}\overline{\overline{\mathsf{M}}}_1 = 0$, while the skewon part satisfies

$$(\overline{\overline{\mathsf{M}}}_2 \lfloor\lfloor \overline{\overline{\mathsf{I}}})^\wedge_\wedge \overline{\overline{\mathsf{I}}}^T = 2\overline{\overline{\mathsf{M}}}_2, \tag{3.79}$$

implying $\mathrm{tr}\overline{\overline{\mathsf{M}}}_2 = 0$ and, the axion part satisfies

$$(\overline{\overline{\mathsf{M}}}_3 \lfloor\lfloor \overline{\overline{\mathsf{I}}})^\wedge_\wedge \overline{\overline{\mathsf{I}}}^T = 6\overline{\overline{\mathsf{M}}}_3. \tag{3.80}$$

The form (3.79) shows us that the skewon part must be of the form

$$\overline{\overline{\mathsf{M}}}_2 = (\overline{\overline{\mathsf{B}}}_o {}^\wedge_\wedge \overline{\overline{\mathsf{I}}})^T, \tag{3.81}$$

where $\overline{\overline{\mathsf{B}}}_o \in \mathbb{E}_1\mathbb{F}_1$ is a trace-free dyadic. The condition (3.78) can be used as a definition of a principal bidyadic, a bidyadic with no skewon and axion parts. In fact, from $\overline{\overline{\mathsf{M}}}\lfloor\lfloor\overline{\overline{\mathsf{I}}} = 0$ it follows that $\mathrm{tr}\overline{\overline{\mathsf{M}}} = 0$ whence $\overline{\overline{\mathsf{M}}} = \overline{\overline{\mathsf{M}}}_1 + \overline{\overline{\mathsf{M}}}_2$. From (3.79) we then obtain $2\overline{\overline{\mathsf{M}}}_2 = (\overline{\overline{\mathsf{M}}}\lfloor\lfloor\overline{\overline{\mathsf{I}}})^{\wedge}_{\wedge}\overline{\overline{\mathsf{I}}}^T = 0$.

For the metric bidyadic $\overline{\overline{\mathsf{M}}}_m \in \mathbb{E}_2\mathbb{E}_2$ mapping two-forms to bivectors and related to the bidyadic $\overline{\overline{\mathsf{M}}}$ by

$$\overline{\overline{\mathsf{M}}}_m = \mathbf{e}_N\lfloor\overline{\overline{\mathsf{M}}}, \quad \overline{\overline{\mathsf{M}}} = \boldsymbol{\varepsilon}_N\lfloor\overline{\overline{\mathsf{M}}}_m, \tag{3.82}$$

the eigenproblem (3.71) corresponds to

$$\overline{\overline{\mathsf{I}}}^{(4)}\lfloor\lfloor\overline{\overline{\mathsf{M}}} = \mathbf{e}_N\boldsymbol{\varepsilon}_N\lfloor\lfloor(\boldsymbol{\varepsilon}_N\lfloor\overline{\overline{\mathsf{M}}}_m) = \overline{\overline{\mathsf{M}}}_m\rfloor\boldsymbol{\varepsilon}_N = \lambda\overline{\overline{\mathsf{M}}}_m^T\rfloor\boldsymbol{\varepsilon}_N, \tag{3.83}$$

or,

$$\overline{\overline{\mathsf{M}}}_m^T = \lambda\overline{\overline{\mathsf{M}}}_m, \quad \lambda = \pm 1. \tag{3.84}$$

Thus, $\overline{\overline{\mathsf{M}}}_{m+}$ and $\overline{\overline{\mathsf{M}}}_{m-}$ corresponding to the eigenbidyadics $\overline{\overline{\mathsf{M}}}_+$ and $\overline{\overline{\mathsf{M}}}_-$ are respectively symmetric and antisymmetric bidyadics. The corresponding Hehl–Obukhov decomposition for the metric bidyadic is, then,

$$\overline{\overline{\mathsf{M}}}_m = \overline{\overline{\mathsf{M}}}_{m1} + \overline{\overline{\mathsf{M}}}_{m2} + \overline{\overline{\mathsf{M}}}_{m3}, \tag{3.85}$$

where $\overline{\overline{\mathsf{M}}}_{m2}$ equals the antisymmetric part of $\overline{\overline{\mathsf{M}}}_m$ while $\overline{\overline{\mathsf{M}}}_{m3}$ is a multiple of the symmetric dyadic $\mathbf{e}_N\lfloor\overline{\overline{\mathsf{I}}}^{(2)T}$ and $\overline{\overline{\mathsf{M}}}_{m1}$ makes the rest of the symmetric part of $\overline{\overline{\mathsf{M}}}_m$ satisfying $\mathrm{tr}(\boldsymbol{\varepsilon}_N\lfloor\overline{\overline{\mathsf{M}}}_{m1}) = 0$.

From (3.59) it follows that any antisymmetric bidyadic $\overline{\overline{\mathsf{M}}}_{m2} \in \mathbb{E}_2\mathbb{E}_2$ can be expressed in terms of a trace-free dyadic $\overline{\overline{\mathsf{B}}}_o \in \mathbb{E}_1\mathbb{F}_1$ in the form

$$\overline{\overline{\mathsf{M}}}_{m2} = \mathbf{e}_N\lfloor(\overline{\overline{\mathsf{B}}}_o{}^{\wedge}_{\wedge}\overline{\overline{\mathsf{I}}})^T, \quad \mathrm{tr}\overline{\overline{\mathsf{B}}}_o = 0. \tag{3.86}$$

Let us check the number of parameters of the Hehl–Obukhov decomposition. Since an antisymmetric bidyadic corresponds to an antisymmetric 6×6 matrix, it has $(36 - 6)/2 = 15$ free parameters. On the other hand, $\overline{\overline{\mathsf{B}}}_o$ corresponds to a trace-free 4×4 matrix whose number of parameters amounts to $16 - 1 = 15$. Since the axion part has one free parameter, the number of parameters of the principal part equals $36 - 15 - 1 = 20$.

The Hehl–Obukhov decomposition plays an essential role in the definition of various classes of electromagnetic media.

3.4 EXAMPLE: SIMPLE ANTISYMMETRIC BIDYADIC

Let us consider a simple antisymmetric bidyadic $\overline{\overline{\mathsf{U}}} \in \mathbb{F}_2\mathbb{F}_2$, mapping bivectors to two-forms, defined in terms of two two-forms $\mathbf{\Phi}, \mathbf{\Psi} \in \mathbb{F}_2$ as

$$\overline{\overline{\mathsf{U}}} = \frac{1}{2}(\mathbf{\Phi}\mathbf{\Psi} - \mathbf{\Psi}\mathbf{\Phi}). \tag{3.87}$$

Such a bidyadic is of importance when considering quantities like electromagnetic stress, energy, and momentum. The two-forms are assumed to be linearly independent, because otherwise the bidyadic $\overline{\overline{\mathsf{U}}}$ would vanish. Let us study some of its properties.

Applying the natural dot product of two-forms, $\mathbf{\Phi} \cdot \mathbf{\Psi} = \mathbf{e}_N\lfloor(\mathbf{\Phi} \wedge \mathbf{\Psi})$, we can write

$$\overline{\overline{\mathsf{U}}}^2 \cdot \mathbf{\Phi} = \Lambda^2\mathbf{\Phi}, \quad \overline{\overline{\mathsf{U}}}^2 \cdot \mathbf{\Psi} = \Lambda^2\mathbf{\Psi}, \tag{3.88}$$

with

$$\Lambda = \frac{1}{2}\sqrt{(\mathbf{\Psi} \cdot \mathbf{\Phi})^2 - (\mathbf{\Psi} \cdot \mathbf{\Psi})(\mathbf{\Phi} \cdot \mathbf{\Phi})}. \tag{3.89}$$

Thus, for $\Lambda \neq 0$ the bidyadic $(\overline{\overline{\mathsf{U}}}/\Lambda)^2$ maps any two-form of the form $A\mathbf{\Phi} + B\mathbf{\Psi}$ to itself in the dot product. Considering the eigenproblem

$$\overline{\overline{\mathsf{U}}} \cdot \mathbf{\Xi} = \lambda\mathbf{\Xi}, \tag{3.90}$$

from

$$\overline{\overline{\mathsf{U}}}^2 \cdot \mathbf{\Xi} = \lambda\overline{\overline{\mathsf{U}}} \cdot \mathbf{\Xi} = \lambda^2\mathbf{\Xi} \tag{3.91}$$

we conclude that the solutions fall in two sets, either

$$\lambda = 0, \quad \mathbf{\Psi} \cdot \mathbf{\Xi} = \mathbf{\Phi} \cdot \mathbf{\Xi} = 0 \tag{3.92}$$

or

$$\lambda = \pm\Lambda, \quad \mathbf{\Xi}_\pm = A_\pm\mathbf{\Phi} + B_\pm\mathbf{\Psi} \tag{3.93}$$

for some $A_\pm, B_\pm$. Considering the latter case, from

$$\begin{aligned}(\overline{\overline{\mathsf{U}}}^2 - \Lambda^2\overline{\overline{\mathsf{E}}}^{-1})|\mathbf{\Pi} &= (\overline{\overline{\mathsf{U}}} - \Lambda\overline{\overline{\mathsf{E}}}^{-1}) \cdot (\overline{\overline{\mathsf{U}}} + \Lambda\overline{\overline{\mathsf{E}}}^{-1}) \cdot \mathbf{\Pi} \\ &= (\overline{\overline{\mathsf{U}}} + \Lambda\overline{\overline{\mathsf{E}}}^{-1}) \cdot (\overline{\overline{\mathsf{U}}} - \Lambda\overline{\overline{\mathsf{E}}}^{-1}) \cdot \mathbf{\Pi} = 0,\end{aligned} \tag{3.94}$$

valid for any two-form $\mathbf{\Pi}$ which is a linear combination of $\mathbf{\Phi}$ and $\mathbf{\Psi}$, we may express the eigen two-forms as

$$\mathbf{\Xi}_\pm = (\overline{\overline{\mathsf{U}}} \mp \Lambda \overline{\overline{\mathsf{E}}}^{-1}) \cdot \mathbf{\Pi}, \tag{3.95}$$

for any such $\mathbf{\Pi}$ yielding $\mathbf{\Xi}_\pm \neq 0$. The eigen two-forms $\mathbf{\Xi}_\pm$ satisfy the properties

$$\begin{aligned} \mathbf{\Xi}_\pm \cdot \mathbf{\Xi}_\pm &= \mathbf{\Pi} \cdot (-\overline{\overline{\mathsf{U}}} \mp \Lambda \overline{\overline{\mathsf{E}}}^{-1}) \cdot (\overline{\overline{\mathsf{U}}} \mp \Lambda \overline{\overline{\mathsf{E}}}^{-1}) \cdot \mathbf{\Pi} \\ &= -\mathbf{\Pi} \cdot (\overline{\overline{\mathsf{U}}}^2 - \Lambda^2 \overline{\overline{\mathsf{E}}}^{-1}) \cdot \mathbf{\Pi} = 0, \end{aligned} \tag{3.96}$$

$$\begin{aligned} \mathbf{\Xi}_\pm \cdot \mathbf{\Xi}_\mp &= \mathbf{\Pi} \cdot (-\overline{\overline{\mathsf{U}}} \mp \Lambda \overline{\overline{\mathsf{E}}}^{-1}) \cdot (\overline{\overline{\mathsf{U}}} \pm \Lambda \overline{\overline{\mathsf{E}}}^{-1}) \cdot \mathbf{\Pi} \\ &= -\mathbf{\Pi} \cdot (\overline{\overline{\mathsf{U}}}^2 + \Lambda^2 \overline{\overline{\mathsf{E}}}^{-1}) \cdot \mathbf{\Pi} = -2\Lambda^2 \mathbf{\Pi} \cdot \mathbf{\Pi}. \end{aligned} \tag{3.97}$$

Here we have applied antisymmetry of $\overline{\overline{\mathsf{U}}}$ and (3.88) for $\mathbf{\Pi}$, which is a linear combination of $\mathbf{\Phi}$ and $\mathbf{\Psi}$. Assuming that the two-form $\mathbf{\Pi}$ is not simple and $\Lambda \neq 0$, we have $\mathbf{\Xi}_+ \cdot \mathbf{\Xi}_- \neq 0$. In this case $\overline{\overline{\mathsf{U}}}$ can be expressed in terms of the eigen two-forms as

$$\overline{\overline{\mathsf{U}}} = \frac{\Lambda}{\mathbf{\Xi}_+ \cdot \mathbf{\Xi}_-} (\mathbf{\Xi}_+ \mathbf{\Xi}_- - \mathbf{\Xi}_- \mathbf{\Xi}_+). \tag{3.98}$$

This can be verified in terms of (3.90). Also, one can show that the bidyadic

$$\overline{\overline{\mathsf{\Pi}}} = \frac{1}{\mathbf{\Xi}_+ \cdot \mathbf{\Xi}_-} \mathbf{e}_N \lfloor (\mathbf{\Xi}_+ \mathbf{\Xi}_- + \mathbf{\Xi}_- \mathbf{\Xi}_+) \in \mathbb{E}_2 \mathbb{F}_2 \tag{3.99}$$

satisfies $\overline{\overline{\mathsf{\Pi}}}^2 = \overline{\overline{\mathsf{\Pi}}}$, whence it can be interpreted as a projection bidyadic.

From (3.96) it follows that each of the eigen two-forms must be simple, whence they can be expressed in terms of four one-forms $\boldsymbol{\alpha}_\pm, \boldsymbol{\beta}_\pm$ as

$$\mathbf{\Xi}_\pm = \boldsymbol{\alpha}_\pm \wedge \boldsymbol{\beta}_\pm \tag{3.100}$$

Because of $\mathbf{\Xi}_+ \cdot \mathbf{\Xi}_- \neq 0$, the four one-forms make a basis. Let us rewrite the basis one-forms as multiples of $\boldsymbol{\varepsilon}_i$ for convenience,

$$\mathbf{\Xi}_+ = \xi_+ \boldsymbol{\varepsilon}_{12}, \quad \mathbf{\Xi}_- = \xi_- \boldsymbol{\varepsilon}_{34}, \quad \mathbf{\Xi}_+ \cdot \mathbf{\Xi}_- = \xi_+ \xi_-. \tag{3.101}$$

The bidyadic (3.90) can now be expressed in the simple form

$$\begin{aligned} \overline{\overline{\mathsf{U}}} &= \frac{\Lambda}{\xi_+ \xi_-} (\mathbf{\Xi}_+ \mathbf{\Xi}_- - \mathbf{\Xi}_- \mathbf{\Xi}_+) \\ &= \Lambda (\boldsymbol{\varepsilon}_{12} \boldsymbol{\varepsilon}_{34} - \boldsymbol{\varepsilon}_{34} \boldsymbol{\varepsilon}_{12}). \end{aligned} \tag{3.102}$$

It is easy to verify that (3.102) satisfies the eigenequation (3.90) for (3.101).

Since $\overline{\overline{\mathsf{U}}}$ is an antisymmetric bidyadic, it can be represented in terms of a trace-free dyadic $\overline{\overline{\mathsf{B}}}_o \in \mathbb{E}_1\mathbb{F}_1$, in a form similar to (3.86), as

$$\overline{\overline{\mathsf{U}}} = \boldsymbol{\varepsilon}_N \lfloor(\overline{\overline{\mathsf{B}}}_o {}^\wedge_\wedge \overline{\overline{\mathsf{I}}}), \qquad \mathrm{tr}\overline{\overline{\mathsf{B}}}_o = 0. \tag{3.103}$$

The dyadic $\overline{\overline{\mathsf{B}}}_o$ can be recovered from $\overline{\overline{\mathsf{U}}}$ as

$$\overline{\overline{\mathsf{B}}}_o = \frac{1}{2}(\mathbf{e}_N \lfloor \overline{\overline{\mathsf{U}}}) \lfloor\lfloor \overline{\overline{\mathsf{I}}}^T. \tag{3.104}$$

Inserting (3.102) yields the representation

$$\begin{aligned}\overline{\overline{\mathsf{B}}}_o &= \frac{\Lambda}{2}(\mathbf{e}_{34}\boldsymbol{\varepsilon}_{34} - \mathbf{e}_{12}\boldsymbol{\varepsilon}_{12}) \lfloor\lfloor \overline{\overline{\mathsf{I}}}^T \\ &= \frac{\Lambda}{2}(\mathbf{e}_3\boldsymbol{\varepsilon}_3 + \mathbf{e}_4\boldsymbol{\varepsilon}_4 - \mathbf{e}_1\boldsymbol{\varepsilon}_1 - \mathbf{e}_2\boldsymbol{\varepsilon}_2),\end{aligned} \tag{3.105}$$

whence $\overline{\overline{\mathsf{B}}}_o$ satisfies the following remarkable property (Minkowski identity)

$$\overline{\overline{\mathsf{B}}}{}_o^2 = \frac{\Lambda^2}{4}\overline{\overline{\mathsf{I}}} = \frac{1}{4}\mathrm{tr}\overline{\overline{\mathsf{B}}}{}_o^2\, \overline{\overline{\mathsf{I}}}, \tag{3.106}$$

independent of any basis. Thus, $\overline{\overline{\mathsf{B}}}_o$ is a multiple of what can be called a unipotent dyadic.

3.5 INVERSE RULES FOR BIDYADICS

To form an analytic expression for the inverse of a dyadic belonging to the space $\mathbb{E}_1\mathbb{F}_1$ we can apply the convenient rule (2.120). Alternatively, one can apply the rule (2.121) obtained from the Cayley–Hamilton equation. For the metric dyadic of the space $\mathbb{E}_1\mathbb{E}_1$, the inverse follows the rule (2.206). There does not seem to exist a simple universal analytic inverse formula for the general bidyadic $\overline{\overline{\mathsf{C}}} \in \mathbb{E}_2\mathbb{F}_2$. However, one can always use the rule (3.34) due to the bidyadic Cayley–Hamilton equation, rewritten here as

$$\overline{\overline{\mathsf{C}}}^{-1} = -\frac{1}{a_6}(a_0\overline{\overline{\mathsf{C}}}^5 + a_1\overline{\overline{\mathsf{C}}}^4 + a_2\overline{\overline{\mathsf{C}}}^3 + a_3\overline{\overline{\mathsf{C}}}^2 + a_4\overline{\overline{\mathsf{C}}} + a_5\overline{\overline{\mathsf{I}}}^{(2)}), \tag{3.107}$$

in which the extensive expressions (3.13)–(3.19) for the coefficients a_i as functions of the bidyadic $\overline{\overline{\mathsf{C}}}$ must be substituted.

For some special bidyadics it is possible to find the inverse in analytic form. For a bidyadic $\overline{\overline{\mathsf{C}}} \in \mathbb{E}_2\mathbb{F}_2$ which can be expressed in terms of a dyadic $\overline{\overline{\mathsf{A}}}$ as $\overline{\overline{\mathsf{C}}} = \overline{\overline{\mathsf{A}}}^{(2)}$ we can apply the rule (2.123) in the form

$$\overline{\overline{\mathsf{C}}}^{-1} = 3\frac{\overline{\overline{\mathsf{I}}}^{(4)T}\lfloor\lfloor\overline{\overline{\mathsf{C}}}^T}{\mathrm{tr}\overline{\overline{\mathsf{C}}}^{(2)}}, \quad \overline{\overline{\mathsf{C}}} = \overline{\overline{\mathsf{A}}}^{(2)}. \tag{3.108}$$

Similarly, for special metric bidyadics of the form $\overline{\overline{\mathsf{F}}} = \overline{\overline{\mathsf{B}}}^{(2)} \in \mathbb{E}_2\mathbb{E}_2$ we can apply the rule (2.207) as

$$\overline{\overline{\mathsf{F}}}^{-1} = 3\frac{\varepsilon_N\varepsilon_N\lfloor\lfloor\overline{\overline{\mathsf{F}}}^T}{\varepsilon_N\varepsilon_N||\overline{\overline{\mathsf{F}}}^{(2)}} \in \mathbb{F}_2\mathbb{F}_2, \quad \overline{\overline{\mathsf{F}}} = \overline{\overline{\mathsf{B}}}^{(2)}. \tag{3.109}$$

Let us consider other special cases of bidyadics for which we can form inverses in simple analytic form.

3.5.1 Skewon Bidyadic

It turns out to be possible to find a simple rule for the inverse of the general skewon (or antisymmetric) bidyadic. The route to this result can be made through dyadic identities as was shown in [10]. A somewhat shorter route can be found based on the fact that the inverse of an antisymmetric metric bidyadic is antisymmetric, from which it follows that the inverse of a skewon bidyadic is also a skewon bidyadic. This is based on the property (2.191) which is valid for bidyadics as well. Thus, the inverse can be expressed in the form

$$(\overline{\overline{\mathsf{B}}}_o{}^\wedge_\wedge\overline{\overline{\mathsf{I}}})^{-1} = \overline{\overline{\mathsf{A}}}_o{}^\wedge_\wedge\overline{\overline{\mathsf{I}}}. \tag{3.110}$$

where $\overline{\overline{\mathsf{A}}}_o \in \mathbb{E}_1\mathbb{F}_1$ is another trace-free dyadic. The problem is to find an analytic expression for the dyadic $\overline{\overline{\mathsf{A}}}_o$ in terms of the given dyadic $\overline{\overline{\mathsf{B}}}_o$.

Expanding the basic equation for $\overline{\overline{\mathsf{A}}}_o$ through (2.41),

$$(\overline{\overline{\mathsf{B}}}_o{}^\wedge_\wedge\overline{\overline{\mathsf{I}}})|(\overline{\overline{\mathsf{A}}}_o{}^\wedge_\wedge\overline{\overline{\mathsf{I}}}) = (\overline{\overline{\mathsf{B}}}_o|\overline{\overline{\mathsf{A}}}_o){}^\wedge_\wedge\overline{\overline{\mathsf{I}}} + \overline{\overline{\mathsf{B}}}_o{}^\wedge_\wedge\overline{\overline{\mathsf{A}}}_o = \overline{\overline{\mathsf{I}}}^{(2)} \tag{3.111}$$

and contracting this by $\lfloor\lfloor\overline{\overline{\mathsf{I}}}^T$ through (2.82) yields

$$\mathrm{tr}(\overline{\overline{\mathsf{B}}}_o|\overline{\overline{\mathsf{A}}}_o)\overline{\overline{\mathsf{I}}} + 2\overline{\overline{\mathsf{B}}}_o|\overline{\overline{\mathsf{A}}}_o - \overline{\overline{\mathsf{B}}}_o|\overline{\overline{\mathsf{A}}}_o - \overline{\overline{\mathsf{A}}}_o|\overline{\overline{\mathsf{B}}}_o = 3\overline{\overline{\mathsf{I}}}. \tag{3.112}$$

Solving the trace as

$$\mathrm{tr}(\overline{\overline{\mathsf{B}}}_o|\overline{\overline{\mathsf{A}}}_o) = \mathrm{tr}(\overline{\overline{\mathsf{A}}}_o|\overline{\overline{\mathsf{B}}}_o) = 3, \tag{3.113}$$

and inserting it in the previous equation yields

$$\overline{\overline{\mathsf{B}}}_o|\overline{\overline{\mathsf{A}}}_o = \overline{\overline{\mathsf{A}}}_o|\overline{\overline{\mathsf{B}}}_o. \tag{3.114}$$

This means that $\overline{\overline{\mathsf{A}}}_o$ and $\overline{\overline{\mathsf{B}}}_o$ must have the same sets of eigenvectors and eigen one-forms.

Assuming the eigenexpansion

$$\overline{\overline{\mathsf{B}}}_o = \sum_{i=1}^{4} B_i \mathbf{e}_i \boldsymbol{\varepsilon}_i, \tag{3.115}$$

the corresponding expansion for $\overline{\overline{\mathsf{A}}}_o$ must be of the form

$$\overline{\overline{\mathsf{A}}}_o = \sum_{i=1}^{4} A_i \mathbf{e}_i \boldsymbol{\varepsilon}_i. \tag{3.116}$$

After some simple steps, the two skewon bidyadics can be expanded as

$$\overline{\overline{\mathsf{B}}}_o {}^\wedge_\wedge \overline{\overline{\mathsf{I}}} = \sum_{i<j} (B_i + B_j) \mathbf{e}_{ij} \boldsymbol{\varepsilon}_{ij}, \tag{3.117}$$

$$\overline{\overline{\mathsf{A}}}_o {}^\wedge_\wedge \overline{\overline{\mathsf{I}}} = \sum_{i<j} (A_i + A_j) \mathbf{e}_{ij} \boldsymbol{\varepsilon}_{ij}. \tag{3.118}$$

Inserted in (3.111) the following relations for the two sets of eigenvalues are obtained:

$$A_i + A_j = \frac{1}{B_i + B_j}, \quad i \neq j. \tag{3.119}$$

At this point we can apply the trace-free properties of $\overline{\overline{\mathsf{B}}}_o$ and $\overline{\overline{\mathsf{A}}}_o$ by substituting

$$B_4 = -(B_1 + B_2 + B_3), \quad A_4 = -(A_1 + A_2 + A_3), \tag{3.120}$$

and solve (3.119) for A_1, A_2 and A_3 in the form

$$A_i = -\frac{B_i^2}{(B_1 + B_2)(B_2 + B_3)(B_3 + B_1)} + A, \quad i = 1, 2, 3, \tag{3.121}$$

with

$$A = \frac{B_1^2 + B_2^2 + B_3^2 + B_1 B_2 + B_2 B_3 + B_3 B_1}{2(B_1 + B_2)(B_2 + B_3)(B_3 + B_1)}. \tag{3.122}$$

Applying these, from (3.120) we obtain the fourth coefficient A_4 in a similar form,

$$A_4 = -\frac{B_4^2}{(B_1 + B_2)(B_2 + B_3)(B_3 + B_1)} + A. \tag{3.123}$$

To interpret the result in coordinate-free form, let us expand

$$\begin{aligned}\mathrm{tr}\overline{\overline{\mathsf{B}}}_o^{(3)} &= B_1B_2B_3 + B_1B_2B_4 + B_1B_3B_4 + B_2B_3B_4 \\ &= -(B_1 + B_2)(B_2 + B_3)(B_3 + B_1),\end{aligned} \tag{3.124}$$

$$\begin{aligned}\mathrm{tr}\overline{\overline{\mathsf{B}}}_o^{(2)} &= B_1B_2 + B_2B_3 + B_3B_1 + B_1B_4 + B_2B_4 + B_3B_4 \\ &= -(B_1^2 + B_2^2 + B_3^2 + B_1B_2 + B_2B_3 + B_3B_1),\end{aligned} \tag{3.125}$$

in terms of which we can write

$$A_i = \frac{B_i^2}{\mathrm{tr}\overline{\overline{\mathsf{B}}}_o^{(3)}} + \frac{\mathrm{tr}\overline{\overline{\mathsf{B}}}_o^{(2)}}{2\mathrm{tr}\overline{\overline{\mathsf{B}}}_o^{(3)}}, \quad i = 1, 2, 3, 4. \tag{3.126}$$

Thus, finally, the dyadic $\overline{\overline{\mathsf{A}}}_o$ can be expressed as

$$\overline{\overline{\mathsf{A}}}_o = \frac{1}{\mathrm{tr}\overline{\overline{\mathsf{B}}}_o^{(3)}}\left(\overline{\overline{\mathsf{B}}}_o^2 + \frac{1}{2}\mathrm{tr}\overline{\overline{\mathsf{B}}}_o^{(2)}\overline{\overline{\mathsf{I}}}\right), \tag{3.127}$$

whence the inverse of the skewon bidyadic has the compact analytic form

$$(\overline{\overline{\mathsf{B}}}_o{}^\wedge_\wedge\overline{\overline{\mathsf{I}}})^{-1} = \overline{\overline{\mathsf{A}}}_o{}^\wedge_\wedge\overline{\overline{\mathsf{I}}} = \frac{1}{\mathrm{tr}\overline{\overline{\mathsf{B}}}_o^{(3)}}(\overline{\overline{\mathsf{B}}}_o^2{}^\wedge_\wedge\overline{\overline{\mathsf{I}}} + \mathrm{tr}\overline{\overline{\mathsf{B}}}_o^{(2)}\overline{\overline{\mathsf{I}}}^{(2)}). \tag{3.128}$$

There is no inverse in the case of $\mathrm{tr}\overline{\overline{\mathsf{B}}}_o^{(3)} = 0$. (3.128) can be checked by starting from the converse relation

$$\overline{\overline{\mathsf{B}}}_o = \frac{1}{\mathrm{tr}\overline{\overline{\mathsf{A}}}_o^{(3)}}\left(\overline{\overline{\mathsf{A}}}_o^2 + \frac{1}{2}\mathrm{tr}\overline{\overline{\mathsf{A}}}_o^{(2)}\overline{\overline{\mathsf{I}}}\right), \tag{3.129}$$

and substituting (3.127) to obtain the result $\overline{\overline{\mathsf{B}}}_o = \overline{\overline{\mathsf{B}}}_o$. Details are left as an exercise.

Another form for the inverse of a skewon bidyadic can be found by starting from the rule (3.54) for $\overline{\overline{\mathsf{C}}} = \overline{\overline{\mathsf{B}}}_o^{(2)T}$ and expanding

$$\begin{aligned}\overline{\overline{\mathsf{I}}}^{(4)}\lfloor\lfloor\overline{\overline{\mathsf{B}}}_o^{(2)T} &= \mathrm{tr}\overline{\overline{\mathsf{B}}}_o^{(2)}\,\overline{\overline{\mathsf{I}}}^{(2)} - (\overline{\overline{\mathsf{B}}}_o^{(2)}\lfloor\lfloor\overline{\overline{\mathsf{I}}}^T){}^\wedge_\wedge\overline{\overline{\mathsf{I}}} + \overline{\overline{\mathsf{B}}}_o^{(2)} \\ &= \mathrm{tr}\overline{\overline{\mathsf{B}}}_o^{(2)}\,\overline{\overline{\mathsf{I}}}^{(2)} + \overline{\overline{\mathsf{B}}}_o^2{}^\wedge_\wedge\overline{\overline{\mathsf{I}}} + \overline{\overline{\mathsf{B}}}_o^{(2)} \\ &= \mathrm{tr}\overline{\overline{\mathsf{B}}}_o^{(3)}(\overline{\overline{\mathsf{B}}}_o{}^\wedge_\wedge\overline{\overline{\mathsf{I}}})^{-1} + \overline{\overline{\mathsf{B}}}_o^{(2)}.\end{aligned} \tag{3.130}$$

At the last stage we have applied (3.128). From this we obtain the expression

$$(\overline{\overline{\mathsf{B}}}_o{}^\wedge_\wedge\overline{\overline{\mathsf{I}}})^{-1} = \frac{1}{\mathrm{tr}\overline{\overline{\mathsf{B}}}_o^{(3)}}(\overline{\overline{\mathsf{I}}}^{(4)}\lfloor\lfloor\overline{\overline{\mathsf{B}}}_o^{(2)T} - \overline{\overline{\mathsf{B}}}_o^{(2)}). \tag{3.131}$$

As a check we can verify that this solution satisfies

$$\overline{\overline{\mathsf{I}}}^{(4)}\lfloor\lfloor(\overline{\overline{\mathsf{B}}}_o{}^\wedge_\wedge\overline{\overline{\mathsf{I}}})^{-1} = -(\overline{\overline{\mathsf{B}}}_o{}^\wedge_\wedge\overline{\overline{\mathsf{I}}})^{-1T}, \tag{3.132}$$

which is the condition for a skewon bidyadic.

A third expression for the inverse of a skewon bidyadic can be found through the special form (3.42) of the Cayley–Hamilton equation where $\overline{\overline{\mathsf{D}}} = (\overline{\overline{\mathsf{B}}}_o{}^\wedge_\wedge\overline{\overline{\mathsf{I}}})^2$. In this case we have

$$\begin{aligned}(\overline{\overline{\mathsf{B}}}_o{}^\wedge_\wedge\overline{\overline{\mathsf{I}}})^{-1} &= (\overline{\overline{\mathsf{B}}}_o{}^\wedge_\wedge\overline{\overline{\mathsf{I}}})|\overline{\overline{\mathsf{D}}}^{-1} = -\frac{1}{a_6}(\overline{\overline{\mathsf{B}}}_o{}^\wedge_\wedge\overline{\overline{\mathsf{I}}})|(\overline{\overline{\mathsf{D}}}^2 + a_2\overline{\overline{\mathsf{D}}} + a_4\overline{\overline{\mathsf{I}}}^{(2)}) \\ &= -\frac{1}{\det(\overline{\overline{\mathsf{B}}}_o{}^\wedge_\wedge\overline{\overline{\mathsf{I}}})}\Big|\Big((\overline{\overline{\mathsf{B}}}_o{}^\wedge_\wedge\overline{\overline{\mathsf{I}}})^5 - (\mathrm{tr}\overline{\overline{\mathsf{B}}}_o^2)(\overline{\overline{\mathsf{B}}}_o{}^\wedge_\wedge\overline{\overline{\mathsf{I}}})^3 \\ &\quad - \left(\frac{1}{4}(\mathrm{tr}\overline{\overline{\mathsf{B}}}_o^2)^2 - \mathrm{tr}\overline{\overline{\mathsf{B}}}_o^4\right)(\overline{\overline{\mathsf{B}}}_o{}^\wedge_\wedge\overline{\overline{\mathsf{I}}})\Big).\end{aligned} \tag{3.133}$$

Applying (3.37), the determinant of the skewon bidyadic can be shown to reduce to

$$\det(\overline{\overline{\mathsf{B}}}_o{}^\wedge_\wedge\overline{\overline{\mathsf{I}}}) = -(\mathrm{tr}\overline{\overline{\mathsf{B}}}_o^{(3)})^2, \tag{3.134}$$

details of which are left as an exercise.

The inverse of an antisymmetric bidyadic $\overline{\overline{\mathsf{D}}} \in \mathbb{E}_2\mathbb{E}_2$ can be found by expressing it in terms of a skewon bidyadic as $\overline{\overline{\mathsf{D}}} = (\overline{\overline{\mathsf{B}}}_o{}^\wedge_\wedge\overline{\overline{\mathsf{I}}})\rfloor\mathbf{e}_N$ and applying the inverse rule (3.128) to

$$\overline{\overline{\mathsf{D}}}^{-1} = ((\overline{\overline{\mathsf{B}}}_o{}^\wedge_\wedge\overline{\overline{\mathsf{I}}})|(\overline{\overline{\mathsf{I}}}^{(2)}\rfloor\mathbf{e}_N))^{-1} = (\overline{\overline{\mathsf{I}}}^{(2)}\rfloor\mathbf{e}_N)^{-1}|(\overline{\overline{\mathsf{B}}}_o{}^\wedge_\wedge\overline{\overline{\mathsf{I}}})|^{-1} = \boldsymbol{\varepsilon}_N\lfloor(\overline{\overline{\mathsf{B}}}_o{}^\wedge_\wedge\overline{\overline{\mathsf{I}}})^{-1}. \tag{3.135}$$

Since it is known that the inverse of a symmetric and antisymmetric bidyadic is respectively symmetric and antisymmetric, it follows that the inverse of a skewon bidyadic is a skewon bidyadic and the inverse of a principal–axion bidyadic is a principal–axion bidyadic. However, the inverse of a principal bidyadic may contain an axion term and the inverse of a skewon–axion bidyadic may contain a principal term.

3.5.2 Extended Bidyadics

If the inverse of a bidyadic $\overline{\overline{\mathsf{C}}} \in \mathbb{E}_2\mathbb{F}_2$ is known, the inverse of the extended bidyadic

$$\overline{\overline{\mathsf{D}}} = \overline{\overline{\mathsf{C}}} + \mathbf{A}\boldsymbol{\Phi} \tag{3.136}$$

can be formed for any bivector $\mathbf{A}$ and two-form $\mathbf{\Phi}$ satisfying $\mathbf{\Phi}|\overline{\overline{\mathsf{C}}}^{-1}|\mathbf{A} \neq -1$. In fact, the expression

$$\overline{\overline{\mathsf{D}}}^{-1} = \overline{\overline{\mathsf{C}}}^{-1} - \frac{\overline{\overline{\mathsf{C}}}^{-1}|\mathbf{A}\mathbf{\Phi}|\overline{\overline{\mathsf{C}}}^{-1}}{1 + \mathbf{\Phi}|\overline{\overline{\mathsf{C}}}^{-1}|\mathbf{A}} \tag{3.137}$$

can be straightforwardly verified when multiplying by $\overline{\overline{\mathsf{D}}}|$. Similarly, for metric bidyadics $\overline{\overline{\mathsf{D}}}, \overline{\overline{\mathsf{C}}} \in \mathbb{E}_2\mathbb{E}_2$ related by

$$\overline{\overline{\mathsf{D}}} = \overline{\overline{\mathsf{C}}} + \mathbf{AB} \tag{3.138}$$

the inverse becomes

$$\overline{\overline{\mathsf{D}}}^{-1} = \overline{\overline{\mathsf{C}}}^{-1} - \frac{\overline{\overline{\mathsf{C}}}^{-1}|\mathbf{AB}|\overline{\overline{\mathsf{C}}}^{-1}}{1 + \mathbf{B}|\overline{\overline{\mathsf{C}}}^{-1}|\mathbf{A}}. \tag{3.139}$$

The previous rule (3.137) can be generalized. To simplify the notation without essential loss of generality, let us consider the rule for the bidyadic

$$\overline{\overline{\mathsf{D}}} = \overline{\overline{\mathsf{I}}}^{(2)} + \overline{\overline{\mathsf{F}}}, \quad \overline{\overline{\mathsf{F}}} = \mathbf{A}\mathbf{\Phi}. \tag{3.140}$$

In this case the inverse can be expressed in the form

$$(\overline{\overline{\mathsf{I}}}^{(2)} + \overline{\overline{\mathsf{F}}})^{-1} = \overline{\overline{\mathsf{I}}}^{(2)} - \frac{\overline{\overline{\mathsf{F}}}}{1 - a_1}, \quad a_1 = -\mathrm{tr}\overline{\overline{\mathsf{F}}}. \tag{3.141}$$

The expression can be verified by applying the equation

$$\overline{\overline{\mathsf{F}}}^2 + a_1\overline{\overline{\mathsf{F}}} = 0 \tag{3.142}$$

satisfied by the bidyadic $\overline{\overline{\mathsf{F}}}$. In fact, we obtain

$$\begin{aligned}\left(\overline{\overline{\mathsf{I}}}^{(2)} - \frac{\overline{\overline{\mathsf{F}}}}{1 - a_1}\right)|(\overline{\overline{\mathsf{I}}}^{(2)} + \overline{\overline{\mathsf{F}}}) &= \overline{\overline{\mathsf{I}}}^{(2)} + \overline{\overline{\mathsf{F}}} - \frac{\overline{\overline{\mathsf{F}}} + \overline{\overline{\mathsf{F}}}^2}{1 - a_1} \\ &= \overline{\overline{\mathsf{I}}}^{(2)} + \overline{\overline{\mathsf{F}}} - \frac{\overline{\overline{\mathsf{F}}} - a_1\overline{\overline{\mathsf{F}}}}{1 - a_1} = \overline{\overline{\mathsf{I}}}^{(2)}.\end{aligned} \tag{3.143}$$

Actually, the inverse is valid for any bidyadic $\overline{\overline{\mathsf{F}}}$ satisfying (3.142), where a_1 may be any scalar.

Applying the rule (3.137) twice we can find the inverse rule for the bidyadic

$$\overline{\overline{\mathsf{D}}} = \overline{\overline{\mathsf{I}}}^{(2)} + \overline{\overline{\mathsf{F}}}, \quad \overline{\overline{\mathsf{F}}} = \mathbf{A}\mathbf{\Phi} + \mathbf{B}\mathbf{\Psi}. \tag{3.144}$$

After some algebraic steps, details of which are left as an exercise, the inverse formula can be expressed as

$$\overline{\overline{\mathsf{D}}}^{-1} = \overline{\overline{\mathrm{I}}}^{(2)} - \frac{\overline{\overline{\mathsf{N}}}}{D}, \tag{3.145}$$

with

$$\overline{\overline{\mathsf{N}}} = \mathbf{A\Phi} + \mathbf{B\Psi} + (\mathbf{A}|\mathbf{\Phi})\mathbf{B\Psi} + (\mathbf{B}|\mathbf{\Psi})\mathbf{A\Phi} - \mathbf{A\Phi}|\mathbf{B\Psi} - \mathbf{B\Psi}|\mathbf{A\Phi} \tag{3.146}$$

and

$$D = 1 + \mathbf{A}|\mathbf{\Phi} + \mathbf{B}|\mathbf{\Psi} + (\mathbf{A}|\mathbf{\Phi})(\mathbf{B}|\mathbf{\Psi}) - (\mathbf{A}|\mathbf{\Psi})(\mathbf{B}|\mathbf{\Phi}). \tag{3.147}$$

However, the rule (3.145) can be written more compactly as

$$(\overline{\overline{\mathrm{I}}}^{(2)} + \overline{\overline{\mathsf{F}}})^{-1} = \overline{\overline{\mathrm{I}}}^{(2)} - \frac{(1 - a_1)\overline{\overline{\mathsf{F}}} - \overline{\overline{\mathsf{F}}}^2}{1 - a_1 + a_2}, \tag{3.148}$$

with a_1 as above and

$$a_2 = \frac{1}{2}((\mathrm{tr}\overline{\overline{\mathsf{F}}})^2 - \mathrm{tr}\overline{\overline{\mathsf{F}}}^2). \tag{3.149}$$

The expression (3.148) can be verified by first showing that, in this case, $\overline{\overline{\mathsf{F}}}$ satisfies the equation

$$\overline{\overline{\mathsf{F}}}^3 + a_1\overline{\overline{\mathsf{F}}}^2 + a_2\overline{\overline{\mathsf{F}}} = 0. \tag{3.150}$$

The process can be continued by considering a bidyadic $\overline{\overline{\mathsf{F}}}$ satisfying

$$\overline{\overline{\mathsf{F}}}^4 + a_1\overline{\overline{\mathsf{F}}}^3 + a_2\overline{\overline{\mathsf{F}}}^2 + a_3\overline{\overline{\mathsf{F}}} = 0, \tag{3.151}$$

with a_1 and a_2 as above and

$$a_3 = -\frac{1}{3!}((\mathrm{tr}\overline{\overline{\mathsf{F}}})^3 - 3(\mathrm{tr}\overline{\overline{\mathsf{F}}})(\mathrm{tr}\overline{\overline{\mathsf{F}}}^2) + 2\mathrm{tr}\overline{\overline{\mathsf{F}}}^3). \tag{3.152}$$

In this case the inverse becomes

$$(\overline{\overline{\mathrm{I}}} + \overline{\overline{\mathsf{F}}})^{-1} = \overline{\overline{\mathrm{I}}}^{(2)} - \frac{(1 - a_1 + a_2)\overline{\overline{\mathsf{F}}} - (1 - a_1)\overline{\overline{\mathsf{F}}}^2 + \overline{\overline{\mathsf{F}}}^3}{1 - a_1 + a_2 - a_3}. \tag{3.153}$$

Again, (3.153) is valid for a bidyadic $\overline{\overline{\mathsf{F}}}$ satisfying (3.151) for any three scalars a_1, a_2, a_3.

Finally, because any bidyadic $\overline{\overline{\mathsf{F}}}$ satisfies the Cayley–Hamilton equation

$$\overline{\overline{\mathsf{F}}}^6 + a_1\overline{\overline{\mathsf{F}}}^5 + a_2\overline{\overline{\mathsf{F}}}^4 + a_3\overline{\overline{\mathsf{F}}}^2 + a_5\overline{\overline{\mathsf{F}}} + a_6\overline{\overline{\mathsf{I}}}^{(2)} = 0, \tag{3.154}$$

the expression for the inverse of $\overline{\overline{\mathsf{I}}}^{(2)} + \overline{\overline{\mathsf{F}}}$ can be formed by continuing the previous pattern as

$$(\overline{\overline{\mathsf{I}}}^{(2)} + \overline{\overline{\mathsf{F}}})^{-1} = \overline{\overline{\mathsf{I}}}^{(2)} + \frac{1}{A_6}\sum_{j=0}^{5} A_i(-\overline{\overline{\mathsf{F}}})^{6-i}, \tag{3.155}$$

with

$$A_i = \sum_{j=0}^{i}(-1)^j a_j = 1 - a_1 + \cdots + (-1)^i a_i. \tag{3.156}$$

As an example let us form the inverse of a skewon–axion bidyadic

$$\overline{\overline{\mathsf{D}}} = \overline{\overline{\mathsf{C}}}_o + \overline{\overline{\mathsf{I}}}, \quad \overline{\overline{\mathsf{C}}}_o = \overline{\overline{\mathsf{B}}}_o {}^\wedge_\wedge \overline{\overline{\mathsf{I}}}. \tag{3.157}$$

The skewon bidyadic satisfies the reduced Cayley–Hamilton equation

$$\overline{\overline{\mathsf{C}}}_o^6 + a_2\overline{\overline{\mathsf{C}}}_o^4 + a_4\overline{\overline{\mathsf{C}}}_o^2 + a_6\overline{\overline{\mathsf{I}}}^{(2)} = 0, \tag{3.158}$$

with coefficients a_i obtained from (3.13) to (3.19) with $\overline{\overline{\mathsf{C}}} = \overline{\overline{\mathsf{C}}}_o$. Applying the inverse rule (3.155) we obtain

$$(\overline{\overline{\mathsf{I}}}^{(2)} + \overline{\overline{\mathsf{C}}}_o)^{-1} = \overline{\overline{\mathsf{I}}}^{(2)} - \frac{\overline{\overline{\mathsf{C}}}_o - \overline{\overline{\mathsf{C}}}_o^2}{1 + a_2 + a_4 + a_6}|((1 + a_2 + a_4)\overline{\overline{\mathsf{I}}}^{(2)}$$
$$+ (1 + a_2)\overline{\overline{\mathsf{C}}}_o^2 + \overline{\overline{\mathsf{C}}}_o^4). \tag{3.159}$$

Verifying this is left as an exercise.

3.5.3 3D Expansions

Expanding a bidyadic $\overline{\overline{\mathsf{D}}} \in \mathbb{E}_2\mathbb{F}_2$ in its spatial and temporal components as

$$\overline{\overline{\mathsf{D}}} = \overline{\overline{\mathsf{A}}}_s + \mathbf{e}_4 \wedge \overline{\overline{\mathsf{B}}}_s + \overline{\overline{\mathsf{C}}}_s \wedge \boldsymbol{\varepsilon}_4 + \mathbf{e}_4 \wedge \overline{\overline{\mathsf{D}}}_s \wedge \boldsymbol{\varepsilon}_4, \tag{3.160}$$

it is possible to find its inverse in analytic form. Applying the corresponding expansion for the inverse bidyadic

$$\overline{\overline{\mathsf{D}}}^{-1} = \overline{\overline{\mathsf{A}}}'_s + \mathbf{e}_4 \wedge \overline{\overline{\mathsf{B}}}'_s + \overline{\overline{\mathsf{C}}}'_s \wedge \boldsymbol{\varepsilon}_4 + \mathbf{e}_4 \wedge \overline{\overline{\mathsf{D}}}'_s \wedge \boldsymbol{\varepsilon}_4, \tag{3.161}$$

we obtain the relations

$$\overline{\overline{\mathsf{A}}}_s|\overline{\overline{\mathsf{A}}}'_s - \overline{\overline{\mathsf{C}}}_s|\overline{\overline{\mathsf{B}}}'_s = \overline{\overline{\mathsf{I}}}^{(2)}_s, \tag{3.162}$$

$$\overline{\overline{\mathsf{A}}}_s|\overline{\overline{\mathsf{C}}}'_s - \overline{\overline{\mathsf{C}}}_s|\overline{\overline{\mathsf{D}}}'_s = 0, \tag{3.163}$$

$$\overline{\overline{\mathsf{B}}}_s|\overline{\overline{\mathsf{A}}}'_s - \overline{\overline{\mathsf{D}}}_s|\overline{\overline{\mathsf{B}}}'_s = 0, \tag{3.164}$$

$$\overline{\overline{\mathsf{B}}}_s|\overline{\overline{\mathsf{C}}}'_s - \overline{\overline{\mathsf{D}}}_s|\overline{\overline{\mathsf{D}}}'_s = \overline{\overline{\mathsf{I}}}_s. \tag{3.165}$$

To solve for the spatial dyadics $\overline{\overline{\mathsf{A}}}'_s..\overline{\overline{\mathsf{D}}}'_s$ requires certain assumptions on the spatial inverses of some of the dyadics $\overline{\overline{\mathsf{A}}}_s..\overline{\overline{\mathsf{D}}}_s$. For example, assuming the existence of $\overline{\overline{\mathsf{A}}}_s^{-1}$ and $\overline{\overline{\mathsf{B}}}_s^{-1}$, we obtain the expressions

$$\overline{\overline{\mathsf{A}}}'_s = \overline{\overline{\mathsf{A}}}_s^{-1} - \overline{\overline{\mathsf{A}}}_s^{-1}|\overline{\overline{\mathsf{C}}}_s|(\overline{\overline{\mathsf{C}}}_s - \overline{\overline{\mathsf{A}}}_s|\overline{\overline{\mathsf{B}}}_s^{-1}|\overline{\overline{\mathsf{D}}}_s)^{-1}, \tag{3.166}$$

$$\overline{\overline{\mathsf{B}}}'_s = -(\overline{\overline{\mathsf{C}}}_s - \overline{\overline{\mathsf{A}}}_s|\overline{\overline{\mathsf{B}}}_s^{-1}|\overline{\overline{\mathsf{D}}}_s)^{-1}, \tag{3.167}$$

$$\overline{\overline{\mathsf{C}}}'_s = -\overline{\overline{\mathsf{A}}}_s^{-1}|\overline{\overline{\mathsf{C}}}_s|(\overline{\overline{\mathsf{D}}}_s - \overline{\overline{\mathsf{B}}}_s|\overline{\overline{\mathsf{A}}}_s^{-1}|\overline{\overline{\mathsf{C}}}_s)^{-1}, \tag{3.168}$$

$$\overline{\overline{\mathsf{D}}}'_s = -(\overline{\overline{\mathsf{D}}}_s - \overline{\overline{\mathsf{B}}}_s|\overline{\overline{\mathsf{A}}}_s^{-1}|\overline{\overline{\mathsf{C}}}_s)^{-1}. \tag{3.169}$$

Also, the two bracketed spatial dyadic expressions are assumed to have inverses. Similar expressions are obtained by starting from other possibilities. If all of the spatial dyadics have inverses, all expressions are simultaneously valid.

PROBLEMS

3.1 Prove that if a bidyadic $\overline{\overline{\mathsf{C}}} \in \mathbb{E}_2\mathbb{F}_2$ satisfies $\overline{\overline{\mathsf{C}}}\lfloor\mathbf{a} = 0$ for any vector $\mathbf{a}$, it must vanish, $\overline{\overline{\mathsf{C}}} = 0$.

3.2 Solve the bidyadic equation

$$\overline{\overline{\mathsf{I}}}{}^\wedge_\wedge\overline{\overline{\mathsf{X}}} = \overline{\overline{\mathsf{Y}}}$$

for the dyadic $\overline{\overline{\mathsf{X}}} \in \mathbb{E}_1\mathbb{F}_1$ and bidyadic $\overline{\overline{\mathsf{Y}}} \in \mathbb{E}_2\mathbb{F}_2$, assuming that such a solution exists. Show that a necessary and sufficient condition for the solution to exist is that $\overline{\overline{\mathsf{Y}}}$ satisfy the bidyadic equation

$$6\overline{\overline{\mathsf{Y}}} + 2\overline{\overline{\mathsf{I}}}^{(2)}\mathrm{tr}\overline{\overline{\mathsf{Y}}} - 3(\overline{\overline{\mathsf{Y}}}\lfloor\lfloor\overline{\overline{\mathsf{I}}}^T){}^\wedge_\wedge\overline{\overline{\mathsf{I}}} = 0.$$

Hint: contract both sides of the equation by $()\lfloor\lfloor\overline{\overline{\mathsf{I}}}^T$.

3.3 An antisymmetric bidyadic $\overline{\overline{\mathsf{C}}} \in \mathbb{E}_2\mathbb{E}_2$ is called simple if it can be expressed in terms of two bivectors as $\mathbf{AB} - \mathbf{BA}$. Show that the most general antisymmetric bidyadic can be expressed as a sum of three simple antisymmetric bidyadics, that is, in terms of six linearly independent bivectors $\mathbf{A}_i$ in the form

$$\overline{\overline{\mathsf{C}}} = (\mathbf{A}_1\mathbf{A}_2 - \mathbf{A}_2\mathbf{A}_1) + (\mathbf{A}_3\mathbf{A}_4 - \mathbf{A}_4\mathbf{A}_3) + (\mathbf{A}_5\mathbf{A}_6 - \mathbf{A}_6\mathbf{A}_5).$$

Hint: show that a sum of four simple antisymmetric bidyadics can always be reduced to a sum of three simple antisymmetric bidyadics.

3.4 Show that if one of the six bivectors in the expansion of the bidyadic in the previous problem is linearly dependent on the other five, the bidyadic can be expressed as a sum of two simple antisymmetric bidyadics.

3.5 Assuming that the six bivectors $\mathbf{A}_i$ of Problem 3.3 form a basis and the two-forms $\mathbf{\Pi}_i$ make the reciprocal two-form basis satisfying $\mathbf{A}_i|\mathbf{\Pi}_j = \delta_{ij}$, find an expression for the inverse of the general antisymmetric bidyadic $\overline{\overline{\mathsf{C}}}$ of Problem 3.3.

3.6 Derive the inverse expression for the skewon–axion metric bidyadic

$$\overline{\overline{\mathsf{D}}} = \overline{\overline{\mathsf{C}}} + \alpha\overline{\overline{\mathsf{E}}}, \quad \overline{\overline{\mathsf{C}}} = \mathbf{AB} - \mathbf{BA},$$

where $\mathbf{A}, \mathbf{B}$ are two bivectors. Hint: start by expanding $\overline{\overline{\mathsf{C}}}^3$ to find a useful condition.

3.7 Derive the inverse of the bidyadic

$$\overline{\overline{\mathsf{D}}} = \overline{\overline{\mathsf{I}}}^{(2)} + \mathbf{A\Phi} + \mathbf{B\Psi}$$

by applying the rule (3.137) twice. Verify the result by forming the dyadic $\overline{\overline{\mathsf{D}}}|\overline{\overline{\mathsf{D}}}^{-1}$.

3.8 Check the expression for the inverse of a skewon bidyadic (3.128) by starting from the relation

$$\overline{\overline{\mathsf{B}}}_o = \frac{1}{\mathrm{tr}\overline{\overline{\mathsf{A}}}_o^{(3)}} \left(\overline{\overline{\mathsf{A}}}_o^2 + \frac{1}{2}\mathrm{tr}\overline{\overline{\mathsf{A}}}_o^{(2)}\overline{\overline{\mathsf{I}}} \right),$$

where the dyadic $\overline{\overline{\mathsf{A}}}_o$ is related to $\overline{\overline{\mathsf{B}}}_o$ by (3.127), by substituting it in the above expression.

3.9 Find the inverse of the bidyadic

$$\overline{\overline{\mathsf{D}}} = \alpha\overline{\overline{\mathsf{I}}}^{(2)} + \overline{\overline{\mathsf{N}}}{}^{\wedge}_{\wedge}\overline{\overline{\mathsf{I}}} \in \mathbb{E}_2\mathbb{F}_2,$$

where $\overline{\mathsf{N}} \in \mathbb{E}_1\mathbb{F}_1$ is a nilpotent dyadic satisfying $\overline{\mathsf{N}}^2 = 0$. Hint: expand first $(\overline{\mathsf{N}}{}^{\wedge}_{\wedge}\overline{\mathsf{I}})^3$.

3.10 Find the inverse of the bidyadic $\overline{\overline{\mathsf{C}}} = \alpha\overline{\overline{\mathsf{I}}}^{(2)} + \mathbf{a}\boldsymbol{\alpha}{}^{\wedge}_{\wedge}\overline{\mathsf{I}}$, $\alpha \neq 0$.

3.11 Prove that if a bidyadic $\overline{\overline{\mathsf{C}}} \in \mathbb{E}_2\mathbb{F}_2$ satisfies (1) $\mathbf{a} \wedge \overline{\overline{\mathsf{C}}} = 0$ or (2) $\overline{\overline{\mathsf{C}}}\lfloor\mathbf{a} = 0$ for any vector $\mathbf{a}$, we must have $\overline{\overline{\mathsf{C}}} = 0$.

3.12 Prove that if a bidyadic $\overline{\overline{\mathsf{M}}} \in \mathbb{F}_2\mathbb{E}_2$ satisfies $\boldsymbol{\alpha} \wedge \overline{\overline{\mathsf{M}}}\lfloor\boldsymbol{\alpha} = 0$ for any one-form $\boldsymbol{\alpha}$, $\overline{\overline{\mathsf{M}}}$ must be a multiple of the unit bidyadic $\overline{\overline{\mathsf{I}}}^{(2)T}$. Show that the same problem can be expressed for a metric bidyadic $\overline{\overline{\mathsf{M}}}_m \in \mathbb{E}_2\mathbb{E}_2$ in the form $\overline{\overline{\mathsf{M}}}_m\lfloor\lfloor\boldsymbol{\alpha}\boldsymbol{\alpha} = 0$ for all $\boldsymbol{\alpha}$ implies $\overline{\overline{\mathsf{M}}}_m = \alpha\mathbf{e}_N\lfloor\overline{\overline{\mathsf{I}}}^{(2)T}$.

3.13 Find the solutions $\overline{\overline{\mathsf{Q}}} \in \mathbb{E}_1\mathbb{E}_1$ to the bidyadic equation

$$\overline{\overline{\mathsf{Q}}}^{(2)} = Q\mathbf{e}_N\lfloor\overline{\overline{\mathsf{I}}}^{(2)T}.$$

Hint: operate both sides by $\rfloor\rfloor\boldsymbol{\alpha}\boldsymbol{\alpha}$.

3.14 (1) Derive from (3.10) the Cayley–Hamilton equation for the bidyadic $\overline{\overline{\mathsf{C}}}^{-1}$, assuming $\det\overline{\overline{\mathsf{C}}} = a_6 \neq 0$.
(2) Show that if the bidyadic $\overline{\overline{\mathsf{C}}}$ satisfies the Cayley–Hamilton equation (3.10), the bidyadic $\overline{\overline{\mathsf{D}}} = \overline{\overline{\mathsf{I}}}^{(4)}\lfloor\lfloor\overline{\overline{\mathsf{C}}}^T$ satisfies the same equation.

3.15 Check that the bidyadic $\overline{\overline{\mathsf{D}}} = \overline{\overline{\mathsf{C}}}^2$ defined by

$$\overline{\overline{\mathsf{C}}} = \overline{\mathsf{B}}_o{}^{\wedge}_{\wedge}\overline{\mathsf{I}}, \qquad \overline{\mathsf{B}}_o = \frac{1}{2}(\overline{\mathsf{I}} - \mathbf{a}\boldsymbol{\alpha}),$$

with $\mathbf{a}|\boldsymbol{\alpha} = 4$ satisfies the equation (3.42)

$$a_0\overline{\overline{\mathsf{D}}}^3 + a_2\overline{\overline{\mathsf{D}}}^2 + A_4\overline{\overline{\mathsf{D}}} + a_6\overline{\overline{\mathsf{I}}}^{(2)} = 0$$

with coefficients defined by (3.43)–(3.45). Check also that $\overline{\overline{\mathsf{C}}}\rfloor\mathbf{e}_N$ is antisymmetric.

3.16 Find the inverse of a bidyadic of the form

$$\overline{\overline{\mathsf{M}}} = \boldsymbol{\Pi}\mathbf{C} + \boldsymbol{\Lambda}\mathbf{D} + \alpha\overline{\overline{\mathsf{I}}}^{(2)} \in \mathbb{F}_2\mathbb{E}_2$$

defined by two two-forms $\boldsymbol{\Pi}, \boldsymbol{\Lambda}$, two bivectors $\mathbf{C}, \mathbf{D}$ and a scalar α, by assuming that the inverse bidyadic has the form

$$\overline{\overline{\mathsf{M}}}^{-1} = \boldsymbol{\Pi}\mathbf{C}' + \boldsymbol{\Lambda}\mathbf{D}' + \alpha'\overline{\overline{\mathsf{I}}}^{(2)}$$

where $\mathbf{C}, \mathbf{D}$ and α are replaced by the unknown quantities $\mathbf{C}', \mathbf{D}', \alpha'$.

3.17 As a verification of the equation (3.42) show that the trace of its left-hand side of vanishes.

3.18 Check the relations (3.31).

3.19 Verify that another form for the inverse of the skewon bidyadic $\overline{\overline{\mathsf{B}}}_{o}{}^{\wedge}_{\wedge}\overline{\overline{\mathsf{I}}}$ with $\mathrm{tr}\overline{\overline{\mathsf{B}}}_{o} = 0$, is

$$(\overline{\overline{\mathsf{B}}}_{o}{}^{\wedge}_{\wedge}\overline{\overline{\mathsf{I}}})^{-1} = \frac{1}{\mathrm{tr}\overline{\overline{\mathsf{B}}}{}^{(3)}_{o}}(\overline{\overline{\mathsf{I}}}{}^{(4)}\lfloor\lfloor\overline{\overline{\mathsf{B}}}{}^{(2)T}_{o} - \overline{\overline{\mathsf{B}}}{}^{(2)}_{o}).$$

3.20 Show that if a bidyadic $\overline{\overline{\mathsf{M}}} \in \mathbb{F}_2\mathbb{E}_2$ is of full rank and satisfies $\boldsymbol{\alpha} \wedge \overline{\overline{\mathsf{M}}}\lfloor\boldsymbol{\alpha} = 0$ for all one-forms $\boldsymbol{\alpha}$, it must be of the form $\overline{\overline{\mathsf{M}}} = \alpha\overline{\overline{\mathsf{I}}}{}^{(2)T}$. As a conclusion show that the condition $\overline{\overline{\mathsf{M}}}_m\lfloor\lfloor\boldsymbol{\alpha\alpha} = 0$ for all $\boldsymbol{\alpha}$ yields $\overline{\overline{\mathsf{M}}}_m = A\overline{\overline{\mathsf{E}}}$ when $\overline{\overline{\mathsf{M}}}_m \in \mathbb{E}_2\mathbb{E}_2$ is a metric bidyadic.

3.21 Find the form of the Cayley–Hamilton equation for the skewon bidyadic $\overline{\overline{\mathsf{C}}} = \overline{\overline{\mathsf{B}}}_{o}{}^{\wedge}_{\wedge}\overline{\overline{\mathsf{I}}}$. Hint: apply the identity

$$(\overline{\overline{\mathsf{A}}}{}^{\wedge}_{\wedge}\overline{\overline{\mathsf{B}}})|(\overline{\overline{\mathsf{C}}}{}^{\wedge}_{\wedge}\overline{\overline{\mathsf{D}}}) = (\overline{\overline{\mathsf{A}}}|\overline{\overline{\mathsf{C}}}){}^{\wedge}_{\wedge}(\overline{\overline{\mathsf{B}}}{}^{\wedge}_{\wedge}\overline{\overline{\mathsf{D}}}) + (\overline{\overline{\mathsf{A}}}|\overline{\overline{\mathsf{D}}}){}^{\wedge}_{\wedge}(\overline{\overline{\mathsf{B}}}|\overline{\overline{\mathsf{C}}})$$

to expanding various powers $\overline{\overline{\mathsf{C}}}{}^{p}$.

3.22 Applying the results of the previous problem, find the determinant of the skewon bidyadic $\overline{\overline{\mathsf{B}}}_{o}{}^{\wedge}_{\wedge}\overline{\overline{\mathsf{I}}}$.

3.23 Show that when a bidyadic $\overline{\overline{\mathsf{C}}} \in \mathbb{E}_2\mathbb{F}_2$ satisfies $\overline{\overline{\mathsf{I}}}{}^{(3)}\lfloor\lfloor\overline{\overline{\mathsf{C}}}{}^{T} = 0$, it is a principal bidyadic, possessing no axion or skewon part. Hint: start from the identity $\overline{\overline{\mathsf{I}}}{}^{(3)}\lfloor\lfloor\boldsymbol{\alpha}\mathbf{a} = (\mathbf{a}|\boldsymbol{\alpha})\overline{\overline{\mathsf{I}}} - \overline{\overline{\mathsf{I}}}{}^{\wedge}_{\wedge}\mathbf{a}\boldsymbol{\alpha}$, valid for any vector $\mathbf{a}$ and one-form $\boldsymbol{\alpha}$.

3.24 Show by applying the result of Problem 3.6 for the metric bidyadic $\overline{\overline{\mathsf{D}}} = \overline{\overline{\mathsf{E}}} + \mathbf{AB} - \mathbf{BA}$ that the inverse of a skewon–axion bidyadic may contain a principal part.

3.25 Verify the rule (3.159) for the inverse of a skewon–axion bidyadic.

CHAPTER 4

Special Dyadics and Bidyadics

4.1 ORTHOGONALITY CONDITIONS

Let us consider conditions between dyadics and bidyadics which may loosely be called orthogonality conditions.

4.1.1 Orthogonality of Dyadics

Let us assume that the two dyadics $\overline{\overline{\mathsf{A}}}, \overline{\overline{\mathsf{B}}} \in \mathbb{E}_1\mathbb{F}_1$ satisfy the condition

$$\overline{\overline{\mathsf{A}}}|\overline{\overline{\mathsf{B}}} = 0. \tag{4.1}$$

It is obvious that, if either one of the dyadics has an inverse, the other one must be zero. Let us ignore that case and assume that the rank of $\overline{\overline{\mathsf{A}}}$ is less than 4 and find the conditions for $\overline{\overline{\mathsf{B}}}$. From symmetry, similar conditions are valid for $\overline{\overline{\mathsf{A}}}$ when starting from the properties of $\overline{\overline{\mathsf{B}}}$.

$\overline{\overline{\mathsf{A}}}$ *of Rank 3*

Assuming that the dyadic $\overline{\overline{\mathsf{A}}}$ is of rank 3 satisfying

$$\overline{\overline{\mathsf{A}}}^{(4)} = 0, \quad \overline{\overline{\mathsf{A}}}^{(3)} \neq 0, \tag{4.2}$$

it can be expressed as

$$\overline{\overline{\mathsf{A}}} = \mathbf{a}_1\boldsymbol{\varepsilon}_i + \mathbf{a}_2\boldsymbol{\varepsilon}_2 + \mathbf{a}_3\boldsymbol{\varepsilon}_3, \quad \mathbf{a}_1 \wedge \mathbf{a}_2 \wedge \mathbf{a}_3 \neq 0, \tag{4.3}$$

Multiforms, Dyadics, and Electromagnetic Media, First Edition. Ismo V. Lindell.

where $\boldsymbol{\varepsilon}_i$, $i = 1 - 4$ is some basis of one-forms. Expanding the dyadic $\overline{\overline{\mathsf{B}}}$ as

$$\overline{\overline{\mathsf{B}}} = \sum \mathbf{e}_i \boldsymbol{\beta}_i, \tag{4.4}$$

where the vectors $\mathbf{e}_j$ make the reciprocal basis, from (4.1) we obtain

$$\mathbf{a}_1 \boldsymbol{\beta}_1 + \mathbf{a}_2 \boldsymbol{\beta}_2 + \mathbf{a}_3 \boldsymbol{\beta}_3 = 0, \tag{4.5}$$

which implies $\boldsymbol{\beta}_1 = \boldsymbol{\beta}_2 = \boldsymbol{\beta}_3 = 0$. Thus, the $\overline{\overline{\mathsf{B}}}$ dyadic must be of the form

$$\overline{\overline{\mathsf{B}}} = \mathbf{e}_4 \boldsymbol{\beta}_4. \tag{4.6}$$

This means that if $\overline{\overline{\mathsf{A}}}$ is of rank 3, $\overline{\overline{\mathsf{B}}}$ must be of rank 1, of the form $\overline{\overline{\mathsf{B}}} = \mathbf{b}\boldsymbol{\beta}$ with $\mathbf{b}$ satisfying $\overline{\overline{\mathsf{A}}}|\mathbf{b} = 0$. Similarly, if $\overline{\overline{\mathsf{B}}}$ is of rank 3, $\overline{\overline{\mathsf{A}}}$ must be of rank 1.

$\overline{\overline{\mathsf{A}}}$ *of Rank 2*

Assuming that the dyadic $\overline{\overline{\mathsf{A}}}$ is of rank 2 satisfying

$$\overline{\overline{\mathsf{A}}}^{(3)} = 0, \quad \overline{\overline{\mathsf{A}}}^{(2)} \neq 0, \tag{4.7}$$

it can be expanded as

$$\overline{\overline{\mathsf{A}}} = \mathbf{a}_1 \boldsymbol{\varepsilon}_1 + \mathbf{a}_2 \boldsymbol{\varepsilon}_2, \quad \overline{\overline{\mathsf{A}}}^{(2)} = \mathbf{a}_1 \wedge \mathbf{a}_2 \boldsymbol{\varepsilon}_{12} \neq 0. \tag{4.8}$$

Expressing the dyadic $\overline{\overline{\mathsf{B}}}$ again as (4.4), we obtain

$$\mathbf{a}_1 \boldsymbol{\beta}_1 + \mathbf{a}_2 \boldsymbol{\beta}_2 = 0, \tag{4.9}$$

which from (4.8) leads to $\boldsymbol{\beta}_1 = \boldsymbol{\beta}_2 = 0$. Thus, $\overline{\overline{\mathsf{B}}}$ must be of the form

$$\overline{\overline{\mathsf{B}}} = \mathbf{e}_3 \boldsymbol{\beta}_3 + \mathbf{e}_4 \boldsymbol{\beta}_4, \tag{4.10}$$

whence it is at most of rank 2. Another way to express the relation is

$$\overline{\overline{\mathsf{B}}} = \overline{\overline{\mathsf{I}}}^{(4)} \lfloor\lfloor (\overline{\overline{\mathsf{A}}}^{(2)} {}^{\wedge}_{\wedge} \overline{\overline{\mathsf{C}}})^T, \tag{4.11}$$

where $\overline{\overline{\mathsf{C}}} \in \mathbb{E}_1\mathbb{F}_1$ is an arbitrary dyadic.

Because of symmetry, if $\overline{\overline{\mathsf{A}}}$ is of rank 1, $\overline{\overline{\mathsf{B}}}$ may be at most of rank 3. As a conclusion, the sum of the ranks of the two dyadics satisfying the orthogonality condition (4.1) must be ≤ 4.

4.1.2 Orthogonality of Bidyadics

Assuming that $\overline{\overline{\mathsf{C}}}, \overline{\overline{\mathsf{D}}} \in \mathbb{E}_2\mathbb{F}_2$ are two bidyadics satisfying

$$\overline{\overline{\mathsf{C}}}|\overline{\overline{\mathsf{D}}} = 0, \tag{4.12}$$

after reasoning similar to that above, the orthogonality condition (4.12) can be shown to require that the sum of the ranks of the two bidyadics may not exceed 6.

Skipping the other possibilities, let us consider as a representative example the case when the bidyadic $\overline{\overline{\mathsf{C}}}$ is of rank 3, that is, of the form

$$\overline{\overline{\mathsf{C}}} = \mathbf{A}_1\mathbf{\Phi}_1 + \mathbf{A}_2\mathbf{\Phi}_2 + \mathbf{A}_3\mathbf{\Phi}_3, \tag{4.13}$$

where both the bivectors $\mathbf{A}_i$ and the two-forms $\mathbf{\Phi}_i$ are linearly independent. If the two-forms $\mathbf{\Phi}_i$ are completed to a basis by adding three two-forms $\mathbf{\Phi}_4, \mathbf{\Phi}_5$ and $\mathbf{\Phi}_6$ and assuming a basis of bivectors $\mathbf{B}_i$, $i = 1, \ldots, 6$ satisfying $\mathbf{\Phi}_i|\mathbf{B}_j = \delta_{i,j}$, expressing the bidyadic $\overline{\overline{\mathsf{D}}}$ in the form

$$\overline{\overline{\mathsf{D}}} = \sum_{j=1}^{6} \mathbf{B}_j\mathbf{\Psi}_j, \tag{4.14}$$

(4.1) leads to

$$\mathbf{A}_1\mathbf{\Psi}_1 + \mathbf{A}_2\mathbf{\Psi}_2 + \mathbf{A}_3\mathbf{\Psi}_3 = 0. \tag{4.15}$$

From linear independence of the bivectors $\mathbf{A}_j$ we now obtain $\mathbf{\Psi}_1 = \mathbf{\Psi}_2 = \mathbf{\Psi}_3 = 0$. Thus, $\overline{\overline{\mathsf{D}}}$ is reduced to

$$\overline{\overline{\mathsf{D}}} = \mathbf{B}_4\mathbf{\Psi}_4 + \mathbf{B}_5\mathbf{\Psi}_5 + \mathbf{B}_6\mathbf{\Psi}_6. \tag{4.16}$$

In conclusion, the case when the two bidyadics satisfy the orthogonality condition (4.12) and $\overline{\overline{\mathsf{C}}}$ is of rank 3, the bidyadic $\overline{\overline{\mathsf{D}}}$ must be of rank 3 or less. The expansion (4.16) is only limited by requiring that the bivectors $\mathbf{B}_4, \mathbf{B}_5$ and $\mathbf{B}_6$ satisfy $\overline{\overline{\mathsf{C}}}|\mathbf{B}_i = 0$, that is, they are eigenbivectors of $\overline{\overline{\mathsf{C}}}$ corresponding to the eigenvalue zero.

4.2 NILPOTENT DYADICS AND BIDYADICS

Dyadics $\overline{\overline{\mathsf{A}}} \in \mathbb{E}_1\mathbb{F}_1$ are called nilpotent of grade p when they satisfy

$$\overline{\overline{\mathsf{A}}}^p = 0, \quad p > 1. \tag{4.17}$$

Let us consider solutions to the grade 2 equation

$$\overline{\overline{\mathsf{A}}}^2 = 0. \tag{4.18}$$

Obviously, any solution to (4.18) satisfies (4.17) for any $p > 2$ as well, but the converse is not true in general. Replacing $\overline{\overline{\mathsf{B}}}$ by $\overline{\overline{\mathsf{A}}}$ in (4.1) we may conclude that the $\overline{\overline{\mathsf{A}}}$ may be at most of rank 2 satisfying

$$\overline{\overline{\mathsf{A}}}^{(3)} = 0, \quad \Rightarrow \quad \overline{\overline{\mathsf{A}}} = \mathbf{a}_1\boldsymbol{\alpha}_1 + \mathbf{a}_2\boldsymbol{\alpha}_2. \tag{4.19}$$

Inserting this in (4.18) requires that the four conditions

$$\boldsymbol{\alpha}_i|\mathbf{a}_j = 0, \quad i,j = 1,2 \tag{4.20}$$

must be satisfied by the two vectors and one-forms. The dyadic

$$\overline{\overline{\mathsf{A}}} = \mathbf{a}\boldsymbol{\alpha}, \quad \mathrm{tr}\overline{\overline{\mathsf{A}}} = \mathbf{a}|\boldsymbol{\alpha} = 0 \tag{4.21}$$

is a simple special case.

Considering the case of nilpotent bidyadics $\overline{\overline{\mathsf{C}}} \in \mathbb{E}_2\mathbb{F}_2$ of grade 2 satisfying

$$\overline{\overline{\mathsf{C}}}^2 = 0, \tag{4.22}$$

the orthogonality condition (4.12) requires that it must be at most of rank 3, whence it can be represented in the form

$$\overline{\overline{\mathsf{C}}} = \mathbf{A}_1\boldsymbol{\Phi}_1 + \mathbf{A}_2\boldsymbol{\Phi}_2 + \mathbf{A}_3\boldsymbol{\Phi}_3. \tag{4.23}$$

Assuming that the rank is 3, the bivector and two-form triples must be linearly independent. Inserted in (4.18), on the left side we obtain a sum of nine terms each of which must vanish individually,

$$\mathbf{A}_i|\boldsymbol{\Phi}_j = 0, \quad i,j = 1,2,3. \tag{4.24}$$

Thus, each term in (4.23) is a nilpotent bidyadic of rank 1 and each two terms are orthogonal bidyadics. As a simple example of a nilpotent bidyadic of grade 2 and rank 3 we may write

$$\overline{\overline{\mathsf{C}}} = A\mathbf{e}_{12}\boldsymbol{\varepsilon}_{34} + B\mathbf{e}_{23}\boldsymbol{\varepsilon}_{14} + C\mathbf{e}_{31}\boldsymbol{\varepsilon}_{24}. \tag{4.25}$$

Lower-rank nilpotent bidyadics can be obtained by omitting one or two terms of the expansion.

4.3 PROJECTION DYADICS AND BIDYADICS

Projection dyadics $\overline{\overline{\Pi}} \in \mathbb{E}_1\mathbb{F}_1$ are defined by the condition

$$\overline{\overline{\Pi}}^2 = \overline{\overline{\Pi}}, \tag{4.26}$$

which imply

$$\overline{\overline{\Pi}}^p = \overline{\overline{\Pi}}, \quad p \geq 1, \tag{4.27}$$

and

$$\mathrm{tr}\overline{\overline{\Pi}}^p = \mathrm{tr}\overline{\overline{\Pi}}, \quad p \geq 1. \tag{4.28}$$

The dyadic defined by

$$\overline{\overline{\Pi}}' = \overline{\overline{I}} - \overline{\overline{\Pi}} \tag{4.29}$$

can be called the complementary projection dyadic because it satisfies similar conditions,

$$\overline{\overline{\Pi}}'^2 = \overline{\overline{I}} - 2\overline{\overline{\Pi}} + \overline{\overline{\Pi}}^2 = \overline{\overline{I}} - \overline{\overline{\Pi}} = \overline{\overline{\Pi}}', \tag{4.30}$$

$$\overline{\overline{\Pi}}'^p = \overline{\overline{\Pi}}', \quad \mathrm{tr}\overline{\overline{\Pi}}'^p = \mathrm{tr}\overline{\overline{\Pi}}' = 4 - \mathrm{tr}\overline{\overline{\Pi}}, \quad p \geq 1, \tag{4.31}$$

and the orthogonality

$$\overline{\overline{\Pi}}|\overline{\overline{\Pi}}' = \overline{\overline{\Pi}}'|\overline{\overline{\Pi}} = 0. \tag{4.32}$$

Applying the identities

$$\mathrm{tr}\overline{\overline{\mathsf{A}}}^{(2)} = \frac{1}{2}((\mathrm{tr}\overline{\overline{\mathsf{A}}})^2 - \mathrm{tr}\overline{\overline{\mathsf{A}}}^2), \tag{4.33}$$

$$\mathrm{tr}\overline{\overline{\mathsf{A}}}^{(3)} = \frac{1}{6}((\mathrm{tr}\overline{\overline{\mathsf{A}}})^3 - 3(\mathrm{tr}\overline{\overline{\mathsf{A}}})(\mathrm{tr}\overline{\overline{\mathsf{A}}}^2) + 2\mathrm{tr}\overline{\overline{\mathsf{A}}}^3), \tag{4.34}$$

$$\mathrm{tr}\overline{\overline{\mathsf{A}}}^{(4)} = \frac{1}{24}((\mathrm{tr}\overline{\overline{\mathsf{A}}})^4 - 6(\mathrm{tr}\overline{\overline{\mathsf{A}}})^2(\mathrm{tr}\overline{\overline{\mathsf{A}}}^2) + 8(\mathrm{tr}\overline{\overline{\mathsf{A}}})(\mathrm{tr}\overline{\overline{\mathsf{A}}}^3) + 3(\mathrm{tr}\overline{\overline{\mathsf{A}}}^2)^2 - 6\mathrm{tr}\overline{\overline{\mathsf{A}}}^4), \tag{4.35}$$

valid for any dyadic $\overline{\overline{\mathsf{A}}} \in \mathbb{E}_1\mathbb{F}_1$, we obtain

$$\mathrm{tr}\overline{\overline{\Pi}}^{(2)} = \frac{1}{2}(\mathrm{tr}\overline{\overline{\Pi}})(\mathrm{tr}\overline{\overline{\Pi}} - 1), \tag{4.36}$$

$$\mathrm{tr}\overline{\overline{\Pi}}^{(3)} = \frac{1}{6}(\mathrm{tr}\overline{\overline{\Pi}})(\mathrm{tr}\overline{\overline{\Pi}} - 1)(\mathrm{tr}\overline{\overline{\Pi}} - 2), \tag{4.37}$$

$$\mathrm{tr}\overline{\overline{\Pi}}^{(4)} = \frac{1}{24}(\mathrm{tr}\overline{\overline{\Pi}})(\mathrm{tr}\overline{\overline{\Pi}} - 1)(\mathrm{tr}\overline{\overline{\Pi}} - 2)(\mathrm{tr}\overline{\overline{\Pi}} - 3). \tag{4.38}$$

Inserting these and (4.27) in the Cayley–Hamilton equation

$$\overline{\overline{\Pi}}^4 - (\mathrm{tr}\overline{\overline{\Pi}})\overline{\overline{\Pi}}^3 + (\mathrm{tr}\overline{\overline{\Pi}}^{(2)})\overline{\overline{\Pi}}^2 - (\mathrm{tr}\overline{\overline{\Pi}}^{(3)})\overline{\overline{\Pi}} + (\mathrm{tr}\overline{\overline{\Pi}}^{(4)})\overline{\overline{I}} = 0, \qquad (4.39)$$

yields the following condition for the projection dyadic,

$$(\mathrm{tr}\overline{\overline{\Pi}} - 1)(\mathrm{tr}\overline{\overline{\Pi}} - 2)(\mathrm{tr}\overline{\overline{\Pi}} - 3)\left(\overline{\overline{\Pi}} - \frac{1}{4}(\mathrm{tr}\overline{\overline{\Pi}})\overline{\overline{I}}\right) = 0. \qquad (4.40)$$

Vanishing of the last bracketed dyadic of (4.40) corresponds to $\overline{\overline{\Pi}}$ equaling zero or the unit dyadic. The other three solutions denoted by $\overline{\overline{\Pi}}_1, \overline{\overline{\Pi}}_2, \overline{\overline{\Pi}}_3$ are restricted by

$$\mathrm{tr}\overline{\overline{\Pi}}_i = i, \quad i = 1, 2, 3. \qquad (4.41)$$

From (4.36)–(4.38), we now obtain

$$\mathrm{tr}\overline{\overline{\Pi}}_1 = 1, \;\Rightarrow\; \mathrm{tr}\overline{\overline{\Pi}}_1^{(2)} = \mathrm{tr}\overline{\overline{\Pi}}_1^{(3)} = \mathrm{tr}\overline{\overline{\Pi}}_1^{(4)} = 0, \qquad (4.42)$$

$$\mathrm{tr}\overline{\overline{\Pi}}_2 = 2, \;\Rightarrow\; \mathrm{tr}\overline{\overline{\Pi}}_2^{(2)} = 1, \;\; \mathrm{tr}\overline{\overline{\Pi}}_2^{(3)} = \mathrm{tr}\overline{\overline{\Pi}}_2^{(4)} = 0, \qquad (4.43)$$

$$\mathrm{tr}\overline{\overline{\Pi}}_3 = 3, \;\Rightarrow\; \mathrm{tr}\overline{\overline{\Pi}}_3^{(2)} = 3, \;\; \mathrm{tr}\overline{\overline{\Pi}}_3^{(3)} = 1, \;\; \mathrm{tr}\overline{\overline{\Pi}}_3^{(4)} = 0. \qquad (4.44)$$

Inserting the projection dyadic $\overline{\overline{A}} = \overline{\overline{\Pi}}$ in the dyadic identities

$$\overline{\overline{A}}^{(4)} \lfloor\lfloor \overline{\overline{I}}^T = (\mathrm{tr}\overline{\overline{A}})\overline{\overline{A}}^{(3)} - \overline{\overline{A}}^{(2)} {}^\wedge_\wedge \overline{\overline{A}}^2 = (\mathrm{tr}\overline{\overline{A}}^{(4)})\overline{\overline{I}}^{(3)}, \qquad (4.45)$$

$$\overline{\overline{A}}^{(3)} \lfloor\lfloor \overline{\overline{I}}^T = (\mathrm{tr}\overline{\overline{A}})\overline{\overline{A}}^{(2)} - \overline{\overline{A}} {}^\wedge_\wedge \overline{\overline{A}}^2, \qquad (4.46)$$

$$\overline{\overline{A}}^{(2)} \lfloor\lfloor \overline{\overline{I}}^T = (\mathrm{tr}\overline{\overline{A}})\overline{\overline{A}} - \overline{\overline{A}}^2, \qquad (4.47)$$

they are respectively reduced to

$$(\mathrm{tr}\overline{\overline{\Pi}}^{(4)})\overline{\overline{I}}^{(3)} = (\mathrm{tr}\overline{\overline{\Pi}} - 3)\overline{\overline{\Pi}}^{(3)}, \qquad (4.48)$$

$$\overline{\overline{\Pi}}^{(3)} \lfloor\lfloor \overline{\overline{I}}^T = (\mathrm{tr}\overline{\overline{\Pi}} - 2)\overline{\overline{\Pi}}^{(2)}, \qquad (4.49)$$

$$\overline{\overline{\Pi}}^{(2)} \lfloor\lfloor \overline{\overline{I}}^T = (\mathrm{tr}\overline{\overline{\Pi}} - 1)\overline{\overline{\Pi}}. \qquad (4.50)$$

Applying (4.42)–(4.44) they are further reduced to

$$\overline{\overline{\Pi}}_1^{(3)} = 0, \;\; \overline{\overline{\Pi}}_1^{(2)} = 0, \;\; \mathrm{tr}\overline{\overline{\Pi}}_1 = 1, \qquad (4.51)$$

$$\overline{\overline{\Pi}}_2^{(3)} = 0, \;\; \mathrm{tr}\overline{\overline{\Pi}}_2^{(2)} = 1, \qquad (4.52)$$

$$\mathrm{tr}\overline{\overline{\Pi}}_3^{(3)} = 1. \qquad (4.53)$$

From these conditions it follows that the different classes of projection dyadics can be represented by dyadic polynomials of the form

$$\overline{\overline{\Pi}}_0 = 0, \quad (4.54)$$
$$\overline{\overline{\Pi}}_1 = \mathbf{p}_1\boldsymbol{\pi}_1, \quad (4.55)$$
$$\overline{\overline{\Pi}}_2 = \mathbf{p}_1\boldsymbol{\pi}_1 + \mathbf{p}_2\boldsymbol{\pi}_2, \quad (4.56)$$
$$\overline{\overline{\Pi}}_3 = \mathbf{p}_1\boldsymbol{\pi}_1 + \mathbf{p}_2\boldsymbol{\pi}_2 + \mathbf{p}_3\boldsymbol{\pi}_3, \quad (4.57)$$
$$\overline{\overline{\Pi}}_4 = \mathbf{p}_1\boldsymbol{\pi}_1 + \mathbf{p}_2\boldsymbol{\pi}_2 + \mathbf{p}_3\boldsymbol{\pi}_3 + \mathbf{p}_4\boldsymbol{\pi}_4 = \overline{\overline{\mathsf{I}}}, \quad (4.58)$$

where the vectors $\mathbf{p}_i$ and one-forms $\boldsymbol{\pi}_j$ are taken from any reciprocal system of basis vectors and one-forms. Actually, the different projection dyadics act as unit dyadics in certain subspaces of $\mathbb{E}_1\mathbb{F}_1$.

Projection bidyadics can be studied along similar lines of analysis. Adopting the same symbol, the projection bidyadic $\overline{\overline{\Pi}} \in \mathbb{E}_2\mathbb{F}_2$ and its complement $\overline{\overline{\Pi}}' = \overline{\overline{\mathsf{I}}}^{(2)} - \overline{\overline{\Pi}}$ satisfy

$$\overline{\overline{\Pi}}^2 = \overline{\overline{\Pi}}, \quad \overline{\overline{\Pi}}'^2 = \overline{\overline{\Pi}}', \quad \overline{\overline{\Pi}}|\overline{\overline{\Pi}}' = \overline{\overline{\Pi}}'|\overline{\overline{\Pi}} = 0, \quad (4.59)$$

Without further proof we can extend the previous result to mappings of the six-dimensional bivector space as

$$\overline{\overline{\Pi}}_0 = 0, \quad (4.60)$$
$$\overline{\overline{\Pi}}_1 = \mathbf{P}_1\mathbf{\Pi}_1, \quad (4.61)$$
$$\overline{\overline{\Pi}}_2 = \mathbf{P}_1\mathbf{\Pi}_1 + \mathbf{P}_2\mathbf{\Pi}_2, \quad (4.62)$$
$$\overline{\overline{\Pi}}_3 = \mathbf{P}_1\mathbf{\Pi}_1 + \mathbf{P}_2\mathbf{\Pi}_2 + \mathbf{P}_3\mathbf{\Pi}_3, \quad (4.63)$$
$$\overline{\overline{\Pi}}_4 = \mathbf{P}_1\mathbf{\Pi}_1 + \mathbf{P}_2\mathbf{\Pi}_2 + \mathbf{P}_3\mathbf{\Pi}_3 + \mathbf{P}_4\mathbf{\Pi}_4, \quad (4.64)$$
$$\overline{\overline{\Pi}}_5 = \mathbf{P}_1\mathbf{\Pi}_1 + \mathbf{P}_2\mathbf{\Pi}_2 + \mathbf{P}_3\mathbf{\Pi}_3 + \mathbf{P}_4\mathbf{\Pi}_4 + \mathbf{P}_5\mathbf{\Pi}_5, \quad (4.65)$$
$$\overline{\overline{\Pi}}_6 = \mathbf{P}_1\mathbf{\Pi}_1 + \mathbf{P}_2\mathbf{\Pi}_2 + \mathbf{P}_3\mathbf{\Pi}_3 + \mathbf{P}_4\mathbf{\Pi}_4 + \mathbf{P}_5\mathbf{\Pi}_5 + \mathbf{P}_6\mathbf{\Pi}_6, \quad (4.66)$$

where the bivectors and two-forms satisfy $\mathbf{P}_i|\mathbf{\Pi}_j = \delta_{i,j}$. The last projection bidyadic equals the unit bidyadic, $\overline{\overline{\Pi}}_6 = \overline{\overline{\mathsf{I}}}^{(2)}$.

4.4 UNIPOTENT DYADICS AND BIDYADICS

The class of unipotent dyadics $\overline{\overline{\mathsf{K}}} \in \mathbb{E}_1\mathbb{F}_1$ is defined by the condition

$$\overline{\overline{\mathsf{K}}}^2 = \overline{\overline{\mathsf{I}}}. \quad (4.67)$$

$\overline{\overline{\mathsf{K}}}$ must be of rank 4 because

$$(\overline{\overline{\mathsf{K}}}^2)^{(4)} = (\overline{\overline{\mathsf{K}}}^{(4)})^2 = (\mathrm{tr}\overline{\overline{\mathsf{K}}}^{(4)})^2\overline{\overline{\mathsf{I}}}^{(4)} = \overline{\overline{\mathsf{I}}}^{(4)} \tag{4.68}$$

implies either $\mathrm{tr}\overline{\overline{\mathsf{K}}}^{(4)} = 1$ or $= -1$. Considering solutions in terms of a dyadic $\overline{\overline{\Pi}}$ in the form

$$\overline{\overline{\mathsf{K}}} = \pm(\overline{\overline{\mathsf{I}}} - 2\overline{\overline{\Pi}}), \tag{4.69}$$

(4.67) is reduced to

$$\overline{\overline{\mathsf{K}}}^2 = \overline{\overline{\mathsf{I}}} - 4\overline{\overline{\Pi}} + 4\overline{\overline{\Pi}}^2 = \overline{\overline{\mathsf{I}}} \quad \Rightarrow \quad \overline{\overline{\Pi}}^2 = \overline{\overline{\Pi}}. \tag{4.70}$$

Because the dyadic $\overline{\overline{\Pi}}$ satisfies (4.26), it must be a projection dyadic. Thus, any unipotent dyadic can be defined in terms of some projection dyadic through the representation (4.69) and, conversely, any projection dyadic can be represented in terms of some unipotent dyadic. The unipotent dyadic $\overline{\overline{\mathsf{K}}}$ can be expressed in a more symmetric form in terms of the complementary projection dyadic $\overline{\overline{\Pi}}'$ defined by (4.29) as

$$\overline{\overline{\mathsf{K}}} = \pm(\overline{\overline{\Pi}}' - \overline{\overline{\Pi}}). \tag{4.71}$$

Because of (4.54)–(4.58), the projection dyadics satisfy

$$\overline{\overline{\Pi}}_i' = \overline{\overline{\mathsf{I}}} - \overline{\overline{\Pi}}_i = \overline{\overline{\Pi}}_{4-i}, \tag{4.72}$$

whence the possible unipotent dyadics have the following three basic forms,

$$\overline{\overline{\mathsf{K}}} = \pm\overline{\overline{\mathsf{I}}}, \tag{4.73}$$

$$\overline{\overline{\mathsf{K}}} = \pm(\mathbf{p}_1\boldsymbol{\pi}_1 - \mathbf{p}_2\boldsymbol{\pi}_2 - \mathbf{p}_3\boldsymbol{\pi}_3 - \mathbf{p}_4\boldsymbol{\pi}_4), \tag{4.74}$$

$$\overline{\overline{\mathsf{K}}} = \pm(\mathbf{p}_1\boldsymbol{\pi}_1 + \mathbf{p}_2\boldsymbol{\pi}_2 - \mathbf{p}_3\boldsymbol{\pi}_3 - \mathbf{p}_4\boldsymbol{\pi}_4). \tag{4.75}$$

The class of unipotent bidyadics $\overline{\overline{\mathsf{K}}} \in \mathbb{E}_2\mathbb{F}_2$ can be defined by the condition

$$\overline{\overline{\mathsf{K}}}^2 = \overline{\overline{\mathsf{I}}}^{(2)}. \tag{4.76}$$

Their classification can be made along the previous pattern in terms of projection bidyadics $\overline{\overline{\Pi}}_i \in \mathbb{E}_2\mathbb{F}_2$ as

$$\overline{\overline{\mathsf{K}}}_i = \pm(\overline{\overline{\mathsf{I}}}^{(2)} - 2\overline{\overline{\Pi}}_i) = \pm(\overline{\overline{\Pi}}_i' - \overline{\overline{\Pi}}_i). \tag{4.77}$$

There are four different types of unipotent bidyadics which can be listed as follows:

$$\overline{\overline{\mathsf{K}}} = \pm \overline{\overline{\mathsf{I}}}^{(2)}, \tag{4.78}$$

$$\overline{\overline{\mathsf{K}}} = \pm(\mathbf{P}_1\mathbf{\Pi}_1 - \mathbf{P}_2\mathbf{\Pi}_2 - \mathbf{P}_3\mathbf{\Pi}_3 - \mathbf{P}_4\mathbf{\Pi}_4 - \mathbf{P}_5\mathbf{\Pi}_5 - \mathbf{P}_6\mathbf{\Pi}_6), \tag{4.79}$$

$$\overline{\overline{\mathsf{K}}} = \pm(\mathbf{P}_1\mathbf{\Pi}_1 + \mathbf{P}_2\mathbf{\Pi}_2 - \mathbf{P}_3\mathbf{\Pi}_3 - \mathbf{P}_4\mathbf{\Pi}_4 - \mathbf{P}_5\mathbf{\Pi}_5 - \mathbf{P}_6\mathbf{\Pi}_6), \tag{4.80}$$

$$\overline{\overline{\mathsf{K}}} = \pm(\mathbf{P}_1\mathbf{\Pi}_1 + \mathbf{P}_2\mathbf{\Pi}_2 + \mathbf{P}_3\mathbf{\Pi}_3 - \mathbf{P}_4\mathbf{\Pi}_4 - \mathbf{P}_5\mathbf{\Pi}_5 - \mathbf{P}_6\mathbf{\Pi}_6). \tag{4.81}$$

4.5 ALMOST-COMPLEX DYADICS

An n-dimensional real-valued vector space is said to possess an almost complex structure [9] (or complex structure [8]) if there exists a real-valued dyadic $\overline{\overline{\mathsf{J}}} \in \mathbb{E}_1\mathbb{F}_1$ satisfying

$$\overline{\overline{\mathsf{J}}}^2 = -\overline{\overline{\mathsf{I}}}. \tag{4.82}$$

Let us call such a dyadic $\overline{\overline{\mathsf{J}}}$ by the name almost-complex (AC) dyadic. Inserting expansions of $\overline{\overline{\mathsf{J}}}$ and $\overline{\overline{\mathsf{I}}}$ in (4.82) as

$$\left(\sum_{i,j} J_{i,j}\mathbf{e}_i\boldsymbol{\varepsilon}_j\right)^2 = \sum_{i,j} \mathbf{e}_i\boldsymbol{\varepsilon}_j\left(\sum_k J_{i,k}J_{k,j}\right) = \sum_{i,j}(-\delta_{i,j})\mathbf{e}_i\boldsymbol{\varepsilon}_j, \tag{4.83}$$

and requiring that the determinant of the coefficient matrices on each side must be the same:

$$(\det[J_{i,j}])^2 = \det[-\delta_{i,j}] = (-1)^n, \tag{4.84}$$

we can conclude that, for $\overline{\overline{\mathsf{J}}}$ to be a real-valued dyadic, n must be an even number. For $n \leq 4$, we may concentrate to vector spaces of dimensions $n = 2$ and $n = 4$.

For $n = 4$, the Cayley–Hamilton equation (2.119) for an AC dyadic $\overline{\overline{\mathsf{J}}}$ is reduced to

$$\overline{\overline{\mathsf{I}}} + \mathrm{tr}\overline{\overline{\mathsf{J}}}\,\overline{\overline{\mathsf{J}}} - \mathrm{tr}\overline{\overline{\mathsf{J}}}^{(2)}\,\overline{\overline{\mathsf{I}}} - \mathrm{tr}\overline{\overline{\mathsf{J}}}^{(3)}\,\overline{\overline{\mathsf{J}}} + \mathrm{tr}\overline{\overline{\mathsf{J}}}^{(4)}\,\overline{\overline{\mathsf{I}}} = 0. \tag{4.85}$$

Since a real AC dyadic $\overline{\overline{\mathsf{J}}}$ obviously cannot be a multiple of $\overline{\overline{\mathsf{I}}}$, the coefficients of $\overline{\overline{\mathsf{J}}}$ and $\overline{\overline{\mathsf{I}}}$ must individually vanish, which leads to the conditions

$$\mathrm{tr}\overline{\overline{\mathsf{J}}}^{(3)} - \mathrm{tr}\overline{\overline{\mathsf{J}}} = 0, \tag{4.86}$$

$$\mathrm{tr}\overline{\overline{\mathsf{J}}}^{(4)} - \mathrm{tr}\overline{\overline{\mathsf{J}}}^{(2)} + 1 = 0. \tag{4.87}$$

Applying the identities (4.33)–(4.35) for $\overline{\overline{\mathsf{A}}}$ replaced by $\overline{\overline{\mathsf{J}}}$ these conditions take the form

$$\mathrm{tr}\overline{\overline{\mathsf{J}}}((\mathrm{tr}\overline{\overline{\mathsf{J}}})^2+4)=0, \tag{4.88}$$
$$(\mathrm{tr}\overline{\overline{\mathsf{J}}})^2((\mathrm{tr}\overline{\overline{\mathsf{J}}})^2+4)=0. \tag{4.89}$$

For real $\overline{\overline{\mathsf{J}}}$ we must have $\mathrm{tr}\overline{\overline{\mathsf{J}}}=0$ which leads to a set of other conditions,

$$\mathrm{tr}\overline{\overline{\mathsf{J}}}=\mathrm{tr}\overline{\overline{\mathsf{J}}}^3=\mathrm{tr}\overline{\overline{\mathsf{J}}}^{(3)}=0, \tag{4.90}$$
$$\mathrm{tr}\overline{\overline{\mathsf{J}}}^2=-4,\quad \mathrm{tr}\overline{\overline{\mathsf{J}}}^{(2)}=2,\quad \mathrm{tr}\overline{\overline{\mathsf{J}}}^4=4,\quad \mathrm{tr}\overline{\overline{\mathsf{J}}}^{(4)}=1. \tag{4.91}$$

This means that all AC dyadics $\overline{\overline{\mathsf{J}}}$ satisfy the same Cayley–Hamilton equation, that is, they have the same set of four eigenvalues λ_i. Actually, the eigenvalues λ_i can be solved by writing (4.82) as

$$(\overline{\overline{\mathsf{J}}}-j\overline{\overline{\mathsf{I}}})|(\overline{\overline{\mathsf{J}}}+j\overline{\overline{\mathsf{I}}})=0, \tag{4.92}$$

whence the eigenproblem

$$\overline{\overline{\mathsf{J}}}|\mathbf{x}_i=\lambda_i\mathbf{x}_i \quad \boldsymbol{\xi}_i|\overline{\overline{\mathsf{J}}}=\lambda_i\boldsymbol{\xi}, \tag{4.93}$$

has four eigenvectors of the form

$$\mathbf{x}_i=(\overline{\overline{\mathsf{J}}}\pm j\overline{\overline{\mathsf{I}}})|\mathbf{a},\quad \lambda_i=\pm j, \tag{4.94}$$

for any vector $\mathbf{a}$ yielding $\mathbf{x}_i\neq 0$. Because $\overline{\overline{\mathsf{J}}}$ is trace free, each of $+j$ and $-j$ is a double eigenvalue. Denoting them by

$$\lambda_1=j,\quad \lambda_2=-j,\quad \lambda_3=j,\quad \lambda_4=-j, \tag{4.95}$$

they can be checked through the expressions (4.90), (4.91) as

$$\mathrm{tr}\overline{\overline{\mathsf{J}}}=\lambda_1+\lambda_2+\lambda_3+\lambda_4=0,$$
$$\mathrm{tr}\overline{\overline{\mathsf{J}}}^{(2)}=\lambda_1\lambda_2+\lambda_1\lambda_3+\lambda_1\lambda_4+\lambda_2\lambda_3+\lambda_3\lambda_4=2,$$
$$\mathrm{tr}\overline{\overline{\mathsf{J}}}^{(3)}=\lambda_1\lambda_2\lambda_3+\lambda_1\lambda_2\lambda_4+\lambda_1\lambda_3\lambda_4+\lambda_2\lambda_3\lambda_4=0,$$
$$\mathrm{tr}\overline{\overline{\mathsf{J}}}^{(4)}=\lambda_1\lambda_2\lambda_3\lambda_4=1.$$

Applying the similarity transformation (3.12) for dyadics satisfying the same Cayley–Hamilton equation, two AC dyadics are similar in the sense that they can be transformed to one another as

$$\overline{\overline{\mathsf{J}}}_2=\overline{\overline{\mathsf{D}}}|\overline{\overline{\mathsf{J}}}_1|\overline{\overline{\mathsf{D}}}^{-1} \tag{4.96}$$

in terms of some dyadic $\overline{\overline{\mathsf{D}}}\in\mathbb{E}_1\mathbb{F}_1$ possessing an inverse. Thus, knowing one AC dyadic $\overline{\overline{\mathsf{J}}}_1$, all other AC dyadics can be obtained through a suitable dyadic $\overline{\overline{\mathsf{D}}}$ from (4.96).

4.5.1 Two-Dimensional AC Dyadics

To find explicit expressions for AC dyadics let us find a basis expansion of the 2D AC dyadic satisfying

$$\overline{\overline{\mathsf{J}}}^2 = -\overline{\overline{\mathsf{I}}}_{12} = -(\mathbf{e}_1\boldsymbol{\varepsilon}_1 + \mathbf{e}_2\boldsymbol{\varepsilon}_2). \tag{4.97}$$

Since in this case $\overline{\overline{\mathsf{J}}}^{(3)} = 0$ and $\overline{\overline{\mathsf{J}}}^{(4)} = 0$, (4.86) and (4.87) are reduced to

$$\mathrm{tr}\overline{\overline{\mathsf{J}}} = 0, \quad \mathrm{tr}\overline{\overline{\mathsf{J}}}^{(2)} = 1, \tag{4.98}$$

the latter of which can be written in the dyadic form as

$$\overline{\overline{\mathsf{J}}}^{(2)} = \mathrm{tr}\overline{\overline{\mathsf{J}}}^{(2)}\,\overline{\overline{\mathsf{I}}}^{(2)}_{12} = \overline{\overline{\mathsf{I}}}^{(2)}_{12} = \mathbf{e}_{12}\boldsymbol{\varepsilon}_{12}. \tag{4.99}$$

Actually, the conditions (4.98) are equivalent to (4.97) in the 2D case, as can be easily verified.

One can show that the most general solution to (4.97) can be expressed in the form

$$\overline{\overline{\mathsf{J}}} = \mathbf{e}_1\boldsymbol{\varepsilon}_2 - \mathbf{e}_2\boldsymbol{\varepsilon}_1, \tag{4.100}$$

in terms of any two 2D vectors $\mathbf{e}_1, \mathbf{e}_2$ and one-forms $\boldsymbol{\varepsilon}_1, \boldsymbol{\varepsilon}_2$ satisfying $\mathbf{e}_i|\boldsymbol{\varepsilon}_j = \delta_{i,j}$. The result of the form (4.100) can be obtained by starting from the expansion

$$\overline{\overline{\mathsf{J}}} = J_{1,1}\mathbf{e}_1\boldsymbol{\varepsilon}_1 + J_{1,2}\mathbf{e}_1\boldsymbol{\varepsilon}_2 + J_{2,1}\mathbf{e}_2\boldsymbol{\varepsilon}_1 + J_{2,2}\mathbf{e}_2\boldsymbol{\varepsilon}_2. \tag{4.101}$$

Details of the proof are left as an exercise. The 2D AC dyadic (4.100) performs the operation

$$\overline{\overline{\mathsf{J}}}|\mathbf{x} = \overline{\overline{\mathsf{J}}}|(x_1\mathbf{e}_1 + x_2\mathbf{e}_2) = -x_1\mathbf{e}_2 + x_2\mathbf{e}_1, \tag{4.102}$$

which equals 90° rotation of vectors in a plane. A more general rotation by an angle θ can be represented by the dyadic

$$\overline{\overline{\mathsf{R}}}(\theta) = \cos\theta\,\overline{\overline{\mathsf{I}}}_{12} + \sin\theta\,\overline{\overline{\mathsf{J}}} = e^{\theta\overline{\overline{\mathsf{J}}}}. \tag{4.103}$$

4.5.2 Four-Dimensional AC Dyadics

To find a 4D solution of (4.82) let us consider the eigenproblem (4.93) whose eigenvalues are defined by (4.95). The corresponding eigenvectors and eigen one-forms $\mathbf{x}_i, \boldsymbol{\xi}_i$ can be assumed to satisfy the orthogonality conditions

$$\mathbf{x}_i|\boldsymbol{\xi}_j = 0 \tag{4.104}$$

for all $i \neq j$ because of linear independence of the eigenvectors. Further, the eigenvectors and eigen one-forms can be assumed normalized as

$$\mathbf{x}_i|\xi_i = 1, \tag{4.105}$$

whence they form reciprocal sets of vector and one-form bases. In this case the unit dyadic can be represented by

$$\overline{\overline{\mathsf{I}}} = \mathbf{x}_1\xi_1 + \mathbf{x}_2\xi_2 + \mathbf{x}_3\xi_3 + \mathbf{x}_4\xi_4, \tag{4.106}$$

and the AC dyadic has the eigenexpansion

$$\overline{\overline{\mathsf{J}}} = j(\mathbf{x}_1\xi_1 - \mathbf{x}_2\xi_2 + \mathbf{x}_3\xi_3 - \mathbf{x}_4\xi_4). \tag{4.107}$$

It can be easily verified that (4.107) satisfies (4.82) and the conditions (4.90), (4.91). For $\overline{\overline{\mathsf{J}}}$ to be real, all eigenvectors and/or eigen one-forms cannot be real valued.

The form of the 4D AC dyadic (4.107) suggests that it can be split in two 2D AC dyadics by defining

$$\overline{\overline{\mathsf{J}}} = \overline{\overline{\mathsf{J}}}_{12} + \overline{\overline{\mathsf{J}}}_{34}, \tag{4.108}$$

$$\overline{\overline{\mathsf{J}}}_{12} = j(\mathbf{x}_1\xi_1 - \mathbf{x}_2\xi_2), \tag{4.109}$$

$$\overline{\overline{\mathsf{J}}}_{34} = j(\mathbf{x}_3\xi_3 - \mathbf{x}_4\xi_4), \tag{4.110}$$

while $\overline{\overline{\mathsf{I}}}$ can be split in two complementary projection dyadics as

$$\overline{\overline{\mathsf{I}}} = \overline{\overline{\mathsf{I}}}_{12} + \overline{\overline{\mathsf{I}}}_{34}, \tag{4.111}$$

$$\overline{\overline{\mathsf{I}}}_{12} = \mathbf{x}_1\xi_1 + \mathbf{x}_2\xi_2, \tag{4.112}$$

$$\overline{\overline{\mathsf{I}}}_{34} = \mathbf{x}_3\xi_3 + \mathbf{x}_4\xi_4. \tag{4.113}$$

Because of the orthogonality conditions (4.104) $\overline{\overline{\mathsf{J}}}_{12}$ and $\overline{\overline{\mathsf{J}}}_{34}$ satisfy

$$\overline{\overline{\mathsf{J}}}_{12}^2 = -\overline{\overline{\mathsf{I}}}_{12} \quad \overline{\overline{\mathsf{J}}}_{34}^2 = -\overline{\overline{\mathsf{I}}}_{34}, \quad \overline{\overline{\mathsf{J}}}_{12}|\overline{\overline{\mathsf{J}}}_{34} = \overline{\overline{\mathsf{J}}}_{34}|\overline{\overline{\mathsf{J}}}_{12} = 0. \tag{4.114}$$

To express $\overline{\overline{\mathsf{J}}}_{12}$ and $\overline{\overline{\mathsf{J}}}_{34}$ in terms of real-valued components in the form of (4.100), let us require that the following vectors and one-forms are real,

$$\mathbf{e}_1 = \frac{1}{\sqrt{2}}(\mathbf{x}_1 - j\mathbf{x}_2) \quad \mathbf{e}_2 = \frac{1}{\sqrt{2}}(\mathbf{x}_2 - j\mathbf{x}_1), \tag{4.115}$$

$$\mathbf{e}_3 = \frac{1}{\sqrt{2}}(\mathbf{x}_3 - j\mathbf{x}_4) \quad \mathbf{e}_4 = \frac{1}{\sqrt{2}}(\mathbf{x}_4 - j\mathbf{x}_3), \tag{4.116}$$

$$\boldsymbol{\varepsilon}_1 = \frac{1}{\sqrt{2}}(\boldsymbol{\xi}_1 + j\boldsymbol{\xi}_2) \quad \boldsymbol{\varepsilon}_2 = \frac{1}{\sqrt{2}}(\boldsymbol{\xi}_2 + j\boldsymbol{\xi}_1), \tag{4.117}$$

$$\boldsymbol{\varepsilon}_3 = \frac{1}{\sqrt{2}}(\boldsymbol{\xi}_3 + j\boldsymbol{\xi}_4) \quad \boldsymbol{\varepsilon}_4 = \frac{1}{\sqrt{2}}(\boldsymbol{\xi}_4 + j\boldsymbol{\xi}_3). \tag{4.118}$$

Actually, these quantities form a reciprocal set of basis vectors and one-forms because they satisfy

$$\mathbf{e}_i|\boldsymbol{\varepsilon}_j = \delta_{i,j}. \tag{4.119}$$

The two AC dyadics have the simple appearance

$$\overline{\overline{\mathsf{J}}}_{12} = \mathbf{e}_1\boldsymbol{\varepsilon}_2 - \mathbf{e}_2\boldsymbol{\varepsilon}_1 \quad \overline{\overline{\mathsf{J}}}_{34} = \mathbf{e}_3\boldsymbol{\varepsilon}_4 - \mathbf{e}_4\boldsymbol{\varepsilon}_3, \tag{4.120}$$

It is left as an exercise to show that the eigenvectors and eigen one-forms must be alternatively real and imaginary valued to obtain the representation (4.108) in terms of real-valued dyadics (4.120).

Applying the similarity transformation (4.96), any AC dyadic can be expressed in the form

$$\overline{\overline{\mathsf{J}}} = \overline{\overline{\mathsf{D}}}|(\overline{\overline{\mathsf{J}}}_{12} + \overline{\overline{\mathsf{J}}}_{34})|\overline{\overline{\mathsf{D}}}^{-1} \tag{4.121}$$

in terms of a suitable dyadic $\overline{\overline{\mathsf{D}}} \in \mathbb{E}_1\mathbb{F}_1$ possessing an inverse [8], p. 156.

4.6 ALMOST-COMPLEX BIDYADICS

AC bidyadics $\overline{\overline{\mathsf{J}}} \in \mathbb{E}_2\mathbb{F}_2$, are defined as solutions of the bidyadic equation

$$\overline{\overline{\mathsf{J}}}^2 = -\overline{\overline{\mathsf{I}}}^{(2)}. \tag{4.122}$$

Since the underlying bivector space is even-dimensional, $n = 6$, one may expect to find real-valued bidyadic solutions for (4.122). The traces of different powers of the bidyadic $\overline{\overline{\mathsf{J}}}$ satisfy

$$\mathrm{tr}\overline{\overline{\mathsf{J}}}^6 = -\mathrm{tr}\overline{\overline{\mathsf{J}}}^4 = \mathrm{tr}\overline{\overline{\mathsf{J}}}^2 = -6, \tag{4.123}$$

$$\mathrm{tr}\overline{\overline{\mathsf{J}}}^5 = -\mathrm{tr}\overline{\overline{\mathsf{J}}}^3 = \mathrm{tr}\overline{\overline{\mathsf{J}}}. \tag{4.124}$$

The equation

$$(\overline{\overline{\mathsf{J}}}^2 + \overline{\overline{\mathsf{I}}}^{(2)})^3 = \overline{\overline{\mathsf{J}}}^6 + 3\overline{\overline{\mathsf{J}}}^4 + 3\overline{\overline{\mathsf{J}}}^2 + \overline{\overline{\mathsf{I}}}^{(2)} = 0, \tag{4.125}$$

implies that the AC bidyadic $\overline{\mathsf{J}}$ has triple eigenvalues of both $+j$ and $-j$, whence its trace vanishes,

$$\mathrm{tr}\overline{\overline{\mathsf{J}}} = \sum \lambda_i = 0. \tag{4.126}$$

Taking this into account, (4.125) can be shown to equal the Cayley–Hamilton equation (3.10).

Because any AC bidyadic can be obtained from a given AC bidyadic through a similarity transformation, let us construct a solution through a spatial dyadic $\overline{\overline{\mathsf{P}}}_s \in \mathbb{E}_2\mathbb{F}_1$ possessing a spatial inverse $\overline{\overline{\mathsf{P}}}_s^{-1} \in \mathbb{E}_1\mathbb{F}_2$ satisfying

$$\overline{\overline{\mathsf{P}}}_s|\overline{\overline{\mathsf{P}}}_s^{-1} = \overline{\overline{\mathsf{I}}}_s^{(2)}, \quad \overline{\overline{\mathsf{P}}}_s^{-1}|\overline{\overline{\mathsf{P}}}_s = \overline{\overline{\mathsf{I}}}_s. \tag{4.127}$$

To simplify the notation, let us define two bidyadics $\overline{\overline{\mathsf{N}}}_a, \overline{\overline{\mathsf{N}}}_b \in \mathbb{E}_2\mathbb{F}_2$ by

$$\overline{\overline{\mathsf{N}}}_a = \overline{\overline{\mathsf{P}}}_s \wedge \boldsymbol{\varepsilon}_4, \quad \overline{\overline{\mathsf{N}}}_b = \mathbf{e}_4 \wedge \overline{\overline{\mathsf{P}}}_s^{-1}. \tag{4.128}$$

They are actually nilpotent,

$$\overline{\overline{\mathsf{N}}}_a^2 = \overline{\overline{\mathsf{N}}}_b^2 = 0, \tag{4.129}$$

and give rise to two complementary projection bidyadics as

$$\begin{aligned}\overline{\overline{\Pi}}_{ab} &= -\overline{\overline{\mathsf{N}}}_a|\overline{\overline{\mathsf{N}}}_b = -(\overline{\overline{\mathsf{P}}}_s \wedge \boldsymbol{\varepsilon}_4)|(\mathbf{e}_4 \wedge \overline{\overline{\mathsf{P}}}_s^{-1}) \\ &= \overline{\overline{\mathsf{P}}}_s|\overline{\overline{\mathsf{P}}}_s^{-1} = \overline{\overline{\mathsf{I}}}_s^{(2)},\end{aligned} \tag{4.130}$$

$$\begin{aligned}\overline{\overline{\Pi}}_{ba} &= -\overline{\overline{\mathsf{N}}}_b|\overline{\overline{\mathsf{N}}}_a = -(\mathbf{e}_4 \wedge \overline{\overline{\mathsf{P}}}_s^{-1})|(\overline{\overline{\mathsf{P}}}_s \wedge \boldsymbol{\varepsilon}_4) \\ &= -\mathbf{e}_4 \wedge \overline{\overline{\mathsf{I}}}_s \wedge \boldsymbol{\varepsilon}_4 = \overline{\overline{\mathsf{I}}}_s {}^{\wedge}_{\wedge} \mathbf{e}_4\boldsymbol{\varepsilon}_4.\end{aligned} \tag{4.131}$$

because they satisfy

$$\overline{\overline{\Pi}}_{ab}^2 = \overline{\overline{\Pi}}_{ab}, \quad \overline{\overline{\Pi}}_{ba}^2 = \overline{\overline{\Pi}}_{ba}, \quad \overline{\overline{\Pi}}_{ab} + \overline{\overline{\Pi}}_{ba} = \overline{\overline{\mathsf{I}}}^{(2)}. \tag{4.132}$$

Now one can show that an AC bidyadic can be expressed as

$$\overline{\overline{\mathsf{J}}} = \overline{\overline{\mathsf{N}}}_a + \overline{\overline{\mathsf{N}}}_b. \tag{4.133}$$

In fact, applying the above properties we have

$$\overline{\overline{\mathsf{J}}}^2 = \overline{\overline{\mathsf{N}}}_a|\overline{\overline{\mathsf{N}}}_b + \overline{\overline{\mathsf{N}}}_b|\overline{\overline{\mathsf{N}}}_a = -\overline{\overline{\Pi}}_{ab} - \overline{\overline{\Pi}}_{ba} = -\overline{\overline{\mathsf{I}}}^{(2)}. \tag{4.134}$$

As an example, let us consider the symmetric spatial dyadics

$$\overline{\overline{\mathsf{G}}}_s = \mathbf{e}_1\mathbf{e}_1 + \mathbf{e}_2\mathbf{e}_2 + \mathbf{e}_3\mathbf{e}_3, \tag{4.135}$$

$$\overline{\overline{\Gamma}}_s = \boldsymbol{\varepsilon}_1\boldsymbol{\varepsilon}_1 + \boldsymbol{\varepsilon}_2\boldsymbol{\varepsilon}_2 + \boldsymbol{\varepsilon}_3\boldsymbol{\varepsilon}_3, \tag{4.136}$$

in terms of which the two nilpotent bidyadics can be represented as

$$\overline{\overline{\mathsf{N}}}_a = \mathbf{e}_{123}\lfloor\overline{\overline{\Gamma}}_s \wedge \boldsymbol{\varepsilon}_4 = \mathbf{e}_{23}\boldsymbol{\varepsilon}_{14} + \mathbf{e}_{31}\boldsymbol{\varepsilon}_{24} + \mathbf{e}_{12}\boldsymbol{\varepsilon}_{34}, \tag{4.137}$$

$$\overline{\overline{\mathsf{N}}}_b = \mathbf{e}_4 \wedge \overline{\overline{\mathsf{G}}}_s\rfloor\boldsymbol{\varepsilon}_{123} = -\mathbf{e}_{14}\boldsymbol{\varepsilon}_{23} - \mathbf{e}_{24}\boldsymbol{\varepsilon}_{31} - \mathbf{e}_{34}\boldsymbol{\varepsilon}_{12}. \tag{4.138}$$

In this case the AC bidyadic becomes

$$\overline{\overline{\mathsf{J}}} = \mathbf{e}_{23}\boldsymbol{\varepsilon}_{14} + \mathbf{e}_{31}\boldsymbol{\varepsilon}_{24} + \mathbf{e}_{12}\boldsymbol{\varepsilon}_{34} - \mathbf{e}_{14}\boldsymbol{\varepsilon}_{23} - \mathbf{e}_{24}\boldsymbol{\varepsilon}_{31} - \mathbf{e}_{34}\boldsymbol{\varepsilon}_{12}. \tag{4.139}$$

4.7 MODIFIED CLOSURE RELATION

A condition of the form

$$\overline{\overline{\mathsf{M}}}|\overline{\overline{\mathsf{M}}} = \alpha\overline{\overline{\mathsf{I}}}^{(2)T}, \tag{4.140}$$

is called the closure relation for bidyadics $\overline{\overline{\mathsf{M}}} \in \mathbb{F}_2\mathbb{E}_2$ [9, 11]. Actually, this requires that $\overline{\overline{\mathsf{M}}}$ is a multiple of a unipotent bidyadic (4.76). A somewhat similar condition either for the bidyadic $\overline{\overline{\mathsf{M}}}$ or for the metric bidyadic $\overline{\overline{\mathsf{M}}}_m = \mathbf{e}_N\lfloor\overline{\overline{\mathsf{M}}} \in \mathbb{E}_2\mathbb{E}_2$

$$\overline{\overline{\mathsf{M}}}^T \cdot \overline{\overline{\mathsf{M}}} = \overline{\overline{\mathsf{M}}}^T|(\mathbf{e}_N\lfloor\overline{\overline{\mathsf{M}}}) = \alpha\overline{\overline{\mathsf{E}}}, \tag{4.141}$$

$$\overline{\overline{\mathsf{M}}}_m^T \cdot \overline{\overline{\mathsf{M}}}_m = \overline{\overline{\mathsf{M}}}_m^T|(\boldsymbol{\varepsilon}_N\lfloor\overline{\overline{\mathsf{M}}}_m) = \alpha\overline{\overline{\mathsf{E}}}. \tag{4.142}$$

can be called the modified closure relation [12] for $\alpha \neq 0$. Equation (4.141) implies that

$$(\overline{\overline{\mathsf{M}}}|\boldsymbol{\Phi}_1) \cdot (\overline{\overline{\mathsf{M}}}|\boldsymbol{\Phi}_2) = \alpha\boldsymbol{\Phi}_1 \cdot \boldsymbol{\Phi}_2, \tag{4.143}$$

is valid for any two-forms $\boldsymbol{\Phi}_1, \boldsymbol{\Phi}_2$. When $\alpha = 1$, the bidyadic $\overline{\overline{\mathsf{M}}}$ resembles a rotation operator in its property of preserving the dot product of two-forms.

One can show through 3D expansions that, to satisfy (4.141), the bidyadic $\overline{\overline{\mathsf{M}}}$ can be expressed in the form

$$\overline{\overline{\mathsf{M}}} = \overline{\overline{\mathsf{P}}}^{(2)T}, \quad \overline{\overline{\mathsf{M}}}_m = \mathbf{e}_N\lfloor\overline{\overline{\mathsf{P}}}^{(2)T}, \tag{4.144}$$

or in the form

$$\overline{\overline{\mathsf{M}}} = \boldsymbol{\varepsilon}_N\lfloor\overline{\overline{\mathsf{Q}}}^{(2)}, \quad \overline{\overline{\mathsf{M}}}_m = \overline{\overline{\mathsf{Q}}}^{(2)}, \tag{4.145}$$

for some dyadic $\overline{\overline{\mathsf{P}}} \in \mathbb{E}_1\mathbb{F}_1$ or metric dyadic $\overline{\overline{\mathsf{Q}}} \in \mathbb{E}_1\mathbb{E}_1$ [12]. Thus, the modified closure condition restricts the number of parameters defining the bidyadic $\overline{\overline{\mathsf{M}}}$ from $6 \times 6 = 36$ to $4 \times 4 = 16$. The two cases (4.144) and (4.145) are called P-solutions and Q-solutions, respectively. Let us consider the problem without 3D expansions.

4.7.1 Equivalent Conditions

Let us start from the result (see Problem 3.20) that if a metric bidyadic $\overline{\overline{\mathsf{X}}} \in \mathbb{E}_2\mathbb{E}_2$ satisfies

$$\overline{\overline{\mathsf{X}}}\lfloor\lfloor\boldsymbol{\nu}\boldsymbol{\nu} = 0 \tag{4.146}$$

for all one-forms $\boldsymbol{\nu}$, it must be of the form

$$\overline{\overline{\mathsf{X}}} = \alpha\overline{\overline{\mathsf{E}}}, \tag{4.147}$$

for some scalar α. Thus, we may state that if the bidyadic $\overline{\overline{\mathsf{M}}}$ satisfies

$$(\overline{\overline{\mathsf{M}}}^T \cdot \overline{\overline{\mathsf{M}}})\lfloor\lfloor\boldsymbol{\nu}\boldsymbol{\nu} = 0 \tag{4.148}$$

for all $\boldsymbol{\nu}$, it must satisfy a modified closure relation of the form (4.141) for some $\boldsymbol{\alpha}$. Since the converse is also true, (4.141) and (4.148) are equivalent.

Now it is also known that if a metric dyadic $\overline{\overline{\mathsf{A}}} \in \mathbb{E}_1\mathbb{E}_1$ satisfies

$$\overline{\overline{\mathsf{A}}}||\boldsymbol{\mu}\boldsymbol{\mu} = 0 \tag{4.149}$$

for all $\boldsymbol{\mu} \in \mathbb{F}_1$, the symmetric part of $\overline{\overline{\mathsf{A}}}$ must vanish. Thus, if the symmetric dyadic $\overline{\overline{\mathsf{A}}} = (\overline{\overline{\mathsf{M}}}^T \cdot \overline{\overline{\mathsf{M}}})\lfloor\lfloor\boldsymbol{\nu}\boldsymbol{\nu} \in \mathbb{E}_1\mathbb{E}_1$ satisfies (4.149), it must vanish. Combining these two conditions, we may state that if $\overline{\overline{\mathsf{M}}}$ satisfies

$$((\overline{\overline{\mathsf{M}}}^T \cdot \overline{\overline{\mathsf{M}}})\lfloor\lfloor\boldsymbol{\nu}\boldsymbol{\nu})||\boldsymbol{\mu}\boldsymbol{\mu} = (\overline{\overline{\mathsf{M}}}|(\boldsymbol{\nu}\wedge\boldsymbol{\mu}))\cdot(\overline{\overline{\mathsf{M}}}|(\boldsymbol{\nu}\wedge\boldsymbol{\mu})) = 0 \tag{4.150}$$

for all one-forms $\boldsymbol{\nu}$ and $\boldsymbol{\mu}$, it must satisfy the modified closure relation (4.141) for some scalar $\boldsymbol{\alpha}$. Since the converse is also true, (4.141) and (4.150) are equivalent. The same condition for the modified bidyadic is

$$((\overline{\overline{\mathsf{M}}}_m^T \cdot \overline{\overline{\mathsf{M}}}_m)\lfloor\lfloor\boldsymbol{\nu}\boldsymbol{\nu})||\boldsymbol{\mu}\boldsymbol{\mu} = (\overline{\overline{\mathsf{M}}}_m|(\boldsymbol{\nu}\wedge\boldsymbol{\mu}))\cdot(\overline{\overline{\mathsf{M}}}_m|(\boldsymbol{\nu}\wedge\boldsymbol{\mu})) = 0. \tag{4.151}$$

4.7.2 Solutions

The condition (4.150) can be interpreted so that a bidyadic $\overline{\overline{\mathsf{M}}}$ satisfying the modified closure condition (4.141) maps any simple two-form $\boldsymbol{\nu}\wedge\boldsymbol{\mu}$ to another simple two-form. A simple two-form can be expressed either in terms of two one-forms $\boldsymbol{\alpha}, \boldsymbol{\beta}$ as $\boldsymbol{\alpha}\wedge\boldsymbol{\beta}$, or in terms of two vectors **a**, **b**, as $\varepsilon_N\lfloor(\mathbf{a}\wedge\mathbf{b})$. These lead to two representations,

$$\overline{\overline{\mathsf{M}}}|(\boldsymbol{\nu}\wedge\boldsymbol{\mu}) = \boldsymbol{\alpha}\wedge\boldsymbol{\beta}, \tag{4.152}$$

$$\overline{\overline{\mathsf{M}}}|(\boldsymbol{\nu}\wedge\boldsymbol{\mu}) = \varepsilon_N\lfloor(\mathbf{a}\wedge\mathbf{b}), \tag{4.153}$$

the latter of which can be more conveniently expressed for the metric bidyadic $\overline{\overline{\mathsf{M}}}_m$ as

$$\overline{\overline{\mathsf{M}}}_m|(\boldsymbol{\nu}\wedge\boldsymbol{\mu}) = \mathbf{a}\wedge\mathbf{b}. \tag{4.154}$$

Let us consider these two cases separately.

P-Solution

Since the left side of (4.152) is linearly dependent on $\boldsymbol{\mu}$, either of the one-forms $\boldsymbol{\alpha}$ or $\boldsymbol{\beta}$ must be a linear function of $\boldsymbol{\mu}$. There is no loss of generality if we express

$$\boldsymbol{\beta} = \overline{\overline{\mathsf{P}}}^T|\boldsymbol{\mu}, \tag{4.155}$$

in terms of some dyadic $\overline{\overline{\mathsf{P}}}\in\mathbb{E}_1\mathbb{F}_1$. From

$$((\overline{\overline{\mathsf{M}}}\lfloor\boldsymbol{\nu}) - \boldsymbol{\alpha}\wedge\overline{\overline{\mathsf{P}}}^T)|\boldsymbol{\mu} = 0, \tag{4.156}$$

which must be valid for any one-form $\boldsymbol{\mu}$, we obtain the condition

$$\overline{\overline{\mathsf{M}}}\lfloor\boldsymbol{\nu} = \boldsymbol{\alpha}\wedge\overline{\overline{\mathsf{P}}}^T. \tag{4.157}$$

Because this implies

$$\boldsymbol{\alpha}\wedge\overline{\overline{\mathsf{P}}}^T|\boldsymbol{\nu} = 0, \tag{4.158}$$

the one-form $\boldsymbol{\alpha}$ must be of the form

$$\boldsymbol{\alpha} = M\overline{\overline{\mathsf{P}}}^T|\boldsymbol{\nu}, \tag{4.159}$$

for some scalar M. The condition (4.157) then becomes

$$\overline{\overline{\mathsf{M}}}\lfloor\boldsymbol{\nu} = M(\overline{\overline{\mathsf{P}}}^T|\boldsymbol{\nu})\wedge\overline{\overline{\mathsf{P}}}^T = M\overline{\overline{\mathsf{P}}}^{(2)T}\lfloor\boldsymbol{\nu}. \tag{4.160}$$

Since this must be valid for all $\boldsymbol{\nu}$, we arrive at one solution to the modified closure relation,

$$\overline{\overline{\mathsf{M}}} = M\overline{\overline{\mathsf{P}}}^{(2)T}. \tag{4.161}$$

When the scalar M is absorbed in the dyadic $\overline{\overline{\mathsf{P}}}$, this equals (4.144), the P-solution.

Q-Solution

Starting from (4.154), we can proceed similarly. Assuming that the vector **b** depends linearly on $\boldsymbol{\mu}$ as

$$\mathbf{b} = \overline{\overline{\mathsf{Q}}}|\boldsymbol{\mu}, \tag{4.162}$$

where $\overline{\overline{\mathsf{Q}}} \in \mathbb{E}_1\mathbb{E}_1$ is some metric dyadic, (4.154) yields

$$\overline{\overline{\mathsf{M}}}_m\lfloor\boldsymbol{\nu} = \mathbf{a}\wedge\overline{\overline{\mathsf{Q}}}. \tag{4.163}$$

This implies

$$\mathbf{a}\wedge\overline{\overline{\mathsf{Q}}}|\boldsymbol{\nu} = 0, \tag{4.164}$$

whence **a** must be of the form

$$\mathbf{a} = M\overline{\overline{\mathsf{Q}}}|\boldsymbol{\nu}. \tag{4.165}$$

(4.163) now becomes

$$\overline{\overline{\mathsf{M}}}_m\lfloor\boldsymbol{\nu} = M(\overline{\overline{\mathsf{Q}}}|\boldsymbol{\nu})\wedge\overline{\overline{\mathsf{Q}}} = M\overline{\overline{\mathsf{Q}}}^{(2)}\lfloor\boldsymbol{\nu}. \tag{4.166}$$

Because this must be valid for all $\boldsymbol{\nu}$, the modified medium bidyadic must be of the form

$$\overline{\overline{\mathsf{M}}}_m = M\overline{\overline{\mathsf{Q}}}^{(2)}. \tag{4.167}$$

Again, if the scalar M is absorbed by the dyadic $\overline{\overline{\mathsf{Q}}}$, this yields (4.145), the Q-solution.

4.7.3 Testing the Two Solutions

Since the modified closure relation (4.141) or (4.142) has two sets of solutions, one may ask how to distinguish whether a given solution $\overline{\overline{\mathsf{M}}}$ or $\overline{\overline{\mathsf{M}}}_m$ is a P-solution or a Q-solution?

Considering the metric bidyadic $\overline{\overline{\mathsf{M}}}_m \in \mathbb{E}_2\mathbb{E}_2$ satisfying the modified closure equation (4.142) let us first study the Q-solution by double contracting by one-forms $\boldsymbol{\alpha}\boldsymbol{\alpha}$ and finding its double-wedge powers as

$$\overline{\overline{\mathsf{M}}}_m\lfloor\lfloor\boldsymbol{\alpha}\boldsymbol{\alpha} = \overline{\overline{\mathsf{Q}}}^{(2)}\lfloor\lfloor\boldsymbol{\alpha}\boldsymbol{\alpha} = (\overline{\overline{\mathsf{Q}}}||\boldsymbol{\alpha}\boldsymbol{\alpha})\overline{\overline{\mathsf{Q}}} - (\overline{\overline{\mathsf{Q}}}|\boldsymbol{\alpha})(\boldsymbol{\alpha}|\overline{\overline{\mathsf{Q}}}), \tag{4.168}$$

$$(\overline{\overline{\mathsf{M}}}_m\lfloor\lfloor\boldsymbol{\alpha}\boldsymbol{\alpha})^{(2)} = (\overline{\overline{\mathsf{Q}}}||\boldsymbol{\alpha}\boldsymbol{\alpha})((\overline{\overline{\mathsf{Q}}}||\boldsymbol{\alpha}\boldsymbol{\alpha})\overline{\overline{\mathsf{Q}}}^{(2)} - \overline{\overline{\mathsf{Q}}}{}^{\wedge}_{\wedge}((\overline{\overline{\mathsf{Q}}}|\boldsymbol{\alpha})(\boldsymbol{\alpha}|\overline{\overline{\mathsf{Q}}}))), \tag{4.169}$$

$$\begin{aligned}(\overline{\overline{\mathsf{M}}}_m\lfloor\lfloor\boldsymbol{\alpha}\boldsymbol{\alpha})^{(3)} &= (\overline{\overline{\mathsf{Q}}}||\boldsymbol{\alpha}\boldsymbol{\alpha})^2((\overline{\overline{\mathsf{Q}}}||\boldsymbol{\alpha}\boldsymbol{\alpha})\overline{\overline{\mathsf{Q}}}^{(3)} - \overline{\overline{\mathsf{Q}}}^{(2)}{}^{\wedge}_{\wedge}((\overline{\overline{\mathsf{Q}}}|\boldsymbol{\alpha})(\boldsymbol{\alpha}|\overline{\overline{\mathsf{Q}}}))) \\ &= (\overline{\overline{\mathsf{Q}}}||\boldsymbol{\alpha}\boldsymbol{\alpha})^2(\boldsymbol{\varepsilon}_N\boldsymbol{\varepsilon}_N||\overline{\overline{\mathsf{Q}}}^{(4)})(\mathbf{e}_N\mathbf{e}_N\lfloor\lfloor\boldsymbol{\alpha}\boldsymbol{\alpha}).\end{aligned} \tag{4.170}$$

Let us now do the same operations with the P-solution.

$$\overline{\overline{\mathsf{M}}}_m\lfloor\lfloor\boldsymbol{\alpha}\boldsymbol{\alpha} = (\mathbf{e}_N\lfloor\overline{\overline{\mathsf{P}}}^{(2)T})\lfloor\lfloor\boldsymbol{\alpha}\boldsymbol{\alpha}$$
$$= \mathbf{e}_N\lfloor(\boldsymbol{\alpha}\wedge(\overline{\overline{\mathsf{P}}}^T|\boldsymbol{\alpha})\wedge\overline{\overline{\mathsf{P}}}^T). \tag{4.171}$$

Denoting the simple bivector by

$$\mathbf{e}_N\lfloor(\boldsymbol{\alpha}\wedge(\overline{\overline{\mathsf{P}}}^T|\boldsymbol{\alpha})) = \mathbf{a}\wedge\mathbf{b}, \tag{4.172}$$

we have

$$\overline{\overline{\mathsf{M}}}_m\lfloor\lfloor\boldsymbol{\alpha}\boldsymbol{\alpha} = (\mathbf{a}\wedge\mathbf{b})\lfloor\overline{\overline{\mathsf{P}}}^T = \mathbf{b}(\mathbf{a}|\overline{\overline{\mathsf{P}}}^T) - \mathbf{a}(\mathbf{b}|\overline{\overline{\mathsf{P}}}^T), \tag{4.173}$$
$$(\overline{\overline{\mathsf{M}}}_m\lfloor\lfloor\boldsymbol{\alpha}\boldsymbol{\alpha})^{(2)} = (\mathbf{a}\wedge\mathbf{b})(\mathbf{a}\wedge\mathbf{b})|\overline{\overline{\mathsf{P}}}^{(2)T}, \tag{4.174}$$
$$(\overline{\overline{\mathsf{M}}}_m\lfloor\lfloor\boldsymbol{\alpha}\boldsymbol{\alpha})^{(3)} = 0. \tag{4.175}$$

The above conditions, valid for any one-form $\boldsymbol{\alpha}$, allow one to test the character of a metric bidyadic $\overline{\overline{\mathsf{M}}}_m$ satisfying the modified closure relation (4.142) in the case when it is of full rank, whence both $\overline{\overline{\mathsf{Q}}}$ and $\overline{\overline{\mathsf{P}}}$ must be of rank 4, satisfying $\overline{\overline{\mathsf{Q}}}^{(4)} \neq 0$ and $\overline{\overline{\mathsf{P}}}^{(4)} \neq 0$. The different possibilities can be listed as based on the rank of the dyadic $\overline{\overline{\mathsf{M}}}_m\lfloor\lfloor\boldsymbol{\alpha}\boldsymbol{\alpha}$ as follows.

- $\overline{\overline{\mathsf{M}}}_m$ satisfies $(\overline{\overline{\mathsf{M}}}_m\lfloor\lfloor\boldsymbol{\alpha}\boldsymbol{\alpha})^{(3)} \neq 0$ for some one-form $\boldsymbol{\alpha}$. Because of (4.175), it cannot correspond to a P-solution, whence it must be a Q-solution.
- $\overline{\overline{\mathsf{M}}}_m$ satisfies $(\overline{\overline{\mathsf{M}}}_m\lfloor\lfloor\boldsymbol{\alpha}\boldsymbol{\alpha})^{(3)} = 0$ for all $\boldsymbol{\alpha}$ and $(\overline{\overline{\mathsf{M}}}_m\lfloor\lfloor\boldsymbol{\alpha}\boldsymbol{\alpha})^{(2)} \neq 0$ for some $\boldsymbol{\alpha}$. From (4.170) a full-rank Q-solution must then satisfy $\overline{\overline{\mathsf{Q}}}||\boldsymbol{\alpha}\boldsymbol{\alpha} = 0$, which from (4.169) is contrary to $(\overline{\overline{\mathsf{M}}}_m\lfloor\lfloor\boldsymbol{\alpha}\boldsymbol{\alpha})^{(2)} \neq 0$. Thus, this case must correspond to a P-solution.
- $\overline{\overline{\mathsf{M}}}_m$ satisfies $(\overline{\overline{\mathsf{M}}}_m\lfloor\lfloor\boldsymbol{\alpha}\boldsymbol{\alpha})^{(2)} = 0$ for all $\boldsymbol{\alpha}$ and $\overline{\overline{\mathsf{M}}}_m\lfloor\lfloor\boldsymbol{\alpha}\boldsymbol{\alpha} \neq 0$ for some $\boldsymbol{\alpha}$. This is satisfied by a Q-solution with antisymmetric $\overline{\overline{\mathsf{Q}}}$ while from (4.174) we have $\mathbf{a}\wedge\mathbf{b} = 0$ whence $\overline{\overline{\mathsf{M}}}_m\lfloor\lfloor\boldsymbol{\alpha}\boldsymbol{\alpha} \neq 0$ cannot be satisfied by any full-rank P-solution.
- $\overline{\overline{\mathsf{M}}}_m$ satisfies $\overline{\overline{\mathsf{M}}}_m\lfloor\lfloor\boldsymbol{\alpha}\boldsymbol{\alpha} = 0$ for all $\boldsymbol{\alpha}$ and $\overline{\overline{\mathsf{M}}}_m \neq 0$. This requires that $\overline{\overline{\mathsf{M}}}_m = M\mathbf{e}_N\lfloor\overline{\overline{\mathsf{I}}}^{(2)T}$ which is again a P-solution.

Thus, the Q- and P-solutions are distinct when $\overline{\overline{\mathsf{M}}}$ is of full rank in which case $\overline{\overline{\mathsf{P}}}$ or $\overline{\overline{\mathsf{Q}}}$ must also be of full rank. Actually, one can show that there are no full-rank solutions for $\overline{\overline{\mathsf{Q}}}$ and $\overline{\overline{\mathsf{P}}}$ satisfying an equation of the form

$$\overline{\overline{\mathsf{Q}}}^{(2)} = \mathbf{e}_N\lfloor\overline{\overline{\mathsf{P}}}^{(2)T}. \tag{4.176}$$

In the case when $\overline{\overline{\mathsf{M}}}$ is not of full rank, the P- and Q-solutions may both be possible. Further considerations are left as exercises.

PROBLEMS

4.1 Show that the condition

$$\overline{\overline{\mathsf{A}}}_s|\overline{\overline{\mathsf{B}}}_s = 0$$

for two spatial dyadics $\overline{\overline{\mathsf{A}}}_s, \overline{\overline{\mathsf{B}}}_s \in \mathbb{E}_1\mathbb{F}_1$ requires that either $\overline{\overline{\mathsf{A}}}_s^{(2)} = 0$ or $\overline{\overline{\mathsf{B}}}_s^{(2)} = 0$ must be satisfied, whence one of the two dyadics must be of the form $\mathbf{a}_s\boldsymbol{\alpha}_s$ for some spatial vector $\mathbf{a}_s$ and spatial one-form $\boldsymbol{\alpha}_s$.

4.2 Show that the dyadic

$$\overline{\overline{\mathsf{K}}} = \overline{\overline{\mathsf{I}}} + \mathbf{e}_3\boldsymbol{\varepsilon}_4 + \mathbf{e}_4\boldsymbol{\varepsilon}_3 - \mathbf{e}_3\boldsymbol{\varepsilon}_3 - \mathbf{e}_4\boldsymbol{\varepsilon}_4$$

is unipotent when $\mathbf{e}_i$ and $\boldsymbol{\varepsilon}_i$ form reciprocal vector and one-form bases. Find its expansion either in the form (4.74) or (4.75).

4.3 Show that a dyadic $\overline{\overline{\mathsf{A}}} \in \mathbb{E}_1\mathbb{F}_1$ satisfying

$$\overline{\overline{\mathsf{A}}}^3 = 0, \quad \overline{\overline{\mathsf{A}}}^2 \neq 0$$

also satisfies $\mathrm{tr}\overline{\overline{\mathsf{A}}}^i = 0$ for $i = 1, \ldots, 4$. Hint: use the Cayley–Hamilton equation.

4.4 Show that the conditions (4.98) for the 2D AC dyadics are equivalent to (4.97). Hint: start from the expression $\overline{\overline{\mathsf{J}}}^{(2)}\lfloor\lfloor\overline{\overline{\mathsf{I}}}^T_{12}$.

4.5 Show that the most general solution to (4.97) can be expressed in a form similar to (4.100) by starting from the most general 2D dyadic expansion $\overline{\overline{\mathsf{J}}} = \sum J_{i,j}\mathbf{e}_i\boldsymbol{\varepsilon}_j$. Hint: It may help to denote without loss of generality $J_{1,1} = \sinh\theta$, $J_{1,2} = A\cosh\theta$ and $J_{2,1} = -(1/A)\cosh\theta$.

4.6 Starting from (4.82) and the definition of the AC dyadic, show that the AC dyadic $\overline{\overline{\mathsf{J}}} \in \mathbb{E}_1\mathbb{F}_1$ satisfies the following properties:

$$\overline{\overline{\mathsf{J}}}^{(2)}\lfloor\lfloor\overline{\overline{\mathsf{I}}}^T = \overline{\overline{\mathsf{I}}},$$
$$\overline{\overline{\mathsf{J}}}^{(3)}\lfloor\lfloor\overline{\overline{\mathsf{I}}}^T = \overline{\overline{\mathsf{J}}}{}^\wedge_\wedge\overline{\overline{\mathsf{I}}},$$
$$\overline{\overline{\mathsf{J}}}^{(3)}\lfloor\lfloor\overline{\overline{\mathsf{I}}}^{(2)T} = \overline{\overline{\mathsf{J}}},$$
$$\overline{\overline{\mathsf{I}}}^{(3)}\lfloor\lfloor\overline{\overline{\mathsf{J}}}^T = -\overline{\overline{\mathsf{I}}}{}^\wedge_\wedge\overline{\overline{\mathsf{J}}},$$
$$\overline{\overline{\mathsf{I}}}^{(4)}\lfloor\lfloor\overline{\overline{\mathsf{J}}}^{(3)T} = -\overline{\overline{\mathsf{J}}},$$
$$\overline{\overline{\mathsf{I}}}^{(3)}\lfloor\lfloor\overline{\overline{\mathsf{J}}}^{(2)T} = \overline{\overline{\mathsf{I}}}.$$

4.7 Show that a bidyadic $\overline{\overline{\mathsf{M}}} \in \mathbb{F}_2\mathbb{E}_2$ consisting of a real-valued AC dyadic $\overline{\overline{\mathsf{J}}}$ defined by $\overline{\overline{\mathsf{M}}} = M\overline{\overline{\mathsf{J}}}^{(2)T}$ does not contain a skewon component. Show also that such a bidyadic contains a nonzero principal part.

4.8 Show that if $\overline{\overline{\mathsf{J}}} \in \mathbb{E}_2\mathbb{F}_2$ is an AC bidyadic, $\overline{\overline{\mathsf{I}}}^{(4)}\lfloor\lfloor\overline{\overline{\mathsf{J}}}^T$ is also an AC bidyadic.

4.9 Derive a general expression for the AC bidyadic $\overline{\overline{\mathsf{J}}}$ by expressing $\overline{\overline{\mathsf{J}}}$ in its spatial and temporal components as

$$\overline{\overline{\mathsf{J}}} = \overline{\overline{\mathsf{A}}}_s + \mathbf{e}_4 \wedge \overline{\overline{\mathsf{B}}}_s + \overline{\overline{\mathsf{C}}}_s \wedge \boldsymbol{\varepsilon}_4 + \mathbf{e}_4 \wedge \overline{\overline{\mathsf{D}}}_s \wedge \boldsymbol{\varepsilon}_4.$$

Assuming that $\overline{\overline{\mathsf{B}}}_s$ has an inverse, show that only two 3D equations are needed to define the problem.

4.10 Show that the eigenvectors $\mathbf{x}_i$ and eigen one-forms $\boldsymbol{\xi}_i$ of the AC dyadic $\overline{\overline{\mathsf{J}}}$ must be alternatively real and imaginary valued to obtain the representation (4.108) in terms of real-valued dyadics (4.120).

4.11 Given two AC dyadics defined by

$$\overline{\overline{\mathsf{J}}} = \mathbf{e}_1\boldsymbol{\varepsilon}_2 - \mathbf{e}_2\boldsymbol{\varepsilon}_1 + \mathbf{e}_3\boldsymbol{\varepsilon}_4 - \mathbf{e}_4\boldsymbol{\varepsilon}_3,$$
$$\overline{\overline{\mathsf{J}}}' = \mathbf{e}_1\boldsymbol{\varepsilon}_3 + \mathbf{e}_2\boldsymbol{\varepsilon}_4 - \mathbf{e}_3\boldsymbol{\varepsilon}_1 - \mathbf{e}_4\boldsymbol{\varepsilon}_2, \qquad (4.177)$$

find a dyadic $\overline{\overline{\mathsf{D}}} \in \mathbb{E}_1\mathbb{F}_1$ yielding the connection $\overline{\overline{\mathsf{J}}}' = \overline{\overline{\mathsf{D}}}|\overline{\overline{\mathsf{J}}}|\overline{\overline{\mathsf{D}}}^{-1}$ between them.

4.12 Find the most general unipotent spatial dyadics $\overline{\overline{\beta}}_s \in \mathbb{F}_1\mathbb{E}_1$ and spatial bidyadics $\overline{\overline{\alpha}}_s \in \mathbb{F}_2\mathbb{E}_2$.

4.13 Show that the equation

$$\overline{\overline{\mathsf{Q}}}^{(2)} = \mathbf{e}_N\lfloor\overline{\overline{\mathsf{I}}}^{(2)T}$$

has no solutions $\overline{\overline{\mathsf{Q}}} \in \mathbb{E}_1\mathbb{E}_1$.

4.14 Study possible solutions $\overline{\overline{\mathsf{Q}}}$ to the equation

$$\overline{\overline{\mathsf{Q}}}^{(2)} = \mathbf{e}_N\lfloor\overline{\overline{\mathsf{P}}}^{(2)T}$$

for different values of the rank of the dyadic $\overline{\overline{\mathsf{P}}} \in \mathbb{E}_1\mathbb{F}_1$.

4.15 Show that, by applying the transformation bidyadic

$$\overline{\overline{\mathsf{D}}} = \overline{\overline{\mathsf{I}}}_s^{(2)} + (\overline{\overline{\mathsf{G}}}'_s|\overline{\overline{\Gamma}}_s)^{\wedge}_{\wedge}\mathbf{e}_4\boldsymbol{\varepsilon}_4,$$

an AC bidyadic defined by a spatial metric dyadic $\overline{\overline{\mathsf{G}}}_s$, which is not symmetric, can be transformed to an AC bidyadic defined by a symmetric spatial dyadic $\overline{\overline{\mathsf{G}}}'_s$.

4.16 Consider the Cayley–Hamilton equation (2.119) for the unipotent dyadic $\overline{\overline{\mathsf{K}}}$ and find the possible values of tr$\overline{\overline{\mathsf{K}}}$.

4.17 Show that for an AC bidyadic satisfying (4.122), the equation (4.125) actually equals the Cayley–Hamilton equation (3.10).

CHAPTER 5

Electromagnetic Fields

The electromagnetic field quantities and their sources are represented by differential forms of different grades. Applying the notation introduced by Deschamps [1, 13], the electromagnetic fields are represented by two-forms $\boldsymbol{\Psi}, \boldsymbol{\Phi} \in \mathbb{F}_2$ and their source by a three-form $\gamma \in \mathbb{F}_3$. The fields and the source can be decomposed in spatial and temporal parts as

$$\boldsymbol{\Psi} = \mathbf{D} - \mathbf{H} \wedge \boldsymbol{\varepsilon}_4, \tag{5.1}$$

$$\boldsymbol{\Phi} = \mathbf{B} + \mathbf{E} \wedge \boldsymbol{\varepsilon}_4, \tag{5.2}$$

$$\gamma = \varrho - \mathbf{J} \wedge \boldsymbol{\varepsilon}_4. \tag{5.3}$$

Applying conventional notation, $\mathbf{D}, \mathbf{B} \in \mathbb{F}_2$ represent spatial electromagnetic two-forms while $\mathbf{H}, \mathbf{E} \in \mathbb{F}_1$ are two spatial electromagnetic one-forms. $\varrho \in \mathbb{F}_3$ denotes the spatial charge three-form and $\mathbf{J} \in \mathbb{F}_2$ is the spatial current two-form. In the following, $\boldsymbol{\varepsilon}_1, \boldsymbol{\varepsilon}_2, \boldsymbol{\varepsilon}_3$ form the spatial basis of one-forms while $\boldsymbol{\varepsilon}_4$ denotes the temporal one-form.

5.1 FIELD EQUATIONS

5.1.1 Differentiation Operator

The spacetime vector $\mathbf{x} \in \mathbb{E}_1$ is defined by

$$\mathbf{x} = \mathbf{r} + \mathbf{e}_4 \tau, \quad \tau = ct, \tag{5.4}$$

Multiforms, Dyadics, and Electromagnetic Media, First Edition. Ismo V. Lindell.

where $\mathbf{r}$ is the spatial part of $\mathbf{x}$ and $\mathbf{e}_4$ denotes the temporal basis vector. $ct = \boldsymbol{\varepsilon}_4|\mathbf{x}$ is the time variable multiplied by the velocity of light c. The differential operator $\mathbf{d} \in \mathbb{F}_1$ can be defined by the dyadic operation

$$\mathbf{dx} = \overline{\overline{\mathsf{I}}}^T. \tag{5.5}$$

Expanding $\mathbf{x}$ in Cartesian coordinates as

$$\mathbf{x} = \mathbf{e}_1 x_1 + \mathbf{e}_2 x_2 + \mathbf{e}_3 x_3 + \mathbf{e}_4 x_4 = \sum_i \mathbf{e}_i x_i, \tag{5.6}$$

where the basis vectors $\mathbf{e}_i$ are assumed independent of $\mathbf{x}$, (5.5) can be written as

$$\mathbf{dx} = \sum_i \mathbf{d}x_i \mathbf{e}_i = \overline{\overline{\mathsf{I}}}^T = \sum_i \boldsymbol{\varepsilon}_i \mathbf{e}_i. \tag{5.7}$$

The one-forms $\boldsymbol{\varepsilon}_i$ defined by

$$\mathbf{d}x_i = \boldsymbol{\varepsilon}_i \tag{5.8}$$

form a basis reciprocal to the vector basis $\{\mathbf{e}_i\}$. The differential operator can be expanded as

$$\mathbf{d} = \sum_i \boldsymbol{\varepsilon}_i \partial_{x_i} = \mathbf{d}_s + \boldsymbol{\varepsilon}_4 \partial_\tau, \tag{5.9}$$

where $\mathbf{d}_s$ denotes differentiation in the spatial coordinates x_1, x_2, x_3. It is often of interest to consider time-harmonic electromagnetic sources and fields in terms of complex time dependence

$$e^{j\omega t} = e^{jk\tau}, \quad k = \omega/c. \tag{5.10}$$

In such a case the differential operator can be replaced by

$$\mathbf{d} = \mathbf{d}_s + jk\boldsymbol{\varepsilon}_4. \tag{5.11}$$

The differential operator $\mathbf{d}$ allows many operations on multivector and multiform functions of $\mathbf{x}$:

- Differentiation of a scalar $\phi(\mathbf{x})$ yields a one-form,

$$\mathbf{d}\phi(\mathbf{x}) \in \mathbb{F}_1. \tag{5.12}$$

- Exterior differentiation of a p-form yields a $p+1$-form,

$$\boldsymbol{\phi}(\mathbf{x}) \in \mathbb{F}_1, \quad \mathbf{d} \wedge \boldsymbol{\phi}(\mathbf{x}) \in \mathbb{F}_2, \tag{5.13}$$

$$\boldsymbol{\Phi}(\mathbf{x}) \in \mathbb{F}_2, \quad \mathbf{d} \wedge \boldsymbol{\Phi}(\mathbf{x}) \in \mathbb{F}_3. \tag{5.14}$$

- Interior differentiation of vector yields a scalar,

$$\mathbf{d}|\mathbf{f}(\mathbf{x}) \in \mathbb{E}_0, \tag{5.15}$$

- Contraction differentiation of a p-vector yields a $p-1$-vector,

$$\mathbf{F}(\mathbf{x}) \in \mathbb{E}_2, \quad \mathbf{d}\rfloor\mathbf{F}(\mathbf{x}) \in \mathbb{E}_1, \tag{5.16}$$

$$\mathbf{k}(\mathbf{x}) \in \mathbb{E}_3, \quad \mathbf{d}\rfloor\mathbf{k}(\mathbf{x}) \in \mathbb{E}_2. \tag{5.17}$$

- "No-sign" differentiation yields a dyadic,

$$\boldsymbol{\phi}(\mathbf{x}) \in \mathbb{F}_1, \quad \mathbf{d}\boldsymbol{\phi}(\mathbf{x}) \in \mathbb{F}_1\mathbb{F}_1, \tag{5.18}$$

$$\mathbf{f}(\mathbf{x}) \in \mathbb{E}_1, \quad \mathbf{df}(\mathbf{x}) \in \mathbb{F}_1\mathbb{E}_1. \tag{5.19}$$

Various algebraic rules can be transformed to differentiation rules when replacing a one-form by the differential operator **d** and taking proper care of the differentiation. For example, the bac-cab rule (1.43) for two vector functions of **x** yields the differentiation rule

$$\mathbf{d}\rfloor(\mathbf{f} \wedge \mathbf{g}) = \mathbf{f}_c(\mathbf{d}|\mathbf{g}) - \mathbf{g}_c(\mathbf{d}|\mathbf{f}) + \mathbf{g}_c|(\mathbf{df}) - \mathbf{f}_c|(\mathbf{dg}), \tag{5.20}$$

where subscript $()_c$ denotes that the quantity is considered constant, that is, independent of **x** in differentiation, while **df** and **dg** are dyadic quantities.

The property $\mathbf{d} \wedge \mathbf{d} \wedge \boldsymbol{\xi}(\mathbf{x}) = 0$ is valid for any p-form function $\boldsymbol{\xi}(\mathbf{x}) \in \mathbb{F}_p$. The inverse of this is known as de Rham's theorem [14]. If $\mathbf{d} \wedge \boldsymbol{\xi} = 0$ for some p-form, there exists a $p-1$-form $\boldsymbol{\zeta}$ such that we can express $\boldsymbol{\xi} = \mathbf{d} \wedge \boldsymbol{\zeta}(\mathbf{x})$. However, $\boldsymbol{\zeta}(\mathbf{x})$ may be multi-valued if it is defined in a multiply connected region such as the doughnut.

5.1.2 Maxwell Equations

In differential-form formalism the Maxwell equations can be given a simple form [1, 4, 9, 13, 15–20] as

$$\mathbf{d} \wedge \boldsymbol{\Phi} = 0, \tag{5.21}$$

$$\mathbf{d} \wedge \boldsymbol{\Psi} = \boldsymbol{\gamma}. \tag{5.22}$$

In some cases magnetic charges and currents may be included as equivalent sources. To distinguish the sources, (5.22) and (5.21) will then be replaced by

$$\mathbf{d} \wedge \mathbf{\Phi} = \gamma_m, \tag{5.23}$$

$$\mathbf{d} \wedge \mathbf{\Psi} = \gamma_e, \tag{5.24}$$

with subscript $()_e$ added to the electric sources,

$$\gamma_e = \varrho_e - \mathbf{J}_e \wedge \varepsilon_4, \tag{5.25}$$

$$\gamma_m = \varrho_m - \mathbf{J}_m \wedge \varepsilon_4. \tag{5.26}$$

When there are no magnetic sources present, the subscript $()_e$ will be omitted. Inserting (5.2) in (5.21) and expanding,

$$\mathbf{d} \wedge \mathbf{\Phi} = \mathbf{d} \wedge \mathbf{B} + \mathbf{d} \wedge \mathbf{E} \wedge \varepsilon_4 \tag{5.27}$$

$$= \mathbf{d}_s \wedge \mathbf{B} + (\mathbf{d}_s \wedge \mathbf{E} + \partial_\tau \mathbf{B}) \wedge \varepsilon_4 = 0, \tag{5.28}$$

after separating the spatial and temporal parts, the corresponding 3D equations for spatial field quantities are obtained,

$$\mathbf{d}_s \wedge \mathbf{B} = 0, \tag{5.29}$$

$$\mathbf{d}_s \wedge \mathbf{E} + \partial_\tau \mathbf{B} = 0. \tag{5.30}$$

Similarly, inserting (5.1) and (5.3) in (5.22) yields

$$\mathbf{d}_s \wedge \mathbf{D} = \varrho, \tag{5.31}$$

$$\mathbf{d}_s \wedge \mathbf{H} - \partial_\tau \mathbf{D} = \mathbf{J}. \tag{5.32}$$

The spatial differential forms $\mathbf{D}, \mathbf{B}, \mathbf{E}, \mathbf{H}$ can be transformed to Gibbsian vector fields $\mathbf{B}_g, \mathbf{E}_g, \mathbf{H}_g, \mathbf{D}_g, \mathbf{J}_g$ according to the rules given in Appendix B by

$$\mathbf{D}_g = \mathbf{e}_{123} \lfloor \mathbf{D} \in \mathbb{E}_1, \tag{5.33}$$

$$\mathbf{B}_g = \mathbf{e}_{123} \lfloor \mathbf{B} \in \mathbb{E}_1, \tag{5.34}$$

$$\mathbf{E}_g = \overline{\overline{\mathsf{G}}}_s | \mathbf{E} \in \mathbb{E}_1, \tag{5.35}$$

$$\mathbf{H}_g = \overline{\overline{\mathsf{G}}}_s | \mathbf{H} \in \mathbb{E}_1, \tag{5.36}$$

where

$$\overline{\overline{\mathsf{G}}}_s = \mathbf{e}_1\mathbf{e}_1 + \mathbf{e}_2\mathbf{e}_2 + \mathbf{e}_3\mathbf{e}_3 \tag{5.37}$$

is a chosen spatial metric dyadic defining the dot product. The Gibbsian sources

$$\varrho_g = \mathbf{e}_{123} | \varrho \in \mathbb{E}_0, \tag{5.38}$$

$$\mathbf{J}_g = \mathbf{e}_{123} \lfloor \mathbf{J} \in \mathbb{E}_1 \tag{5.39}$$

are respectively scalar and vector valued. Applying these rules and defining the vector operator $\nabla = \overline{\overline{\mathsf{G}}}_s|\mathbf{d}_s$, the Maxwell equations can be written for the Gibbsian vector fields and sources as (details are left as an exercise)

$$\nabla \cdot \mathbf{B}_g = 0, \tag{5.40}$$

$$\nabla \times \mathbf{E}_g + \partial_t \mathbf{B}_g = 0, \tag{5.41}$$

$$\nabla \cdot \mathbf{D}_g = \varrho_g, \tag{5.42}$$

$$\nabla \times \mathbf{H}_g - \partial_t \mathbf{D}_g = \mathbf{J}_g. \tag{5.43}$$

In their original form, the Maxwell equations consisted of 20 scalar equations [21], which were later expressed by Heaviside [22] and Hertz [23] in a more compact form equivalent to (5.40)–(5.43).

5.1.3 Potential One-Form

With no magnetic sources, $\boldsymbol{\gamma}_m = 0$, de Rham's theorem and (5.21) imply that the two-form $\boldsymbol{\Phi}$ can be expressed in terms of a one-form, the potential $\boldsymbol{\phi}$, as

$$\boldsymbol{\Phi} = \mathbf{d} \wedge \boldsymbol{\phi}. \tag{5.44}$$

Expanding $\boldsymbol{\phi}$ in its spatial and temporal parts as

$$\boldsymbol{\phi} = \mathbf{A} - \varphi \boldsymbol{\varepsilon}_4, \tag{5.45}$$

where $\mathbf{A} \in \mathbb{F}_1$ is the spatial one-form and φ is the scalar potential, we obtain

$$\boldsymbol{\Phi} = \mathbf{d}_s \wedge \mathbf{A} + (\partial_\tau \mathbf{A} - \mathbf{d}_s \varphi) \wedge \boldsymbol{\varepsilon}_4, \tag{5.46}$$

which can be split as

$$\mathbf{B} = \mathbf{d}_s \wedge \mathbf{A}, \tag{5.47}$$

$$\mathbf{E} = \partial_\tau \mathbf{A} - \mathbf{d}_s \varphi. \tag{5.48}$$

The corresponding Gibbsian equations for the scalar potential φ and the vector potential $\mathbf{A}_g = \overline{\overline{\mathsf{G}}}_s|\mathbf{A}$ are

$$\mathbf{B}_g = \nabla \cdot \mathbf{A}_g, \tag{5.49}$$

$$\mathbf{E}_g = \partial_\tau \mathbf{A}_g - \nabla \varphi. \tag{5.50}$$

5.2 MEDIUM EQUATIONS

In macroscopic consideration, physical materials can be represented by a set of medium parameters which define relations between the electromagnetic field quantities.

5.2.1 Medium Bidyadics

Assuming a simple homogeneous, time-invariant and linear medium, the electromagnetic two-forms are related by an algebraic equation of the form

$$\boldsymbol{\Psi} = \overline{\overline{\mathsf{M}}}|\boldsymbol{\Phi}, \tag{5.51}$$

where $\overline{\overline{\mathsf{M}}} \in \mathbb{F}_2\mathbb{E}_2$ is the medium bidyadic mapping two-forms to two-forms. Since $\overline{\overline{\mathsf{M}}}$ corresponds to a 6×6 matrix, it can be represented by 36 parameters when expanded in some basis. Another way to define the medium equation is

$$\boldsymbol{\Phi} = \overline{\overline{\mathsf{N}}}|\boldsymbol{\Psi}, \tag{5.52}$$

and there is basically no reason to prefer one to the other one, although (5.51) will mostly be applied in the sequel. When the bidyadic $\overline{\overline{\mathsf{M}}}$ has an inverse, we have the relations

$$\overline{\overline{\mathsf{N}}} = \overline{\overline{\mathsf{M}}}^{-1}, \quad \overline{\overline{\mathsf{M}}} = \overline{\overline{\mathsf{N}}}^{-1}. \tag{5.53}$$

However, if there is no inverse for $\overline{\overline{\mathsf{M}}}$ or for $\overline{\overline{\mathsf{N}}}$, only one of the equations (5.51), (5.52) makes sense.

Another useful form for the medium equation, equivalent to (5.51), is

$$\mathbf{e}_N\lfloor\boldsymbol{\Psi} = \overline{\overline{\mathsf{M}}}_m|\boldsymbol{\Phi} \in \mathbb{E}_2, \tag{5.54}$$

where

$$\overline{\overline{\mathsf{M}}}_m = \mathbf{e}_N\lfloor\overline{\overline{\mathsf{M}}} \in \mathbb{E}_2\mathbb{E}_2 \tag{5.55}$$

is a metric bidyadic called the modified medium bidyadic (in [1] the modified medium bidyadic was denoted by $\overline{\overline{\mathsf{M}}}_g$). Similarly, the modified version of the medium bidyadic $\overline{\overline{\mathsf{N}}}$ is defined by

$$\mathbf{e}_N\lfloor\boldsymbol{\Phi} = \overline{\overline{\mathsf{N}}}_m|\boldsymbol{\Psi}, \quad \overline{\overline{\mathsf{N}}}_m = \mathbf{e}_N\lfloor\overline{\overline{\mathsf{N}}}. \tag{5.56}$$

While the two medium bidyadics are related by (5.53), the two modified bidyadics satisfy the more complicated relations

$$\overline{\overline{\mathsf{N}}}_m = \mathbf{e}_N\mathbf{e}_N\lfloor\lfloor\overline{\overline{\mathsf{M}}}_m^{-1}, \quad \overline{\overline{\mathsf{M}}}_m = \mathbf{e}_N\mathbf{e}_N\lfloor\lfloor\overline{\overline{\mathsf{N}}}_m^{-1}. \tag{5.57}$$

5.2.2 Potential Equation

Knowledge of the medium bidyadic $\overline{\overline{\mathsf{M}}}$ allows one to form an equation for the potential one-form $\boldsymbol{\phi}$. For a homogeneous and time-invariant medium $\overline{\overline{\mathsf{M}}}$ and $\overline{\overline{\mathsf{M}}}_m$ have no functional dependence on $\mathbf{x}$, whence inserting (5.51) and (5.44) in (5.22) yields

$$\mathbf{d}\wedge\boldsymbol{\Psi} = \mathbf{d}\wedge\overline{\overline{\mathsf{M}}}|\boldsymbol{\Phi} = \mathbf{d}\wedge\overline{\overline{\mathsf{M}}}|(\mathbf{d}\wedge\boldsymbol{\phi}) = \boldsymbol{\gamma}. \tag{5.58}$$

Written in the form

$$(\mathbf{d}\wedge\overline{\overline{\mathsf{M}}}\lfloor\mathbf{d})|\boldsymbol{\phi} = \boldsymbol{\gamma}, \tag{5.59}$$

it represents a mapping from a one-form $\boldsymbol{\phi}$ to a three-form $\boldsymbol{\gamma}$. Contracting by $\mathbf{e}_N\lfloor$ and applying the rule

$$\mathbf{e}_N\lfloor(\boldsymbol{\beta}\wedge\boldsymbol{\gamma}) = -\boldsymbol{\beta}\rfloor(\mathbf{e}_N\lfloor\boldsymbol{\gamma}), \tag{5.60}$$

valid for any one-form $\boldsymbol{\beta}$ and two-form $\boldsymbol{\gamma}$, yields

$$\mathbf{e}_N\lfloor(\mathbf{d}\wedge\overline{\overline{\mathsf{M}}}\lfloor\mathbf{d})|\boldsymbol{\alpha} = -(\mathbf{d}\rfloor\overline{\overline{\mathsf{M}}}_m\lfloor\mathbf{d})|\boldsymbol{\alpha} = \mathbf{e}_N\lfloor\boldsymbol{\gamma}. \tag{5.61}$$

This is of the form

$$\overline{\overline{\mathsf{D}}}(\mathbf{d})|\boldsymbol{\phi} = \mathbf{g}, \tag{5.62}$$

where the second-order dyadic operator is defined by

$$\overline{\overline{\mathsf{D}}}(\mathbf{d}) = -\mathbf{d}\rfloor\overline{\overline{\mathsf{M}}}_m\lfloor\mathbf{d} = \overline{\overline{\mathsf{M}}}_m\lfloor\lfloor\mathbf{dd} \in \mathbb{E}_1\mathbb{F}_1, \tag{5.63}$$

and the source vector by

$$\mathbf{g} = \mathbf{e}_N\lfloor\boldsymbol{\gamma} \in \mathbb{E}_1. \tag{5.64}$$

5.2.3 Expansions of Medium Bidyadics

The medium bidyadic can be expanded in its spatial and temporal parts as

$$\overline{\overline{\mathsf{M}}} = \overline{\overline{\alpha}} + \overline{\overline{\epsilon}}'\wedge\mathbf{e}_4 + \boldsymbol{\varepsilon}_4\wedge\overline{\overline{\mu}}^{-1} + \boldsymbol{\varepsilon}_4\wedge\overline{\overline{\beta}}\wedge\mathbf{e}_4, \tag{5.65}$$

where the four spatial dyadics belong to different spaces as

$$\overline{\overline{\alpha}} \in \mathbb{F}_2\mathbb{E}_2, \quad \overline{\overline{\epsilon}}' \in \mathbb{F}_2\mathbb{E}_1, \quad \overline{\overline{\mu}}^{-1} \in \mathbb{F}_1\mathbb{E}_2, \quad \overline{\overline{\beta}} \in \mathbb{F}_1\mathbb{E}_1. \tag{5.66}$$

Because the dyadics $\overline{\overline{\alpha}}, \overline{\overline{\epsilon}}', \overline{\overline{\mu}}^{-1}, \overline{\overline{\beta}}$ map spatial one-forms or two-forms to spatial one-forms or two-forms, each of them can be represented by 9 scalar parameters in a given basis. Inserting (5.65) in (5.51) we have

$$\boldsymbol{\Psi} = \mathbf{D} - \mathbf{H} \wedge \boldsymbol{\varepsilon}_4 = (\overline{\overline{\alpha}} + \boldsymbol{\varepsilon}_4 \wedge \overline{\overline{\mu}}^{-1})|\mathbf{B} + (\overline{\overline{\epsilon}}' + \boldsymbol{\varepsilon}_4 \wedge \overline{\overline{\beta}})|\mathbf{E}, \tag{5.67}$$

which corresponds to the spatial medium equations

$$\mathbf{D} = \overline{\overline{\alpha}}|\mathbf{B} + \overline{\overline{\epsilon}}'|\mathbf{E}, \tag{5.68}$$

$$\mathbf{H} = \overline{\overline{\mu}}^{-1}|\mathbf{B} + \overline{\overline{\beta}}|\mathbf{E}. \tag{5.69}$$

In electrical engineering it is customary to express the Gibbsian field relations between other field quantities [24, 25]. In differential-form formalism the equations take the form

$$\mathbf{D} = \overline{\overline{\epsilon}}|\mathbf{E} + \overline{\overline{\xi}}|\mathbf{H}, \tag{5.70}$$

$$\mathbf{B} = \overline{\overline{\zeta}}|\mathbf{E} + \overline{\overline{\mu}}|\mathbf{H}. \tag{5.71}$$

In this notation all four dyadics are members of the same space, $\overline{\overline{\epsilon}}, \overline{\overline{\xi}}, \overline{\overline{\zeta}}, \overline{\overline{\mu}} \in \mathbb{F}_2\mathbb{E}_1$. One must note that $\overline{\overline{\epsilon}}$ and $\overline{\overline{\epsilon}}'$ are different dyadics in general. They are the same when $\overline{\overline{\alpha}} = 0$ and $\overline{\overline{\beta}} = 0$ or, equivalently, when $\overline{\overline{\xi}} = \overline{\overline{\zeta}} = 0$. However, $\overline{\overline{\mu}}$ is the same dyadic in both representations.

One can define media which cannot be represented in the form (5.68), (5.69) or in the form (5.70), (5.71). For example, if the dyadic $\overline{\overline{\mu}}$ does not have an inverse, (5.69) is obviously not a good choice. On the other hand, one can find examples of media for which $\overline{\overline{\mu}}^{-1}$ exists but $\overline{\overline{\mu}}$ does not, in which case (5.70) and (5.71) cannot be applied.

When (5.68), (5.69) and (5.70), (5.71) are both valid representations of the medium conditions, relations between the two sets of medium dyadics can be expressed as

$$\overline{\overline{\epsilon}}' = \overline{\overline{\epsilon}} - \overline{\overline{\xi}}|\overline{\overline{\mu}}^{-1}|\overline{\overline{\zeta}}, \quad \overline{\overline{\alpha}} = \overline{\overline{\xi}}|\overline{\overline{\mu}}^{-1}, \quad \overline{\overline{\beta}} = -\overline{\overline{\mu}}^{-1}|\overline{\overline{\zeta}}, \tag{5.72}$$

and

$$\overline{\overline{\epsilon}} = \overline{\overline{\epsilon}}' - \overline{\overline{\alpha}}|\overline{\overline{\mu}}|\overline{\overline{\beta}}, \quad \overline{\overline{\xi}} = \overline{\overline{\alpha}}|\overline{\overline{\mu}}, \quad \overline{\overline{\zeta}} = -\overline{\overline{\mu}}|\overline{\overline{\beta}}. \tag{5.73}$$

Inserting (5.72) in (5.67) as

$$\mathbf{D} - \mathbf{H} \wedge \boldsymbol{\varepsilon}_4 = (\overline{\overline{\xi}}|\overline{\overline{\mu}}^{-1} + \boldsymbol{\varepsilon}_4 \wedge \overline{\overline{\mu}}^{-1})|\mathbf{B} + (\overline{\overline{\epsilon}} - \overline{\overline{\xi}}|\overline{\overline{\mu}}^{-1}|\overline{\overline{\zeta}} - \boldsymbol{\varepsilon}_4 \wedge \overline{\overline{\mu}}^{-1}|\overline{\overline{\zeta}})|\mathbf{E}, \tag{5.74}$$

leads to another expansion for the medium bidyadic,

$$\overline{\overline{\mathsf{M}}} = \overline{\overline{\epsilon}} \wedge \mathbf{e}_4 + (\varepsilon_4 \wedge \overline{\overline{\mathrm{I}}}{}_s^T + \overline{\overline{\xi}})|\overline{\overline{\mu}}^{-1}|(\overline{\overline{\mathrm{I}}}{}_s^{(2)T} - \overline{\overline{\zeta}} \wedge \mathbf{e}_4). \tag{5.75}$$

On the other hand, starting from the representation

$$\mathbf{B} + \mathbf{E} \wedge \varepsilon_4 = (\overline{\overline{\zeta}}|\overline{\overline{\epsilon}}^{-1} - \varepsilon_4 \wedge \overline{\overline{\epsilon}}^{-1})|\mathbf{D} + (\overline{\overline{\mu}} + \varepsilon_4 \wedge \overline{\overline{\epsilon}}^{-1}|\overline{\overline{\xi}} - \overline{\overline{\zeta}}|\overline{\overline{\epsilon}}^{-1}|\overline{\overline{\xi}})|\mathbf{H}, \tag{5.76}$$

yields the following expansion for the medium bidyadic $\overline{\overline{\mathsf{N}}}$,

$$\overline{\overline{\mathsf{N}}} = \overline{\overline{\mathsf{M}}}^{-1} = -\overline{\overline{\mu}} \wedge \mathbf{e}_4 - (\varepsilon_4 \wedge \overline{\overline{\mathrm{I}}}{}_s^T - \overline{\overline{\zeta}})|\overline{\overline{\epsilon}}^{-1}|(\overline{\overline{\mathrm{I}}}{}_s^{(2)T} + \overline{\overline{\xi}} \wedge \mathbf{e}_4). \tag{5.77}$$

One can verify that (5.75) and (5.77) satisfy

$$\overline{\overline{\mathsf{M}}}|\overline{\overline{\mathsf{N}}} = \overline{\overline{\mathsf{N}}}|\overline{\overline{\mathsf{M}}} = \overline{\overline{\mathrm{I}}}^{(2)T}, \tag{5.78}$$

details of which are left as an exercise.

Actually, there exist six possibilities to present the medium equations between two pairs of the four spatial fields $\mathbf{E}, \mathbf{H}, \mathbf{B}, \mathbf{D}$. Equations (5.68) and (5.69) could be inverted and presented as $\mathbf{B}, \mathbf{E}$ in terms of $\mathbf{D}, \mathbf{H}$. Similarly, (5.70) and (5.71) could be written as $\mathbf{E}, \mathbf{H}$ in terms of $\mathbf{D}, \mathbf{B}$. Yet another, equally valid, set of equations is

$$\mathbf{D} = \overline{\overline{\mathsf{A}}}|\mathbf{B} + \overline{\overline{\mathsf{B}}}|\mathbf{H}, \tag{5.79}$$

$$\mathbf{E} = \overline{\overline{\mathsf{C}}}|\mathbf{B} + \overline{\overline{\mathsf{D}}}|\mathbf{H}, \tag{5.80}$$

and, conversely, we can express $\mathbf{B}, \mathbf{H}$ in terms of $\mathbf{D}, \mathbf{E}$. These last two possibilities have been ignored in the literature. However, there may well exist media whose simplest representation requires the form (5.79), (5.80) or its inverse.

5.2.4 Gibbsian Representation

Let us now consider the connection of the medium dyadics and the corresponding Gibbsian dyadics. The Gibbsian vector fields $\mathbf{D}_g, \mathbf{B}_g, \mathbf{E}_g, \mathbf{H}_g$ defined by (5.33)–(5.36) are related by Gibbsian medium dyadics $\overline{\overline{\epsilon}}_g, \overline{\overline{\xi}}_g, \overline{\overline{\zeta}}_g, \overline{\overline{\mu}}_g$, which all belong to the same space $\mathbb{E}_1\mathbb{E}_1$,

$$\mathbf{D}_g = \overline{\overline{\epsilon}}_g \cdot \mathbf{E}_g + \overline{\overline{\xi}}_g \cdot \mathbf{H}_g = \overline{\overline{\epsilon}}_g|\mathbf{E} + \overline{\overline{\xi}}_g|\mathbf{H}, \tag{5.81}$$

$$\mathbf{B}_g = \overline{\overline{\zeta}}_g \cdot \mathbf{E}_g + \overline{\overline{\mu}}_g \cdot \mathbf{H}_g = \overline{\overline{\zeta}}_g|\mathbf{E} + \overline{\overline{\mu}}_g|\mathbf{H}. \tag{5.82}$$

Here we apply rules given in Appendix B. Relations between the two sets of medium dyadics can be represented as

$$\overline{\overline{\epsilon}}' = \varepsilon_{123}\lfloor(\overline{\overline{\epsilon}}_g - \overline{\overline{\xi}}_g|\overline{\overline{\mu}}_g^{-1}|\overline{\overline{\zeta}}_g), \tag{5.83}$$

$$\overline{\overline{\mu}} = \varepsilon_{123}\lfloor\overline{\overline{\mu}}_g, \tag{5.84}$$

$$\overline{\overline{\alpha}} = \varepsilon_{123}\lfloor(\overline{\overline{\xi}}_g|\overline{\overline{\mu}}_g^{-1})\rfloor\mathbf{e}_{123}, \tag{5.85}$$

$$\overline{\overline{\beta}} = -\overline{\overline{\mu}}_g^{-1}|\overline{\overline{\zeta}}_g. \tag{5.86}$$

Inserting the latter set in (5.65) produces an expansion of the medium bidyadic in terms of the Gibbsian dyadics,

$$\overline{\overline{\mathsf{M}}} = \varepsilon_{123}\lfloor\overline{\overline{\epsilon}}_g \wedge \mathbf{e}_4 + (\varepsilon_{123}\lfloor\overline{\overline{\xi}}_g + \varepsilon_4 \wedge \overline{\overline{\mathsf{I}}}^T)|\overline{\overline{\mu}}_g^{-1}|(\overline{\overline{\mathsf{I}}}\rfloor\mathbf{e}_{123} - \overline{\overline{\zeta}}_g \wedge \mathbf{e}_4), \tag{5.87}$$

and the corresponding expansion for the modified medium bidyadic is obtained from $\overline{\overline{\mathsf{M}}}_m = \mathbf{e}_N\lfloor\overline{\overline{\mathsf{M}}}$ as

$$\overline{\overline{\mathsf{M}}}_m = \overline{\overline{\epsilon}}_g{}^{\wedge}_{\wedge}\mathbf{e}_4\mathbf{e}_4 - (\mathbf{e}_{123}\lfloor\overline{\overline{\mathsf{I}}}^T + \mathbf{e}_4 \wedge \overline{\overline{\xi}}_g)|\overline{\overline{\mu}}_g^{-1}|(\overline{\overline{\mathsf{I}}}\rfloor\mathbf{e}_{123} - \overline{\overline{\zeta}}_g \wedge \mathbf{e}_4). \tag{5.88}$$

Its inverse can be expressed following the form of (5.77) as

$$\overline{\overline{\mathsf{M}}}_m^{-1} = -\varepsilon_{123}\varepsilon_{123}\lfloor\lfloor\overline{\overline{\mu}}_g - (\varepsilon_4 \wedge \overline{\overline{\mathsf{I}}}^T - \varepsilon_{123}\lfloor\overline{\overline{\zeta}}_g)|\overline{\overline{\epsilon}}_g^{-1}|(\overline{\overline{\mathsf{I}}} \wedge \varepsilon_4 + \overline{\overline{\xi}}_g\rfloor\varepsilon_{123}). \tag{5.89}$$

One can verify that the two dyadics satisfy the relations

$$\overline{\overline{\mathsf{M}}}_m|\overline{\overline{\mathsf{M}}}_m^{-1} = \overline{\overline{\mathsf{I}}}^{(2)}, \quad \overline{\overline{\mathsf{M}}}_m^{-1}|\overline{\overline{\mathsf{M}}}_m = \overline{\overline{\mathsf{I}}}^{(2)T}, \tag{5.90}$$

details of which are left as an exercise.

5.3 BASIC CLASSES OF MEDIA

5.3.1 Hehl–Obukhov Decomposition

Based on the medium bidyadic $\overline{\overline{\mathsf{M}}}$, the Hehl–Obukhov decomposition (3.74)

$$\overline{\overline{\mathsf{M}}} = \overline{\overline{\mathsf{M}}}_1 + \overline{\overline{\mathsf{M}}}_2 + \overline{\overline{\mathsf{M}}}_3, \tag{5.91}$$

and the corresponding modified medium bidyadic decomposition (3.85),

$$\overline{\overline{\mathsf{M}}}_m = \overline{\overline{\mathsf{M}}}_{m1} + \overline{\overline{\mathsf{M}}}_{m2} + \overline{\overline{\mathsf{M}}}_{m3}, \tag{5.92}$$

allows one to define three basic classes of electromagnetic media. Properties of the different components in the decomposition as defined in Section 3.3 can be summarized as follows.

- **The axion part** $\overline{\overline{\mathsf{M}}}_3$ consists of the trace of the medium bidyadic $\overline{\overline{\mathsf{M}}}$,

$$\overline{\overline{\mathsf{M}}}_3 = \frac{1}{6}\mathrm{tr}\overline{\overline{\mathsf{M}}}\,\overline{\overline{\mathsf{I}}}^{(2)T}. \tag{5.93}$$

- **The skewon part** $\overline{\overline{\mathsf{M}}}_2$ satisfies

$$\overline{\overline{\mathsf{I}}}^{(4)}\lfloor\lfloor\overline{\overline{\mathsf{M}}}_2 = -\overline{\overline{\mathsf{M}}}_2^T, \tag{5.94}$$

whence it is trace free, $\mathrm{tr}\overline{\overline{\mathsf{M}}}_2 = 0$. The modified medium bidyadic $\overline{\overline{\mathsf{M}}}_{m2}$ consists of the antisymmetric part of $\overline{\overline{\mathsf{M}}}_m$,

$$\overline{\overline{\mathsf{M}}}_{m2} = \frac{1}{2}(\overline{\overline{\mathsf{M}}}_m - \overline{\overline{\mathsf{M}}}_m^T). \tag{5.95}$$

As was shown by (3.86), the skewon part of $\overline{\overline{\mathsf{M}}}$ can be expressed in terms of a trace-free dyadic $\overline{\overline{\mathsf{B}}}_o \in \mathbb{E}_1\mathbb{F}_1$ in the form

$$\overline{\overline{\mathsf{M}}}_2 = \varepsilon_N\lfloor\overline{\overline{\mathsf{M}}}_{m2} = (\overline{\overline{\mathsf{B}}}_o{}^\wedge_\wedge\overline{\overline{\mathsf{I}}})^T \tag{5.96}$$

and the dyadic $\overline{\overline{\mathsf{B}}}_o$ is obtained from $\overline{\overline{\mathsf{M}}}_2$ through the operation

$$\overline{\overline{\mathsf{B}}}_o = \frac{1}{2}(\overline{\overline{\mathsf{M}}}_2\lfloor\lfloor\overline{\overline{\mathsf{I}}})^T. \tag{5.97}$$

- **The principal part** is defined by $\overline{\overline{\mathsf{M}}}_1 = \overline{\overline{\mathsf{M}}} - \overline{\overline{\mathsf{M}}}_2 - \overline{\overline{\mathsf{M}}}_3$. It satisfies

$$\overline{\overline{\mathsf{I}}}^{(4)}\lfloor\lfloor\overline{\overline{\mathsf{M}}}_1 = \overline{\overline{\mathsf{M}}}_1^T, \quad \mathrm{tr}\overline{\overline{\mathsf{M}}}_1 = 0, \tag{5.98}$$

whence the modified medium bidyadic $\overline{\overline{\mathsf{M}}}_{m1}$ is symmetric. It equals the symmetric part of $\overline{\overline{\mathsf{M}}}_m$ without the axion part,

$$\overline{\overline{\mathsf{M}}}_{m1} = \frac{1}{2}(\overline{\overline{\mathsf{M}}}_m + \overline{\overline{\mathsf{M}}}_m^T) - \frac{1}{6}\mathrm{tr}\overline{\overline{\mathsf{M}}}\,\mathbf{e}_N\lfloor\overline{\overline{\mathsf{I}}}^{(2)T}. \tag{5.99}$$

In (3.78) it was shown that that if the medium bidyadic satisfies the condition

$$\overline{\overline{\mathsf{M}}}\lfloor\lfloor\overline{\overline{\mathsf{I}}} = 0, \tag{5.100}$$

it equals its principal part, $\overline{\overline{\mathsf{M}}} = \overline{\overline{\mathsf{M}}}_1$. Thus, (5.100) represents the definition of a medium with no skewon and axion components.

The Hehl–Obukhov decomposition of the medium bidyadic can be applied to define basic classes of media, independent of any basis systems by requiring that the medium bidyadic $\overline{\overline{\mathsf{M}}}$ does not contain all of the three Hehl–Obukhov parts $\overline{\overline{\mathsf{M}}}_i$. There are six different possibilities given in the following table.

$\overline{\overline{\mathsf{M}}}$	Medium class
$\overline{\overline{\mathsf{M}}}_1$	Principal media
$\overline{\overline{\mathsf{M}}}_2$	Skewon media
$\overline{\overline{\mathsf{M}}}_3$	Axion media
$\overline{\overline{\mathsf{M}}}_1 + \overline{\overline{\mathsf{M}}}_2$	Axion-free media
$\overline{\overline{\mathsf{M}}}_1 + \overline{\overline{\mathsf{M}}}_3$	Principal–axion media
$\overline{\overline{\mathsf{M}}}_2 + \overline{\overline{\mathsf{M}}}_3$	Skewon–axion media

The classification could also be made through the medium bidyadic $\overline{\overline{\mathsf{N}}}$ defined in (5.52). However, one must note that the medium classes do not coincide with those defined through $\overline{\overline{\mathsf{M}}}$ in all cases. Because the inverse of a symmetric and an antisymmetric bidyadic is respectively symmetric and antisymmetric, the classes of skewon media and principal–axion media are the same in both definitions and so is the class of axion media. However, in the other cases the two definitions do not lead to the same classes in all cases.

As an example, a principal medium defined through the bidyadic $\overline{\overline{\mathsf{M}}}$ may contain an axion component when defined in terms of the bidyadic $\overline{\overline{\mathsf{N}}}$. In fact, one can show that $\overline{\overline{\mathsf{M}}}\lfloor\lfloor\overline{\overline{\mathsf{I}}} = 0$ does not imply the condition $\overline{\overline{\mathsf{M}}}^{-1}\lfloor\lfloor\overline{\overline{\mathsf{I}}} = 0$. However, there does exist media satisfying both conditions. As an example we may consider $\overline{\overline{\mathsf{M}}} = \varepsilon_N\lfloor\overline{\overline{\mathsf{Q}}}^{(2)}$ with a symmetric dyadic $\overline{\overline{\mathsf{Q}}} \in \mathbb{E}_1\mathbb{E}_1$. In this case one can show that $\mathrm{tr}\overline{\overline{\mathsf{M}}} = 0$ implies $\mathrm{tr}\overline{\overline{\mathsf{M}}}^{-1} = 0$, whence both $\overline{\overline{\mathsf{M}}}$ and $\overline{\overline{\mathsf{N}}} = \overline{\overline{\mathsf{M}}}^{-1}$ are principal bidyadics. Similarly, for the class of skewon–axion media, from $\overline{\overline{\mathsf{M}}} = (\overline{\overline{\mathsf{B}}}{}^\wedge_\wedge\overline{\overline{\mathsf{I}}})^T$ it does not follow that the inverse bidyadic has a similar form $\overline{\overline{\mathsf{M}}}^{-1} = (\overline{\overline{\mathsf{A}}}{}^\wedge_\wedge\overline{\overline{\mathsf{I}}})^T$, unless $\overline{\overline{\mathsf{B}}} = \overline{\overline{\mathsf{B}}}_o$ happens to be trace free. In the following we will define the medium classes through the medium bidyadic $\overline{\overline{\mathsf{M}}}$ whenever it exists.

5.3.2 3D Expansions

The two solutions of the bidyadic eigenproblem (3.48) rewritten here as

$$\overline{\overline{\mathsf{I}}}^{(4)}\lfloor\lfloor\overline{\overline{\mathsf{M}}}_\pm = \pm\overline{\overline{\mathsf{M}}}_\pm^T, \qquad (5.101)$$

coincide with the principal–axion bidyadic $\overline{\overline{\mathsf{M}}}_+ = \overline{\overline{\mathsf{M}}}_1 + \overline{\overline{\mathsf{M}}}_3$ and the skewon bidyadic $\overline{\overline{\mathsf{M}}}_- = \overline{\overline{\mathsf{M}}}_2$. Let us study the properties of the corresponding spatial medium dyadics by inserting the expansion (5.65) in (5.101) and equating similar terms on both sides. This leads to the set of double relations

$$\overline{\overline{\alpha}}_\pm = \mp\varepsilon_{123}\mathbf{e}_{123}\lfloor\lfloor\overline{\overline{\beta}}{}^T_\pm, \tag{5.102}$$

$$\overline{\overline{\epsilon}}{}'_\pm = \pm\varepsilon_{123}\mathbf{e}_{123}\lfloor\lfloor\overline{\overline{\epsilon}}{}'^T_\pm, \tag{5.103}$$

$$\overline{\overline{\mu}}{}^{-1}_\pm = \pm\varepsilon_{123}\mathbf{e}_{123}\lfloor\lfloor\overline{\overline{\mu}}{}^{-1T}_\pm, \tag{5.104}$$

$$\overline{\overline{\beta}}_\pm = \mp\varepsilon_{123}\mathbf{e}_{123}\lfloor\lfloor\overline{\overline{\alpha}}{}^T_\pm, \tag{5.105}$$

for the principal–axion (+) and skewon (–) dyadics. They can be written more compactly as

$$\overline{\overline{\mathsf{I}}}{}^{(3)}_s\lfloor\lfloor\begin{pmatrix}\overline{\overline{\epsilon}}{}' & \overline{\overline{\alpha}} \\ \overline{\overline{\beta}} & \overline{\overline{\mu}}{}^{-1}\end{pmatrix}_\pm = \pm\begin{pmatrix}\overline{\overline{\epsilon}}{}' & -\overline{\overline{\alpha}} \\ -\overline{\overline{\beta}} & \overline{\overline{\mu}}{}^{-1}\end{pmatrix}^T_\pm. \tag{5.106}$$

The relations (5.102) and (5.105) are actually equivalent. Expressing (5.103) and (5.105) as

$$\mathbf{e}_{123}\lfloor\overline{\overline{\epsilon}}{}'_\pm = \pm(\mathbf{e}_{123}\lfloor\overline{\overline{\epsilon}}{}'_\pm)^T \in \mathbb{E}_1\mathbb{E}_1, \tag{5.107}$$

$$\mathbf{e}_{123}\lfloor\overline{\overline{\mu}}{}^{-1}_\pm = \pm(\mathbf{e}_{123}\lfloor\overline{\overline{\mu}}{}^{-1}_\pm)^T \in \mathbb{E}_2\mathbb{E}_2, \tag{5.108}$$

shows us that the dyadics $\mathbf{e}_{123}\lfloor\overline{\overline{\epsilon}}{}'_+$ and $\mathbf{e}_{123}\lfloor\overline{\overline{\mu}}{}^{-1}_+$ must be symmetric and the dyadics $\mathbf{e}_{123}\lfloor\overline{\overline{\epsilon}}{}'_-$ and $\mathbf{e}_{123}\lfloor\overline{\overline{\mu}}{}^{-1}_-$ must be antisymmetric. Thus, for a skewon medium [26], neither $\overline{\overline{\epsilon}}{}'$ nor $\overline{\overline{\mu}}{}^{-1}$ possesses a spatial inverse. In particular, since a dyadic $\overline{\overline{\mu}}_-$ does not exist, there is no Gibbsian representation of the form (5.81), (5.82) for the skewon medium. In contrast, a Gibbsian representation exists for the principal–axion medium in which case the medium dyadic matrix is symmetric,

$$\begin{pmatrix}\overline{\overline{\epsilon}}_g & \overline{\overline{\xi}}_g \\ \overline{\overline{\zeta}}_g & \overline{\overline{\mu}}_g\end{pmatrix}^T_+ = \begin{pmatrix}\overline{\overline{\epsilon}}{}^T_g & \overline{\overline{\zeta}}{}^T_g \\ \overline{\overline{\xi}}{}^T_g & \overline{\overline{\mu}}{}^T_g\end{pmatrix}_+ = \begin{pmatrix}\overline{\overline{\epsilon}}_g & \overline{\overline{\xi}}_g \\ \overline{\overline{\zeta}}_g & \overline{\overline{\mu}}_g\end{pmatrix}_+. \tag{5.109}$$

It is known that symmetric medium dyadics $\overline{\overline{\epsilon}}_g$ and $\overline{\overline{\mu}}_g$ can be realized by dielectric and magnetic crystals while the symmetric part of the dyadic $\overline{\overline{\xi}}_g + \overline{\overline{\zeta}}{}^T_g$ exists in a Tellegen medium [27] and the antisymmetric part effectively arises when a medium is in uniform motion [28].

Considering the pure principal medium, $\overline{\overline{\mathsf{M}}} = \overline{\overline{\mathsf{M}}}_+$ with $\mathrm{tr}\overline{\overline{\mathsf{M}}} = 0$, from (5.65) the latter condition becomes

$$\mathrm{tr}\overline{\overline{\alpha}} = \mathrm{tr}\overline{\overline{\beta}}. \tag{5.110}$$

Expressing $\overline{\overline{\alpha}}$ and $\overline{\overline{\beta}}$ in terms of the Gibbsian medium dyadics as

$$\overline{\overline{\alpha}} = \boldsymbol{\varepsilon}_{123}\mathbf{e}_{123}\lfloor\lfloor(\overline{\overline{\xi}}_g|\overline{\overline{\mu}}_g^{-1}), \quad \overline{\overline{\beta}} = -\overline{\overline{\mu}}_g^{-1}|\overline{\overline{\zeta}}_g, \tag{5.111}$$

(5.110) can be written as

$$\mathrm{tr}(\overline{\overline{\xi}}_g|\overline{\overline{\mu}}_g^{-1}) + \mathrm{tr}(\overline{\overline{\mu}}_g^{-1}|\overline{\overline{\zeta}}_g) = 0. \tag{5.112}$$

Since from (5.109) we obtain $\overline{\overline{\zeta}}_g = \overline{\overline{\xi}}_g^T$ and $\overline{\overline{\mu}}_g^{-1} = \overline{\overline{\mu}}_g^{-1T}$, the axion-free condition (5.110) for a principal medium is further reduced to

$$\overline{\overline{\xi}}_g||\overline{\overline{\mu}}_g^{-1} = 0. \tag{5.113}$$

Because $\overline{\overline{\mu}}_g^{-1}$ is symmetric, this does not restrict the choice of the antisymmetric part of $\overline{\overline{\xi}}_g$. Equations (5.109) and (5.113) define the principal medium in terms of Gibbsian dyadics. The number of free parameters of a principal medium bidyadic equals 36 − 15 − 1 = 20.

5.3.3 Simple Principal Medium

As an example of a simple electromagnetic medium let us consider one depending on a spatial dyadic $\overline{\overline{\mathsf{C}}}_s \in \mathbb{E}_2\mathbb{F}_1$ possessing a spatial inverse $\overline{\overline{\mathsf{C}}}_s^{-1} \in \mathbb{E}_1\mathbb{F}_2$, defined by

$$\overline{\overline{\alpha}} = 0, \quad \overline{\overline{\epsilon}}' = \epsilon\overline{\overline{\mathsf{C}}}_s, \quad \overline{\overline{\mu}}^{-1} = (\mu\overline{\overline{\mathsf{C}}}_s)^{-1}, \quad \overline{\overline{\beta}} = 0. \tag{5.114}$$

The corresponding medium bidyadic has the form

$$\overline{\overline{\mathsf{M}}} = \epsilon\overline{\overline{\mathsf{C}}}_s \wedge \mathbf{e}_4 + \boldsymbol{\varepsilon}_4 \wedge (\mu\overline{\overline{\mathsf{C}}}_s)^{-1}. \tag{5.115}$$

Because of the form (5.115), $\overline{\overline{\mathsf{M}}}$ is trace free, whence the medium has no axion component. The modified medium bidyadic corresponding to (5.115) can be expanded as

$$\overline{\overline{\mathsf{M}}}_m = \mathbf{e}_N\lfloor(\epsilon\overline{\overline{\mathsf{C}}}_s \wedge \mathbf{e}_4 + \boldsymbol{\varepsilon}_4 \wedge (\mu\overline{\overline{\mathsf{C}}}_s)^{-1}) \tag{5.116}$$

$$= -\epsilon\mathbf{e}_4 \wedge (\mathbf{e}_{123}\lfloor\overline{\overline{\mathsf{C}}}_s) \wedge \mathbf{e}_4 - \frac{1}{\mu}\mathbf{e}_{123}\lfloor\overline{\overline{\mathsf{C}}}_s^{-1} \tag{5.117}$$

$$= \frac{1}{\mu}(\mu\epsilon(\mathbf{e}_{123}\lfloor\overline{\overline{\mathsf{C}}}_s)^{\wedge}_{\wedge}\mathbf{e}_4\mathbf{e}_4 - \mathbf{e}_{123}\mathbf{e}_{123}\lfloor\lfloor(\mathbf{e}_{123}\lfloor\overline{\overline{\mathsf{C}}}_s)^{-1}). \tag{5.118}$$

Because of the scalar factors ϵ and μ, the dyadic $\overline{\overline{\mathsf{C}}}_s$ can be normalized without losing the generality. Let us define a spatial metric dyadic $\overline{\overline{\mathsf{G}}}_s \in \mathbb{E}_1\mathbb{E}_1$ with the normalization,

$$\overline{\overline{\mathsf{G}}}_s = \mathbf{e}_{123}\rfloor\overline{\overline{\mathsf{C}}}_s, \quad \boldsymbol{\varepsilon}_{123}\boldsymbol{\varepsilon}_{123}||\overline{\overline{\mathsf{G}}}_s^{(3)} = 1. \tag{5.119}$$

The spatial inverse is obtained through (2.211) as

$$\overline{\overline{\mathsf{G}}}_s^{-1} = \frac{\boldsymbol{\varepsilon}_{123}\boldsymbol{\varepsilon}_{123}\lfloor\lfloor\overline{\overline{\mathsf{G}}}_s^{(2)T}}{\boldsymbol{\varepsilon}_{123}\boldsymbol{\varepsilon}_{123}||\overline{\overline{\mathsf{G}}}_s^{(3)}} = \boldsymbol{\varepsilon}_{123}\boldsymbol{\varepsilon}_{123}\lfloor\lfloor\overline{\overline{\mathsf{G}}}_s^{(2)T}. \tag{5.120}$$

Thus, we obtain the representation,

$$\overline{\overline{\mathsf{M}}}_m = \frac{1}{\mu}(\mu\epsilon\overline{\overline{\mathsf{G}}}_s{}^\wedge_\wedge\mathbf{e}_4\mathbf{e}_4 - \overline{\overline{\mathsf{G}}}_s^{(2)T}). \tag{5.121}$$

To be a principal medium $\overline{\overline{\mathsf{M}}}_m$ must be symmetric, whence $\overline{\overline{\mathsf{G}}}_s$ must be a symmetric dyadic, $\overline{\overline{\mathsf{G}}}_s^T = \overline{\overline{\mathsf{G}}}_s$. In this case the modified medium bidyadic takes the simple form

$$\overline{\overline{\mathsf{M}}}_m = -\frac{1}{\mu}(\overline{\overline{\mathsf{G}}}_s^{(2)} - \mu\epsilon\overline{\overline{\mathsf{G}}}_s{}^\wedge_\wedge\mathbf{e}_4\mathbf{e}_4) = -\frac{1}{\mu}(\overline{\overline{\mathsf{G}}}_s - \mu\epsilon\mathbf{e}_4\mathbf{e}_4)^{(2)}. \tag{5.122}$$

It was pointed out in the previous section that a medium defined by a bidyadic of a form $\overline{\overline{\mathsf{M}}} = \boldsymbol{\varepsilon}_N\lfloor\overline{\overline{\mathsf{Q}}}^{(2)}$ for any symmetric $\overline{\overline{\mathsf{Q}}}$ belongs to the class of principal media. A medium defined by (5.122) with symmetric spatial dyadic $\overline{\overline{\mathsf{G}}}_s$ will be called by the name simple principal medium.

Comparison with the expansions (5.75), (5.87), (5.88) leads to the following sets of 3D medium dyadics describing the simple principal medium

$$\overline{\overline{\alpha}} = 0, \quad \overline{\overline{\epsilon}}' = \epsilon\overline{\overline{\mathsf{C}}}_s, \quad \overline{\overline{\mu}}^{-1} = (\mu\overline{\overline{\mathsf{C}}}_s)^{-1}, \quad \overline{\overline{\beta}} = 0, \tag{5.123}$$

$$\overline{\overline{\epsilon}} = \epsilon\overline{\overline{\mathsf{C}}}_s, \quad \overline{\overline{\mu}} = \mu\overline{\overline{\mathsf{C}}}_s, \quad \overline{\overline{\xi}} = \overline{\overline{\zeta}} = 0, \tag{5.124}$$

$$\overline{\overline{\epsilon}}_g = \epsilon\overline{\overline{\mathsf{G}}}_s, \quad \overline{\overline{\mu}}_g = \mu\overline{\overline{\mathsf{G}}}_s, \quad \overline{\overline{\xi}}_g = \overline{\overline{\zeta}}_g = 0. \tag{5.125}$$

Choosing a vector basis properly, we can express the symmetric metric dyadic by

$$\overline{\overline{\mathsf{G}}}_s = \mathbf{e}_1\mathbf{e}_1 + \mathbf{e}_2\mathbf{e}_2 + \mathbf{e}_3\mathbf{e}_3, \tag{5.126}$$

whence, identifying $\overline{\overline{\mathsf{G}}}_s$ with the Gibbsian unit dyadic $\overline{\overline{\mathsf{I}}}_g$, the simple principal medium actually corresponds to an isotropic medium in Gibbsian formalism. In terms of (5.126), the modified medium bidyadic of the simple principal medium has the expansion

$$\overline{\overline{\mathsf{M}}}_m = \sqrt{\frac{\epsilon}{\mu}}(\mathbf{e}_{14}\mathbf{e}_{14} + \mathbf{e}_{24}\mathbf{e}_{24} + \mathbf{e}_{34}\mathbf{e}_{34} - \mathbf{e}_{12}\mathbf{e}_{12} - \mathbf{e}_{23}\mathbf{e}_{23} - \mathbf{e}_{31}\mathbf{e}_{31}). \tag{5.127}$$

Because the dyadic $\overline{\overline{\mathsf{C}}}_s$ and its inverse satisfy the relations

$$\overline{\overline{\mathsf{C}}}_s|\overline{\overline{\mathsf{C}}}_s^{-1} = \overline{\overline{\mathsf{I}}}_s^{(2)T}, \quad \overline{\overline{\mathsf{C}}}_s^{-1}|\overline{\overline{\mathsf{C}}}_s = \overline{\overline{\mathsf{I}}}_s^T, \tag{5.128}$$

we obtain

$$\begin{aligned}\overline{\overline{\mathsf{M}}}^2 &= (\epsilon\overline{\overline{\mathsf{C}}}_s \wedge \mathbf{e}_4)|(\boldsymbol{\varepsilon}_4 \wedge (\mu\overline{\overline{\mathsf{C}}}_s)^{-1}) + (\boldsymbol{\varepsilon}_4 \wedge (\mu\overline{\overline{\mathsf{C}}}_s)^{-1})|(\epsilon\overline{\overline{\mathsf{C}}}_s \wedge \mathbf{e}_4) \\ &= -\frac{\epsilon}{\mu}\overline{\overline{\mathsf{I}}}_s^{(2)T} + \frac{\epsilon}{\mu}\boldsymbol{\varepsilon}_4 \wedge \overline{\overline{\mathsf{I}}}_s^T \wedge \mathbf{e}_4 = -\frac{\epsilon}{\mu}\overline{\overline{\mathsf{I}}}^{(2)T}.\end{aligned} \tag{5.129}$$

From this we conclude that the medium bidyadic is a multiple of an AC bidyadic or a unipotent bidyadic depending on the sign of ϵ/μ. For complex medium parameters these two cases cannot be distinguished.

Let us consider the equation (5.62) for the potential one-form $\boldsymbol{\phi}$ in the simple principal medium defined by (5.122). Applying the rule

$$\overline{\overline{\mathsf{A}}}^{(2)}\lfloor\lfloor\overline{\overline{\mathsf{B}}}^T = (\overline{\overline{\mathsf{A}}}||\overline{\overline{\mathsf{B}}})\overline{\overline{\mathsf{A}}} - \overline{\overline{\mathsf{A}}}|\overline{\overline{\mathsf{B}}}|\overline{\overline{\mathsf{A}}}, \tag{5.130}$$

valid for $\overline{\overline{\mathsf{A}}} \in \mathbb{E}_1\mathbb{E}_1$ and $\overline{\overline{\mathsf{B}}} \in \mathbb{F}_1\mathbb{F}_1$, the dyadic differential operator can be expanded as

$$\begin{aligned}\overline{\overline{\mathsf{D}}}(\mathbf{d}) &= -\frac{1}{\mu}(\overline{\overline{\mathsf{G}}}_s - \mu\epsilon\mathbf{e}_4\mathbf{e}_4)^{(2)}\lfloor\lfloor\mathbf{dd} \\ &= -\frac{1}{\mu}(\overline{\overline{\mathsf{G}}}_s||\mathbf{dd} - \mu\epsilon\partial_\tau^2)(\overline{\overline{\mathsf{G}}}_s - \mu\epsilon\mathbf{e}_4\mathbf{e}_4) \\ &\quad + \frac{1}{\mu}(\overline{\overline{\mathsf{G}}}_s|\mathbf{d} - \mu\epsilon\mathbf{e}_4\partial_\tau)(\overline{\overline{\mathsf{G}}}_s|\mathbf{d} - \mu\epsilon\mathbf{e}_4\partial_\tau).\end{aligned} \tag{5.131}$$

The potential is not unique because $\boldsymbol{\phi}$ can be replaced by $\boldsymbol{\phi} + \mathbf{d}\lambda$ for any scalar function λ without changing the field two-form $\boldsymbol{\Phi} = \mathbf{d} \wedge \boldsymbol{\phi}$. For uniqueness we can restrict the potential by requiring that it satisfy an additional scalar-valued condition. Choosing the condition

$$(\overline{\overline{\mathsf{G}}}_s|\mathbf{d} - \mu\epsilon\mathbf{e}_4\partial_\tau)|\boldsymbol{\phi} = 0, \tag{5.132}$$

(the Lorenz condition), which can be expressed for the spatial potential one-form $\mathbf{A}$ and scalar φ as

$$\mathbf{d}|\overline{\overline{\mathsf{G}}}_s|\mathbf{A} + \mu\epsilon\partial_\tau\varphi = 0, \tag{5.133}$$

(5.62) is reduced to an equation with a scalar operator,

$$(\overline{\overline{\mathsf{G}}}_s\lfloor\lfloor\mathbf{dd} - \mu\epsilon\partial_\tau^2)(\overline{\overline{\mathsf{G}}}_s - \mu\epsilon\mathbf{e}_4\mathbf{e}_4)|\boldsymbol{\phi} = -\mu\mathbf{e}_N\lfloor\boldsymbol{\gamma}. \tag{5.134}$$

This is equivalent with

$$(\overline{\overline{\mathsf{G}}}_s\lfloor\lfloor\mathbf{dd} - \mu\epsilon\partial_\tau^2)\boldsymbol{\phi} = -\left(\mu\overline{\overline{\mathsf{G}}}_s^{-1} - \frac{1}{\epsilon}\mathbf{e}_4\mathbf{e}_4\right)|(\mathbf{e}_N\lfloor\boldsymbol{\gamma}). \tag{5.135}$$

where we can write $\overline{\overline{\mathsf{G}}}_s\lfloor\lfloor\mathbf{dd} = \partial_{x_1}^2 + \partial_{x_2}^2 + \partial_{x_3}^2$. This is a second-order hyperbolic equation (wave equation) for $\mu\epsilon > 0$ and and elliptic equation for $\mu\epsilon < 0$.

5.4 INTERFACES AND BOUNDARIES

An interface is a surface which separates two media while a boundary is a mathematical concept, a surface where the region of interest ends. At an interface the fields on one side of the surface depend on the fields on the other side of the surface through continuity conditions. At a boundary the fields satisfy certain conditions and the question what are the fields behind the boundary is irrelevant. As is known from a medium called perfect electric conductor (PEC), a medium interface may act as a boundary. To show some examples of media with a similar property, let us consider fields at a planar interface of two media.

5.4.1 Interface Conditions

Let us assume that two media $\overline{\overline{\mathsf{M}}}_a$ and $\overline{\overline{\mathsf{M}}}_b$ are separated by a planar interface S defined by $x_3 = 0$ in terms of a Cartesian coordinate function x_3. Because of the discontinuity in the medium, the fields $\boldsymbol{\Phi}, \boldsymbol{\Psi}$ may have a discontinuity at the interface. Denoting the fields on each side of the interface by subscripts a and b, the total field two-forms can be expressed as

$$\boldsymbol{\Phi} = \boldsymbol{\Phi}_a + \boldsymbol{\Phi}_b, \quad \boldsymbol{\Psi} = \boldsymbol{\Psi}_a + \boldsymbol{\Psi}_b, \tag{5.136}$$

with

$$\begin{pmatrix}\boldsymbol{\Phi}_a(\mathbf{x})\\ \boldsymbol{\Psi}_a(\mathbf{x})\end{pmatrix} = \begin{pmatrix}\boldsymbol{\Phi}'_a(\mathbf{x})\\ \boldsymbol{\Psi}'_a(\mathbf{x})\end{pmatrix} P_a(\mathbf{r}), \tag{5.137}$$

$$\begin{pmatrix}\boldsymbol{\Phi}_b(\mathbf{x})\\ \boldsymbol{\Psi}_b(\mathbf{x})\end{pmatrix} = \begin{pmatrix}\boldsymbol{\Phi}'_b(\mathbf{x})\\ \boldsymbol{\Psi}'_b(\mathbf{x})\end{pmatrix} P_b(\mathbf{r}). \tag{5.138}$$

The primed fields are assumed to be analytical continuations of the fields without primes, extending over the surface S with no discontinuity in medium, while the characteristic functions P_a and P_b defined by

$$P_a(\mathbf{r}) = \theta(x_3), \qquad P_b(\mathbf{r}) = \theta(-x_3) = 1 - P_a(x_3), \tag{5.139}$$

extract the unphysical parts of the primed fields. Here, $\theta(x)$ denotes the Heaviside unit step function satisfying

$$\theta(x) = 1, \;\; x > 0, \qquad \theta(x) = 0, \;\; x < 0, \tag{5.140}$$

$$\partial_x \theta(x) = \delta(x). \tag{5.141}$$

Assuming no sources at the interface, the Maxwell equations around the surface S can be expanded as

$$\begin{aligned}
\mathbf{d} \wedge \begin{pmatrix} \mathbf{\Phi}(\mathbf{x}) \\ \mathbf{\Psi}(\mathbf{x}) \end{pmatrix} &= \mathbf{d} \wedge \begin{pmatrix} \mathbf{\Phi}_a(\mathbf{x}) \\ \mathbf{\Psi}_a(\mathbf{x}) \end{pmatrix} + \mathbf{d} \wedge \begin{pmatrix} \mathbf{\Phi}_b(\mathbf{x}) \\ \mathbf{\Psi}_b(\mathbf{x}) \end{pmatrix} \\
&= \mathbf{d}P_a(\mathbf{r}) \wedge \begin{pmatrix} \mathbf{\Phi}'_a(\mathbf{x}) \\ \mathbf{\Psi}'_a(\mathbf{x}) \end{pmatrix} + \mathbf{d}P_b(\mathbf{r}) \wedge \begin{pmatrix} \mathbf{\Phi}'_b(\mathbf{x}) \\ \mathbf{\Psi}'_b(\mathbf{x}) \end{pmatrix} \\
&= \mathbf{d}P_a(\mathbf{r}) \wedge \begin{pmatrix} \mathbf{\Phi}_a(\mathbf{x}) - \mathbf{\Phi}_b(\mathbf{x}) \\ \mathbf{\Psi}_a(\mathbf{x}) - \mathbf{\Psi}_b(\mathbf{x}) \end{pmatrix} = \begin{pmatrix} 0 \\ 0 \end{pmatrix}.
\end{aligned} \tag{5.142}$$

Inserting

$$\mathbf{d}P_a(\mathbf{r}) = \mathbf{d}\theta(x_3) = \mathbf{d}x_3 \partial_{x_3} \theta(x_3) = \varepsilon_3 \delta(x_3), \tag{5.143}$$

we obtain conditions for the fields at the interface $x_3 = 0$:

$$\varepsilon_3 \wedge \mathbf{\Phi}_a = \varepsilon_3 \wedge \mathbf{\Phi}_b, \qquad \varepsilon_3 \wedge \mathbf{\Psi}_a = \varepsilon_3 \wedge \mathbf{\Psi}_b. \tag{5.144}$$

Expanded in spatial field components each of the interface conditions can be split in two conditions,

$$\varepsilon_3 \wedge \mathbf{B}_a = \varepsilon_3 \wedge \mathbf{B}_b, \qquad \varepsilon_3 \wedge \mathbf{E}_a = \varepsilon_3 \wedge \mathbf{E}_b, \tag{5.145}$$

$$\varepsilon_3 \wedge \mathbf{D}_a = \varepsilon_3 \wedge \mathbf{D}_b, \qquad \varepsilon_3 \wedge \mathbf{H}_a = \varepsilon_3 \wedge \mathbf{H}_b, \tag{5.146}$$

which correspond to the well-known conditions for the Gibbsian fields,

$$\mathbf{e}_3 \cdot \mathbf{B}_{ga} = \mathbf{e}_3 \cdot \mathbf{B}_{gb}, \qquad \mathbf{e}_3 \times \mathbf{E}_{ga} = \mathbf{e}_3 \times \mathbf{E}_{gb}, \tag{5.147}$$

$$\mathbf{e}_3 \cdot \mathbf{D}_{ga} = \mathbf{e}_3 \cdot \mathbf{D}_{gb}, \qquad \mathbf{e}_3 \times \mathbf{H}_{ga} = \mathbf{e}_3 \times \mathbf{H}_{gb}, \tag{5.148}$$

where $\mathbf{e}_3$ denotes the unit vector normal to the interface. Being local conditions, (5.144) are actually valid for any smooth surface making the interface.

5.4.2 Boundary Conditions

Since the interface conditions (5.144) were derived from the Maxwell equations, they are valid for any two media defined by the medium bidyadics $\overline{\overline{\mathsf{M}}}_a$ and $\overline{\overline{\mathsf{M}}}_b$ or through other representations like (5.52) or (5.70) and (5.71) in each of the two media. For certain media b the fields in medium a can be shown to satisfy a set of conditions at the interface depending only on the medium bidyadic $\overline{\overline{\mathsf{M}}}_b$ or its inverse bidyadic $\overline{\overline{\mathsf{N}}}_b$ no matter what the fields are in the medium b. Let us consider some examples of boundary conditions defined by media b.

PMC Boundary

The perfect magnetic conductor (PMC) medium in the region b corresponds to the medium bidyadic

$$\overline{\overline{\mathsf{M}}}_b = 0. \tag{5.149}$$

Since this requires that the field in medium b satisfy

$$\mathbf{\Psi}_b(\mathbf{x}) = 0, \tag{5.150}$$

from (5.144) we obtain for the field $\mathbf{\Psi}_a$ at $x_3 = 0$ the boundary condition

$$\varepsilon_3 \wedge \mathbf{\Psi}_a = 0. \tag{5.151}$$

This corresponds to the conditions

$$\varepsilon_3 \wedge \mathbf{D}_a = 0, \quad \varepsilon_3 \wedge \mathbf{H}_a = 0, \tag{5.152}$$

and the Gibbsian conditions

$$\mathbf{e}_3 \cdot \mathbf{D}_{ga} = 0, \quad \mathbf{e}_3 \times \mathbf{H}_{ga} = 0. \tag{5.153}$$

The same PMC boundary conditions (5.151) can be obtained for the less restrictive medium condition

$$\varepsilon_3 \wedge \overline{\overline{\mathsf{M}}}_b = 0, \tag{5.154}$$

which allows the bidyadic $\overline{\overline{\mathsf{M}}}_b$ to have some nonzero components. In fact, (5.154) requires that the fields in medium b satisfy

$$\varepsilon_3 \wedge \mathbf{\Psi}_b = 0, \quad \Rightarrow \quad \varepsilon_3 \wedge \mathbf{D}_b = 0, \quad \varepsilon_3 \wedge \mathbf{H}_b = 0, \tag{5.155}$$

whence the PMC boundary conditions (5.151), (5.152) are also obtained for this more general medium. One should notice that if the one-form

ε_3 is not independent of $\mathbf{x}$ (nonplanar boundary), (5.154) requires that the medium b must be inhomogeneous in a certain way.

PMC boundary has obtained an important role in antenna synthesis because, unlike the PEC boundary, it does not cancel the radiation of a current element parallel to it but, rather, enhances it. Various methods to realize the PMC condition by metasurfaces under the name "High-impedance electromagnetic surface" have been suggested since 1999 [29].

PEC Boundary

The PEC medium is defined by

$$\overline{\overline{\mathsf{N}}}_b = 0 \tag{5.156}$$

in the representation (5.52). In this case the field condition is

$$\mathbf{\Phi}_b(\mathbf{x}) = 0. \tag{5.157}$$

and we obtain at $x_3 = 0$ the boundary condition

$$\varepsilon_3 \wedge \mathbf{\Phi}_a = 0. \tag{5.158}$$

For the spatial field components this can be expressed as

$$\varepsilon_3 \wedge \mathbf{B}_a = 0, \quad \varepsilon_3 \wedge \mathbf{E}_a = 0, \tag{5.159}$$

and, for the Gibbsian vector fields,

$$\mathbf{e}_3 \cdot \mathbf{B}_{ga} = 0, \quad \mathbf{e}_3 \times \mathbf{E}_{ga} = 0. \tag{5.160}$$

As in the PMC case, the medium condition can be actually relaxed to

$$\varepsilon_3 \wedge \overline{\overline{\mathsf{N}}}_b = 0 \tag{5.161}$$

for the same PEC boundary conditions.

PEMC Boundary

The previous two boundary conditions can be generalized by considering the axion medium defined by

$$\overline{\overline{\mathsf{M}}}_b = M\overline{\overline{\mathsf{I}}}^{(2)T}, \tag{5.162}$$

which corresponds to the field condition

$$\mathbf{\Psi}_b(\mathbf{x}) - M\mathbf{\Phi}_b(\mathbf{x}) = 0. \tag{5.163}$$

From (5.144) we then arrive at the boundary condition

$$\varepsilon_3 \wedge (\boldsymbol{\Psi}_a - M\boldsymbol{\Phi}_a) = 0, \qquad (5.164)$$

which can be split into

$$\varepsilon_3 \wedge (\mathbf{D}_a - M\mathbf{B}_a) = 0, \quad \varepsilon_3 \wedge (\mathbf{H}_a + M\mathbf{E}_a) = 0. \qquad (5.165)$$

Since this generalizes the PMC ($M = 0$) and PEC ($1/M = 0$) conditions, the medium has been dubbed as the perfect electromagnetic conductor (PEMC) [30]. The Gibbsian form for the PEMC boundary conditions is

$$\mathbf{e}_3 \cdot (\mathbf{D}_{ga} - M\mathbf{B}_{ga}) = 0, \quad \mathbf{e}_3 \times (\mathbf{H}_{ga} + M\mathbf{E}_{ga}) = 0. \qquad (5.166)$$

Following the pattern of both of the previous examples, the PEMC conditions can also be obtained by a medium satisfying the less restricting medium condition

$$\varepsilon_3 \wedge (\overline{\overline{\mathsf{M}}}_b - M\overline{\overline{\mathsf{I}}}^{(2)T}) = 0. \qquad (5.167)$$

The PEMC boundary has the useful property of changing the polarization of a plane wave in reflection, whence some effort has been given to realize such a boundary by a metasurface [31, 32].

DB Boundary

As another restriction to the medium b we may require a set of two scalar conditions

$$\varepsilon_3 \wedge \varepsilon_4 \wedge \overline{\overline{\mathsf{M}}}_b = 0, \quad \varepsilon_3 \wedge \varepsilon_4 \wedge \overline{\overline{\mathsf{N}}}_b = 0, \qquad (5.168)$$

which imply the field conditions

$$\varepsilon_3 \wedge \varepsilon_4 \wedge \boldsymbol{\Psi}_b = \varepsilon_3 \wedge \varepsilon_4 \wedge \mathbf{D}_b = 0, \quad \Rightarrow \quad \varepsilon_3 \wedge \mathbf{D}_b = 0 \qquad (5.169)$$

$$\varepsilon_3 \wedge \varepsilon_4 \wedge \boldsymbol{\Phi}_b = \varepsilon_3 \wedge \varepsilon_4 \wedge \mathbf{B}_b = 0, \quad \Rightarrow \quad \varepsilon_3 \wedge \mathbf{B}_b = 0. \qquad (5.170)$$

These give rise to the boundary conditions

$$\varepsilon_3 \wedge \mathbf{D}_a = 0, \quad \varepsilon_3 \wedge \mathbf{B}_a = 0. \qquad (5.171)$$

Written in Gibbsian form,

$$\mathbf{e}_3 \cdot \mathbf{D}_{ga} = 0, \quad \mathbf{e}_3 \cdot \mathbf{B}_{ga} = 0. \qquad (5.172)$$

they are known as DB boundary conditions [33, 34]. Inserting the expansions (5.75) and (5.77) in (5.168) we obtain

$$\varepsilon_3 \wedge \varepsilon_4 \wedge \overline{\overline{\mathsf{M}}}_b = \varepsilon_3 \wedge \varepsilon_4 \wedge \overline{\overline{\epsilon}}_b \wedge \mathbf{e}_4 + \varepsilon_3 \wedge \varepsilon_4 \wedge \overline{\overline{\xi}}_b | \overline{\overline{\mu}}_b^{-1} | \times (\overline{\overline{\mathsf{I}}}_s^{(2)T} - \overline{\overline{\zeta}}_b \wedge \mathbf{e}_4) = 0, \tag{5.173}$$

$$\varepsilon_3 \wedge \varepsilon_4 \wedge \overline{\overline{\mathsf{N}}}_b = -\varepsilon_3 \wedge \varepsilon_4 \wedge \overline{\overline{\mu}}_b \wedge \mathbf{e}_4 + \varepsilon_3 \wedge \varepsilon_4 \wedge \overline{\overline{\zeta}}_b | \overline{\overline{\epsilon}}_b^{-1} | \times (\overline{\overline{\mathsf{I}}}_s^{(2)T} + \overline{\overline{\xi}}_b \wedge \mathbf{e}_4) = 0, \tag{5.174}$$

which can be split in four conditions for the spatial medium dyadics as

$$\varepsilon_3 \wedge \overline{\overline{\xi}}_b | \overline{\overline{\mu}}_b^{-1} = 0, \qquad \varepsilon_3 \wedge (\overline{\overline{\epsilon}}_b - \overline{\overline{\xi}}_b | \overline{\overline{\mu}}_b^{-1} | \overline{\overline{\zeta}}_b) = 0, \tag{5.175}$$

$$\varepsilon_3 \wedge \overline{\overline{\zeta}}_b | \overline{\overline{\epsilon}}_b^{-1} = 0, \qquad \varepsilon_3 \wedge (\overline{\overline{\mu}}_b - \overline{\overline{\zeta}}_b | \overline{\overline{\epsilon}}_b^{-1} | \overline{\overline{\xi}}_b) = 0. \tag{5.176}$$

Considering a medium with no magnetoelectric dyadics, $\overline{\overline{\xi}}_b = \overline{\overline{\zeta}}_b = 0$, these reduce to

$$\varepsilon_3 \wedge \overline{\overline{\epsilon}}_b = 0, \quad \varepsilon_3 \wedge \overline{\overline{\mu}}_b = 0, \tag{5.177}$$

which corresponds to a medium restricted by Gibbsian dyadics satisfying

$$\mathbf{e}_3 \cdot \overline{\overline{\epsilon}}_{gb} = 0, \quad \overline{\overline{\xi}}_{gb} = 0, \quad \overline{\overline{\zeta}}_{gb} = 0, \quad \mathbf{e}_3 \cdot \overline{\overline{\mu}}_{gb} = 0. \tag{5.178}$$

The DB boundary conditions were introduced already in 1959 [35]. Novel interest followed when they were discovered to have application in electromagnetic invisibility cloaking problems in the 2000s [36, 37]. Realization of the DB boundary by a metasurface was subsequently suggested in [38].

SH Boundary

A case similar to the previous one is obtained by starting from medium b restricted by conditions of the form

$$\varepsilon_2 \wedge \varepsilon_3 \wedge \overline{\overline{\mathsf{M}}}_b = 0, \quad \varepsilon_2 \wedge \varepsilon_3 \wedge \overline{\overline{\mathsf{N}}}_b = 0. \tag{5.179}$$

These imply conditions for the fields

$$\varepsilon_2 \wedge \varepsilon_3 \wedge \mathbf{\Psi}_b = \varepsilon_2 \wedge \varepsilon_3 \wedge \varepsilon_4 \wedge \mathbf{H}_b = 0, \quad \Rightarrow \quad \varepsilon_2 \wedge \varepsilon_3 \wedge \mathbf{H}_b = 0 \tag{5.180}$$

$$\varepsilon_2 \wedge \varepsilon_3 \wedge \mathbf{\Phi}_b = -\varepsilon_2 \wedge \varepsilon_3 \wedge \varepsilon_4 \wedge \mathbf{E}_b = 0, \quad \Rightarrow \quad \varepsilon_2 \wedge \varepsilon_3 \wedge \mathbf{E}_b = 0, \tag{5.181}$$

and, consequently, yield the boundary conditions

$$\varepsilon_2 \wedge \varepsilon_3 \wedge \mathbf{E}_a = 0, \quad \varepsilon_2 \wedge \varepsilon_3 \wedge \mathbf{H}_a = 0. \tag{5.182}$$

In Gibbsian form the conditions are

$$\mathbf{e}_2 \times \mathbf{e}_3 \cdot \mathbf{E}_{ga} = \mathbf{e}_1 \cdot \mathbf{E}_{ga} = 0, \quad \mathbf{e}_2 \times \mathbf{e}_3 \cdot \mathbf{H}_{ga} = \mathbf{e}_1 \cdot \mathbf{H}_{ga} = 0, \tag{5.183}$$

where the unit vector $\mathbf{e}_1$ is parallel to the boundary surface. These conditions have been labeled as the soft-and-hard boundary conditions or SH conditions in the past [39]. SH boundary conditions have been realized by corrugated conducting surfaces since the 1940s, for example, to obtain rotationally symmetric radiation patterns in horn antennas.

In all of the previous examples the boundary conditions for fields in medium a were obtained directly from the medium conditions of medium b without having to consider further restrictions to the fields imposed by the Maxwell equations. Examples of other cases will be considered in subsequent chapters.

5.5 POWER AND ENERGY

5.5.1 Bilinear Invariants

Let us consider quantities which are bilinear in the field two-forms $\mathbf{\Phi}$ and $\mathbf{\Psi}$ or quadratic in either of them. The four-forms

$$\mathbf{\Phi} \wedge \mathbf{\Phi}, \quad \mathbf{\Psi} \wedge \mathbf{\Psi}, \quad \mathbf{\Phi} \wedge \mathbf{\Psi} \tag{5.184}$$

are examples of such quantities. Expanded in terms of 3D fields they become

$$\begin{aligned}\mathbf{\Phi} \wedge \mathbf{\Phi} &= (\mathbf{B} + \mathbf{E} \wedge \varepsilon_4) \wedge (\mathbf{B} + \mathbf{E} \wedge \varepsilon_4) \\ &= 2\mathbf{B} \wedge \mathbf{E} \wedge \varepsilon_4,\end{aligned} \tag{5.185}$$

$$\begin{aligned}\mathbf{\Psi} \wedge \mathbf{\Psi} &= (\mathbf{D} - \mathbf{H} \wedge \varepsilon_4) \wedge (\mathbf{D} - \mathbf{H} \wedge \varepsilon_4) \\ &= -2\mathbf{D} \wedge \mathbf{H} \wedge \varepsilon_4,\end{aligned} \tag{5.186}$$

$$\begin{aligned}\mathbf{\Phi} \wedge \mathbf{\Psi} &= (\mathbf{B} + \mathbf{E} \wedge \varepsilon_4) \wedge (\mathbf{D} - \mathbf{H} \wedge \varepsilon_4) \\ &= (\mathbf{D} \wedge \mathbf{E} - \mathbf{B} \wedge \mathbf{H}) \wedge \varepsilon_4.\end{aligned} \tag{5.187}$$

These quantities are invariants, that is, independent of any basis systems.

The dyadic product of field two-forms $\mathbf{\Phi\Psi}$ is another example of a bilinear quantity. Its antisymmetric part denoted by

$$\overline{\overline{\mathsf{U}}} = \frac{1}{2}(\mathbf{\Phi\Psi} - \mathbf{\Psi\Phi}) = -\overline{\overline{\mathsf{U}}}^T \in \mathbb{F}_2\mathbb{F}_2 \tag{5.188}$$

is of special interest. Properties of such a simple bidyadic (3.87) were already studied in Chapter 3. It can be expanded as

$$\overline{\overline{\mathsf{U}}} = \frac{1}{2}(\mathbf{BD} - \mathbf{DB} - \boldsymbol{\varepsilon}_4 \wedge (\mathbf{ED} + \mathbf{HB}) - (\mathbf{DE} + \mathbf{BH}) \wedge \boldsymbol{\varepsilon}_4 + \boldsymbol{\varepsilon}_4 \wedge (\mathbf{EH} - \mathbf{HE}) \wedge \boldsymbol{\varepsilon}_4). \tag{5.189}$$

Being antisymmetric, the bidyadic $\overline{\overline{\mathsf{U}}}$ can be represented in terms of a certain trace-free dyadic $\overline{\overline{\mathsf{B}}}_o \in \mathbb{E}_1\mathbb{F}_1$ as

$$\overline{\overline{\mathsf{U}}} = \boldsymbol{\varepsilon}_N\lfloor(\overline{\overline{\mathsf{B}}}_o{}^\wedge_\wedge\overline{\overline{\mathsf{I}}}). \tag{5.190}$$

$\overline{\overline{\mathsf{B}}}_o$ serves as another electromagnetic bilinear invariant. Applying (5.97) to the the skewon bidyadic $\mathbf{e}_N\lfloor\overline{\overline{\mathsf{U}}}$, it can be extracted as

$$\overline{\overline{\mathsf{B}}}_o = \frac{1}{2}(\mathbf{e}_N\lfloor\overline{\overline{\mathsf{U}}})\lfloor\lfloor\overline{\overline{\mathsf{I}}}^T. \tag{5.191}$$

$$= \frac{1}{4}\sum_i((\mathbf{e}_N\lfloor\mathbf{\Psi})\lfloor\boldsymbol{\varepsilon}_i(\mathbf{\Phi}\lfloor\mathbf{e}_i) - (\mathbf{e}_N\lfloor\mathbf{\Phi})\lfloor\boldsymbol{\varepsilon}_i(\mathbf{\Psi}\lfloor\mathbf{e}_i))$$

$$= -\frac{1}{4}\mathbf{e}_N\lfloor(\mathbf{\Psi}\wedge\overline{\overline{\mathsf{I}}}^T\rfloor\mathbf{\Phi} - \mathbf{\Phi}\wedge\overline{\overline{\mathsf{I}}}^T\rfloor\mathbf{\Psi})$$

$$= -\frac{1}{2}\mathbf{e}_N\lfloor\overline{\overline{\mathsf{T}}}. \tag{5.192}$$

The dyadic $\overline{\overline{\mathsf{T}}} = 2\boldsymbol{\varepsilon}_N\lfloor\overline{\overline{\mathsf{B}}}_o \in \mathbb{F}_3\mathbb{F}_1$ is yet another bilinear invariant, known as the stress–energy dyadic [1, 40]. Applying (3.106), we can write

$$\overline{\overline{\mathsf{B}}}{}_o^2 = \frac{1}{4}(\mathbf{e}_N\lfloor\overline{\overline{\mathsf{T}}})^2 = -\frac{1}{4}(\mathbf{e}_N\mathbf{e}_N\lfloor\lfloor\overline{\overline{\mathsf{T}}})|\overline{\overline{\mathsf{T}}}$$

$$= \frac{1}{16}\mathrm{tr}((\mathbf{e}_N\lfloor\overline{\overline{\mathsf{T}}})^2)\,\overline{\overline{\mathsf{I}}}, \tag{5.193}$$

which shows us that $\overline{\overline{\mathsf{B}}}_o$ and, hence $\mathbf{e}_N\lfloor\overline{\overline{\mathsf{T}}}$, is a multiple of an unipotent dyadic. Thus, the inverse of the stress–energy dyadic $\overline{\overline{\mathsf{T}}}$ can be expressed as

$$\overline{\overline{\mathsf{T}}}^{-1} = -4\frac{\mathbf{e}_N\mathbf{e}_N\lfloor\lfloor\overline{\overline{\mathsf{T}}}}{\mathrm{tr}((\mathbf{e}_N\lfloor\overline{\overline{\mathsf{T}}})^2)}, \tag{5.194}$$

when $\mathrm{tr}((\mathbf{e}_N\lfloor\overline{\overline{\mathsf{T}}})^2)$ does not vanish.

5.5.2 The Stress–Energy Dyadic

The stress–energy density dyadic $\overline{\overline{\mathsf{T}}}$ was obtained above as arising from bilinear invariants. It can be given a physical meaning by starting from the Gibbsian expression for energy density, represented by the scalar quantity

$$w_g = \frac{1}{2}(\mathbf{D}_g \cdot \mathbf{E}_g + \mathbf{B}_g \cdot \mathbf{H}_g). \tag{5.195}$$

The corresponding three-form energy-density is obtained through the rules of Appendix B as

$$\mathbf{w} = \frac{1}{2}(\mathbf{D} \wedge \mathbf{E} + \mathbf{B} \wedge \mathbf{H}). \tag{5.196}$$

To express this in terms of the 4D electromagnetic two-forms we start by substituting

$$\mathbf{B} = \boldsymbol{\Phi} - \mathbf{E} \wedge \boldsymbol{\varepsilon}_4 \tag{5.197}$$
$$\mathbf{D} = \boldsymbol{\Psi} + \mathbf{H} \wedge \boldsymbol{\varepsilon}_4 \tag{5.198}$$
$$\mathbf{E} = -(\mathbf{E} \wedge \boldsymbol{\varepsilon}_4)\rfloor\mathbf{e}_4 = -\boldsymbol{\Phi}\rfloor\mathbf{e}_4, \tag{5.199}$$
$$\mathbf{H} = -(\mathbf{H} \wedge \boldsymbol{\varepsilon}_4)\rfloor\mathbf{e}_4 = \boldsymbol{\Psi}\rfloor\mathbf{e}_4, \tag{5.200}$$

whence we arrive at the expression

$$\mathbf{w} = \frac{1}{2}(-\boldsymbol{\Psi} \wedge (\boldsymbol{\Phi}\rfloor\mathbf{e}_4) + \boldsymbol{\Phi} \wedge (\boldsymbol{\Psi}\rfloor\mathbf{e}_4) + 2\mathbf{E} \wedge \mathbf{H} \wedge \boldsymbol{\varepsilon}_4). \tag{5.201}$$

This suggests that we should rather concentrate on the 4D three-form

$$\begin{aligned}\mathbf{W} &= \frac{1}{2}(-\boldsymbol{\Psi} \wedge (\boldsymbol{\Phi}\rfloor\mathbf{e}_4) + \boldsymbol{\Phi} \wedge (\boldsymbol{\Psi}\rfloor\mathbf{e}_4)) \\ &= \mathbf{w} - \mathbf{E} \wedge \mathbf{H} \wedge \boldsymbol{\varepsilon}_4.\end{aligned} \tag{5.202}$$

The term $\mathbf{E} \wedge \mathbf{H}$ is the Poynting two-form which corresponds to the Poynting vector $\mathbf{E}_g \times \mathbf{H}_g$ in the Gibbsian sense. Thus, $\mathbf{W}$ can be called as the energy–power density three-form. Its expression still depends on the basis through the vector $\mathbf{e}_4$. To find an expression independent of any basis, let us first multiply $\mathbf{W}$ dyadically by $\boldsymbol{\varepsilon}_4$ from the right as

$$\mathbf{W}\boldsymbol{\varepsilon}_4 = \frac{1}{2}(\boldsymbol{\Phi} \wedge (\boldsymbol{\Psi}\rfloor\mathbf{e}_4\boldsymbol{\varepsilon}_4) - \boldsymbol{\Psi} \wedge (\boldsymbol{\Phi}\rfloor\mathbf{e}_4\boldsymbol{\varepsilon}_4)), \tag{5.203}$$

and generalize it by replacing $\mathbf{e}_4\boldsymbol{\varepsilon}_4$ by $\overline{\overline{\mathsf{I}}} = \sum \mathbf{e}_i\boldsymbol{\varepsilon}_i$. In this way we arrive at what is called the Maxwell stress–energy dyadic $\overline{\overline{\mathsf{T}}} \in \mathbb{F}_3\mathbb{F}_1$,

$$\overline{\overline{\mathsf{T}}} = \frac{1}{2}(\boldsymbol{\Phi} \wedge (\boldsymbol{\Psi}\lfloor\overline{\overline{\mathsf{I}}}) - \boldsymbol{\Psi} \wedge (\boldsymbol{\Phi}\lfloor\overline{\overline{\mathsf{I}}})). \tag{5.204}$$

Applying the rule $\boldsymbol{\Phi}\lfloor\overline{\overline{\mathsf{I}}} = -\overline{\overline{\mathsf{I}}}^T\rfloor\boldsymbol{\Phi}$ we obtain the form

$$\overline{\overline{\mathsf{T}}} = \frac{1}{2}(\boldsymbol{\Psi} \wedge (\overline{\overline{\mathsf{I}}}^T\rfloor\boldsymbol{\Phi}) - \boldsymbol{\Phi} \wedge (\overline{\overline{\mathsf{I}}}^T\rfloor\boldsymbol{\Psi})). \tag{5.205}$$

This derivation gives a meaning to the dyadic which was obtained through algebraic manipulation of quadratic invariants, in (5.192). The converse relation between the Maxwell stress–energy dyadic $\overline{\overline{\mathsf{T}}}$ and the dyadic invariant $\overline{\overline{\mathsf{U}}}$ can be found as

$$\begin{aligned}\overline{\overline{\mathsf{T}}} &= \frac{1}{2}\sum_i(\boldsymbol{\Psi} \wedge \boldsymbol{\varepsilon}_i\mathbf{e}_i\rfloor\boldsymbol{\Phi} - \boldsymbol{\Phi} \wedge \boldsymbol{\varepsilon}_i\mathbf{e}_i\rfloor\boldsymbol{\Psi}) \\ &= -\frac{1}{2}\sum_i \boldsymbol{\varepsilon}_i \wedge (\boldsymbol{\Psi}\boldsymbol{\Phi} - \boldsymbol{\Phi}\boldsymbol{\Psi})\lfloor\mathbf{e}_i = \sum_i \boldsymbol{\varepsilon}_i \wedge \overline{\overline{\mathsf{U}}}\lfloor\mathbf{e}_i.\end{aligned} \tag{5.206}$$

The stress–energy dyadic can be written in yet another form as

$$\begin{aligned}\overline{\overline{\mathsf{T}}} &= \boldsymbol{\Psi} \wedge (\overline{\overline{\mathsf{I}}}^T\rfloor\boldsymbol{\Phi}) - T\boldsymbol{\varepsilon}_N\lfloor\overline{\overline{\mathsf{I}}} \\ &= -\boldsymbol{\Phi} \wedge (\overline{\overline{\mathsf{I}}}^T\rfloor\boldsymbol{\Psi})) + T\boldsymbol{\varepsilon}_N\lfloor\overline{\overline{\mathsf{I}}},\end{aligned} \tag{5.207}$$

with the scalar

$$T = \frac{1}{2}(\boldsymbol{\Psi} \wedge \boldsymbol{\Phi})|\mathbf{e}_N. \tag{5.208}$$

Let us study the components of the stress–energy dyadic (5.205). The spatial energy density three-form can be obtained as its spatial-temporal part, which yields the scalar

$$\mathbf{e}_{123}|\overline{\overline{\mathsf{T}}}|\mathbf{e}_4 = \mathbf{e}_{123}|\mathbf{w}. \tag{5.209}$$

On the other hand, the temporal–temporal part yields the Poynting two-form as

$$\mathbf{e}_4\rfloor\overline{\overline{\mathsf{T}}}|\mathbf{e}_4 = \mathbf{E} \wedge \mathbf{H}. \tag{5.210}$$

The remaining components are the spatial–spatial part which can be expressed as the momentum two-form

$$\mathbf{e}_{123}|\overline{\overline{\mathsf{T}}}\rfloor\mathbf{e}_{123} = \mathbf{e}_{123}|(\mathbf{D} \wedge \overline{\overline{\mathsf{I}}}^T\rfloor\mathbf{B})\rfloor\mathbf{e}_{123}, \tag{5.211}$$

and the temporal–spatial part which corresponds to the stress dyadic

$$\mathbf{e}_4\rfloor\overline{\overline{\mathsf{T}}}\rfloor\mathbf{e}_{123} = \frac{1}{2}(\mathbf{DE} + \mathbf{BH} + \mathbf{H}\wedge\overline{\overline{\mathsf{I}}}^T\rfloor\mathbf{B} + \mathbf{E}\wedge\overline{\overline{\mathsf{I}}}^T\rfloor\mathbf{D})\rfloor\mathbf{e}_{123}. \tag{5.212}$$

5.5.3 Differentiation Rule

Differentiating the stress–energy dyadic leads to the expression

$$\begin{aligned} 2\mathbf{d}\wedge\overline{\overline{\mathsf{T}}} &= \mathbf{d}\wedge\mathbf{\Psi}\wedge\overline{\overline{\mathsf{I}}}^T\rfloor\mathbf{\Phi}_c + \mathbf{d}\wedge\mathbf{\Psi}_c\wedge\overline{\overline{\mathsf{I}}}^T\rfloor\mathbf{\Phi} \\ &\quad - \mathbf{d}\wedge\mathbf{\Phi}\wedge\overline{\overline{\mathsf{I}}}^T\rfloor\mathbf{\Psi}_c - \mathbf{d}\wedge\mathbf{\Phi}_c\wedge\overline{\overline{\mathsf{I}}}^T\rfloor\mathbf{\Psi} \\ &= 2\mathbf{d}\wedge\mathbf{\Psi}\wedge\overline{\overline{\mathsf{I}}}^T\rfloor\mathbf{\Phi}_c - 2\mathbf{d}\wedge\mathbf{\Phi}\wedge\overline{\overline{\mathsf{I}}}^T\rfloor\mathbf{\Psi}_c \\ &\quad + \mathbf{d}\wedge((\mathbf{\Psi}_c\wedge\mathbf{\Phi} - \mathbf{\Phi}_c\wedge\mathbf{\Psi})\lfloor\overline{\overline{\mathsf{I}}}), \end{aligned} \tag{5.213}$$

where the subscript c marks a quantity which is kept constant in differentiation. Here we have applied the identity

$$(\mathbf{\Psi}\wedge\mathbf{\Phi})\lfloor\overline{\overline{\mathsf{I}}} = (\mathbf{\Phi}\wedge\mathbf{\Psi})\lfloor\overline{\overline{\mathsf{I}}} = \mathbf{\Psi}\wedge\overline{\overline{\mathsf{I}}}^T\rfloor\mathbf{\Phi} + \mathbf{\Phi}\wedge\overline{\overline{\mathsf{I}}}^T\rfloor\mathbf{\Psi}. \tag{5.214}$$

Invoking the Maxwell equations we obtain the relation

$$\mathbf{d}\wedge\overline{\overline{\mathsf{T}}} = \boldsymbol{\gamma}_e\wedge\overline{\overline{\mathsf{I}}}^T\rfloor\mathbf{\Phi} - \boldsymbol{\gamma}_m\wedge\overline{\overline{\mathsf{I}}}^T\rfloor\mathbf{\Psi} + \frac{1}{2}\mathbf{d}\wedge((\mathbf{\Psi}_c\wedge\mathbf{\Phi} - \mathbf{\Phi}_c\wedge\mathbf{\Psi})\lfloor\overline{\overline{\mathsf{I}}}), \tag{5.215}$$

where $\boldsymbol{\gamma}_e$ and $\boldsymbol{\gamma}_m$ denote the respective electric and magnetic source three-forms.

Let us consider the case when the last term of (5.215) vanishes for all possible fields, which is equivalent to

$$\mathbf{d}(\mathbf{\Phi}\cdot\mathbf{\Psi}_c - \mathbf{\Psi}\cdot\mathbf{\Phi}_c) = \mathbf{d}\mathbf{\Phi}|(\overline{\overline{\mathsf{M}}}_m - \overline{\overline{\mathsf{M}}}_m^T)|\mathbf{\Phi}_c = 0. \tag{5.216}$$

This requires that the quantity behind $\mathbf{d}$ must be constant for any $\mathbf{\Phi}(\mathbf{x})$, which is possible only when the modified medium bidyadic $\overline{\overline{\mathsf{M}}}_m$ is symmetric,

$$\overline{\overline{\mathsf{M}}}_m^T = \overline{\overline{\mathsf{M}}}_m, \tag{5.217}$$

that is, there is no skewon component in the medium. Thus, the differentiation rule

$$\mathbf{d}\wedge\overline{\overline{\mathsf{T}}} = \boldsymbol{\gamma}_e\wedge(\overline{\overline{\mathsf{I}}}^T\rfloor\mathbf{\Phi}) - \boldsymbol{\gamma}_m\wedge(\overline{\overline{\mathsf{I}}}^T\rfloor\mathbf{\Psi}), \tag{5.218}$$

is valid for the principal–axion medium.

To interpret the right side of (5.218), we can write

$$\gamma_e \wedge (\bar{\bar{\mathsf{I}}}^T \rfloor \mathbf{\Phi}) = -\varepsilon_4 \wedge (\varrho_e \mathbf{E} + \mathbf{J}_e \wedge \bar{\bar{\mathsf{I}}}_s^T \rfloor \mathbf{B} - (\mathbf{J}_e \wedge \mathbf{E})\varepsilon_4). \tag{5.219}$$

The Gibbsian counterparts of the consecutive terms in the brackets are the electric force on electric charge, $\varrho_{eg}\mathbf{E}_g$, the magnetic force on the electric current, $\mathbf{J}_{eg} \times \mathbf{B}_g$ and the Joule loss $\mathbf{J}_{eg} \cdot \mathbf{E}_g$. In (5.218) they appear as dyadic density quantities of the space $\mathbb{F}_4\mathbb{F}_1$ which can be integrated over a 4D volume quadrivector to yield a one-form force quantity. The last term in (5.218) can be interpreted likewise as forces on magnetic sources.

5.6 PLANE WAVES

Assuming a medium which is space and time invariant, as defined by a constant bidyadic $\overline{\overline{\mathsf{M}}}$ or modified bidyadic $\overline{\overline{\mathsf{M}}}_m$, the simplest solutions to the Maxwell equations appear to be plane waves whose sources may be ignored since they can be assumed to be outside the region under interest.

5.6.1 Basic Equations

In time-harmonic Gibbsian analysis of complex fields the plane wave is described by an exponential function

$$\mathbf{E}_g(\mathbf{r}) = \mathbf{E}_g e^{-j\mathbf{k}\cdot\mathbf{r}}, \tag{5.220}$$

where $\mathbf{k}$ is the (Gibbsian) wave vector. The corresponding 4D representation is

$$\mathbf{\Phi}(\mathbf{x}) = \mathbf{\Phi} \exp(\boldsymbol{\nu}|\mathbf{x}), \tag{5.221}$$

where $\mathbf{\Phi}$ is an amplitude two-form. The wave one-form $\boldsymbol{\nu}$ can be expanded as

$$\boldsymbol{\nu} = \boldsymbol{\beta} - k\varepsilon_4, \tag{5.222}$$

where the spatial wave one-form $\boldsymbol{\beta}$ corresponds to the Gibbsian $\mathbf{k}$ vector and the scalar k corresponds to the variable ω/c in the time-harmonic case. Equation (5.221) represents a plane wave because the field is constant on spatial planes $\mathbf{r}(\tau)$ satisfying

$$\boldsymbol{\beta}|\mathbf{r} = k\tau. \tag{5.223}$$

Applying the medium equation (5.51), the field two-form $\mathbf{\Psi}(\mathbf{x})$ has the same exponential dependence on $\mathbf{x}$,

$$\mathbf{\Psi}(\mathbf{x}) = \overline{\overline{\mathsf{M}}}|\mathbf{\Phi}(\mathbf{x}) = \mathbf{\Psi} \exp(\boldsymbol{\nu}|\mathbf{x}). \tag{5.224}$$

The Maxwell equations become algebraic equations for the field two-forms

$$\boldsymbol{\nu} \wedge \boldsymbol{\Phi} = 0, \tag{5.225}$$

$$\boldsymbol{\nu} \wedge \boldsymbol{\Psi} = 0. \tag{5.226}$$

From the form of these equations it follows that there must exist potential one-forms $\boldsymbol{\phi}$ and $\boldsymbol{\psi}$ in terms of which the field two-forms can be expressed as

$$\boldsymbol{\Phi} = \boldsymbol{\nu} \wedge \boldsymbol{\phi}, \tag{5.227}$$

$$\boldsymbol{\Psi} = \boldsymbol{\nu} \wedge \boldsymbol{\psi}. \tag{5.228}$$

In fact, applying the rule

$$\mathbf{h}\rfloor(\nu \wedge \boldsymbol{\Phi}) = \nu \wedge (\mathbf{h}\rfloor\boldsymbol{\Phi}) + (\mathbf{h}|\nu)\boldsymbol{\Phi} = 0 \tag{5.229}$$

we can write

$$\boldsymbol{\phi} = -\frac{\mathbf{h}\rfloor\boldsymbol{\Phi}}{\mathbf{h}|\boldsymbol{\nu}}, \tag{5.230}$$

for any vector $\mathbf{h}$ satisfying $\mathbf{h}|\boldsymbol{\nu} \neq 0$. From (5.230) it can be seen that the potential $\boldsymbol{\phi}$ is not unique, since it depends on the choice of the vector $\mathbf{h}$. Once $\mathbf{h}$ is chosen, the potential satisfies the additional condition

$$\mathbf{h}|\boldsymbol{\phi} = 0. \tag{5.231}$$

Because of (5.227) and (5.228), the field two-forms $\boldsymbol{\Phi}$ and $\boldsymbol{\Psi}$ of a plane wave must be simple and they satisfy three orthogonality conditions

$$\boldsymbol{\Phi} \wedge \boldsymbol{\Phi} = 0, \quad \boldsymbol{\Phi} \wedge \boldsymbol{\Psi} = 0, \quad \boldsymbol{\Psi} \wedge \boldsymbol{\Psi} = 0, \tag{5.232}$$

or, equivalently,

$$\boldsymbol{\Phi} \cdot \boldsymbol{\Phi} = 0, \quad \boldsymbol{\Phi} \cdot \boldsymbol{\Psi} = 0, \quad \boldsymbol{\Psi} \cdot \boldsymbol{\Psi} = 0. \tag{5.233}$$

From (5.226) we obtain a three-form equation for the potential one-form $\boldsymbol{\phi}$,

$$\boldsymbol{\nu} \wedge \boldsymbol{\Psi} = \boldsymbol{\nu} \wedge \overline{\overline{\mathsf{M}}}|\boldsymbol{\Phi} = \boldsymbol{\nu} \wedge \overline{\overline{\mathsf{M}}}|(\boldsymbol{\nu} \wedge \boldsymbol{\phi}) = 0. \tag{5.234}$$

Contracted by $\mathbf{e}_N\lfloor$ this becomes a vector equation,

$$\mathbf{e}_N\lfloor(\boldsymbol{\nu} \wedge \overline{\overline{\mathsf{M}}}\lfloor\boldsymbol{\nu})|\boldsymbol{\phi} = -\boldsymbol{\nu}\rfloor(\overline{\overline{\mathsf{M}}}_m\lfloor\boldsymbol{\nu})|\boldsymbol{\phi} = 0, \tag{5.235}$$

which can be written as

$$\overline{\overline{\mathsf{D}}}(\boldsymbol{\nu})|\boldsymbol{\phi} = 0. \tag{5.236}$$

The dyadic $\overline{\overline{\mathsf{D}}}(\boldsymbol{\nu}) \in \mathbb{E}_1\mathbb{E}_1$ defined by

$$\overline{\overline{\mathsf{D}}}(\boldsymbol{\nu}) = -\boldsymbol{\nu}\rfloor\overline{\overline{\mathsf{M}}}_m\lfloor\boldsymbol{\nu} = \overline{\overline{\mathsf{M}}}_m\lfloor\lfloor\boldsymbol{\nu}\boldsymbol{\nu} \tag{5.237}$$

is called the dispersion dyadic. One may note that (5.237) coincides with the dyadic differential operator (5.63) when $\mathbf{d}$ is replaced by $\boldsymbol{\nu}$. Because of

$$\begin{aligned}(\mathbf{e}_N\lfloor\overline{\overline{\mathsf{I}}}^{(2)T})\lfloor\lfloor\boldsymbol{\nu}\boldsymbol{\nu} &= -\boldsymbol{\nu}\rfloor(\mathbf{e}_N\lfloor\overline{\overline{\mathsf{I}}}^{(2)})\lfloor\boldsymbol{\nu} = -\boldsymbol{\nu}\rfloor(\mathbf{e}_N\lfloor(\boldsymbol{\nu}\wedge\overline{\overline{\mathsf{I}}}^T))\\ &= \mathbf{e}_N\lfloor(\boldsymbol{\nu}\wedge\boldsymbol{\nu}\wedge\overline{\overline{\mathsf{I}}}^T) = 0,\end{aligned} \tag{5.238}$$

the axion part of the medium bidyadic $\overline{\overline{\mathsf{M}}}$ can be ignored when forming the dispersion dyadic and it does not have any effect on a plane wave propagating in a homogeneous medium.

5.6.2 Dispersion Equation

Since the dispersion dyadic satisfies (5.236) it has no inverse whence its rank can be at most 3. However, because it also satisfies

$$\overline{\overline{\mathsf{D}}}(\boldsymbol{\nu})|\boldsymbol{\nu} = -\boldsymbol{\nu}\rfloor\overline{\overline{\mathsf{M}}}_m|(\boldsymbol{\nu}\wedge\boldsymbol{\nu}) = 0, \tag{5.239}$$

while for $\boldsymbol{\Phi} = \boldsymbol{\nu}\wedge\boldsymbol{\phi} \neq 0$, $\boldsymbol{\nu}$ and $\boldsymbol{\phi}$ are linearly independent, the rank of the dispersion dyadic $\overline{\overline{\mathsf{D}}}(\boldsymbol{\nu})$ can be at most 2. This means that the dispersion dyadic satisfies

$$\overline{\overline{\mathsf{D}}}^{(3)}(\boldsymbol{\nu}) = 0, \tag{5.240}$$

which is a dyadic equation for the wave one-form $\boldsymbol{\nu}$. Actually, (5.240) is equivalent to a scalar equation called the dispersion equation. To find the scalar equation let us form the double-wedge square and cube of the dispersion dyadic,

$$\begin{aligned}\overline{\overline{\mathsf{D}}}^{(2)}(\boldsymbol{\nu}) &= \frac{1}{2}\overline{\overline{\mathsf{D}}}(\boldsymbol{\nu})^{\wedge}_{\wedge}\overline{\overline{\mathsf{D}}}(\boldsymbol{\nu})\\ &= \frac{1}{2}\boldsymbol{\nu}\boldsymbol{\nu}\rfloor\rfloor(\overline{\overline{\mathsf{M}}}_m{}^{\wedge}_{\wedge}\overline{\overline{\mathsf{D}}}(\boldsymbol{\nu}))\\ &= \frac{1}{2}\boldsymbol{\nu}\boldsymbol{\nu}\rfloor\rfloor(\overline{\overline{\mathsf{M}}}_m{}^{\wedge}_{\wedge}(\boldsymbol{\nu}\boldsymbol{\nu}\rfloor\rfloor\overline{\overline{\mathsf{M}}}_m)),\end{aligned} \tag{5.241}$$

$$\begin{aligned}\overline{\overline{\mathsf{D}}}^{(3)}(\boldsymbol{\nu}) &= \frac{1}{3}\overline{\overline{\mathsf{D}}}(\boldsymbol{\nu})^{\wedge}_{\wedge}\overline{\overline{\mathsf{D}}}^{(2)}(\boldsymbol{\nu})\\ &= \frac{1}{3}\boldsymbol{\nu}\boldsymbol{\nu}\rfloor\rfloor(\overline{\overline{\mathsf{M}}}_m{}^{\wedge}_{\wedge}\overline{\overline{\mathsf{D}}}^{(2)}(\boldsymbol{\nu})).\end{aligned} \tag{5.242}$$

Because the last dyadic in brackets belongs to the space $\mathbb{E}_4\mathbb{E}_4$, it must be a multiple of $\mathbf{e}_N\mathbf{e}_N$. Thus, we can express

$$\overline{\overline{\mathsf{D}}}^{(3)}(\boldsymbol{\nu}) = (\boldsymbol{\nu}\boldsymbol{\nu}\rfloor\rfloor\mathbf{e}_N\mathbf{e}_N)D(\boldsymbol{\nu}), \tag{5.243}$$

where the bracketed dyadic does not vanish for $\boldsymbol{\nu} \neq 0$. The scalar dispersion function $D(\boldsymbol{\nu})$ depends on the medium through the modified medium bidyadic $\overline{\overline{\mathsf{M}}}_m$. It has the following analytic form:

$$\begin{aligned} D(\boldsymbol{\nu}) &= \frac{1}{3}\boldsymbol{\varepsilon}_N\boldsymbol{\varepsilon}_N||(\overline{\overline{\mathsf{M}}}_m{}^{\wedge}_{\wedge}\overline{\overline{\mathsf{D}}}^{(2)}(\boldsymbol{\nu})) \\ &= \frac{1}{6}\boldsymbol{\varepsilon}_N\boldsymbol{\varepsilon}_N||(\overline{\overline{\mathsf{M}}}_m{}^{\wedge}_{\wedge}(\boldsymbol{\nu}\boldsymbol{\nu}\rfloor\rfloor(\overline{\overline{\mathsf{M}}}_m{}^{\wedge}_{\wedge}(\boldsymbol{\nu}\boldsymbol{\nu}\rfloor\rfloor\overline{\overline{\mathsf{M}}}_m)))). \end{aligned} \tag{5.244}$$

Thus, for $\boldsymbol{\nu} \neq 0$, we conclude that the dyadic dispersion equation (5.240) is equivalent to the scalar dispersion equation

$$D(\boldsymbol{\nu}) = 0. \tag{5.245}$$

Substituting (5.222), the dispersion equation has the form $f(k, \boldsymbol{\beta}) = 0$. Assuming that the scalar k is given and $\boldsymbol{\beta} = \beta\boldsymbol{\varepsilon}$ where $\boldsymbol{\varepsilon}$ is a given spatial one-form, the dispersion equation becomes an algebraic equation for the scalar magnitude $\beta = \beta(\boldsymbol{\varepsilon}, k)$ of the one-form $\boldsymbol{\beta}$. Because the dispersion function $D(\boldsymbol{\nu})$ is a fourth-order polynomial in $\boldsymbol{\nu}$, (5.245) represents an algebraic equation of the fourth order for $\boldsymbol{\nu}$. For a given value of k, it can be pictured as defining a surface of the fourth order in the 3D space of $\boldsymbol{\beta}$.

The dispersion equation could be alternatively obtained by starting from the medium equation (5.52) instead of (5.51). In fact, expressing the field two-form $\boldsymbol{\Psi}$ as (5.228) and following similar steps, the equation has the form

$$\overline{\overline{\mathsf{D}}}'(\boldsymbol{\nu})|\boldsymbol{\psi} = 0, \quad \overline{\overline{\mathsf{D}}}'(\boldsymbol{\nu}) = \overline{\overline{\mathsf{N}}}_m\lfloor\lfloor\boldsymbol{\nu}\boldsymbol{\nu}, \tag{5.246}$$

and the dyadic dispersion equation becomes

$$\overline{\overline{\mathsf{D}}}'^{(3)}(\boldsymbol{\nu}) = (\mathbf{e}_N\mathbf{e}_N\lfloor\lfloor\boldsymbol{\nu}\boldsymbol{\nu})D'(\boldsymbol{\nu}) = 0. \tag{5.247}$$

Now it is obvious that $D'(\boldsymbol{\nu}) = 0$ represents the same dispersion equation as (5.245) because they both define the wave one-form $\boldsymbol{\nu}$ of the same plane wave in the same medium. Of course, when one of the bidyadics $\overline{\overline{\mathsf{M}}}_m$ and $\overline{\overline{\mathsf{N}}}_m$ does not exist, the corresponding dispersion equation does not exist.

5.6.3 Special Cases

There are various special cases of the dispersion equation depending on the medium in question.

1. $D(\boldsymbol{\nu}) = 0$ is a quartic equation for the general medium.
2. For some media the dispersion function can be factorized as a product of two second-order functions $D(\boldsymbol{\nu}) = D_1(\boldsymbol{\nu})D_2(\boldsymbol{\nu})$, whence the quartic equation can be decomposed in two quadratic equations $D_1(\boldsymbol{\nu}) = 0$ and $D_2(\boldsymbol{\nu}) = 0$. The corresponding media can be called decomposable media (DC media).
3. For some DC media the two dispersion functions coincide, $D(\boldsymbol{\nu}) = (D_1(\boldsymbol{\nu}))^2$, whence there is only one quadratic equation $D_1(\boldsymbol{\nu}) = 0$. The corresponding media can be called nonbirefringent (NB media).
4. For some media the dispersion equation $D(\boldsymbol{\nu}) = 0$ is satisfied identically for any $\boldsymbol{\nu}$, whence the choice of the one-form $\boldsymbol{\nu}$ is not restricted by the medium. Such media can be called media with no dispersion equation (NDE media).

Examples of media with different dispersion equation properties will come up in subsequent chapters.

5.6.4 Plane-Wave Fields

Let us assume that $\boldsymbol{\nu}$ is a known solution for the dispersion equation of the form (5.245) or (5.240). Since the potential $\boldsymbol{\phi}$ and the wave one-form $\boldsymbol{\nu}$ satisfy $\overline{\overline{\mathsf{D}}}(\boldsymbol{\nu})|\boldsymbol{\phi} = 0$ and $\overline{\overline{\mathsf{D}}}(\boldsymbol{\nu})|\boldsymbol{\nu} = 0$, the field two-form $\boldsymbol{\Phi} = \boldsymbol{\nu} \wedge \boldsymbol{\phi}$ satisfies the dyadic equation

$$\overline{\overline{\mathsf{D}}}(\boldsymbol{\nu})\rfloor\boldsymbol{\Phi} = (\overline{\overline{\mathsf{D}}}(\boldsymbol{\nu})|\boldsymbol{\phi})\boldsymbol{\nu} - (\overline{\overline{\mathsf{D}}}(\boldsymbol{\nu})|\boldsymbol{\nu})\boldsymbol{\phi} = 0. \tag{5.248}$$

To find the field two-form, we can start from the condition

$$\overline{\overline{\mathsf{D}}}\rfloor(\boldsymbol{\varepsilon}_N\lfloor\overline{\overline{\mathsf{D}}}^{(2)T}) = 0, \tag{5.249}$$

which can be shown to be valid for any metric dyadic $\overline{\overline{\mathsf{D}}} \in \mathbb{E}_1\mathbb{E}_1$ satisfying $\overline{\overline{\mathsf{D}}}^{(3)} = 0$. Choosing $\overline{\overline{\mathsf{D}}}$ equal to the dispersion dyadic $\overline{\overline{\mathsf{D}}}(\boldsymbol{\nu})$ of rank 2, satisfying $\overline{\overline{\mathsf{D}}}^{(2)}(\boldsymbol{\nu}) \neq 0$, we can write a solution to (5.248) as

$$\boldsymbol{\Phi} = \boldsymbol{\varepsilon}_N\lfloor\overline{\overline{\mathsf{D}}}^{(2)T}|\boldsymbol{\Xi} = \boldsymbol{\varepsilon}_N\lfloor(\boldsymbol{\Xi}|\overline{\overline{\mathsf{D}}}^{(2)}), \tag{5.250}$$

where $\boldsymbol{\Xi}$ may be any two-form yielding a nonzero result. As a check we can expand

$$\begin{aligned}\mathbf{e}_N\lfloor(\boldsymbol{\Phi}\wedge\boldsymbol{\nu}) &= \mathbf{e}_N\lfloor((\boldsymbol{\varepsilon}_N\lfloor(\boldsymbol{\Xi}|\overline{\overline{\mathsf{D}}}^{(2)}))\wedge\boldsymbol{\nu}) \\ &= (\mathbf{e}_N\lfloor(\boldsymbol{\varepsilon}_N\lfloor(\boldsymbol{\Xi}|\overline{\overline{\mathsf{D}}}^{(2)})))\lfloor\boldsymbol{\nu} \\ &= (\boldsymbol{\Xi}|\overline{\overline{\mathsf{D}}}^{(2)})\lfloor\boldsymbol{\nu} \\ &= \boldsymbol{\Xi}|((\overline{\overline{\mathsf{D}}}|\boldsymbol{\nu})\wedge\overline{\overline{\mathsf{D}}}) = 0,\end{aligned} \tag{5.251}$$

which implies $\boldsymbol{\Phi}\wedge\boldsymbol{\nu} = 0$ because of (5.239). Thus, (5.250) satisfies (5.225).

The corresponding potential solution is obtained from (5.230). Assuming that the chosen vector $\mathbf{h}$ satisfies $\mathbf{h}|\boldsymbol{\nu} = 1$, we obtain

$$\boldsymbol{\phi} = -\mathbf{h}\rfloor(\boldsymbol{\varepsilon}_N\lfloor(\boldsymbol{\Xi}|\overline{\overline{\mathsf{D}}}^{(2)})) = \boldsymbol{\varepsilon}_N\lfloor(\mathbf{h}\wedge(\boldsymbol{\Xi}|\overline{\overline{\mathsf{D}}}^{(2)})). \tag{5.252}$$

The potential is not unique because of the vector $\mathbf{h}$. The polarization of the field two-form is unique because the dispersion dyadic is of rank 2 and can be expressed in the form $\overline{\overline{\mathsf{D}}}(\boldsymbol{\nu}) = \mathbf{ac} + \mathbf{bd}$ for some vectors $\mathbf{a}-\mathbf{d}$. Thus, $\boldsymbol{\Phi}$ is a multiple of $\boldsymbol{\varepsilon}_N\lfloor(\mathbf{c}\wedge\mathbf{d})$ for any two-form $\boldsymbol{\Xi}$.

The rule (5.250) obviously fails for media with wave one-form $\boldsymbol{\nu}$ satisfying $\overline{\overline{\mathsf{D}}}^{(2)}(\boldsymbol{\nu}) = 0$, in which case the dispersion dyadic $\overline{\overline{\mathsf{D}}}(\boldsymbol{\nu})$ is of rank 1. Because in this case we can write $\overline{\overline{\mathsf{D}}} = \mathbf{ac}$, any two-form satisfying $\mathbf{c}\rfloor\boldsymbol{\Phi} = 0$ may represent a valid electromagnetic field. The two-form can be expressed in the form

$$\boldsymbol{\Phi} = \boldsymbol{\varepsilon}_N\lfloor(\mathbf{c}\wedge\mathbf{d}), \tag{5.253}$$

where $\mathbf{d}$ may be any vector yielding a nonzero result. An equivalent form can be written as

$$\boldsymbol{\Phi} = \boldsymbol{\varepsilon}_N\lfloor((\boldsymbol{\alpha}|\overline{\overline{\mathsf{D}}})\wedge\mathbf{d}), \tag{5.254}$$

which depends on a one-form $\boldsymbol{\alpha}$ and a vector $\mathbf{d}$. To check the validity of (5.225), we expand

$$\begin{aligned}\mathbf{e}_N\lfloor(\boldsymbol{\Phi}\wedge\boldsymbol{\nu}) &= \mathbf{e}_N\lfloor(\boldsymbol{\varepsilon}_N\lfloor((\boldsymbol{\alpha}|\overline{\overline{\mathsf{D}}})\wedge\mathbf{d})\wedge\boldsymbol{\nu}) \\ &= (\mathbf{e}_N\lfloor(\boldsymbol{\varepsilon}_N\lfloor((\boldsymbol{\alpha}|\overline{\overline{\mathsf{D}}})\wedge\mathbf{d})))\lfloor\boldsymbol{\nu} \\ &= ((\boldsymbol{\alpha}|\overline{\overline{\mathsf{D}}})\wedge\mathbf{d})\lfloor\boldsymbol{\nu} \\ &= \mathbf{d}(\boldsymbol{\alpha}|\overline{\overline{\mathsf{D}}}|\boldsymbol{\nu}) - (\mathbf{d}|\boldsymbol{\nu})(\boldsymbol{\alpha}|\overline{\overline{\mathsf{D}}}) \\ &= -(\mathbf{d}|\boldsymbol{\nu})(\boldsymbol{\alpha}|\overline{\overline{\mathsf{D}}}),\end{aligned}$$

which requires that the vector $\mathbf{d}$ must satisfy $\mathbf{d}|\boldsymbol{\nu} = 0$ for the given one-form $\boldsymbol{\nu}$. With this, (5.254) satisfies (5.225) and represents a valid polarization of the plane wave field for any one-form $\boldsymbol{\alpha}$.

5.6.5 Simple Principal Medium

As an example let us consider plane-wave propagation in the simple principal medium defined by the modified medium bidyadic (5.122),

$$\overline{\overline{\mathsf{M}}}_m = -\frac{1}{\mu}\overline{\overline{\mathsf{G}}}^{(2)}, \tag{5.255}$$

with the symmetric dyadic $\overline{\overline{\mathsf{G}}} \in \mathbb{E}_1\mathbb{E}_1$ defined by

$$\overline{\overline{\mathsf{G}}} = \overline{\overline{\mathsf{G}}}_s - \mu\epsilon\mathbf{e}_4\mathbf{e}_4, \quad \overline{\overline{\mathsf{G}}}_s = \mathbf{e}_1\mathbf{e}_1 + \mathbf{e}_2\mathbf{e}_2 + \mathbf{e}_3\mathbf{e}_3, \tag{5.256}$$

and satisfying

$$\overline{\overline{\mathsf{G}}}_s^{(3)} = \mathbf{e}_{123}\mathbf{e}_{123}, \quad \overline{\overline{\mathsf{G}}}^{(4)} = -\mu\epsilon\mathbf{e}_N\mathbf{e}_N. \tag{5.257}$$

The dispersion dyadic and its double-wedge powers can now be expanded as

$$\mu\overline{\overline{\mathsf{D}}}(\boldsymbol{\nu}) = -\overline{\overline{\mathsf{G}}}^{(2)}\lfloor\lfloor\boldsymbol{\nu}\boldsymbol{\nu} = -(\overline{\overline{\mathsf{G}}}||\boldsymbol{\nu}\boldsymbol{\nu})\overline{\overline{\mathsf{G}}} + (\overline{\overline{\mathsf{G}}}|\boldsymbol{\nu})(\overline{\overline{\mathsf{G}}}|\boldsymbol{\nu}), \tag{5.258}$$

$$\mu^2\overline{\overline{\mathsf{D}}}^{(2)}(\boldsymbol{\nu}) = (\overline{\overline{\mathsf{G}}}||\boldsymbol{\nu}\boldsymbol{\nu})(\overline{\overline{\mathsf{G}}}^{(2)}(\overline{\overline{\mathsf{G}}}||\boldsymbol{\nu}\boldsymbol{\nu}) - \overline{\overline{\mathsf{G}}}{}^\wedge_\wedge(\overline{\overline{\mathsf{G}}}|\boldsymbol{\nu})(\overline{\overline{\mathsf{G}}}|\boldsymbol{\nu})) \tag{5.259}$$

$$= (\overline{\overline{\mathsf{G}}}||\boldsymbol{\nu}\boldsymbol{\nu})\overline{\overline{\mathsf{G}}}^{(3)}\lfloor\lfloor\boldsymbol{\nu}\boldsymbol{\nu}, \tag{5.260}$$

$$\begin{aligned}\mu^3\overline{\overline{\mathsf{D}}}^{(3)}(\boldsymbol{\nu}) &= -(\overline{\overline{\mathsf{G}}}||\boldsymbol{\nu}\boldsymbol{\nu})^2((\overline{\overline{\mathsf{G}}}||\boldsymbol{\nu}\boldsymbol{\nu})\overline{\overline{\mathsf{G}}}^{(3)} - \overline{\overline{\mathsf{G}}}^{(2)}{}^\wedge_\wedge(\overline{\overline{\mathsf{G}}}|\boldsymbol{\nu})(\overline{\overline{\mathsf{G}}}|\boldsymbol{\nu})) \\ &= -(\overline{\overline{\mathsf{G}}}||\boldsymbol{\nu}\boldsymbol{\nu})^2(\overline{\overline{\mathsf{G}}}^{(4)}\lfloor\lfloor\boldsymbol{\nu}\boldsymbol{\nu}) \\ &= \mu\epsilon(\overline{\overline{\mathsf{G}}}||\boldsymbol{\nu}\boldsymbol{\nu})^2(\mathbf{e}_N\mathbf{e}_N\lfloor\lfloor\boldsymbol{\nu}\boldsymbol{\nu}).\end{aligned} \tag{5.261}$$

Here we have applied the following identity, valid for any metric dyadics $\overline{\overline{\mathsf{G}}} \in \mathbb{E}_1\mathbb{E}_1$ and $\overline{\overline{\Theta}} \in \mathbb{F}_1\mathbb{F}_1$,

$$\overline{\overline{\mathsf{G}}}^{(p)}\lfloor\lfloor\overline{\overline{\Theta}}^T = (\overline{\overline{\mathsf{G}}}||\overline{\overline{\Theta}}^T)\overline{\overline{\mathsf{G}}}^{(p-1)} - \overline{\overline{\mathsf{G}}}^{(p-2)}{}^\wedge_\wedge(\overline{\overline{\mathsf{G}}}|\overline{\overline{\Theta}}|\overline{\overline{\mathsf{G}}}), \quad p = 4, 3. \tag{5.262}$$

From (5.261) the dispersion equation (5.240) is seen to reduce to

$$\overline{\overline{\mathsf{G}}}||\boldsymbol{\nu}\boldsymbol{\nu} = 0, \tag{5.263}$$

or, in terms of spatial quantities, to

$$\boldsymbol{\beta}|\overline{\overline{\mathsf{G}}}_s|\boldsymbol{\beta} - \mu\epsilon k^2 = 0. \tag{5.264}$$

This corresponds to a special case when the dispersion surface in the spatial $\boldsymbol{\beta}$ space becomes a single quadric. This means that the simple principal medium is a nonbirefringent medium.

When $\boldsymbol{\nu}$ is a solution to (5.263), from (5.259) it follows that the dyadic dispersion equation is of the more restricted form $\overline{\overline{\mathsf{D}}}^{(2)}(\boldsymbol{\nu}) = 0$ and the dispersion dyadic (5.258) is of rank 1 as

$$\overline{\overline{\mathsf{D}}}(\boldsymbol{\nu}) = \frac{1}{\mu}(\overline{\overline{\mathsf{G}}}|\boldsymbol{\nu})(\overline{\overline{\mathsf{G}}}|\boldsymbol{\nu}). \tag{5.265}$$

The equation (5.236) for the corresponding potential one-form $\boldsymbol{\phi}$ has now the simple scalar form

$$\boldsymbol{\nu}|\overline{\overline{\mathsf{G}}}|\boldsymbol{\phi} = 0, \tag{5.266}$$

whence $\boldsymbol{\phi}$ can be expressed in the form

$$\boldsymbol{\phi} = \boldsymbol{\nu}|\overline{\overline{\mathsf{G}}}\rfloor\boldsymbol{\Xi} \tag{5.267}$$

for any two-form $\boldsymbol{\Xi}$ yielding a nonzero result. The corresponding field two-form becomes

$$\boldsymbol{\Phi} = \boldsymbol{\nu} \wedge (\boldsymbol{\nu}|\overline{\overline{\mathsf{G}}}\rfloor\boldsymbol{\Xi}). \tag{5.268}$$

It is left as an exercise to show that (5.268) corresponds to the previously obtained expression (5.254).

5.6.6 Handedness of Plane Wave

Spatial 3D trivectors $\boldsymbol{\kappa}_s \in \mathbb{F}_3$ are similar to scalars, because they are multiples of a given spatial trivector, say $\boldsymbol{\varepsilon}_{123}$,

$$\boldsymbol{\kappa}_s = \kappa\boldsymbol{\varepsilon}_{123}, \qquad \kappa = \boldsymbol{\kappa}_s|\mathbf{e}_{123}. \tag{5.269}$$

If κ is positive, $\boldsymbol{\kappa}_s$ and $\boldsymbol{\varepsilon}_{123}$ have the same handedness, if negative, they have opposite handedness. A plane wave has a special spatial three-form, $\boldsymbol{\beta} \wedge \mathbf{E} \wedge \mathbf{H}$, where $\boldsymbol{\beta}$ is the spatial part of $\boldsymbol{\nu}$. Let us assume that $\mathbf{e}_{123}$ is a right-handed trivector. It turns out that the handedness of the spatial trivector $\boldsymbol{\beta} \wedge \mathbf{E} \wedge \mathbf{H}$ depends on the medium.

For plane-wave fields, the energy–power density three-form (5.202) satisfies

$$\begin{aligned}\boldsymbol{\nu} \wedge \mathbf{W} &= \frac{1}{2}\boldsymbol{\nu} \wedge (-\boldsymbol{\Psi} \wedge (\boldsymbol{\Phi}\lfloor\mathbf{e}_4) + \boldsymbol{\Phi} \wedge (\boldsymbol{\Psi}\lfloor\mathbf{e}_4))\\ &= \frac{1}{2}\boldsymbol{\nu} \wedge (-\boldsymbol{\nu} \wedge \boldsymbol{\psi} \wedge (\boldsymbol{\Phi}\lfloor\mathbf{e}_4) + \boldsymbol{\nu} \wedge \boldsymbol{\phi} \wedge (\boldsymbol{\Psi}\lfloor\mathbf{e}_4)) = 0. \end{aligned} \tag{5.270}$$

On the other hand, since it consists of the energy density three-form $\mathbf{w}$ and the Poynting two-form $\mathbf{E} \wedge \mathbf{H}$ as

$$\mathbf{W} = \mathbf{w} - \mathbf{E} \wedge \mathbf{H} \wedge \varepsilon_4, \tag{5.271}$$

we have

$$\mathbf{e}_N \lfloor (\boldsymbol{\nu} \wedge \mathbf{w}) = \mathbf{e}_N \lfloor (\boldsymbol{\nu} \wedge \mathbf{E} \wedge \mathbf{H} \wedge \varepsilon_4). \tag{5.272}$$

Expanding both sides of (5.272) as

$$\begin{aligned} \mathbf{e}_N \lfloor (\boldsymbol{\nu} \wedge \mathbf{w}) &= \mathbf{e}_N \lfloor (-k\varepsilon_4 \wedge (\mathbf{D} \wedge \mathbf{E} + \mathbf{B} \wedge \mathbf{H})) \\ &= k\mathbf{e}_{123} \lfloor (\mathbf{D} \wedge \mathbf{E} + \mathbf{B} \wedge \mathbf{E}) \\ &= k(\mathbf{D}_g \cdot \mathbf{E}_g + \mathbf{B}_g \cdot \mathbf{H}_g) \end{aligned} \tag{5.273}$$

$$\begin{aligned} \mathbf{e}_N \lfloor (\boldsymbol{\nu} \wedge \mathbf{E} \wedge \mathbf{H} \wedge \varepsilon_4) &= \mathbf{e}_N \lfloor (\boldsymbol{\beta} \wedge \mathbf{E} \wedge \mathbf{H} \wedge \varepsilon_4) \\ &= \mathbf{e}_{123} \lfloor (\boldsymbol{\beta} \wedge \mathbf{E} \wedge \mathbf{H}), \end{aligned} \tag{5.274}$$

the condition (5.272) can be written for Gibbsian fields as

$$\mathbf{e}_{123} \lfloor (\boldsymbol{\beta} \wedge \mathbf{E} \wedge \mathbf{H}) = k(\mathbf{E}_g \ \mathbf{H}_g) \cdot \begin{pmatrix} \overline{\overline{\epsilon}}_g & \overline{\overline{\xi}}_g \\ \overline{\overline{\zeta}}_g & \overline{\overline{\mu}}_g \end{pmatrix} \cdot \begin{pmatrix} \mathbf{E}_g \\ \mathbf{H}_g \end{pmatrix}. \tag{5.275}$$

Because this is valid for any plane-wave fields, it links together the handedness of the three-form $\boldsymbol{\beta} \wedge \mathbf{E} \wedge \mathbf{H}$ when compared with the trivector $\mathbf{e}_{123}$ and the definiteness of the matrix of Gibbsian medium dyadics. As an example, let us consider the simple principal medium with

$$\overline{\overline{\epsilon}}_g = \epsilon \overline{\overline{\mathsf{G}}}_s, \quad \overline{\overline{\xi}}_g = 0, \quad \overline{\overline{\zeta}}_g = 0, \quad \overline{\overline{\mu}}_g = \mu \overline{\overline{\mathsf{G}}}_s, \tag{5.276}$$

in which case we have

$$\mathbf{e}_{123} \lfloor (\boldsymbol{\beta} \wedge \mathbf{E} \wedge \mathbf{H}) = k(\epsilon \mathbf{E}_g \cdot \mathbf{E}_g + \mu \mathbf{H}_g \cdot \mathbf{H}_g), \tag{5.277}$$

where $k = \mathbf{e}_4 | \boldsymbol{\nu}$ equals ω/c for a time-harmonic field. Thus, for positive ϵ and μ the three-form $\boldsymbol{\beta} \wedge \mathbf{E} \wedge \mathbf{H}$ is right handed while for negative ϵ and μ it is left handed. This is the reason for the concept of "left-handed medium" launched by Veselago [41] for media with negative ϵ and μ.

PROBLEMS

5.1 Show that the Gibbsian form of Maxwell equations (5.40)–(5.42) can be obtained from (5.21) and (5.22) through the rules (5.33)–(5.36).

5.2 Find the equation for the potential representation of the field $\boldsymbol{\Psi}$ outside the source region, corresponding to (5.62), (5.63).

5.3 Derive the expansion (5.87) for the medium bidyadic in terms of Gibbsian medium dyadics, starting from (5.83)–(5.86).

5.4 Prove (5.97).

5.5 Verify that the bidyadic $\overline{\overline{\mathsf{N}}}$ defined by the 3D expansion (5.77) is the inverse of the bidyadic $\overline{\overline{\mathsf{M}}}$ defined by (5.75) and that (5.89) is the inverse of the modified medium bidyadic (5.88).

5.6 Show that a medium defined by the modified medium bidyadic $\overline{\overline{\mathsf{M}}}_m = \overline{\overline{\mathsf{Q}}}^{(2)}$ with symmetric dyadic $\overline{\overline{\mathsf{Q}}} \in \mathbb{E}_1\mathbb{E}_1$ is a principal medium.

5.7 Find the Hehl–Obukhov decomposition of the medium bidyadic $\overline{\overline{\mathsf{M}}} = \varepsilon_N \lfloor \overline{\overline{\mathsf{Q}}}^{(2)}$ where the dyadic $\overline{\overline{\mathsf{Q}}}$ is antisymmetric, of the form $\overline{\overline{\mathsf{Q}}} = \mathbf{Q}\lfloor \overline{\overline{\mathsf{I}}}^T$ where $\mathbf{Q} \in \mathbb{E}_2$ is some bidyadic.

5.8 Derive conditions for the spatial medium dyadics for the SH conditions starting from (5.179) and proceeding like for the DB conditions, (5.173) and (5.174).

5.9 Verify by basis expansion that the simple principal medium bidyadic

$$\overline{\overline{\mathsf{M}}}_m = -\frac{1}{\mu}(\overline{\overline{\mathsf{G}}}_s - \mu\epsilon \mathbf{e}_4\mathbf{e}_4)^{(2)}$$

does not have an axion component.

5.10 Starting from medium conditions slightly different from those of (5.179),

$$\varepsilon_2 \wedge \varepsilon_3 \wedge \mathbf{E}_b = 0, \qquad \varepsilon_1 \wedge \varepsilon_3 \wedge \mathbf{H}_b = 0,$$

study the boundary conditions at the interface $x_3 = 0$ defined by such a medium. Interpret the conditions in terms of Gibbsian vector fields.

5.11 Verify that the trace-free dyadic $\overline{\overline{\mathsf{B}}}_o$ defined by the simple antisymmetric bidyadic $\overline{\overline{\mathsf{U}}}$ by (5.191) satisfies $\overline{\overline{\mathsf{B}}}_o^2 = \alpha \overline{\overline{\mathsf{I}}}$ for some scalar α. Hint: show that, for any vector $\mathbf{a}$, $\overline{\overline{\mathsf{B}}}_o^2|\mathbf{a}$ is a scalar multiple of $\mathbf{a}$.

5.12 Show that for a skewon–axion medium defined by $\overline{\overline{\mathsf{M}}} = (\overline{\overline{\mathsf{B}}}{}^{\wedge}_{\wedge}\overline{\overline{\mathsf{I}}})^T$ where $\overline{\overline{\mathsf{B}}} \in \mathbb{E}_1\mathbb{F}_1$ is any dyadic, the dyadic dispersion equation (5.243) is satisfied identically.

5.13 Derive the expression (5.219) in detail.

5.14 Prove that the Maxwell energy–stress dyadic vanishes for any fields if and only if the medium bidyadic has only the axion term, that is, if it is PEMC.

5.15 Show that when the field two-form $\boldsymbol{\Phi}$ is simple, that is, satisfies $\boldsymbol{\Phi} \wedge \boldsymbol{\Phi} = 0$, there must exist a spatial one-form $\boldsymbol{\alpha}$ such that $\mathbf{B} = \boldsymbol{\alpha} \wedge \mathbf{E}$.

5.16 Show that the condition for the modified medium bidyadic

$$A\overline{\overline{\mathsf{M}}}_m + B\overline{\overline{\mathsf{M}}}_m^{-1T} \rfloor\rfloor \mathbf{e}_N\mathbf{e}_N = 0$$

for some scalars A, B implies the following conditions for the Gibbsian dyadics:

$$A\overline{\overline{\epsilon}}_g^T = B\overline{\overline{\mu}}_g^T, \quad \overline{\overline{\xi}}_g^T = -\overline{\overline{\xi}}_g, \quad \overline{\overline{\zeta}}_g^T = -\overline{\overline{\zeta}}_g,$$

provided the 3D inverses of the Gibbsian dyadics $\overline{\overline{\epsilon}}_g, \overline{\overline{\mu}}_g$ exist.

5.17 Derive the dispersion equation for a plane wave in a medium defined by the modified medium bidyadic

$$\overline{\overline{\mathsf{M}}}_m = \overline{\overline{\mathsf{Q}}}^{(2)} + \mathbf{A}\mathbf{B},$$

where $\overline{\overline{\mathsf{Q}}} \in \mathbb{E}_1\mathbb{E}_1$ is a metric dyadic and $\mathbf{A}, \mathbf{B}$ are two bivectors.

5.18 Find the dispersion equation for a medium defined by the modified medium bidyadic

$$\overline{\overline{\mathsf{M}}}_m = M\overline{\overline{\mathsf{E}}} + \mathbf{A}_1\mathbf{C}_1 + \mathbf{A}_2\mathbf{C}_2, \quad \overline{\overline{\mathsf{E}}} = \mathbf{e}_N \lfloor \overline{\overline{\mathsf{I}}}^{(2)T}$$

where $\mathbf{A}_i$, $\mathbf{C}_i$ are two bivectors.

5.19 Find the dispersion equation for a medium defined by the modified medium bidyadic

$$\overline{\overline{\mathsf{M}}}_m = M\overline{\overline{\mathsf{E}}} + \mathbf{A}_1\mathbf{C}_1 + \mathbf{A}_2\mathbf{C}_2 + \mathbf{A}_3\mathbf{C}_3, \quad \overline{\overline{\mathsf{E}}} = \mathbf{e}_N \lfloor \overline{\overline{\mathsf{I}}}^{(2)T}$$

where $\mathbf{A}_i$, $\mathbf{C}_i$ are six bivectors. Show that the dispersion equation is split in two quadratic equations.

5.20 Show that the expressions (5.268) and (5.254) correspond to the same field two-form of a plane wave in a simple principal medium.

5.21 Prove that (5.249) is satisfied for any metric dyadic $\overline{\overline{\mathsf{C}}} \in \mathbb{E}_1\mathbb{E}_1$ of rank 2.

CHAPTER 6

Transformation of Fields and Media

Applying various transformations of fields, sources, media and boundary conditions allows one to produce solutions for new problems from those of old problems without having to go through the solution process.

6.1 AFFINE TRANSFORMATION

Affine transformation is a linear transformation of spacetime which affects fields, sources, and media.

6.1.1 Transformation of Fields

Assuming a dyadic $\overline{\overline{\mathsf{A}}} \in \mathbb{E}_1\mathbb{F}_1$ of rank 4, satisfying $\mathrm{tr}\overline{\overline{\mathsf{A}}}^{(4)} \neq 0$, vectors can be mapped to other vectors as

$$\mathbf{x} \rightarrow \mathbf{x}_a = \overline{\overline{\mathsf{A}}}|\mathbf{x}, \tag{6.1}$$

which is called affine transformation [1, 25]. In its more general form affine transformation also involves translation by a vector [42], which is omitted here. The differential operator $\mathbf{d}$ is simultaneously transformed as

$$\mathbf{d}_a = \overline{\overline{\mathsf{A}}}^{-1T}|\mathbf{d}, \tag{6.2}$$

because it satisfies

$$\mathbf{d}_a\mathbf{x}_a = \overline{\overline{\mathsf{A}}}^{-1T}|\mathbf{d}\mathbf{x}|\overline{\overline{\mathsf{A}}}^T = \overline{\overline{\mathsf{I}}}^T, \tag{6.3}$$

Multiforms, Dyadics, and Electromagnetic Media, First Edition. Ismo V. Lindell.

when applying the property (5.5)

$$\mathbf{dx} = \overline{\overline{\mathsf{I}}}^T. \tag{6.4}$$

The wedge product of two transformed vectors,

$$\mathbf{x}_a \wedge \mathbf{y}_a = (\overline{\overline{\mathsf{A}}}|\mathbf{x}) \wedge (\overline{\overline{\mathsf{A}}}|\mathbf{y}) = \overline{\overline{\mathsf{A}}}^{(2)}|(\mathbf{x} \wedge \mathbf{y}), \tag{6.5}$$

yields a transformation rule for any bivector $\mathbf{A}$,

$$\mathbf{A}_a = \overline{\overline{\mathsf{A}}}^{(2)}|\mathbf{A}. \tag{6.6}$$

Similarly, trivectors $\mathbf{k} \in \mathbb{E}_3$ and quadrivectors $\mathbf{q}_N \in \mathbb{E}_4$ are transformed as

$$\mathbf{k}_a = \overline{\overline{\mathsf{A}}}^{(3)}|\mathbf{k}, \tag{6.7}$$

$$\mathbf{q}_{Na} = \mathrm{tr}\overline{\overline{\mathsf{A}}}^{(4)}\ \mathbf{q}_N. \tag{6.8}$$

Transformation rule of a one-form $\boldsymbol{\phi}$ follows that of the differential operator $\mathbf{d}$, while that of a two-form $\boldsymbol{\Phi}$, a three-form $\boldsymbol{\gamma}$ and a four-form $\boldsymbol{\kappa}_N$ can be expressed as

$$\boldsymbol{\phi}_a = \overline{\overline{\mathsf{A}}}^{-1T}|\boldsymbol{\phi}, \tag{6.9}$$

$$\boldsymbol{\Phi}_a = \overline{\overline{\mathsf{A}}}^{(-2)T}|\boldsymbol{\Phi}, \tag{6.10}$$

$$\boldsymbol{\gamma}_a = \overline{\overline{\mathsf{A}}}^{(-3)T}|\boldsymbol{\gamma}, \tag{6.11}$$

$$\boldsymbol{\kappa}_{Na} = \mathrm{tr}\overline{\overline{\mathsf{A}}}^{(-1)}\ \boldsymbol{\kappa}_N. \tag{6.12}$$

The Maxwell equations can be shown to be invariant in form in the affine transformation:

$$\begin{aligned}\mathbf{d}_a \wedge \boldsymbol{\Psi}_a - \boldsymbol{\gamma}_{ea} &= (\overline{\overline{\mathsf{A}}}^{-1T}|\mathbf{d}) \wedge (\overline{\overline{\mathsf{A}}}^{(-2)T}|\boldsymbol{\Psi}) - \overline{\overline{\mathsf{A}}}^{(-3)T}|\boldsymbol{\gamma}_e \\ &= \overline{\overline{\mathsf{A}}}^{(-3)T}|(\mathbf{d} \wedge \boldsymbol{\Psi} - \boldsymbol{\gamma}_e) = 0,\end{aligned} \tag{6.13}$$

$$\begin{aligned}\mathbf{d}_a \wedge \boldsymbol{\Psi}_a - \boldsymbol{\gamma}_{ea} &= (\overline{\overline{\mathsf{A}}}^{-1T}|\mathbf{d}) \wedge (\overline{\overline{\mathsf{A}}}^{(-2)T}|\boldsymbol{\Psi}) - \overline{\overline{\mathsf{A}}}^{(-3)T}|\boldsymbol{\gamma}_e \\ &= \overline{\overline{\mathsf{A}}}^{(-3)T}|(\mathbf{d} \wedge \boldsymbol{\Phi} - \boldsymbol{\gamma}_m) = 0.\end{aligned} \tag{6.14}$$

6.1.2 Transformation of Media

Expanding

$$\begin{aligned}\boldsymbol{\Psi}_a &= \overline{\overline{\mathsf{A}}}^{(-2)T}|\boldsymbol{\Psi} = \overline{\overline{\mathsf{A}}}^{(-2)T}|\overline{\overline{\mathsf{M}}}|\boldsymbol{\Phi} \\ &= \overline{\overline{\mathsf{A}}}^{(-2)T}|\overline{\overline{\mathsf{M}}}|\overline{\overline{\mathsf{A}}}^{(2)T}|\overline{\overline{\mathsf{A}}}^{(-2)T}|\boldsymbol{\Phi}, \\ &= \overline{\overline{\mathsf{A}}}^{(-2)T}|\overline{\overline{\mathsf{M}}}|\overline{\overline{\mathsf{A}}}^{(2)T}|\boldsymbol{\Phi}_a \\ &= \overline{\overline{\mathsf{M}}}_a|\boldsymbol{\Phi}_a,\end{aligned} \tag{6.15}$$

we obtain the transformation rule for the medium bidyadic $\overline{\overline{\mathsf{M}}}$,

$$\overline{\overline{\mathsf{M}}}_a = \overline{\overline{\mathsf{A}}}^{(-2)T}|\overline{\overline{\mathsf{M}}}|\overline{\overline{\mathsf{A}}}^{(2)T}. \tag{6.16}$$

From (6.16) it follows that the trace of the medium bidyadic is invariant in the affine transformation,

$$\mathrm{tr}\overline{\overline{\mathsf{M}}}_a = \mathrm{tr}\overline{\overline{\mathsf{M}}}. \tag{6.17}$$

The transformed modified medium bidyadic is defined as the modified transformed medium bidyadic,

$$\overline{\overline{\mathsf{M}}}_{ma} = (\overline{\overline{\mathsf{M}}}_a)_m = \mathbf{e}_{Na}\lfloor\overline{\overline{\mathsf{M}}}_a, \tag{6.18}$$

with

$$\mathbf{e}_{Na} = \mathrm{tr}\overline{\overline{\mathsf{A}}}^{(4)}\,\mathbf{e}_N. \tag{6.19}$$

The transformation rule can be found from

$$\begin{aligned}\overline{\overline{\mathsf{M}}}_{ma} &= \mathrm{tr}\overline{\overline{\mathsf{A}}}^{(4)}\mathbf{e}_N\lfloor\overline{\overline{\mathsf{A}}}^{(-2)T}|\overline{\overline{\mathsf{M}}}|\overline{\overline{\mathsf{A}}}^{(2)T}\\ &= \mathrm{tr}\overline{\overline{\mathsf{A}}}^{(4)}\mathbf{e}_N\lfloor\overline{\overline{\mathsf{A}}}^{(-2)T}\rfloor\boldsymbol{\varepsilon}_N|\overline{\overline{\mathsf{M}}}_m|\overline{\overline{\mathsf{A}}}^{(2)T}\\ &= \overline{\overline{\mathsf{A}}}^{(2)}|\overline{\overline{\mathsf{M}}}_m|\overline{\overline{\mathsf{A}}}^{(2)T},\end{aligned} \tag{6.20}$$

where we have applied the inverse rule (2.123) as

$$\overline{\overline{\mathsf{A}}}^{(2)} = \mathrm{tr}\overline{\overline{\mathsf{A}}}^{(4)}\overline{\overline{\mathsf{I}}}^{(4)}\lfloor\lfloor\overline{\overline{\mathsf{A}}}^{(-2)T} = \mathrm{tr}\overline{\overline{\mathsf{A}}}^{(4)}\mathbf{e}_N\lfloor\overline{\overline{\mathsf{A}}}^{(-2)T}\rfloor\boldsymbol{\varepsilon}_N. \tag{6.21}$$

From (6.20) it follows that symmetric and antisymmetric bidyadics $\overline{\overline{\mathsf{M}}}_m$ are respectively mapped to symmetric and antisymmetric bidyadics $\overline{\overline{\mathsf{M}}}_{ma}$ in an affine transformation. Thus, in the Hehl–Obukhov decomposition (3.85) of the modified medium

$$\overline{\overline{\mathsf{M}}}_m = \overline{\overline{\mathsf{M}}}_{m1} + \overline{\overline{\mathsf{M}}}_{m2} + \overline{\overline{\mathsf{M}}}_{m3}, \tag{6.22}$$

the skewon part $\overline{\overline{\mathsf{M}}}_{m2}$ is mapped to the skewon part of $\overline{\overline{\mathsf{M}}}_{ma}$. Since from (6.17) it follows that the axion part is invariant in the transformation, also the principal part is invariant. Thus, in conclusion, the principal, skewon, and axion parts of $\overline{\overline{\mathsf{M}}}$ are transformed individually to the principal, skewon, and axion parts of $\overline{\overline{\mathsf{M}}}_a$.

To find the transformation rule of a skewon medium bidyadic we apply the representation (3.86)

$$\overline{\overline{\mathsf{M}}} = (\overline{\overline{\mathsf{B}}}_o{}^{\wedge}_{\wedge}\overline{\overline{\mathsf{I}}})^T \tag{6.23}$$

in terms of a trace-free dyadic $\overline{\overline{\mathsf{B}}}_o \in \mathbb{E}_1\mathbb{F}_1$. Invoking the rule (2.41) in the form

$$\overline{\overline{\mathsf{A}}}^{(2)}|(\overline{\overline{\mathsf{B}}}_o{}^\wedge_\wedge\overline{\overline{\mathsf{I}}}) = (\overline{\overline{\mathsf{A}}}|\overline{\overline{\mathsf{B}}}_o)^\wedge_\wedge\overline{\overline{\mathsf{A}}}, \tag{6.24}$$

we can form the transformed medium bidyadic as

$$\begin{aligned}\overline{\overline{\mathsf{M}}}_a &= (\overline{\overline{\mathsf{A}}}^{(2)}|(\overline{\overline{\mathsf{B}}}_o{}^\wedge_\wedge\overline{\overline{\mathsf{I}}})|\overline{\overline{\mathsf{A}}}^{(-2)})^T \\ &= ((\overline{\overline{\mathsf{A}}}|\overline{\overline{\mathsf{B}}}_o|\overline{\overline{\mathsf{A}}}^{-1})^\wedge_\wedge(\overline{\overline{\mathsf{A}}}|\overline{\overline{\mathsf{A}}}^{-1}))^T \\ &= ((\overline{\overline{\mathsf{A}}}|\overline{\overline{\mathsf{B}}}_o|\overline{\overline{\mathsf{A}}}^{-1})^\wedge_\wedge\overline{\overline{\mathsf{I}}})^T,\end{aligned} \tag{6.25}$$

which has the skewon form,

$$\overline{\overline{\mathsf{M}}}_a = (\overline{\overline{\mathsf{B}}}_{oa}{}^\wedge_\wedge\overline{\overline{\mathsf{I}}})^T, \quad \overline{\overline{\mathsf{B}}}_{oa} = \overline{\overline{\mathsf{A}}}|\overline{\overline{\mathsf{B}}}_o|\overline{\overline{\mathsf{A}}}^{-1}. \tag{6.26}$$

6.1.3 Dispersion Equation

Considering a plane wave in the transformed medium, the dispersion dyadic (5.237) can be expanded as

$$\begin{aligned}\overline{\overline{\mathsf{D}}}_a(\boldsymbol{\nu}) &= \overline{\overline{\mathsf{M}}}_{ma}\lfloor\lfloor\boldsymbol{\nu}\boldsymbol{\nu} \\ &= -\boldsymbol{\nu}\rfloor(\overline{\overline{\mathsf{A}}}^{(2)}|\overline{\overline{\mathsf{M}}}_m|\overline{\overline{\mathsf{A}}}^{(2)T})\lfloor\boldsymbol{\nu} \\ &= -(\overline{\overline{\mathsf{A}}}\wedge(\boldsymbol{\nu}|\overline{\overline{\mathsf{A}}}))|\overline{\overline{\mathsf{M}}}_m|((\overline{\overline{\mathsf{A}}}^T|\boldsymbol{\nu})\wedge\overline{\overline{\mathsf{A}}}^T) \\ &= -\overline{\overline{\mathsf{A}}}|((\boldsymbol{\nu}|\overline{\overline{\mathsf{A}}})\rfloor\overline{\overline{\mathsf{M}}}_m\lfloor(\overline{\overline{\mathsf{A}}}^T|\boldsymbol{\nu}))|\overline{\overline{\mathsf{A}}}^T \\ &= \overline{\overline{\mathsf{A}}}|(\overline{\overline{\mathsf{M}}}_m\lfloor\lfloor\boldsymbol{\nu}_a\boldsymbol{\nu}_a)|\overline{\overline{\mathsf{A}}}^T \\ &= \overline{\overline{\mathsf{A}}}|\overline{\overline{\mathsf{D}}}(\boldsymbol{\nu}_a)|\overline{\overline{\mathsf{A}}}^T,\end{aligned} \tag{6.27}$$

where we denote

$$\boldsymbol{\nu}_a = \overline{\overline{\mathsf{A}}}^T|\boldsymbol{\nu}. \tag{6.28}$$

The dyadic dispersion equation (5.240) for the transformed medium can be expressed as

$$\overline{\overline{\mathsf{D}}}_a^{(3)}(\boldsymbol{\nu}) = \overline{\overline{\mathsf{A}}}^{(3)}|\overline{\overline{\mathsf{D}}}^{(3)}(\boldsymbol{\nu}_a)|\overline{\overline{\mathsf{A}}}^{(3)T} = 0, \tag{6.29}$$

which implies

$$\overline{\overline{\mathsf{D}}}^{(3)}(\boldsymbol{\nu}_a) = 0. \tag{6.30}$$

Thus, the scalar dispersion equation (5.245) for the transformed medium becomes

$$D(\boldsymbol{\nu}_a) = D(\overline{\overline{\mathsf{A}}}^T|\boldsymbol{\nu}) = 0, \tag{6.31}$$

where the dispersion function D involves the original medium bidyadic $\overline{\overline{\mathsf{M}}}$.

One may conclude that the affine transformation does not change the character of the dispersion equation. In particular, if the original dispersion equation can be reduced to two quadratic equations, so can the transformed equation. If the medium is nonbirefringent, so is the transformed medium. If there is no dispersion equation for the original medium, there is no dispersion equation for the transformed medium.

6.1.4 Simple Principal Medium

As an example, let us consider affine transformation of the simple principal medium defined by (5.115). Applying the spatial medium dyadics

$$\overline{\overline{\alpha}} = 0, \quad \overline{\overline{\epsilon}}' = \epsilon \varepsilon_{123} \lfloor \overline{\overline{\mathsf{G}}}_s, \quad \overline{\overline{\mu}} = \mu \varepsilon_{123} \lfloor \overline{\overline{\mathsf{G}}}_s, \quad \overline{\overline{\beta}} = 0, \tag{6.32}$$

the modified medium bidyadic has the form (5.122)

$$\overline{\overline{\mathsf{M}}}_m = -\frac{1}{\mu}(\overline{\overline{\mathsf{G}}}_s - \mu\epsilon \mathbf{e}_4\mathbf{e}_4)^{(2)}. \tag{6.33}$$

The affine-transformed bidyadic now becomes

$$\begin{aligned} \overline{\overline{\mathsf{M}}}_{ma} &= \overline{\overline{\mathsf{A}}}^{(2)}|\overline{\overline{\mathsf{M}}}_m|\overline{\overline{\mathsf{A}}}^{(2)T} \\ &= -\frac{1}{\mu}(\overline{\overline{\mathsf{A}}}|\overline{\overline{\mathsf{G}}}_s|\overline{\overline{\mathsf{A}}}^T - \mu\epsilon(\overline{\overline{\mathsf{A}}}|\mathbf{e}_4)(\overline{\overline{\mathsf{A}}}|\mathbf{e}_4))^{(2)} \\ &= -\frac{1}{\mu}(\overline{\overline{\mathsf{G}}}_{sa} - \mu\epsilon \mathbf{e}_{4a}\mathbf{e}_{4a})^{(2)}, \end{aligned} \tag{6.34}$$

with

$$\mathbf{e}_{ia} = \overline{\overline{\mathsf{A}}}|\mathbf{e}_i, \quad \overline{\overline{\mathsf{G}}}_{sa} = \sum_1^3 \mathbf{e}_{ia}\mathbf{e}_{ia}. \tag{6.35}$$

Thus, the transformed simple principal medium is another simple principal medium.

In terms of Gibbsian dyadics, the simple principal medium defined by $\overline{\overline{\epsilon}}_g = A\overline{\overline{\mu}}_g^T$ and $\overline{\overline{\xi}}_g = \overline{\overline{\zeta}}_g = 0$ has been called affine-isotropic medium because one can find an affine transformation in terms of which it can be transformed to an isotropic medium [25].

6.2 DUALITY TRANSFORMATION

Duality transformation is based on the symmetry of electric and magnetic quantities in the Maxwell equations. By changing symbols of fields and sources the Maxwell equations remain the same in form which leads to corresponding similarity in different electromagnetic problems.

6.2.1 Transformation of Fields

Let us consider transformation of fields $\mathbf{\Psi}, \mathbf{\Phi}$ to dual fields $\mathbf{\Psi}_d, \mathbf{\Phi}_d$ as defined by the linear relation [1]

$$\begin{pmatrix} Z_d\mathbf{\Psi}_d \\ \mathbf{\Phi}_d \end{pmatrix} = \begin{pmatrix} \cos\theta & \sin\theta \\ -\sin\theta & \cos\theta \end{pmatrix} \begin{pmatrix} Z\mathbf{\Psi} \\ \mathbf{\Phi} \end{pmatrix}. \tag{6.36}$$

More explicitly, the transformation is represented by

$$\mathbf{\Psi}_d = \frac{1}{Z_d}(\cos\theta\, Z\mathbf{\Psi} + \sin\theta\, \mathbf{\Phi}), \tag{6.37}$$

$$\mathbf{\Phi}_d = -\sin\theta\, Z\mathbf{\Psi} + \cos\theta\, \mathbf{\Phi}, \tag{6.38}$$

as defined by three scalar parameters Z, Z_d, and θ. The trivial case $Z_d = Z$, $\cos\theta = 1$ leads to the identity transformation, $\mathbf{\Psi}_d = \mathbf{\Psi}$, $\mathbf{\Phi}_d = \mathbf{\Phi}$, which will be omitted in the sequel. The inverse of the duality transformation is

$$\begin{pmatrix} Z\mathbf{\Psi} \\ \mathbf{\Phi} \end{pmatrix} = \begin{pmatrix} \cos\theta & -\sin\theta \\ \sin\theta & \cos\theta \end{pmatrix} \begin{pmatrix} Z_d\mathbf{\Psi}_d \\ \mathbf{\Phi}_d \end{pmatrix}. \tag{6.39}$$

The corresponding transformation of the electric and magnetic source three-forms follows a similar rule,

$$\begin{pmatrix} Z_d\gamma_{ed} \\ \gamma_{md} \end{pmatrix} = \begin{pmatrix} \cos\theta & \sin\theta \\ -\sin\theta & \cos\theta \end{pmatrix} \begin{pmatrix} Z\gamma_e \\ \gamma_m \end{pmatrix}. \tag{6.40}$$

Considering the antisymmetric bidyadic function of fields (5.188),

$$\overline{\overline{\mathsf{U}}} = \frac{1}{2}(\mathbf{\Phi}\mathbf{\Psi} - \mathbf{\Psi}\mathbf{\Phi}) \in \mathbb{F}_2\mathbb{F}_2, \tag{6.41}$$

its transformation can be written after substitution of (6.37) and (6.38) as

$$\begin{aligned} \overline{\overline{\mathsf{U}}}_d &= \frac{1}{2}(\mathbf{\Phi}_d\mathbf{\Psi}_d - \mathbf{\Psi}_d\mathbf{\Phi}_d) \\ &= \frac{Z}{2Z_d}(\mathbf{\Phi}\mathbf{\Psi} - \mathbf{\Psi}\mathbf{\Phi}) \\ &= \frac{Z}{Z_d}\overline{\overline{\mathsf{U}}}. \end{aligned} \tag{6.42}$$

Because $\overline{\overline{\mathsf{U}}}$ has a linear relation to the stress–energy dyadic $\overline{\overline{\mathsf{T}}}$ (5.205) as (5.206), its dual is similarly obtained as

$$\overline{\overline{\mathsf{T}}}_d = \frac{Z}{Z_d}\overline{\overline{\mathsf{T}}}. \tag{6.43}$$

The same is also valid for all components of the stress–energy dyadic. For example, the dual of the Poynting two-form $\mathbf{S} = \mathbf{E} \wedge \mathbf{H}$ is then

$$\mathbf{S}_d = \frac{Z}{Z_d}\mathbf{S}. \tag{6.44}$$

For the special choice of transformation parameters $Z_d = Z$, these quantities are actually invariant in the duality transformation.

The duality transformation rules (6.37), (6.38) can be decomposed in their spatial and temporal parts as

$$\begin{pmatrix} Z_d\mathbf{D}_d \\ \mathbf{B}_d \end{pmatrix} = \begin{pmatrix} \cos\theta & \sin\theta \\ -\sin\theta & \cos\theta \end{pmatrix} \begin{pmatrix} Z\mathbf{D} \\ \mathbf{B} \end{pmatrix}, \tag{6.45}$$

$$\begin{pmatrix} \mathbf{E} \\ Z_d\mathbf{H}_d \end{pmatrix} = \begin{pmatrix} \cos\theta & \sin\theta \\ -\sin\theta & \cos\theta \end{pmatrix} \begin{pmatrix} \mathbf{E} \\ Z\mathbf{H} \end{pmatrix}. \tag{6.46}$$

6.2.2 Involutionary Duality Transformation

Let us require that the transformation (6.36) be an involution, that is, that it for all $\boldsymbol{\Psi}, \boldsymbol{\Phi}$ the conditions

$$(\boldsymbol{\Psi}_d)_d = \frac{1}{Z_d}(\cos\theta\, Z\boldsymbol{\Psi}_d + \sin\theta\boldsymbol{\Phi}_d) = \boldsymbol{\Psi}, \tag{6.47}$$

$$(\boldsymbol{\Phi}_d)_d = -\sin\theta\, Z\boldsymbol{\Psi}_d + \cos\theta\, \boldsymbol{\Phi}_d = \boldsymbol{\Phi} \tag{6.48}$$

are valid. Substituting from (6.37) and (6.38) we obtain two conditions for the three transformation parameters,

$$\frac{Z + Z_d}{Z_d^2}[(Z_d \sin^2\theta + (Z - Z_d)\cos^2\theta)\boldsymbol{\Psi} + \sin\theta\cos\theta\boldsymbol{\Phi}] = 0, \tag{6.49}$$

$$\frac{Z + Z_d}{Z_d}(\sin\theta\cos\theta Z\boldsymbol{\Psi} + \sin^2\theta\boldsymbol{\Phi}) = 0. \tag{6.50}$$

Leaving θ as a free parameter, these conditions require $Z_d = -Z$, whence the two-parameter involutionary transformation is defined by

$$\begin{pmatrix} Z\mathbf{\Psi}_d \\ \mathbf{\Phi}_d \end{pmatrix} = -\begin{pmatrix} \cos\theta & \sin\theta \\ \sin\theta & -\cos\theta \end{pmatrix}\begin{pmatrix} Z\mathbf{\Psi} \\ \mathbf{\Phi} \end{pmatrix}. \tag{6.51}$$

Two simple special cases of (6.51) are obtained by choosing $\theta = \pi/2$ which reduces the transformation to

$$\begin{pmatrix} \mathbf{\Psi}_d \\ \mathbf{\Phi}_d \end{pmatrix} = -\begin{pmatrix} 0 & 1/Z \\ Z & 0 \end{pmatrix}\begin{pmatrix} \mathbf{\Psi} \\ \mathbf{\Phi} \end{pmatrix}, \tag{6.52}$$

or $\theta = -\pi/2$, which changes the sign in (6.52). One should note that (6.51) does not contain the identity transformation as a special case since $\theta = 0$ yields $\mathbf{\Psi}_d = -\mathbf{\Psi}$, $\mathbf{\Phi}_d = \mathbf{\Phi}$. For $\theta = \pi/2$ the spatial fields transform as

$$\mathbf{B}_d = -Z\mathbf{D}, \quad \mathbf{D}_d = -\mathbf{B}/Z, \quad \mathbf{E}_d = Z\mathbf{H}, \quad \mathbf{H}_d = \mathbf{E}/Z, \tag{6.53}$$

which show the essence of the electric–magnetic duality.

The stress–energy dyadic is transformed as $\overline{\overline{\mathsf{T}}}_d = -\overline{\overline{\mathsf{T}}}$ in the involutionary transformation. The Poynting two-form is also reversed in the transformation, $\mathbf{S}_d = -\mathbf{S}$, which corresponds to changing $\mathbf{E}_g \rightleftarrows \mathbf{H}_g$ in the Gibbsian Poynting vector $\mathbf{E}_g \times \mathbf{H}_g$.

Any field can be split in two parts with respect to a given involutionary duality transformation:

$$\begin{pmatrix} \mathbf{\Psi} \\ \mathbf{\Phi} \end{pmatrix} = \begin{pmatrix} \mathbf{\Psi} \\ \mathbf{\Phi} \end{pmatrix}_{sd} + \begin{pmatrix} \mathbf{\Psi} \\ \mathbf{\Phi} \end{pmatrix}_{ad}, \tag{6.54}$$

so that the field $\mathbf{\Psi}_{sd}, \mathbf{\Phi}_{sd}$ is invariant ("self-dual"), while the field $\mathbf{\Psi}_{ad}, \mathbf{\Phi}_{ad}$ changes sign ("anti-self-dual") in this duality transformation. The two partial fields are

$$\begin{pmatrix} \mathbf{\Psi} \\ \mathbf{\Phi} \end{pmatrix}_{sd} = \frac{1}{2}\begin{pmatrix} 1 & -1/Z \\ -Z & 1 \end{pmatrix}\begin{pmatrix} \mathbf{\Psi} \\ \mathbf{\Phi} \end{pmatrix}, \tag{6.55}$$

$$\begin{pmatrix} \mathbf{\Psi} \\ \mathbf{\Phi} \end{pmatrix}_{ad} = \frac{1}{2}\begin{pmatrix} 1 & 1/Z \\ Z & 1 \end{pmatrix}\begin{pmatrix} \mathbf{\Psi} \\ \mathbf{\Phi} \end{pmatrix}. \tag{6.56}$$

6.2.3 Transformation of Media

Inserting the dual fields from (6.36) in the transformed medium equation,

$$\mathbf{\Psi}_d = \overline{\overline{\mathsf{M}}}_d|\mathbf{\Phi}_d \tag{6.57}$$

yields the relation

$$\cos\theta\, Z\overline{\overline{\mathsf{M}}}|\mathbf{\Phi} + \sin\theta\, \mathbf{\Phi} = Z_d\overline{\overline{\mathsf{M}}}_d|(-\sin\theta\, Z\overline{\overline{\mathsf{M}}}|\mathbf{\Phi} + \cos\theta\, \mathbf{\Phi}), \tag{6.58}$$

which must be valid for any $\mathbf{\Phi}$. The transformation rule for the medium bidyadic can be extracted in the form

$$Z_d\overline{\overline{\mathsf{M}}}_d = (\sin\theta\, \overline{\overline{\mathsf{I}}}^{(2)T} + \cos\theta\, Z\overline{\overline{\mathsf{M}}})|(\cos\theta\, \overline{\overline{\mathsf{I}}}^{(2)T} - \sin\theta\, Z\overline{\overline{\mathsf{M}}})^{-1}, \tag{6.59}$$

or, more symmetrically, as

$$(\cos\theta \overline{\overline{\mathsf{I}}}^{(2)T} + \sin\theta\, Z_d\overline{\overline{\mathsf{M}}}_d)|(\cos\theta \overline{\overline{\mathsf{I}}}^{(2)T} - \sin\theta\, Z\overline{\overline{\mathsf{M}}}) = \overline{\overline{\mathsf{I}}}^{(2)T}. \tag{6.60}$$

From this one can show that, in the general case, the medium bidyadics $\overline{\overline{\mathsf{M}}}$ and $\overline{\overline{\mathsf{M}}}_d$ commute, $\overline{\overline{\mathsf{M}}}|\overline{\overline{\mathsf{M}}}_d = \overline{\overline{\mathsf{M}}}_d|\overline{\overline{\mathsf{M}}}$, whence they have the same set of eigen two-forms. Details are left as an exercise.

A medium is called self-dual if one can find a nontrivial duality transformation in which the medium bidyadic is invariant, $\overline{\overline{\mathsf{M}}}_d = \overline{\overline{\mathsf{M}}}$. In a self-dual medium electromagnetic fields and sources can be combined with transformed fields and sources because the medium for both sets of fields and sources is the same.

The transformation rule (6.59) can be expanded in terms of spatial medium dyadics as [1]

$$\overline{\overline{\epsilon}}_d Z_d + \overline{\overline{\mu}}_d/Z_d = \overline{\overline{\epsilon}} Z + \overline{\overline{\mu}}/Z, \tag{6.61}$$

$$\overline{\overline{\xi}}_d - \overline{\overline{\zeta}}_d = \overline{\overline{\xi}} - \overline{\overline{\zeta}}, \tag{6.62}$$

$$\overline{\overline{\epsilon}}_d Z_d - \overline{\overline{\mu}}_d/Z_d = (\overline{\overline{\epsilon}} Z - \overline{\overline{\mu}}/Z)\cos 2\theta + (\overline{\overline{\xi}} + \overline{\overline{\zeta}})\sin 2\theta, \tag{6.63}$$

$$\overline{\overline{\xi}}_d + \overline{\overline{\zeta}}_d = -(\overline{\overline{\epsilon}} Z - \overline{\overline{\mu}}/Z)\sin 2\theta + (\overline{\overline{\xi}} + \overline{\overline{\zeta}})\cos 2\theta. \tag{6.64}$$

It is worth noting that the dyadic $\overline{\overline{\xi}} - \overline{\overline{\zeta}}$ is invariant in any duality transformation.

For the special involutionary duality transformation defined by $\theta = \pi/2$ and $Z_d = -Z$, the medium bidyadic obeys the simple transformation rule

$$\overline{\overline{\mathsf{M}}}_d = (Z^2\overline{\overline{\mathsf{M}}})^{-1} = \frac{1}{Z^2}\overline{\overline{\mathsf{N}}}, \tag{6.65}$$

with $\overline{\overline{\mathsf{N}}}$ defined as in (5.52). This corresponds to

$$\overline{\overline{\epsilon}}_d = -\overline{\overline{\mu}}/Z^2, \qquad \overline{\overline{\mu}}_d = -\overline{\overline{\epsilon}}Z^2, \tag{6.66}$$

$$\overline{\overline{\xi}}_d = -\overline{\overline{\zeta}}, \qquad \overline{\overline{\zeta}}_d = -\overline{\overline{\xi}}. \tag{6.67}$$

The modified medium bidyadic is transformed as

$$\overline{\overline{\mathsf{M}}}_{md} = \frac{1}{Z^2}\mathbf{e}_N\lfloor(\boldsymbol{\varepsilon}_N\lfloor\overline{\overline{\mathsf{M}}}_m)^{-1} = \frac{1}{Z^2}\mathbf{e}_N\mathbf{e}_N\lfloor\lfloor\overline{\overline{\mathsf{M}}}_m^{-1} = \frac{1}{Z^2}\overline{\overline{\mathsf{N}}}_m, \tag{6.68}$$

with $\overline{\overline{\mathsf{N}}}_m = \mathbf{e}_N\lfloor\overline{\overline{\mathsf{N}}}$. Equations (6.65) and (6.68) show us that inverse medium bidyadics have a close relation to duality-transformed medium bidyadics.

The transformation rule (6.65) can actually be applied to finding the analytic expression of the inverse of a given medium bidyadic. In fact, inserting the spatial expansion (5.75) for the dual medium on the right side of (6.65) and substituting from (6.66), (6.67) yields

$$\overline{\overline{\mathsf{M}}}^{-1} = Z^2\overline{\overline{\mathsf{M}}}_d \tag{6.69}$$

$$= Z^2\overline{\overline{\epsilon}}_d \wedge \mathbf{e}_4 + Z^2(\boldsymbol{\varepsilon}_4 \wedge \overline{\overline{\mathsf{I}}}^T + \overline{\overline{\xi}}_d)\lfloor\overline{\overline{\mu}}_d^{-1}\rfloor(\overline{\overline{\mathsf{I}}}^{(2)T} - \overline{\overline{\zeta}}_d \wedge \mathbf{e}_4)$$

$$= -\overline{\overline{\mu}} \wedge \mathbf{e}_4 - (\boldsymbol{\varepsilon}_4 \wedge \overline{\overline{\mathsf{I}}}^T - \overline{\overline{\zeta}})\lfloor\overline{\overline{\epsilon}}^{-1}\rfloor(\overline{\overline{\mathsf{I}}}^{(2)T} + \overline{\overline{\xi}} \wedge \mathbf{e}_4), \tag{6.70}$$

which coincides with (5.77).

6.3 TRANSFORMATION OF BOUNDARY CONDITIONS

Let us consider the set of boundary conditions of Section 5.4.2 in the duality transformation.

- **PMC Boundary**
  Starting from the condition of the PMC boundary,

$$\boldsymbol{\varepsilon}_3 \wedge \boldsymbol{\Psi} = 0, \tag{6.71}$$

  the three-parameter inverse duality transformation (6.39) yields a condition of the form

$$\boldsymbol{\varepsilon}_3 \wedge (A\boldsymbol{\Psi}_d + B\boldsymbol{\Phi}_d) = 0, \tag{6.72}$$

  with

$$A = (Z_d/Z)\cos\theta, \quad B = -(1/Z)\sin\theta. \tag{6.73}$$

In general case, the PMC boundary is transformed to a PEMC boundary. In particular, for $\theta = \pm\pi/2$ the PMC boundary is transformed to the PEC boundary.

- **PEC Boundary**
  Applying similar equations, the PEC boundary satisfying

$$\varepsilon_3 \wedge \mathbf{\Phi} = 0 \tag{6.74}$$

  can also be transformed to a PEMC boundary.

- **PEMC Boundary**
  Inverting the previous transformations, the perfect electromagnetic boundary with fields satisfying

$$\varepsilon_3 \wedge (\mathbf{\Psi} - M\mathbf{\Phi}) = 0 \tag{6.75}$$

  can be transformed to the PMC or the PEC boundary. More generally, it can be transformed to another PEMC boundary satisfying a condition of the form (6.72) with

$$A = \frac{Z_d}{Z}(\cos\theta - MZ\sin\theta), \quad B = -\frac{1}{Z}(\sin\theta + MZ\cos\theta). \tag{6.76}$$

  Any PEMC boundary can be self-dual for a certain choice of transformation parameters. Details are left as an exercise.

- **DB Boundary**
  The DB boundary conditions expressed as

$$\varepsilon_3 \wedge \varepsilon_4 \wedge \begin{pmatrix} \mathbf{\Psi} \\ \mathbf{\Phi} \end{pmatrix} = \begin{pmatrix} 0 \\ 0 \end{pmatrix} \tag{6.77}$$

  are obviously self-dual in any duality transformations, as can be seen from (6.36).

- **SH Boundary**
  Similarly, the SH boundary conditions

$$\varepsilon_2 \wedge \varepsilon_3 \wedge \begin{pmatrix} \mathbf{\Psi} \\ \mathbf{\Phi} \end{pmatrix} = \begin{pmatrix} 0 \\ 0 \end{pmatrix} \tag{6.78}$$

  are also self-dual for any transformation parameters.

It has been shown that an object with a self-dual boundary and certain symmetry in its geometry has no back scattering for an incident plane wave [43]. Thus, such an object appears invisible for the radar.

6.3.1 Simple Principal Medium

As an example, let us consider the simple principal medium defined by the medium bidyadic

$$\overline{\overline{\mathsf{M}}} = \epsilon\overline{\overline{\mathsf{C}}}_s \wedge \mathbf{e}_4 + \boldsymbol{\varepsilon}_4 \wedge (\mu\overline{\overline{\mathsf{C}}}_s)^{-1}, \tag{6.79}$$

where the spatial dyadic $\overline{\overline{\mathsf{C}}}_s \in \mathbb{F}_2\mathbb{E}_1$ can be expressed as $\overline{\overline{\mathsf{C}}}_s = \boldsymbol{\varepsilon}_{123}\lfloor\overline{\overline{\mathsf{G}}}_s$ in terms of a spatial symmetric dyadic $\overline{\overline{\mathsf{G}}}_s$. Because (6.79) satisfies (5.129) as

$$\overline{\overline{\mathsf{M}}}^2 = -M^2\overline{\overline{\mathsf{I}}}^{(2)T}, \quad M^2 = \epsilon/\mu, \tag{6.80}$$

it also satisfies

$$(\cos\theta\overline{\overline{\mathsf{I}}}^{(2)T} + \sin\theta\, Z\overline{\overline{\mathsf{M}}})|(\cos\theta\overline{\overline{\mathsf{I}}}^{(2)T} - \sin\theta\, Z\overline{\overline{\mathsf{M}}})$$
$$= (\cos^2\theta + \sin^2\theta M^2Z^2)\overline{\overline{\mathsf{I}}}^{(2)T}, \tag{6.81}$$

as can be verified after expanding the left side. Now one can show that the simple principal medium is actually self-dual in terms of a suitable transformation. In fact, choosing

$$Z = Z_d = 1/M = \sqrt{\mu/\epsilon}, \tag{6.82}$$

and comparing (6.81) to (6.60), we can identify from the two equations the condition

$$\overline{\overline{\mathsf{M}}}_d = \overline{\overline{\mathsf{M}}}. \tag{6.83}$$

6.3.2 Plane Wave

A plane wave in a medium defined by a bidyadic $\overline{\overline{\mathsf{M}}}$ is governed by a wave one-form $\boldsymbol{\nu}$ satisfying the dispersion equation (5.245). Making the duality transformation changes the medium and the fields but, unlike for the affine transformation, it does not change the wave one-form $\boldsymbol{\nu}$. In fact, assuming that $\boldsymbol{\nu}$ satisfies both $\boldsymbol{\nu} \wedge \boldsymbol{\Psi} = 0$ and $\boldsymbol{\nu} \wedge \boldsymbol{\Phi} = 0$, from (6.39) we have

$$\boldsymbol{\nu} \wedge \begin{pmatrix} Z\boldsymbol{\Psi} \\ \boldsymbol{\Phi} \end{pmatrix} = \begin{pmatrix} \cos\theta & -\sin\theta \\ \sin\theta & \cos\theta \end{pmatrix} \boldsymbol{\nu} \wedge \begin{pmatrix} Z_d\boldsymbol{\Psi}_d \\ \boldsymbol{\Phi}_d \end{pmatrix} = \begin{pmatrix} 0 \\ 0 \end{pmatrix}, \tag{6.84}$$

whence the same $\boldsymbol{\nu}$ also satisfies $\boldsymbol{\nu} \wedge \boldsymbol{\Psi}_d = 0$ and $\boldsymbol{\nu} \wedge \boldsymbol{\Phi}_d = 0$. Thus, we can express the transformed field two-forms in terms of potential one-forms as

$$\begin{pmatrix} \boldsymbol{\Psi}_d \\ \boldsymbol{\Phi}_d \end{pmatrix} = \nu \wedge \begin{pmatrix} \psi_d \\ \phi_d \end{pmatrix}, \tag{6.85}$$

which satisfy the transformation rules

$$\begin{pmatrix} Z_d \psi_d \\ \phi_d \end{pmatrix} = \begin{pmatrix} \cos\theta & \sin\theta \\ -\sin\theta & \cos\theta \end{pmatrix} \begin{pmatrix} Z\psi \\ \phi \end{pmatrix}. \tag{6.86}$$

In particular, considering the special involutory duality transformation defined by the parameters $\theta = \pi/2$ and $Z_d = -Z$ implying the relation (6.68) between the modified bidyadics $\overline{\overline{\mathsf{M}}}_{md}$ and $\overline{\overline{\mathsf{N}}}_m$, indicates that the dispersion equation (5.245) must have the same solutions as the dispersion equation (5.246).

6.4 RECIPROCITY TRANSFORMATION

Reciprocity (also called as Lorentz reciprocity [44]) in electromagnetics is related to the question whether it is possible to swap the positions of a transmitter and a receiver without a change in the received signal [45, 46]. This is only possible if the medium satisfies a certain property which is called the reciprocity condition.

6.4.1 Medium Transformation

One can define a reciprocal medium by requiring that the medium bidyadic be invariant in a transformation called the reciprocity transformation, which can be defined as

$$\overline{\overline{\mathsf{M}}} \rightarrow \overline{\overline{\mathsf{M}}}_r = -\overline{\overline{\mathsf{I}}}^{(4)T} \lfloor\lfloor \overline{\overline{\mathsf{M}}}^{*T}. \tag{6.87}$$

Here $()^*$ denotes a special affine transformation called time reversion. It is defined for vectors and one-forms by the unipotent dyadic $\overline{\overline{\mathsf{T}}} \in \mathbb{E}_1\mathbb{F}_1$

$$\overline{\overline{\mathsf{T}}} = \overline{\overline{\mathsf{I}}}_s - \mathbf{e}_4\boldsymbol{\varepsilon}_4, \tag{6.88}$$

$$\overline{\overline{\mathsf{T}}}^2 = \overline{\overline{\mathsf{I}}}, \tag{6.89}$$

as

$$\mathbf{a}^* = \overline{\overline{\mathsf{T}}}|\mathbf{a}, \quad \boldsymbol{\alpha}^* = \boldsymbol{\alpha}|\overline{\overline{\mathsf{T}}}. \tag{6.90}$$

For bivectors and two-forms the time reversion bidyadic becomes

$$\overline{\overline{\mathsf{T}}}^{(2)} = \overline{\overline{\mathsf{I}}}^{(2)}_s - \overline{\mathsf{I}}_s {}^{\wedge}_{\wedge} \mathbf{e}_4 \boldsymbol{\varepsilon}_4$$
$$= \mathbf{e}_{12}\boldsymbol{\varepsilon}_{12} + \mathbf{e}_{23}\boldsymbol{\varepsilon}_{23} + \mathbf{e}_{31}\boldsymbol{\varepsilon}_{31}$$
$$- \mathbf{e}_{14}\boldsymbol{\varepsilon}_{14} - \mathbf{e}_{24}\boldsymbol{\varepsilon}_{24} - \mathbf{e}_{34}\boldsymbol{\varepsilon}_{34}, \tag{6.91}$$
$$(\overline{\overline{\mathsf{T}}}^{(2)})^2 = \overline{\overline{\mathsf{I}}}^{(2)}, \tag{6.92}$$

and it operates as

$$\mathbf{A}^* = \overline{\overline{\mathsf{T}}}^{(2)}|\mathbf{A}, \quad \boldsymbol{\Phi}^* = \boldsymbol{\Phi}|\overline{\overline{\mathsf{T}}}^{(2)}. \tag{6.93}$$

For medium bidyadics, the time reversion operation is defined by

$$\overline{\overline{\mathsf{M}}}^* = \overline{\overline{\mathsf{T}}}^{(2)T}|\overline{\overline{\mathsf{M}}}|\overline{\overline{\mathsf{T}}}^{(2)T}, \quad \overline{\overline{\mathsf{M}}}^*_m = \overline{\overline{\mathsf{T}}}^{(2)}|\overline{\overline{\mathsf{M}}}_m|\overline{\overline{\mathsf{T}}}^{(2)T}. \tag{6.94}$$

Inserting the 3D expansion of the medium bidyadic we obtain

$$\overline{\overline{\mathsf{M}}}^* = (\overline{\overline{\alpha}} + \overline{\overline{\epsilon}}' \wedge \mathbf{e}_4 + \boldsymbol{\varepsilon}_4 \wedge \overline{\overline{\mu}}^{-1} + \boldsymbol{\varepsilon}_4 \wedge \overline{\overline{\beta}} \wedge \mathbf{e}_4)^*$$
$$= \overline{\overline{\alpha}} - \overline{\overline{\epsilon}}' \wedge \mathbf{e}_4 - \boldsymbol{\varepsilon}_4 \wedge \overline{\overline{\mu}}^{-1} + \boldsymbol{\varepsilon}_4 \wedge \overline{\overline{\beta}} \wedge \mathbf{e}_4. \tag{6.95}$$

Thus, the effect of time reversion to a medium is changing the sign of the dyadics $\overline{\overline{\epsilon}}'$ and $\overline{\overline{\mu}}^{-1}$.

The reciprocity-transformed medium bidyadic (6.87) can be expanded as

$$\overline{\overline{\mathsf{M}}}_r = \overline{\overline{\alpha}}_r + \overline{\overline{\epsilon}}'_r \wedge \mathbf{e}_4 + \boldsymbol{\varepsilon}_4 \wedge \overline{\overline{\mu}}_r^{-1} + \boldsymbol{\varepsilon}_4 \wedge \overline{\overline{\beta}}_r \wedge \mathbf{e}_4$$
$$= -\boldsymbol{\varepsilon}_N \lfloor \overline{\overline{\mathsf{M}}}^{*T} \rfloor \mathbf{e}_N$$
$$= -\boldsymbol{\varepsilon}_N \lfloor (\overline{\overline{\alpha}}^T + \mathbf{e}_4 \wedge \overline{\overline{\epsilon}}'^T + \overline{\overline{\mu}}^{-1T} \wedge \boldsymbol{\varepsilon}_4 + \mathbf{e}_4 \wedge \overline{\overline{\beta}}^T \wedge \boldsymbol{\varepsilon}_4) \rfloor \mathbf{e}_N$$
$$= \boldsymbol{\varepsilon}_4 \wedge (\boldsymbol{\varepsilon}_{123}\mathbf{e}_{123} \lfloor\lfloor \overline{\overline{\alpha}}^T) \wedge \mathbf{e}_4 + (\boldsymbol{\varepsilon}_{123}\mathbf{e}_{123} \lfloor\lfloor \overline{\overline{\epsilon}}'^T) \wedge \mathbf{e}_4$$
$$+ \boldsymbol{\varepsilon}_4 \wedge (\boldsymbol{\varepsilon}_{123}\mathbf{e}_{123} \lfloor\lfloor \overline{\overline{\mu}}^{-1T}) + \boldsymbol{\varepsilon}_{123}\mathbf{e}_{123} \lfloor\lfloor \overline{\overline{\beta}}^T, \tag{6.96}$$

where we have applied the rules

$$\mathbf{e}_N \lfloor \boldsymbol{\alpha}_s = \mathbf{e}_4 \wedge (\mathbf{e}_{123} \lfloor \boldsymbol{\alpha}_s), \quad \boldsymbol{\alpha}_s \rfloor \mathbf{e}_N = -(\boldsymbol{\alpha}_s \rfloor \mathbf{e}_{123}) \wedge \mathbf{e}_4, \tag{6.97}$$
$$\mathbf{e}_N \lfloor \boldsymbol{\Phi}_s = -\mathbf{e}_4 \wedge (\mathbf{e}_{123} \lfloor \boldsymbol{\Phi}_s), \quad \boldsymbol{\Phi}_s \rfloor \mathbf{e}_N = (\boldsymbol{\Phi}_s \rfloor \mathbf{e}_{123}) \wedge \mathbf{e}_4, \tag{6.98}$$

valid for spatial one-forms $\boldsymbol{\alpha}_s$ and two-forms $\boldsymbol{\Phi}_s$. In terms of these rules, reciprocity transformations of the different spatial medium dyadics can be compactly expressed as

$$\begin{pmatrix} \overline{\overline{\alpha}}_r & \overline{\overline{\epsilon}}'_r \\ \overline{\overline{\mu}}_r^{-1} & \overline{\overline{\beta}}_r \end{pmatrix} = \boldsymbol{\varepsilon}_{123}\mathbf{e}_{123} \lfloor\lfloor \begin{pmatrix} \overline{\overline{\beta}}^T & \overline{\overline{\epsilon}}'^T \\ \overline{\overline{\mu}}^{-1T} & \overline{\overline{\alpha}}^T \end{pmatrix}. \tag{6.99}$$

From

$$\mathrm{tr}\overline{\overline{\alpha}}_r = \mathrm{tr}(\boldsymbol{\varepsilon}_{123}\mathbf{e}_{123}\lfloor\lfloor\overline{\overline{\beta}}^T) = \mathrm{tr}\overline{\overline{\beta}}, \qquad (6.100)$$

$$\mathrm{tr}\overline{\overline{\beta}}_r = \mathrm{tr}(\boldsymbol{\varepsilon}_{123}\mathbf{e}_{123}\lfloor\lfloor\overline{\overline{\alpha}}^T) = \mathrm{tr}\overline{\overline{\alpha}}, \qquad (6.101)$$

we obtain the relation

$$\mathrm{tr}\overline{\overline{\mathsf{M}}}_r = \mathrm{tr}\overline{\overline{\alpha}}_r - \mathrm{tr}\overline{\overline{\beta}}_r = \mathrm{tr}\overline{\overline{\beta}} - \mathrm{tr}\overline{\overline{\alpha}} = -\mathrm{tr}\overline{\overline{\mathsf{M}}}. \qquad (6.102)$$

This means that the axion part of the medium bidyadic changes sign in the reciprocity transformation.

For the modified medium bidyadic the reciprocity transformation is obtained from (6.87) as

$$\overline{\overline{\mathsf{M}}}_{mr} = \overline{\overline{\mathsf{M}}}_{rm} = \mathbf{e}_N\lfloor\overline{\overline{\mathsf{M}}}_r = -\overline{\overline{\mathsf{M}}}^{*T}\rfloor\mathbf{e}_N = -(\mathbf{e}_N^*\lfloor\overline{\overline{\mathsf{M}}})^{*T} = \overline{\overline{\mathsf{M}}}_m^{*T}, \qquad (6.103)$$

with $\mathbf{e}_N^* = -\mathbf{e}_N$. For the Gibbsian medium parameters the transformation becomes

$$\begin{pmatrix} \overline{\overline{\epsilon}}_g & \overline{\overline{\xi}}_g \\ \overline{\overline{\zeta}}_g & \overline{\overline{\mu}}_g \end{pmatrix}_r = \begin{pmatrix} \overline{\overline{\epsilon}}_g^T & -\overline{\overline{\zeta}}_g^T \\ -\overline{\overline{\xi}}_g^T & \overline{\overline{\mu}}_g^T \end{pmatrix}. \qquad (6.104)$$

6.4.2 Reciprocity Conditions

A reciprocal medium is defined by requiring invariance of the medium bidyadic in the the reciprocity transformation,

$$\overline{\overline{\mathsf{M}}} = \overline{\overline{\mathsf{M}}}_r = -\overline{\overline{\mathsf{I}}}^{(4)T}\lfloor\lfloor\overline{\overline{\mathsf{M}}}^{T*} = -(\overline{\overline{\mathsf{I}}}^{(4)}\lfloor\lfloor\overline{\overline{\mathsf{M}}})^{T*}, \qquad (6.105)$$

or,

$$\overline{\overline{\mathsf{M}}}_m = \overline{\overline{\mathsf{M}}}_{mr} = \overline{\overline{\mathsf{M}}}_m^{*T}. \qquad (6.106)$$

A medium which fails to satisfy these conditions is called nonreciprocal. Media satisfying

$$\overline{\overline{\mathsf{M}}} = -\overline{\overline{\mathsf{M}}}_r = (\overline{\overline{\mathsf{I}}}^{(4)}\lfloor\lfloor\overline{\overline{\mathsf{M}}})^{T*}, \qquad (6.107)$$

or

$$\overline{\overline{\mathsf{M}}}_m = -\overline{\overline{\mathsf{M}}}_{mr} = -\overline{\overline{\mathsf{M}}}_m^{*T}, \qquad (6.108)$$

may be called anti-reciprocal. Because the reciprocity transformation is involutory,

$$(\overline{\overline{\mathsf{M}}}_r)_r = -(\overline{\overline{\mathsf{I}}}^{(4)}\lfloor\lfloor(-\overline{\overline{\mathsf{I}}}^{(4)}\lfloor\lfloor\overline{\overline{\mathsf{M}}})^{T*})^{T*} = \overline{\overline{\mathsf{I}}}^{(4)T}\lfloor\lfloor(\overline{\overline{\mathsf{I}}}^{(4)}\lfloor\lfloor\overline{\overline{\mathsf{M}}}) = \overline{\overline{\mathsf{M}}}, \qquad (6.109)$$

$$(\overline{\overline{\mathsf{M}}}_{mr})_r = (\overline{\overline{\mathsf{M}}}_m^{*T})_r = (\overline{\overline{\mathsf{M}}}_m^{*T})^{*T} = \overline{\overline{\mathsf{M}}}_m, \qquad (6.110)$$

any medium bidyadic can be decomposed in a sum of reciprocal and anti-reciprocal parts as

$$\overline{\overline{\mathsf{M}}} = \frac{1}{2}(\overline{\overline{\mathsf{M}}} - (\overline{\overline{\mathsf{I}}}^{(4)} \lfloor\lfloor \overline{\overline{\mathsf{M}}})^{*T}) + \frac{1}{2}(\overline{\overline{\mathsf{M}}}_m + (\overline{\overline{\mathsf{I}}}^{(4)T} \lfloor\lfloor \overline{\overline{\mathsf{M}}})^{*T}), \quad (6.111)$$

$$\overline{\overline{\mathsf{M}}}_m = \frac{1}{2}(\overline{\overline{\mathsf{M}}}_m + \overline{\overline{\mathsf{M}}}_m^{*T}) + \frac{1}{2}(\overline{\overline{\mathsf{M}}}_m - \overline{\overline{\mathsf{M}}}_m^{*T}). \quad (6.112)$$

Because of (6.102), any reciprocal medium satisfies $\mathrm{tr}\overline{\overline{\mathsf{M}}} = 0$. Thus, it does not contain an axion part and the axion medium is manifestly anti-reciprocal. The skewon part $\overline{\overline{\mathsf{M}}}_2$ in the Hehl–Obukhov decomposition (5.91), satisfying $\overline{\overline{\mathsf{I}}}^{(4)} \lfloor\lfloor \overline{\overline{\mathsf{M}}}_2 = -\overline{\overline{\mathsf{M}}}_2^T$, has the respective reciprocal and anti-reciprocal parts

$$\overline{\overline{\mathsf{M}}}_2 = \frac{1}{2}(\overline{\overline{\mathsf{M}}}_2 + \overline{\overline{\mathsf{M}}}_2^{*}) + \frac{1}{2}(\overline{\overline{\mathsf{M}}}_2 - \overline{\overline{\mathsf{M}}}_2^{*}), \quad (6.113)$$

while the principal part $\overline{\overline{\mathsf{M}}}_1$ has the corresponding decomposition

$$\overline{\overline{\mathsf{M}}}_1 = \frac{1}{2}(\overline{\overline{\mathsf{M}}}_1 - \overline{\overline{\mathsf{M}}}_1^{*}) + \frac{1}{2}(\overline{\overline{\mathsf{M}}}_1 + \overline{\overline{\mathsf{M}}}_1^{*}). \quad (6.114)$$

Thus, if the skewon part is reciprocal, it must satisfy $\overline{\overline{\mathsf{M}}}_2 = \overline{\overline{\mathsf{M}}}_2^{*}$, whence the dyadics $\overline{\overline{\epsilon}}_2'$ and $\overline{\overline{\mu}}_2^{-1}$ must vanish. Correspondingly, if the principal part is reciprocal, it must satisfy $\overline{\overline{\mathsf{M}}}_1 = -\overline{\overline{\mathsf{M}}}_1^{*}$, whence the dyadics $\overline{\overline{\alpha}}_1$ and $\overline{\overline{\beta}}_1$ must vanish.

The reciprocity conditions for the spatial medium dyadics are obtained from (6.99) as

$$\overline{\overline{\alpha}} = \varepsilon_{123}\mathbf{e}_{123} \lfloor\lfloor \overline{\overline{\beta}}^T, \quad (6.115)$$

$$\overline{\overline{\epsilon}}' = \varepsilon_{123}\mathbf{e}_{123} \lfloor\lfloor \overline{\overline{\epsilon}}'^T, \quad (6.116)$$

$$\overline{\overline{\mu}}^{-1} = \varepsilon_{123}\mathbf{e}_{123} \lfloor\lfloor \overline{\overline{\mu}}^{-1T}, \quad (6.117)$$

$$\overline{\overline{\beta}} = \varepsilon_{123}\mathbf{e}_{123} \lfloor\lfloor \overline{\overline{\alpha}}^T, \quad (6.118)$$

or

$$\mathbf{e}_{123} \lfloor \overline{\overline{\alpha}} = (\mathbf{e}_{123} \lfloor \overline{\overline{\beta}})^T, \quad \mathbf{e}_{123} \lfloor \overline{\overline{\epsilon}}' = (\mathbf{e}_{123} \lfloor \overline{\overline{\epsilon}}')^T, \quad \mathbf{e}_{123} \lfloor \overline{\overline{\mu}}^{-1} = (\mathbf{e}_{123} \lfloor \overline{\overline{\mu}}^{-1})^T. \quad (6.119)$$

For the Gibbsian medium dyadics, the reciprocity conditions can be expressed as [24]

$$\overline{\overline{\epsilon}}_g = \overline{\overline{\epsilon}}_g^T, \quad \overline{\overline{\xi}}_g = -\overline{\overline{\zeta}}_g^T, \quad \overline{\overline{\zeta}}_g = -\overline{\overline{\xi}}_g^T, \quad \overline{\overline{\mu}}_g = \overline{\overline{\mu}}_g^T. \quad (6.120)$$

Thus, $\overline{\overline{\epsilon}}_g$ and $\overline{\overline{\mu}}_g$ must be symmetric dyadics for a reciprocal medium.

As an example, the modified bidyadic of the simple principal medium (5.122) is transformed as

$$\overline{\overline{\mathsf{M}}}_{mr} = -\frac{1}{\mu}(\overline{\overline{\mathsf{G}}}_s - \mu\epsilon \mathbf{e}_4\mathbf{e}_4)^{*T}$$
$$= -\frac{1}{\mu}(\overline{\overline{\mathsf{G}}}_s - \mu\epsilon \mathbf{e}_4\mathbf{e}_4), \tag{6.121}$$

which means that the simple principal medium is invariant in the reciprocity transformation, that is, it is reciprocal. This can also be seen from the Gibbsian representation (6.120) because $\overline{\overline{\epsilon}}_g$ and $\overline{\overline{\mu}}_g$ are symmetric dyadics and $\overline{\overline{\xi}}_g = \overline{\overline{\zeta}}_g = 0$. It is left as an exercise to prove that axion, skewon, and principal parts of a medium bidyadic are transformed individually to axion, skewon, and principal parts of the transformed medium bidyadic.

6.4.3 Field Relations

The effect of medium reciprocity to electromagnetic fields can be elucidated by considering two media which are reciprocal to one another, that is, whose medium bidyadics $\overline{\overline{\mathsf{M}}}_1$ and $\overline{\overline{\mathsf{M}}}_2$ are related through the reciprocity transformation. Assuming that the modified bidyadics satisfy

$$\overline{\overline{\mathsf{M}}}_{1m} = \overline{\overline{\mathsf{M}}}_{2mr} = \overline{\overline{\mathsf{M}}}_{2m}^{T*}, \tag{6.122}$$

the fields 1 and 2 are related by

$$\mathbf{e}_N|(\mathbf{\Phi}_2^* \wedge \mathbf{\Psi}_1) = \mathbf{\Phi}_2^*|(\mathbf{e}_N\lfloor\mathbf{\Psi}_1) = \mathbf{\Phi}_2^*|(\overline{\overline{\mathsf{M}}}_{1m}|\mathbf{\Psi}_1)$$
$$= \mathbf{\Phi}_2^*|(\overline{\overline{\mathsf{M}}}_{2m}^{T*}|\mathbf{\Phi}_1) = (\overline{\overline{\mathsf{M}}}_{2m}|\mathbf{\Phi}_2)^*|\mathbf{\Phi}_1$$
$$= (\mathbf{e}_N\lfloor\mathbf{\Psi}_2)^*|\mathbf{\Phi}_1 = \mathbf{e}_N^*|(\mathbf{\Psi}_2^* \wedge \mathbf{\Phi}_1)$$
$$= -\mathbf{e}_N|(\mathbf{\Psi}_2^* \wedge \mathbf{\Phi}_1), \tag{6.123}$$

which equals the condition

$$\mathbf{\Phi}_2^* \wedge \mathbf{\Psi}_1 + \mathbf{\Phi}_1 \wedge \mathbf{\Psi}_2^* = 0. \tag{6.124}$$

In terms of spatial fields this becomes

$$-\mathbf{B}_2 \wedge \mathbf{H}_1 \wedge \varepsilon_4 - \mathbf{E}_2 \wedge \mathbf{D}_1 \wedge \varepsilon_4 + \mathbf{B}_1 \wedge \mathbf{H}_2 \wedge \varepsilon_4 + \mathbf{E}_1 \wedge \mathbf{D}_2 \wedge \varepsilon_4 = 0, \tag{6.125}$$

which equals the condition

$$\mathbf{B}_2 \wedge \mathbf{H}_1 + \mathbf{E}_2 \wedge \mathbf{D}_1 = \mathbf{B}_1 \wedge \mathbf{H}_2 + \mathbf{E}_1 \wedge \mathbf{D}_2. \tag{6.126}$$

Working backwards, we can state that if (6.126) is satisfied for any fields 1 in medium 1 and any fields 2 in medium 2, the two medium bidyadics must be related by (6.122). If the two fields are in the same medium, this requires that the medium bidyadic satisfy the reciprocity condition (6.105) or (6.106).

6.4.4 Time-Harmonic Fields

Let us finally interpret (6.126) in terms of time-harmonic sources and fields with time dependence $\exp(j\omega t) = \exp(jk\tau)$. Assuming electric and magnetic sources γ_{e1}, γ_{m1} in medium 1, and a second set of sources γ_{e2}, γ_{m2} in medium 2, which produce the respective fields $\mathbf{\Phi}_1, \mathbf{\Psi}_1$ and $\mathbf{\Phi}_2, \mathbf{\Phi}_2$, the condition (6.126) can be expanded as

$$\begin{aligned}
0 &= jk(\mathbf{B}_2 \wedge \mathbf{H}_1 + \mathbf{E}_2 \wedge \mathbf{D}_1 - \mathbf{B}_1 \wedge \mathbf{H}_2 - \mathbf{E}_1 \wedge \mathbf{D}_2)\\
&= (\partial_\tau \mathbf{B}_2) \wedge \mathbf{H}_1 + \mathbf{E}_2 \wedge (\partial_\tau \mathbf{D}_1) - (\partial_\tau \mathbf{B}_1) \wedge \mathbf{H}_2 - \mathbf{E}_1 \wedge (\partial_\tau \mathbf{D}_2)\\
&= -(\mathbf{d} \wedge \mathbf{E}_2 + \mathbf{J}_{m2}) \wedge \mathbf{H}_1 + \mathbf{E}_2 \wedge (\mathbf{d} \wedge \mathbf{H}_1 - \mathbf{J}_{e1})\\
&\quad + (\mathbf{d} \wedge \mathbf{E}_1 + \mathbf{J}_{m1}) \wedge \mathbf{H}_2 - \mathbf{E}_1 \wedge (\mathbf{d} \wedge \mathbf{H}_2 - \mathbf{J}_{e2})\\
&= -(\mathbf{d} \wedge \mathbf{E}_2) \wedge \mathbf{H}_1 + (\mathbf{d} \wedge \mathbf{H}_1) \wedge \mathbf{E}_2 + (\mathbf{d} \wedge \mathbf{E}_1) \wedge \mathbf{H}_2 - (\mathbf{d} \wedge \mathbf{H}_2) \wedge \mathbf{E}_1\\
&\quad - \mathbf{J}_{m2} \wedge \mathbf{H}_1 - \mathbf{E}_2 \wedge \mathbf{J}_{e1} + \mathbf{J}_{m1} \wedge \mathbf{H}_2 + \mathbf{E}_1 \wedge \mathbf{J}_{e2}.
\end{aligned} \tag{6.127}$$

The resulting relation between the sources and fields can be rewritten in the form

$$\begin{aligned}
&(\mathbf{J}_{e1} \wedge \mathbf{E}_2 - \mathbf{J}_{m1} \wedge \mathbf{H}_2) - (\mathbf{J}_{e2} \wedge \mathbf{E}_1 - \mathbf{J}_{m2} \wedge \mathbf{H}_1)\\
&\quad = \mathbf{d} \wedge (\mathbf{E}_1 \wedge \mathbf{H}_2 - \mathbf{E}_2 \wedge \mathbf{H}_1).
\end{aligned} \tag{6.128}$$

Integrating over the whole 3D space and assuming vanishing of the term on the right side expressed as a surface integral over a surface in infinity (the sources are assumed to be concentrated in a finite region), the left side yields

$$\int (\mathbf{J}_{e1} \wedge \mathbf{E}_2 - \mathbf{J}_{m1} \wedge \mathbf{H}_2) dV = \int (\mathbf{J}_{e2} \wedge \mathbf{E}_1 - \mathbf{J}_{m2} \wedge \mathbf{H}_1) dV. \tag{6.129}$$

In classical terminology, this can be interpreted as the reaction of field 2 to source 1 being equal to the reaction of field 1 to source 2, which forms the essence of reciprocity in electromagnetics [46].

6.5 CONFORMAL TRANSFORMATION

After Lorentz and Einstein demonstrated that the Maxwell equations remain form-invariant in the Lorentz transformation, another transformation defined by Bateman and Cunningham in 1909 [47, 48], was also shown to leave the Maxwell equations invariant. Labeled as the conformal transformation it was based on the inversion transformation, introduced in 1845 to 3D electrostatics by William Thomson (Kelvin).

The inversion transformation for a four-vector $\mathbf{x}$ is defined by

$$\mathbf{x} \rightarrow \mathbf{I}(\mathbf{x}) = \frac{\mathbf{x}}{\mathbf{x} \cdot \mathbf{x}}, \tag{6.130}$$

where the dot product $\mathbf{x} \cdot \mathbf{x} = \mathbf{x}|\overline{\overline{\Gamma}}|\mathbf{x}$ is associated with the Minkowskian metric dyadic

$$\overline{\overline{\Gamma}} = \varepsilon_1\varepsilon_1 + \varepsilon_2\varepsilon_2 + \varepsilon_3\varepsilon_3 - \varepsilon_4\varepsilon_4 = \overline{\overline{\mathsf{G}}}^{-1}. \tag{6.131}$$

The conformal transformation $\mathbf{x} \rightarrow \mathbf{x}'$ is defined in terms of the inversion transformation by

$$\mathbf{x}' = \mathbf{C}(\mathbf{x}, \mathbf{a}) = \mathbf{I}(\mathbf{a} + \mathbf{I}(\mathbf{x})) \tag{6.132}$$

$$= \frac{\mathbf{a} + \frac{\mathbf{x}}{\mathbf{x}\cdot\mathbf{x}}}{\left(\mathbf{a} + \frac{\mathbf{x}}{\mathbf{x}\cdot\mathbf{x}}\right) \cdot \left(\mathbf{a} + \frac{\mathbf{x}}{\mathbf{x}\cdot\mathbf{x}}\right)} = \frac{\mathbf{x} + \mathbf{a}(\mathbf{x} \cdot \mathbf{x})}{\sigma(\mathbf{x}, \mathbf{a})}, \tag{6.133}$$

with

$$\sigma(\mathbf{x}, \mathbf{a}) = 1 + 2\mathbf{a} \cdot \mathbf{x} + (\mathbf{a} \cdot \mathbf{a})(\mathbf{x} \cdot \mathbf{x}). \tag{6.134}$$

The transformation has a vector parameter $\mathbf{a}$ ("the conformal acceleration") and depends on the metric dyadic $\overline{\overline{\Gamma}} = \overline{\overline{\mathsf{G}}}^{-1}$ defining the dot product. Cunningham's paper from 1909 [48] contained a demonstration that the Maxwell equations retain their form under the conformal transformation. Thus, any solution to a 4D field problem could be transformed to produce a solution to the transformed problem.

If we express the $\mathbf{a}$ and $\mathbf{x}$ vectors in their respective spatial and temporal components as

$$\mathbf{a} = \mathbf{a}_s + a\mathbf{e}_4, \quad \mathbf{x} = \mathbf{r} + \tau\mathbf{e}_4, \quad \mathbf{e}_4 \cdot \mathbf{e}_4 = -1, \tag{6.135}$$

the transformation can be expressed as

$$\mathbf{r}' = \frac{1}{\sigma(\mathbf{x},\mathbf{a})}(\mathbf{r} + \mathbf{a}_s(r^2 - \tau^2)), \quad (6.136)$$

$$\tau' = \frac{1}{\sigma(\mathbf{x},\mathbf{a})}(\tau + a(r^2 - \tau^2)), \quad (6.137)$$

$$\sigma(\mathbf{x},\mathbf{a}) = 1 + 2(\mathbf{r}\cdot\mathbf{a}_s - a\tau) + (r^2 - \tau^2)(\mathbf{a}_s\cdot\mathbf{a}_s - a^2). \quad (6.138)$$

Because of its nonlinearity, the transformation of a time-harmonic field is no longer time-harmonic in general but obeys a more complicated time dependence. Also, a plane wave is transformed to a field of more general space dependence. Simple boundary surfaces constant in time are transformed to surfaces moving and deforming in time.

After its introduction, the conformal transformation was applied to different branches of physics [49]. In electrical engineering the transformation appears lesser known, an exception being its use in studying certain beam-like forms of electromagnetic radiation (focus-wave modes) [50]. In addition to changing fields and their sources, the transformation has also the property of changing the electromagnetic media and boundaries. In the following just a few basic properties will be considered to open up the topic [51].

6.5.1 Properties of the Conformal Transformation

Transformation of Vectors

Let us start by considering the basic properties of the conformal transformation $\mathbf{x} \rightarrow \mathbf{x}'$. Equation (6.133) can be interpreted by the procedure "invert $\mathbf{x}$, add $\mathbf{a}$ and invert the result." For $\mathbf{a} = 0$ the transformation reduces to two inversions which returns the original vector,

$$\mathbf{x}' = \mathbf{C}(\mathbf{x}, 0) = \mathbf{I}(\mathbf{I}(\mathbf{x})) = \mathbf{x}. \quad (6.139)$$

One can readily show that consecutive transformations with vectors $\mathbf{a}_1, \mathbf{a}_2, \ldots$ can be reduced to one transformation with the vector $\mathbf{a}_1 + \mathbf{a}_2 + \cdots$. In fact, a transformation with $\mathbf{a}_1$ followed by a transformation with $\mathbf{a}_2$ results in

$$\mathbf{C}(\mathbf{C}(\mathbf{x},\mathbf{a}_1),\mathbf{a}_2) = \mathbf{I}(\mathbf{a}_2 + \mathbf{I}(\mathbf{I}(\mathbf{a}_1 + \mathbf{I}(\mathbf{x})))) = \mathbf{I}(\mathbf{a}_2 + \mathbf{a}_1 + \mathbf{I}(\mathbf{x}))$$
$$= \mathbf{C}(\mathbf{x}, \mathbf{a}_1 + \mathbf{a}_2). \quad (6.140)$$

From this it follows that the inverse of a transformation with the vector $\mathbf{a}$ equals the transformation with the vector $-\mathbf{a}$:

$$\mathbf{C}(\mathbf{x}, \mathbf{a}) = \mathbf{y}, \quad \Rightarrow \quad \mathbf{x} = \mathbf{C}(\mathbf{y}, -\mathbf{a}). \tag{6.141}$$

The transformed vector satisfies

$$\mathbf{x}' \cdot \mathbf{x}' = \frac{\mathbf{x} \cdot \mathbf{x}}{\sigma(\mathbf{x}, \mathbf{a})} = \frac{1}{\left(\mathbf{a} + \frac{\mathbf{x}}{\mathbf{x}\cdot\mathbf{x}}\right) \cdot \left(\mathbf{a} + \frac{\mathbf{x}}{\mathbf{x}\cdot\mathbf{x}}\right)}, \tag{6.142}$$

whence $\mathbf{x} \cdot \mathbf{x} = 0$ implies $\mathbf{x}' \cdot \mathbf{x}' = 0$. For $\sigma(\mathbf{x}, \mathbf{a}) < 0$ spatial vectors satisfying $\mathbf{x} \cdot \mathbf{x} > 0$ are transformed to temporal vectors satisfying $\mathbf{x}' \cdot \mathbf{x}' < 0$ and vice versa, which may violate causality. Vectors on the light cone $(\mathbf{x} \cdot \mathbf{x} = 0)$ are transformed to vectors

$$\mathbf{x}' = \frac{\mathbf{x}}{1 + 2\mathbf{x} \cdot \mathbf{a}}, \tag{6.143}$$

which also lie on the light cone. The null vector is mapped to the null vector:

$$\mathbf{x}' = \mathbf{C}(0, \mathbf{a}) = 0. \tag{6.144}$$

Substituting from (6.133) we can prove the relation

$$(\mathbf{z}' - \mathbf{y}') \cdot (\mathbf{z}' - \mathbf{y}') = \frac{(\mathbf{z} - \mathbf{y}) \cdot (\mathbf{z} - \mathbf{y})}{\sigma(\mathbf{z}, \mathbf{a})\sigma(\mathbf{y}, \mathbf{a})}, \tag{6.145}$$

for the transformation of any two vectors $\mathbf{z}, \mathbf{y}$. Setting now

$$\mathbf{z} = \mathbf{x} + d\mathbf{x}, \quad \mathbf{y} = \mathbf{x} + d\mathbf{y}, \tag{6.146}$$

where $d\mathbf{x}$ and $d\mathbf{y}$ are small and denoting the transformed vectors by

$$\mathbf{z}' = \mathbf{x}' + d\mathbf{x}', \quad \mathbf{y}' = \mathbf{x}' + d\mathbf{y}', \tag{6.147}$$

the relation (6.145) becomes

$$(d\mathbf{x}' - d\mathbf{y}') \cdot (d\mathbf{x}' - d\mathbf{y}') = \frac{(d\mathbf{x} - d\mathbf{y}) \cdot (d\mathbf{x} - d\mathbf{y})}{\sigma^2(\mathbf{x}, \mathbf{a})}. \tag{6.148}$$

In the two special cases with $d\mathbf{y} = 0$ or $d\mathbf{x} = 0$, we have, respectively,

$$d\mathbf{x}' \cdot d\mathbf{x}' = \frac{d\mathbf{x} \cdot d\mathbf{x}}{\sigma^2(\mathbf{x}, \mathbf{a})}, \quad d\mathbf{y}' \cdot d\mathbf{y}' = \frac{d\mathbf{y} \cdot d\mathbf{y}}{\sigma^2(\mathbf{x}, \mathbf{a})}, \tag{6.149}$$

which inserted to (6.148) yield

$$d\mathbf{x}' \cdot d\mathbf{y}' = \frac{d\mathbf{x} \cdot d\mathbf{y}}{\sigma^2(\mathbf{x}, \mathbf{a})}. \tag{6.150}$$

In analogy to 3D vectors we can define the angles φ', φ as

$$\cos\varphi' = \frac{d\mathbf{x}'\cdot d\mathbf{y}'}{\sqrt{d\mathbf{x}'\cdot d\mathbf{x}'}\sqrt{d\mathbf{y}'\cdot d\mathbf{y}'}} = \frac{d\mathbf{x}\cdot d\mathbf{y}}{\sqrt{d\mathbf{x}\cdot d\mathbf{x}}\sqrt{d\mathbf{y}\cdot d\mathbf{y}}} = \cos\varphi. \tag{6.151}$$

This relation serves as the reason why the transformation is called conformal: "the angle between $d\mathbf{x}$ and $d\mathbf{y}$" is invariant in the transformation.

Differential Operations

From the basic differentiation rule we can write

$$\mathbf{d}'\mathbf{x}' = \overline{\overline{\mathsf{I}}}^T. \tag{6.152}$$

To find the relation between the operators $\mathbf{d}'$ and $\mathbf{d}$, let us first express the relation between $\mathbf{x}$ and $\mathbf{x}'$ as

$$\frac{\mathbf{x}'}{\mathbf{x}'\cdot\mathbf{x}'} = \mathbf{a} + \frac{\mathbf{x}}{\mathbf{x}\cdot\mathbf{x}}. \tag{6.153}$$

This can be differentiated as

$$\mathbf{d}\frac{\mathbf{x}'}{\mathbf{x}'\cdot\mathbf{x}'} = \mathbf{d}\frac{\mathbf{x}}{\mathbf{x}\cdot\mathbf{x}}. \tag{6.154}$$

Applying the rules

$$\mathbf{d}(\mathbf{x}\cdot\mathbf{x}) = 2(\mathbf{d}\mathbf{x})|\overline{\overline{\Gamma}}|\mathbf{x} = 2\overline{\overline{\Gamma}}|\mathbf{x}, \tag{6.155}$$

$$\mathbf{d}(\mathbf{x}'\cdot\mathbf{x}') = 2(\mathbf{d}\mathbf{x}')\cdot\mathbf{x}' = 2(\mathbf{d}\mathbf{x}')|\overline{\overline{\Gamma}}|\mathbf{x}', \tag{6.156}$$

(6.154) is transformed to

$$(\mathbf{x}\cdot\mathbf{x})\mathbf{d}\mathbf{x}'|\left(\overline{\overline{\mathsf{I}}}^T - 2\frac{(\overline{\overline{\Gamma}}|\mathbf{x}')\mathbf{x}'}{\mathbf{x}'\cdot\mathbf{x}'}\right) = (\mathbf{x}'\cdot\mathbf{x}')\left(\overline{\overline{\mathsf{I}}}^T - 2\frac{(\overline{\overline{\Gamma}}|\mathbf{x})\mathbf{x}}{\mathbf{x}\cdot\mathbf{x}}\right). \tag{6.157}$$

The dyadics in brackets can be shown to be unipotent, whence they are equal to their inverses. Writing $\sigma(\mathbf{x},\mathbf{a}) = \sigma$ for brevity and applying (6.142), (6.157) can be expressed as

$$\sigma\left(\overline{\overline{\mathsf{I}}}^T - 2\frac{(\overline{\overline{\Gamma}}|\mathbf{x}')\mathbf{x}'}{\mathbf{x}'\cdot\mathbf{x}'}\right)|\left(\overline{\overline{\mathsf{I}}}^T - 2\frac{(\overline{\overline{\Gamma}}|\mathbf{x})\mathbf{x}}{\mathbf{x}\cdot\mathbf{x}}\right)|\mathbf{d}\mathbf{x}' = \overline{\overline{\mathsf{I}}}^T. \tag{6.158}$$

Comparing with (6.152), we can now identify the transformation law for the differential operator $\mathbf{d}$ as

$$\mathbf{d} \rightarrow \mathbf{d}' = \overline{\overline{\mathsf{C}}}|\mathbf{d}, \tag{6.159}$$

where the dyadic $\overline{\overline{\mathsf{C}}}$ is a function of $\mathbf{x}$ defined by

$$\overline{\overline{\mathsf{C}}} = \sigma\left(\overline{\overline{\mathsf{I}}}^T - 2\frac{(\overline{\overline{\Gamma}}|\mathbf{x}')\mathbf{x}'}{\mathbf{x}'\cdot\mathbf{x}'}\right)\Big|\left(\overline{\overline{\mathsf{I}}}^T - 2\frac{(\overline{\overline{\Gamma}}|\mathbf{x})\mathbf{x}}{\mathbf{x}\cdot\mathbf{x}}\right). \tag{6.160}$$

We can also define

$$\overline{\overline{\mathsf{C}}} = \overline{\overline{\mathsf{C}}}|(\mathbf{dx}) = (\overline{\overline{\mathsf{C}}}|\mathbf{d})\mathbf{x} = \mathbf{d}'\mathbf{x}. \tag{6.161}$$

while from (6.159) we have

$$\mathbf{dx}' = \overline{\overline{\mathsf{C}}}^{-1}|\mathbf{d}'\mathbf{x}' = \overline{\overline{\mathsf{C}}}^{-1}. \tag{6.162}$$

Properties of the Dyadic $\overline{\overline{\mathsf{C}}}$

The dyadic $\overline{\overline{\mathsf{C}}}$ satisfies the differentiation rules

$$\mathbf{d}'\wedge\overline{\overline{\mathsf{C}}} = \mathbf{d}'\wedge\mathbf{d}'\mathbf{x} = 0, \qquad \mathbf{d}'\wedge\overline{\overline{\mathsf{C}}}^{(2)} = (\mathbf{d}'\wedge\overline{\overline{\mathsf{C}}})^{\wedge}_{\wedge}\overline{\overline{\mathsf{C}}}_c = 0, \tag{6.163}$$

and

$$\mathbf{d}\wedge\overline{\overline{\mathsf{C}}}^{-1} = \mathbf{d}\wedge\mathbf{dx}' = 0, \qquad \mathbf{d}\wedge\overline{\overline{\mathsf{C}}}^{(-2)} = (\mathbf{d}\wedge\overline{\overline{\mathsf{C}}}^{-1})^{\wedge}_{\wedge}\overline{\overline{\mathsf{C}}}_c^{-1} = 0, \tag{6.164}$$

where the subscript $_c$ again marks a quantity considered constant in differentiation.

The inverse of the dyadic $\overline{\overline{\mathsf{C}}}$ can be expressed as

$$\overline{\overline{\mathsf{C}}}^{-1} = \frac{1}{\sigma}\left(\overline{\overline{\mathsf{I}}}^T - 2\frac{(\overline{\overline{\Gamma}}|\mathbf{x})\mathbf{x}}{\mathbf{x}\cdot\mathbf{x}}\right)\Big|\left(\overline{\overline{\mathsf{I}}}^T - 2\frac{(\overline{\overline{\Gamma}}|\mathbf{x}')\mathbf{x}'}{\mathbf{x}'\cdot\mathbf{x}'}\right) = \frac{1}{\sigma^2}\overline{\overline{\Gamma}}|\overline{\overline{\mathsf{C}}}^T|\overline{\overline{\mathsf{G}}}. \tag{6.165}$$

Thus, the $\overline{\overline{\mathsf{C}}}$ dyadic satisfies the quadratic equations

$$\overline{\overline{\mathsf{C}}}|\overline{\overline{\Gamma}}|\overline{\overline{\mathsf{C}}}^T = \sigma^2\overline{\overline{\Gamma}}, \quad \overline{\overline{\mathsf{C}}}^T|\overline{\overline{\mathsf{G}}}|\overline{\overline{\mathsf{C}}} = \sigma^2\overline{\overline{\mathsf{G}}}. \tag{6.166}$$

Obviously, we have

$$\overline{\overline{\mathsf{C}}}^{(p)} = \sigma^p\left(\overline{\overline{\mathsf{I}}}^T - 2\frac{(\overline{\overline{\Gamma}}|\mathbf{x}')\mathbf{x}'}{\mathbf{x}'\cdot\mathbf{x}'}\right)^{(p)}\Big|\left(\overline{\overline{\mathsf{I}}}^T - 2\frac{(\overline{\overline{\Gamma}}|\mathbf{x})\mathbf{x}}{\mathbf{x}\cdot\mathbf{x}}\right)^{(p)}, \quad -4\le p\le 4. \tag{6.167}$$

Applying

$$(\overline{\overline{\mathsf{I}}} + \mathbf{a}\boldsymbol{\alpha})^{(4)} = \overline{\overline{\mathsf{I}}}^{(4)} + \overline{\overline{\mathsf{I}}}^{(3)}{}^{\wedge}_{\wedge}\mathbf{a}\boldsymbol{\alpha} = \overline{\overline{\mathsf{I}}}^{(4)}(1 + \mathrm{tr}(\overline{\overline{\mathsf{I}}}^{(3)}{}^{\wedge}_{\wedge}\mathbf{a}\boldsymbol{\alpha})) = (1 + \mathbf{a}|\boldsymbol{\alpha})\overline{\overline{\mathsf{I}}}^{(4)}, \tag{6.168}$$

we obtain

$$\left(\overline{\overline{\mathsf{I}}}^T - 2\frac{(\overline{\overline{\Gamma}}|\mathbf{x})\mathbf{x}}{\mathbf{x}\cdot\mathbf{x}}\right)^{(4)} = \left(\overline{\overline{\mathsf{I}}}^T - 2\frac{(\overline{\overline{\Gamma}}|\mathbf{x}')\mathbf{x}'}{\mathbf{x}'\cdot\mathbf{x}'}\right)^{(4)} = -\overline{\overline{\mathsf{I}}}^{(4)T}, \qquad (6.169)$$

and

$$\overline{\overline{\mathsf{C}}}^{(4)} = \sigma^4\overline{\overline{\mathsf{I}}}^{(4)T}, \quad \sigma^4 = \mathrm{tr}\overline{\overline{\mathsf{C}}}^{(4)}. \qquad (6.170)$$

Applying the rule of the inverse dyadic gives us

$$\overline{\overline{\mathsf{C}}}^{(-2)} = \frac{\overline{\overline{\mathsf{I}}}^{(4)T}\lfloor\lfloor\overline{\overline{\mathsf{C}}}^{(2)T}}{\mathrm{tr}\overline{\overline{\mathsf{C}}}^{(4)}} = \frac{1}{\sigma^4}\overline{\overline{\mathsf{I}}}^{(4)T}\lfloor\lfloor\overline{\overline{\mathsf{C}}}^{(2)T}, \qquad (6.171)$$

which combined with (6.165) yields the relation

$$\overline{\overline{\mathsf{I}}}^{(4)}\lfloor\lfloor\overline{\overline{\mathsf{C}}}^{(2)} = \overline{\overline{\mathsf{G}}}^{(2)}|\overline{\overline{\mathsf{C}}}^{(2)}|\overline{\overline{\Gamma}}^{(2)}. \qquad (6.172)$$

Further, we have

$$\mathrm{tr}\overline{\overline{\mathsf{C}}} = 4\sigma\frac{(\mathbf{x}\cdot\mathbf{x}')^2}{(\mathbf{x}\cdot\mathbf{x})(\mathbf{x}'\cdot\mathbf{x}')} = 4(1+\mathbf{a}\cdot\mathbf{x})^2 = \sigma^2\mathrm{tr}\overline{\overline{\mathsf{C}}}^{-1}, \qquad (6.173)$$

6.5.2 Field Transformation

Because the one-forms are transformed in the same manner as the operator **d**,

$$\boldsymbol{\alpha} \to \boldsymbol{\alpha}'(\mathbf{x}') = \overline{\overline{\mathsf{C}}}|\boldsymbol{\alpha}(\mathbf{x}), \qquad (6.174)$$

the electromagnetic field two-forms and the source three-forms are transformed as

$$\boldsymbol{\Phi} \to \boldsymbol{\Phi}'(\mathbf{x}') = \overline{\overline{\mathsf{C}}}^{(2)}|\boldsymbol{\Phi}(\mathbf{x}), \qquad (6.175)$$

$$\boldsymbol{\Psi} \to \boldsymbol{\Psi}'(\mathbf{x}') = \overline{\overline{\mathsf{C}}}^{(2)}|\boldsymbol{\Psi}(\mathbf{x}), \qquad (6.176)$$

$$\boldsymbol{\gamma} \to \boldsymbol{\gamma}'(\mathbf{x}') = \overline{\overline{\mathsf{C}}}^{(3)}|\boldsymbol{\gamma}(\mathbf{x}). \qquad (6.177)$$

Since the transformation is not a linear one, the invariance of the Maxwell equations in the transformation is not obvious. However, taking into account the relation (6.163), we can expand

$$\begin{aligned}\mathbf{d}'\wedge\boldsymbol{\Phi}' &= \mathbf{d}'\wedge(\overline{\overline{\mathsf{C}}}^{(2)}|\boldsymbol{\Phi})\\ &= (\mathbf{d}'\wedge\overline{\overline{\mathsf{C}}}^{(2)})|\boldsymbol{\Phi}_c + (\overline{\overline{\mathsf{C}}}_c|\mathbf{d})\wedge\overline{\overline{\mathsf{C}}}_c^{(2)}|\boldsymbol{\Phi}\\ &= \overline{\overline{\mathsf{C}}}_c^{(3)}|(\mathbf{d}\wedge\boldsymbol{\Phi}) = 0,\end{aligned} \qquad (6.178)$$

and, similarly, the second one becomes

$$\mathbf{d}' \wedge \boldsymbol{\Psi}' = \overline{\overline{\mathsf{C}}}_c^{(3)}|(\mathbf{d} \wedge \boldsymbol{\Psi}) = \overline{\overline{\mathsf{C}}}^{(3)}|\gamma = \gamma'. \tag{6.179}$$

This proves that the Maxwell equations appear form-invariant in the conformal transformation. As a consequence, solutions to the Maxwell equations (electromagnetic fields) remain solutions to the Maxwell equations after the conformal transformation (transformed electromagnetic fields).

6.5.3 Medium Transformation

From the medium equation manipulated as

$$\boldsymbol{\Psi}' = \overline{\overline{\mathsf{C}}}^{(2)}|\boldsymbol{\Psi} = \overline{\overline{\mathsf{C}}}^{(2)}|\overline{\overline{\mathsf{M}}}|\boldsymbol{\Phi} = \overline{\overline{\mathsf{M}}}'|\boldsymbol{\Phi}' = \overline{\overline{\mathsf{M}}}'|\overline{\overline{\mathsf{C}}}^{(2)}|\boldsymbol{\Phi}, \tag{6.180}$$

we obtain the transformation rule for the medium bidyadic,

$$\overline{\overline{\mathsf{M}}}' = \overline{\overline{\mathsf{C}}}^{(2)}|\overline{\overline{\mathsf{M}}}|\overline{\overline{\mathsf{C}}}^{(-2)}. \tag{6.181}$$

This resembles the rule for the affine transformation (6.16), but one must remember that $\overline{\overline{\mathsf{C}}}$ is not a constant dyadic whence a medium loses the homogeneous and time-invariant character in the transformation in the general case. Because (6.181) implies

$$\overline{\overline{\mathsf{M}}}'^p = \overline{\overline{\mathsf{C}}}^{(2)}|\overline{\overline{\mathsf{M}}}^p|\overline{\overline{\mathsf{C}}}^{(-2)}, \tag{6.182}$$

for all integer powers p, a medium whose medium bidyadic $\overline{\overline{\mathsf{M}}}$ satisfies an algebraic equation with constant coefficients is transformed to a medium satisfying the same equation.

The modified medium bidyadic is transformed as

$$\overline{\overline{\mathsf{M}}}'_m = \mathbf{e}_N\lfloor\overline{\overline{\mathsf{M}}}' = \mathbf{e}_N\lfloor\overline{\overline{\mathsf{C}}}^{(2)}|(\varepsilon_N\lfloor\lfloor\overline{\overline{\mathsf{M}}}_m)|\overline{\overline{\mathsf{C}}}^{(-2)} = (\overline{\overline{\mathsf{I}}}^{(4)}\lfloor\lfloor\overline{\overline{\mathsf{C}}}^{(2)})|\overline{\overline{\mathsf{M}}}_m|\overline{\overline{\mathsf{C}}}^{(-2)}. \tag{6.183}$$

Applying (6.171) we can write

$$\begin{aligned}\overline{\overline{\mathsf{M}}}'_m &= \sigma^4\overline{\overline{\mathsf{C}}}^{(-2)T}|\overline{\overline{\mathsf{M}}}_m|\overline{\overline{\mathsf{C}}}^{(-2)} \\ &= \left(\overline{\overline{\mathsf{I}}} - 2\frac{\mathbf{x}'(\overline{\overline{\Gamma}}|\mathbf{x}')}{\mathbf{x}'\cdot\mathbf{x}'}\right)^{(2)}\left|\left(\overline{\overline{\mathsf{I}}} - 2\frac{\mathbf{x}(\overline{\overline{\Gamma}}|\mathbf{x})}{\mathbf{x}\cdot\mathbf{x}}\right)^{(2)}\right|\overline{\overline{\mathsf{M}}}_m\left|\left(\overline{\overline{\mathsf{I}}}^T - 2\frac{(\overline{\overline{\Gamma}}|\mathbf{x})\mathbf{x}}{\mathbf{x}\cdot\mathbf{x}}\right)^{(2)}\right| \\ &\quad\times\left(\overline{\overline{\mathsf{I}}}^T - 2\frac{(\overline{\overline{\Gamma}}|\mathbf{x}')\mathbf{x}'}{\mathbf{x}'\cdot\mathbf{x}'}\right)^{(2)}.\end{aligned} \tag{6.184}$$

From (6.181) we have

$$\mathrm{tr}\overline{\overline{\mathsf{M}}}' = \mathrm{tr}\overline{\overline{\mathsf{M}}}. \tag{6.185}$$

PROBLEMS

6.1 Find the relation between the bidyadics $\overline{\overline{\mathsf{I}}}^{(4)}\lfloor\lfloor\overline{\overline{\mathsf{M}}}$ and $\overline{\overline{\mathsf{I}}}^{(4)}\lfloor\lfloor\overline{\overline{\mathsf{M}}}_a$, when $\overline{\overline{\mathsf{M}}}_a$ is the affine transformation of the medium bidyadic $\overline{\overline{\mathsf{M}}}$ in terms of a given transformation dyadic $\overline{\overline{\mathsf{A}}}$.

6.2 Find the affine-transformation rules for the 3D medium dyadics $\overline{\overline{\alpha}}, \overline{\overline{\epsilon}}', \overline{\overline{\mu}}^{-1}, \overline{\overline{\beta}}$ and $\overline{\overline{\epsilon}}_g, \overline{\overline{\xi}}_g, \overline{\overline{\zeta}}_g, \overline{\overline{\mu}}_g$ corresponding to the time reversion transformation defined by the dyadic

$$\overline{\overline{\mathsf{T}}} = \overline{\overline{\mathsf{I}}}_s - \mathbf{e}_4\boldsymbol{\varepsilon}_4.$$

What media are invariant in time reversion? What about the Gibbsian representation?

6.3 Defining the dyadic quotient function by [52]

$$\overline{\overline{\mathsf{Q}}}(\overline{\overline{\mathsf{X}}}) = (\overline{\overline{\mathsf{I}}} - \overline{\overline{\mathsf{X}}})|(\overline{\overline{\mathsf{I}}} + \overline{\overline{\mathsf{X}}})^{-1} = (\overline{\overline{\mathsf{I}}} + \overline{\overline{\mathsf{X}}})^{-1}|(\overline{\overline{\mathsf{I}}} - \overline{\overline{\mathsf{X}}}),$$

prove the following properties

$$\overline{\overline{\mathsf{Q}}}(\overline{\overline{\mathsf{Q}}}(\overline{\overline{\mathsf{X}}})) = \overline{\overline{\mathsf{X}}}, \quad \overline{\overline{\mathsf{Q}}}(0) = \overline{\overline{\mathsf{I}}}, \quad \overline{\overline{\mathsf{Q}}}(\overline{\overline{\mathsf{I}}}) = 0$$

$$\overline{\overline{\mathsf{Q}}}(-\overline{\overline{\mathsf{X}}}) = (\overline{\overline{\mathsf{Q}}}(\overline{\overline{\mathsf{X}}}))^{-1}, \quad \overline{\overline{\mathsf{Q}}}(\overline{\overline{\mathsf{X}}}^{-1}) = -\overline{\overline{\mathsf{Q}}}(\overline{\overline{\mathsf{X}}})$$

$$\overline{\overline{\mathsf{Q}}}(\overline{\overline{\mathsf{G}}}|\overline{\overline{\mathsf{X}}}^T|\overline{\overline{\Gamma}}) = \overline{\overline{\mathsf{G}}}|\overline{\overline{\mathsf{Q}}}^T(\overline{\overline{\mathsf{X}}})|\overline{\overline{\Gamma}}$$

where $\overline{\overline{\mathsf{G}}} \in \mathbb{E}_1\mathbb{E}_1$ is a symmetric dyadic and $\overline{\overline{\Gamma}}$ is its inverse.

6.4 Defining the dot product of two vectors $\mathbf{a} \cdot \mathbf{b} = \mathbf{a}|\overline{\overline{\Gamma}}|\mathbf{b}$ in terms of a symmetric dyadic $\overline{\overline{\Gamma}} = \overline{\overline{\mathsf{G}}}^{-1} \in \mathbb{F}_1\mathbb{F}_1$, show that an affine transformation (rotation transformation) defined by the dyadic $\overline{\overline{\mathsf{R}}} \in \mathbb{E}_1\mathbb{F}_1$ satisfying the property

$$(\overline{\overline{\mathsf{R}}}|\mathbf{a}) \cdot (\overline{\overline{\mathsf{R}}}|\mathbf{a}) = \mathbf{a} \cdot \mathbf{a},$$

for any vector $\mathbf{a}$, can be represented by

$$\overline{\overline{\mathsf{R}}} = \overline{\overline{\mathsf{Q}}}(\overline{\overline{\mathsf{G}}}\rfloor\boldsymbol{\Pi}),$$

where $\overline{\overline{\mathsf{Q}}}(\overline{\overline{\mathsf{X}}})$ is the dyadic quotient function of Problem 6.3 while $\boldsymbol{\Pi} \in \mathbb{F}_2$ is some two-form.

6.5 Derive the duality transformation conditions for the medium dyadics (6.61)–(6.64).

6.6 Show that the Hehl–Obukhov decomposed parts of the general medium bidyadic are transformed individually in the reciprocity transformation.

6.7 Derive the transformation rule

$$\overline{\overline{\mathsf{T}}}' = \overline{\overline{\mathsf{C}}}^{(3)}|\overline{\overline{\mathsf{T}}}|\overline{\overline{\mathsf{C}}}^T$$

for the stress–energy dyadic $\overline{\overline{\mathsf{T}}}$ in the conformal transformation.

6.8 Show that the Hehl–Obukhov decomposed parts of the general medium bidyadic are transformed individually in the conformal transformation.

6.9 Expressing the relation between the four medium dyadics and their dual medium dyadics as

$$\begin{pmatrix} \overline{\overline{\epsilon}}_d Z_d \\ \overline{\overline{\mu}}_d/Z_d \\ \overline{\overline{\xi}}_d \\ \overline{\overline{\zeta}}_d \end{pmatrix} = \mathcal{R}(\theta) \begin{pmatrix} \overline{\overline{\epsilon}} Z \\ \overline{\overline{\mu}}/Z \\ \overline{\overline{\xi}} \\ \overline{\overline{\zeta}} \end{pmatrix},$$

show that the 4×4 matrix $\mathcal{R}(\theta)$ is a rotation matrix, that is, it satisfies $\mathcal{R}(\theta)\mathcal{R}^T(\theta) = \mathcal{I}$, where $\mathcal{I}$ is the unit 4×4 matrix.

6.10 A transformation of medium bidyadics $\overline{\overline{\mathsf{M}}} \to \overline{\overline{\mathsf{M}}}'$ can be defined as

$$\overline{\overline{\mathsf{M}}}' = Y^2 \overline{\overline{\mathsf{I}}}^{(4)T} \lfloor\lfloor \overline{\overline{\mathsf{M}}}^{-1T},$$

where Y is a scalar parameter and $\overline{\overline{\mathsf{M}}}$ is a full-rank bidyadic. Find the corresponding transformation for the Gibbsian medium dyadics $\overline{\overline{\epsilon}}_g, \overline{\overline{\xi}}_g, \overline{\overline{\zeta}}_g, \overline{\overline{\mu}}_g$.

6.11 Show that a modified medium bidyadic of the form

$$\overline{\overline{\mathsf{M}}}_m = \alpha \overline{\overline{\mathsf{E}}} + \overline{\overline{\mathsf{Q}}}^{(2)}, \quad \overline{\overline{\mathsf{Q}}} \in \mathbb{E}_1\mathbb{E}_1$$

can be transformed to a bidyadic of similar form

$$\overline{\overline{\mathsf{M}}}_{md} = \alpha_d \overline{\overline{\mathsf{E}}} + \overline{\overline{\mathsf{Q}}}_d^{(2)}$$

in some duality transformation.

6.12 Show that from (6.60) it follows that $\overline{\overline{\mathsf{M}}}$ and $\overline{\overline{\mathsf{M}}}_d$ satisfy $\overline{\overline{\mathsf{M}}}|\overline{\overline{\mathsf{M}}}_d = \overline{\overline{\mathsf{M}}}_d|\overline{\overline{\mathsf{M}}}$.

6.13 Show that the parameters of the duality transformation can be chosen so that the PEMC boundary condition (6.75) appears self-dual.

CHAPTER 7

Basic Classes of Electromagnetic Media

7.1 GIBBSIAN ISOTROPY

The simplest electromagnetic media in terms of Gibbsian representation are isotropic media, defined by equations of the form

$$\mathbf{D}_g = \epsilon \mathbf{E}_g, \quad \mathbf{B}_g = \mu \mathbf{H}_g. \tag{7.1}$$

In fact, they do not depend on any special spatial direction and they involve just two scalar medium parameters ϵ and μ. From a more general point of view such media can be called spatially isotropic, or invariant in spatial rotations.

7.1.1 Gibbsian Isotropic Medium

In 4D electromagnetics, **E** and **H** are one-forms and **D** and **B** are two-forms whence the scalars ϵ and μ in (7.1) must be replaced by dyadics $\overline{\overline{\epsilon}}$ and $\overline{\overline{\mu}}$ mapping one-forms to two-forms. Thus, in the present formalism, (7.1) should be expressed as (see Appendix B)

$$\mathbf{D} = \overline{\overline{\epsilon}}|\mathbf{E} = \epsilon\overline{\overline{\mathsf{C}}}_s|\mathbf{E}, \quad \mathbf{B} = \overline{\overline{\mu}}|\mathbf{H} = \mu\overline{\overline{\mathsf{C}}}_s|\mathbf{H}, \tag{7.2}$$

with

$$\overline{\overline{\mathsf{C}}}_s = \boldsymbol{\varepsilon}_{123}\lfloor\overline{\overline{\mathsf{G}}}_s = \boldsymbol{\varepsilon}_{12}\mathbf{e}_3 + \boldsymbol{\varepsilon}_{23}\mathbf{e}_1 + \boldsymbol{\varepsilon}_{31}\mathbf{e}_2 \in \mathbb{F}_2\mathbb{E}_1, \tag{7.3}$$

Multiforms, Dyadics, and Electromagnetic Media, First Edition. Ismo V. Lindell.

in terms of a symmetric spatial metric dyadic $\overline{\overline{\mathsf{G}}}_s = \mathbf{e}_1\mathbf{e}_1 + \mathbf{e}_2\mathbf{e}_2 + \mathbf{e}_3\mathbf{e}_3$. The dyadic $\overline{\overline{\mathsf{C}}}_s$ has the spatial inverse

$$\overline{\overline{\mathsf{C}}}_s^{-1} = \overline{\overline{\mathsf{G}}}_s^{-1}\rfloor \mathbf{e}_{123} = \boldsymbol{\varepsilon}_1\mathbf{e}_{23} + \boldsymbol{\varepsilon}_2\mathbf{e}_{31} + \boldsymbol{\varepsilon}_3\mathbf{e}_{12} \in \mathbb{F}_1\mathbb{E}_2. \tag{7.4}$$

The medium bidyadic corresponding to (7.1) has the form

$$\begin{aligned}\overline{\overline{\mathsf{M}}} &= \overline{\overline{\epsilon}} \wedge \mathbf{e}_4 + \boldsymbol{\varepsilon}_4 \wedge \overline{\overline{\mu}}^{-1} \\ &= \epsilon\overline{\overline{\mathsf{C}}}_s \wedge \mathbf{e}_4 + \boldsymbol{\varepsilon}_4 \wedge (\mu\overline{\overline{\mathsf{C}}}_s)^{-1},\end{aligned} \tag{7.5}$$

and the medium can be called Gibbsian isotropic medium. Actually, the Gibbsian isotropic medium equals the simple principal medium defined by (5.115). The modified medium bidyadic has the compact form (5.122)

$$\overline{\overline{\mathsf{M}}}_m = -\frac{1}{\mu}(\overline{\overline{\mathsf{G}}}_s - \mu\epsilon\mathbf{e}_4\mathbf{e}_4)^{(2)} = -\frac{1}{\mu}\overline{\overline{\mathsf{G}}}^{(2)}. \tag{7.6}$$

Making an affine transformation with the dyadic

$$\overline{\overline{\mathsf{A}}} = \overline{\overline{\mathsf{I}}}_s + \frac{1}{\sqrt{\mu\epsilon}}\mathbf{e}_4\boldsymbol{\varepsilon}_4, \tag{7.7}$$

the medium bidyadic (7.5) expressed as

$$\overline{\overline{\mathsf{M}}} = \sqrt{\epsilon/\mu}\,((\sqrt{\mu\epsilon}\,\overline{\overline{\mathsf{C}}}_s) \wedge \mathbf{e}_4 + \boldsymbol{\varepsilon}_4 \wedge (\sqrt{\mu\epsilon}\,\overline{\overline{\mathsf{C}}}_s)^{-1}), \tag{7.8}$$

is transformed to

$$\overline{\overline{\mathsf{M}}}_a = \overline{\overline{\mathsf{A}}}^{(-2)T}|\overline{\overline{\mathsf{M}}}|\overline{\overline{\mathsf{A}}}^{(2)T} = \sqrt{\frac{\epsilon}{\mu}}(\overline{\overline{\mathsf{C}}}_s \wedge \mathbf{e}_4 + \boldsymbol{\varepsilon}_4 \wedge \overline{\overline{\mathsf{C}}}_s^{-1}), \tag{7.9}$$

and the modified medium bidyadic is transformed to

$$\overline{\overline{\mathsf{M}}}_{ma} = -\sqrt{\frac{\epsilon}{\mu}}(\overline{\overline{\mathsf{G}}}_s - \mathbf{e}_4\boldsymbol{\varepsilon}_4)^{(2)}, \tag{7.10}$$

with transformed parameters satisfying $\epsilon\mu = 1$.

7.1.2 Gibbsian Bi-isotropic Medium

In the Gibbsian formalism, the bi-isotropic medium is defined by conditions of the form [28]

$$\mathbf{D}_g = \epsilon\mathbf{E}_g + \xi\mathbf{H}_g, \quad \mathbf{B}_g = \zeta\mathbf{E}_g + \mu\mathbf{H}_g. \tag{7.11}$$

They correspond to the conditions

$$\begin{pmatrix}\mathbf{D}\\ \mathbf{B}\end{pmatrix} = \begin{pmatrix}\epsilon\overline{\overline{\mathsf{C}}}_s & \xi\overline{\overline{\mathsf{C}}}_s\\ \zeta\overline{\overline{\mathsf{C}}}_s & \mu\overline{\overline{\mathsf{C}}}_s\end{pmatrix} | \begin{pmatrix}\mathbf{E}\\ \mathbf{H}\end{pmatrix}, \tag{7.12}$$

defining a class of what can be called Gibbsian bi-isotropic (GBI) media. Writing these relations in the form

$$\begin{pmatrix}\mathbf{D}\\ \mathbf{H}\end{pmatrix} = \begin{pmatrix}(\xi/\mu)\overline{\overline{\mathsf{I}}}_s^{(2)T} & (\epsilon - \xi\zeta/\mu)\overline{\overline{\mathsf{C}}}_s\\ (\mu\overline{\overline{\mathsf{C}}}_s)^{-1} & -(\zeta/\mu)\overline{\overline{\mathsf{I}}}_s^{T}\end{pmatrix} | \begin{pmatrix}\mathbf{B}\\ \mathbf{E}\end{pmatrix}, \tag{7.13}$$

the medium bidyadic of the general GBI medium can be expressed as

$$\begin{aligned}\overline{\overline{\mathsf{M}}} &= \overline{\overline{\alpha}} + \overline{\overline{\epsilon}}' \wedge \mathbf{e}_4 + \boldsymbol{\varepsilon}_4 \wedge \overline{\overline{\mu}}^{-1} + \boldsymbol{\varepsilon}_4 \wedge \overline{\overline{\beta}} \wedge \mathbf{e}_4\\ &= \alpha\overline{\overline{\mathsf{I}}}_s^{(2)T} + \epsilon'\overline{\overline{\mathsf{C}}}_s \wedge \mathbf{e}_4 + \boldsymbol{\varepsilon}_4 \wedge (\mu\overline{\overline{\mathsf{C}}}_s)^{-1} + \beta\boldsymbol{\varepsilon}_4 \wedge \overline{\overline{\mathsf{I}}}_s^{T} \wedge \mathbf{e}_4.\end{aligned} \tag{7.14}$$

The two sets of medium parameters are related as

$$\begin{pmatrix}\alpha & \epsilon'\\ 1/\mu & \beta\end{pmatrix} = \begin{pmatrix}\xi/\mu & \epsilon - \xi\zeta/\mu\\ 1/\mu & -\zeta/\mu\end{pmatrix}. \tag{7.15}$$

For $\xi = \zeta = 0$, or $\alpha = \beta = 0$, the GBI medium (7.14) reduces to the Gibbsian isotropic medium (7.5).

7.1.3 Decomposition of GBI Medium

Let us find the Hehl–Obukhov decomposition (5.91) of the general GBI medium. Denoting

$$\alpha = A + B, \quad \beta = B - A, \tag{7.16}$$

(7.14) can be rewritten as

$$\overline{\overline{\mathsf{M}}} = A\overline{\overline{\mathsf{I}}}^{(2)T} + B(\overline{\overline{\mathsf{I}}}_s - \mathbf{e}_4\boldsymbol{\varepsilon}_4)^{(2)T} + \boldsymbol{\varepsilon}_{123}\lfloor \epsilon'\overline{\overline{\mathsf{G}}}_s \wedge \mathbf{e}_4 + \boldsymbol{\varepsilon}_4 \wedge (\mu\overline{\overline{\mathsf{G}}}_s)^{-1}\rfloor \mathbf{e}_{123}. \tag{7.17}$$

Because the trace of each of the last three terms vanishes, the first term equals the axion part of the medium bidyadic. The coefficient

$$A = \frac{\alpha - \beta}{2} = \frac{\xi + \zeta}{2\mu} \tag{7.18}$$

is proportional to what is known as the Tellegen parameter of the medium, while the coefficient

$$B = \frac{\alpha + \beta}{2} = \frac{\xi - \zeta}{2\mu} \tag{7.19}$$

is proportional to the chirality parameter of the medium [28]. After some steps of algebra, the medium bidyadic (7.17) can be expressed in the form

$$\overline{\overline{\mathsf{M}}} = A\overline{\overline{\mathrm{I}}}^{(2)T} + B\overline{\overline{\mathsf{T}}}^{(2)T} - \frac{1}{\mu}\varepsilon_N\lfloor\overline{\overline{\mathsf{G}}}^{(2)}, \tag{7.20}$$

where $\overline{\overline{\mathsf{T}}} = \overline{\overline{\mathrm{I}}}_s - \mathbf{e}_4\varepsilon_4$ is the time reversion dyadic and the symmetric dyadic $\overline{\overline{\mathsf{G}}}$ is here defined by

$$\overline{\overline{\mathsf{G}}} = \overline{\overline{\mathsf{G}}}_s - \mu\epsilon'\mathbf{e}_4\mathbf{e}_4, \quad \overline{\overline{\mathsf{G}}}_s = \mathbf{e}_1\mathbf{e}_1 + \mathbf{e}_2\mathbf{e}_2 + \mathbf{e}_3\mathbf{e}_3, \tag{7.21}$$

satisfying $\overline{\overline{\mathsf{G}}}^{(4)} = -\mu\epsilon'\mathbf{e}_N\mathbf{e}_N$. Since the metric bidyadic

$$\begin{aligned}\mathbf{e}_N\lfloor\overline{\overline{\mathsf{T}}}^{(2)T} &= \mathbf{e}_{34}\mathbf{e}_{12} + \mathbf{e}_{14}\mathbf{e}_{23} + \mathbf{e}_{24}\mathbf{e}_{31} - \mathbf{e}_{12}\mathbf{e}_{34} - \mathbf{e}_{23}\mathbf{e}_{14} - \mathbf{e}_{31}\mathbf{e}_{24} \\ &= -(\mathbf{e}_N\lfloor\overline{\overline{\mathsf{T}}}^{(2)T})^T,\end{aligned} \tag{7.22}$$

appears antisymmetric, the middle term in (7.20) can be identified as a skewon bidyadic. Finally, since $\overline{\overline{\mathsf{G}}}^{(2)}$ is symmetric and $\mathrm{tr}(\varepsilon_N\lfloor\overline{\overline{\mathsf{G}}}^{(2)}) = 0$, as is seen from (7.17), the last term in (7.20) makes the principal part of the medium bidyadic. Thus, the expansion (7.20) actually represents the Hehl–Obukhov decomposition of the modified GBI medium bidyadic, in terms of its respective axion, skewon, and principal parts.

One can show that the GBI medium is reciprocal exactly when the axion term vanishes, that is, for $\mathrm{tr}\overline{\overline{\mathsf{M}}} = 0$. In fact, applying the time reversion (6.94) to the GBI medium bidyadic,

$$\begin{aligned}\overline{\overline{\mathsf{M}}}^* &= \overline{\overline{\mathsf{T}}}^{(2)T}|\overline{\overline{\mathsf{M}}}|\overline{\overline{\mathsf{T}}}^{(2)T} \\ &= A(\overline{\overline{\mathrm{I}}}^{(2)T})^* + B(\overline{\overline{\mathsf{T}}}^{(2)T})^* - \frac{1}{\mu}(\varepsilon_N\lfloor\overline{\overline{\mathsf{G}}}^{(2)})^* \\ &= A\overline{\overline{\mathrm{I}}}^{(2)T} + B\overline{\overline{\mathsf{T}}}^{(2)T} + \frac{1}{\mu}\varepsilon_N\lfloor\overline{\overline{\mathsf{G}}}^{(2)},\end{aligned} \tag{7.23}$$

the reciprocity transformation (6.96) yields

$$\begin{aligned}\overline{\overline{\mathsf{M}}}_r &= -\overline{\overline{\mathrm{I}}}^{(4)T}\lfloor\lfloor\overline{\overline{\mathsf{M}}}^{T*} \\ &= -A\overline{\overline{\mathrm{I}}}^{(4)T}\lfloor\lfloor\overline{\overline{\mathrm{I}}}^{(2)} - B\overline{\overline{\mathrm{I}}}^{(4)T}\lfloor\lfloor\overline{\overline{\mathsf{T}}}^{(2)} - \frac{1}{\mu}\overline{\overline{\mathrm{I}}}^{(4)T}\lfloor\lfloor(\varepsilon_N\lfloor\overline{\overline{\mathsf{G}}}^{(2)T}) \\ &= -A\overline{\overline{\mathrm{I}}}^{(2)T} + B\overline{\overline{\mathsf{T}}}^{(2)T} - \frac{1}{\mu}\varepsilon_N\lfloor\overline{\overline{\mathsf{G}}}^{(2)}.\end{aligned} \tag{7.24}$$

Comparing with (7.20), the transformation is seen to reverse the sign of the axion term. Thus, the GBI medium is reciprocal whenever $A = 0$, that is, when the Tellegen parameter vanishes [28].

7.1.4 Affine Transformation

Let us consider the affine-transformed GBI medium bidyadic

$$\begin{aligned}\overline{\overline{\mathsf{M}}}_a &= \overline{\overline{\mathsf{A}}}^{(-2)T}|\overline{\overline{\mathsf{M}}}|\overline{\overline{\mathsf{A}}}^{(2)T} \\ &= \overline{\overline{\mathsf{A}}}^{(-2)T}|(A\overline{\overline{\mathsf{I}}}^{(2)T} + B\overline{\overline{\mathsf{T}}}^{(2)T} - \frac{1}{\mu}\varepsilon_N\lfloor\overline{\overline{\mathsf{G}}}^{(2)})|\overline{\overline{\mathsf{A}}}^{(2)T}.\end{aligned} \tag{7.25}$$

The first (axion) term is seen to be invariant. The second (skewon) term becomes

$$B\overline{\overline{\mathsf{A}}}^{(-2)T}|\overline{\overline{\mathsf{T}}}^{(2)T}|\overline{\overline{\mathsf{A}}}^{(2)T} = B(\overline{\overline{\mathsf{A}}}|\overline{\overline{\mathsf{T}}}|\overline{\overline{\mathsf{A}}}^{-1})^{(2)T} = B\overline{\overline{\mathsf{T}}}_a^{(2)T}, \tag{7.26}$$

or it is another skewon term with $\overline{\overline{\mathsf{T}}}$ replaced by

$$\overline{\overline{\mathsf{T}}}_a = \sum_1^3 \mathbf{e}_{ai}\boldsymbol{\varepsilon}_{ai} - \mathbf{e}_{a4}\boldsymbol{\varepsilon}_{a4}, \tag{7.27}$$

as defined by the transformed basis vectors and one-forms

$$\mathbf{e}_{ai} = \overline{\overline{\mathsf{A}}}|\mathbf{e}_i, \quad \boldsymbol{\varepsilon}_{ai} = \overline{\overline{\mathsf{A}}}^{-1T}|\boldsymbol{\varepsilon}_i. \tag{7.28}$$

Finally, the third (principal) term is transformed as

$$\begin{aligned}-\frac{1}{\mu}\overline{\overline{\mathsf{A}}}^{(-2)T}|(\varepsilon_N\lfloor\overline{\overline{\mathsf{G}}}^{(2)})|\overline{\overline{\mathsf{A}}}^{(2)T} &= -\frac{1}{\mu\mathrm{tr}\overline{\overline{\mathsf{A}}}^{(4)}}\varepsilon_N\lfloor\overline{\overline{\mathsf{A}}}^{(2)}|\overline{\overline{\mathsf{G}}}^{(2)}|\overline{\overline{\mathsf{A}}}^{(2)T} \\ &= -\frac{1}{\mu}\varepsilon_{aN}\lfloor\overline{\overline{\mathsf{G}}}_a^{(2)},\end{aligned} \tag{7.29}$$

with

$$\overline{\overline{\mathsf{G}}}_a = \sum_1^3 \mathbf{e}_{ai}\mathbf{e}_{ai} - \mu\epsilon'\mathbf{e}_{a4}\mathbf{e}_{a4}. \tag{7.30}$$

Thus, the GBI medium is transformed to another GBI medium with axion, skewon, and principal parts transformed individually to respective axion, skewon, and principal parts. This means, for example, that there is no affine transformation which would map the general GBI medium to a simpler Gibbsian isotropic medium with no axion and skewon components.

7.1.5 Eigenfields in GBI Medium

The square of the GBI medium bidyadic can be expanded as

$$\begin{aligned}\overline{\overline{\mathsf{M}}}^2 &= (A\overline{\overline{\mathsf{I}}}^{(2)T} + B\overline{\overline{\mathsf{T}}}^{(2)T} - \frac{1}{\mu}\varepsilon_N\lfloor\overline{\overline{\mathsf{G}}}^{(2)})^2 \\ &= \left(A^2 + B^2 - \frac{\epsilon'}{\mu}\right)\overline{\overline{\mathsf{I}}}^{(2)T} + 2AB\overline{\overline{\mathsf{T}}}^{(2)T} - \frac{2A}{\mu}\varepsilon_N\lfloor\overline{\overline{\mathsf{G}}}^{(2)T} \\ &= 2A\overline{\overline{\mathsf{M}}} + \left(-A^2 + B^2 - \frac{\epsilon'}{\mu}\right)\overline{\overline{\mathsf{I}}}^{(2)T} \end{aligned} \qquad (7.31)$$

by applying the following properties, which can be easily checked,

$$(\overline{\overline{\mathsf{T}}}^{(2)T})^2 = \overline{\overline{\mathsf{I}}}^{(2)T} \qquad (7.32)$$

$$\begin{aligned}(\varepsilon_N\lfloor\overline{\overline{\mathsf{G}}}^{(2)})^2 &= (\varepsilon_N\varepsilon_N\lfloor\lfloor\overline{\overline{\mathsf{G}}}^{(2)})|\overline{\overline{\mathsf{G}}}^{(2)} = (\varepsilon_N\varepsilon_N||\overline{\overline{\mathsf{G}}}^{(4)})\overline{\overline{\mathsf{I}}}^{(2)T} \\ &= -\mu\epsilon'\overline{\overline{\mathsf{I}}}^{(2)T}\end{aligned} \qquad (7.33)$$

$$\overline{\overline{\mathsf{T}}}^{(2)T}|(\varepsilon_N\lfloor\overline{\overline{\mathsf{G}}}^{(2)}) = -(\varepsilon_N\lfloor\overline{\overline{\mathsf{G}}}^{(2)})|\overline{\overline{\mathsf{T}}}^{(2)T}. \qquad (7.34)$$

From (7.31) it follows that $\overline{\overline{\mathsf{M}}}$ satisfies an algebraic equation of the second order. It can be written in the form

$$(\overline{\overline{\mathsf{M}}} - M_+\overline{\overline{\mathsf{I}}}^{(2)T})|(\overline{\overline{\mathsf{M}}} - M_-\overline{\overline{\mathsf{I}}}^{(2)T}) = 0, \qquad (7.35)$$

with

$$M_\pm = A \pm jS, \quad S = \sqrt{(\epsilon'/\mu) - B^2}, \qquad (7.36)$$

where A and B are defined by (7.18) and (7.19). M_+ and M_- equal the two eigenvalues in the problem

$$\overline{\overline{\mathsf{M}}}|\mathbf{\Phi}_\pm = M_\pm\mathbf{\Phi}_\pm. \qquad (7.37)$$

From (7.35) it follows that the two eigen two-forms $\mathbf{\Phi}_\pm$ can be expressed as

$$\mathbf{\Phi}_\pm = (\overline{\overline{\mathsf{M}}} - M_\mp\overline{\overline{\mathsf{I}}}^{(2)T})|\mathbf{\Theta}, \qquad (7.38)$$

where $\mathbf{\Theta}$ may be any two-form yielding $\mathbf{\Phi}_\pm \neq 0$. Excluding the case $S = 0$, that is, assuming $M_+ \neq M_-$, we can express

$$\mathbf{\Theta} = \pm\frac{1}{M_+ - M_-}\mathbf{\Phi}, \qquad (7.39)$$

in terms of some two-form $\mathbf{\Phi}$, whence from (7.38) we obtain

$$\mathbf{\Phi}_\pm = \frac{\pm 1}{M_+ - M_-}(\overline{\overline{\mathsf{M}}} - M_\mp \overline{\overline{\mathsf{I}}}^{(2)T})|\mathbf{\Phi}. \tag{7.40}$$

Because the eigenfields satisfy

$$\mathbf{\Phi}_+ + \mathbf{\Phi}_- = \mathbf{\Phi}, \tag{7.41}$$

(7.40) actually represents a rule how the two eigen two-forms components are obtained for a given two-form $\mathbf{\Phi}$ after solving for the eigenvalues $M_\pm$. The sum rule (7.41) is more obvious through the representation

$$\mathbf{\Phi}_\pm = \frac{1}{2}\mathbf{\Phi} \pm \frac{1}{2jS}\left(B\overline{\overline{\mathsf{T}}}^{(2)T} - \frac{1}{\mu}\boldsymbol{\varepsilon}_N \lfloor \overline{\overline{\mathsf{G}}}^{(2)}\right)|\mathbf{\Phi}. \tag{7.42}$$

The excluded case $S = 0$ with coinciding eigenvalues arises when the medium bidyadic is of the special form $\overline{\overline{\mathsf{M}}} = A\overline{\overline{\mathsf{I}}}^{(2)T} + \overline{\overline{\mathsf{N}}}$ where $\overline{\overline{\mathsf{N}}}$ is any nilpotent bidyadic satisfying $\overline{\overline{\mathsf{N}}}^2 = 0$. This can be directly seen from (7.31) written in the form

$$(\overline{\overline{\mathsf{M}}} - A\overline{\overline{\mathsf{I}}}^{(2)T})^2 = -S^2\overline{\overline{\mathsf{I}}}^{(2)}. \tag{7.43}$$

For any homogeneous and time-invariant GBI medium possessing constant eigenvalues $M_+ \neq M_-$, the two Maxwell equations

$$\mathbf{d} \wedge \mathbf{\Phi} = \mathbf{d} \wedge \mathbf{\Phi}_+ + \mathbf{d} \wedge \mathbf{\Phi}_- = \gamma_m, \tag{7.44}$$

$$\mathbf{d} \wedge \mathbf{\Psi} = M_+\mathbf{d} \wedge \mathbf{\Phi}_+ + M_-\mathbf{d} \wedge \mathbf{\Phi}_- = \gamma_e, \tag{7.45}$$

can be decomposed in two noninteracting parts. In fact, defining two source three-forms as

$$\gamma_\pm = \frac{1}{M_+ - M_-}(\gamma_e - M_\mp \gamma_m), \tag{7.46}$$

the two eigenfields satisfy individually the Maxwell equations

$$\mathbf{d} \wedge \mathbf{\Phi}_+ = \gamma_+, \qquad \mathbf{d} \wedge \mathbf{\Psi}_+ = M_+\gamma_+, \tag{7.47}$$

$$\mathbf{d} \wedge \mathbf{\Phi}_- = \gamma_-, \qquad \mathbf{d} \wedge \mathbf{\Psi}_- = M_-\gamma_-, \tag{7.48}$$

whence the problem can be split in two separate parts.

One can show that, for each of the two eigenfields $\mathbf{\Phi}_+$ and $\mathbf{\Phi}_-$ the original GBI medium can be replaced by an equivalent Gibbsian isotropic (simple principal) medium. The medium equations for the eigenfields can be written as

$$\mathbf{\Psi}_\pm = \overline{\overline{\mathsf{M}}}_\pm|\mathbf{\Phi}_\pm, \tag{7.49}$$

where the equivalent medium bidyadics have a simpler form corresponding to Gibbsian isotropic media,

$$\overline{\overline{\mathsf{M}}}_\pm = \epsilon_\pm \overline{\overline{\mathsf{C}}}_s \wedge \mathbf{e}_4 + \boldsymbol{\varepsilon}_4 \wedge (\mu_\pm \overline{\overline{\mathsf{C}}}_s)^{-1}. \tag{7.50}$$

The medium parameters of (7.50) depend on the original medium parameters as

$$\epsilon_\pm = \epsilon' \frac{M_\pm}{M_\pm - A - B} = \epsilon' \frac{A \pm \sqrt{B^2 - \epsilon'/\mu}}{-B \pm \sqrt{B^2 - \epsilon'/\mu}} \tag{7.51}$$

$$\mu_\pm = \mu \frac{M_\pm - A + B}{M_\pm} = \mu \frac{B \pm \sqrt{B^2 - \epsilon'/\mu}}{A \pm \sqrt{B^2 - \epsilon'/\mu}}. \tag{7.52}$$

Derivation of these expressions is left as an exercise.

From the expression (5.205) one can notice that, for the eigenfields $\boldsymbol{\Psi}_\pm, \boldsymbol{\Phi}_\pm$ the stress–energy dyadic vanishes,

$$\begin{aligned}\overline{\overline{\mathsf{T}}}_\pm &= \frac{1}{2}\boldsymbol{\Psi}_\pm \wedge \overline{\overline{\mathsf{I}}}^T \rfloor \boldsymbol{\Phi}_\pm - \frac{1}{2}\boldsymbol{\Phi}_\pm \wedge \overline{\overline{\mathsf{I}}}^T \rfloor \boldsymbol{\Psi}_\pm \\ &= \frac{1}{2}M_\pm \boldsymbol{\Phi}_\pm \wedge \overline{\overline{\mathsf{I}}}^T \rfloor \boldsymbol{\Phi}_\pm - \frac{1}{2}M_\pm \boldsymbol{\Phi}_\pm \wedge \overline{\overline{\mathsf{I}}}^T \rfloor \boldsymbol{\Phi}_\pm = 0. \end{aligned} \tag{7.53}$$

Thus, to carry energy, the field must contain components of both eigenfields.

7.1.6 Plane Wave in GBI Medium

The dispersion dyadic of a plane wave in a GBI medium, defined by the medium bidyadic (7.20), or the modified medium bidyadic

$$\overline{\overline{\mathsf{M}}}_m = \mathbf{e}_N \lfloor (A\overline{\overline{\mathsf{I}}}^{(2)T} + B\overline{\overline{\mathsf{T}}}^{(2)T}) - \frac{1}{\mu}\overline{\overline{\mathsf{G}}}^{(2)}, \tag{7.54}$$

can be expanded as

$$\begin{aligned}\mu\overline{\overline{\mathsf{D}}}(\boldsymbol{\nu}) &= \mu\overline{\overline{\mathsf{M}}}_m \lfloor\lfloor \boldsymbol{\nu}\boldsymbol{\nu} \\ &= \mu B(\mathbf{e}_N \lfloor \overline{\overline{\mathsf{T}}}^{(2)T}) \lfloor\lfloor \boldsymbol{\nu}\boldsymbol{\nu} - \overline{\overline{\mathsf{G}}}^{(2)} \lfloor\lfloor \boldsymbol{\nu}\boldsymbol{\nu} \\ &= -2\mu B(\mathbf{e}_N \lfloor (\overline{\overline{\mathsf{I}}}^T {}^\wedge_\wedge \boldsymbol{\varepsilon}_4 \mathbf{e}_4)) \lfloor\lfloor \boldsymbol{\nu}\boldsymbol{\nu} - \overline{\overline{\mathsf{G}}}^{(2)} \lfloor\lfloor \boldsymbol{\nu}\boldsymbol{\nu}. \end{aligned} \tag{7.55}$$

The axion term has no effect on plane-wave propagation due to (5.238). Applying the identity

$$(\overline{\overline{\mathsf{A}}} {}^\wedge_\wedge \overline{\overline{\mathsf{B}}})^T \lfloor \boldsymbol{\alpha} = (\overline{\overline{\mathsf{A}}}^T | \boldsymbol{\alpha}) \wedge \overline{\overline{\mathsf{B}}}^T + (\overline{\overline{\mathsf{B}}}^T | \boldsymbol{\alpha}) \wedge \overline{\overline{\mathsf{A}}}^T, \tag{7.56}$$

we have

$$
\begin{aligned}
-2\mu B(\mathbf{e}_N\lfloor(\overline{\overline{\mathsf{I}}}^T{}^\wedge_\wedge\varepsilon_4\mathbf{e}_4))\lfloor\lfloor\boldsymbol{\nu}\boldsymbol{\nu} &= 2\mu B\boldsymbol{\nu}\rfloor(\mathbf{e}_N\lfloor(\overline{\overline{\mathsf{I}}}^T{}^\wedge_\wedge\varepsilon_4\mathbf{e}_4))\lfloor\boldsymbol{\nu} \\
&= -2\mu B\mathbf{e}_N\lfloor(\boldsymbol{\nu}\wedge(\boldsymbol{\nu}\wedge\varepsilon_4\mathbf{e}_4+(\varepsilon_4\mathbf{e}_4|\boldsymbol{\nu})\wedge\overline{\overline{\mathsf{I}}}^T)) \\
&= -2\mu B(\mathbf{e}_4|\boldsymbol{\nu})\mathbf{e}_N\lfloor(\boldsymbol{\nu}\wedge\varepsilon_4\wedge\overline{\overline{\mathsf{I}}}^T) \\
&= \mathbf{A}_s\lfloor\overline{\overline{\mathsf{I}}}^T. \qquad (7.57)
\end{aligned}
$$

The resulting dyadic is of antisymmetric form depending on the spatial bivector $\mathbf{A}_s$ defined by

$$
\mathbf{A}_s = \alpha\mathbf{e}_N\lfloor(\boldsymbol{\nu}\wedge\varepsilon_4) = \alpha\mathbf{e}_{123}\lfloor\boldsymbol{\nu}_s, \quad \alpha = -2\mu B(\mathbf{e}_4|\boldsymbol{\nu}), \qquad (7.58)
$$

and satisfying

$$
\mathbf{A}_s\wedge\mathbf{A}_s = 0, \quad \mathbf{A}_s\lfloor\varepsilon_4 = 0, \quad \mathbf{A}_s\lfloor\boldsymbol{\nu} = 0, \qquad (7.59)
$$

$$
(\mathbf{A}_s\lfloor\overline{\overline{\mathsf{I}}}^T)^{(2)} = \mathbf{A}_s\mathbf{A}_s, \quad (\mathbf{A}_s\lfloor\overline{\overline{\mathsf{I}}}^T)^{(3)} = 0. \qquad (7.60)
$$

The dyadic dispersion equation (5.240) of the GBI medium can now be expanded as

$$
\begin{aligned}
\mu^3\overline{\overline{\mathsf{D}}}^{(3)}(\boldsymbol{\nu}) &= (\mathbf{A}_s\lfloor\overline{\overline{\mathsf{I}}}^T - \overline{\overline{\mathsf{G}}}^{(2)}\lfloor\lfloor\boldsymbol{\nu}\boldsymbol{\nu})^{(3)} \\
&= (\mathbf{A}_s\lfloor\overline{\overline{\mathsf{I}}}^T)^{(3)} - (\mathbf{A}_s\lfloor\overline{\overline{\mathsf{I}}}^T)^{(2)}{}^\wedge_\wedge(\overline{\overline{\mathsf{G}}}^{(2)}\lfloor\lfloor\boldsymbol{\nu}\boldsymbol{\nu}) \\
&\quad + (\mathbf{A}_s\lfloor\overline{\overline{\mathsf{I}}}^T){}^\wedge_\wedge(\overline{\overline{\mathsf{G}}}^{(2)}\lfloor\lfloor\boldsymbol{\nu}\boldsymbol{\nu})^{(2)} - (\overline{\overline{\mathsf{G}}}^{(2)}\lfloor\lfloor\boldsymbol{\nu}\boldsymbol{\nu})^{(3)} \\
&= -\mathbf{A}_s\mathbf{A}_s{}^\wedge_\wedge(\overline{\overline{\mathsf{G}}}^{(2)}\lfloor\lfloor\boldsymbol{\nu}\boldsymbol{\nu}) + (\mathbf{A}_s\lfloor\lfloor\overline{\overline{\mathsf{I}}}^T){}^\wedge_\wedge(\overline{\overline{\mathsf{G}}}^{(2)}\lfloor\lfloor\boldsymbol{\nu}\boldsymbol{\nu})^{(2)} \\
&\quad - (\overline{\overline{\mathsf{G}}}^{(2)}\lfloor\lfloor\boldsymbol{\nu}\boldsymbol{\nu})^{(3)} \\
&= 0. \qquad (7.61)
\end{aligned}
$$

Since this is a dyadic equation, its symmetric and antisymmetric parts must vanish individually. Now one can show that the antisymmetric equation is identically valid. Writing the symmetric dyadic equation in the form

$$
\overline{\overline{\mathsf{D}}}^{(3)}(\boldsymbol{\nu}) = D(\boldsymbol{\nu})(\mathbf{e}_N\mathbf{e}_N\lfloor\lfloor\boldsymbol{\nu}\boldsymbol{\nu}) = 0, \qquad (7.62)
$$

it is reduced to a scalar equation. Leaving the details of the analysis as an exercise, the resulting equation can be reduced to

$$
\mu^3 D(\boldsymbol{\nu}) = \mu\epsilon' D_+(\boldsymbol{\nu})D_-(\boldsymbol{\nu}) = 0. \qquad (7.63)
$$

The GBI medium is an example of a decomposable (DC) medium since its quartic dispersion equation can be decomposed in two quadratic dispersion equations,

$$
D_+(\boldsymbol{\nu}) = 0, \quad D_-(\boldsymbol{\nu}) = 0, \qquad (7.64)
$$

defined by two dispersion functions which can be given the form of the Gibbsian isotropic medium,

$$D_\pm(\nu) = \overline{\overline{\mathsf{G}}}_\pm || \nu\nu \tag{7.65}$$

$$\overline{\overline{\mathsf{G}}}_\pm = \overline{\overline{\mathsf{G}}}_s - \mu_\pm \epsilon_\pm \mathbf{e}_4 \mathbf{e}_4, \tag{7.66}$$

$$\mu_\pm \epsilon_\pm = -(\mu B \pm \sqrt{(\mu B)^2 - \mu\epsilon'})^2. \tag{7.67}$$

As a check, setting $B = 0$ for vanishing chirality, we obtain $\mu_\pm \epsilon_\pm = \mu\epsilon'$ and, hence, $\overline{\overline{\mathsf{G}}}_\pm = \overline{\overline{\mathsf{G}}}$. This corresponds to the Gibbsian isotropic (simple principal) medium case (5.263), for which the two dispersion equations coincide to $\overline{\overline{\mathsf{G}}}||\nu\nu = 0$.

Actually, the previous results can be more directly obtained through the eigenfield expansions. Because for each eigenfield the GBI medium can be replaced by an equivalent Gibbsian isotropic medium with medium bidyadics $\overline{\overline{\mathsf{M}}}_\pm$ defined by (7.50), the dispersion equations can be directly found from (7.65) and the relation (7.67) results from combining (7.51) and (7.52). It is interesting to note that the axion parameter A, which has an effect on the parameters $\epsilon_\pm$ and $\mu_\pm$, has no effect on the product $\epsilon_\pm \mu_\pm$ which governs the wave propagation.

7.2 THE AXION MEDIUM

The GBI medium (7.20) consisting of its axion part only,

$$\overline{\overline{\mathsf{M}}} = M \overline{\overline{\mathsf{I}}}^{(2)T}, \tag{7.68}$$

can be conceived as the simplest electromagnetic medium since it involves just a single parameter M [53]. Because the axion medium bidyadic is invariant in the affine transformation

$$\overline{\overline{\mathsf{M}}}_a = \overline{\overline{\mathsf{A}}}^{(-2)T} | \overline{\overline{\mathsf{M}}} | \overline{\overline{\mathsf{A}}}^{(2)T} = \overline{\overline{\mathsf{M}}} \tag{7.69}$$

for any dyadic $\overline{\overline{\mathsf{A}}} \in \mathbb{E}_1\mathbb{F}_1$ possessing an inverse (including all 4D rotations) the axion media could also be called 4D isotropic media. In contrast, the special GBI media with only the skewon term,

$$\overline{\overline{\mathsf{M}}} = B \overline{\overline{\mathsf{T}}}^{(2)T} = B(\overline{\overline{\mathsf{I}}}^{(2)T} - 2\overline{\overline{\mathsf{I}}}^T {}^\wedge_\wedge \boldsymbol{\varepsilon}_4 \mathbf{e}_4), \tag{7.70}$$

also labeled as simple-skewon (SS) media [54], depend on the choice of the temporal basis vector $\mathbf{e}_4$ and one-form $\boldsymbol{\varepsilon}_4$ as well as the coefficient B.

The medium equation of axion media,

$$\mathbf{\Psi} = \overline{\overline{\mathsf{M}}}|\mathbf{\Phi} = M\mathbf{\Phi}, \tag{7.71}$$

written for spatial fields as

$$\mathbf{D} = M\mathbf{B}, \quad \mathbf{H} = -M\mathbf{E}, \tag{7.72}$$

corresponds to spatial medium dyadics of the form

$$\overline{\overline{\alpha}} = M\overline{\overline{\mathrm{I}}}_s^{(2)T}, \quad \overline{\overline{\epsilon}}' = 0, \quad \overline{\overline{\mu}}^{-1} = 0, \quad \overline{\overline{\beta}} = -M\overline{\overline{\mathrm{I}}}_s^T. \tag{7.73}$$

Because the dyadic $\overline{\overline{\mu}}$ does not exist, the axion medium cannot be represented in terms of the four Gibbsian medium dyadics $\overline{\overline{\epsilon}}_g, \overline{\overline{\mu}}_g, \overline{\overline{\xi}}_g, \overline{\overline{\zeta}}_g$, at least not in a straightforward manner. In comparison, the medium equations for the SS medium (7.70) differ from (7.72) just for a sign [54].

$$\mathbf{D} = B\mathbf{B}, \quad \mathbf{H} = B\mathbf{E}. \tag{7.74}$$

7.2.1 Perfect Electromagnetic Conductor

Because for $M \to 0$ the medium conditions (7.71) and (7.72) of the axion medium reduce to those of the PMC

$$\mathbf{\Psi} = 0, \quad \Rightarrow \quad \mathbf{D} = 0, \quad \mathbf{H} = 0, \tag{7.75}$$

and for $1/M \to 0$ to those of the PEC,

$$\mathbf{\Phi} = 0, \quad \Rightarrow \quad \mathbf{B} = 0, \quad \mathbf{E} = 0, \tag{7.76}$$

being a generalization of these, the axion medium has been given the name PEMC [30, 55, 56]. The coefficient M can then be called the PEMC admittance.

Assuming both electric and magnetic sources γ_e, γ_m in a PEMC medium with constant and finite M, the Maxwell equations written as

$$\mathbf{d} \wedge \mathbf{\Phi} = \gamma_e/M, \quad \mathbf{d} \wedge \mathbf{\Phi} = \gamma_m, \tag{7.77}$$

impose a condition for the sources,

$$\gamma_e = M\gamma_m, \tag{7.78}$$

restricting the choice of possible sources in a PEMC medium. Such a restriction is not too extraordinary. In fact, for the same reason, the sources in the PEC and PMC media must satisfy the respective conditions $\gamma_m = 0$ and $\gamma_e = 0$.

PEMC is an example of a medium for which the dispersion equation of a plane wave $D(\mathbf{\nu}) = 0$ is identically satisfied for any one-form $\mathbf{\nu}$ due to (5.238). In fact, since the plane-wave field $\mathbf{\Phi}$ satisfies the equations

$$\mathbf{\nu} \wedge \mathbf{\Phi} = 0, \quad M(\mathbf{\nu} \wedge \mathbf{\Phi}) = 0, \tag{7.79}$$

which for finite M is the same equation, a solution can be found for any choice of $\mathbf{\nu}$. Of course, at the interface of the PEMC medium, the interface conditions impose further restrictions for the one-form $\mathbf{\nu}$ and the two-form $\mathbf{\Phi}$ of a plane wave transmitted into the medium. As another property, the antisymmetric bidyadic $\mathbf{\Psi\Phi} - \mathbf{\Phi\Psi}$ vanishes identically for the PEMC medium whence, for example, the energy density and Poynting two-form in the medium are zero. Since PEMC has appeared most interesting when defining electromagnetic boundary conditions at its interface, in which case the exterior fields can be computed without knowing the interior fields, the question of interior fields can be left to various realizations of the PEMC medium in terms of more conventional media.

7.2.2 PEMC as Limiting Case of GBI Medium

Although it is not possible to define the PEMC medium in terms of Gibbsian dyadics $\overline{\overline{\epsilon}}_g, \overline{\overline{\mu}}_g, \overline{\overline{\xi}}_g, \overline{\overline{\zeta}}_g$ in a simple manner, we may consider PEMC as a limiting case of skewon-free GBI medium by letting the principal part tend to zero. For that purpose, let us start from (7.20) by setting $B = 0$ and inserting an extra parameter q as

$$\begin{aligned}\overline{\overline{\mathsf{M}}} &= M\overline{\overline{\mathrm{I}}}^{(2)T} - \frac{1}{q\mu}(\boldsymbol{\varepsilon}_N \lfloor \overline{\overline{\mathsf{G}}}^{(2)}) \\ &= M\left(\overline{\overline{\mathrm{I}}}_s^{(2)T} + \frac{1}{q}(\epsilon\overline{\overline{\mathsf{C}}}_s \wedge \mathbf{e}_4 + \boldsymbol{\varepsilon}_4 \wedge (\mu\overline{\overline{\mathsf{C}}}_s)^{-1}) + \overline{\overline{\mathrm{I}}}_s^T {}^\wedge_\wedge \boldsymbol{\varepsilon}_4\mathbf{e}_4\right).\end{aligned} \tag{7.80}$$

Obviously, for $q \to \infty$, the expression (7.80) approaches the medium bidyadic of a PEMC medium.

The corresponding Gibbsian medium dyadics can be found after comparing (7.80) term wise with the expansion (5.87), written here as

$$\overline{\overline{\mathsf{M}}} = \boldsymbol{\varepsilon}_{123}\lfloor\overline{\overline{\epsilon}}_g \wedge \mathbf{e}_4 + (\boldsymbol{\varepsilon}_{123}\lfloor\overline{\overline{\xi}}_g + \boldsymbol{\varepsilon}_4 \wedge \overline{\overline{\mathrm{I}}}_s^T)|\overline{\overline{\mu}}_g^{-1}|(\overline{\overline{\mathrm{I}}}\rfloor\mathbf{e}_{123} - \overline{\overline{\zeta}}_g \wedge \mathbf{e}_4). \tag{7.81}$$

This leads to the following Gibbsian representation for the PEMC medium

$$\begin{pmatrix} \overline{\overline{\epsilon}}_g & \overline{\overline{\xi}}_g \\ \overline{\overline{\zeta}}_g & \overline{\overline{\mu}}_g \end{pmatrix} = \lim_{q\to\infty} q \begin{pmatrix} M(1+(\mu\epsilon/q^2)) & 1 \\ 1 & 1/M \end{pmatrix} \overline{\overline{\mathsf{G}}}_s, \tag{7.82}$$

It is noteworthy that in this representation all four Gibbsian medium dyadics obtain infinite magnitudes in the limit. However, in spite of this, the determinant of the matrix multiplying the dyadic $\overline{\overline{\mathsf{G}}}_s$ has the finite value $\mu\epsilon/q$ and approaches zero in the limit [57]. Also, the dyadic product $\overline{\overline{\epsilon}}_g \cdot \overline{\overline{\mu}}_g^{-1} = M^2\overline{\overline{\mathsf{I}}}_s$ remains finite. Applying the representation (7.82), computer codes designed for media expressed in terms of Gibbsian dyadics can be used for problems involving PEMC structures by letting q assume large values.

7.2.3 PEMC Boundary Problems

PEC and PMC have been useful concepts because they define ideal boundary conditions at the interface. The same can be anticipated to be the case for the more general PEMC medium. Basic problems involving PEC boundaries concern closed regions like waveguides and cavity resonators and open problems like scattering from PEC objects. As was shown in Chapter 6, there exist duality transformations capable of transforming a PEMC boundary to a PEC boundary. Actually, it is possible to choose the transformation so that PEMC boundaries can be transformed to similar PEC boundaries without changing the adjacent medium when it is simple enough [55].

To show this, let us invoke the three-parameter duality transformation rule (6.59) which transforms a medium bidyadic $\overline{\overline{\mathsf{M}}}$ to $\overline{\overline{\mathsf{M}}}_d$ as

$$Z_d\overline{\overline{\mathsf{M}}}_d = (\sin\theta\, \overline{\overline{\mathsf{I}}}^{(2)T} + \cos\theta\, Z\overline{\overline{\mathsf{M}}})|(\cos\theta\, \overline{\overline{\mathsf{I}}}^{(2)T} - \sin\theta\, Z\overline{\overline{\mathsf{M}}})^{-1}. \tag{7.83}$$

Inserting the PEMC medium bidyadic $\overline{\overline{\mathsf{M}}} = M\overline{\overline{\mathsf{I}}}^{(2)T}$ in (7.83) we obtain another PEMC medium bidyadic $\overline{\overline{\mathsf{M}}}_d = M_d\overline{\overline{\mathsf{I}}}^{(2)T}$ obeying the following dependence on M and the transformation parameters Z, Z_d, θ,

$$Z_dM_d = \frac{\sin\theta + \cos\theta\, ZM}{\cos\theta - \sin\theta\, ZM}. \tag{7.84}$$

Thus, if we wish to transform PEMC to PEC, that is, require that the magnitude of $\overline{\overline{\mathsf{M}}}_d$ be infinitely large, we must choose θ to satisfy

$$\cot\theta = ZM. \tag{7.85}$$

The transformation rule (7.83) then becomes

$$Z_d\overline{\overline{\mathsf{M}}}_d = (\overline{\overline{\mathsf{I}}}^{(2)T} + Z^2 M\overline{\overline{\mathsf{M}}})|(ZM\,\overline{\overline{\mathsf{I}}}^{(2)T} - Z\overline{\overline{\mathsf{M}}})^{-1}, \tag{7.86}$$

where the parameters Z and Z_d may still be arbitrary at this stage.

To define these two parameters we may require that the medium outside the PEMC boundary be the same after the duality transformation. This is not possible for just any given medium, actually the medium must belong to the class of self-dual media. In Section 6.3.1, it was shown that the simple principal (Gibbsian isotropic) medium represented by the medium bidyadic

$$\overline{\overline{\mathsf{M}}} = -\frac{1}{\mu}\epsilon_N\lfloor\overline{\overline{\mathsf{G}}}^{(2)}, \quad \overline{\overline{\mathsf{G}}} = \overline{\overline{\mathsf{G}}}_s - \mu\epsilon\mathbf{e}_4\mathbf{e}_4 \tag{7.87}$$

is invariant if the parameters are chosen as

$$Z = Z_d = \sqrt{\mu/\epsilon}. \tag{7.88}$$

In conclusion, the three-parameter duality transformation defined by parameters Z, Z_d satisfying (7.88) and θ satisfying $\cot\theta = M\sqrt{\mu/\epsilon}$ performs two tasks: changes PEMC boundaries to PEC boundaries and keeps the Gibbsian isotropic medium unchanged. After solving the transformed problem with the PEC boundary, the solution can be transformed back through the inverse duality transformation to yield the solution for the corresponding problem associated with the PEMC boundary.

7.3 SKEWON–AXION MEDIA

As was shown by (5.96), a medium defined by a bidyadic of the form

$$\overline{\overline{\mathsf{M}}} = (\overline{\overline{\mathsf{B}}}_o{}^\wedge_\wedge\overline{\overline{\mathsf{I}}})^T, \qquad \mathrm{tr}\overline{\overline{\mathsf{B}}}_o = 0, \tag{7.89}$$

where $\overline{\overline{\mathsf{B}}}_o \in \mathbb{E}_1\mathbb{F}_1$ is a trace-free dyadic, has only a skewon part. More generally, media with no principal part can be represented by bidyadics of the form

$$\overline{\overline{\mathsf{M}}} = (\overline{\overline{\mathsf{B}}}{}^\wedge_\wedge\overline{\overline{\mathsf{I}}})^T = (\overline{\overline{\mathsf{I}}}{}^\wedge_\wedge\overline{\overline{\mathsf{B}}})^T, \tag{7.90}$$

in terms of a dyadic $\overline{\overline{\mathsf{B}}}$ which may have nonzero trace. Such media are called skewon–axion media [9] (or "IB-media" [58]). Because the dyadic $\overline{\overline{\mathsf{B}}}$ has $4 \times 4 = 16$ free parameters, the bidyadic $\overline{\overline{\mathsf{M}}}$ is defined by the same number of parameters instead of $6 \times 6 = 36$ parameters corresponding to the most general medium bidyadic. Here we must (again) emphasize that the inverse bidyadic $\overline{\overline{\mathsf{N}}} = \overline{\overline{\mathsf{M}}}^{-1}$ is not generally of the form $(\overline{\overline{\mathsf{A}}}{}^{\wedge}_{\wedge}\overline{\mathsf{I}})^T$. However, the inverse of a pure skewon $\overline{\overline{\mathsf{M}}}$ is of the skewon form $\overline{\overline{\mathsf{N}}} = (\overline{\overline{\mathsf{A}}}_o{}^{\wedge}_{\wedge}\overline{\mathsf{I}})^T$ as shown by (3.110).

To be of the form (7.90), the medium bidyadic $\overline{\overline{\mathsf{M}}}$ must satisfy a certain condition. It can be found by contracting (7.90) as

$$\overline{\overline{\mathsf{M}}}\lfloor\lfloor\overline{\mathsf{I}} = (\overline{\overline{\mathsf{B}}}{}^{\wedge}_{\wedge}\overline{\mathsf{I}})^T\lfloor\lfloor\overline{\mathsf{I}} = (\mathrm{tr}\overline{\mathsf{I}})\overline{\overline{\mathsf{B}}}^T + (\mathrm{tr}\overline{\overline{\mathsf{B}}})\overline{\mathsf{I}}^T - 2\overline{\overline{\mathsf{B}}}^T \tag{7.91}$$

$$= 2\overline{\overline{\mathsf{B}}}^T + (\mathrm{tr}\overline{\overline{\mathsf{B}}})\overline{\mathsf{I}}^T. \tag{7.92}$$

The trace of the left side is $\mathrm{tr}(\overline{\overline{\mathsf{M}}}\lfloor\lfloor\overline{\mathsf{I}}) = 2\mathrm{tr}\overline{\overline{\mathsf{M}}}$ while the right side yields $6\mathrm{tr}\overline{\overline{\mathsf{B}}}$, whence

$$\mathrm{tr}\overline{\overline{\mathsf{M}}} = 3\mathrm{tr}\overline{\overline{\mathsf{B}}}, \tag{7.93}$$

which allows us to solve $\overline{\overline{\mathsf{B}}}$ from (7.92) as

$$\overline{\overline{\mathsf{B}}} = \frac{1}{2}(\overline{\overline{\mathsf{M}}}\lfloor\lfloor\overline{\mathsf{I}})^T - \frac{1}{6}\overline{\mathsf{I}}(\mathrm{tr}\overline{\overline{\mathsf{M}}}). \tag{7.94}$$

Substituting (7.94) in (7.90) we obtain a condition for the medium bidyadic $\overline{\overline{\mathsf{M}}}$ to be of the skewon–axion form (7.90),

$$6\overline{\overline{\mathsf{M}}} + 2(\mathrm{tr}\overline{\overline{\mathsf{M}}})\overline{\overline{\mathsf{I}}}^{(2)T} - 3(\overline{\overline{\mathsf{M}}}\lfloor\lfloor\overline{\mathsf{I}}){}^{\wedge}_{\wedge}\overline{\mathsf{I}}^T = 0. \tag{7.95}$$

Because (7.95) follows from (7.90), it is a necessary condition. It is also sufficient because the form of (7.95) proves that $\overline{\overline{\mathsf{M}}}$ can be expressed as $\overline{\mathsf{I}}{}^{\wedge}_{\wedge}\overline{\overline{\mathsf{B}}}$ with $\overline{\overline{\mathsf{B}}}$ defined by (7.94).

The PEMC or axion medium bidyadic is a special case of (7.90) with

$$\overline{\overline{\mathsf{M}}} = M\overline{\overline{\mathsf{I}}}^{(2)T} = (\overline{\overline{\mathsf{B}}}{}^{\wedge}_{\wedge}\overline{\mathsf{I}})^T, \quad \overline{\overline{\mathsf{B}}} = \frac{M}{2}\overline{\mathsf{I}}, \quad M = \frac{1}{2}\mathrm{tr}\overline{\overline{\mathsf{B}}}. \tag{7.96}$$

The axion and skewon parts of the skewon–axion medium can be separated as

$$\overline{\overline{\mathsf{M}}} = M\overline{\overline{\mathsf{I}}}^{(2)T} + \overline{\overline{\mathsf{M}}}_o, \quad M = \frac{1}{6}\mathrm{tr}\overline{\overline{\mathsf{M}}}, \tag{7.97}$$

with $\mathrm{tr}\overline{\overline{\mathsf{M}}}_o = 0$.

7.3.1 Plane Wave in Skewon–Axion Medium

Let us consider a plane wave in the skewon–axion medium defined by (7.90). Applying the identity

$$(\overline{\overline{\mathsf{A}}}{}^{\wedge}_{\wedge}\overline{\overline{\mathsf{B}}})^T|(\boldsymbol{\alpha}\wedge\boldsymbol{\beta}) = (\overline{\overline{\mathsf{A}}}{}^T|\boldsymbol{\alpha})\wedge(\overline{\overline{\mathsf{B}}}{}^T|\boldsymbol{\beta}) + (\overline{\overline{\mathsf{B}}}{}^T|\boldsymbol{\alpha})\wedge(\overline{\overline{\mathsf{A}}}{}^T|\boldsymbol{\beta}), \tag{7.98}$$

valid for any dyadics $\overline{\overline{\mathsf{A}}}, \overline{\overline{\mathsf{B}}} \in \mathbb{E}_1\mathbb{F}_1$ and one-forms $\boldsymbol{\alpha}, \boldsymbol{\beta}$, the dispersion equation for the potential one-form $\boldsymbol{\phi}$ can be derived as

$$\begin{aligned} \boldsymbol{\nu}\wedge\boldsymbol{\Psi} &= \boldsymbol{\nu}\wedge\overline{\overline{\mathsf{M}}}|(\boldsymbol{\nu}\wedge\boldsymbol{\phi}) \\ &= \boldsymbol{\nu}\wedge(\overline{\overline{\mathsf{B}}}{}^T{}^{\wedge}_{\wedge}\overline{\overline{\mathsf{I}}}{}^T)|(\boldsymbol{\nu}\wedge\boldsymbol{\phi}) \\ &= \boldsymbol{\nu}\wedge(\overline{\overline{\mathsf{B}}}{}^T|\boldsymbol{\nu})\wedge\boldsymbol{\phi} + \boldsymbol{\nu}\wedge\boldsymbol{\nu}\wedge(\overline{\overline{\mathsf{B}}}{}^T|\boldsymbol{\phi}) \\ &= \boldsymbol{\nu}\wedge(\boldsymbol{\nu}|\overline{\overline{\mathsf{B}}})\wedge\boldsymbol{\phi} = 0. \end{aligned} \tag{7.99}$$

Assuming that $\boldsymbol{\nu}$ and $\boldsymbol{\nu}|\overline{\overline{\mathsf{B}}}$ are linearly independent: $\boldsymbol{\nu}\wedge(\boldsymbol{\nu}|\overline{\overline{\mathsf{B}}}) \neq 0$, that is, $\boldsymbol{\nu}$ is not a left eigen one-form of the dyadic $\overline{\overline{\mathsf{B}}}$, $\boldsymbol{\phi}$ must be linearly dependent on the two one-forms $\boldsymbol{\nu}$ and $\boldsymbol{\nu}|\overline{\overline{\mathsf{B}}}$ as

$$\boldsymbol{\phi} = c_1\boldsymbol{\nu} + c_2\boldsymbol{\nu}|\overline{\overline{\mathsf{B}}}. \tag{7.100}$$

In this case the field two-forms can be expressed as

$$\boldsymbol{\Phi} = \boldsymbol{\nu}\wedge\boldsymbol{\phi} = c_2\boldsymbol{\nu}\wedge(\boldsymbol{\nu}|\overline{\overline{\mathsf{B}}}), \tag{7.101}$$

$$\boldsymbol{\Psi} = (\overline{\overline{\mathsf{B}}}{}^{\wedge}_{\wedge}\overline{\overline{\mathsf{I}}})^T|(\boldsymbol{\nu}\wedge\boldsymbol{\phi}) = c_2\boldsymbol{\nu}\wedge(\boldsymbol{\nu}|\overline{\overline{\mathsf{B}}}{}^2). \tag{7.102}$$

There is no restriction for the wave one-form $\boldsymbol{\nu}$ except $\boldsymbol{\nu}\wedge(\boldsymbol{\nu}|\overline{\overline{\mathsf{B}}}) \neq 0$ which excludes the pure axion medium. Actually, the wave one-form $\boldsymbol{\nu}$ in the skewon–axion medium can be freely chosen, after which the potential one-form can be determined by (7.99). On the other hand, if $\boldsymbol{\nu}$ is chosen to be an eigen one-form of $\overline{\overline{\mathsf{B}}}$, satisfying

$$\boldsymbol{\nu}|\overline{\overline{\mathsf{B}}} = \lambda\boldsymbol{\nu}, \quad \lambda \neq 0, \tag{7.103}$$

(7.99) is valid for any one-form $\boldsymbol{\phi}$. In this case (7.101) and (7.102) are no longer valid. In such a case there is a lot of freedom for $\boldsymbol{\Phi}$ to choose from, because it is only required to satisfy $\boldsymbol{\nu}\wedge\boldsymbol{\Phi} = 0$.

The previous analysis shows that skewon–axion media fall in the class of NDE media, media with no dispersion equation. This can be shown more directly by expanding the dispersion dyadic (5.237) as

$$\begin{aligned}\overline{\overline{\mathsf{D}}}(\boldsymbol{\nu}) &= \overline{\overline{\mathsf{M}}}_m\lfloor\lfloor\boldsymbol{\nu}\boldsymbol{\nu} = -\boldsymbol{\nu}\rfloor(\mathbf{e}_N\lfloor(\overline{\overline{\mathsf{B}}}{}^\wedge_\wedge\overline{\overline{\mathsf{I}}})^T)\lfloor\boldsymbol{\nu}\\ &= -\boldsymbol{\nu}\rfloor(\mathbf{e}_N\lfloor((\overline{\overline{\mathsf{B}}}{}^T|\boldsymbol{\nu})\wedge\overline{\overline{\mathsf{I}}}{}^T+\boldsymbol{\nu}\wedge\overline{\overline{\mathsf{B}}}{}^T))\\ &= \mathbf{e}_N\lfloor(\boldsymbol{\nu}\wedge(\boldsymbol{\nu}|\overline{\overline{\mathsf{B}}})\wedge\overline{\overline{\mathsf{I}}}{}^T)\\ &= \mathbf{F}\lfloor\overline{\overline{\mathsf{I}}}{}^T, \end{aligned} \tag{7.104}$$

where the bivector $\mathbf{F}$ is defined by

$$\mathbf{F} = \mathbf{e}_N\lfloor(\boldsymbol{\nu}\wedge(\boldsymbol{\nu}|\overline{\overline{\mathsf{B}}})). \tag{7.105}$$

Because of the form (7.104), the dispersion dyadic of a skewon–axion medium is antisymmetric. From

$$\mathbf{F}\wedge\mathbf{F} = \mathbf{e}_N\boldsymbol{\varepsilon}_N|(\boldsymbol{\nu}\wedge(\boldsymbol{\nu}|\overline{\overline{\mathsf{B}}})\wedge\boldsymbol{\nu}\wedge(\boldsymbol{\nu}|\overline{\overline{\mathsf{B}}})) = 0, \tag{7.106}$$

it follows that $\mathbf{F}$ is a simple bivector. Writing $\mathbf{F} = \mathbf{a}\wedge\mathbf{b}$, the dyadic dispersion equation can be shown to be satisfied identically for any $\boldsymbol{\nu}$,

$$\overline{\overline{\mathsf{D}}}{}^{(3)}(\boldsymbol{\nu}) = (\mathbf{F}\lfloor\overline{\overline{\mathsf{I}}}{}^T)^{(3)} = (\mathbf{ba}-\mathbf{ab})^{(3)} = 0. \tag{7.107}$$

7.3.2 Gibbsian Representation

To define a skewon–axion medium (7.90) in terms of Gibbsian dyadics, let us start by expanding the dyadic $\overline{\overline{\mathsf{B}}}$ as

$$\overline{\overline{\mathsf{B}}} = \overline{\overline{\mathsf{B}}}_s + \mathbf{b}_s\boldsymbol{\varepsilon}_4 + \mathbf{e}_4\boldsymbol{\beta}_s + b\mathbf{e}_4\boldsymbol{\varepsilon}_4, \tag{7.108}$$

whence the medium bidyadic has the form

$$\overline{\overline{\mathsf{M}}} = \overline{\overline{\mathsf{I}}}{}^T_s{}^\wedge_\wedge\overline{\overline{\mathsf{B}}}{}^T_s - \boldsymbol{\varepsilon}_4\wedge\overline{\overline{\mathsf{I}}}{}^T_s\wedge\mathbf{b}_s - \boldsymbol{\beta}_s\wedge\overline{\overline{\mathsf{I}}}{}^T_s\wedge\mathbf{e}_4 - \boldsymbol{\varepsilon}_4\wedge(\overline{\overline{\mathsf{B}}}{}^T_s + b\overline{\overline{\mathsf{I}}}{}^T_s)\wedge\mathbf{e}_4. \tag{7.109}$$

Comparing with (5.65), we obtain expressions for the spatial medium dyadics

$$\overline{\overline{\alpha}} = \overline{\overline{\mathsf{I}}}{}^T_s{}^\wedge_\wedge\overline{\overline{\mathsf{B}}}{}^T_s = \mathrm{tr}\overline{\overline{\mathsf{B}}}_s\,\overline{\overline{\mathsf{I}}}{}^{(2)T}_s - \boldsymbol{\varepsilon}_{123}\mathbf{e}_{123}\lfloor\lfloor\overline{\overline{\mathsf{B}}}_s, \tag{7.110}$$

$$\overline{\overline{\epsilon}}' = -\boldsymbol{\beta}_s\wedge\overline{\overline{\mathsf{I}}}{}^T_s = -\overline{\overline{\mathsf{I}}}{}^{(2)}\lfloor\boldsymbol{\beta}_s, \tag{7.111}$$

$$\overline{\overline{\mu}}{}^{-1} = -\mathbf{b}_s\rfloor\overline{\overline{\mathsf{I}}}{}^{(2)T}_s = -\overline{\overline{\mathsf{I}}}{}^T_s\wedge\mathbf{b}_s, \tag{7.112}$$

$$\overline{\overline{\beta}} = -(\overline{\overline{\mathsf{B}}}{}^T_s + b\overline{\overline{\mathsf{I}}}{}^T_s). \tag{7.113}$$

Because of $\overline{\overline{\epsilon}}'|\boldsymbol{\beta}_s = 0$ and $\mathbf{b}_s|\overline{\overline{\mu}}^{-1} = 0$, $\overline{\overline{\epsilon}}'$ and $\overline{\overline{\mu}}^{-1}$ do not possess spatial inverses. The medium equations can be written as

$$\mathbf{D} = \mathrm{tr}\overline{\overline{\mathsf{B}}}_s\,\mathbf{B} - \boldsymbol{\varepsilon}_{123}\lfloor\overline{\overline{\mathsf{B}}}_s|(\mathbf{e}_{123}\lfloor\mathbf{B}) - \boldsymbol{\beta}_s \wedge \mathbf{E}, \tag{7.114}$$

$$\mathbf{H} = -\mathbf{b}_s\rfloor\mathbf{B} - \overline{\overline{\mathsf{B}}}_s^T|\mathbf{E} - b\mathbf{E}. \tag{7.115}$$

In terms of Gibbsian field vectors, (7.114) and (7.115) take the form (see Appendix B)

$$\mathbf{D}_g = (\mathrm{tr}\overline{\overline{\mathsf{B}}}_s)\mathbf{B}_g - \overline{\overline{\mathsf{B}}}_g \cdot \mathbf{B}_g - \boldsymbol{\beta}_g \times \mathbf{E}_g, \tag{7.116}$$

$$\mathbf{H}_g = -\mathbf{b}_s \times \mathbf{B}_g - \overline{\overline{\mathsf{B}}}_g^T \cdot \mathbf{E}_g - b\mathbf{E}_g, \tag{7.117}$$

where we define

$$\overline{\overline{\mathsf{B}}}_g = \overline{\overline{\mathsf{B}}}_s|\overline{\overline{\mathsf{G}}}_s, \quad \boldsymbol{\beta}_g = \overline{\overline{\mathsf{G}}}_s|\boldsymbol{\beta}_s. \tag{7.118}$$

It appears instructive to check the previously derived plane-wave properties also through the Gibbsian representation of the skewon–axion medium, (7.116), (7.117). Assuming a time-harmonic plane wave with $\exp(-j\mathbf{k}\cdot\mathbf{r})$ dependence and denoting $\mathbf{p} = \mathbf{k}/\omega$, the Maxwell equations become

$$\mathbf{p} \times \mathbf{E}_g = \mathbf{B}_g, \tag{7.119}$$

$$\mathbf{p} \times \mathbf{H}_g = -\mathbf{D}_g. \tag{7.120}$$

Substituting $\mathbf{B}_g$ in terms of $\mathbf{E}_g$ from (7.119) in (7.116) and (7.117) and substituting $\mathbf{D}_g$ and $\mathbf{H}_g$ in (7.120) leaves us with the following equation,

$$(\mathrm{tr}\overline{\overline{\mathsf{B}}}_s + \mathbf{b}_s \cdot \mathbf{p} - b)\mathbf{p} \times \mathbf{E}_g - \boldsymbol{\beta}_g \times \mathbf{E}_g - (\mathbf{p} \times \overline{\overline{\mathsf{B}}}_g^T + \overline{\overline{\mathsf{B}}}_g \times \mathbf{p}) \cdot \mathbf{E}_g = 0. \tag{7.121}$$

The dyadic in brackets is actually antisymmetric and can be expressed in the form

$$\mathbf{p} \times \overline{\overline{\mathsf{B}}}_g^T + \overline{\overline{\mathsf{B}}}_g \times \mathbf{p} = (\mathrm{tr}\overline{\overline{\mathsf{B}}}_s\,\mathbf{p} - \mathbf{p} \cdot \overline{\overline{\mathsf{B}}}_g) \times \overline{\overline{\mathsf{I}}}_g, \tag{7.122}$$

where the Gibbsian unit dyadic equals the spatial metric dyadic, $\overline{\overline{\mathsf{I}}}_g = \overline{\overline{\mathsf{G}}}_s$. Derivation of (7.122) is left as an exercise. Inserting (7.122) in (7.121) yields an equation of the form

$$\overline{\overline{\mathsf{D}}}(\mathbf{p}) \cdot \mathbf{E}_g = \mathbf{q}(\mathbf{p}) \times \mathbf{E}_g = 0, \tag{7.123}$$

where the vector function of $\mathbf{p}$ is defined by

$$\mathbf{q}(\mathbf{p}) = (\mathbf{b}_s \cdot \mathbf{p} - b)\mathbf{p} - \boldsymbol{\beta}_g + \mathbf{p} \cdot \overline{\overline{\mathsf{B}}}_g. \tag{7.124}$$

In an analysis of more conventional media we proceed from here by requiring $\det\overline{\overline{\mathsf{D}}}(\mathbf{p}) = 0$, which is called the dispersion equation, yielding the possible values for the normalized propagation vector $\mathbf{p}$ as its solutions. In the present case $\overline{\overline{\mathsf{D}}}(\mathbf{p}) = \mathbf{q}(\mathbf{p}) \times \overline{\overline{\mathsf{I}}}_g$ is an antisymmetric dyadic, whence $\det\overline{\overline{\mathsf{D}}}(\mathbf{p}) = 0$ is identically satisfied for any $\mathbf{p}$, which means that a plane wave is possible for any chosen vector $\mathbf{p}$. From (7.123) we see that the corresponding electric field vector must be parallel to the vector $\mathbf{q}(\mathbf{p})$. Also, the field vector $\mathbf{B}_g$ must be parallel to $\mathbf{p} \times \mathbf{q}(\mathbf{p}) = \boldsymbol{\beta}_g \times \mathbf{p} - \mathbf{p} \cdot \overline{\overline{\mathsf{B}}}_g \times \mathbf{p}$. However, if $\mathbf{p}$ is chosen so that $\mathbf{q}(\mathbf{p}) = 0$, there is no restriction to the field vector $\mathbf{E}_g$ due to (7.123).

It is also interesting to study the converse, what media would make the dispersion dyadic $\overline{\overline{\mathsf{D}}}(\mathbf{p})$ antisymmetric for any Gibbsian vector $\mathbf{p}$. Inserting the general medium equations in (7.120) and applying (7.119), we obtain

$$\overline{\overline{\mathsf{D}}}(\mathbf{p}) = \mathbf{p} \times \overline{\overline{\mu}}_g^{-1} \times \mathbf{p} + \mathbf{p} \times \overline{\overline{\beta}}_g + \overline{\overline{\alpha}}_g \times \mathbf{p} + \overline{\overline{\epsilon}}'_g = 0. \qquad (7.125)$$

Requiring $\overline{\overline{\mathsf{D}}}(\mathbf{p})$ to be antisymmetric for any vector $\mathbf{p}$, terms with different orders of $\mathbf{p}$ must be individually antisymmetric. This implies that the dyadics $\overline{\overline{\mu}}_g^{-1}$ and $\overline{\overline{\epsilon}}'_g$ must be antisymmetric and $\overline{\overline{\alpha}}_g^T = \overline{\overline{\beta}}_g$, which requires that the principal part of the medium bidyadic must vanish.

7.3.3 Boundary Conditions

Like the pure axion medium, the skewon–axion media may also appear somewhat strange as electromagnetic media and one has to wait for engineering applications to change this attitude. However, the skewon–axion media do have importance in creating useful boundary conditions at the medium interface. This was already demonstrated by PEC, PMC and PEMC media, which are all special cases of the axion medium. In Chapter 5, it was shown that boundary conditions can be defined at the interface of media defined by certain medium bidyadics without having to know about the fields in the medium. In the present case we must know some properties of the fields in the medium to be able to define boundary conditions at the interface.

Let us study conditions for the plane-wave fields at the boundary $\boldsymbol{\varepsilon}_3|\mathbf{x} = 0$ of certain skewon–axion media occupying the region $\boldsymbol{\varepsilon}_3|\mathbf{x} < 0$. The medium is defined by the choice of the dyadic $\overline{\overline{\mathsf{B}}}$. The pure axion medium is known to produce the PEMC boundary condition. Let us consider a few cases with simple extensions to the axion medium.

Case 1

The axion medium can be extended by adding a simple dyadic term $\mathbf{a}\boldsymbol{\alpha}$ to the defining dyadic $\overline{\overline{\mathsf{B}}}$ as

$$\overline{\overline{\mathsf{B}}} = B\overline{\overline{\mathsf{I}}} + \mathbf{a}\boldsymbol{\alpha}, \tag{7.126}$$

$$\overline{\overline{\mathsf{B}}}^2 = B^2\overline{\overline{\mathsf{I}}} + (2B + \mathbf{a}|\boldsymbol{\alpha})\mathbf{a}\boldsymbol{\alpha}, \tag{7.127}$$

where $\boldsymbol{\alpha}$ is a one-form and $\mathbf{a}$ is a vector. In this case the field two-forms of a plane wave in the medium (7.101) and (7.102) can be expressed as

$$\boldsymbol{\Phi} = \boldsymbol{\nu} \wedge \boldsymbol{\alpha}(\mathbf{a}|\boldsymbol{\nu}), \tag{7.128}$$

$$\boldsymbol{\Psi} = (2B + \mathbf{a}|\boldsymbol{\alpha})\boldsymbol{\nu} \wedge \boldsymbol{\alpha}(\mathbf{a}|\boldsymbol{\nu}), \tag{7.129}$$

whence they satisfy the conditions

$$\boldsymbol{\alpha} \wedge \boldsymbol{\Phi} = 0, \qquad \boldsymbol{\alpha} \wedge \boldsymbol{\Psi} = 0 \tag{7.130}$$

for any chosen one-form $\boldsymbol{\nu}$. Let us first assume that $\boldsymbol{\alpha}$ is a temporal one-form $\boldsymbol{\alpha} = \boldsymbol{\varepsilon}_4$. In this case the conditions (7.130) yield

$$\boldsymbol{\varepsilon}_4 \wedge \boldsymbol{\Phi} = \boldsymbol{\varepsilon}_4 \wedge \mathbf{B} = 0, \quad \Rightarrow \quad \mathbf{B} = 0, \tag{7.131}$$

$$\boldsymbol{\varepsilon}_4 \wedge \boldsymbol{\Psi} = \boldsymbol{\varepsilon}_4 \wedge \mathbf{D} = 0, \quad \Rightarrow \quad \mathbf{D} = 0. \tag{7.132}$$

Since the fields $\mathbf{B}$ and $\mathbf{D}$ of the plane wave must vanish in the medium, from continuity at an interface defined by $\boldsymbol{\varepsilon}_3|\mathbf{x} = 0$, the fields must satisfy the boundary conditions

$$\boldsymbol{\varepsilon}_3 \wedge \mathbf{D} = 0, \qquad \boldsymbol{\varepsilon}_3 \wedge \mathbf{B} = 0, \tag{7.133}$$

known as the DB conditions. The corresponding Gibbsian conditions are (5.172)

$$\mathbf{e}_3 \cdot \mathbf{D}_g = 0, \quad \mathbf{e}_3 \cdot \mathbf{B}_g = 0, \tag{7.134}$$

for the Gibbsian vector fields.

Because (7.133) are linear conditions for the fields and independent of the one-form $\boldsymbol{\nu}$, they are valid for any sum or integral of plane waves and, consequently, for fields created by any sources. Actually, they could have been derived for general fields without the plane-wave assumption [59].

Case 2

As another special case of (7.126), (7.127), let us assume that $\boldsymbol{\alpha}$ is a spatial one-form, $\boldsymbol{\alpha} = \boldsymbol{\varepsilon}_1$. In this case the fields in the medium satisfy

$$\boldsymbol{\varepsilon}_1 \wedge \boldsymbol{\Phi} = 0, \quad \Rightarrow \quad \boldsymbol{\varepsilon}_1 \wedge \mathbf{B} = 0, \quad \boldsymbol{\varepsilon}_1 \wedge \mathbf{E} = 0 \tag{7.135}$$

$$\boldsymbol{\varepsilon}_1 \wedge \boldsymbol{\Psi} = 0, \quad \Rightarrow \quad \boldsymbol{\varepsilon}_1 \wedge \mathbf{D} = 0, \quad \boldsymbol{\varepsilon}_1 \wedge \mathbf{H} = 0. \tag{7.136}$$

In this case the conditions for **B** and **D** are not continuous through the interface while those for **E** and **H** yield the boundary conditions

$$\boldsymbol{\varepsilon}_3 \wedge \boldsymbol{\varepsilon}_1 \wedge \mathbf{E} = 0, \tag{7.137}$$

$$\boldsymbol{\varepsilon}_3 \wedge \boldsymbol{\varepsilon}_1 \wedge \mathbf{H} = 0. \tag{7.138}$$

For the Gibbsian fields they correspond to (5.183)

$$(\mathbf{e}_3 \times \mathbf{e}_1) \cdot \mathbf{E}_g = \mathbf{e}_2 \cdot \mathbf{E}_g = 0, \tag{7.139}$$

$$(\mathbf{e}_3 \times \mathbf{e}_1) \cdot \mathbf{H}_g = \mathbf{e}_2 \cdot \mathbf{H}_g = 0, \tag{7.140}$$

and can be identified as the SH boundary conditions [39]. Here, $\mathbf{e}_2$ is a unit vector tangential to the interface $\mathbf{e}_3 \cdot \mathbf{r} = 0$.

Case 3

As the next possibility let us extend the previous dyadic $\overline{\overline{\mathsf{B}}}$ by adding a second term $\mathbf{b}\boldsymbol{\beta}$ as

$$\overline{\overline{\mathsf{B}}} = B\overline{\overline{\mathsf{I}}} + \mathbf{a}\boldsymbol{\alpha} + \mathbf{b}\boldsymbol{\beta}, \tag{7.141}$$

$$\overline{\overline{\mathsf{B}}}^2 = B^2\overline{\overline{\mathsf{I}}} + \mathbf{a}'\boldsymbol{\alpha} + \mathbf{b}'\boldsymbol{\beta}, \tag{7.142}$$

with

$$\mathbf{a}' = 2B\mathbf{a} + \mathbf{a}|(\boldsymbol{\alpha}\mathbf{a} + \boldsymbol{\beta}\mathbf{b}), \tag{7.143}$$

$$\mathbf{b}' = 2B\mathbf{b} + \mathbf{b}|(\boldsymbol{\alpha}\mathbf{a} + \boldsymbol{\beta}\mathbf{b}). \tag{7.144}$$

Because the field two-forms in such a medium (7.101) and (7.102) have the form

$$\boldsymbol{\Phi} = \boldsymbol{\nu} \wedge (\boldsymbol{\alpha}\mathbf{a} + \boldsymbol{\beta}\mathbf{b})|\boldsymbol{\nu}, \tag{7.145}$$

$$\boldsymbol{\Psi} = \boldsymbol{\nu} \wedge (\boldsymbol{\alpha}\mathbf{a}' + \boldsymbol{\beta}\mathbf{b}')|\boldsymbol{\nu}, \tag{7.146}$$

they must satisfy the conditions

$$\boldsymbol{\alpha} \wedge \boldsymbol{\beta} \wedge \boldsymbol{\Phi} = 0, \tag{7.147}$$

$$\boldsymbol{\alpha} \wedge \boldsymbol{\beta} \wedge \boldsymbol{\Psi} = 0. \tag{7.148}$$

Let us first assume that one of the two one-forms is temporal and the other one spatial, $\boldsymbol{\alpha} = \boldsymbol{\varepsilon}_3$ and $\boldsymbol{\beta} = \boldsymbol{\varepsilon}_4$. In this case we obtain the DB conditions

$$\boldsymbol{\varepsilon}_3 \wedge \mathbf{B} = 0, \quad \boldsymbol{\varepsilon}_3 \wedge \mathbf{D} = 0. \tag{7.149}$$

As a second possibility we assume that both one-forms are spatial, $\boldsymbol{\alpha} = \boldsymbol{\varepsilon}_3$ and $\boldsymbol{\beta} = \boldsymbol{\varepsilon}_1$. In this case the fields at the interface $\boldsymbol{\varepsilon}_3|\mathbf{x} = 0$ satisfy

$$\boldsymbol{\varepsilon}_3 \wedge \boldsymbol{\varepsilon}_1 \wedge \mathbf{E} = 0, \quad \boldsymbol{\varepsilon}_3 \wedge \boldsymbol{\varepsilon}_1 \wedge \mathbf{H} = 0, \tag{7.150}$$

which yield the SH conditions (7.139), (7.140).

Case 4

As a further generalization of Case 3, let us choose $\boldsymbol{\alpha} = \boldsymbol{\varepsilon}_3$ and $\boldsymbol{\beta} = \boldsymbol{\varepsilon}_1 + A\boldsymbol{\varepsilon}_4$, where A is a scalar. The conditions (7.147), (7.148), for the fields in the skewon–axion medium now become

$$\boldsymbol{\varepsilon}_3 \wedge (A\boldsymbol{\varepsilon}_4 \wedge \mathbf{B} + \boldsymbol{\varepsilon}_1 \wedge \mathbf{E} \wedge \boldsymbol{\varepsilon}_4) = 0, \tag{7.151}$$

$$\boldsymbol{\varepsilon}_3 \wedge (A\boldsymbol{\varepsilon}_4 \wedge \mathbf{D} - \boldsymbol{\varepsilon}_1 \wedge \mathbf{H} \wedge \boldsymbol{\varepsilon}_4) = 0, \tag{7.152}$$

and they are equivalent to the spatial conditions

$$\boldsymbol{\varepsilon}_3 \wedge (A\mathbf{B} + \boldsymbol{\varepsilon}_1 \wedge \mathbf{E}) = 0, \tag{7.153}$$

$$\boldsymbol{\varepsilon}_3 \wedge (A\mathbf{D} - \boldsymbol{\varepsilon}_1 \wedge \mathbf{H}) = 0, \tag{7.154}$$

which, because of continuity, again serve as boundary conditions at the interface $\boldsymbol{\varepsilon}_3|\mathbf{x} = 0$. The corresponding Gibbsian conditions are of the form

$$A\mathbf{e}_3 \cdot \mathbf{B}_g + \mathbf{e}_2 \cdot \mathbf{E}_g = 0, \tag{7.155}$$

$$A\mathbf{e}_3 \cdot \mathbf{D}_g - \mathbf{e}_2 \cdot \mathbf{H}_g = 0. \tag{7.156}$$

Since these conditions generalize the SH conditions (obtained for $A = 0$) and the DB conditions (obtained for $|A| \to \infty$), they have been dubbed as SHDB conditions [60]. It is left as an exercise to show that SHDB conditions can be also obtained through other definitions of skewon–axion media.

Case 5

The electromagnetic medium can also be defined through the bidyadic $\overline{\overline{\mathsf{N}}}$ in the alternative way of expressing the medium equation (5.52),

$$\mathbf{\Phi} = \overline{\overline{\mathsf{N}}}|\mathbf{\Psi}. \tag{7.157}$$

When $\overline{\overline{\mathsf{M}}}$ has an inverse, we have $\overline{\overline{\mathsf{N}}} = \overline{\overline{\mathsf{M}}}^{-1}$. The previous analysis could have made by defining $\overline{\overline{\mathsf{N}}} = (\overline{\overline{\mathsf{A}}}{}^{\wedge}_{\wedge}\overline{\overline{\mathsf{I}}})^T$ and all of the results corresponding to the cases 1–5 could have arrived at through this route as well. However in the case when $\overline{\overline{\mathsf{N}}}^{-1}$ does not exist, there is no bidyadic $\overline{\overline{\mathsf{M}}}$.

As an example of such a case let us start by considering a skewon–axion medium defined by the bidyadic $\overline{\overline{\mathsf{M}}}$ with a unipotent dyadic $\overline{\overline{\mathsf{B}}}$ satisfying

$$\overline{\overline{\mathsf{B}}}^2 = \overline{\overline{\mathsf{I}}}. \tag{7.158}$$

From (7.102) we then obtain

$$\Psi = \boldsymbol{\nu} \wedge (\boldsymbol{\nu}|\overline{\overline{\mathsf{B}}}^2) = \boldsymbol{\nu} \wedge \boldsymbol{\nu} = 0, \tag{7.159}$$

whence the plane wave satisfies the PMC conditions

$$\mathbf{D} = 0, \quad \mathbf{H} = 0, \tag{7.160}$$

in such a medium. This means that at any interface the Gibbsian PMC conditions

$$\mathbf{n} \cdot \mathbf{D}_g = 0, \quad \mathbf{n} \times \mathbf{H}_g = 0 \tag{7.161}$$

are valid. The corresponding medium yielding the PEC condition in the medium, $\mathbf{\Phi} = 0$, and at the interface,

$$\mathbf{n} \cdot \mathbf{B}_g = 0, \quad \mathbf{n} \times \mathbf{E}_g = 0 \tag{7.162}$$

are obtained by defining a skewon–axion medium through a bidyadic $\overline{\overline{\mathsf{N}}}$ in terms of a unipotent dyadic $\overline{\overline{\mathsf{A}}}$ as

$$\overline{\overline{\mathsf{N}}} = (\overline{\overline{\mathsf{A}}}{}^{\wedge}_{\wedge}\overline{\overline{\mathsf{I}}})^T, \quad \overline{\overline{\mathsf{A}}}^2 = \overline{\overline{\mathsf{I}}}. \tag{7.163}$$

As a conclusion, it was seen that same boundary conditions can be obtained at the interface of many different media. A more recent addition is the SHDB boundary generalizing the SH and DB boundary conditions which have been known since the 1950s. One can show that the SHDB conditions share the property of both SH and DB conditions as being self-dual, that is, invariant in any three-parameter duality transformation of the form (6.36) or

$$\begin{pmatrix} Z_d\boldsymbol{\Psi}_d \\ \boldsymbol{\Phi}_d \end{pmatrix} = \begin{pmatrix} \cos\theta & \sin\theta \\ -\sin\theta & \cos\theta \end{pmatrix} \begin{pmatrix} Z\boldsymbol{\Psi} \\ \boldsymbol{\Phi} \end{pmatrix}. \tag{7.164}$$

In fact, expressing (7.153) and (7.154) as

$$\mathbf{f}(\mathbf{E}, \mathbf{B}) = \varepsilon_3 \wedge (A\mathbf{B} + \varepsilon_1 \wedge \mathbf{E}) = 0, \tag{7.165}$$

$$\mathbf{f}(-\mathbf{H}, \mathbf{D}) = \varepsilon_3 \wedge (A\mathbf{D} - \varepsilon_1 \wedge \mathbf{H}) = 0, \tag{7.166}$$

we can write the relation

$$\begin{pmatrix} \mathbf{f}(\mathbf{E}_d, \mathbf{B}_d) \\ \mathbf{f}(-\mathbf{H}_d, \mathbf{D}_d) \end{pmatrix} = \begin{pmatrix} \cos\theta & -Z\sin\theta \\ (1/Z_d)\sin\theta & (Z/Z_d)\cos\theta \end{pmatrix} \begin{pmatrix} \mathbf{f}(\mathbf{E}, \mathbf{B}) \\ \mathbf{f}(-\mathbf{H}, \mathbf{D}) \end{pmatrix}, \tag{7.167}$$

from which we obtain

$$\mathbf{f}(\mathbf{E}_d, \mathbf{B}_d) = 0, \quad \mathbf{f}(-\mathbf{H}_d, \mathbf{D}_d) = 0. \tag{7.168}$$

Thus, an object with an SHDB boundary and suitable symmetry has zero backscattering, that is, it is invisible for the radar [43].

7.4 EXTENDED SKEWON–AXION MEDIA

The skewon–axion medium can be extended by adding a simple term to the medium bidyadic as

$$\overline{\overline{\mathsf{M}}} = (\overline{\overline{\mathsf{B}}}{}^{\wedge}_{\wedge}\overline{\overline{\mathsf{I}}})^T + \boldsymbol{\Delta}\mathbf{B}, \tag{7.169}$$

where $\boldsymbol{\Delta}$ is some two-form and $\mathbf{B}$ some bivector. The corresponding modified medium bidyadic has the form

$$\overline{\overline{\mathsf{M}}}_m = \mathbf{e}_N \lfloor (\overline{\overline{\mathsf{B}}}{}^{\wedge}_{\wedge}\overline{\overline{\mathsf{I}}})^T + \mathbf{A}\mathbf{B}, \tag{7.170}$$

where $\mathbf{A} = \mathbf{e}_N \lfloor \boldsymbol{\Delta}$ is a bivector. Let us consider the effect of the added term to plane-wave properties. The dispersion dyadic (7.104) is extended correspondingly as

$$\overline{\overline{\mathsf{D}}}(\boldsymbol{\nu}) = \mathbf{F} \lfloor \overline{\overline{\mathsf{I}}}^T + \mathbf{a}\mathbf{b}, \tag{7.171}$$

with

$$\mathbf{a} = \mathbf{A}\lfloor \boldsymbol{\nu}, \quad \mathbf{b} = \mathbf{B}\lfloor \boldsymbol{\nu}, \quad \mathbf{F} = \mathbf{e}_N \lfloor (\boldsymbol{\nu} \wedge (\boldsymbol{\nu}|\overline{\overline{\mathsf{B}}})). \tag{7.172}$$

Because the bivector $\mathbf{F}$ is simple, it satisfies

$$(\mathbf{F}\lfloor\overline{\overline{\mathsf{I}}}^T)^{(2)} = \mathbf{FF}, \quad (\mathbf{F}\lfloor\overline{\overline{\mathsf{I}}}^T)^{(3)} = 0, \tag{7.173}$$

whence the dyadic dispersion equation (5.240) becomes

$$\begin{aligned}\overline{\overline{\mathsf{D}}}^{(3)}(\boldsymbol{\nu}) &= (\mathbf{F}\lfloor\overline{\overline{\mathsf{I}}}^T)^{(3)} + (\mathbf{F}\lfloor\overline{\overline{\mathsf{I}}}^T)^{(2)}{}^{\wedge}_{\wedge}\mathbf{ab}\\ &= \mathbf{FF}{}^{\wedge}_{\wedge}\mathbf{ab} = 0.\end{aligned} \tag{7.174}$$

Expanding

$$\begin{aligned}\boldsymbol{\varepsilon}_N\boldsymbol{\varepsilon}_N\lfloor\lfloor(\mathbf{FF}{}^{\wedge}_{\wedge}\mathbf{ab}) &= (\boldsymbol{\varepsilon}_N\boldsymbol{\varepsilon}_N\lfloor\lfloor\mathbf{FF})\lfloor\lfloor\mathbf{ab}\\ &= ((\boldsymbol{\nu}\wedge(\boldsymbol{\nu}|\overline{\overline{\mathsf{B}}}))\lfloor\mathbf{a})((\boldsymbol{\nu}\wedge(\boldsymbol{\nu}|\overline{\overline{\mathsf{B}}}))\lfloor\mathbf{b})\\ &= \boldsymbol{\nu\nu}(\boldsymbol{\nu}|\overline{\overline{\mathsf{B}}}|\mathbf{a})(\boldsymbol{\nu}|\overline{\overline{\mathsf{B}}}|\mathbf{b}),\end{aligned} \tag{7.175}$$

where we have applied $\boldsymbol{\nu}|\mathbf{a} = \boldsymbol{\nu}|\mathbf{b} = 0$ in the bac-cab rule, the quartic dispersion equation is decomposed in two scalar dispersion equations of the second order,

$$\boldsymbol{\nu}|\overline{\overline{\mathsf{B}}}|\mathbf{a} = \boldsymbol{\nu}|(\overline{\overline{\mathsf{B}}}\rfloor\mathbf{A})|\boldsymbol{\nu} = 0, \tag{7.176}$$

$$\boldsymbol{\nu}|\overline{\overline{\mathsf{B}}}|\mathbf{b} = \boldsymbol{\nu}|(\overline{\overline{\mathsf{B}}}\rfloor\mathbf{B})|\boldsymbol{\nu} = 0, \tag{7.177}$$

whose solutions define two characteristic waves which may respectively be called by the names A wave and B wave. Thus, the extended skewon–axion medium is an example of a DC medium.

The previous analysis shows us that, extending the skewon–axion medium bidyadic by a term as in (7.170), the one-form $\boldsymbol{\nu}$ can no longer be freely chosen but, instead, must satisfy either of the two conditions (7.176), (7.177). This is because the added term contains a nonzero principal component. In the general case, the A wave and the B wave obey different dispersion equations, whence such a medium is birefringent. The two equations become the same when the symmetric parts of the two dyadics $\overline{\overline{\mathsf{B}}}\rfloor\mathbf{A}$ and $\overline{\overline{\mathsf{B}}}\rfloor\mathbf{B}$ are multiples of each other.

Inserting (7.171) in (5.236), the equation for the potential one-form $\boldsymbol{\phi}$ of the plane wave in the extended skewon–axion medium becomes

$$\begin{aligned}\overline{\overline{\mathsf{D}}}(\boldsymbol{\nu})|\boldsymbol{\phi} &= (\mathbf{F}\lfloor\overline{\overline{\mathsf{I}}}^T + \mathbf{ab})|\boldsymbol{\phi}\\ &= \mathbf{e}_N\lfloor(\boldsymbol{\nu}\wedge(\boldsymbol{\nu}|\overline{\overline{\mathsf{B}}})\wedge\boldsymbol{\phi}) + \mathbf{a}(\mathbf{b}|\boldsymbol{\phi}) = 0.\end{aligned} \tag{7.178}$$

Multiplying this by $\boldsymbol{\phi}|$ we obtain

$$\begin{aligned}\boldsymbol{\phi}|\overline{\overline{\mathsf{D}}}|\boldsymbol{\phi} &= \mathbf{F}|(\boldsymbol{\phi}\wedge\boldsymbol{\phi}) + (\mathbf{a}|\boldsymbol{\phi})(\mathbf{b}|\boldsymbol{\phi})\\ &= (\mathbf{a}|\boldsymbol{\phi})(\mathbf{b}|\boldsymbol{\phi}) = 0,\end{aligned} \tag{7.179}$$

whence the potential one-form of the plane wave must satisfy one of the two polarization conditions

$$\mathbf{a}|\boldsymbol{\phi} = \mathbf{A}|(\boldsymbol{\nu} \wedge \boldsymbol{\phi}) = \mathbf{A}|\boldsymbol{\Phi} = 0, \tag{7.180}$$

$$\mathbf{b}|\boldsymbol{\phi} = \mathbf{B}|(\boldsymbol{\nu} \wedge \boldsymbol{\phi}) = \mathbf{B}|\boldsymbol{\Phi} = 0. \tag{7.181}$$

To find the connection between the two dispersion equations (7.176), (7.177) and the two polarization conditions (7.180), (7.181), we multiply (7.178) by $(\boldsymbol{\nu}|\overline{\overline{\mathsf{B}}})|$ which yields

$$(\boldsymbol{\nu}|\overline{\overline{\mathsf{B}}}|\mathbf{a})(\mathbf{b}|\boldsymbol{\phi}) = 0. \tag{7.182}$$

Assuming $\mathbf{b}|\boldsymbol{\phi} \neq 0$, whence $\mathbf{a}|\boldsymbol{\phi} = 0$, we must have $\boldsymbol{\nu}|\overline{\overline{\mathsf{B}}}|\mathbf{a} = 0$, whence (7.180) corresponds to (7.176) of the A-wave. Thus, we may expect that (7.181) must correspond to (7.177) of the B-wave. Details of the proof are left as an exercise. This reasoning fails when (7.176) and (7.177) are both simultaneously valid, in which case the medium is nonbirefringent and the polarization is not restricted by either of the conditions.

PROBLEMS

7.1 Show that if the Gibbsian isotropic medium bidyadic (7.5) has only a principal component, by requiring the condition $\overline{\overline{\mathsf{M}}}\lfloor\lfloor\overline{\overline{\mathsf{I}}} = 0$. The metric dyadic $\mathbf{e}_{123}\lfloor\overline{\overline{\mathsf{C}}}_s = \overline{\overline{\mathsf{G}}}_s \in \mathbb{E}_1\mathbb{E}_1$ must be symmetric.

7.2 Starting from the medium bidyadic (7.17) of the GBI medium, derive the expression (7.20).

7.3 Do the details in the derivation of (7.33) and (7.34) to arrive at (7.31).

7.4 Show that the bidyadic $\overline{\overline{\mathsf{T}}}^{(2)} = (\overline{\overline{\mathsf{I}}}_s - \mathbf{e}_4\boldsymbol{\varepsilon}_4)^{(2)}$ is a skewon bidyadic and that it can be expressed in the form $\overline{\overline{\mathsf{B}}}_o{}^\wedge_\wedge\overline{\overline{\mathsf{I}}}$, in terms of a trace-free dyadic $\overline{\overline{\mathsf{B}}}_o$.

7.5 Derive the expressions for equivalent medium parameters (7.51) and (7.52) of two Gibbsian isotropic medium bidyadics

$$\overline{\overline{\mathsf{M}}}_\pm = \epsilon_\pm\overline{\overline{\mathsf{C}}}_s \wedge \mathbf{e}_4 + \frac{1}{\mu_\pm}\boldsymbol{\varepsilon}_4 \wedge \overline{\overline{\mathsf{C}}}_s^{-1}$$

replacing a given GBI medium bidyadic

$$\overline{\overline{\mathsf{M}}} = A\overline{\overline{\mathsf{I}}}^{(2)T} + B\overline{\overline{\mathsf{T}}}^{(2)T} + \epsilon'\overline{\overline{\mathsf{C}}}_s \wedge \mathbf{e}_4 + \frac{1}{\mu}\boldsymbol{\varepsilon}_4 \wedge \overline{\overline{\mathsf{C}}}_s^{-1}$$

for eigenfields as $\overline{\overline{\mathsf{M}}}_\pm|\boldsymbol{\Phi}_\pm = M_\pm\boldsymbol{\Phi}_\pm$, by applying 3D decompositions $\boldsymbol{\Phi} = \mathbf{B} + \mathbf{E} \wedge \boldsymbol{\varepsilon}_4$.

7.6 Show that the antisymmetric part of the dyadic dispersion equation (7.61) vanishes identically for all $\boldsymbol{\nu}$. The identity of Problem 2.22 may prove useful in the analysis.

7.7 Expand the symmetric part of the dyadic dispersion equation (7.61) in detail and verify that the dispersion functions are of the form (7.65)– (7.67).

7.8 Show that the general GBI-medium bidyadic (7.20) can be expressed as

$$\overline{\overline{\mathsf{M}}} = A\overline{\overline{\mathsf{I}}}^{(2)T} + B\overline{\overline{\mathsf{U}}} + C\overline{\overline{\mathsf{J}}},$$

where $\overline{\overline{\mathsf{U}}}$ is a unipotent bidyadic and $\overline{\overline{\mathsf{J}}}$ is an AC bidyadic satisfying $\overline{\overline{\mathsf{U}}}^2 = -\overline{\overline{\mathsf{J}}}^2 = \overline{\overline{\mathsf{I}}}^{(2)T}$. Show also that the three terms are mapped to similar terms in the similarity transformation $\overline{\overline{\mathsf{D}}}|\overline{\overline{\mathsf{M}}}|\overline{\overline{\mathsf{D}}}^{-1}$.

7.9 Show that a duality transformation can be defined so that a given GBI medium is self-dual, that is, invariant in the transformation.

7.10 Study the fields in the simple skewon–axion medium defined by the medium bidyadic

$$\overline{\overline{\mathsf{M}}} = A\overline{\overline{\mathsf{I}}}^{(2)T} + B\overline{\overline{\mathsf{T}}}^{(2)T} = \alpha\overline{\overline{\mathsf{I}}}_s^{(2)T} - \beta\overline{\overline{\mathsf{I}}}_s^{T} {}^\wedge_\wedge \boldsymbol{\varepsilon}_4\mathbf{e}_4,$$

with $B \neq 0$. Consider solutions to the Maxwell equations in a homogeneous medium and show that such a medium produces DB boundary conditions (5.171) at its interface.

7.11 Derive the Gibbsian dyadic rule (7.122).

7.12 Derive the Gibbsian medium equations (7.116) and (7.117) for the skewon–axion medium.

7.13 Show that if the dispersion dyadic corresponding to the medium bidyadic $\overline{\overline{\mathsf{M}}}$ can be expressed in the form

$$\overline{\overline{\mathsf{D}}}(\boldsymbol{\nu}) = \mathbf{e}_N\lfloor(\boldsymbol{\alpha}(\boldsymbol{\nu}) \wedge \boldsymbol{\beta}(\boldsymbol{\nu}) \wedge \overline{\overline{\mathsf{I}}}^T),$$

there is no dispersion equation, that is, $D(\boldsymbol{\nu}) = 0$ is satisfied for any $\boldsymbol{\nu}$. Here the one-forms $\boldsymbol{\alpha}$ and $\boldsymbol{\beta}$ are functions of $\boldsymbol{\nu}$. Show also that in such a case the medium cannot have a principal component. Hint: apply the rule $\overline{\overline{\mathsf{C}}}\lfloor\lfloor\boldsymbol{\gamma}\boldsymbol{\gamma} = 0$ for all $\boldsymbol{\gamma} \in \mathbb{F}_1$ implies that $\overline{\overline{\mathsf{C}}} \in \mathbb{E}_2\mathbb{E}_2$ is of the form $\alpha\overline{\overline{\mathsf{E}}}$.

7.14 Study the possibility of defining boundary conditions at an interface of a skewon–axion medium $\overline{\overline{\mathsf{M}}} = (\overline{\overline{\mathsf{B}}}{}^\wedge_\wedge\overline{\overline{\mathsf{I}}})^T$ by defining $\boldsymbol{\alpha} = A\boldsymbol{\varepsilon}_4 + \boldsymbol{\varepsilon}_1$ in (7.126) and (7.127).

7.15 Derive the condition (7.95) by requiring that the trace-free part $\overline{\overline{\mathsf{M}}}_o$ of the medium bidyadic $\overline{\overline{\mathsf{M}}}$ is an eigendyadic of $\overline{\overline{\mathsf{I}}}^{(4)T}\lfloor\lfloor\overline{\overline{\mathsf{M}}}_o^T = \lambda\overline{\overline{\mathsf{M}}}_o$ corresponding to the eigenvalue $\lambda = -1$.

7.16 Find the inverse of the extended skewon-medium bidyadic (7.169).

7.17 Find the conditions for the Gibbsian medium dyadics corresponding to a skewon medium by starting from the relations (5.102)–(5.105).

7.18 Show that a principal–axion medium satisfying $\overline{\overline{\epsilon}}' = 0$ and $\overline{\overline{\mu}}^{-1} = 0$ is defined by medium equations of the form

$$\mathbf{D} = -(\boldsymbol{\varepsilon}_{123}\lfloor\overline{\overline{\beta}}^T\rfloor\mathbf{e}_{123})|\mathbf{B}, \quad \mathbf{H} = \overline{\overline{\beta}}|\mathbf{E}.$$

Find the dispersion equation for a plane wave in such a medium.

7.19 Consider the special principal–axion medium of the previous problem in terms of Gibbsian symbols and find the dispersion equation of a plane wave.

7.20 Show in detail that the dispersion equation for a plane wave in an extended skewon–axion medium defined by (7.170) corresponding to the polarization condition $\mathbf{b}|\boldsymbol{\phi} = 0$ is $\boldsymbol{\nu}|(\overline{\overline{\mathsf{B}}}\rfloor\mathbf{B})|\boldsymbol{\nu} = 0$.

CHAPTER 8

Quadratic Media

In this chapter we consider media whose medium bidyadic or modified medium bidyadic can be expressed as a double-wedge square of some dyadic. In both of these cases the the number of free parameters is 16 instead of 36. In addition, simple extensions of these media will be covered.

8.1 P MEDIA AND Q MEDIA

There are two possibilities in defining media in terms of bidyadics obeying a double-wedge square law. If $\overline{\overline{\mathsf{P}}} \in \mathbb{E}_1\mathbb{F}_1$ is a dyadic mapping vectors to vectors, a medium bidyadic can be defined as [61]

$$\overline{\overline{\mathsf{M}}} = \overline{\overline{\mathsf{P}}}^{(2)T} \in \mathbb{F}_2\mathbb{E}_2. \tag{8.1}$$

On the other hand, if $\overline{\overline{\mathsf{Q}}} \in \mathbb{E}_1\mathbb{E}_1$ is a dyadic mapping one-forms to vectors, a modified medium bidyadic can be defined as [1, 62]

$$\overline{\overline{\mathsf{M}}}_m = \overline{\overline{\mathsf{Q}}}^{(2)} \in \mathbb{E}_2\mathbb{E}_2. \tag{8.2}$$

Both (8.1) and (8.2) define certain classes of electromagnetic media which have been called by the respective names P media and Q media in the past. Alternatively, it is possible to define

$$\overline{\overline{\mathsf{M}}} = M\overline{\overline{\mathsf{P}}}^{(2)T}, \tag{8.3}$$

$$\overline{\overline{\mathsf{M}}}_m = M\overline{\overline{\mathsf{Q}}}^{(2)}, \tag{8.4}$$

Multiforms, Dyadics, and Electromagnetic Media, First Edition. Ismo V. Lindell.

by adding a coefficient M in which case we can impose a normalizing condition on the dyadics $\overline{\overline{\mathsf{P}}}$ and $\overline{\overline{\mathsf{Q}}}$. For example, if the dyadics are of full rank, we could choose

$$\Delta_P = \mathrm{tr}\overline{\overline{\mathsf{P}}}^{(4)} = 1, \tag{8.5}$$

$$\Delta_Q = \boldsymbol{\varepsilon}_N \boldsymbol{\varepsilon}_N || \overline{\overline{\mathsf{Q}}}^{(4)} = 1, \tag{8.6}$$

or -1 instead of 1. However, since (8.1) and (8.2) appear simpler and are not limited by the full-rank requirement, they will be applied in the sequel even if the dyadics $\overline{\overline{\mathsf{Q}}}$ and $\overline{\overline{\mathsf{P}}}$ may involve complex components.

As an example, the simple principal medium (5.122) (Gibbsian isotropic medium), defined by the modified medium bidyadic of the form

$$\overline{\overline{\mathsf{M}}}_m = -\frac{1}{\mu}\overline{\overline{\mathsf{G}}}^{(2)}, \quad \overline{\overline{\mathsf{G}}} = \overline{\overline{\mathsf{G}}}_s - \mu\epsilon \mathbf{e}_4\mathbf{e}_4, \tag{8.7}$$

belongs to a subclass of Q media with a symmetric bidyadic $\overline{\overline{\mathsf{Q}}}^{(2)}$ because we can set $\overline{\overline{\mathsf{Q}}} = \pm\overline{\overline{\mathsf{G}}}/\sqrt{-\mu}$ with either sign.

One can show that any spatial bidyadic or a modified spatial bidyadic possessing a spatial inverse can be uniquely expressed in terms of a spatial dyadic $\overline{\overline{\mathsf{P}}}$ or a metric dyadic $\overline{\overline{\mathsf{Q}}}$ in the form (8.1) or (8.2). The proof of this property, which is known for Gibbsian 3D dyadics [25], is left as an exercise. However, since this is not valid in the general 4D case, bidyadics and modified bidyadics satisfying (8.1) or (8.2) form two definite classes of media.

One can also show that P media and Q media do not have common elements when $\overline{\overline{\mathsf{P}}}$ and $\overline{\overline{\mathsf{Q}}}$ are full-rank dyadics. To prove this, we may consider the equation

$$\overline{\overline{\mathsf{Q}}}^{(2)} = \mathbf{e}_N \lfloor \overline{\overline{\mathsf{P}}}^{(2)T} \tag{8.8}$$

and show that there is no solution $\overline{\overline{\mathsf{Q}}}$ for any given $\overline{\overline{\mathsf{P}}}$ satisfying $\Delta_P \neq 0$. An equivalent equation is

$$\overline{\overline{\mathsf{X}}}^{(2)} = \mathbf{e}_N \lfloor \overline{\overline{\mathsf{I}}}^{(2)T}, \tag{8.9}$$

for the metric dyadic $\overline{\overline{\mathsf{X}}} = \overline{\overline{\mathsf{Q}}}|\overline{\overline{\mathsf{P}}}^{-1T}$. Details of the proof are left as an exercise. However, when the bidyadics are not restricted by the full-rank requirement, there exist solutions to (8.8). For example, choosing $\overline{\overline{\mathsf{P}}} = \mathbf{e}_1\boldsymbol{\varepsilon}_3 + \mathbf{e}_2\boldsymbol{\varepsilon}_4$ we have a solution $\overline{\overline{\mathsf{Q}}} = \mathbf{e}_1\mathbf{e}_1 + \mathbf{e}_2\mathbf{e}_2$.

For full-rank dyadics $\overline{\overline{\mathsf{P}}}$ and $\overline{\overline{\mathsf{Q}}}$, the corresponding medium bidyadics have inverses in both cases and their respective expressions can be obtained from the rules (2.123) and (2.207) as

$$\overline{\overline{\mathsf{M}}}^{-1} = \overline{\overline{\mathsf{P}}}^{(-2)T} = \frac{1}{\Delta_P}\overline{\overline{\mathrm{I}}}^{(4)T}\lfloor\lfloor\overline{\overline{\mathsf{P}}}^{(2)} = \frac{1}{\Delta_P}\overline{\overline{\mathrm{I}}}^{(4)T}\lfloor\lfloor\overline{\overline{\mathsf{M}}}^{T}, \tag{8.10}$$

$$\overline{\overline{\mathsf{M}}}_m^{-1} = \overline{\overline{\mathsf{Q}}}^{(-2)} = \frac{1}{\Delta_Q}\boldsymbol{\varepsilon}_N\boldsymbol{\varepsilon}_N\lfloor\lfloor\overline{\overline{\mathsf{Q}}}^{(2)T} = \frac{1}{\Delta_Q}\boldsymbol{\varepsilon}_N\boldsymbol{\varepsilon}_N\lfloor\lfloor\overline{\overline{\mathsf{M}}}_m^{T}. \tag{8.11}$$

Considering the representation (5.52), the $\overline{\overline{\mathsf{N}}}$ and $\overline{\overline{\mathsf{N}}}_m$ bidyadics can be expressed for the P medium as

$$\overline{\overline{\mathsf{N}}} = \overline{\overline{\mathsf{M}}}^{-1} = (\overline{\overline{\mathsf{P}}}^{-1T})^{(2)}, \tag{8.12}$$

and for the Q medium as

$$\begin{aligned}\overline{\overline{\mathsf{N}}}_m &= \mathbf{e}_N\lfloor\overline{\overline{\mathsf{N}}} = \mathbf{e}_N\lfloor\overline{\overline{\mathsf{M}}}^{-1}\\ &= \mathbf{e}_N\lfloor(\boldsymbol{\varepsilon}_N\lfloor\overline{\overline{\mathsf{M}}}_m)^{-1} = \mathbf{e}_N\lfloor\overline{\overline{\mathsf{M}}}_m^{-1}\rfloor\mathbf{e}_N\\ &= \frac{1}{\Delta_Q}\overline{\overline{\mathsf{Q}}}^{(2)T} = (\overline{\overline{\mathsf{Q}}}^{T}/\sqrt{\Delta_Q})^{(2)}.\end{aligned} \tag{8.13}$$

Consequently, the classes of P media and Q media can equally well defined through the $\overline{\overline{\mathsf{N}}}$ bidyadic representation (5.52) in the case when the dyadics are of full rank.

Because of the expansions

$$\overline{\overline{\mathsf{P}}}^{(2)}\cdot\overline{\overline{\mathsf{P}}}^{(2)T} = \overline{\overline{\mathsf{P}}}^{(2)}\lfloor\mathbf{e}_N\lfloor\overline{\overline{\mathsf{P}}}^{(2)T} = \Delta_P\overline{\overline{\mathrm{I}}}^{(2)}\rfloor\mathbf{e}_N, \tag{8.14}$$

and

$$\overline{\overline{\mathsf{Q}}}^{(2)T}\cdot\overline{\overline{\mathsf{Q}}}^{(2)} = \overline{\overline{\mathsf{Q}}}^{(2)T}\rfloor\boldsymbol{\varepsilon}_N\lfloor\overline{\overline{\mathsf{Q}}}^{(2)} = \Delta_Q\mathbf{e}_N\lfloor\overline{\overline{\mathrm{I}}}^{(2)T}, \tag{8.15}$$

the medium bidyadics of P media and modified medium bidyadics of Q media satisfy the modified closure conditions (4.141) and (4.142) as

$$\overline{\overline{\mathsf{M}}}^{T}\cdot\overline{\overline{\mathsf{M}}} = \alpha\mathbf{e}_N\lfloor\overline{\overline{\mathrm{I}}}^{(2)T}, \tag{8.16}$$

$$\overline{\overline{\mathsf{M}}}_m^{T}\cdot\overline{\overline{\mathsf{M}}}_m = \alpha\mathbf{e}_N\lfloor\overline{\overline{\mathrm{I}}}^{(2)T}, \tag{8.17}$$

where α equals respectively Δ_P or Δ_Q. The converse is also true. In fact, solutions of the equation

$$\overline{\overline{\mathsf{M}}}^{T}\cdot\overline{\overline{\mathsf{M}}} = \alpha\mathbf{e}_N\lfloor\overline{\overline{\mathrm{I}}}^{(2)T} \tag{8.18}$$

for $\alpha \neq 0$ have been shown to be either of the P medium form (4.144) or the Q medium form (4.145).

8.2 TRANSFORMATIONS

An affine transformation (6.16) or (6.20) in terms of a dyadic $\overline{\overline{\mathsf{A}}} \in \mathbb{E}_1\mathbb{F}_1$ yields for the P medium

$$\begin{aligned}\overline{\overline{\mathsf{M}}}_a &= \overline{\overline{\mathsf{A}}}^{(-2)}|\overline{\overline{\mathsf{M}}}|\overline{\overline{\mathsf{A}}}^{(2)T} = \overline{\overline{\mathsf{A}}}^{(-2)}|\overline{\overline{\mathsf{P}}}^{(2)T}|\overline{\overline{\mathsf{A}}}^{(2)T} \\ &= (\overline{\overline{\mathsf{A}}}^{-1}|\overline{\overline{\mathsf{P}}}^{T}|\overline{\overline{\mathsf{A}}}^{T})^{(2)} = \overline{\overline{\mathsf{P}}}_a^{(2)T},\end{aligned} \tag{8.19}$$

and, for the Q medium,

$$\begin{aligned}\overline{\overline{\mathsf{M}}}_{ma} &= \overline{\overline{\mathsf{A}}}^{(2)}|\overline{\overline{\mathsf{M}}}_m|\overline{\overline{\mathsf{A}}}^{(2)T} = \overline{\overline{\mathsf{A}}}^{(2)}|\overline{\overline{\mathsf{Q}}}^{(2)}|\overline{\overline{\mathsf{A}}}^{(2)T} \\ &= (\overline{\overline{\mathsf{A}}}|\overline{\overline{\mathsf{Q}}}|\overline{\overline{\mathsf{A}}}^{T})^{(2)} = \overline{\overline{\mathsf{Q}}}_a^{(2)}.\end{aligned} \tag{8.20}$$

Thus, a P medium is transformed to another P medium and a Q medium is transformed to another Q medium as defined by the respective dyadics

$$\overline{\overline{\mathsf{P}}}_a = \overline{\overline{\mathsf{A}}}|\overline{\overline{\mathsf{P}}}|\overline{\overline{\mathsf{A}}}^{-1T}, \quad \overline{\overline{\mathsf{Q}}}_a = \overline{\overline{\mathsf{A}}}|\overline{\overline{\mathsf{Q}}}|\overline{\overline{\mathsf{A}}}^{T}. \tag{8.21}$$

In particular, if $\overline{\overline{\mathsf{Q}}}$ is symmetric or antisymmetric, so is $\overline{\overline{\mathsf{Q}}}_a$. Also, if $\overline{\overline{\mathsf{Q}}}$ is not symmetric but $\overline{\overline{\mathsf{Q}}}^{(2)}$ is symmetric, $\overline{\overline{\mathsf{Q}}}_a$ has the same property.

The three-parameter involutionary duality transformation of the medium bidyadic can be studied through the rule (6.65) as

$$\overline{\overline{\mathsf{M}}}_d = \frac{1}{Z^2}\overline{\overline{\mathsf{M}}}^{-1} = \frac{1}{Z^2}\overline{\overline{\mathsf{N}}}, \tag{8.22}$$

whence the P medium and the Q medium bidyadics are respectively transformed to

$$\overline{\overline{\mathsf{M}}}_d = \frac{1}{Z^2}(\overline{\overline{\mathsf{P}}}^{(2)T})^{-1} = ((Z\overline{\overline{\mathsf{P}}}^{T})^{-1})^{(2)T}, \tag{8.23}$$

$$\overline{\overline{\mathsf{M}}}_d = \frac{1}{Z^2}(\boldsymbol{\varepsilon}_N\lfloor\overline{\overline{\mathsf{Q}}}^{(2)})^{-1} = (Z\overline{\overline{\mathsf{Q}}})^{(-2)}\rfloor\mathbf{e}_N. \tag{8.24}$$

Thus, the P medium is transformed to another P medium with $\overline{\overline{\mathsf{P}}}_d = (Z\overline{\overline{\mathsf{P}}})^{-1}$, while the Q medium is transformed to another Q medium defined by

$$\begin{aligned}\overline{\overline{\mathsf{M}}}_{md} &= \mathbf{e}_N\lfloor\overline{\overline{\mathsf{M}}}_d = \mathbf{e}_N\mathbf{e}_N\lfloor\lfloor(Z\overline{\overline{\mathsf{Q}}})^{(-2)} \\ &= \frac{1}{Z^2\Delta_Q}\overline{\overline{\mathsf{Q}}}^{(2)T} = \overline{\overline{\mathsf{Q}}}_d^{(2)},\end{aligned} \tag{8.25}$$

$$\overline{\overline{\mathsf{Q}}}_d = \frac{1}{Z\sqrt{\Delta_Q}}\overline{\overline{\mathsf{Q}}}^{T}. \tag{8.26}$$

For symmetric dyadics $\overline{\overline{\mathsf{Q}}}$ the Q medium is self-dual, that is, there exists an involutionary duality transformation defined by $Z = 1/\sqrt{\Delta_Q}$ which leaves the Q medium invariant.

8.3 SPATIAL EXPANSIONS

8.3.1 Spatial Expansion of Q Media

Let us find the spatial medium dyadics of a Q medium in the expansion

$$\overline{\overline{\mathsf{M}}} = \overline{\overline{\alpha}} + \overline{\overline{\epsilon}}' \wedge \mathbf{e}_4 + \boldsymbol{\varepsilon}_4 \wedge \overline{\overline{\mu}}^{-1} + \boldsymbol{\varepsilon}_4 \wedge \overline{\overline{\beta}} \wedge \mathbf{e}_4, \tag{8.27}$$

by first expanding the metric dyadic $\overline{\overline{\mathsf{Q}}}$ as

$$\overline{\overline{\mathsf{Q}}} = \overline{\overline{\mathsf{Q}}}_s + \mathbf{e}_4\mathbf{a}_s + \mathbf{b}_s\mathbf{e}_4 + c\mathbf{e}_4\mathbf{e}_4, \tag{8.28}$$

where $\overline{\overline{\mathsf{Q}}}_s \in \mathbb{E}_1\mathbb{E}_1$ is a spatial dyadic, $\mathbf{a}_s, \mathbf{b}_s \in \mathbb{E}_1$ are two spatial vectors and c is a scalar. The corresponding expansion for the double-wedge square is

$$\overline{\overline{\mathsf{Q}}}^{(2)} = \overline{\overline{\mathsf{Q}}}_s^{(2)} - \mathbf{e}_4 \wedge \overline{\overline{\mathsf{Q}}}_s \wedge \mathbf{a}_s - \mathbf{b}_s \wedge \overline{\overline{\mathsf{Q}}}_s \wedge \mathbf{e}_4 - \boldsymbol{\varepsilon}_4 \wedge (c\overline{\overline{\mathsf{Q}}}_s - \mathbf{b}_s\mathbf{a}_s) \wedge \mathbf{e}_4, \tag{8.29}$$

which yields

$$\begin{aligned}\overline{\overline{\mathsf{M}}} &= \boldsymbol{\varepsilon}_N \lfloor \overline{\overline{\mathsf{Q}}}^{(2)} \\ &= -\boldsymbol{\varepsilon}_4 \wedge (\boldsymbol{\varepsilon}_{123} \lfloor \overline{\overline{\mathsf{Q}}}_s^{(2)}) + \boldsymbol{\varepsilon}_{123} \lfloor \overline{\overline{\mathsf{Q}}}_s \wedge \mathbf{a}_s \\ &\quad + \boldsymbol{\varepsilon}_4 \wedge (\boldsymbol{\varepsilon}_{123} \lfloor (\mathbf{b}_s \wedge \overline{\overline{\mathsf{Q}}}_s)) \wedge \boldsymbol{\varepsilon}_4 + \boldsymbol{\varepsilon}_{123} \lfloor (c\overline{\overline{\mathsf{Q}}}_s - \mathbf{b}_s\mathbf{a}_s) \wedge \mathbf{e}_4. \end{aligned} \tag{8.30}$$

Comparing with (8.27) and equating the respective terms, leads to the following set of relations,

$$\overline{\overline{\alpha}} = \boldsymbol{\varepsilon}_{123} \lfloor \overline{\overline{\mathsf{Q}}}_s \wedge \mathbf{a}_s, \tag{8.31}$$

$$\overline{\overline{\epsilon}}' = \boldsymbol{\varepsilon}_{123} \lfloor (c\overline{\overline{\mathsf{Q}}}_s - \mathbf{b}_s\mathbf{a}_s), \tag{8.32}$$

$$\overline{\overline{\mu}}^{-1} = -\boldsymbol{\varepsilon}_{123} \lfloor \overline{\overline{\mathsf{Q}}}_s^{(2)}, \tag{8.33}$$

$$\overline{\overline{\beta}} = \boldsymbol{\varepsilon}_{123} \lfloor (\mathbf{b}_s \wedge \overline{\overline{\mathsf{Q}}}_s). \tag{8.34}$$

One should note that the dyadics $\overline{\overline{\alpha}}$ and $\overline{\overline{\beta}}$ of a Q medium do not have spatial inverses because they satisfy

$$\overline{\overline{\alpha}} \wedge \mathbf{a}_s = \boldsymbol{\varepsilon}_{123} \lfloor \overline{\overline{\mathsf{Q}}}_s \wedge \mathbf{a}_s \wedge \mathbf{a}_s = 0, \tag{8.35}$$

$$\mathbf{b}_s | \overline{\overline{\beta}} = \boldsymbol{\varepsilon}_{123} | (\mathbf{b}_s \wedge \mathbf{b}_s \wedge \overline{\overline{\mathsf{Q}}}_s) = 0. \tag{8.36}$$

Assuming that $\overline{\overline{\mathsf{Q}}}_s \in \mathbb{E}_1\mathbb{E}_1$ is of full rank 3 with $\Delta_{Qs} = \overline{\overline{\mathsf{Q}}}_s^{(3)} || \boldsymbol{\varepsilon}_{123}\boldsymbol{\varepsilon}_{123} \neq 0$, it has the spatial inverse

$$\overline{\overline{\mathsf{Q}}}_s^{-1} = \frac{1}{\Delta_{Qs}} \boldsymbol{\varepsilon}_{123}\boldsymbol{\varepsilon}_{123} \lfloor\lfloor \overline{\overline{\mathsf{Q}}}_s^{(2)T}. \tag{8.37}$$

Applying the expansion (5.87), the Gibbsian medium dyadics of the Q medium can be identified as

$$\begin{pmatrix} \overline{\overline{\epsilon}}_g & \overline{\overline{\xi}}_g \\ \overline{\overline{\zeta}}_g & \overline{\overline{\mu}}_g \end{pmatrix} = \begin{pmatrix} \epsilon \overline{\overline{\mathsf{Q}}}_s & \overline{\overline{\mathsf{I}}}_s \rfloor \mathbf{X}_s \\ \overline{\overline{\mathsf{I}}}_s \rfloor \mathbf{Z}_s & \mu \overline{\overline{\mathsf{Q}}}_s^T \end{pmatrix}, \tag{8.38}$$

with the scalars

$$\epsilon = c - \mathbf{a}_s | \overline{\overline{\mathsf{Q}}}_s^{-1} | \mathbf{b}_s, \quad \mu = -1/\Delta_{Qs} \tag{8.39}$$

and the spatial bivectors

$$\mathbf{X}_s = -\mathbf{e}_{123} \lfloor \overline{\overline{\mathsf{Q}}}_s^{-1T} | \mathbf{a}_s, \quad \mathbf{Z}_s = \mathbf{e}_{123} \lfloor \overline{\overline{\mathsf{Q}}}_s^{-1} | \mathbf{b}_s. \tag{8.40}$$

From (8.38) we conclude that the Gibbsian medium dyadics $\overline{\overline{\epsilon}}_g$ and $\overline{\overline{\mu}}_g$ of a Q medium must be related through a condition of the form

$$\overline{\overline{\epsilon}}_g = a \overline{\overline{\mu}}_g^T, \tag{8.41}$$

while the Gibbsian medium dyadics $\overline{\overline{\xi}}_g$ and $\overline{\overline{\zeta}}_g$ may be any antisymmetric dyadics.

Let us compare the number of free parameters in the different definitions of the Q medium. In the Gibbsian representation, the permittivity and permeability are defined by $9 + 1 = 10$ parameters because a dyadic $\overline{\overline{\mu}}_g$ and a scalar a can be chosen freely. The antisymmetric dyadics $\overline{\overline{\xi}}_g$ and $\overline{\overline{\zeta}}_g$ involve two spatial bivectors $\mathbf{X}_s, \mathbf{Z}_s$ which correspond to $3 + 3 = 6$ parameters. The total of $10 + 6 = 16$ parameters equals 4×4 of the 4D dyadic $\overline{\overline{\mathsf{Q}}}$.

8.3.2 Spatial Expansion of P Media

To find the expansion of the medium bidyadic (8.27) for the P medium, let us expand

$$\overline{\overline{\mathsf{P}}} = \overline{\overline{\mathsf{P}}}_s + \mathbf{e}_4\boldsymbol{\pi}_s + \mathbf{p}_s\boldsymbol{\varepsilon}_4 + p\mathbf{e}_4\boldsymbol{\varepsilon}_4, \tag{8.42}$$

where the dyadic $\overline{\overline{\mathsf{P}}}_s$, vector $\mathbf{p}_s$ and one-form $\boldsymbol{\pi}_s$ are spatial quantities, while p is a scalar. Inserting (8.42) in (8.1) we obtain

$$\overline{\overline{\mathsf{M}}} = \overline{\overline{\mathsf{P}}}_s^{(2)T} + \overline{\overline{\mathsf{P}}}_s^T{}^{\wedge}_{\wedge}(\boldsymbol{\varepsilon}_4\mathbf{p}_s + \boldsymbol{\pi}_s\mathbf{e}_4) + (p\overline{\overline{\mathsf{P}}}_s^T - \boldsymbol{\pi}_s\mathbf{p}_s)^{\wedge}_{\wedge}\boldsymbol{\varepsilon}_4\mathbf{e}_4. \tag{8.43}$$

Comparing termwise with (8.27), the spatial medium dyadics can be identified as

$$\overline{\overline{\alpha}} = \overline{\overline{\mathsf{P}}}_s^{(2)T}, \tag{8.44}$$

$$\overline{\overline{\epsilon}}' = -\boldsymbol{\pi}_s \wedge \overline{\overline{\mathsf{P}}}_s^T, \tag{8.45}$$

$$\overline{\overline{\mu}}^{-1} = -\overline{\overline{\mathsf{P}}}_s^T \wedge \mathbf{p}_s, \tag{8.46}$$

$$\overline{\overline{\beta}} = \boldsymbol{\pi}_s\mathbf{p}_s - p\overline{\overline{\mathsf{P}}}_s^T. \tag{8.47}$$

The expansion parameters of a P medium can be shown to depend on each other through the relation

$$3\overline{\overline{\beta}}^{(3)} = p^2(\overline{\overline{\alpha}}{}^{\wedge}_{\wedge}\overline{\overline{\beta}} + \overline{\overline{\epsilon}}'{}^{\wedge}_{\wedge}\overline{\overline{\mu}}^{-1}). \tag{8.48}$$

Actually, this is equivalent to a scalar relation obtained by taking the trace operation of both sides of (8.48).

From (8.44)–(8.47) it follows that the dyadic $\overline{\overline{\beta}}$ is essential for the existence of a P medium. In fact, assuming $\overline{\overline{\beta}} = 0$ and $p \neq 0$, we have $\overline{\overline{\mathsf{P}}}_s = \mathbf{p}_s\boldsymbol{\pi}_s/p$, and, consequently, $\overline{\overline{\alpha}} = 0$, $\overline{\overline{\epsilon}}' = 0$ and $\overline{\overline{\mu}}^{-1} = 0$, which leads to vanishing of the whole medium bidyadic, $\overline{\overline{\mathsf{M}}} = 0$.

Because (8.46) implies $\overline{\overline{\mu}}^{-1} \wedge \mathbf{p}_s = 0$, the dyadic $\overline{\overline{\mu}}^{-1}$ does not have a spatial inverse. Thus, unlike in the case of Q media, there is no simple way to express the medium equations in the form (5.70), (5.71) with **D**, **B** in terms of **E**, **H**. However, there exists another, less conventional, representation in terms of Gibbsian field vectors as

$$\begin{pmatrix} \mathbf{B}_g \\ \mathbf{H}_g \end{pmatrix} = \begin{pmatrix} \overline{\overline{\mathsf{A}}}_g & \overline{\overline{\mathsf{B}}}_g \\ \overline{\overline{\mathsf{C}}}_g & \overline{\overline{\mathsf{D}}}_g \end{pmatrix} \cdot \begin{pmatrix} \mathbf{D}_g \\ \mathbf{E}_g \end{pmatrix}. \tag{8.49}$$

In fact, the four Gibbsian medium dyadics turn out to take the quite simple appearance

$$\overline{\overline{\mathsf{A}}}_g = (\mathbf{e}_{123}\boldsymbol{\varepsilon}_{123}\lfloor\lfloor\overline{\overline{\alpha}}^{-1})|\overline{\overline{\mathsf{G}}}_s = \overline{\overline{\mathsf{P}}}_s|\overline{\overline{\mathsf{G}}}_s/\mathrm{tr}\overline{\overline{\mathsf{P}}}_s^{(3)}, \tag{8.50}$$

$$\overline{\overline{\mathsf{B}}}_g = -\mathbf{e}_{123}\lfloor\overline{\overline{\alpha}}^{-1}|\overline{\overline{\epsilon}}' = \mathbf{a}_g \times \overline{\overline{\mathsf{I}}}_g, \tag{8.51}$$

$$\overline{\overline{\mathsf{C}}}_g = (\overline{\overline{\mathsf{G}}}_s|(\overline{\overline{\mu}}^{-1}|\overline{\overline{\alpha}}^{-1})\rfloor\boldsymbol{\varepsilon}_{123})|\overline{\overline{\mathsf{G}}}_s = \overline{\overline{\mathsf{I}}}_g \times \mathbf{b}_g, \tag{8.52}$$

$$\overline{\overline{\mathsf{D}}}_g = \overline{\overline{\mathsf{G}}}_s|(\overline{\overline{\beta}} - \overline{\overline{\mu}}^{-1}|\overline{\overline{\alpha}}^{-1}|\overline{\overline{\epsilon}}') = a\overline{\overline{\mathsf{G}}}_s|\overline{\overline{\mathsf{P}}}_s^T. \tag{8.53}$$

Here we have applied the Gibbsian quantities of Appendix B, $\overline{\overline{\mathsf{I}}}_g = \overline{\overline{\mathsf{G}}}_s = \sum \mathbf{e}_i\mathbf{e}_i$ and

$$\mathbf{a}_g = \boldsymbol{\pi}_s|\overline{\overline{\mathsf{P}}}_s^{-1}|\overline{\overline{\mathsf{G}}}_s, \quad \mathbf{b}_g = -\overline{\overline{\mathsf{P}}}_s^{-1}|\mathbf{p}_s, \quad a = \boldsymbol{\pi}_s|\overline{\overline{\mathsf{P}}}_s^{-1}|\mathbf{p}_s - p. \tag{8.54}$$

Details of the derivation are left as an exercise. From (8.50) and (8.53) it follows that, for the P medium, the dyadics $\overline{\overline{\mathsf{A}}}_g$ and $\overline{\overline{\mathsf{D}}}_g$ must satisfy the relation

$$\overline{\overline{\mathsf{A}}}_g = \lambda\overline{\overline{\mathsf{D}}}_g^T, \tag{8.55}$$

while $\overline{\overline{\mathsf{B}}}_g$ and $\overline{\overline{\mathsf{C}}}_g$ may be any antisymmetric Gibbsian dyadics. This representation of the P medium bears close similarity to the Gibbsian representation (8.38) of the Q medium. In fact, the relation (8.55) appears similar to (8.41) while the off-diagonal dyadics may be arbitrary antisymmetric dyadics in both cases. As a check, the number of free parameters in the representation (8.49) is $9(\overline{\overline{\mathsf{D}}}_g) + 1(\lambda) + 3(\mathbf{a}_g) + 3(\mathbf{b}_g) = 16$ equals that of the dyadic $\overline{\overline{\mathsf{P}}}$.

8.3.3 Relation Between P Media and Q Media

Comparison of the spatial expressions of Q media and P media gives an impression of certain complementarity of the two medium classes. In fact, while for Q media the dyadics $\overline{\overline{\epsilon}}', \overline{\overline{\mu}}^{-1}$ can be of full rank and $\overline{\overline{\alpha}}, \overline{\overline{\beta}}$ do not have inverses, the opposite is true for P media. Actually, a relation between the spatial dyadic expressions (8.44)–(8.47) and (8.31)–(8.34) can be obtained through a connection between a P medium bidyadic $\overline{\overline{\mathsf{M}}}_P$ and a modified Q medium bidyadic $\overline{\overline{\mathsf{M}}}_{mQ}$ in the form

$$\overline{\overline{\mathsf{M}}}_{mQ} = \overline{\overline{\mathsf{G}}}^{(2)}|\overline{\overline{\mathsf{M}}}_P, \qquad \overline{\overline{\mathsf{M}}}_P = \overline{\overline{\Gamma}}^{(2)}|\overline{\overline{\mathsf{M}}}_{mQ}, \tag{8.56}$$

by applying a symmetric metric dyadic of full rank, $\overline{\overline{\mathsf{G}}} = \overline{\overline{\Gamma}}^{-1} \in \mathbb{E}_1\mathbb{E}_1$. In fact, (8.56) is equivalent with the relations

$$\overline{\overline{\mathsf{Q}}} = \overline{\overline{\mathsf{G}}}|\overline{\overline{\mathsf{P}}}^T, \quad \overline{\overline{\mathsf{P}}} = \overline{\overline{\mathsf{Q}}}^T|\overline{\overline{\Gamma}}. \tag{8.57}$$

The spatial dyadic expressions (8.31)–(8.34) for the Q medium are related to those of the P medium, (8.44)–(8.47), as

$$\overline{\overline{\alpha}}_Q = -\varepsilon_{123}\lfloor\overline{\overline{\mathsf{G}}}_s|\overline{\overline{\mu}}_P^{-1}, \tag{8.58}$$
$$\overline{\overline{\epsilon}}'_Q = -\varepsilon_{123}\lfloor\overline{\overline{\mathsf{G}}}_s|\overline{\overline{\beta}}_P, \tag{8.59}$$
$$\overline{\overline{\mu}}_Q^{-1} = -\varepsilon_{123}\lfloor\overline{\overline{\mathsf{G}}}_s^{(2)}|\overline{\overline{\alpha}}_P, \tag{8.60}$$
$$\overline{\overline{\beta}}_Q = -\varepsilon_{123}\lfloor\overline{\overline{\mathsf{G}}}_s^{(2)}|\overline{\overline{\epsilon}}'_P, \tag{8.61}$$

where $\overline{\overline{\mathsf{G}}}_s$ is the spatial part of $\overline{\overline{\mathsf{G}}}$. Thus, it appears that $\overline{\overline{\mu}}^{-1}$ and $\overline{\overline{\alpha}}$ on one hand, and $\overline{\overline{\epsilon}}'$ and $\overline{\overline{\beta}}$ on the other hand, are related through this approach.

8.4 PLANE WAVES

8.4.1 Plane Waves in Q Media

The condition (5.236) for the potential one-form $\boldsymbol{\phi}$ of a plane wave in a Q medium is

$$\overline{\overline{\mathsf{D}}}(\boldsymbol{\nu})|\boldsymbol{\phi} = (\overline{\overline{\mathsf{Q}}}^{(2)}\lfloor\lfloor\boldsymbol{\nu}\boldsymbol{\nu})|\boldsymbol{\phi} = 0. \tag{8.62}$$

The dispersion dyadic and its double-wedge powers can be expanded as

$$\overline{\overline{\mathsf{D}}}(\boldsymbol{\nu}) = (\overline{\overline{\mathsf{Q}}}||\boldsymbol{\nu}\boldsymbol{\nu})\overline{\overline{\mathsf{Q}}} - (\overline{\overline{\mathsf{Q}}}|\boldsymbol{\nu})(\boldsymbol{\nu}|\overline{\overline{\mathsf{Q}}}), \tag{8.63}$$
$$\overline{\overline{\mathsf{D}}}^{(2)}(\boldsymbol{\nu}) = (\overline{\overline{\mathsf{Q}}}||\boldsymbol{\nu}\boldsymbol{\nu})((\overline{\overline{\mathsf{Q}}}||\boldsymbol{\nu}\boldsymbol{\nu})\overline{\overline{\mathsf{Q}}}^{(2)} - \overline{\overline{\mathsf{Q}}}{}^{\wedge}_{\wedge}(\overline{\overline{\mathsf{Q}}}|\boldsymbol{\nu})(\boldsymbol{\nu}|\overline{\overline{\mathsf{Q}}})), \tag{8.64}$$
$$\overline{\overline{\mathsf{D}}}^{(3)}(\boldsymbol{\nu}) = (\overline{\overline{\mathsf{Q}}}||\boldsymbol{\nu}\boldsymbol{\nu})^2((\overline{\overline{\mathsf{Q}}}||\boldsymbol{\nu}\boldsymbol{\nu})\overline{\overline{\mathsf{Q}}}^{(3)} - \overline{\overline{\mathsf{Q}}}^{(2)}{}^{\wedge}_{\wedge}(\overline{\overline{\mathsf{Q}}}|\boldsymbol{\nu})(\boldsymbol{\nu}|\overline{\overline{\mathsf{Q}}})). \tag{8.65}$$

Applying the identities

$$\overline{\overline{\mathsf{Q}}}^{(3)}\lfloor\lfloor\boldsymbol{\nu}\boldsymbol{\nu} = (\overline{\overline{\mathsf{Q}}}||\boldsymbol{\nu}\boldsymbol{\nu})\overline{\overline{\mathsf{Q}}}^{(2)} - \overline{\overline{\mathsf{Q}}}{}^{\wedge}_{\wedge}(\overline{\overline{\mathsf{Q}}}|\boldsymbol{\nu})(\boldsymbol{\nu}|\overline{\overline{\mathsf{Q}}}), \tag{8.66}$$
$$\overline{\overline{\mathsf{Q}}}^{(4)}\lfloor\lfloor\boldsymbol{\nu}\boldsymbol{\nu} = (\overline{\overline{\mathsf{Q}}}||\boldsymbol{\nu}\boldsymbol{\nu})\overline{\overline{\mathsf{Q}}}^{(3)} - \overline{\overline{\mathsf{Q}}}^{(2)}{}^{\wedge}_{\wedge}(\overline{\overline{\mathsf{Q}}}|\boldsymbol{\nu})(\boldsymbol{\nu}|\overline{\overline{\mathsf{Q}}}), \tag{8.67}$$

we obtain the alternative representations

$$\overline{\overline{\mathsf{D}}}^{(2)}(\boldsymbol{\nu}) = (\overline{\overline{\mathsf{Q}}}||\boldsymbol{\nu}\boldsymbol{\nu})(\overline{\overline{\mathsf{Q}}}^{(3)}\lfloor\lfloor\boldsymbol{\nu}\boldsymbol{\nu}), \tag{8.68}$$
$$\overline{\overline{\mathsf{D}}}^{(3)}(\boldsymbol{\nu}) = (\overline{\overline{\mathsf{Q}}}||\boldsymbol{\nu}\boldsymbol{\nu})^2(\overline{\overline{\mathsf{Q}}}^{(4)}\lfloor\lfloor\boldsymbol{\nu}\boldsymbol{\nu}) \tag{8.69}$$
$$= (\overline{\overline{\mathsf{Q}}}||\boldsymbol{\nu}\boldsymbol{\nu})^2\Delta_Q(\mathbf{e}_N\mathbf{e}_N\lfloor\lfloor\boldsymbol{\nu}\boldsymbol{\nu}). \tag{8.70}$$

The dyadic dispersion equation (5.240),

$$\overline{\overline{\mathsf{D}}}^{(3)}(\boldsymbol{\nu}) = 0, \tag{8.71}$$

can now be reduced for Q media to the scalar equation

$$(\overline{\overline{\mathsf{Q}}}||\boldsymbol{\nu}\boldsymbol{\nu})^2 \Delta_Q = 0. \tag{8.72}$$

Here we can separate two cases. For Q media with less than full-rank dyadics $\overline{\overline{\mathsf{Q}}}$, satisfying

$$\Delta_Q = \boldsymbol{\varepsilon}_N \boldsymbol{\varepsilon}_N || \overline{\overline{\mathsf{Q}}}^{(4)} = 0, \tag{8.73}$$

the dispersion equation (8.71) is satisfied identically for any one-form $\boldsymbol{\nu}$. Thus, a Q medium defined by a dyadic $\overline{\overline{\mathsf{Q}}}$ of rank 3 or lower, has no dispersion equation. This means that it belongs to the class of NDE media for which the one-form $\boldsymbol{\nu}$ of the plane wave can be arbitrarily chosen.

In the second case, assuming $\overline{\overline{\mathsf{Q}}}$ is a full-rank dyadic satisfying $\Delta_Q \neq 0$, the dispersion equation has the form

$$\overline{\overline{\mathsf{Q}}}||\boldsymbol{\nu}\boldsymbol{\nu} = 0. \tag{8.74}$$

Since (8.74) is a second-order equation, such a Q medium is nonbirefringent. Here one should note that the antisymmetric part of the dyadic $\overline{\overline{\mathsf{Q}}}$ does not affect the solution of (8.74). In particular, if $\overline{\overline{\mathsf{Q}}}$ is antisymmetric, the dispersion equation (8.71) is again satisfied identically.

For $\Delta_Q \neq 0$ we obtain a condition for the potential one-form by inserting (8.74) in (8.63), which yields

$$\overline{\overline{\mathsf{D}}}(\boldsymbol{\nu})|\boldsymbol{\phi} = -(\overline{\overline{\mathsf{Q}}}|\boldsymbol{\nu})(\boldsymbol{\nu}|\overline{\overline{\mathsf{Q}}})|\boldsymbol{\phi} = 0, \quad \Rightarrow \quad \boldsymbol{\nu}|\overline{\overline{\mathsf{Q}}}|\boldsymbol{\phi} = 0. \tag{8.75}$$

Actually, any potential satisfying this condition is a valid solution. The general form for the potential of the plane wave can be expressed in the form

$$\boldsymbol{\phi} = (\boldsymbol{\nu}|\overline{\overline{\mathsf{Q}}})\rfloor\boldsymbol{\Gamma}, \tag{8.76}$$

where $\boldsymbol{\Gamma} \in \mathbb{F}_2$ may be any two-form yielding a nonzero result. It is left as an exercise to verify that (8.76) satisfies (8.62).

8.4.2 Plane Waves in P Media

The equation for the potential one-form of a plane wave in a P medium is obtained directly from the Maxwell equation as

$$\boldsymbol{\nu}\wedge\boldsymbol{\Psi} = \boldsymbol{\nu}\wedge\overline{\overline{\mathsf{P}}}^{(2)T}|(\boldsymbol{\nu}\wedge\boldsymbol{\phi}) = \boldsymbol{\nu}\wedge(\overline{\overline{\mathsf{P}}}^{T}|\boldsymbol{\nu})\wedge(\overline{\overline{\mathsf{P}}}^{T}|\boldsymbol{\phi}) = 0. \tag{8.77}$$

Assuming that $\boldsymbol{\nu}$ is not an eigen one-form of the dyadic $\overline{\overline{\mathsf{P}}}^{T}$, that is, $\boldsymbol{\nu}\wedge(\overline{\overline{\mathsf{P}}}^{T}|\boldsymbol{\nu})\neq 0$, from (8.77) it follows that the one-form $\overline{\overline{\mathsf{P}}}^{T}|\boldsymbol{\phi}$ must be a linear combination of $\boldsymbol{\nu}$ and $\overline{\overline{\mathsf{P}}}^{T}|\boldsymbol{\nu}$. Thus, for some parameters A, B we can write

$$\boldsymbol{\phi} = A\overline{\overline{\mathsf{P}}}^{-1T}|\boldsymbol{\nu} + B\boldsymbol{\nu} \tag{8.78}$$

The field two-forms then become

$$\boldsymbol{\Phi} = \boldsymbol{\nu}\wedge\boldsymbol{\phi} = A\boldsymbol{\nu}\wedge(\overline{\overline{\mathsf{P}}}^{-1T}|\boldsymbol{\nu}), \tag{8.79}$$

$$\boldsymbol{\Psi} = \overline{\overline{\mathsf{P}}}^{(2)T}|\boldsymbol{\Phi} = A(\overline{\overline{\mathsf{P}}}^{T}|\boldsymbol{\nu})\wedge\boldsymbol{\nu}. \tag{8.80}$$

No condition restricting the choice of the wave one-form $\boldsymbol{\nu}$ appears here, whence any P medium belongs to the class of NDE media. Thus, a plane wave in the P medium can be defined for any chosen $\boldsymbol{\nu}$ and the corresponding field two-forms are obtained from (8.79) and (8.80). A similar property was shown to be valid for skewon–axion media. To verify that the dispersion equation is identically valid, let us form the dispersion dyadic by

$$\begin{aligned}\overline{\overline{\mathsf{D}}}(\boldsymbol{\nu}) &= \overline{\overline{\mathsf{M}}}_m\lfloor\lfloor\boldsymbol{\nu}\boldsymbol{\nu} = -\mathbf{e}_N\lfloor(\boldsymbol{\nu}\wedge\overline{\overline{\mathsf{P}}}^{(2)T}\lfloor\boldsymbol{\nu})\\ &= -\mathbf{e}_N\lfloor(\boldsymbol{\nu}\wedge(\overline{\overline{\mathsf{P}}}^{T}|\boldsymbol{\nu})\wedge\overline{\overline{\mathsf{P}}}^{T}) = (\mathbf{F}\lfloor\overline{\overline{\mathsf{I}}}^{T})|\overline{\overline{\mathsf{P}}}^{T},\end{aligned} \tag{8.81}$$

where the bivector $\mathbf{F}$ is defined by

$$\mathbf{F}(\boldsymbol{\nu}) = \mathbf{e}_N\lfloor(\boldsymbol{\nu}\wedge(\overline{\overline{\mathsf{P}}}^{T}|\boldsymbol{\nu})). \tag{8.82}$$

$\mathbf{F}$ is a simple bivector because it satisfies

$$\begin{aligned}\mathbf{F}\cdot\mathbf{F} &= \mathbf{F}|(\boldsymbol{\varepsilon}_N\lfloor\mathbf{F}) = \mathbf{F}|(\boldsymbol{\nu}\wedge(\overline{\overline{\mathsf{P}}}^{T}|\boldsymbol{\nu}))\\ &= \mathbf{e}_N|(\boldsymbol{\nu}\wedge(\overline{\overline{\mathsf{P}}}^{T}|\boldsymbol{\nu})\wedge\boldsymbol{\nu}\wedge(\overline{\overline{\mathsf{P}}}^{T}|\boldsymbol{\nu})) = 0.\end{aligned} \tag{8.83}$$

Applying the rule (see Problem 2.8)

$$(\mathbf{A}\lfloor\overline{\overline{\mathsf{I}}}^{T})^2 = \mathbf{AA} - \frac{1}{2}(\mathbf{A}\wedge\mathbf{A})\lfloor\overline{\overline{\mathsf{I}}}^{(2)T} \tag{8.84}$$

we have

$$(\mathbf{F}\lfloor\overline{\overline{\mathsf{I}}}^{T})^{(2)} = \mathbf{FF}, \quad (\mathbf{F}\lfloor\overline{\overline{\mathsf{I}}}^{T})^{(3)} = 0. \tag{8.85}$$

Because of this, the dyadic dispersion equation is satisfied identically,

$$\overline{\overline{\mathsf{D}}}^{(3)}(\boldsymbol{\nu}) = (\mathbf{F}(\boldsymbol{\nu})\lfloor\overline{\overline{\mathsf{I}}}^T)^{(3)}|\overline{\overline{\mathsf{P}}}^{(3)T} = 0, \tag{8.86}$$

for any one-form $\boldsymbol{\nu}$ of a plane wave in a P medium.

8.4.3 P Medium as Boundary Material

Properties of plane waves in P media and skewon–axion media appear somewhat similar because for both of them the dispersion equation is identically satisfied for any wave one-form $\boldsymbol{\nu}$. Also, one can show that an interface of a P medium can produce boundary conditions similar to those of a skewon–axion medium. As an example, consider a P medium defined by

$$\overline{\overline{\mathsf{P}}} = \overline{\overline{\mathsf{I}}} + \mathbf{a}\boldsymbol{\alpha}, \quad \mathbf{a}|\boldsymbol{\alpha} \neq -1, \tag{8.87}$$

with

$$\overline{\overline{\mathsf{P}}}^{-1} = \overline{\overline{\mathsf{I}}} - \frac{\mathbf{a}\boldsymbol{\alpha}}{1 + \mathbf{a}|\boldsymbol{\alpha}}. \tag{8.88}$$

From (8.79) and (8.80) the field two-forms of a plane wave in such a P medium can be expressed as

$$\boldsymbol{\Phi} = -\frac{A(\mathbf{a}|\boldsymbol{\alpha})}{1 + \mathbf{a}|\boldsymbol{\alpha}}\boldsymbol{\nu} \wedge \boldsymbol{\alpha}, \tag{8.89}$$

$$\boldsymbol{\Psi} = -A(\mathbf{a}|\boldsymbol{\alpha})\boldsymbol{\nu} \wedge \boldsymbol{\alpha}, \tag{8.90}$$

whence they satisfy the conditions

$$\boldsymbol{\alpha} \wedge \boldsymbol{\Phi} = 0, \quad \boldsymbol{\alpha} \wedge \boldsymbol{\Psi} = 0 \tag{8.91}$$

in this medium. Now these conditions coincide with (7.130) obtained for a special skewon–axion medium, whence after similar conclusions, DB boundary conditions (7.133) or SH boundary conditions (7.137) and (7.138) can be obtained at the interface $\boldsymbol{\varepsilon}_3|\mathbf{x} = 0$. Actually, this is no wonder since for the choice (8.87) the P medium bidyadic can be written as

$$\overline{\overline{\mathsf{M}}} = \overline{\overline{\mathsf{P}}}^{(2)T} = (\overline{\overline{\mathsf{I}}}{}^{\wedge}_{\wedge}\overline{\overline{\mathsf{B}}})^T, \quad \overline{\overline{\mathsf{B}}} = \frac{1}{2}\overline{\overline{\mathsf{I}}} + \mathbf{a}\boldsymbol{\alpha}, \tag{8.92}$$

which is of the skewon–axion form (7.90).

Because a nonzero principal part would distinguish a P medium from a skewon–axion medium, let us study the Hehl–Obukhov expansion of the general P medium bidyadic. Applying (3.73), the skewon part can be expressed as

$$\overline{\overline{\mathsf{M}}}_2 = \overline{\overline{\mathsf{M}}}_- = \frac{1}{2}(\overline{\overline{\mathsf{P}}}^{(2)T} - \overline{\overline{\mathsf{I}}}^{(4)T}\lfloor\lfloor\overline{\overline{\mathsf{P}}}^{(2)}). \tag{8.93}$$

Because the axion part has the form (3.77),

$$\overline{\overline{\mathsf{M}}}_3 = \frac{1}{6}(\mathrm{tr}\overline{\overline{\mathsf{P}}}^{(2)})\overline{\overline{\mathsf{I}}}^{(2)T}, \tag{8.94}$$

the principal part of the P-medium bidyadic can be written as

$$\begin{aligned}\overline{\overline{\mathsf{M}}}_1 &= \overline{\overline{\mathsf{M}}} - \overline{\overline{\mathsf{M}}}_2 - \overline{\overline{\mathsf{M}}}_3 \\ &= \overline{\overline{\mathsf{P}}}^{(2)T} + \frac{1}{3}(\mathrm{tr}\overline{\overline{\mathsf{P}}}^{(2)T})\overline{\overline{\mathsf{I}}}^{(2)T} - \frac{1}{2}(\overline{\overline{\mathsf{P}}}^{(2)T}\lfloor\lfloor\overline{\overline{\mathsf{I}}})^{\wedge}_{\wedge}\overline{\overline{\mathsf{I}}}^T.\end{aligned} \tag{8.95}$$

Since this does not vanish identically, the class of P media does not coincide with the class of skewon–axion media. Actually, one can show that it is possible to define a P-medium bidyadic consisting just of its principal part, that is, $\overline{\overline{\mathsf{M}}} = \overline{\overline{\mathsf{P}}}^{(3)T}$ satisfying $\overline{\overline{\mathsf{M}}}\lfloor\lfloor\overline{\overline{\mathsf{I}}} = 0$. However, in such a case the medium bidyadic turns out to be of rank 1. Details are left as an exercise.

8.5 P-AXION AND Q-AXION MEDIA

A straightforward generalization of the class of P media and Q media can be obtained by adding an axion term to the definitions (8.1) and (8.2),

$$\overline{\overline{\mathsf{M}}} = \overline{\overline{\mathsf{P}}}^{(2)T} + \alpha\overline{\overline{\mathsf{I}}}^{(2)T}, \tag{8.96}$$

$$\overline{\overline{\mathsf{M}}}_m = \overline{\overline{\mathsf{Q}}}^{(2)} + \alpha\overline{\overline{\mathsf{E}}}, \tag{8.97}$$

with $\overline{\overline{\mathsf{E}}} = \mathbf{e}_N\lfloor\overline{\overline{\mathsf{I}}}^{(2)T}$. The number of free parameters is 17 in both cases. Since the axion term does not have an effect on the dispersion equation of the plane wave, there is no dispersion equation for the P-axion medium while (8.74) remains valid for the Q-axion medium. The effect of the axion term can be detected at an interface where the medium parameters change, as a change in reflection and transmission properties of plane waves.

Certain special P-axion and Q-axion media have an indirect relation to skewon–axion media. The relation can be detected through the modified closure conditions (8.16) and (8.17). Let us consider the condition (8.17) for the modified Q-axion medium bidyadic (8.97) written as

$$(\overline{\overline{\mathsf{M}}}_m - \alpha\overline{\overline{\mathsf{E}}})^T \cdot (\overline{\overline{\mathsf{M}}}_m - \alpha\overline{\overline{\mathsf{E}}}) = \overline{\overline{\mathsf{Q}}}^{(2)T} \cdot \overline{\overline{\mathsf{Q}}}^{(2)} = \Delta_Q \overline{\overline{\mathsf{E}}}, \tag{8.98}$$

or

$$\overline{\overline{\mathsf{M}}}_m^T \cdot \overline{\overline{\mathsf{M}}}_m - \alpha(\overline{\overline{\mathsf{M}}}_m^T + \overline{\overline{\mathsf{M}}}_m) + (\alpha^2 - \Delta_Q)\overline{\overline{\mathsf{E}}} = 0. \tag{8.99}$$

Considering now a special Q-axion medium satisfying

$$\alpha = \sqrt{\Delta_Q}, \tag{8.100}$$

with either branch of the square root, whence the number of free parameters is reduced to 16. The closure condition becomes

$$\overline{\overline{\mathsf{M}}}_m^T \cdot \overline{\overline{\mathsf{M}}}_m - \alpha(\overline{\overline{\mathsf{M}}}_m^T + \overline{\overline{\mathsf{M}}}_m) = 0. \tag{8.101}$$

Assuming that the inverse bidyadic $\overline{\overline{\mathsf{M}}}_m^{-1}$ exists, multiplying (8.101) by $\overline{\overline{\mathsf{M}}}_m^{-1T}|$ from the left and by $|\overline{\overline{\mathsf{M}}}_m^{-1}$ from the right yields the condition

$$\overline{\overline{\mathsf{M}}}_m^{-1T} + \overline{\overline{\mathsf{M}}}_m^{-1} - \frac{1}{\alpha}\overline{\overline{\mathsf{E}}} = 0, \tag{8.102}$$

which can be rewritten as

$$(\overline{\overline{\mathsf{M}}}_m^{-1} - \frac{1}{2\alpha}\overline{\overline{\mathsf{E}}})^T + \left(\overline{\overline{\mathsf{M}}}_m^{-1} - \frac{1}{2\alpha}\overline{\overline{\mathsf{E}}}\right) = 0. \tag{8.103}$$

This condition requires that the bidyadic in brackets must be antisymmetric, that is, there must exist an antisymmetric bidyadic $\overline{\overline{\mathsf{X}}} \in \mathbb{F}_2\mathbb{F}_2$ in terms of which we can write

$$\overline{\overline{\mathsf{M}}}_m^{-1} = \overline{\overline{\mathsf{X}}} + \frac{1}{2\alpha}\overline{\overline{\mathsf{E}}}. \tag{8.104}$$

Expressing this in terms of the medium bidyadic $\overline{\overline{\mathsf{N}}}$ as defined by (5.52), we have

$$\overline{\overline{\mathsf{N}}}_m = \mathbf{e}_N\mathbf{e}_N \lfloor\lfloor \overline{\overline{\mathsf{M}}}_m^{-1} = \overline{\overline{\mathsf{Y}}} + \frac{1}{2\alpha}\overline{\overline{\mathsf{E}}}, \tag{8.105}$$

where $\overline{\overline{\mathsf{Y}}} = \mathbf{e}_N\mathbf{e}_N \lfloor\lfloor \overline{\overline{\mathsf{X}}} \in \mathbb{E}_2\mathbb{E}_2$ is another antisymmetric bidyadic. Obviously, the right side of (8.105) is of the skewon–axion form. Since the skewon–axion medium has been defined through the bidyadic $\overline{\overline{\mathsf{M}}}$ and not $\overline{\overline{\mathsf{N}}}$, it is not, however, proper to talk about skewon–axion medium in this case. Actually this is a special Q-axion medium. A similar relation can be obtained by starting from a special P-axion medium bidyadic.

Here one must note that a relation of the form (8.104) requires that $\overline{\overline{\mathsf{M}}}_m$ must have a nonzero antisymmetric part. In fact, assuming that $\overline{\overline{\mathsf{M}}}_m$ is symmetric, so must be its inverse $\overline{\overline{\mathsf{M}}}_m^{-1}$, in which case (8.104) would require $\overline{\overline{\mathsf{X}}} = 0$, whence $\overline{\overline{\mathsf{M}}}_m^{-1}$ would be a multiple of $\overline{\overline{\mathsf{E}}}$, which would lead to a contradiction. Actually, it can be shown that if $\overline{\overline{\mathsf{M}}}_m$ defined by (8.97) with (8.100) is symmetric, it does not have an inverse. Details are left as an exercise.

8.6 EXTENDED Q MEDIA

The class of Q media can be extended in the same way as the class of skewon–axion media, by adding a simple dyadic term as

$$\overline{\overline{\mathsf{M}}} = \varepsilon_N \lfloor \overline{\overline{\mathsf{Q}}}^{(2)} + \mathbf{\Delta B}, \tag{8.106}$$

$$\overline{\overline{\mathsf{M}}}_m = \overline{\overline{\mathsf{Q}}}^{(2)} + \mathbf{AB}, \tag{8.107}$$

where $\mathbf{A}$ and $\mathbf{B}$ are two bivectors and $\mathbf{\Delta} = \varepsilon_N \lfloor \mathbf{A}$. The corresponding class of media can be called that of extended Q media (or generalized Q media [1, 63]). When $\overline{\overline{\mathsf{Q}}}$ is of full rank and $1 + \mathbf{B}|\overline{\overline{\mathsf{Q}}}^{(-2)}|\mathbf{A}$ does not vanish, the bidyadic $\overline{\overline{\mathsf{M}}}_m$ has an inverse of the the same form as (8.107),

$$\overline{\overline{\mathsf{M}}}_m^{-1} = \overline{\overline{\mathsf{Q}}}^{(-2)} - \frac{(\overline{\overline{\mathsf{Q}}}^{(-2)}|\mathbf{A})(\mathbf{B}|\overline{\overline{\mathsf{Q}}}^{(-2)})}{1 + \mathbf{B}|\overline{\overline{\mathsf{Q}}}^{(-2)}|\mathbf{A}}, \tag{8.108}$$

as can be easily verified by multiplication. Applying (5.57) we obtain

$$\begin{aligned} \overline{\overline{\mathsf{N}}}_m &= \mathbf{e}_N \lfloor \overline{\overline{\mathsf{M}}}_m^{-1} \rfloor \mathbf{e}_N \\ &= \frac{1}{\Delta_Q} \overline{\overline{\mathsf{Q}}}^{(2)T} - \frac{(\mathbf{e}_N \lfloor \overline{\overline{\mathsf{Q}}}^{(-2)}|\mathbf{A})(\mathbf{B}|\overline{\overline{\mathsf{Q}}}^{(-2)} \rfloor \mathbf{e}_N)}{1 + \mathbf{B}|\overline{\overline{\mathsf{Q}}}^{(-2)}|\mathbf{A}}, \end{aligned} \tag{8.109}$$

which is, again, of the similar form. Thus, the class of extended Q media can be defined through either the bidyadic $\overline{\overline{\mathsf{M}}}$ or $\overline{\overline{\mathsf{N}}}$.

8.6.1 Gibbsian Representation

Expanding the two bivectors as

$$\mathbf{A} = \mathbf{e}_{123} \lfloor \boldsymbol{\alpha}_1 + \mathbf{a}_2 \wedge \mathbf{e}_4, \quad \mathbf{B} = \mathbf{e}_{123} \lfloor \boldsymbol{\beta}_1 + \mathbf{b}_2 \wedge \mathbf{e}_4, \tag{8.110}$$

where $\boldsymbol{\alpha}_1, \boldsymbol{\beta}_1$ are spatial one-forms and $\mathbf{a}_2, \mathbf{b}_2$ spatial vectors, we can write

$$\mathbf{AB} = (\mathbf{e}_{123}\lfloor\boldsymbol{\alpha}_1)(\mathbf{e}_{123}\lfloor\boldsymbol{\beta}_1) - \mathbf{e}_4 \wedge \mathbf{a}_2(\boldsymbol{\beta}_1\rfloor\mathbf{e}_{123}) + (\mathbf{e}_{123}\lfloor\boldsymbol{\alpha}_1)\mathbf{b}_2 \wedge \mathbf{e}_4 - \mathbf{e}_4 \wedge (\mathbf{a}_2\mathbf{b}_2) \wedge \mathbf{e}_4, \tag{8.111}$$

and

$$\boldsymbol{\varepsilon}_N\lfloor\mathbf{AB} = -\boldsymbol{\varepsilon}_4 \wedge \boldsymbol{\alpha}_1(\boldsymbol{\beta}_1\rfloor\mathbf{e}_{123}) + (\boldsymbol{\varepsilon}_{123}\lfloor\mathbf{a}_2)(\boldsymbol{\beta}_1\rfloor\mathbf{e}_{123}) - \boldsymbol{\varepsilon}_4 \wedge \boldsymbol{\alpha}_1\mathbf{b}_2 \wedge \mathbf{e}_4 + (\boldsymbol{\varepsilon}_{123}\lfloor\mathbf{a}_2)\mathbf{b}_2 \wedge \mathbf{e}_4. \tag{8.112}$$

Combined with the expansion (8.28) we obtain expressions of the spatial medium dyadics, extended from those of the Q medium (8.31)–(8.34), as

$$\overline{\overline{\alpha}} = \boldsymbol{\varepsilon}_{123}\lfloor\overline{\overline{\mathsf{Q}}}_s \wedge \mathbf{a}_s + (\boldsymbol{\varepsilon}_{123}\lfloor\mathbf{a}_2)(\boldsymbol{\beta}_1\rfloor\mathbf{e}_{123}), \tag{8.113}$$

$$\overline{\overline{\epsilon}}' = \boldsymbol{\varepsilon}_{123}\lfloor(c\overline{\overline{\mathsf{Q}}}_s - \mathbf{b}_s\mathbf{a}_s) + (\boldsymbol{\varepsilon}_{123}\lfloor\mathbf{a}_2)\mathbf{b}_2, \tag{8.114}$$

$$\overline{\overline{\mu}}^{-1} = -\boldsymbol{\varepsilon}_{123}\lfloor\overline{\overline{\mathsf{Q}}}_s^{(2)} - \boldsymbol{\alpha}_1(\boldsymbol{\beta}_1\rfloor\mathbf{e}_{123}), \tag{8.115}$$

$$\overline{\overline{\beta}} = \boldsymbol{\varepsilon}_{123}\lfloor(\mathbf{b}_s \wedge \overline{\overline{\mathsf{Q}}}_s) - \boldsymbol{\alpha}_1\mathbf{b}_2. \tag{8.116}$$

To find the corresponding Gibbsian medium dyadics we can start from

$$\overline{\overline{\mu}}_g^{-1} = \overline{\overline{\mu}}^{-1}\rfloor\boldsymbol{\varepsilon}_{123} = -\boldsymbol{\varepsilon}_{123}\boldsymbol{\varepsilon}_{123}\lfloor\lfloor\overline{\overline{\mathsf{Q}}}_s^{(2)} - \boldsymbol{\alpha}_1\boldsymbol{\beta}_1 = -\Delta_{Qs}\overline{\overline{\mathsf{Q}}}_s^{-1T} - \boldsymbol{\alpha}_1\boldsymbol{\beta}_1. \tag{8.117}$$

Assuming $\Delta_{Q_s} = \boldsymbol{\varepsilon}_{123}\boldsymbol{\varepsilon}_{123}||\overline{\overline{\mu}}^{(-3)} \neq 0$ and $\Delta_{Q_s} \neq -\boldsymbol{\alpha}_1|\overline{\overline{\mathsf{Q}}}_s|\boldsymbol{\beta}_1$, its inverse can be formed as

$$\overline{\overline{\mu}}_g = -\frac{1}{\Delta_{Qs}}\left(\overline{\overline{\mathsf{Q}}}_s^T - \frac{(\boldsymbol{\alpha}_1|\overline{\overline{\mathsf{Q}}}_s)(\overline{\overline{\mathsf{Q}}}_s|\boldsymbol{\beta}_1)}{\Delta_{Qs} + \boldsymbol{\alpha}_1|\overline{\overline{\mathsf{Q}}}_s|\boldsymbol{\beta}_1}\right). \tag{8.118}$$

The other Gibbsian dyadics can now be found from

$$\overline{\overline{\xi}}_g = (\mathbf{e}_{123}\boldsymbol{\varepsilon}_{123}\lfloor\lfloor\overline{\overline{\alpha}})|\overline{\overline{\mu}}_g, \tag{8.119}$$

$$\overline{\overline{\zeta}}_g = -\overline{\overline{\mu}}_g|\overline{\overline{\beta}}, \tag{8.120}$$

$$\overline{\overline{\epsilon}}_g = \mathbf{e}_{123}\lfloor\overline{\overline{\epsilon}}' + \overline{\overline{\xi}}_g|\overline{\overline{\mu}}_g^{-1}|\overline{\overline{\zeta}}_g. \tag{8.121}$$

Leaving the details as exercises, the results can be shown to take the form

$$\overline{\overline{\epsilon}}_g = \epsilon\overline{\overline{\mathsf{Q}}}_s + \mathbf{p}_2\mathbf{q}_1, \tag{8.122}$$

$$\overline{\overline{\xi}}_g = \overline{\overline{\mathsf{I}}}\rfloor\mathbf{X}_s + \mathbf{p}_2\mathbf{q}_2, \tag{8.123}$$

$$\overline{\overline{\zeta}}_g = \overline{\overline{\mathsf{I}}}\rfloor\mathbf{Z}_s + \mathbf{p}_1\mathbf{q}_1, \tag{8.124}$$

$$\overline{\overline{\mu}}_g = \mu\overline{\overline{\mathsf{Q}}}_s^T + \mathbf{p}_1\mathbf{q}_2, \tag{8.125}$$

where the scalars ϵ, μ and the spatial bivectors $\mathbf{X}_s, \mathbf{Z}_s$ are the same as in (8.38). Actually, the Gibbsian medium dyadics of the extended Q medium equal those of the Q medium (8.123)–(8.125) extended by terms involving four spatial vectors $\mathbf{p}_1, \mathbf{p}_2, \mathbf{q}_1, \mathbf{q}_2$ in the dyadic products

$$\mathbf{p}_1\mathbf{q}_1 = -\frac{\Delta_{Qs}(\boldsymbol{\alpha}_1|\overline{\overline{\mathsf{Q}}}_s)(\boldsymbol{\beta}_1\rfloor\mathbf{Z}_s + \mathbf{b}_2)}{\Delta_{Qs} + \boldsymbol{\alpha}_1|\overline{\overline{\mathsf{Q}}}_s|\boldsymbol{\beta}_1}, \tag{8.126}$$

$$\mathbf{p}_1\mathbf{q}_2 = \frac{(\boldsymbol{\alpha}_1|\overline{\overline{\mathsf{Q}}}_s)(\overline{\overline{\mathsf{Q}}}_s|\boldsymbol{\beta}_1)}{\Delta_{Qs} + \boldsymbol{\alpha}_1|\overline{\overline{\mathsf{Q}}}_s|\boldsymbol{\beta}_1}, \tag{8.127}$$

$$\mathbf{p}_2\mathbf{q}_1 = \frac{\Delta_{Qs}(\mathbf{X}_s\lfloor\boldsymbol{\alpha}_1 + \mathbf{a}_2)(\boldsymbol{\beta}_1\rfloor\mathbf{Z}_s + \mathbf{b}_2)}{\epsilon(\Delta_{Qs} + \boldsymbol{\alpha}_1|\overline{\overline{\mathsf{Q}}}_s|\boldsymbol{\beta}_1)}, \tag{8.128}$$

$$\mathbf{p}_2\mathbf{q}_2 = -\frac{(\mathbf{X}_s\lfloor\boldsymbol{\alpha}_1 + \mathbf{a}_2)(\overline{\overline{\mathsf{Q}}}_s|\boldsymbol{\beta}_1)}{\epsilon(\Delta_{Qs} + \boldsymbol{\alpha}_1|\overline{\overline{\mathsf{Q}}}_s|\boldsymbol{\beta}_1)}. \tag{8.129}$$

The expressions for the four vectors can be extracted as

$$\mathbf{p}_1 = -\boldsymbol{\alpha}_1|\overline{\overline{\mathsf{Q}}}_s, \tag{8.130}$$

$$\mathbf{p}_2 = \frac{1}{\epsilon}(\mathbf{X}_s\lfloor\boldsymbol{\alpha}_1 + \mathbf{a}_2), \tag{8.131}$$

$$\mathbf{q}_1 = \frac{\Delta_{Qs}(\boldsymbol{\beta}_1\rfloor\mathbf{Z}_s + \mathbf{b}_2)}{\Delta_{Qs} + \boldsymbol{\alpha}_1|\overline{\overline{\mathsf{Q}}}_s|\boldsymbol{\beta}_1}, \tag{8.132}$$

$$\mathbf{q}_2 = -\frac{\overline{\overline{\mathsf{Q}}}_s|\boldsymbol{\beta}_1}{\Delta_{Qs} + \boldsymbol{\alpha}_1|\overline{\overline{\mathsf{Q}}}_s|\boldsymbol{\beta}_1}. \tag{8.133}$$

The economy in 4D notation (8.107) over (8.122)–(8.125), with (8.126)–(8.129) substituted, appears striking.

8.6.2 Field Decomposition

It will turn out that plane-wave fields in an extended Q medium can be decomposed in two sets each obeying a polarization condition of its own. To see this, let us apply the conditions (5.233) satisfied by the fields of any plane wave in any medium,

$$\mathbf{\Phi} \cdot \mathbf{\Phi} = 0, \quad \mathbf{\Phi} \cdot \mathbf{\Psi} = 0, \quad \mathbf{\Psi} \cdot \mathbf{\Psi} = 0. \tag{8.134}$$

Starting from the condition

$$(\overline{\overline{\mathsf{M}}}_m - \mathbf{AB})^T \cdot (\overline{\overline{\mathsf{M}}}_m - \mathbf{AB}) = \overline{\overline{\mathsf{Q}}}^{(2)T} \cdot \overline{\overline{\mathsf{Q}}}^{(2)} = \Delta_Q \overline{\overline{\mathsf{E}}}, \tag{8.135}$$

and multiplying this from both sides as $\mathbf{\Phi}|()|\mathbf{\Phi}$, we can expand

$$\begin{aligned}
\Delta_Q \mathbf{\Phi}|\overline{\overline{\mathsf{E}}}|\mathbf{\Phi} &= \Delta_Q \mathbf{\Phi} \cdot \mathbf{\Phi} = 0 \\
&= (\overline{\overline{\mathsf{M}}}_m|\mathbf{\Phi} - (\mathbf{B}|\mathbf{\Phi})\mathbf{A}) \cdot (\overline{\overline{\mathsf{M}}}_m|\mathbf{\Phi} - \mathbf{A}(\mathbf{B}|\mathbf{\Phi})) \\
&= (\mathbf{e}_N \lfloor \mathbf{\Psi} - (\mathbf{B}|\mathbf{\Phi})\mathbf{A}) \cdot (\mathbf{e}_N \lfloor \mathbf{\Psi} - \mathbf{A}(\mathbf{B}|\mathbf{\Phi})) \\
&= \mathbf{\Psi} \cdot \mathbf{\Psi} - 2(\mathbf{A}|\mathbf{\Psi})(\mathbf{B}|\mathbf{\Phi}) + (\mathbf{A} \cdot \mathbf{A})(\mathbf{B}|\mathbf{\Phi})^2 \\
&= -(2\mathbf{A}|\mathbf{\Psi} - \mathbf{A} \cdot \mathbf{A}(\mathbf{B}|\mathbf{\Phi}))(\mathbf{B}|\mathbf{\Phi}) = 0.
\end{aligned}$$

This condition is valid for any plane wave in the extended Q medium. It follows that any given plane wave must satisfy one of the two polarization conditions

$$\mathbf{B}|\mathbf{\Phi} = 0, \tag{8.136}$$

$$2\mathbf{A}|\mathbf{\Psi} - \mathbf{A} \cdot \mathbf{A}(\mathbf{B}|\mathbf{\Phi}) = 0. \tag{8.137}$$

If $\mathbf{A}$ is a simple bivector, the latter condition reduces to $\mathbf{A}|\mathbf{\Psi} = 0$. In the general case it can be rewritten in the alternative form

$$\mathbf{A} \cdot (2\overline{\overline{\mathsf{Q}}}^{(2)} + \mathbf{AB}) \cdot \mathbf{\Phi} = 0. \tag{8.138}$$

To conclude, all plane waves in an extended Q medium can be decomposed in two sets, satisfying either the first one or the second one of the two orthogonality conditions (8.136), (8.138). Since the conditions are independent of the wave one-form $\boldsymbol{\nu}$ of the plane wave and linear in the field two-form $\mathbf{\Phi}$, actually any field which can be expressed as a sum or integral of plane waves can be decomposed in two fields satisfying these conditions. This is why such a medium has been called by the name decomposable (DC) medium in the past [64]. Because an axion term does not affect a plane wave in a homogeneous medium, the same field decomposition is valid in more general media defined by

$$\overline{\overline{\mathsf{M}}}_m = \alpha \mathbf{e}_N \lfloor \overline{\overline{\mathsf{I}}}^{(2)T} + \overline{\overline{\mathsf{Q}}}^{(2)} + \mathbf{AB}, \tag{8.139}$$

$$\overline{\overline{\mathsf{M}}} = \alpha \overline{\overline{\mathsf{I}}}^{(2)T} + \boldsymbol{\varepsilon}_N \lfloor \overline{\overline{\mathsf{Q}}}^{(2)} + \boldsymbol{\Delta}\mathbf{B}. \tag{8.140}$$

8.6.3 Transformations

The extended Q medium shares the property of the Q medium of being form-invariant in the affine transformation:

$$\overline{\overline{\mathsf{M}}}_{ma} = \overline{\overline{\mathsf{A}}}^{(2)}|\overline{\overline{\mathsf{M}}}_m|\overline{\overline{\mathsf{A}}}^{(2)T} = \overline{\overline{\mathsf{Q}}}_a^{(2)} + \mathbf{A}_a\mathbf{B}_a, \tag{8.141}$$

when the transformed quantities are defined by

$$\overline{\overline{\mathsf{Q}}}_a = \overline{\overline{\mathsf{A}}}|\overline{\overline{\mathsf{Q}}}|\overline{\overline{\mathsf{A}}}^T, \quad \mathbf{A}_a = \overline{\overline{\mathsf{A}}}^{(2)}|\mathbf{A}, \quad \overline{\overline{\mathsf{B}}}_a = \overline{\overline{\mathsf{A}}}^{(2)}|\mathbf{B}. \tag{8.142}$$

Considering the special involutionary duality transformation (6.65), defined by the parameters $Z_d = -Z$ and $\theta = \pi/2$, the transformed extended Q medium bidyadic becomes

$$\overline{\overline{\mathsf{M}}}_d = \frac{1}{Z^2}\overline{\overline{\mathsf{M}}}^{-1} = \frac{1}{Z^2}(\boldsymbol{\varepsilon}_N \lfloor (\overline{\overline{\mathsf{Q}}}^{(2)} + \mathbf{AB}))^{-1}. \tag{8.143}$$

Invoking (8.108), the result can be written in the form

$$\overline{\overline{\mathsf{M}}}_{md} = \overline{\overline{\mathsf{Q}}}_d^{(2)} + \mathbf{A}_d\mathbf{B}_d, \tag{8.144}$$

with

$$\overline{\overline{\mathsf{Q}}}_d = \frac{1}{Z\sqrt{\Delta_Q}}\overline{\overline{\mathsf{Q}}}^T, \tag{8.145}$$

and

$$\mathbf{A}_d\mathbf{B}_d = -\frac{(\mathbf{A}\cdot\overline{\overline{\mathsf{Q}}}^{(2)})(\overline{\overline{\mathsf{Q}}}^{(2)}\cdot\mathbf{B})}{Z^2\Delta_Q(\Delta_Q + \mathbf{A}\cdot\overline{\overline{\mathsf{Q}}}^{(2)}\cdot\mathbf{B})}. \tag{8.146}$$

Thus, the extended Q medium is form-invariant also in the involutionary duality transformation.

8.6.4 Plane Waves in Extended Q Media

The dispersion dyadic corresponding to a plane wave in the extended Q medium can be expressed as

$$\overline{\overline{\mathsf{D}}}(\boldsymbol{\nu}) = \boldsymbol{\nu}\boldsymbol{\nu}\rfloor\rfloor\overline{\overline{\mathsf{M}}}_m = \boldsymbol{\nu}\boldsymbol{\nu}\rfloor\rfloor\overline{\overline{\mathsf{Q}}}^{(2)} + (\boldsymbol{\nu}\rfloor\mathbf{A})(\boldsymbol{\nu}\rfloor\mathbf{B}). \tag{8.147}$$

Applying the identities (8.66) and (8.67), the double-wedge powers can be expanded as

$$\overline{\overline{\mathsf{D}}}^{(2)}(\boldsymbol{\nu}) = (\boldsymbol{\nu}\boldsymbol{\nu}||\overline{\overline{\mathsf{Q}}})(\boldsymbol{\nu}\boldsymbol{\nu}\rfloor\rfloor\overline{\overline{\mathsf{Q}}}^{(3)}) + (\boldsymbol{\nu}\boldsymbol{\nu}\rfloor\rfloor\overline{\overline{\mathsf{Q}}}^{(2)})^{\wedge}_{\wedge}(\boldsymbol{\nu}\rfloor\mathbf{A})(\boldsymbol{\nu}\rfloor\mathbf{B}), \quad (8.148)$$

$$\overline{\overline{\mathsf{D}}}^{(3)}(\boldsymbol{\nu}) = (\boldsymbol{\nu}\boldsymbol{\nu}||\overline{\overline{\mathsf{Q}}})^2(\boldsymbol{\nu}\boldsymbol{\nu}\rfloor\rfloor\overline{\overline{\mathsf{Q}}}^{(4)}) + (\boldsymbol{\nu}\boldsymbol{\nu}||\overline{\overline{\mathsf{Q}}})(\boldsymbol{\nu}\boldsymbol{\nu}\rfloor\rfloor\overline{\overline{\mathsf{Q}}}^{(3)})^{\wedge}_{\wedge}(\boldsymbol{\nu}\rfloor\mathbf{A})(\boldsymbol{\nu}\rfloor\mathbf{B}). \quad (8.149)$$

Assuming that $\overline{\overline{\mathsf{Q}}}$ is of full rank, we can apply the rules

$$\overline{\overline{\mathsf{Q}}}^{(3)} = \mathbf{e}_N\mathbf{e}_N\lfloor\lfloor\overline{\overline{\mathsf{Q}}}^{-1T}\Delta_Q = \overline{\overline{\mathsf{Q}}}^{(4)}\lfloor\lfloor\overline{\overline{\mathsf{Q}}}^{-1T} \quad (8.150)$$

and

$$(\boldsymbol{\alpha}\rfloor(\mathbf{e}_N\lfloor\boldsymbol{\beta}))\wedge\mathbf{a} = \mathbf{e}_N\lfloor((\boldsymbol{\alpha}\wedge\boldsymbol{\beta})\lfloor\mathbf{a}), \quad (8.151)$$

valid for any one-forms $\boldsymbol{\alpha}, \boldsymbol{\beta}$ and a vector $\mathbf{a}$. With these, we can further expand

$$\begin{aligned}(\boldsymbol{\nu}\boldsymbol{\nu}\rfloor\rfloor\overline{\overline{\mathsf{Q}}}^{(3)})^{\wedge}_{\wedge}(\boldsymbol{\nu}\rfloor\mathbf{A})(\boldsymbol{\nu}\rfloor\mathbf{B}) &= \Delta_Q(\boldsymbol{\nu}\boldsymbol{\nu}\rfloor\rfloor(\mathbf{e}_N\mathbf{e}_N\lfloor\lfloor\overline{\overline{\mathsf{Q}}}^{-1T}))^{\wedge}_{\wedge}(\boldsymbol{\nu}\rfloor\mathbf{A})(\boldsymbol{\nu}\rfloor\mathbf{B})\\ &= \Delta_Q\mathbf{e}_N\mathbf{e}_N\lfloor\lfloor((\boldsymbol{\nu}\boldsymbol{\nu}^{\wedge}_{\wedge}\overline{\overline{\mathsf{Q}}}^{-1T})^{\wedge}_{\wedge}(\boldsymbol{\nu}\rfloor\mathbf{A})(\boldsymbol{\nu}\rfloor\mathbf{B}))\\ &= (\boldsymbol{\nu}\boldsymbol{\nu}\rfloor\rfloor\overline{\overline{\mathsf{Q}}}^{(4)})(\overline{\overline{\mathsf{Q}}}^{-1T}||(\boldsymbol{\nu}\rfloor\mathbf{A})(\boldsymbol{\nu}\rfloor\mathbf{B}))\\ &= \Delta_Q(\boldsymbol{\nu}\boldsymbol{\nu}\rfloor\rfloor\mathbf{e}_N\mathbf{e}_N)\boldsymbol{\nu}\boldsymbol{\nu}||(\overline{\overline{\mathsf{Q}}}^{-1T}\rfloor\rfloor\mathbf{AB}). \quad (8.152)\end{aligned}$$

Inserting this in (8.149) we obtain

$$\begin{aligned}\overline{\overline{\mathsf{D}}}^{(3)}(\boldsymbol{\nu}) &= (\boldsymbol{\nu}\boldsymbol{\nu}\rfloor\rfloor\overline{\overline{\mathsf{Q}}}^{(4)})(\boldsymbol{\nu}\boldsymbol{\nu}||\overline{\overline{\mathsf{Q}}})(\boldsymbol{\nu}\boldsymbol{\nu}||(\overline{\overline{\mathsf{Q}}}+\overline{\overline{\mathsf{Q}}}^{-1T}\rfloor\rfloor\mathbf{AB}))\\ &= \Delta_Q(\boldsymbol{\nu}\boldsymbol{\nu}\rfloor\rfloor\mathbf{e}_N\mathbf{e}_N)D(\boldsymbol{\nu}). \quad (8.153)\end{aligned}$$

The dispersion function defined by

$$D(\boldsymbol{\nu}) = (\boldsymbol{\nu}\boldsymbol{\nu}||\overline{\overline{\mathsf{Q}}})\ \boldsymbol{\nu}\boldsymbol{\nu}||(\overline{\overline{\mathsf{Q}}}+\overline{\overline{\mathsf{Q}}}^{-1T}\rfloor\rfloor\mathbf{AB}) = D_1(\boldsymbol{\nu})D_2(\boldsymbol{\nu}), \quad (8.154)$$

has a factorized form. Thus, for a full-rank dyadic $\overline{\overline{\mathsf{Q}}}$, the dispersion equation $D(\boldsymbol{\nu}) = 0$ can be split in two second-order equations as

$$D_1(\boldsymbol{\nu}) = \boldsymbol{\nu}\boldsymbol{\nu}||\overline{\overline{\mathsf{Q}}} = 0, \quad (8.155)$$

$$D_2(\boldsymbol{\nu}) = \boldsymbol{\nu}\boldsymbol{\nu}||(\overline{\overline{\mathsf{Q}}}+\overline{\overline{\mathsf{Q}}}^{-1T}\rfloor\rfloor\mathbf{AB}) = 0. \quad (8.156)$$

The former of these coincides with the dispersion equation of the pure Q medium (8.74), and the corresponding potential one-form $\boldsymbol{\phi}_1$ satisfies

$$\overline{\overline{\mathsf{D}}}(\boldsymbol{\nu}_1)|\boldsymbol{\phi}_1 = -(\overline{\overline{\mathsf{Q}}}|\boldsymbol{\nu}_1)(\boldsymbol{\nu}_1|\overline{\overline{\mathsf{Q}}}|\boldsymbol{\phi}_1) - (\boldsymbol{\nu}_1\rfloor\mathbf{A})(\boldsymbol{\nu}_1\rfloor\mathbf{B})|\boldsymbol{\phi}_1 = 0, \quad (8.157)$$

when $\boldsymbol{\nu}_1$ is a solution of (8.155). Choosing

$$\boldsymbol{\nu}_1|\overline{\overline{\mathsf{Q}}}|\boldsymbol{\phi}_1 = 0, \qquad (8.158)$$

as the gauge condition for the potential, we obtain

$$-(\boldsymbol{\nu}_1\rfloor\mathbf{B})|\boldsymbol{\phi}_1 = \mathbf{B}|(\boldsymbol{\nu}_1 \wedge \boldsymbol{\phi}_1) = \mathbf{B}|\boldsymbol{\Phi}_1 = 0, \qquad (8.159)$$

which represents the polarization condition for the field two-form. The medium condition for the fields of the plane wave 1

$$\boldsymbol{\Psi}_1 = \overline{\overline{\mathsf{M}}}|\boldsymbol{\Phi}_1 = \epsilon_N \lfloor(\overline{\overline{\mathsf{Q}}}^{(2)} + \mathbf{AB})|\boldsymbol{\Phi}_1 = \epsilon_N \lfloor\overline{\overline{\mathsf{Q}}}^{(2)}|\boldsymbol{\Phi}_1, \qquad (8.160)$$

shows us that the the extended Q medium can be replaced by a simpler Q medium without changing the wave 1.

The second solution $\boldsymbol{\nu} = \boldsymbol{\nu}_2$ satisfies the dispersion equation (8.156), or

$$(\boldsymbol{\nu}_2\boldsymbol{\nu}_2||\overline{\overline{\mathsf{Q}}}) = -(\boldsymbol{\nu}_2\boldsymbol{\nu}_2\rfloor\rfloor\mathbf{AB})||\overline{\overline{\mathsf{Q}}}^{-1T}, \qquad (8.161)$$

whence the corresponding potential one-form satisfies

$$\begin{aligned}\overline{\overline{\mathsf{D}}}(\boldsymbol{\nu}_2)|\boldsymbol{\phi}_2 &= (\boldsymbol{\nu}_2\boldsymbol{\nu}_2\rfloor\rfloor\overline{\overline{\mathsf{Q}}}^{(2)} + (\boldsymbol{\nu}_2\boldsymbol{\nu}_2\rfloor\rfloor\mathbf{AB}))|\boldsymbol{\phi}_2\\ &= ((\boldsymbol{\nu}_2\boldsymbol{\nu}_2||\overline{\overline{\mathsf{Q}}})\overline{\overline{\mathsf{Q}}} - \overline{\overline{\mathsf{Q}}}|\boldsymbol{\nu}_2\boldsymbol{\nu}_2|\overline{\overline{\mathsf{Q}}})|\boldsymbol{\phi}_2 + (\boldsymbol{\nu}_2\boldsymbol{\nu}_2\rfloor\rfloor\mathbf{AB})|\boldsymbol{\phi}_2\\ &= -((\boldsymbol{\nu}_2\boldsymbol{\nu}_2\rfloor\rfloor\mathbf{AB})||\overline{\overline{\mathsf{Q}}}^{-1T})\overline{\overline{\mathsf{Q}}}|\boldsymbol{\phi}_2 - (\overline{\overline{\mathsf{Q}}}|\boldsymbol{\nu}_2)(\boldsymbol{\nu}_2|\overline{\overline{\mathsf{Q}}}|\boldsymbol{\phi}_2)\\ &\quad + (\boldsymbol{\nu}_2\boldsymbol{\nu}_2\rfloor\rfloor\mathbf{AB})|\boldsymbol{\phi}_2 = 0. \end{aligned} \qquad (8.162)$$

Assuming now the same gauge condition (8.158) in the form

$$\boldsymbol{\nu}_2|\overline{\overline{\mathsf{Q}}}|\boldsymbol{\phi}_2 = 0, \qquad (8.163)$$

the potential equation (8.156) is reduced to

$$((\boldsymbol{\nu}_2\boldsymbol{\nu}_2\rfloor\rfloor\mathbf{AB})||\overline{\overline{\mathsf{Q}}}^{-1T})\,\overline{\overline{\mathsf{Q}}}|\boldsymbol{\phi}_2 = (\boldsymbol{\nu}_2\rfloor\mathbf{A})(\boldsymbol{\nu}_2\rfloor\mathbf{B})|\boldsymbol{\phi}_2. \qquad (8.164)$$

The first bracketed scalar factor can be assumed nonzero because otherwise (8.156) would reduce to (8.155). Thus, the vector $\overline{\overline{\mathsf{Q}}}|\boldsymbol{\phi}$ must be a multiple of $\boldsymbol{\nu}_2\rfloor\mathbf{A}$. This implies that the potential one-form $\boldsymbol{\phi}_2$ must be of the form

$$\boldsymbol{\phi}_2 = \lambda\overline{\overline{\mathsf{Q}}}^{-1}|(\boldsymbol{\nu}_2\rfloor\mathbf{A}) = \lambda(\overline{\overline{\mathsf{Q}}}^{-1} \wedge \boldsymbol{\nu}_2)|\mathbf{A}, \qquad (8.165)$$

for some scalar factor λ and, the two-form $\boldsymbol{\Phi}_2$ must be of the form

$$\boldsymbol{\Phi}_2 = \boldsymbol{\nu}_2 \wedge \boldsymbol{\phi}_2 = -\lambda(\boldsymbol{\nu}_2\boldsymbol{\nu}_2{}^\wedge_\wedge\overline{\overline{\mathsf{Q}}}^{-1})|\mathbf{A}. \qquad (8.166)$$

The polarization condition for $\mathbf{\Phi}_2$ is somewhat more tricky than (8.159). After some analysis one can show that the condition

$$\left(\varepsilon_N\lfloor(\mathbf{A}\wedge\overline{\overline{\mathsf{Q}}}^{(2)}) + \frac{1}{2}(\mathbf{A}\cdot\mathbf{A})\mathbf{B}\right)|\mathbf{\Phi}_2 = 0 \tag{8.167}$$

is satisfied by the two-form defined by (8.166) when ν_2 satisfies (8.156). Writing (8.167) in the form

$$\mathbf{A}|\mathbf{\Psi}_2 - \frac{1}{2}(\mathbf{A}\cdot\mathbf{A})\mathbf{B}|\mathbf{\Phi}_2 = 0, \tag{8.168}$$

one can note that for the special case when $\mathbf{A}$ is a simple bivector, that is, for $\mathbf{A}\cdot\mathbf{A} = 0$, the second orthogonality condition becomes similar in form to (8.159),

$$\mathbf{A}|\mathbf{\Psi}_2 = 0. \tag{8.169}$$

8.7 EXTENDED P MEDIA

8.7.1 Medium Conditions

The class of P media can be extended following the pattern of the previous section by adding a term involving a two-form $\mathbf{\Delta}$ and a bivector $\mathbf{C}$ as

$$\overline{\overline{\mathsf{M}}} = \overline{\overline{\mathsf{P}}}^{(2)T} + \mathbf{\Delta}\mathbf{C}, \tag{8.170}$$

$$\overline{\overline{\mathsf{M}}}_m = \mathbf{e}_N\lfloor\overline{\overline{\mathsf{P}}}^{(2)T} + \mathbf{D}\mathbf{C}, \tag{8.171}$$

with $\mathbf{D} = \mathbf{e}_N\lfloor\mathbf{\Delta}$. Media defined in this way are called extended P media (or generalized P media [61]). Considering the relation (8.56) connecting P media and Q media, one can show that extended P media and extended Q media obey a similar relation.

To find the spatial expansion of the medium bidyadic, let us start by expressing the two bivectors as

$$\mathbf{D} = \mathbf{e}_{123}\lfloor\boldsymbol{\delta}_s + \mathbf{d}_s\wedge\mathbf{e}_4, \quad \mathbf{C} = \mathbf{e}_{123}\lfloor\boldsymbol{\gamma}_s + \mathbf{c}_s\wedge\mathbf{e}_4, \tag{8.172}$$

in terms of spatial one-forms $\boldsymbol{\gamma}_s$, $\boldsymbol{\delta}_s$ and spatial vectors $\mathbf{c}_s$, $\mathbf{d}_s$. Expanding

$$\varepsilon_N\lfloor\mathbf{D}\mathbf{C} = (-\boldsymbol{\varepsilon}_4\wedge\boldsymbol{\delta}_s + \boldsymbol{\varepsilon}_{123}\lfloor\mathbf{d}_s)(\mathbf{e}_{123}\lfloor\boldsymbol{\gamma}_s + \mathbf{c}_3\wedge\mathbf{e}_4), \tag{8.173}$$

the spatial P medium parameter dyadics can be extended from those of the P medium (8.44)–(8.47) to

$$\overline{\overline{\alpha}} = \overline{\overline{\mathsf{P}}}_s^{(2)T} + \varepsilon_{123}\mathbf{e}_{123}\rfloor\rfloor\mathbf{d}_s\gamma_s, \tag{8.174}$$

$$\overline{\overline{\epsilon}}' = -\boldsymbol{\pi}_s \wedge \overline{\overline{\mathsf{P}}}_s^T + \varepsilon_{123}\rfloor\mathbf{d}_s\mathbf{c}_s, \tag{8.175}$$

$$\overline{\overline{\mu}}^{-1} = -\overline{\overline{\mathsf{P}}}_s^T \wedge \mathbf{p}_s - \delta_s\gamma_s\lfloor\mathbf{e}_{123}, \tag{8.176}$$

$$\overline{\overline{\beta}} = (\boldsymbol{\pi}_s\mathbf{p}_s - p\overline{\overline{\mathsf{P}}}_s^T) - \delta_s\mathbf{c}_s. \tag{8.177}$$

The extension makes it possible for the spatial dyadics $\overline{\overline{\epsilon}}'$ and $\overline{\overline{\mu}}^{-1}$ to have spatial inverses.

Following steps similar to those leading to the polarization conditions (8.136) and (8.138) for a plane wave in an extended Q medium, the conditions for the extended P medium become

$$\mathbf{C}|\boldsymbol{\Phi} = 0, \tag{8.178}$$

$$2\boldsymbol{\Delta}\cdot\boldsymbol{\Psi} - (\boldsymbol{\Delta}\cdot\boldsymbol{\Delta})(\mathbf{C}|\boldsymbol{\Phi}) = 0, \tag{8.179}$$

where the latter can be replaced by

$$\boldsymbol{\Delta}\cdot(2\overline{\overline{\mathsf{P}}}^{(2)T} + \boldsymbol{\Delta}\mathbf{C})|\boldsymbol{\Phi} = 0. \tag{8.180}$$

Again, being linear in $\boldsymbol{\Phi}$ and independent of $\boldsymbol{\nu}$, these conditions are valid for any fields which can be composed of plane waves.

8.7.2 Plane Waves in Extended P Media

Considering a plane wave in the extended P medium, the dispersion dyadic (8.81) is extended to

$$\begin{aligned}\overline{\overline{\mathsf{D}}}(\boldsymbol{\nu}) &= -\mathbf{e}_N\lfloor(\boldsymbol{\nu}\wedge(\overline{\overline{\mathsf{P}}}^T|\boldsymbol{\nu})\wedge\overline{\overline{\mathsf{P}}}^T) + (\mathbf{D}\lfloor\boldsymbol{\nu})(\mathbf{C}\lfloor\boldsymbol{\nu})\\ &= (\mathbf{F}(\boldsymbol{\nu})\lfloor\overline{\overline{\mathsf{I}}}^T)\lfloor\overline{\overline{\mathsf{P}}}^T + (\mathbf{D}\lfloor\boldsymbol{\nu})(\mathbf{C}\lfloor\boldsymbol{\nu}),\end{aligned} \tag{8.181}$$

where the bivector function $\mathbf{F}(\boldsymbol{\nu})$ equals that introduced in (8.82). Applying (8.85), we can write

$$\overline{\overline{\mathsf{D}}}^{(2)}(\boldsymbol{\nu}) = \mathbf{FF}|\overline{\overline{\mathsf{P}}}^{(2)T} + ((\mathbf{F}\lfloor\overline{\overline{\mathsf{I}}}^T)|\overline{\overline{\mathsf{P}}}^T)^{\wedge}_{\wedge}(\mathbf{D}\lfloor\boldsymbol{\nu})(\mathbf{C}\lfloor\boldsymbol{\nu}), \tag{8.182}$$

and

$$\begin{aligned}\overline{\overline{\mathsf{D}}}^{(3)}(\boldsymbol{\nu}) &= (\mathbf{FF}|\overline{\overline{\mathsf{P}}}^{(2)T})^{\wedge}_{\wedge}(\mathbf{D}\lfloor\boldsymbol{\nu})(\mathbf{C}\lfloor\boldsymbol{\nu})\\ &= (\mathbf{F}\wedge(\mathbf{D}\lfloor\boldsymbol{\nu}))(\mathbf{F}|\overline{\overline{\mathsf{P}}}^{(2)T}\wedge(\mathbf{C}\lfloor\boldsymbol{\nu})).\end{aligned} \tag{8.183}$$

Since the last dyadic appears factorized, the dyadic dispersion equation $\overline{\overline{\mathsf{D}}}^{(3)}(\boldsymbol{\nu}) = 0$ can be split in two trivector equations as

$$\mathbf{F}_1 \wedge (\mathbf{D}\lfloor\boldsymbol{\nu}_1) = 0 \tag{8.184}$$

$$(\overline{\overline{\mathsf{P}}}^{(2)}|\mathbf{F}_2) \wedge (\mathbf{C}\lfloor\boldsymbol{\nu}_2) = 0, \tag{8.185}$$

where we denote $\mathbf{F}_1 = \mathbf{F}(\boldsymbol{\nu}_1)$ and $\mathbf{F}_2 = \mathbf{F}(\boldsymbol{\nu}_2)$. The first equation (8.184) can be expanded as

$$\begin{aligned} 0 &= \boldsymbol{\varepsilon}_N\lfloor(\mathbf{F}_1 \wedge (\mathbf{D}\lfloor\boldsymbol{\nu}_1)) = (\boldsymbol{\varepsilon}_N\lfloor\mathbf{F}_1)\lfloor(\mathbf{D}\lfloor\boldsymbol{\nu}_1) \\ &= (\boldsymbol{\nu}_1 \wedge (\overline{\overline{\mathsf{P}}}^T|\boldsymbol{\nu}_1))\lfloor(\mathbf{D}\lfloor\boldsymbol{\nu}_1) \\ &= (\overline{\overline{\mathsf{P}}}^T|\boldsymbol{\nu}_1)(\mathbf{D}\lfloor\boldsymbol{\nu}_1)|\boldsymbol{\nu}_1 - \boldsymbol{\nu}_1(\mathbf{D}\lfloor\boldsymbol{\nu}_1)|(\overline{\overline{\mathsf{P}}}^T|\boldsymbol{\nu}_1) \\ &= \boldsymbol{\nu}_1(\mathbf{D}\lfloor\overline{\overline{\mathsf{P}}}^T|\boldsymbol{\nu}_1)|\boldsymbol{\nu}_1, \end{aligned} \tag{8.186}$$

which yields the scalar dispersion equation

$$D_1(\boldsymbol{\nu}_1) = \boldsymbol{\nu}_1|(\mathbf{D}\lfloor\overline{\overline{\mathsf{P}}}^T)|\boldsymbol{\nu}_1 = 0. \tag{8.187}$$

The second equation (8.185) can be expanded as

$$\begin{aligned} 0 &= \boldsymbol{\varepsilon}_N\lfloor(\overline{\overline{\mathsf{P}}}^{(-3)}|((\overline{\overline{\mathsf{P}}}^{(2)}|\mathbf{F}_2) \wedge (\mathbf{C}\lfloor\boldsymbol{\nu}_2))) \\ &= \boldsymbol{\varepsilon}_N\lfloor(\mathbf{F}_2 \wedge (\overline{\overline{\mathsf{P}}}^{-1}|(\mathbf{C}\lfloor\boldsymbol{\nu}_2))) \\ &= (\boldsymbol{\varepsilon}_N\lfloor\mathbf{F}_2)\lfloor(\overline{\overline{\mathsf{P}}}^{-1}|(\mathbf{C}\lfloor\boldsymbol{\nu}_2))) \\ &= (\boldsymbol{\nu}_2 \wedge (\overline{\overline{\mathsf{P}}}^T|\boldsymbol{\nu}_2))\lfloor(\overline{\overline{\mathsf{P}}}^{-1}|(\mathbf{C}\lfloor\boldsymbol{\nu}_2))) \\ &= (\overline{\overline{\mathsf{P}}}^T|\boldsymbol{\nu}_2)(\boldsymbol{\nu}_2|\overline{\overline{\mathsf{P}}}^{-1}|(\mathbf{C}\lfloor\boldsymbol{\nu}_2)) - \boldsymbol{\nu}_2(\mathbf{C}\lfloor\boldsymbol{\nu}_2)|\boldsymbol{\nu}_2 \\ &= -(\overline{\overline{\mathsf{P}}}^T|\boldsymbol{\nu}_2)\boldsymbol{\nu}_2|(\mathbf{C}\lfloor\overline{\overline{\mathsf{P}}}^{-1T}|\boldsymbol{\nu}_2), \end{aligned} \tag{8.188}$$

whence the second scalar dispersion equation becomes

$$D_2(\boldsymbol{\nu}_2) = \boldsymbol{\nu}_2|(\mathbf{C}\lfloor\overline{\overline{\mathsf{P}}}^{-1T})|\boldsymbol{\nu}_2 = 0. \tag{8.189}$$

Here we have assumed the existence of $\overline{\overline{\mathsf{P}}}^{-1}$, that is, $\mathrm{tr}\overline{\overline{\mathsf{P}}}^{(4)} \neq 0$. in the converse case we must replace $\overline{\overline{\mathsf{P}}}^{-1T}$ in (8.189) by $\boldsymbol{\varepsilon}_N\mathbf{e}_N\lfloor\lfloor\overline{\overline{\mathsf{P}}}^{(3)}$. For $\overline{\overline{\mathsf{P}}}^{(3)} = 0$ (8.189) is an identity. As a check, in the case of the plain P medium with $\mathbf{C} = \mathbf{D} = 0$, both of the dispersion equations (8.187), (8.189) become identities, valid for any $\boldsymbol{\nu}$.

8.7.3 Field Conditions

Because the fields of any plane wave satisfy the conditions $\boldsymbol{\Phi} \cdot \boldsymbol{\Phi} = 0$ and $\boldsymbol{\Psi} \cdot \boldsymbol{\Psi} = 0$, applying the rule $\overline{\overline{\mathsf{P}}}^{(2)T} \cdot \overline{\overline{\mathsf{P}}}^{(2)} = \mathrm{tr}\overline{\overline{\mathsf{P}}}^{(4)}\mathbf{e}_N\lfloor\overline{\overline{\mathsf{I}}}^{(2)}$, we obtain the following condition for fields in a extended P medium

$$
\begin{aligned}
0 &= \boldsymbol{\Psi}\cdot\boldsymbol{\Psi} - \mathrm{tr}\overline{\overline{\mathsf{P}}}^{(4)}\boldsymbol{\Phi}\cdot\boldsymbol{\Phi} \\
&= \boldsymbol{\Phi}|(\overline{\overline{\mathsf{M}}}^T\cdot\overline{\overline{\mathsf{M}}} - \mathrm{tr}\overline{\overline{\mathsf{P}}}^{(4)}\mathbf{e}_N\lfloor\overline{\overline{\mathsf{I}}}^{(2)T})|\boldsymbol{\Phi} \\
&= \boldsymbol{\Phi}|(\overline{\overline{\mathsf{P}}}^{(2)}\cdot\overline{\overline{\mathsf{P}}}^{(2)T} + \mathbf{C}(\mathbf{D}|\overline{\overline{\mathsf{P}}}^{(2)T}) + (\mathbf{D}|\overline{\overline{\mathsf{P}}}^{(2)T})\mathbf{C} \\
&\quad +(\mathbf{D}\cdot\mathbf{D})\mathbf{C}\mathbf{C} - \mathrm{tr}\overline{\overline{\mathsf{P}}}^{(4)}\mathbf{e}_N\lfloor\overline{\overline{\mathsf{I}}}^{(2)T})|\boldsymbol{\Phi} \\
&= (\boldsymbol{\Phi}|\mathbf{C})(2\mathbf{D}|\overline{\overline{\mathsf{P}}}^{(2)T} + (\mathbf{D}\cdot\mathbf{D})\mathbf{C})|\boldsymbol{\Phi}.
\end{aligned} \tag{8.190}
$$

From this we conclude that the field must satisfy either of the two orthogonality conditions:

$$\mathbf{C}|\boldsymbol{\Phi} = 0, \tag{8.191}$$

$$\left(\mathbf{D}|\overline{\overline{\mathsf{P}}}^{(2)} + \frac{1}{2}(\mathbf{D}\cdot\mathbf{D})\mathbf{C}\right)|\boldsymbol{\Phi} = 0. \tag{8.192}$$

In the special case when the bivector $\mathbf{D}$ is simple, (8.192) written as

$$\left(\mathbf{D}|\overline{\overline{\mathsf{M}}} - \frac{1}{2}(\mathbf{D}\cdot\mathbf{D})\mathbf{C}\right)|\boldsymbol{\Phi} = 0, \tag{8.193}$$

can be seen to reduce to

$$\mathbf{D}|\boldsymbol{\Psi} = 0. \tag{8.194}$$

Obviously, the two orthogonality conditions (8.191) and (8.192) are associated with the two dispersion equations (8.187), (8.189). The question remains which one corresponds to which one. Let us consider the wave associated to (8.191). Because for a field satisfying this condition the extended P medium can be replaced by that of the P medium (8.1), the field two-form can be expressed as (8.79). Requiring that the field satisfy (8.191),

$$\mathbf{C}|\boldsymbol{\Phi} = A\mathbf{C}|(\boldsymbol{\nu}\wedge\overline{\overline{\mathsf{P}}}^{-1T}|\boldsymbol{\nu}) = -A\boldsymbol{\nu}|(\mathbf{C}\lfloor\overline{\overline{\mathsf{P}}}^{-1}T)|\boldsymbol{\nu} = 0, \tag{8.195}$$

we arrive at the dispersion equation (8.189). Thus, (8.189) corresponds to (8.191) and, consequently, (8.187) must correspond to (8.192).

PROBLEMS

8.1 Prove that any spatial bidyadic $\overline{\overline{\mathsf{D}}}_s \in \mathbb{E}_2\mathbb{E}_2$ possessing a spatial inverse can be uniquely expressed as $\overline{\overline{\mathsf{D}}}_s = \overline{\overline{\mathsf{A}}}_s^{(2)}$ in terms of some spatial dyadic $\overline{\overline{\mathsf{A}}}_s \in \mathbb{E}_1\mathbb{E}_1$.

8.2 Derive the expressions in (8.38) for the Gibbsian medium dyadics of a Q medium.

8.3 Show that one can construct the Q medium bidyadic $\overline{\overline{\mathsf{M}}}$ by starting from given antisymmetric Gibbsian dyadics $\overline{\overline{\xi}}_g, \overline{\overline{\zeta}}_g$ and $\overline{\overline{\epsilon}}_g, \overline{\overline{\mu}}_g$ satisfying (8.41). Hint: use the expansion (5.88).

8.4 Derive the dispersion equation for the Q medium starting from

$$D(\boldsymbol{\nu}) = \frac{1}{6}\boldsymbol{\varepsilon}_N\boldsymbol{\varepsilon}_N||(\overline{\overline{\mathsf{M}}}_m{}^{\wedge}_{\wedge}(\boldsymbol{\nu}\boldsymbol{\nu}\rfloor\rfloor(\overline{\overline{\mathsf{M}}}_m{}^{\wedge}_{\wedge}(\boldsymbol{\nu}\boldsymbol{\nu}\rfloor\rfloor\overline{\overline{\mathsf{M}}}_m)) = 0.$$

8.5 Verify that the potential $\boldsymbol{\phi}$ defined by (8.76) satisfies the dispersion equation (8.62). The bac-cab rule

$$\mathbf{A}\lfloor(\boldsymbol{\Gamma}\lfloor\mathbf{a}) = (\mathbf{A}\wedge\mathbf{a})\lfloor\boldsymbol{\Gamma} - (\mathbf{A}|\boldsymbol{\Gamma})\mathbf{a},$$

where $\mathbf{A}$ is a bivector, $\boldsymbol{\Gamma}$ is a two-form and $\mathbf{a}$ is a vector, may appear useful here.

8.6 Show that if the modified medium bidyadic $\overline{\overline{\mathsf{M}}}_m$ of a special Q-axion medium, defined by (8.97) with (8.100), is symmetric, it does not have an inverse.

8.7 Find the conditions for the Gibbsian field vectors $\mathbf{E}_g, \mathbf{H}_g$ corresponding to the conditions (8.136) and (8.138) for the extended Q medium.

8.8 Find the most general P medium consisting of only the principal part, by requiring $\overline{\overline{\mathsf{M}}}\lfloor\lfloor\overline{\overline{\mathsf{I}}} = \overline{\overline{\mathsf{P}}}^{(2)T}\lfloor\lfloor\overline{\overline{\mathsf{I}}} = 0$ and $\overline{\overline{\mathsf{M}}} \neq 0$.

8.9 Consider a Q-axion medium defined by

$$\overline{\overline{\mathsf{M}}} = A\overline{\overline{\mathsf{I}}}^{(2)T} + B\boldsymbol{\varepsilon}_N\lfloor\overline{\overline{\mathsf{Q}}}^{(2)},$$

where $\overline{\overline{\mathsf{Q}}}^{(2)} \in \mathbb{E}_2\mathbb{E}_2$ is a symmetric bidyadic normalized as $\Delta_Q = \boldsymbol{\varepsilon}_N\boldsymbol{\varepsilon}_N||\overline{\overline{\mathsf{Q}}}^{(4)} = -1$. Show that $\overline{\overline{\mathsf{J}}}^T = \boldsymbol{\varepsilon}_N\lfloor\overline{\overline{\mathsf{Q}}}^{(2)} \in \mathbb{F}_2\mathbb{E}_2$ is an AC bidyadic and the medium bidyadic can be expressed in the form

$$\overline{\overline{\mathsf{M}}} = Me^{\psi\overline{\overline{\mathsf{J}}}^T}.$$

8.10 Find the eigenvalues and eigen two-forms corresponding to the equation

$$\overline{\overline{\mathsf{M}}}|\boldsymbol{\Phi}_i = M_i\boldsymbol{\Phi}_i$$

for the symmetric Q-axion medium bidyadic of the previous problem. Denoting the eigen two-forms by $\mathbf{\Phi}_\pm$, show that that any given two-form can be decomposed as $\mathbf{\Phi} = \mathbf{\Phi}_+ + \mathbf{\Phi}_-$ and that they satisfy the orthogonality conditions $\mathbf{\Phi}_+ \cdot \mathbf{\Phi}_- = 0$.

8.11 Show that the decomposition of the electromagnetic two-form $\mathbf{\Phi}$ for the symmetric Q-axion medium of the previous problem gives rise to decoupled Maxwell equations. Find the eigenfield expansion of the bidyadic $\overline{\overline{\mathsf{U}}}(\mathbf{\Psi}, \mathbf{\Phi})$ defined by (5.188), assuming $M_+ \neq M_-$.

8.12 Derive an expression for the dispersion function $D(\boldsymbol{\nu})$ for the extended Q medium starting from the general form (5.244).

8.13 Verify the orthogonality condition (8.167) by substituting the field two-form (8.166) when the wave one-form $\boldsymbol{\nu}_2$ is assumed to satisfy the dispersion equation (8.156) in an extended Q medium.

8.14 Derive the Gibbsian medium dyadic expressions (8.50)–(8.53) for the P medium.

8.15 Express the orthogonality conditions (8.191) and (8.192) for the plane-wave fields in the extended P medium in terms of the spatial field components **D**, **B**, **E**, **H**.

8.16 Derive the duality transformation rules for the extended Q medium (8.145), (8.146), and consider the possibility of defining a transformation in which the medium is self-dual.

8.17 Consider a P medium defined by the dyadic

$$\overline{\overline{\mathsf{P}}} = P\overline{\overline{\mathsf{I}}} + \overline{\overline{\mathsf{P}}}_{34}, \quad \overline{\overline{\mathsf{P}}}_{34} = P_{3,3}\mathbf{e}_3\boldsymbol{\varepsilon}_3 + P_{3,4}\mathbf{e}_3\boldsymbol{\varepsilon}_4 + P_{4,3}\mathbf{e}_4\boldsymbol{\varepsilon}_3 + P_{4,4}\mathbf{e}_4\boldsymbol{\varepsilon}_4.$$

Show that plane-wave fields at an interface $\boldsymbol{\varepsilon}_3|\mathbf{x} = 0$ of the medium satisfy the DB boundary conditions.

CHAPTER 9

Media Defined by Bidyadic Equations

It is known that each medium bidyadic $\overline{\overline{\mathsf{M}}} \in \mathbb{F}_2\mathbb{E}_2$ satisfies an algebraic equation of the sixth order, the Cayley–Hamilton equation (3.9). Equations of lower order,

$$\overline{\overline{\mathsf{M}}}^p + A_1\overline{\overline{\mathsf{M}}}^{p-1} + \cdots + A_p\overline{\overline{\mathsf{I}}}^{(2)T} = 0, \quad p < 6, \tag{9.1}$$

restrict the generality of the medium bidyadic and, thus, define certain classes of media. As the simplest example, medium bidyadics satisfying the first-order equation,

$$\overline{\overline{\mathsf{M}}} - M\overline{\overline{\mathsf{I}}}^{(2)T} = 0, \tag{9.2}$$

form the class of axion (PEMC) media.

There are some properties arising from (9.1). If $A_p \neq 0$, the inverse bidyadic can be expressed as

$$\overline{\overline{\mathsf{M}}}^{-1} = -\frac{1}{A_p}(\overline{\overline{\mathsf{M}}}^{p-1} + A_1\overline{\overline{\mathsf{M}}}^{p-2} + \cdots + A_{p-1}\overline{\overline{\mathsf{I}}}^{(2)T}), \tag{9.3}$$

and it is a member of the same class. This is seen if (9.1) is multiplied by $\overline{\overline{\mathsf{M}}}^{-p}|$. If $A_p = 0$, either there is no inverse or $\overline{\overline{\mathsf{M}}}$ belongs to a class defined by an order lower than p.

If $\overline{\overline{\mathsf{M}}}_1$ is a solution of (9.1), so is

$$\overline{\overline{\mathsf{M}}} = \overline{\overline{\mathsf{D}}}|\overline{\overline{\mathsf{M}}}_1|\overline{\overline{\mathsf{D}}}^{-1} \tag{9.4}$$

Multiforms, Dyadics, and Electromagnetic Media, First Edition. Ismo V. Lindell.

for any bidyadic $\overline{\overline{\mathsf{D}}} \in \mathbb{F}_2\mathbb{E}_2$ possessing an inverse. Thus, one solution generates a family of solutions through this kind of similarity transformation. From

$$\mathrm{tr}(\overline{\overline{\mathsf{D}}}|\overline{\overline{\mathsf{M}}}_1|\overline{\overline{\mathsf{D}}}^{-1}) = \mathrm{tr}(\overline{\overline{\mathsf{D}}}^{-1}|\overline{\overline{\mathsf{D}}}|\overline{\overline{\mathsf{M}}}_1) = \mathrm{tr}\overline{\overline{\mathsf{M}}}_1 \tag{9.5}$$

it follows that the axion part is invariant in this transformation.

If $\overline{\overline{\mathsf{M}}}$ belongs to a class defined by p, the bidyadic $\overline{\overline{\mathsf{M}}} + \alpha\overline{\overline{\mathrm{I}}}^{(2)T}$ belongs to the same class. Thus, the axion part of $\overline{\overline{\mathsf{M}}}$ does not affect the class defined by the order of the equation.

It is the purpose of this chapter to study the some properties of the simplest cases, medium classes defined by quadratic, cubic, and biquadratic equations, with no claim of completeness on these topics.

9.1 QUADRATIC EQUATION

The class of media defined by medium bidyadics satisfying an equation of the second order,

$$\overline{\overline{\mathsf{M}}}^2 - 2A\overline{\overline{\mathsf{M}}} + B\overline{\overline{\mathrm{I}}}^{(2)T} = 0, \tag{9.6}$$

will be called the class of SD media [59]. The equation involves two parameters A and B.

As we have already seen, GBI media, defined by (7.17) as

$$\overline{\overline{\mathsf{M}}} = A\overline{\overline{\mathrm{I}}}^{(2)T} + B\overline{\overline{\mathsf{T}}}^{(2)T} - \frac{1}{\mu}\varepsilon_N\lfloor(\overline{\overline{\mathsf{G}}}_s - \epsilon'\mu\mathbf{e}_4\mathbf{e}_4)^{(2)}, \tag{9.7}$$

where $\overline{\overline{\mathsf{T}}} = \overline{\overline{\mathrm{I}}}_s - \mathbf{e}_4\boldsymbol{\varepsilon}_4$ is the time reversion dyadic, satisfy the second-order equation

$$\overline{\overline{\mathsf{M}}}^2 - 2A\overline{\overline{\mathsf{M}}} + \left(A^2 - B^2 + \frac{\epsilon'}{\mu}\right)\overline{\overline{\mathrm{I}}}^{(2)T} = 0. \tag{9.8}$$

Also, Q-axion media with a symmetric bidyadic $\overline{\overline{\mathsf{Q}}}^{(2)}$, defined by

$$\overline{\overline{\mathsf{M}}} = A\overline{\overline{\mathrm{I}}}^{(2)T} + B\varepsilon_N\lfloor\overline{\overline{\mathsf{Q}}}^{(2)}, \tag{9.9}$$

satisfy

$$\overline{\overline{\mathsf{M}}}^2 - 2A\overline{\overline{\mathsf{M}}} + (A^2 - B^2\Delta_Q)\overline{\overline{\mathrm{I}}}^{(2)T} = 0, \tag{9.10}$$

because of

$$(\mathbf{e}_N\rfloor\overline{\overline{\mathsf{Q}}}^{(2)})^2 = (\mathbf{e}_N\mathbf{e}_N\rfloor\rfloor\overline{\overline{\mathsf{Q}}}^{(2)})|\overline{\overline{\mathsf{Q}}}^{(2)} = \Delta_Q\overline{\overline{\mathsf{Q}}}^{(-2)T}|\overline{\overline{\mathsf{Q}}}^{(2)} = \Delta_Q\overline{\overline{\mathsf{I}}}^{(2)T}. \qquad (9.11)$$

Thus, symmetric Q-axion media belong to the class of SD media.

9.1.1 SD Media

Rewriting (9.6) as

$$(\overline{\overline{\mathsf{M}}} - A\overline{\overline{\mathsf{I}}}^{(2)T})^2 = (A^2 - B)\overline{\overline{\mathsf{I}}}^{(2)T}, \qquad (9.12)$$

we can consider three cases depending on the factor $A^2 - B$.

- For $A^2 - B > 0$, the bidyadic $\overline{\overline{\mathsf{M}}} - A\overline{\overline{\mathsf{I}}}^{(2)T}$ is a multiple of a unipotent bidyadic $\overline{\overline{\mathsf{K}}}^T$,

$$\overline{\overline{\mathsf{M}}} = A\overline{\overline{\mathsf{I}}}^{(2)T} + C\overline{\overline{\mathsf{K}}}^T, \quad \overline{\overline{\mathsf{K}}}^2 = \overline{\overline{\mathsf{I}}}^{(2)}, \qquad (9.13)$$

  with real $C = \sqrt{A^2 - B}$.
- For $A^2 - B < 0$, the bidyadic $\overline{\overline{\mathsf{M}}} - A\overline{\overline{\mathsf{I}}}^{(2)T}$ is a multiple of an AC bidyadic $\overline{\overline{\mathsf{J}}}^T$,

$$\overline{\overline{\mathsf{M}}} = A\overline{\overline{\mathsf{I}}}^{(2)T} + D\overline{\overline{\mathsf{J}}}^T, \quad \overline{\overline{\mathsf{J}}}^2 = -\overline{\overline{\mathsf{I}}}^{(2)}, \qquad (9.14)$$

  with real $D = \sqrt{B - A^2}$.
- For $A^2 - B = 0$, the bidyadic $\overline{\overline{\mathsf{M}}} - A\overline{\overline{\mathsf{I}}}^{(2)T}$ is a nilpotent bidyadic $\overline{\overline{\mathsf{N}}}^T$,

$$\overline{\overline{\mathsf{M}}} = A\overline{\overline{\mathsf{I}}}^{(2)T} + \overline{\overline{\mathsf{N}}}^T, \quad \overline{\overline{\mathsf{N}}}^2 = 0. \qquad (9.15)$$

We could call bidyadics of the form (9.13), (9.14), and (9.15) respectively as SDK, SDJ, and SDN bidyadics. The pure axion medium $\overline{\overline{\mathsf{M}}} = M\overline{\overline{\mathsf{I}}}^{(2)T}$ is ruled out of the SDK and SDJ classes because of the assumption $A^2 \neq B$. One can verify that the inverse of a bidyadic belonging to each of these classes, if it exists, is a member of the same class. In fact, we have

$$(A\overline{\overline{\mathsf{I}}}^{(2)T} + C\overline{\overline{\mathsf{K}}}^T)^{-1} = \frac{1}{B}(A\overline{\overline{\mathsf{I}}}^{(2)T} - C\overline{\overline{\mathsf{K}}}^T), \qquad (9.16)$$

$$(A\overline{\overline{\mathsf{I}}}^{(2)T} + D\overline{\overline{\mathsf{J}}}^T)^{-1} = \frac{1}{B}(A\overline{\overline{\mathsf{I}}}^{(2)T} - D\overline{\overline{\mathsf{J}}}^T), \qquad (9.17)$$

$$(A\overline{\overline{\mathsf{I}}}^{(2)T} + \overline{\overline{\mathsf{N}}}^T)^{-1} = \frac{1}{A^2}(A\overline{\overline{\mathsf{I}}}^{(2)T} - \overline{\overline{\mathsf{N}}}^T). \qquad (9.18)$$

Also, multiplication of two bidyadics of a subclass corresponding to a certain bidyadic $\overline{\overline{\mathsf{K}}}$, $\overline{\overline{\mathsf{J}}}$, or $\overline{\overline{\mathsf{N}}}$ leaves the bidyadic in the same subclass and so does adding an axion term. Finally, it is easy to verify that the similarity transformation (9.4) does not map media out of these three subclasses for any bidyadic $\overline{\overline{\mathsf{D}}}$ possessing an inverse, although the bidyadics $\overline{\overline{\mathsf{K}}}, \overline{\overline{\mathsf{J}}}$, or $\overline{\overline{\mathsf{N}}}$ may change in the transformation.

As an example, the GBI medium defined by the bidyadic (9.7) belongs to the SDK class for $B^2 > \epsilon'/\mu$ and to the SDJ class for $B^2 < \epsilon'/\mu$ while for $B^2 = \epsilon'/\mu$ it belongs to the class SCN. Assuming time-harmonic fields with complex-valued medium parameters, SDK and SDJ classes cannot, however, be distinguished.

9.1.2 Eigenexpansions

To simplify the analysis, let us consider only two subclasses, that of the (proper) SD media and that of SDN media by adopting the definition of SDJ media for SD media, in which case (9.13) will be written as (9.14) with imaginary-valued coefficient D. Defining

$$M = \sqrt{A^2 + D^2}, \quad \tan\psi = D/A, \tag{9.19}$$

(9.14) can be written in the compact form

$$\overline{\overline{\mathsf{M}}} = M(\cos\psi \overline{\overline{\mathrm{I}}}^{(2)T} + \sin\psi \overline{\overline{\mathsf{J}}}^T) = Me^{\psi \overline{\overline{\mathsf{J}}}^T}. \tag{9.20}$$

When the medium bidyadic equation (9.12) is factorized as

$$(\overline{\overline{\mathsf{M}}} - M_+\overline{\overline{\mathrm{I}}}^{(2)T})|(\overline{\overline{\mathsf{M}}} - M_-\overline{\overline{\mathrm{I}}}^{(2)T}) = 0, \tag{9.21}$$

the eigenvalues $M_\pm$ can be identified as

$$M_\pm = A \pm jD = Me^{\pm j\psi}. \tag{9.22}$$

From the assumption $B - A^2 = D^2 \neq 0$ for the SD case it follows that the two eigenvalues are distinct, $M_+ \neq M_-$. In the SDN case the eigenvalues coincide. For the special case $A = 0$ we have $M_+ + M_- = 0$, whence $\mathrm{tr}\overline{\overline{\mathsf{M}}} = 0$. Thus, the cases when $\overline{\overline{\mathsf{M}}}$ is a multiple of either $\overline{\overline{\mathsf{J}}}^T$ or $\overline{\overline{\mathsf{N}}}^T$, there is no axion component.

The eigen two-forms satisfying

$$\overline{\overline{\mathsf{M}}}|\mathbf{\Phi}_\pm = M_\pm \mathbf{\Phi}_\pm \tag{9.23}$$

can be expressed from (9.21) as

$$\begin{aligned}\mathbf{\Phi}_\pm &= \alpha_\pm(\overline{\overline{\mathsf{M}}} - M_\mp \overline{\overline{\mathsf{I}}}^{(2)T})|\mathbf{\Phi},\\ &= \alpha_\pm(Me^{\psi \overline{\overline{\mathsf{J}}}^T} - Me^{\mp j\psi}\overline{\overline{\mathsf{I}}}^{(2)T})|\mathbf{\Phi}\\ &= \alpha_\pm M(\sin\psi\, \overline{\overline{\mathsf{J}}}^T \pm j\sin\psi \overline{\overline{\mathsf{I}}}^{(2)T})|\mathbf{\Phi}\\ &= \pm j\alpha_\pm M\sin\psi(\overline{\overline{\mathsf{I}}}^{(2)T} \mp j\overline{\overline{\mathsf{J}}}^T)|\mathbf{\Phi}.\end{aligned}$$

Here $\mathbf{\Phi}$ is any two-form yielding nonzero results. Choosing the coefficients as

$$\alpha_\pm = \pm\frac{1}{2jM\sin\psi} = \pm\frac{1}{M_+ - M_-}, \tag{9.24}$$

the eigen two-forms are obtained by

$$\mathbf{\Phi}_\pm = \frac{1}{2}(\overline{\overline{\mathsf{I}}}^{(2)T} \mp j\overline{\overline{\mathsf{J}}}^T)|\mathbf{\Phi}. \tag{9.25}$$

Because of the relation

$$\mathbf{\Phi} = \mathbf{\Phi}_+ + \mathbf{\Phi}_-, \tag{9.26}$$

any given field two-form $\mathbf{\Phi}$ can be decomposed as a sum of eigen two-forms $\mathbf{\Phi}_+, \mathbf{\Phi}_-$ which are obtained from (9.25) or, more compactly, from

$$\mathbf{\Phi}_\pm = \overline{\overline{\Pi}}^T_\pm|\mathbf{\Phi}, \tag{9.27}$$

where $\overline{\overline{\Pi}}_+, \overline{\overline{\Pi}}_-$ are two complementary projection bidyadics defined by

$$\overline{\overline{\Pi}}_\pm = \frac{1}{2}(\overline{\overline{\mathsf{I}}}^{(2)} \mp j\overline{\overline{\mathsf{J}}}), \tag{9.28}$$

and obeying the properties

$$\overline{\overline{\Pi}}^2_\pm = \overline{\overline{\Pi}}_\pm, \quad \overline{\overline{\Pi}}_+ + \overline{\overline{\Pi}}_- = \overline{\overline{\mathsf{I}}}^{(2)}, \quad \overline{\overline{\Pi}}_+|\overline{\overline{\Pi}}_- = \overline{\overline{\Pi}}_-|\overline{\overline{\Pi}}_+ = 0. \tag{9.29}$$

9.1.3 Duality Transformation

The three-parameter (Z_d, Z, θ) duality transformation rule (6.59) for the medium bidyadic,

$$Z_d\overline{\overline{\mathsf{M}}}_d = (\sin\theta\, \overline{\overline{\mathsf{I}}}^{(2)T} + \cos\theta\, Z\overline{\overline{\mathsf{M}}})|(\cos\theta\, \overline{\overline{\mathsf{I}}}^{(2)T} - \sin\theta\, Z\overline{\overline{\mathsf{M}}})^{-1}, \tag{9.30}$$

can be shown to map SDK, SDJ, and SDN bidyadics to respective SDK, SDJ, and SDN bidyadics because (9.30) involves additions of axion terms, the inverse operation and multiplication of bidyadics.

An SD medium is self-dual if there exists a duality transformation (excluding the trivial identity transformation), for which the medium bidyadic is invariant, $\overline{\overline{\mathsf{M}}}_d = \overline{\overline{\mathsf{M}}}$. From (9.30) we obtain an equation for the medium bidyadic defining a self-dual medium:

$$\overline{\overline{\mathsf{M}}}^2 - \frac{Z_d - Z}{Z_d Z} \cot\theta\, \overline{\overline{\mathsf{M}}} + \frac{1}{Z_d Z} \overline{\overline{\mathrm{I}}}^{(2)T} = 0, \tag{9.31}$$

which must be valid for some nontrivial parameters Z_d, Z, θ. This shows us that any self-dual medium belongs to the class of SD media.

On the other hand, one can show that any SD medium is self-dual. For this we must find the parameters Z, Z_d, θ of the duality transformation corresponding to a given SD medium bidyadic. Since the SD medium equation involves two parameters and the transformation has three parameters, this leaves us with some freedom. Actually, we can consider the two-parameter involutory transformation (6.51) with $Z_d = -Z$. In this case (9.31) is simplified to

$$\begin{aligned} 0 &= \overline{\overline{\mathsf{M}}}^2 - \frac{2}{Z} \cot\theta\, \overline{\overline{\mathsf{M}}} - \frac{1}{Z^2} \overline{\overline{\mathrm{I}}}^{(2)T}, \\ &= \left(\overline{\overline{\mathsf{M}}} + \frac{1}{Z} \tan(\theta/2) \overline{\overline{\mathrm{I}}}^{(2)T}\right) | \left(\overline{\overline{\mathsf{M}}} - \frac{1}{Z} \cot(\theta/2) \overline{\overline{\mathrm{I}}}^{(2)T}\right), \end{aligned} \tag{9.32}$$

from which we can choose

$$M_+ = -\frac{1}{Z} \tan(\theta/2), \qquad M_- = \frac{1}{Z} \cot(\theta/2). \tag{9.33}$$

Comparing with (9.22), we now have a relation between the parameters of the transformation and the SD medium. Thus, "SD" in the name of the medium class may refer as well to "second" order equation as to "self-dual" medium.

Applying the duality transformation just defined to the eigenfields (9.25) of the SD medium yields

$$\begin{aligned} \begin{pmatrix} \mathbf{\Psi}_{\pm d} \\ \mathbf{\Phi}_{\pm d} \end{pmatrix} &= \begin{pmatrix} (Z/Z_d)\cos\theta & (1/Z_d)\sin\theta \\ -Z\sin\theta & \cos\theta \end{pmatrix} \begin{pmatrix} \mathbf{\Psi}_\pm \\ \mathbf{\Phi}_\pm \end{pmatrix} \\ &= \begin{pmatrix} -\cos\theta & -(1/Z)\sin\theta \\ -Z\sin\theta & \cos\theta \end{pmatrix} \begin{pmatrix} M_\pm \\ 1 \end{pmatrix} \mathbf{\Phi}_\pm \\ &= \pm \begin{pmatrix} M_\pm \\ 1 \end{pmatrix} \mathbf{\Phi}_\pm = \pm \begin{pmatrix} \mathbf{\Psi}_\pm \\ \mathbf{\Phi}_\pm \end{pmatrix}, \end{aligned} \tag{9.34}$$

details of which are left as an exercise. This can be interpreted so that the eigenfields $\boldsymbol{\Psi}_+, \boldsymbol{\Phi}_+$ are self-dual and the eigenfields $\boldsymbol{\Psi}_-, \boldsymbol{\Phi}_-$ are anti-self-dual with respect to the transformation whose parameters Z, θ are related to the parameters A, B of the original medium equation (9.6) as [59]

$$A = \frac{1}{Z}\cot\theta, \quad B = -\frac{1}{Z^2}. \tag{9.35}$$

9.1.4 3D Representations

To define SD media in terms of spatial dyadics, let us start from the axion-free case, restricted by $\mathrm{tr}\overline{\overline{\mathsf{M}}} = 0$, and add the axion part at a later stage for more complete solutions. The axion-free medium bidyadic $\overline{\overline{\mathsf{M}}} = M\overline{\overline{\mathsf{J}}}^T$ satisfies

$$\overline{\overline{\mathsf{M}}}^2 = -M^2\overline{\overline{\mathsf{I}}}^{(2)T}. \tag{9.36}$$

Thus, $\overline{\overline{\mathsf{M}}}$ can be considered as a multiple of an AC bidyadic [12]. Substituting the expansion (5.65)

$$\overline{\overline{\mathsf{M}}} = \overline{\overline{\alpha}} + \overline{\overline{\epsilon}}' \wedge \mathbf{e}_4 + \boldsymbol{\varepsilon}_4 \wedge \overline{\overline{\mu}}^{-1} + \boldsymbol{\varepsilon}_4 \wedge \overline{\overline{\beta}} \wedge \mathbf{e}_4, \tag{9.37}$$

in (9.36) yields four 3D equations,

$$\overline{\overline{\alpha}}^2 - \overline{\overline{\epsilon}}'|\overline{\overline{\mu}}^{-1} = -M^2\overline{\overline{\mathsf{I}}}_s^{(2)T}, \tag{9.38}$$

$$\overline{\overline{\alpha}}|\overline{\overline{\epsilon}}' - \overline{\overline{\epsilon}}'|\overline{\overline{\beta}} = 0, \tag{9.39}$$

$$\overline{\overline{\mu}}^{-1}|\overline{\overline{\alpha}} - \overline{\overline{\beta}}|\overline{\overline{\mu}}^{-1} = 0, \tag{9.40}$$

$$-\overline{\overline{\beta}}^2 + \overline{\overline{\mu}}^{-1}|\overline{\overline{\epsilon}}' = M^2\overline{\overline{\mathsf{I}}}_s^T. \tag{9.41}$$

The trace-free condition requires

$$\mathrm{tr}\overline{\overline{\mathsf{M}}} = \mathrm{tr}\overline{\overline{\alpha}} - \mathrm{tr}\overline{\overline{\beta}} = 0. \tag{9.42}$$

To be able to proceed with the equations (9.38)–(9.41), we must make some assumption. Let us assume that the spatial dyadic $\overline{\overline{\mu}}^{-1}$ has a spatial inverse $\overline{\overline{\mu}}$. A similar result will be obtained by assuming that the spatial dyadic $\overline{\overline{\epsilon}}'$ is invertible.

Let us apply the transformation (9.4) to represent the medium bidyadic in terms of a simpler bidyadic $\overline{\overline{\mathsf{M}}}_1$ by considering the transformation bidyadic

$$\overline{\overline{\mathsf{D}}} = \overline{\overline{\mathsf{I}}}^{(2)T} - \overline{\overline{\alpha}}|\overline{\overline{\mu}} \wedge \mathbf{e}_4, \tag{9.43}$$

whose inverse is

$$\overline{\overline{\mathsf{D}}}^{-1} = \overline{\overline{\mathrm{I}}}^{(2)T} + \overline{\overline{\alpha}}|\overline{\overline{\mu}} \wedge \mathbf{e}_4, \tag{9.44}$$

as can be easily verified. Inserting (9.37) in the inverse of the transformation (9.4) and applying (9.38) and (9.40) we now obtain

$$\overline{\overline{\mathsf{M}}}_1 = \overline{\overline{\mathsf{D}}}^{-1}|\overline{\overline{\mathsf{M}}}|\overline{\overline{\mathsf{D}}} = M^2\overline{\overline{\mu}} \wedge \mathbf{e}_4 + \boldsymbol{\varepsilon}_4 \wedge \overline{\overline{\mu}}^{-1}, \tag{9.45}$$

which is of the form

$$\overline{\overline{\mathsf{M}}}_1 = \overline{\overline{\epsilon}}'_1 \wedge \mathbf{e}_4 + \boldsymbol{\varepsilon}_4 \wedge \overline{\overline{\mu}}_1^{-1}, \tag{9.46}$$

$$\overline{\overline{\epsilon}}'_1 = M^2\overline{\overline{\mu}}, \quad \overline{\overline{\mu}}_1 = \overline{\overline{\mu}}. \tag{9.47}$$

This is a simpler SD medium and can be shown to equal the Gibbsian isotropic medium (7.5) when $\mathbf{e}_{123}\lfloor\overline{\overline{\mu}} \in \mathbb{E}_1\mathbb{E}_1$ is a symmetric dyadic. One can also note that it corresponds to the representation (4.133) of an AC bidyadic.

Thus, to define a SD medium we may take any spatial bidyadic $\overline{\overline{\mathsf{M}}}_1 \in \mathbb{F}_2\mathbb{E}_2$ of the form (9.46) with spatial dyadics related as $\overline{\overline{\epsilon}}'_1 = M^2\overline{\overline{\mu}}_1$ and insert it with (9.47) in (9.4), from which we obtain a more general form for a SD medium bidyadic,

$$\begin{aligned}\overline{\overline{\mathsf{M}}} &= \overline{\overline{\mathsf{D}}}|\overline{\overline{\mathsf{M}}}_1|\overline{\overline{\mathsf{D}}}^{-1} \\ &= (\overline{\overline{\mathrm{I}}}^{(2)T} - \overline{\overline{\alpha}}|\overline{\overline{\mu}} \wedge \mathbf{e}_4)|(M^2\overline{\overline{\mu}} \wedge \mathbf{e}_4 + \boldsymbol{\varepsilon}_4 \wedge \overline{\overline{\mu}}^{-1})|(\overline{\overline{\mathrm{I}}}^{(2)T} + \overline{\overline{\alpha}}|\overline{\overline{\mu}} \wedge \mathbf{e}_4) \\ &= \overline{\overline{\alpha}} + M^2\overline{\overline{\mu}} \wedge \mathbf{e}_4 + \boldsymbol{\varepsilon}_4 \wedge \overline{\overline{\mu}}^{-1} + \boldsymbol{\varepsilon}_4 \wedge \overline{\overline{\mu}}^{-1}|\overline{\overline{\alpha}}|\overline{\overline{\mu}} \wedge \mathbf{e}_4 + \overline{\overline{\alpha}}^2|\overline{\overline{\mu}} \wedge \mathbf{e}_4.\end{aligned} \tag{9.48}$$

Thus, under the assumption that $\overline{\overline{\mu}}$ exists, the axion-free SD media can be defined by conditions of the form

$$\mathbf{D} = \overline{\overline{\alpha}}|\mathbf{B} + (\overline{\overline{\alpha}}^2|\overline{\overline{\mu}} + M^2\overline{\overline{\mu}})|\mathbf{E}, \tag{9.49}$$

$$\mathbf{H} = \overline{\overline{\mu}}^{-1}|\mathbf{B} + (\overline{\overline{\mu}}^{-1}|\overline{\overline{\alpha}}|\overline{\overline{\mu}})|\mathbf{E}. \tag{9.50}$$

The two spatial dyadics $\overline{\overline{\alpha}}$ and $\overline{\overline{\mu}}$ are arbitrary except that $\overline{\overline{\mu}}$ must be invertible. The other two spatial dyadics are obtained from

$$\overline{\overline{\epsilon}}' = \overline{\overline{\alpha}}^2|\overline{\overline{\mu}} + M^2\overline{\overline{\mu}}, \quad \overline{\overline{\beta}} = \overline{\overline{\mu}}^{-1}|\overline{\overline{\alpha}}|\overline{\overline{\mu}}. \tag{9.51}$$

In the other representation of medium equations,

$$\mathbf{D} = \overline{\overline{\epsilon}}|\mathbf{E} + \overline{\overline{\xi}}|\mathbf{H} = M^2\overline{\overline{\mu}}|\mathbf{E} + (\overline{\overline{\alpha}}|\overline{\overline{\mu}})|\mathbf{H}, \tag{9.52}$$

$$\mathbf{B} = \overline{\overline{\zeta}}|\mathbf{E} + \overline{\overline{\mu}}|\mathbf{H} = -(\overline{\overline{\alpha}}|\overline{\overline{\mu}})|\mathbf{E} + \overline{\overline{\mu}}|\mathbf{H}, \tag{9.53}$$

the spatial medium dyadics are seen to obey the relations

$$\overline{\overline{\epsilon}} = M^2\overline{\overline{\mu}}, \quad \overline{\overline{\xi}} = -\overline{\overline{\zeta}} \tag{9.54}$$

for the axion-free SD media.

The bidyadic (9.48) can now be generalized by adding an axion term $A\overline{\overline{\mathsf{I}}}^{(2)T}$. Because this corresponds to replacing $\overline{\overline{\alpha}}$ by $\overline{\overline{\alpha}} + A\overline{\overline{\mathsf{I}}}_s^{(2)T}$ and $\overline{\overline{\beta}}$ by $\overline{\overline{\beta}} - A\overline{\overline{\mathsf{I}}}_s^T$, the medium conditions (9.49) and (9.50) become

$$\mathbf{D} = (\overline{\overline{\alpha}} + A\overline{\overline{\mathsf{I}}}_s^{(2)T})|\mathbf{B} + (\overline{\overline{\alpha}}^2|\overline{\overline{\mu}} + M^2\overline{\overline{\mu}})|\mathbf{E}, \tag{9.55}$$

$$\mathbf{H} = \overline{\overline{\mu}}^{-1}|\mathbf{B} + (\overline{\overline{\mu}}^{-1}|\overline{\overline{\alpha}}|\overline{\overline{\mu}} - A\overline{\overline{\mathsf{I}}}_s^{(2)T})|\mathbf{E}. \tag{9.56}$$

In another representation the medium conditions can be written as

$$\mathbf{D} = (A^2M^2\overline{\overline{\mu}})|\mathbf{E} + (A\overline{\overline{\mu}} + \overline{\overline{\alpha}}|\overline{\overline{\mu}})|\mathbf{H}, \tag{9.57}$$

$$\mathbf{B} = (A\overline{\overline{\mu}} - \overline{\overline{\alpha}}|\overline{\overline{\mu}})|\mathbf{E} + \overline{\overline{\mu}}|\mathbf{H}, \tag{9.58}$$

which correspond to the relations

$$\overline{\overline{\epsilon}} = (A^2 + M^2)\overline{\overline{\mu}}, \quad \overline{\overline{\xi}} + \overline{\overline{\zeta}} = 2A\overline{\overline{\mu}}, \quad \overline{\overline{\xi}} - \overline{\overline{\zeta}} = 2\overline{\overline{\alpha}}|\overline{\overline{\mu}}. \tag{9.59}$$

Thus, for general SD media, the spatial dyadics $\overline{\overline{\epsilon}}, \overline{\overline{\mu}}$, and $\overline{\overline{\xi}} + \overline{\overline{\zeta}}$ must be multiples of the same dyadic while $\overline{\overline{\xi}} - \overline{\overline{\zeta}}$ may be any other dyadic. This result coincides with that for self-dual media in terms of Gibbsian vectors and dyadics [60]. The axion part is unchanged in the similarity transformation (9.4).

The previous medium bidyadic solutions for the quadratic equation were based on the assumption of invertible spatial dyadic $\overline{\overline{\mu}}^{-1}$ or $\overline{\overline{\epsilon}}'$. There are also other possibilities. For example, $\overline{\overline{\mathsf{M}}}$ defined as a multiple of the time reversion bidyadic,

$$\overline{\overline{\mathsf{M}}} = jM\overline{\overline{\mathsf{T}}}^{(2)T} = jM(\overline{\overline{\mathsf{I}}}_s^{(2)T} - \overline{\overline{\mathsf{I}}}_s^T{}^\wedge_\wedge\boldsymbol{\varepsilon}_4\mathbf{e}_4), \tag{9.60}$$

with the spatial medium dyadics

$$\overline{\overline{\alpha}} = jM\overline{\overline{\mathsf{I}}}_s^{(2)T}, \quad \overline{\overline{\epsilon}}' = 0, \quad \overline{\overline{\mu}}^{-1} = 0, \quad \overline{\overline{\beta}} = -jM\overline{\overline{\mathsf{I}}}_s^T. \tag{9.61}$$

obviously satisfies (9.36), whence it serves as another candidate for an SD medium bidyadic. Actually, the corresponding medium conditions

$$\mathbf{D} = jM\mathbf{B}, \quad \mathbf{H} = jM\mathbf{E} \tag{9.62}$$

coincide with those of the SS medium (7.74). Other solutions can be found through any bidyadic $\overline{\overline{\mathsf{D}}}$.

To demonstrate the property that any AC bidyadic can be obtained from any other AC bidyadic through the similarity transformation (9.4) involving a suitable bidyadic $\overline{\overline{\mathsf{D}}}$ ([77], p.156), let us find a bidyadic $\overline{\overline{\mathsf{D}}}$ mapping the bidyadic $\overline{\overline{\mathsf{M}}}_1$ of (9.46) to $\overline{\overline{\mathsf{M}}}$ of (9.60). Leaving the details as an exercise, one can show that, for the bidyadics $\overline{\overline{\mathsf{M}}}$ and $\overline{\overline{\mathsf{M}}}_1$ so defined, the bidyadic $\overline{\overline{\mathsf{D}}} = (\overline{\overline{\mathsf{M}}} + \overline{\overline{\mathsf{M}}}_1)/M$ satisfies

$$\overline{\overline{\mathsf{D}}}|\overline{\overline{\mathsf{M}}}_1|\overline{\overline{\mathsf{D}}}^{-1} = \overline{\overline{\mathsf{M}}}. \tag{9.63}$$

This actually demonstrates that an SD-medium bidyadic $\overline{\overline{\mathsf{M}}}_1$ corresponding to a medium with $\overline{\overline{\alpha}} = 0$ and $\overline{\overline{\beta}} = 0$ can be transformed to another SD-medium bidyadic $\overline{\overline{\mathsf{M}}}$ with $\overline{\overline{\epsilon}}' = 0$, $\overline{\overline{\mu}}^{-1} = 0$.

9.1.5 SDN Media

The SDN medium bidyadics are solutions for (9.6) with $A^2 - B = 0$. Writing the quadratic equation in the form

$$\overline{\overline{\mathsf{M}}}^2 - 2M\overline{\overline{\mathsf{M}}} + M^2\overline{\overline{\mathsf{I}}}^{(2)T} = (\overline{\overline{\mathsf{M}}} - M\overline{\overline{\mathsf{I}}}^{(2)T})^2 = 0, \tag{9.64}$$

its solution can be written as

$$\overline{\overline{\mathsf{M}}} = M\overline{\overline{\mathsf{I}}}^{(2)T} + \overline{\overline{\mathsf{N}}}, \tag{9.65}$$

where, $\overline{\overline{\mathsf{N}}} \in \mathbb{F}_2\mathbb{E}_2$ (not to be mistaken as $\overline{\overline{\mathsf{N}}}$ in (5.52)) may be any nilpotent bidyadic of grade 2.

Let us expand the general nilpotent bidyadic $\overline{\overline{\mathsf{N}}}$ in its spatiotemporal components (9.37)

$$\overline{\overline{\mathsf{N}}} = \overline{\overline{\alpha}} + \overline{\overline{\epsilon}}' \wedge \mathbf{e}_4 + \boldsymbol{\varepsilon}_4 \wedge \overline{\overline{\mu}}^{-1} + \boldsymbol{\varepsilon}_4 \wedge \overline{\overline{\beta}} \wedge \mathbf{e}_4. \tag{9.66}$$

After squaring and requiring that all of its four components vanish, the following set of relations is obtained:

$$\overline{\overline{\alpha}}|\overline{\overline{\epsilon}}' = \overline{\overline{\epsilon}}'|\overline{\overline{\beta}}, \tag{9.67}$$

$$\overline{\overline{\mu}}^{-1}|\overline{\overline{\alpha}} = \overline{\overline{\beta}}|\overline{\overline{\mu}}^{-1}, \tag{9.68}$$

$$\overline{\overline{\alpha}}^2 = \overline{\overline{\epsilon}}'|\overline{\overline{\mu}}^{-1}, \tag{9.69}$$

$$\overline{\overline{\beta}}^2 = \overline{\overline{\mu}}^{-1}|\overline{\overline{\epsilon}}'. \tag{9.70}$$

Assuming again that $\overline{\overline{\mu}}^{-1}$ has an inverse, from (9.68) and (9.69) we can express $\overline{\overline{\beta}}$ and $\overline{\overline{\epsilon}}'$ as

$$\overline{\overline{\beta}} = \overline{\overline{\mu}}^{-1}|\overline{\overline{\alpha}}|\overline{\overline{\mu}}, \tag{9.71}$$

$$\overline{\overline{\epsilon}}' = \overline{\overline{\alpha}}^2|\overline{\overline{\mu}}. \tag{9.72}$$

Inserting these in (9.67) and (9.70) one can show that they are identically valid, which means that two of the medium dyadics can be expressed in terms of the other two which can be freely chosen. The nilpotent bidyadic $\overline{\overline{\mathsf{N}}}$ can be expressed as

$$\begin{aligned}\overline{\overline{\mathsf{N}}} &= \overline{\overline{\alpha}} + \overline{\overline{\alpha}}^2|\overline{\overline{\mu}} \wedge \mathbf{e}_4 + \boldsymbol{\varepsilon}_4 \wedge \overline{\overline{\mu}}^{-1} + \boldsymbol{\varepsilon}_4 \wedge \overline{\overline{\mu}}^{-1}|\overline{\overline{\alpha}}|\overline{\overline{\mu}} \wedge \mathbf{e}_4 \\ &= (\overline{\overline{\alpha}}|\overline{\overline{\mu}} + \boldsymbol{\varepsilon}_4 \wedge \overline{\overline{\mathsf{I}}}_s^T)|\overline{\overline{\mu}}^{-1}|(\overline{\overline{\mathsf{I}}}_s^{(2)T} + \overline{\overline{\alpha}}|\overline{\overline{\mu}} \wedge \mathbf{e}_4),\end{aligned} \tag{9.73}$$

and its square can be easily shown to vanish. The medium equations corresponding to the nilpotent medium bidyadic can now be expressed as

$$\mathbf{D} = \overline{\overline{\alpha}}|(\mathbf{B} + (\overline{\overline{\alpha}}|\overline{\overline{\mu}})|\mathbf{E}) \tag{9.74}$$

$$\mathbf{H} = \overline{\overline{\mu}}^{-1}|(\mathbf{B} + (\overline{\overline{\alpha}}|\overline{\overline{\mu}})|\mathbf{E}), \tag{9.75}$$

or

$$\mathbf{D} = (\overline{\overline{\alpha}}|\overline{\overline{\mu}})|\mathbf{H} \tag{9.76}$$

$$\mathbf{B} = -(\overline{\overline{\alpha}}|\overline{\overline{\mu}})\mathbf{E} + \overline{\overline{\mu}}|\mathbf{H}. \tag{9.77}$$

It is noteworthy that, while the dyadic $\overline{\overline{\epsilon}}$ vanishes, the dyadic $\overline{\overline{\epsilon}}' = \overline{\overline{\alpha}}^2|\overline{\overline{\mu}}$ does not vanish in general for this representation of the nilpotent bidyadic. For the more general SDN medium bidyadic (9.65) an axion term must be added. Again, other possible $\overline{\overline{\mathsf{N}}}$-dyadics are obtained through the similarity transformation.

9.2 CUBIC EQUATION

In this section we consider solutions to the bidyadic cubic equation

$$\overline{\overline{\mathsf{M}}}^3 + A\overline{\overline{\mathsf{M}}}^2 + B\overline{\overline{\mathsf{M}}} + C\overline{\overline{\mathsf{I}}}^{(2)T} = 0, \tag{9.78}$$

which defines the class of media which can be called CU media for brevity.

9.2.1 CU Media

As a first try one can study a medium bidyadic of the form

$$\overline{\overline{\mathsf{M}}} = \overline{\overline{\epsilon}} \wedge \mathbf{e}_4 + \boldsymbol{\varepsilon}_4 \wedge \overline{\overline{\mu}}^{-1} \tag{9.79}$$

as a candidate of a CU medium. However, as one can show (details left as an exercise), assuming full-rank spatial dyadics $\overline{\overline{\epsilon}}, \overline{\overline{\mu}}^{-1}$, any such $\overline{\overline{\mathsf{M}}}$ satisfying the cubic equation (9.78), must actually satisfy a quadratic equation. Thus, a proper CU-medium bidyadic must have terms with nonzero $\overline{\overline{\alpha}}$ and/or $\overline{\overline{\beta}}$ dyadics added to (9.79).

As an example of a true CU medium, let us consider a bidyadic of the form

$$\overline{\overline{\mathsf{M}}} = A\overline{\overline{\mathsf{I}}}^{(2)T} + \overline{\overline{\mathsf{C}}}_o, \tag{9.80}$$

where $\overline{\overline{\mathsf{C}}}_o$ is a simple trace-free bidyadic defined by

$$\overline{\overline{\mathsf{C}}}_o = \boldsymbol{\varepsilon}_N \lfloor (\mathbf{AB} - \mathbf{BA}), \quad \mathrm{tr}\overline{\overline{\mathsf{C}}}_o = 0, \tag{9.81}$$

$\mathbf{A}$ and $\mathbf{B}$ being two bivectors. Thus, (9.80) corresponds to a certain skewon–axion medium. It is not difficult to show that the bidyadic $\overline{\overline{\mathsf{C}}}_o$ satisfies the cubic equation

$$\begin{aligned} \overline{\overline{\mathsf{C}}}_o^{\,3} &= ((\mathbf{A}\cdot\mathbf{B})^2 - (\mathbf{A}\cdot\mathbf{A})(\mathbf{B}\cdot\mathbf{B}))\overline{\overline{\mathsf{C}}}_o \\ &= \frac{1}{2}(\mathrm{tr}\overline{\overline{\mathsf{C}}}_o^{\,2})\overline{\overline{\mathsf{C}}}_o. \end{aligned} \tag{9.82}$$

When substituting $\overline{\overline{\mathsf{C}}}_o = \overline{\overline{\mathsf{M}}} - A\overline{\overline{\mathsf{I}}}^{(2)T}$ in (9.82), we find that the medium bidyadic $\overline{\overline{\mathsf{M}}}$ satisfies the cubic equation

$$(\overline{\overline{\mathsf{M}}} - A\overline{\overline{\mathsf{I}}}^{(2)T})^3 - \frac{1}{2}\mathrm{tr}(\overline{\overline{\mathsf{M}}} - A\overline{\overline{\mathsf{I}}}^{(2)T})(\overline{\overline{\mathsf{M}}} - A\overline{\overline{\mathsf{I}}}^{(2)T}) = 0. \tag{9.83}$$

9.2.2 Eigenexpansions

Expressing (9.78) in factorized form as

$$(\overline{\overline{\mathsf{M}}} - M_a\overline{\overline{\mathsf{I}}}^{(2)T})|(\overline{\overline{\mathsf{M}}} - M_b\overline{\overline{\mathsf{I}}}^{(2)T})|(\overline{\overline{\mathsf{M}}} - M_c\overline{\overline{\mathsf{I}}}^{(2)T}) = 0, \tag{9.84}$$

where the bracketed bidyadics commute. The scalars M_a, M_b, and M_c are related to the coefficients of (9.78) as

$$A = -(M_a + M_b + M_c), \tag{9.85}$$

$$B = M_aM_b + M_bM_c + M_cM_a, \tag{9.86}$$

$$C = -M_aM_bM_c. \tag{9.87}$$

Following the pattern of the previous section for SD media, let us define the following bidyadics in the case $M_a \neq M_b \neq M_c \neq M_a$:

$$\overline{\overline{\Pi}}_a = \frac{(\overline{\overline{\mathsf{M}}} - M_b\overline{\overline{\mathrm{I}}}^{(2)T})|(\overline{\overline{\mathsf{M}}} - M_c\overline{\overline{\mathrm{I}}}^{(2)T})}{(M_a - M_b)(M_a - M_c)}, \tag{9.88}$$

$$\overline{\overline{\Pi}}_b = \frac{(\overline{\overline{\mathsf{M}}} - M_c\overline{\overline{\mathrm{I}}}^{(2)T})|(\overline{\overline{\mathsf{M}}} - M_a\overline{\overline{\mathrm{I}}}^{(2)T})}{(M_b - M_c)(M_b - M_a)}, \tag{9.89}$$

$$\overline{\overline{\Pi}}_c = \frac{(\overline{\overline{\mathsf{M}}} - M_a\overline{\overline{\mathrm{I}}}^{(2)T})|(\overline{\overline{\mathsf{M}}} - M_b\overline{\overline{\mathrm{I}}}^{(2)T})}{(M_c - M_a)(M_c - M_b)}. \tag{9.90}$$

Because of (9.84) they satisfy the eigenequation

$$(\overline{\overline{\mathsf{M}}} - M_i\overline{\overline{\mathrm{I}}}^{(2)T})|\overline{\overline{\Pi}}_i = 0, \tag{9.91}$$

for $i = a, b, c$. Thus, the eigenvalue equation for the CU-medium bidyadic

$$\overline{\overline{\mathsf{M}}}|\mathbf{\Phi}_i = M_i\mathbf{\Phi}_i \tag{9.92}$$

has three solutions and the eigen two-forms can be represented in the form

$$\mathbf{\Phi}_i = \overline{\overline{\Pi}}_i|\mathbf{\Phi}, \tag{9.93}$$

for any two-form $\mathbf{\Phi}$ producing $\mathbf{\Phi}_i \neq 0$. The three bidyadics $\overline{\overline{\Pi}}_i$ can be shown to satisfy

$$\overline{\overline{\Pi}}_a + \overline{\overline{\Pi}}_b + \overline{\overline{\Pi}}_c = \overline{\overline{\mathrm{I}}}^{(2)T}, \tag{9.94}$$

which allows one to expand any two-form in its eigen two-forms as

$$\mathbf{\Phi} = (\overline{\overline{\Pi}}_a + \overline{\overline{\Pi}}_b + \overline{\overline{\Pi}}_c)|\mathbf{\Phi} = \mathbf{\Phi}_a + \mathbf{\Phi}_b + \mathbf{\Phi}_c. \tag{9.95}$$

From (9.84) we obtain the orthogonality conditions

$$\overline{\overline{\Pi}}_i|\overline{\overline{\Pi}}_j = 0, \quad i \neq j, \tag{9.96}$$

while

$$\overline{\overline{\Pi}}_i^2 = \overline{\overline{\Pi}}_i, \tag{9.97}$$

follows from (9.94) and orthogonality. This means that the bidyadics $\overline{\overline{\Pi}}_i$ are actually projection bidyadics in the space of two-forms. Because of the property (9.91) we can write

$$\begin{aligned}\overline{\overline{\mathsf{M}}} &= \overline{\overline{\mathsf{M}}}|(\overline{\overline{\Pi}}_a + \overline{\overline{\Pi}}_b + \overline{\overline{\Pi}}_c) \\ &= M_a\overline{\overline{\Pi}}_a + M_b\overline{\overline{\Pi}}_b + M_c\overline{\overline{\Pi}}_c.\end{aligned} \tag{9.98}$$

This can be generalized to

$$\overline{\overline{\mathsf{M}}}^m = M_a^m \overline{\overline{\Pi}}_a + M_b^m \overline{\overline{\Pi}}_b + M_c^m \overline{\overline{\Pi}}_c, \tag{9.99}$$

where m can be any integer, positive, or negative.

As a check we can substitute the expansions of $\overline{\overline{\mathsf{M}}}^3, \overline{\overline{\mathsf{M}}}^2, \overline{\overline{\mathsf{M}}}$, and $\overline{\overline{\mathsf{I}}}^{(2)T}$ in (9.84) and find that the coefficient expressions of each bidyadic $\overline{\overline{\Pi}}_i$ vanish identically.

For the special CU medium defined by (9.80) the cubic equation (9.83) can be written in the form

$$(\overline{\overline{\mathsf{M}}} - A\overline{\overline{\mathsf{I}}}^{(2)T})| \left[(\overline{\overline{\mathsf{M}}} - A\overline{\overline{\mathsf{I}}}^{(2)T})^2 - \frac{1}{2}(\mathrm{tr}\overline{\overline{\mathsf{C}}}_o{}^2)\overline{\overline{\mathsf{I}}}^{(2)T}\right] = 0, \tag{9.100}$$

from which the eigenvalues of the bidyadic $\overline{\overline{\mathsf{M}}}$ are obtained as

$$M_a = A, \quad M_{b,c} = A \pm \sqrt{\frac{1}{2}\mathrm{tr}\overline{\overline{\mathsf{C}}}_o{}^2}. \tag{9.101}$$

Being a skewon–axion medium, there is no dispersion equation for a plane wave in such a medium.

9.2.3 Examples of CU Media

Let us consider some further examples of medium bidyadics satisfying a cubic equation.

As a symmetric example let us consider projection bidyadics defined by

$$\overline{\overline{\Pi}}_a = \boldsymbol{\varepsilon}_{12}\mathbf{e}_{12} + \boldsymbol{\varepsilon}_{34}\mathbf{e}_{34}, \tag{9.102}$$
$$\overline{\overline{\Pi}}_b = \boldsymbol{\varepsilon}_{23}\mathbf{e}_{23} + \boldsymbol{\varepsilon}_{14}\mathbf{e}_{14}, \tag{9.103}$$
$$\overline{\overline{\Pi}}_c = \boldsymbol{\varepsilon}_{31}\mathbf{e}_{31} + \boldsymbol{\varepsilon}_{24}\mathbf{e}_{24}, \tag{9.104}$$

in terms of which we can construct the medium bidyadic as (9.98) by introducing any three coefficients M_a, M_b, M_c. From (9.99) we obtain

$$\overline{\overline{\mathsf{M}}}^2 = M_a^2 \overline{\overline{\Pi}}_a + M_b^2 \overline{\overline{\Pi}}_b + M_c^2 \overline{\overline{\Pi}}_c, \tag{9.105}$$
$$\overline{\overline{\mathsf{M}}}^3 = M_a^3 \overline{\overline{\Pi}}_a + M_b^3 \overline{\overline{\Pi}}_b + M_c^3 \overline{\overline{\Pi}}_c. \tag{9.106}$$

Eliminating the projection dyadics from the expressions (9.98), (9.105), and (9.106) we find that the medium bidyadic $\overline{\overline{\mathsf{M}}}$ really satisfies a cubic equation of the form (9.78) with coefficients defined by (9.85)–(9.87). In this case the three eigenvalues are double eigenvalues.

As another, less symmetric, example let us consider the projection bidyadics

$$\overline{\overline{\Pi}}_a = \varepsilon_{12}\mathbf{e}_{12}, \tag{9.107}$$

$$\overline{\overline{\Pi}}_b = \varepsilon_{13}\mathbf{e}_{13} + \varepsilon_{23}\mathbf{e}_{23}, \tag{9.108}$$

$$\overline{\overline{\Pi}}_c = \varepsilon_{14}\mathbf{e}_{14} + \varepsilon_{24}\mathbf{e}_{24} + \varepsilon_{34}\mathbf{e}_{34}, \tag{9.109}$$

which satisfy the same orthogonality conditions (9.96). The medium bidyadic defined by (9.80) satisfies also (9.105) and (9.106), whence the medium bidyadic obeys the same cubic equation (9.78) with the same coefficients as in the previous example. However, in this case M_a is a single eigenvalue, M_b is a double eigenvalue, and M_c is a triple eigenvalue. Of course, there are many possibilities for choosing the projection bidyadics.

As a second example, let us consider a generalization of the medium defined by (9.80), a medium bidyadic of the form

$$\overline{\overline{\mathsf{M}}} = \alpha\overline{\overline{\mathrm{I}}}^{(2)T} + \overline{\overline{\mathsf{C}}}, \tag{9.110}$$

where $\overline{\overline{\mathsf{C}}}$ is the bidyadic

$$\overline{\overline{\mathsf{C}}} = \varepsilon_N \lfloor(\mathbf{AC} + \mathbf{BD}), \tag{9.111}$$

defined by four bivectors $\mathbf{A}, \mathbf{B}, \mathbf{C}, \mathbf{D}$. One can show that the bidyadic $\overline{\overline{\mathsf{C}}}$ defined by (9.111) satisfies a cubic equation of the form

$$\overline{\overline{\mathsf{C}}}^3 - (\mathrm{tr}\overline{\overline{\mathsf{C}}})\overline{\overline{\mathsf{C}}}^2 + \frac{1}{2}((\mathrm{tr}\overline{\overline{\mathsf{C}}})^2 - \mathrm{tr}\overline{\overline{\mathsf{C}}}^2)\overline{\overline{\mathsf{C}}} = 0. \tag{9.112}$$

The proof is left as an exercise. From this it follows that the medium bidyadic (9.110) satisfies a cubic equation and, thus, defines a CU medium.

A third example is based on the fact that an antisymmetric bidyadic satisfies a Cayley–Hamilton equation containing only even-powered terms (3.39). Thus, the Cayley–Hamilton equation of a skewon bidyadic can be written in the bicubic form

$$(\overline{\overline{\mathsf{B}}}_o{}^{\wedge}_{\wedge}\overline{\overline{\mathrm{I}}})^6 + a_2(\overline{\overline{\mathsf{B}}}_o{}^{\wedge}_{\wedge}\overline{\overline{\mathrm{I}}})^4 + a_4(\overline{\overline{\mathsf{B}}}_o{}^{\wedge}_{\wedge}\overline{\overline{\mathrm{I}}})^2 + a_6\overline{\overline{\mathrm{I}}}^{(2)} = 0, \quad \mathrm{tr}\overline{\overline{\mathsf{B}}}_o = 0. \tag{9.113}$$

Defining a medium bidyadic as

$$\overline{\overline{\mathsf{M}}} = (\overline{\overline{\mathsf{B}}}_o{}^{\wedge}_{\wedge}\overline{\overline{\mathrm{I}}})^{2T}, \tag{9.114}$$

it satisfies the cubic equation (9.78) for any trace-free dyadic $\overline{\overline{\mathsf{B}}}_o$. It turns out that such a medium bidyadic has no skewon component. In fact, we can expand

$$
\begin{aligned}
\overline{\overline{\mathsf{I}}}^{(4)}\lfloor\lfloor\overline{\overline{\mathsf{M}}} &= (\mathbf{e}_N\lfloor((\overline{\overline{\mathsf{B}}}_o{}^\wedge_\wedge\overline{\overline{\mathsf{I}}})^T)|((\overline{\overline{\mathsf{I}}}{}^\wedge_\wedge\overline{\overline{\mathsf{B}}}_o)^T)\rfloor\boldsymbol{\varepsilon}_N) \\
&= (\mathbf{e}_N\lfloor(\overline{\overline{\mathsf{B}}}_o{}^\wedge_\wedge\overline{\overline{\mathsf{I}}})^T\rfloor\boldsymbol{\varepsilon}_N)|(\mathbf{e}_N\lfloor(\overline{\overline{\mathsf{B}}}_o{}^\wedge_\wedge\overline{\overline{\mathsf{I}}})^T\rfloor\boldsymbol{\varepsilon}_N) \qquad (9.115)\\
&= (\overline{\overline{\mathsf{I}}}^{(4)}\lfloor\lfloor(\overline{\overline{\mathsf{B}}}_o{}^\wedge_\wedge\overline{\overline{\mathsf{I}}})^T)^2 = (-\overline{\overline{\mathsf{B}}}_o{}^\wedge_\wedge\overline{\overline{\mathsf{I}}})^2 = \overline{\overline{\mathsf{M}}}^T
\end{aligned}
$$

by applying the property (3.62),

$$\overline{\overline{\mathsf{I}}}^{(4)}\lfloor\lfloor(\overline{\overline{\mathsf{I}}}{}^\wedge_\wedge\overline{\overline{\mathsf{B}}}_o)^T = -\overline{\overline{\mathsf{I}}}{}^\wedge_\wedge\overline{\overline{\mathsf{B}}}_o. \qquad (9.116)$$

The condition (9.115) means that $\overline{\overline{\mathsf{M}}}$ is an eigenbidyadic of the eigenproblem (3.48) corresponding to the eigenvalue +1, which excludes the skewon component. From

$$
\begin{aligned}
\overline{\overline{\mathsf{M}}}\lfloor\lfloor\overline{\overline{\mathsf{I}}} &= (\overline{\overline{\mathsf{B}}}_o^{2T}{}^\wedge_\wedge\overline{\overline{\mathsf{I}}}^T)\lfloor\lfloor\overline{\overline{\mathsf{I}}} + 2\overline{\overline{\mathsf{B}}}_o^{(2)T}\lfloor\lfloor\overline{\overline{\mathsf{I}}} \\
&= (\mathrm{tr}\overline{\overline{\mathsf{B}}}_o^2)\overline{\overline{\mathsf{I}}} + 2\overline{\overline{\mathsf{B}}}_o^2 - 2(\mathrm{tr}\overline{\overline{\mathsf{B}}}_o\,\overline{\overline{\mathsf{B}}}_o - \overline{\overline{\mathsf{B}}}_o^2) \qquad (9.117)\\
&= \mathrm{tr}\overline{\overline{\mathsf{B}}}_o^2\,\overline{\overline{\mathsf{I}}},
\end{aligned}
$$

we may conclude that $\overline{\overline{\mathsf{M}}}$ defines a pure principal CU medium whenever the trace-free dyadic $\overline{\overline{\mathsf{B}}}_o$ satisfies the additional condition $\mathrm{tr}\overline{\overline{\mathsf{B}}}_o^2 = 0$. Finding the dispersion equation of a plane wave in such a medium is left as an exercise.

9.3 BI-QUADRATIC EQUATION

Instead of the general quartic equation, let us consider media defined by the bi-quadratic equation

$$\overline{\overline{\mathsf{M}}}^4 - 2A\overline{\overline{\mathsf{M}}}^2 + B\overline{\overline{\mathsf{I}}}^{(2)T} = 0, \qquad (9.118)$$

dubbed as BQ media. Comparing (9.6) and (9.118) shows us that the square of the medium bidyadic of any BQ medium serves as a possible medium bidyadic of a SD medium, $\overline{\overline{\mathsf{M}}}_{SD} = \overline{\overline{\mathsf{M}}}_{SQ}^2$. On the other hand, forming the square of the left-hand side of (9.6) shows us that $\overline{\overline{\mathsf{M}}}_{SD}$ is a solution of a more general quartic equation than the biquadratic equation considered here.

9.3.1 BQ Media

As an attempt to define a simple example of a BQ medium let us consider a medium bidyadic of the restricted form

$$\overline{\overline{\mathsf{M}}} = \overline{\overline{\epsilon}} \wedge \mathbf{e}_4 + \varepsilon_4 \wedge \overline{\overline{\mu}}^{-1}. \tag{9.119}$$

Expanding

$$\overline{\overline{\mathsf{M}}}^2 = -\overline{\overline{\epsilon}}|\overline{\overline{\mu}}^{-1} + \varepsilon_4 \wedge \overline{\overline{\mu}}^{-1}|\overline{\overline{\epsilon}} \wedge \mathbf{e}_4, \tag{9.120}$$

$$\overline{\overline{\mathsf{M}}}^4 = (\overline{\overline{\epsilon}}|\overline{\overline{\mu}}^{-1})^2 - \varepsilon_4 \wedge (\overline{\overline{\mu}}^{-1}|\overline{\overline{\epsilon}})^2 \wedge \mathbf{e}_4, \tag{9.121}$$

(9.118) is reduced to two spatial dyadic equations

$$(\overline{\overline{\epsilon}}|\overline{\overline{\mu}}^{-1})^2 + 2A\overline{\overline{\epsilon}}|\overline{\overline{\mu}}^{-1} + B\overline{\overline{\mathsf{I}}}_s^{(2)T} = 0, \tag{9.122}$$

$$(\overline{\overline{\mu}}^{-1}|\overline{\overline{\epsilon}})^2 + 2A\overline{\overline{\mu}}^{-1}|\overline{\overline{\epsilon}} + B\overline{\overline{\mathsf{I}}}_s^T = 0. \tag{9.123}$$

Multiplying (9.122) as $\overline{\overline{\mu}}^{-1}|()|\overline{\overline{\mu}}$ we obtain (9.123), whence they are actually the same equation when $\overline{\overline{\mu}}$ (or $\overline{\overline{\epsilon}}^{-1}$) exists, which is assumed here. Expressing (9.123) as

$$(\overline{\overline{\mu}}^{-1}|\overline{\overline{\epsilon}} + A\overline{\overline{\mathsf{I}}}_s^T)^2 = (A^2 - B)\overline{\overline{\mathsf{I}}}_s^T, \tag{9.124}$$

the dyadic in brackets can be expressed as a multiple of a spatial unipotent dyadic $\overline{\overline{\mathsf{U}}}_s^T \in \mathbb{F}_1\mathbb{E}_1$ satisfying

$$\overline{\overline{\mathsf{U}}}_s^2 = \overline{\overline{\mathsf{I}}}_s, \quad (\overline{\overline{\mathsf{U}}}_s - \overline{\overline{\mathsf{I}}}_s)|(\overline{\overline{\mathsf{U}}}_s + \overline{\overline{\mathsf{I}}}_s) = 0. \tag{9.125}$$

Because one of the dyadics $\overline{\overline{\mathsf{U}}}_s - \overline{\overline{\mathsf{I}}}_s$ and $\overline{\overline{\mathsf{U}}}_s + \overline{\overline{\mathsf{I}}}_s$ must be of rank 1 (see Problem 4.1), we can set

$$\overline{\overline{\mathsf{U}}}_s = \pm(\overline{\overline{\mathsf{I}}}_s^T - \boldsymbol{\alpha}_s\mathbf{a}_s), \quad \mathbf{a}_s|\boldsymbol{\alpha}_s = 1, \tag{9.126}$$

where $\mathbf{a}_s$ and $\boldsymbol{\alpha}_s$ are spatial vector and one-form, with either sign. From (9.124) it follows that the medium dyadics $\overline{\overline{\epsilon}}$ and $\overline{\overline{\mu}}$ must obey a relation of the form

$$\overline{\overline{\epsilon}} = c\overline{\overline{\mu}} + \boldsymbol{\Delta}_s\mathbf{a}_s, \tag{9.127}$$

where c is a scalar and $\boldsymbol{\Delta}_s$ is a spatial two-form. One can readily show that (9.119) with (9.127) inserted satisfies the bi-quadratic equation (9.118) with coefficients

$$A = -c - \frac{1}{2}\mathbf{a}_s|\overline{\overline{\mu}}^{-1}|\boldsymbol{\Delta}_s, \tag{9.128}$$

$$B = c^2 + c\mathbf{a}_s|\overline{\overline{\mu}}^{-1}|\boldsymbol{\Delta}_s, \tag{9.129}$$

whence the bidyadic (9.119), with spatial dyadics related by (9.127), defines a CU medium. One can verify that a medium defined by (9.119) cannot satisfy a second-order equation. Of course, a medium bidyadic $\overline{\overline{\mathsf{M}}}_1 = \overline{\overline{\mathsf{M}}}^2$ as defined by (9.120) satisfies the quadratic equation (9.6).

9.3.2 Eigenexpansions

The eigenvalues M_i of the BQ medium bidyadic $\overline{\overline{\mathsf{M}}}$ satisfying

$$M_i^4 - 2AM_i^2 + B = 0, \tag{9.130}$$

have the analytic form

$$M_i = \pm\sqrt{A \pm \sqrt{A^2 - B}}, \tag{9.131}$$

where the two double signs are independent and responsible for four eigenvalues. The biquadratic equation (9.118) can be written in factorized form as

$$(\overline{\overline{\mathsf{M}}} - M_a\overline{\overline{\mathrm{I}}}^{(2)T})|(\overline{\overline{\mathsf{M}}} + M_a\overline{\overline{\mathrm{I}}}^{(2)T})|(\overline{\overline{\mathsf{M}}} - M_b\overline{\overline{\mathrm{I}}}^{(2)T})|(\overline{\overline{\mathsf{M}}} + M_b\overline{\overline{\mathrm{I}}}^{(2)T}) = 0, \tag{9.132}$$

in which all of the bracketed bidyadics commute. The four eigenvalues denoted by $\pm M_a, \pm M_b$ are related to the coefficients of (9.118) as

$$2A = M_a^2 + M_b^2, \qquad B = M_a^2 M_b^2. \tag{9.133}$$

Since the general medium bidyadic has six eigenvalues, some of the four eigenvalues $\pm M_i$ must be multiple ones and the eigen two-forms $\mathbf{\Phi}_i$ corresponding to those eigenvalues must form subspaces in the six-dimensional space $\mathbb{F}_2$ of two-forms. Let us consider the case when the four eigenvalues are different, which also excludes the possibility of zero eigenvalues.

When the eigenvalues M_i have been solved from (9.130), the corresponding eigen two-forms can be found in terms of the following four bidyadics:

$$\overline{\overline{\Pi}}_{a\pm} = \pm\frac{(\overline{\overline{\mathsf{M}}}^2 - M_b^2\overline{\overline{\mathrm{I}}}^{(2)T})|(\overline{\overline{\mathsf{M}}} \pm M_a\overline{\overline{\mathrm{I}}}^{(2)T})}{2M_a(M_a^2 - M_b^2)}, \tag{9.134}$$

$$\overline{\overline{\Pi}}_{b\pm} = \pm\frac{(\overline{\overline{\mathsf{M}}}^2 - M_a^2\overline{\overline{\mathrm{I}}}^{(2)T})|(\overline{\overline{\mathsf{M}}} \pm M_b\overline{\overline{\mathrm{I}}}^{(2)T})}{2M_b(M_b^2 - M_a^2)}, \tag{9.135}$$

which commute with the medium bidyadic $\overline{\overline{\mathsf{M}}}$. Because of (9.132) the bidyadics satisfy

$$(\overline{\overline{\mathsf{M}}} \mp M_a \overline{\overline{\mathrm{I}}}^{(2)T})|\overline{\overline{\Pi}}_{a\pm} = 0 \quad \Rightarrow \quad \overline{\overline{\mathsf{M}}}|\overline{\overline{\Pi}}_{a\pm} = \pm M_a \overline{\overline{\Pi}}_{a\pm}, \tag{9.136}$$

$$(\overline{\overline{\mathsf{M}}} \mp M_b \overline{\overline{\mathrm{I}}}^{(2)T})|\overline{\overline{\Pi}}_{b\pm} = 0 \quad \Rightarrow \quad \overline{\overline{\mathsf{M}}}|\overline{\overline{\Pi}}_{b\pm} = \pm M_b \overline{\overline{\Pi}}_{b\pm}, \tag{9.137}$$

whence the eigen two-forms can be represented as

$$\mathbf{\Phi}_i = \overline{\overline{\Pi}}_i|\mathbf{\Phi} \tag{9.138}$$

in terms of any two-form $\mathbf{\Phi}$ producing $\mathbf{\Phi}_i \neq 0$.

The four bidyadics $\overline{\overline{\Pi}}_i$ satisfy

$$\overline{\overline{\Pi}}_{a+} + \overline{\overline{\Pi}}_{a-} = \frac{\overline{\overline{\mathsf{M}}}^2 - M_b^2 \overline{\overline{\mathrm{I}}}^{(2)T}}{M_a^2 - M_b^2}, \tag{9.139}$$

$$\overline{\overline{\Pi}}_{b+} + \overline{\overline{\Pi}}_{b-} = \frac{\overline{\overline{\mathsf{M}}}^2 - M_a^2 \overline{\overline{\mathrm{I}}}^{(2)T}}{M_b^2 - M_a^2}, \tag{9.140}$$

and

$$\overline{\overline{\Pi}}_{a+} + \overline{\overline{\Pi}}_{a-} + \overline{\overline{\Pi}}_{b+} + \overline{\overline{\Pi}}_{b-} = \overline{\overline{\mathrm{I}}}^{(2)T}. \tag{9.141}$$

From (9.141) and (9.138) any given two-form $\mathbf{\Phi}$ can be expanded in terms of the eigen two-forms $\mathbf{\Phi}_i$ as

$$\begin{aligned} \mathbf{\Phi} &= (\overline{\overline{\Pi}}_{a+} + \overline{\overline{\Pi}}_{a-} + \overline{\overline{\Pi}}_{b+} + \overline{\overline{\Pi}}_{b-})|\mathbf{\Phi} \\ &= \mathbf{\Phi}_{a+} + \mathbf{\Phi}_{a-} + \mathbf{\Phi}_{b+} + \mathbf{\Phi}_{b-}. \end{aligned} \tag{9.142}$$

Because the bidyadics $\overline{\overline{\Pi}}_i$ satisfy the conditions

$$\overline{\overline{\Pi}}_i|\overline{\overline{\Pi}}_j = 0, \quad i \neq j, \tag{9.143}$$

$$\overline{\overline{\Pi}}_i^2 = \overline{\overline{\Pi}}_i, \tag{9.144}$$

they are projection bidyadics. Applying the expansions

$$\overline{\overline{\mathsf{M}}} = M_a \overline{\overline{\Pi}}_{a+} - M_a \overline{\overline{\Pi}}_{a-} + M_b \overline{\overline{\Pi}}_{b+} - M_b \overline{\overline{\Pi}}_{b-}, \tag{9.145}$$

$$\overline{\overline{\mathsf{M}}}^m = M_a^m \overline{\overline{\Pi}}_{a+} + (-M_a)^m \overline{\overline{\Pi}}_{a-} + M_b^m \overline{\overline{\Pi}}_{b+} + (-M_b)^m \overline{\overline{\Pi}}_{b-}, \tag{9.146}$$

for $m = 2, 4$ and (9.141) on the left-hand side of (9.132), the coefficients of each bidyadic $\overline{\overline{\Pi}}_i$ can be shown to vanish identically.

The expansion (9.145) allows one to construct medium bidyadics satisfying the bi-quadratic equation (9.118) for any choice of two scalars M_a, M_b and four projection bidyadics $\overline{\overline{\Pi}}_i$ satisfying (9.143) and (9.141). Applying the similarity transformation (9.4), all possible medium dyadics $\overline{\overline{\mathsf{M}}}$ satisfying (9.118) for given A and B can be obtained from a given solution $\overline{\overline{\mathsf{M}}}_1$ by applying different bidyadics $\overline{\overline{\mathsf{D}}}$.

9.3.3 3D Representation

Substituting the spatiotemporal expansion (9.37) of the medium bidyadic $\overline{\overline{\mathsf{M}}}$ in

$$\overline{\overline{\mathsf{M}}}^2 = \overline{\overline{\mathsf{A}}}_s + \overline{\overline{\mathsf{B}}}_s \wedge \mathbf{e}_4 + \boldsymbol{\varepsilon}_4 \wedge \overline{\overline{\mathsf{C}}}_s + \boldsymbol{\varepsilon}_4 \wedge \overline{\overline{\mathsf{D}}}_s \wedge \mathbf{e}_4, \tag{9.147}$$

we can identify the four spatial dyadics as

$$\overline{\overline{\mathsf{A}}}_s = \overline{\overline{\alpha}}^2 - \overline{\overline{\epsilon}}'|\overline{\overline{\mu}}^{-1}, \tag{9.148}$$

$$\overline{\overline{\mathsf{B}}}_s = \overline{\overline{\alpha}}|\overline{\overline{\epsilon}}' - \overline{\overline{\epsilon}}'|\overline{\overline{\beta}}, \tag{9.149}$$

$$\overline{\overline{\mathsf{C}}}_s = \overline{\overline{\mu}}^{-1}|\overline{\overline{\alpha}} - \overline{\overline{\beta}}|\overline{\overline{\mu}}^{-1}, \tag{9.150}$$

$$\overline{\overline{\mathsf{D}}}_s = \overline{\overline{\mu}}^{-1}|\overline{\overline{\epsilon}}' - \overline{\overline{\beta}}^2, \tag{9.151}$$

Inserting (9.147) in (9.132), or the equivalent equation

$$(\overline{\overline{\mathsf{M}}}^2 - M_a^2\overline{\overline{\mathrm{I}}}^{(2)T})|(\overline{\overline{\mathsf{M}}}^2 - M_b^2\overline{\overline{\mathrm{I}}}^{(2)T}) = 0, \tag{9.152}$$

and separating spatial and temporal components, the BQ medium conditions take the form of four 3D equations as

$$(\overline{\overline{\mathsf{A}}}_s - M_a^2\overline{\overline{\mathrm{I}}}_s^{(2)T})|(\overline{\overline{\mathsf{A}}}_s - M_b^2\overline{\overline{\mathrm{I}}}_s^{(2)T}) = \overline{\overline{\mathsf{B}}}_s|\overline{\overline{\mathsf{C}}}_s, \tag{9.153}$$

$$(\overline{\overline{\mathsf{A}}}_s - M_a^2\overline{\overline{\mathrm{I}}}_s^{(2)T})|\overline{\overline{\mathsf{B}}}_s = \overline{\overline{\mathsf{B}}}_s|(\overline{\overline{\mathsf{D}}}_s + M_b^2\overline{\overline{\mathrm{I}}}_s^T), \tag{9.154}$$

$$\overline{\overline{\mathsf{C}}}_s|(\overline{\overline{\mathsf{A}}}_s - M_b^2\overline{\overline{\mathrm{I}}}_s^{(2)T}) = (\overline{\overline{\mathsf{D}}}_s + M_a^2\overline{\overline{\mathrm{I}}}_s^T)|\overline{\overline{\mathsf{C}}}_s, \tag{9.155}$$

$$(\overline{\overline{\mathsf{D}}}_s + M_a^2\overline{\overline{\mathrm{I}}}_s^T)|(\overline{\overline{\mathsf{D}}}_s + M_b^2\overline{\overline{\mathrm{I}}}_s^T) = \overline{\overline{\mathsf{C}}}_s|\overline{\overline{\mathsf{B}}}_s. \tag{9.156}$$

These equations are not independent in the general case. For example, if the inverse dyadic $\overline{\overline{\mathsf{B}}}_s^{-1}$ exists, one can show that (9.153) and (9.154) imply (9.155) and (9.156), whence the latter ones can be omitted.

At this instant it appears simpler to shift to the $\overline{\overline{\epsilon}}, \overline{\overline{\xi}}, \overline{\overline{\zeta}}, \overline{\overline{\mu}}$ representation (which presumes the existence of the dyadic $\overline{\overline{\mu}}$) by applying the relations

$$\overline{\overline{\epsilon}}' = \overline{\overline{\epsilon}} - \overline{\overline{\xi}}|\overline{\overline{\mu}}^{-1}|\overline{\overline{\zeta}}, \quad \overline{\overline{\alpha}} = \overline{\overline{\xi}}|\overline{\overline{\mu}}^{-1}, \quad \overline{\overline{\beta}} = -\overline{\overline{\mu}}^{-1}|\overline{\overline{\zeta}}, \tag{9.157}$$

in terms of which we can write

$$\overline{\overline{\mathsf{A}}}_s = \overline{\overline{\xi}}|\overline{\overline{\mu}}^{-1}|(\overline{\overline{\xi}} + \overline{\overline{\zeta}})|\overline{\overline{\mu}}^{-1} - \overline{\overline{\epsilon}}|\overline{\overline{\mu}}^{-1}, \tag{9.158}$$

$$\overline{\overline{\mathsf{B}}}_s = \overline{\overline{\xi}}|\overline{\overline{\mu}}^{-1}|\overline{\overline{\epsilon}} - \overline{\overline{\xi}}|\overline{\overline{\mu}}^{-1}|(\overline{\overline{\xi}} + \overline{\overline{\zeta}})|\overline{\overline{\mu}}^{-1}|\overline{\overline{\zeta}} + \overline{\overline{\epsilon}}|\overline{\overline{\mu}}^{-1}|\overline{\overline{\zeta}}, \tag{9.159}$$

$$\overline{\overline{\mathsf{C}}}_s = \overline{\overline{\mu}}^{-1}|(\overline{\overline{\xi}} + \overline{\overline{\zeta}})|\overline{\overline{\mu}}^{-1}, \tag{9.160}$$

$$\overline{\overline{\mathsf{D}}}_s = \overline{\overline{\mu}}^{-1}|\overline{\overline{\epsilon}} - \overline{\overline{\mu}}^{-1}|(\overline{\overline{\xi}} + \overline{\overline{\zeta}})|\overline{\overline{\mu}}^{-1}|\overline{\overline{\zeta}}. \tag{9.161}$$

When substituted in (9.153)–(9.156), the BQ medium conditions can be finally obtained for the medium dyadics $\overline{\overline{\epsilon}}, \overline{\overline{\xi}}, \overline{\overline{\zeta}}, \overline{\overline{\mu}}$, which appears quite an involved task. The result

$$\overline{\overline{\mu}}^{-1}|\overline{\overline{\epsilon}}|\overline{\overline{\mu}}^{-1}|(\overline{\overline{\xi}} + \overline{\overline{\zeta}})|\overline{\overline{\mu}}^{-1}|\overline{\overline{\epsilon}} = M_a^2 M_b^2 \overline{\overline{\mu}}^{-1}|(\overline{\overline{\xi}} + \overline{\overline{\zeta}}), \tag{9.162}$$

$$\overline{\overline{\mu}}^{-1}|\overline{\overline{\epsilon}}|(\overline{\overline{\mu}}^{-1}|(\overline{\overline{\xi}} + \overline{\overline{\zeta}}))^2 = (\overline{\overline{\mu}}^{-1}|\overline{\overline{\epsilon}} + M_a^2 \overline{\overline{\mathsf{I}}}_s^T)|(\overline{\overline{\mu}}^{-1}|\overline{\overline{\epsilon}} + M_b^2 \overline{\overline{\mathsf{I}}}_s^T) \tag{9.163}$$

has been derived in [44]. Assuming that $\overline{\overline{\mu}}$ and $\overline{\overline{\epsilon}}^{-1}$ exist, the conditions can be given a more symmetric form as

$$\overline{\overline{\mu}}^{-1}|(\overline{\overline{\xi}} + \overline{\overline{\zeta}})|\overline{\overline{\mu}}^{-1} = M_a^2 M_b^2 \overline{\overline{\epsilon}}^{-1}|(\overline{\overline{\xi}} + \overline{\overline{\zeta}})|\overline{\overline{\epsilon}}^{-1} \tag{9.164}$$

$$(\overline{\overline{\xi}} + \overline{\overline{\zeta}})|\overline{\overline{\mu}}^{-1}|(\overline{\overline{\xi}} + \overline{\overline{\zeta}}) = (\overline{\overline{\epsilon}} + M_a^2 \overline{\overline{\mu}})|\overline{\overline{\epsilon}}^{-1}|(\overline{\overline{\epsilon}} + M_b^2 \overline{\overline{\mu}}). \tag{9.165}$$

There is a double redundancy in the conditions (9.153)–(9.156). Actually, two of the dyadic conditions are sufficient to define a BQ medium, provided the dyadic $\overline{\overline{\mu}}^{-1}|\overline{\overline{\epsilon}}$ has an inverse which is assumed here. It is worth noting that, in the conditions (9.162) and (9.163), the four medium dyadics appear through just two dyadics, $\overline{\overline{\mu}}^{-1}|\overline{\overline{\epsilon}}$ and $\overline{\overline{\epsilon}}^{-1}|(\overline{\overline{\xi}} + \overline{\overline{\zeta}})$. This means that the dyadic $\overline{\overline{\xi}} - \overline{\overline{\zeta}}$ is not restricted by the BQ medium conditions.

9.3.4 Special Case

The problem is simplified by considering the special case of medium parameters satisfying the restriction

$$\overline{\overline{\xi}} + \overline{\overline{\zeta}} = 0. \tag{9.166}$$

Since $\overline{\overline{\xi}} - \overline{\overline{\zeta}}$ may be any dyadic, the BQ conditions should actually be independent of the dyadic $\overline{\overline{\xi}} = -\overline{\overline{\zeta}}$. Because (9.158)–(9.161) are now reduced to

$$\overline{\overline{\mathsf{A}}}_s = -\overline{\overline{\epsilon}}|\overline{\overline{\mu}}^{-1}, \tag{9.167}$$

$$\overline{\overline{\mathsf{B}}}_s = \overline{\overline{\xi}}|\overline{\overline{\mu}}^{-1}|\overline{\overline{\epsilon}} - \overline{\overline{\epsilon}}|\overline{\overline{\mu}}^{-1}|\overline{\overline{\xi}}, \tag{9.168}$$

$$\overline{\overline{\mathsf{C}}}_s = 0, \tag{9.169}$$

$$\overline{\overline{\mathsf{D}}}_s = \overline{\overline{\mu}}^{-1}|\overline{\overline{\epsilon}}, \tag{9.170}$$

from (9.147) we can expand

$$\overline{\overline{\mathsf{M}}}^2 = -(\overline{\overline{\epsilon}}|\overline{\overline{\mu}}^{-1})|(\overline{\overline{\mathbf{I}}}_s^{(2)T} + \overline{\overline{\xi}} \wedge \mathbf{e}_4) + (\overline{\overline{\xi}} + \boldsymbol{\varepsilon}_4 \wedge \overline{\overline{\mathbf{I}}}_s^T)|(\overline{\overline{\mu}}^{-1}|\overline{\overline{\epsilon}}) \wedge \mathbf{e}_4, \quad (9.171)$$

$$\overline{\overline{\mathsf{M}}}^4 = (\overline{\overline{\epsilon}}|\overline{\overline{\mu}}^{-1})^2|(\overline{\overline{\mathbf{I}}}_s^{(2)T} + \overline{\overline{\xi}} \wedge \mathbf{e}_4) - (\overline{\overline{\xi}} + \boldsymbol{\varepsilon}_4 \wedge \overline{\overline{\mathbf{I}}}_s^T)|(\overline{\overline{\mu}}^{-1}|\overline{\overline{\epsilon}})^2 \wedge \mathbf{e}_4. \quad (9.172)$$

Inserting these expressions in (9.118) and operating by $\mathbf{e}_4\boldsymbol{\varepsilon}_4\lfloor\lfloor$ the equation is reduced to

$$(\overline{\overline{\mu}}^{-1}|\overline{\overline{\epsilon}})^2 + 2A(\overline{\overline{\mu}}^{-1}|\overline{\overline{\epsilon}}) + B\overline{\overline{\mathbf{I}}}_s^T = 0. \quad (9.173)$$

Similarly, multiplying both sides of (9.118), with (9.171) and (9.172) inserted, by $\overline{\overline{\mathbf{I}}}_s^{(2)T}|$, yields

$$(\overline{\overline{\epsilon}}|\overline{\overline{\mu}}^{-1})^2 + 2A(\overline{\overline{\epsilon}}|\overline{\overline{\mu}}^{-1}) + B\overline{\overline{\mathbf{I}}}_s^{(2)T} = 0. \quad (9.174)$$

Actually, (9.173) and (9.174) represent the same equation when either $\overline{\overline{\mu}}$ or $\overline{\overline{\epsilon}}^{-1}$ exists. Also, (9.173) coincides with (9.163) for this special case. Finally, inserting (9.171) and (9.172) in the original biquadratic equation (9.118), with (9.174) taken into account, can be shown to lead to the identity $0 = 0$. This means that, for $\overline{\overline{\xi}} + \overline{\overline{\zeta}} = 0$, there are no other restrictions to the dyadics $\overline{\overline{\xi}}$ and $\overline{\overline{\zeta}}$ by a BQ medium. Comparing (9.173) with (9.123) and (9.174) with (9.122) we conclude that the special BQ medium bidyadic expressed by (9.119) can actually be generalized by introducing the dyadic $\overline{\overline{\xi}}$ as

$$\overline{\overline{\mathsf{M}}} = \overline{\overline{\epsilon}} \wedge \mathbf{e}_4 + (\overline{\overline{\xi}} + \boldsymbol{\varepsilon}_4 \wedge \overline{\overline{\mathbf{I}}}_s^T)|\overline{\overline{\mu}}^{-1}|(\overline{\overline{\mathbf{I}}}_s^{(2)T} + \overline{\overline{\xi}} \wedge \mathbf{e}_4), \quad (9.175)$$

and $\overline{\overline{\epsilon}}$ and $\overline{\overline{\mu}}$ are still related by (9.127).

PROBLEMS

9.1 Show that the extended Q medium defined by a medium bidyadic of the form

$$\overline{\overline{\mathsf{M}}} = M\overline{\overline{\mathbf{I}}}^{(2)T} + \boldsymbol{\varepsilon}_N\lfloor\mathbf{AB},$$

where **A** and **B** are two bivectors, is an SD medium.

9.2 Do the details in proving that the SD-medium bidyadic $\overline{\overline{\mathsf{M}}}$ (9.37) can be transformed to the form (9.45) when it satisfies (9.36).

9.3 Show that, by assuming that the spatial dyadic $\overline{\overline{\epsilon}}'$ has an inverse, the SD-medium bidyadic (9.37) satisfying (9.36) can be transformed to the form (9.46).

9.4 Check that the representation (9.32) with (9.33) corresponds to (9.31) for $Z_d = -Z$ and find the relations of the transformation parameters Z, θ to those of the SD medium, (9.22).

9.5 Starting from the assumption $\overline{\overline{\epsilon}}' = 0$ and $\overline{\overline{\mu}}^{-1} = 0$, find solutions to SD medium bidyadics $\overline{\overline{\mathsf{M}}}$ satisfying (9.36).

9.6 Show that the similarity transformation rule implies the rule

$$\overline{\overline{\mathsf{M}}}' = \overline{\overline{\mathsf{D}}}|\overline{\overline{\mathsf{M}}}|\overline{\overline{\mathsf{D}}}^{-1} \quad \Rightarrow \quad \overline{\overline{\mathrm{I}}}^{(4)}\lfloor\lfloor\overline{\overline{\mathsf{M}}}' = (\overline{\overline{\mathrm{I}}}^{(4)}\lfloor\lfloor\overline{\overline{\mathsf{D}}})|(\overline{\overline{\mathrm{I}}}^{(4)}\lfloor\lfloor\overline{\overline{\mathsf{M}}})|(\overline{\overline{\mathrm{I}}}^{(4)}\lfloor\lfloor\overline{\overline{\mathsf{D}}})^{-1}.$$

9.7 Show that a similarity transformation (9.4) with the bidyadic $\overline{\overline{\mathsf{D}}} = (\overline{\overline{\mathsf{M}}} + \overline{\overline{\mathsf{M}}}_1)/M$ can be applied to map the SD-medium bidyadic $\overline{\overline{\mathsf{M}}}_1$ of (9.46) to the medium bidyadic $\overline{\overline{\mathsf{M}}}$ corresponding to time reversion,

$$\overline{\overline{\mathsf{M}}} = jM\overline{\overline{\mathrm{T}}}^{(2)T} = jM(\overline{\overline{\mathrm{I}}}_s^{(2)T} - \overline{\overline{\mathrm{I}}}_s^T{}^\wedge_\wedge\varepsilon_4\mathbf{e}_4),$$

and satisfying the condition (9.36) of an SD medium.

9.8 Study the possibility of defining a CU medium with medium bidyadic of the form

$$\overline{\overline{\mathsf{M}}} = \overline{\overline{\epsilon}} \wedge \mathbf{e}_4 + \varepsilon_4 \wedge \overline{\overline{\mu}}^{-1},$$

where the spatial dyadics $\overline{\overline{\epsilon}}, \overline{\overline{\mu}}^{-1}$ are assumed to be of full rank.

9.9 Find the eigenvalues and eigen two-forms corresponding to the CU-medium bidyadic

$$\overline{\overline{\mathsf{M}}} = A\overline{\overline{\mathrm{I}}}^{(2)T} + \varepsilon_N\lfloor(\mathbf{AB} - \mathbf{BA}),$$

where $\mathbf{A}$ and $\mathbf{B}$ are assumed to be linearly independent bivectors.

9.10 Show that the skewon–axion medium bidyadic defined by

$$\overline{\overline{\mathsf{M}}} = A\overline{\overline{\mathrm{I}}}^{(2)T} + \overline{\overline{\mathsf{C}}}_o, \quad \overline{\overline{\mathsf{C}}}_o = \varepsilon_N\lfloor(\mathbf{AB} - \mathbf{BA})$$

satisfies a cubic equation.

9.11 Verify the property (9.94).

9.12 Derive the cubic equation (9.112).

9.13 Verify that the medium defined by a medium bidyadic of the form (9.119) with the relation (9.127) satisfies the bi-quadratic equation (9.118) when the parameters are chosen as (9.128) and (9.129).

9.14 Show that the eigenfields $\mathbf{\Phi}_i$ satisfy the orthogonality condition $\mathbf{\Phi}_i \cdot \mathbf{\Phi}_j = 0$ for $i \neq j$ in the case of a principal BQ medium.

9.15 Starting from the representation (9.148)–(9.151) find the 3D conditions for the BQ medium corresponding to (9.162) and (9.163).

9.16 Derive the expressions (9.171) and (9.172) and show that when they are inserted in (9.118), the biquadratic equation does not give any restriction to the dyadic $\overline{\overline{\xi}}$ when $\overline{\overline{\xi}} + \overline{\overline{\zeta}} = 0$ is valid.

9.17 Find a bidyadic $\overline{\overline{\mathsf{D}}} \in \mathbb{F}_2\mathbb{E}_2$ for a similarity transformation mapping a medium satisfying the BQ medium equation (9.118) to a medium with $\overline{\overline{\xi}} + \overline{\overline{\zeta}} = 0$. Hint: start from $\overline{\overline{\mathsf{D}}} = \overline{\overline{\mathsf{I}}}^{(2)T} - \overline{\overline{\mathsf{F}}}_s \wedge \mathbf{e}_4$, $\overline{\overline{\mathsf{D}}}^{-1} = \overline{\overline{\mathsf{I}}}^{(2)T} + \overline{\overline{\mathsf{F}}}_s \wedge \mathbf{e}_4$, where $\overline{\overline{\mathsf{F}}}_s \in \mathbb{F}_2\mathbb{E}_1$ is a spatial dyadic.

9.18 Find the dispersion equation for a plane wave in a CU medium defined by the medium bidyadic (9.114).

CHAPTER 10

Media Defined by Plane-Wave Properties

Classes of media can also be defined by requiring that plane waves in such media are restricted by certain conditions. Here we only consider two examples: first, media in which the wave vector **k** or the corresponding wave one-form $\boldsymbol{\nu}$ of a plane wave can be freely chosen without being restricted by a dispersion equation and, second, media in which plane waves are required to be decomposable in two sets in terms of their polarization properties. The media thus defined are respectively called "media with no dispersion equation" (NDE media) and "decomposable media" (DC media). Various subclasses of NDE media and DC media are presented in this chapter.

10.1 MEDIA WITH NO DISPERSION EQUATION (NDE MEDIA)

In the previous chapters, several cases of media have emerged with the property that a plane wave is not restricted by a dispersion equation, either in dyadic form $\overline{\overline{\mathsf{D}}}^{(3)}(\boldsymbol{\nu}) = 0$ or in scalar form $D(\boldsymbol{\nu}) = 0$, because the equations are valid for any wave one-form $\boldsymbol{\nu}$. For example, in Section 7.3 it was shown that a plane wave in a skewon medium is not restricted by a dispersion equation. The same was observed in Section 8.1 for Q media when the dyadic $\overline{\overline{\mathsf{Q}}}$ is not of full rank and in Section 8.3 for P media. Since an axion term can be added to the medium bidyadic

Multiforms, Dyadics, and Electromagnetic Media, First Edition. Ismo V. Lindell.

without changing propagation properties of a plane-wave, the previous media can be extended to skewon–axion media, Q-axion media with low-rank dyadic $\overline{\overline{\mathsf{Q}}}$, and P-axion media. Thus, these serve as examples of NDE media.

Here one must note that being an NDE medium does not mean that the medium is not dispersive, quite the opposite. Considering a time-harmonic plane wave propagating in the direction of a unit vector $\mathbf{u}$ with $\mathbf{k} = \mathbf{u}k$ means that the magnitude of k is not limited by a dispersion equation and can be chosen. For a nondispersive NDE medium the group velocity equals the phase velocity ω/k and would exceed the velocity of light for $k < \omega/c$. To avoid this, the medium must be dispersive with function $k = k(\omega)$ satisfying $\partial\omega/\partial k < c$.

10.1.1 Two Cases of Solutions

Because the dyadic dispersion equation requires vanishing of $\overline{\overline{\mathsf{D}}}^{(3)}(\boldsymbol{\nu})$, the class of NDE media is defined by requiring that the dispersion dyadic be of rank less than three for all possible one-forms $\boldsymbol{\nu}$. From this it follows that there must exist four vectors in terms of which we can write

$$\overline{\overline{\mathsf{D}}}(\boldsymbol{\nu}) = \overline{\overline{\mathsf{M}}}_m \lfloor\lfloor \boldsymbol{\nu}\boldsymbol{\nu} = \mathbf{ac} + \mathbf{bd}. \tag{10.1}$$

Different cases can be considered by recognizing that since the dispersion dyadic is a quadratic function of $\boldsymbol{\nu}$, so must be the right side of (10.1).

Let us first assume that the vectors $\mathbf{a} - \mathbf{d}$ may be linear or quadratic functions of $\boldsymbol{\nu}$, or independent of $\boldsymbol{\nu}$ [48]. This gives us a few possibilities.

1. Each of the four vectors is a linear function of $\boldsymbol{\nu}$
2. $\mathbf{a}$ and $\mathbf{b}$ are quadratic functions while $\mathbf{c}$ and $\mathbf{d}$ do not depend on $\boldsymbol{\nu}$
3. $\mathbf{a}$ and $\mathbf{d}$ are quadratic functions while $\mathbf{b}$ and $\mathbf{c}$ do not depend on $\boldsymbol{\nu}$
4. $\mathbf{a}$ is a quadratic function while $\mathbf{b}$ and $\mathbf{d}$ are linear functions and $\mathbf{c}$ does not depend on $\boldsymbol{\nu}$

Other similar possibilities do not seem to bring any new solutions because the dispersion equation (5.245) is invariant to replacing the modified medium bidyadic $\overline{\overline{\mathsf{M}}}_m$ by its transpose $\overline{\overline{\mathsf{M}}}_m^T$ which implies replacing the dispersion dyadic $\overline{\overline{\mathsf{D}}}(\boldsymbol{\nu})$ by its transpose $\overline{\overline{\mathsf{D}}}^T(\boldsymbol{\nu})$ and $\overline{\overline{\mathsf{D}}}^{(3)}(\boldsymbol{\nu})$

by $\overline{\overline{\mathsf{D}}}^{(3)T}(\boldsymbol{\nu})$. Let us consider here only Cases 1 and 2 because 3 and 4 can be reduced to them. Verifying this is left as an exercise. However, these cases do not contain all solutions. For example, some of the vectors $\mathbf{a}-\mathbf{d}$ may be cubic or higher-order functions of $\boldsymbol{\nu}$ divided by linear or higher-order scalar functions of $\boldsymbol{\nu}$ as long as the right side of (10.1) is a quadratic function. These other possibilities will be only covered by an example at the end of this chapter.

Case 1

Because the dispersion dyadic satisfies $\overline{\overline{\mathsf{D}}}(\boldsymbol{\nu})|\boldsymbol{\nu} = \boldsymbol{\nu}|\overline{\overline{\mathsf{D}}}(\boldsymbol{\nu}) = 0$ for all $\boldsymbol{\nu}$, the four vectors must satisfy

$$\mathbf{a}(\mathbf{c}|\boldsymbol{\nu}) + \mathbf{b}(\mathbf{d}|\boldsymbol{\nu}) = 0, \quad (\boldsymbol{\nu}|\mathbf{a})\mathbf{c} + (\boldsymbol{\nu}|\mathbf{b})\mathbf{d} = 0. \tag{10.2}$$

Assuming

$$\overline{\overline{\mathsf{D}}}^{(2)} = (\mathbf{a}\wedge\mathbf{b})(\mathbf{c}\wedge\mathbf{d}) \neq 0, \tag{10.3}$$

for all $\boldsymbol{\nu} \neq 0$, that is, that the dispersion dyadic be of rank 2, the vectors $\mathbf{a}, \mathbf{b}$ must be linearly independent and so must $\mathbf{c}, \mathbf{d}$, whence we obtain

$$\mathbf{a}|\boldsymbol{\nu} = \mathbf{b}|\boldsymbol{\nu} = \mathbf{c}|\boldsymbol{\nu} = \mathbf{d}|\boldsymbol{\nu} = 0, \tag{10.4}$$

for all $\boldsymbol{\nu}$. This requires that there must exist bivectors $\mathbf{A}, \mathbf{B}, \mathbf{C}, \mathbf{D}$ such that we can express

$$\mathbf{a} = \mathbf{A}\lfloor\boldsymbol{\nu}, \quad \mathbf{b} = \mathbf{B}\lfloor\boldsymbol{\nu}, \quad \mathbf{c} = \mathbf{C}\lfloor\boldsymbol{\nu}, \quad \mathbf{d} = \mathbf{D}\lfloor\boldsymbol{\nu}, \tag{10.5}$$

and

$$\overline{\overline{\mathsf{D}}}(\boldsymbol{\nu}) = \overline{\overline{\mathsf{M}}}_m\lfloor\lfloor\boldsymbol{\nu}\boldsymbol{\nu} = (\mathbf{AC}+\mathbf{BD})\lfloor\lfloor\boldsymbol{\nu}\boldsymbol{\nu}. \tag{10.6}$$

Because

$$(\overline{\overline{\mathsf{M}}}_m - (\mathbf{AC}+\mathbf{BD}))\lfloor\lfloor\boldsymbol{\nu}\boldsymbol{\nu} = 0 \tag{10.7}$$

must be valid for all $\boldsymbol{\nu}$, applying the result of Problem 3.20, the modified medium bidyadic $\overline{\overline{\mathsf{M}}}_m$ must be of the form

$$\overline{\overline{\mathsf{M}}}_m = \mathbf{AC}+\mathbf{BD}+\alpha\mathbf{e}_N\lfloor\overline{\overline{\mathsf{I}}}^{(2)T}, \tag{10.8}$$

for some scalar α. Equivalently, the medium bidyadic $\overline{\overline{\mathsf{M}}}$ must be of the form

$$\overline{\overline{\mathsf{M}}} = \boldsymbol{\Pi}\mathbf{C}+\boldsymbol{\Lambda}\mathbf{D}+\alpha\overline{\overline{\mathsf{I}}}^{(2)T}, \tag{10.9}$$

with two-forms $\mathbf{\Pi} = \varepsilon_N \lfloor \mathbf{A}$ and $\mathbf{\Lambda} = \varepsilon_N \lfloor \mathbf{B}$. The expression (10.9) defines in terms of two bivectors, two two-forms and a scalar one possible class of media, for which the dispersion equation is satisfied identically. When the medium bidyadic $\overline{\overline{\mathsf{M}}}$ has an inverse, the dispersion equation can also be defined through the bidyadic $\overline{\overline{\mathsf{N}}} = \overline{\overline{\mathsf{M}}}^{-1}$ of (5.52). One can, however, show that $\overline{\overline{\mathsf{N}}} = \overline{\overline{\mathsf{M}}}^{-1}$ has a form similar to that of (10.9).

In the special case of $\overline{\overline{\mathsf{D}}}^{(2)}(\boldsymbol{\nu}) = 0$ for all $\boldsymbol{\nu}$, (10.8) is reduced to the form

$$\overline{\overline{\mathsf{M}}}_m = \mathbf{AC} + \alpha \mathbf{e}_N \lfloor \overline{\overline{\mathsf{I}}}^{(2)T}, \tag{10.10}$$

while requiring $\overline{\overline{\mathsf{D}}}(\boldsymbol{\nu}) = 0$ for all $\boldsymbol{\nu}$ leads to $\overline{\overline{\mathsf{M}}} = M\overline{\overline{\mathsf{I}}}^{(2)T}$, that is, the axion medium. One may note that the medium defined by (10.10) includes Q-axion media defined by

$$\overline{\overline{\mathsf{M}}}_m = M\overline{\overline{\mathsf{Q}}}^{(2)} + \beta \mathbf{e}_N \lfloor \overline{\overline{\mathsf{I}}}^{(2)T}, \tag{10.11}$$

in the special case when the dyadic $\overline{\overline{\mathsf{Q}}} \in \mathbb{E}_1\mathbb{E}_1$ is antisymmetric. In fact, because an antisymmetric dyadic can be expressed in terms of some bivector $\mathbf{A}$ as $\overline{\overline{\mathsf{Q}}} = \mathbf{A}\lfloor \overline{\overline{\mathsf{I}}}^T$, expanding

$$\begin{aligned}\overline{\overline{\mathsf{M}}}_m &= M(\mathbf{A}\lfloor \overline{\overline{\mathsf{I}}}^T)^{(2)} + \beta \mathbf{e}_N \lfloor \overline{\overline{\mathsf{I}}}^{(2)T} \\ &= M\left(\mathbf{AA} - \frac{1}{2}\varepsilon_N \rfloor (\mathbf{A} \wedge \mathbf{A})\overline{\overline{\mathsf{I}}}^{(2)T}\right) + \beta \mathbf{e}_N \lfloor \overline{\overline{\mathsf{I}}}^{(2)T},\end{aligned} \tag{10.12}$$

we arrive at a special case of (10.10).

It has been here tacitly assumed that none of the four bivectors in (10.8) is simple because, otherwise, the condition (10.3) cannot be valid for all $\boldsymbol{\nu}$. Actually, in the converse case it turns out that $\overline{\overline{\mathsf{D}}}$ has variable rank depending on $\boldsymbol{\nu}$. Details of this are left as an exercise. Since media defined by (10.8) have no dispersion equation regardless of one or more of the bivectors being simple, dispersion dyadics $\overline{\overline{\mathsf{D}}}(\boldsymbol{\nu})$ of variable rank are included in addition to those of rank 2 or rank 1 for all $\boldsymbol{\nu}$.

Case 2

As a second case let us now assume that in (10.1) the vectors $\mathbf{c}$ and $\mathbf{d}$ are independent of $\boldsymbol{\nu}$, whence $\mathbf{a}$ and $\mathbf{b}$ must be quadratic functions of $\boldsymbol{\nu}$. In this case it appears preferable to start from the representation

$$\boldsymbol{\nu} \wedge \overline{\overline{\mathsf{M}}}\lfloor \boldsymbol{\nu} = -\varepsilon_N \lfloor \overline{\overline{\mathsf{D}}}(\boldsymbol{\nu}) = \mathbf{\Gamma c} + \mathbf{\Sigma d}, \tag{10.13}$$

where the two three-forms

$$\mathbf{\Gamma} = -\boldsymbol{\varepsilon}_N \lfloor \mathbf{a}, \quad \mathbf{\Sigma} = -\boldsymbol{\varepsilon}_N \lfloor \mathbf{b}, \quad \in \mathbb{F}_3 \tag{10.14}$$

are now quadratic functions of $\boldsymbol{v}$. Since any two linearly independent three-forms can be interpreted as belonging to a basis of three-forms, which can be constructed in terms of a basis of one-forms, say $\boldsymbol{\alpha}, \boldsymbol{\beta}, \boldsymbol{\gamma}, \boldsymbol{\delta}$, we can express

$$\mathbf{\Gamma} = \boldsymbol{\alpha} \wedge \boldsymbol{\beta} \wedge \boldsymbol{\gamma}, \quad \mathbf{\Sigma} = \boldsymbol{\alpha} \wedge \boldsymbol{\beta} \wedge \boldsymbol{\delta}. \tag{10.15}$$

Thus, (10.13) can be written as

$$\boldsymbol{v} \wedge \overline{\overline{\mathsf{M}}} \lfloor \boldsymbol{v} = \boldsymbol{\alpha} \wedge \boldsymbol{\beta} \wedge (\boldsymbol{\gamma}\mathbf{c} + \boldsymbol{\delta}\mathbf{d}), \tag{10.16}$$

or, more generally, as

$$\boldsymbol{v} \wedge \overline{\overline{\mathsf{M}}} \lfloor \boldsymbol{v} = \boldsymbol{\alpha} \wedge \boldsymbol{\beta} \wedge \overline{\overline{\mathsf{P}}}^T. \tag{10.17}$$

Since terms of the form $\boldsymbol{\alpha}\mathbf{a}' + \boldsymbol{\beta}\mathbf{b}'$ can be added to the dyadic in brackets in (10.16), we may well assume that the dyadic $\overline{\overline{\mathsf{P}}} \in \mathbb{E}_1\mathbb{F}_1$ in (10.17) is of full rank. The special case corresponding to $\overline{\overline{\mathsf{D}}}^{(2)}(\boldsymbol{v}) = 0$, that is, when the bracketed dyadic is reduced by assuming $\boldsymbol{\delta}\mathbf{d} = 0$, will be considered separately. It is left as an exercise to verify that when the medium bidyadic satisfies (10.17) for any $\boldsymbol{v}$, the dispersion equation is satisfied identically.

Considering (10.17), two conditions for the quantities on its right-hand side are obtained by noting that the left-hand side vanishes when multiplying by $|\boldsymbol{v}$ from the right or by $\boldsymbol{v}\wedge$ from the left. These lead to the respective conditions

$$\boldsymbol{\alpha} \wedge \boldsymbol{\beta} \wedge (\overline{\overline{\mathsf{P}}}^T|\boldsymbol{v}) = 0, \tag{10.18}$$

$$\boldsymbol{v} \wedge \boldsymbol{\alpha} \wedge \boldsymbol{\beta} \wedge \overline{\overline{\mathsf{P}}}^T = 0. \tag{10.19}$$

Equation (10.18) requires that the three one-forms $\boldsymbol{\alpha}, \boldsymbol{\beta}$, and $\overline{\overline{\mathsf{P}}}^T|\boldsymbol{v}$ be linearly dependent, whence they must satisfy a relation of the form

$$A\boldsymbol{\alpha} + B\boldsymbol{\beta} + C(\overline{\overline{\mathsf{P}}}^T|\boldsymbol{v}) = 0. \tag{10.20}$$

Considering different possibilities for the coefficients A, B, and C, one can show that (10.17) can be reduced to the form

$$\boldsymbol{v} \wedge \overline{\overline{\mathsf{M}}} \lfloor \boldsymbol{v} = \boldsymbol{\alpha} \wedge (\overline{\overline{\mathsf{P}}}^T|\boldsymbol{v}) \wedge \overline{\overline{\mathsf{P}}}^T, \tag{10.21}$$

details of which are left as an exercise. Since the right-hand side of (10.21) must be a quadratic function of $\boldsymbol{\nu}$, it appears that $\overline{\overline{\mathsf{P}}}$ must be independent of $\boldsymbol{\nu}$. Thus, $\boldsymbol{\alpha}$ must be a linear function of $\boldsymbol{\nu}$ whence it can be expressed as

$$\boldsymbol{\alpha} = \overline{\overline{\mathsf{B}}}^T|\boldsymbol{\nu} \tag{10.22}$$

in terms of some dyadic $\overline{\overline{\mathsf{B}}} \in \mathbb{E}_1\mathbb{F}_1$. Inserting this in (10.21) yields the representation

$$\boldsymbol{\nu} \wedge \overline{\overline{\mathsf{M}}}\lfloor\boldsymbol{\nu} = (\overline{\overline{\mathsf{B}}}^T|\boldsymbol{\nu}) \wedge (\overline{\overline{\mathsf{P}}}^T|\boldsymbol{\nu}) \wedge \overline{\overline{\mathsf{P}}}^T, \tag{10.23}$$

which leads to

$$\boldsymbol{\nu} \wedge (\overline{\overline{\mathsf{B}}}^T|\boldsymbol{\nu}) \wedge (\overline{\overline{\mathsf{P}}}^T|\boldsymbol{\nu}) \wedge \overline{\overline{\mathsf{P}}}^T = 0, \tag{10.24}$$

and it must be valid for any one-form $\boldsymbol{\nu}$. Recalling that $\overline{\overline{\mathsf{P}}}$ was assumed to be a dyadic of full rank, we can multiply (10.24) by $|\overline{\overline{\mathsf{P}}}^{-1T}$ from the right. The condition is then reduced to

$$\boldsymbol{\nu} \wedge (\overline{\overline{\mathsf{B}}}^T|\boldsymbol{\nu}) \wedge (\overline{\overline{\mathsf{P}}}^T|\boldsymbol{\nu}) \wedge \overline{\overline{\mathsf{I}}}^T = 0 \quad \Rightarrow \quad \boldsymbol{\nu} \wedge (\overline{\overline{\mathsf{B}}}^T|\boldsymbol{\nu}) \wedge (\overline{\overline{\mathsf{P}}}^T|\boldsymbol{\nu}) = 0. \tag{10.25}$$

Since the three one-forms are linearly dependent for all $\boldsymbol{\nu}$, the corresponding dyadics must be related as

$$A\overline{\overline{\mathsf{I}}} + B\overline{\overline{\mathsf{B}}} + C\overline{\overline{\mathsf{P}}} = 0, \tag{10.26}$$

for some scalars A, B, C. This leaves us the following three possibilities:

1. $A = 0$ which implies either $\overline{\overline{\mathsf{P}}} = 0$ or $\overline{\overline{\mathsf{B}}}$ is a multiple of $\overline{\overline{\mathsf{P}}}$. In both cases from (10.23) we have $\boldsymbol{\nu} \wedge \overline{\overline{\mathsf{M}}}\lfloor\boldsymbol{\nu} = 0$ for all $\boldsymbol{\nu}$, whence

$$\overline{\overline{\mathsf{M}}} = \alpha\overline{\overline{\mathsf{I}}}^{(2)T}. \tag{10.27}$$

2. $A \neq 0$ and $B = 0$ implies $\overline{\overline{\mathsf{P}}} = -(A/C)\overline{\overline{\mathsf{I}}}$. In this case we have from (10.23)

$$\boldsymbol{\nu} \wedge \overline{\overline{\mathsf{M}}}\lfloor\boldsymbol{\nu} = (A/C)^2(\overline{\overline{\mathsf{B}}}^T|\boldsymbol{\nu}) \wedge \boldsymbol{\nu} \wedge \overline{\overline{\mathsf{I}}}^T = -(A/C)^2\boldsymbol{\nu} \wedge (\overline{\overline{\mathsf{B}}}{}^\wedge_\wedge\overline{\overline{\mathsf{I}}})^T\lfloor\boldsymbol{\nu}, \tag{10.28}$$

which further implies

$$\overline{\overline{\mathsf{M}}} = (\overline{\overline{\mathsf{B}}}{}^\wedge_\wedge\overline{\overline{\mathsf{I}}})^T + \alpha\overline{\overline{\mathsf{I}}}^{(2)T}. \tag{10.29}$$

3. $A \neq 0$ and $B \neq 0$ implies $\overline{\overline{\mathsf{B}}} = -(A/B)\overline{\overline{\mathsf{I}}} - (C/B)\overline{\overline{\mathsf{P}}}$. In this case (10.23) yields

$$\nu \wedge \overline{\overline{\mathsf{M}}}\lfloor \nu = -(A/B)\nu \wedge (\overline{\overline{\mathsf{P}}}^T|\nu) \wedge \overline{\overline{\mathsf{P}}}^T = -(C/B)\nu \wedge \overline{\overline{\mathsf{P}}}^{(2)T}\lfloor \nu, \tag{10.30}$$

whence

$$\overline{\overline{\mathsf{M}}} = \overline{\overline{\mathsf{P}}}^{(2)T} + \alpha \overline{\overline{\mathsf{I}}}^{(2)T}. \tag{10.31}$$

In (10.29) and (10.31) the coefficients A, B, C have been absorbed in the dyadics $\overline{\overline{\mathsf{B}}}$ or $\overline{\overline{\mathsf{P}}}$. Thus, in Case 2 the media with no dispersion equation fall into two classes, those of skewon–axion media and P-axion media while axion media are special cases of both of the two.

The previous Case 2 analysis was based on the assumption $\overline{\overline{\mathsf{D}}}^{(2)}(\nu) \neq 0$ for all ν. The converse case can be handled by replacing (10.1) by

$$\overline{\overline{\mathsf{D}}}(\nu) = \mathbf{ac}, \tag{10.32}$$

where $\mathbf{a}$ is a quadratic function of ν and $\mathbf{c}$ is independent of ν. From

$$\overline{\overline{\mathsf{D}}}(\nu)|\nu = \mathbf{a}(\mathbf{c}|\nu) = 0, \tag{10.33}$$

satisfied for all ν, we must have either $\mathbf{a} = 0$ or $\mathbf{c} = 0$, both of which lead to $\overline{\overline{\mathsf{D}}}(\nu) = 0$ which corresponds to the axion medium case.

One can show that both Case 1 and Case 2 classes of NDE media are closed in affine transformations, details of which are left as an exercise.

10.1.2 Plane-Wave Fields in NDE Media

Let us consider the polarization of plane-wave fields in NDE media defined above under Case 1 and Case 2 for a given one-form ν.

Case 1 Media

Assuming that the dispersion dyadic $\overline{\overline{\mathsf{D}}}(\nu)$ corresponding to the Case 1 medium (10.8) is of rank 2 for any $\nu \neq 0$, the vectors $\mathbf{a} = \mathbf{A}\lfloor \nu$ and $\mathbf{b} = \mathbf{B}\lfloor \nu$ must be linearly independent, in which case the potential must satisfy the two conditions

$$\mathbf{c}|\boldsymbol{\phi} = (\mathbf{C}\lfloor \nu)|\boldsymbol{\phi} = \mathbf{C}|(\nu \wedge \boldsymbol{\phi}) = 0, \tag{10.34}$$

$$\mathbf{d}|\boldsymbol{\phi} = (\mathbf{D}\lfloor \nu)|\boldsymbol{\phi} = \mathbf{D}|(\nu \wedge \boldsymbol{\phi}) = 0, \tag{10.35}$$

which for the field two-form become

$$\mathbf{C}|\boldsymbol{\Phi} = 0, \quad \mathbf{D}|\boldsymbol{\Phi} = 0. \tag{10.36}$$

Because the potential one-form can be expressed as

$$\boldsymbol{\phi} = \boldsymbol{\varepsilon}_N \lfloor (\mathbf{p} \wedge \mathbf{c} \wedge \mathbf{d}), \tag{10.37}$$

for any vector $\mathbf{p}$ yielding a nonzero result, this corresponds to choosing an additional condition

$$\mathbf{p}|\boldsymbol{\phi} = 0 \tag{10.38}$$

to make $\boldsymbol{\phi}$ unique. Thus, the field two-form has the form

$$\boldsymbol{\Phi} = \boldsymbol{\nu} \wedge \boldsymbol{\phi} = \boldsymbol{\nu} \wedge \boldsymbol{\varepsilon}_N \lfloor (\mathbf{p} \wedge \mathbf{c} \wedge \mathbf{d}). \tag{10.39}$$

Both equations (10.36) follow from (10.39), which is seen from

$$\mathbf{C}|(\boldsymbol{\nu} \wedge \boldsymbol{\varepsilon}_N \lfloor (\mathbf{p} \wedge \mathbf{c} \wedge \mathbf{d})) = \mathbf{c}|(\boldsymbol{\varepsilon}_N \lfloor (\mathbf{p} \wedge \mathbf{c} \wedge \mathbf{d})) = \boldsymbol{\varepsilon}_N|(\mathbf{p} \wedge \mathbf{c} \wedge \mathbf{d} \wedge \mathbf{c}) = 0. \tag{10.40}$$

Since C and $\mathbf{D}$ are independent of $\boldsymbol{\nu}$, the orthogonality conditions (10.36) are valid for all choices of the one-form $\boldsymbol{\nu}$. Thus, because of linearity, the conditions are valid not just for a single plane-wave field but for the field of any linear combination of plane waves in such a medium.

Case 2 Media

Considering the fields corresponding to the two main classes of Case 2 NDE media, that of skewon–axion media and P-axion media, we can refer to results obtained in Chapters 7 and 8. In fact, fields in a skewon–axion medium were obtained in the form (7.101) and (7.102) rewritten here as

$$\boldsymbol{\Phi} = A\boldsymbol{\nu} \wedge \overline{\overline{\mathsf{B}}}^T|\boldsymbol{\nu}, \tag{10.41}$$

$$\boldsymbol{\Psi} = A\boldsymbol{\nu} \wedge \overline{\overline{\mathsf{B}}}^{2T}|\boldsymbol{\nu}. \tag{10.42}$$

In this case the polarization of the field depends on the chosen one-form $\boldsymbol{\nu}$. If $\boldsymbol{\nu}$ is chosen as an eigen one-form of $\overline{\overline{\mathsf{B}}}$, the above expressions are not valid. Actually, in such a case $\boldsymbol{\Phi}$ can be any two-form satisfying $\boldsymbol{\nu} \wedge \boldsymbol{\Phi} = 0$.

For the P-axion medium defined by (10.31) the potential one-form $\boldsymbol{\phi}$ obeys the same equation as for the P medium whence the field two-form $\boldsymbol{\Phi}$ is obtained from (8.79) rewritten here as

$$\boldsymbol{\Phi} = A\boldsymbol{\nu} \wedge \overline{\overline{\mathsf{P}}}^{-1T}|\boldsymbol{\nu}. \tag{10.43}$$

However, since the medium bidyadic has an extra axion term, the other field two-form $\boldsymbol{\Psi}$ is different from (8.80) and can be expressed as

$$\begin{aligned}\boldsymbol{\Psi} &= (\overline{\overline{\mathsf{P}}}^{(2)T} + \alpha\overline{\overline{\mathsf{I}}}^{(2)T})|\boldsymbol{\Phi} \\ &= A((\overline{\overline{\mathsf{P}}}^{T}|\boldsymbol{\nu}) \wedge \boldsymbol{\nu} + \alpha\boldsymbol{\nu} \wedge (\overline{\overline{\mathsf{P}}}^{-1T}|\boldsymbol{\nu})).\end{aligned} \tag{10.44}$$

Also in this case the polarization of the field two-forms are uniquely determined by the choice of the wave one-form $\boldsymbol{\nu}$.

10.1.3 Other Possible NDE Media

Let us try to define other media satisfying the NDE condition outside the Case 1 and Case 2 definitions. Recalling that the medium equation could alternatively be defined in terms of a bidyadic $\overline{\overline{\mathsf{N}}}$ in the form $\boldsymbol{\Phi} = \overline{\overline{\mathsf{N}}}|\boldsymbol{\Psi}$, one may wonder whether other NDE media could be found through this route. Starting from the alternate dispersion dyadic $\overline{\overline{\mathsf{D}}}'(\boldsymbol{\nu}) = \overline{\overline{\mathsf{N}}}_m\lfloor\lfloor\boldsymbol{\nu}\boldsymbol{\nu}$ we can work similarly and end up in Case 1 and Case 2 solutions based on the bidyadic $\overline{\overline{\mathsf{N}}}$. To determine whether these differ from those obtained above we must find out what they correspond to in terms of the medium bidyadic $\overline{\overline{\mathsf{M}}}$.

Assuming full-rank bidyadics, we have the relation $\overline{\overline{\mathsf{N}}} = \overline{\overline{\mathsf{M}}}^{-1}$, in which case the problem is reduced to whether the inverse of Case 1 and Case 2 medium bidyadics $\overline{\overline{\mathsf{N}}}$ falls under Case 1 and Case 2 form of bidyadics $\overline{\overline{\mathsf{M}}}$. The answer turns out to be affirmative whence novel medium classes cannot be obtained through this way. Actually, the inverse of a Case 1 bidyadic $\overline{\overline{\mathsf{N}}}$ can be shown to be a Case 1 bidyadic $\overline{\overline{\mathsf{M}}}$. For Case 2, the situation is a bit more complicated. One can show that the inverse of the general P-axion bidyadic $\overline{\overline{\mathsf{N}}}$ similar to (10.31), involving 17 scalar parameters, is a P-axion bidyadic $\overline{\overline{\mathsf{M}}}$ but the inverse of a skewon–axion bidyadic $\overline{\overline{\mathsf{N}}}$, which involves 16 scalar parameters, turns out to be a special P-axion bidyadic $\overline{\overline{\mathsf{M}}}$ with coefficient α satisfying $\mathrm{tr}\overline{\overline{\mathsf{P}}}^{(4)} = \alpha^2$. Also, the inverse of this kind of special P-axion bidyadic $\overline{\overline{\mathsf{N}}}$ turns out to be a skewon–axion bidyadic $\overline{\overline{\mathsf{M}}}$. This was shown in Section 8.4. Applying another result from Section 4.7, one can conclude

Table 10.1.
Relations Between Various Case 2 Medium Bidyadics

$\overline{\overline{\mathsf{M}}}$	$\overline{\overline{\mathsf{N}}}$
Axion	Axion
Skewon	Skewon
P medium	P medium
Skewon–axion	Special P-axion
Special P-axion	Skewon–axion
General P-axion	General P-axion

that if a medium bidyadic $\overline{\overline{\mathsf{M}}}_m$ is of full rank and satisfies the NDE condition $(\overline{\overline{\mathsf{M}}}_m \lfloor\lfloor \boldsymbol{\nu}\boldsymbol{\nu})^{(3)} = \overline{\overline{\mathsf{D}}}^{(3)}(\boldsymbol{\nu}) = 0$ for all $\boldsymbol{\nu}$, it cannot be a Q medium. Because an axion term does not change this conclusion, the inverse of a skewon–axion medium bidyadic must be a P-axion bidyadic and not a Q-axion bidyadic which is not associated to an NDE medium anyway for full-rank dyadic $\overline{\overline{\mathsf{Q}}}$. Details are left as an exercise. The relations between different types of Case 2 NDE-medium bidyadics are shown in Table 10.1.

Even if the consideration of Case 1 and Case 2 solutions through the bidyadic $\overline{\overline{\mathsf{N}}}$ does not reveal new NDE media, the above media are not exhaustive as can be shown by an example. In fact, it was already observed by considering the expression (8.70) that a Q medium with dyadic $\overline{\overline{\mathsf{Q}}}$ of less than full rank satisfies the dyadic dispersion equation (5.240) for any one-form $\boldsymbol{\nu}$. Obviously, such a Q medium can be completed to a Q-axion medium with the same property. The reason why it was not obtained from the previous analysis comes clear when assuming a dyadic $\overline{\overline{\mathsf{Q}}}$ of rank 3 for which we can write

$$\overline{\overline{\mathsf{Q}}} = \mathbf{e}_1\mathbf{q}_1 + \mathbf{e}_2\mathbf{q}_2 + \mathbf{e}_3\mathbf{q}_3, \tag{10.45}$$

$$\overline{\overline{\mathsf{Q}}}^{(2)} = \mathbf{e}_{12}\mathbf{q}_{12} + \mathbf{e}_{23}\mathbf{q}_{23} + \mathbf{e}_{31}\mathbf{q}_{31}. \tag{10.46}$$

Assuming now

$$\overline{\overline{\mathsf{M}}}_m = \overline{\overline{\mathsf{Q}}}^{(2)} + \alpha\mathbf{e}_N \lfloor \overline{\overline{\mathsf{I}}}^{(2)T}, \tag{10.47}$$

and $\mathbf{e}_3|\nu \neq 0$, after some steps (left as an exercise), the dispersion dyadic can be expressed in the form (10.1) with

$$\mathbf{ac} = \frac{1}{\mathbf{e}_3|\nu}((\mathbf{e}_3 \wedge \mathbf{e}_1)\rfloor\nu)((\nu|\overline{\overline{\mathsf{Q}}} \wedge \mathbf{q}_1)\rfloor\nu), \tag{10.48}$$

$$\mathbf{bd} = \frac{1}{\mathbf{e}_3|\nu}((\mathbf{e}_3 \wedge \mathbf{e}_2)\rfloor\nu)((\nu|\overline{\overline{\mathsf{Q}}} \wedge \mathbf{q}_2)\rfloor\nu). \tag{10.49}$$

These can be further combined to

$$\overline{\overline{\mathsf{D}}}(\nu) = \frac{1}{\mathbf{e}_3|\nu}((\mathbf{e}_3(\nu|\overline{\overline{\mathsf{Q}}}))^{\wedge}_{\wedge}\overline{\overline{\mathsf{Q}}})\rfloor\rfloor\nu\nu. \tag{10.50}$$

Obviously, the expansion of the dispersion dyadic (10.1) with (10.48) and (10.49) inserted corresponds to neither Case 1 nor Case 2 because of the divisor $\mathbf{e}_3|\nu$ and cubic dependence on ν of both **ac** and **bd**.

10.2 DECOMPOSABLE MEDIA

The class of DC media is defined by the property that any field in a source-free region of the medium can be split in two independent partial fields each obeying a certain restriction on its polarization. The idea can be understood by considering first TE/TM decomposition of plane-wave fields in a uniaxially anisotropic medium in terms of Gibbsian formalism.

10.2.1 Special Cases

TE/TM Decomposition

Let us consider time-harmonic fields with time dependence $e^{j\omega t}$ in a uniaxially anisotropic medium defined by the Gibbsian medium dyadics

$$\overline{\overline{\epsilon}}_g = \epsilon_t\overline{\overline{\mathsf{I}}}_t + \epsilon_z\mathbf{u}_z\mathbf{u}_z, \tag{10.51}$$

$$\overline{\overline{\mu}}_g = \mu_t\overline{\overline{\mathsf{I}}}_t + \mu_z\mathbf{u}_z\mathbf{u}_z, \tag{10.52}$$

where $\overline{\overline{\mathsf{I}}}_t = \mathbf{u}_x\mathbf{u}_x + \mathbf{u}_y\mathbf{u}_y$. Now, one can show that any plane wave in such a medium can be decomposed in two parts, called transverse electric (TE) and transverse magnetic (TM), as

$$\mathbf{E}_g = \mathbf{E}_{TE} + \mathbf{E}_{TM}, \tag{10.53}$$

$$\mathbf{H}_g = \mathbf{H}_{TE} + \mathbf{H}_{TM}, \tag{10.54}$$

and restricted by the conditions

$$\mathbf{u}_z \cdot \mathbf{E}_{TE} = 0, \tag{10.55}$$
$$\mathbf{u}_z \cdot \mathbf{H}_{TM} = 0. \tag{10.56}$$

Such a property was first published by Clemmow in 1963 [69]. The decomposition is unique when the medium parameters satisfy

$$\epsilon_t \mu_z - \mu_t \epsilon_z \neq 0, \tag{10.57}$$

while in the converse case the decomposition is still valid but not unique. Actually, in the latter case the medium can be transformed to a Gibbsian isotropic medium through a suitable affine transformation [5].

To demonstrate the decomposition, we recall that the fields of a plane wave with space dependence $\exp(-j\mathbf{k} \cdot \mathbf{r})$ in any linear medium satisfy

$$\omega \mathbf{E}_g \cdot \mathbf{B}_g = \mathbf{E}_g \cdot (\mathbf{k} \times \mathbf{E}_g) = 0, \tag{10.58}$$
$$\omega \mathbf{H}_g \cdot \mathbf{D}_g = -\mathbf{H}_g \cdot (\mathbf{k} \times \mathbf{H}_g) = 0, \tag{10.59}$$

whence in the uniaxial medium they must satisfy

$$\epsilon_t \mathbf{E}_g \cdot \mathbf{B}_g - \mu_t \mathbf{H}_g \cdot \mathbf{D}_g = (\epsilon_t \mu_z - \mu_t \epsilon_z)(\mathbf{u}_z \cdot \mathbf{E}_g)(\mathbf{u}_z \cdot \mathbf{H}_g) = 0. \tag{10.60}$$

Thus, assuming (10.57), a plane wave in the uniaxial medium must either be a TE wave or a TM wave with respect to the axial direction defined by the unit vector $\mathbf{u}_z$. From this it follows that any electromagnetic field which can be decomposed in a spectrum of plane waves [37], can be uniquely decomposed in TE and TM parts in the uniaxial medium.

More General Decompositions

The TE/TM decomposition theory can be generalized to certain media where the fields are decomposable to two parts satisfying either $\mathbf{a} \cdot \mathbf{E}_g = 0$ or $\mathbf{b} \cdot \mathbf{H}_g = 0$ where $\mathbf{a}$ and $\mathbf{b}$ are two given (possibly complex) Gibbsian vectors. Even more generally, a decomposition by two polarization conditions

$$\mathbf{a}_1 \cdot \mathbf{E}_g + \mathbf{a}_2 \cdot \mathbf{H}_g = 0, \quad \mathbf{b}_1 \cdot \mathbf{E}_g + \mathbf{b}_2 \cdot \mathbf{H}_g = 0, \tag{10.61}$$

in terms of four Gibbsian vectors $\mathbf{a}_1 - \mathbf{b}_2$ can be made in bi-anisotropic media with Gibbsian medium dyadics defined by [54]

$$\overline{\overline{\epsilon}}_g = \frac{1}{2\tau}\left(-\overline{\overline{\mathsf{B}}}_g^T + \mathbf{a}_2\mathbf{b}_1 + \mathbf{b}_2\mathbf{a}_1\right), \tag{10.62}$$

$$\overline{\overline{\xi}}_g = \mathbf{x} \times \overline{\overline{\mathsf{I}}}_g + \frac{1}{2\tau}(\mathbf{a}_2\mathbf{b}_2 + \mathbf{b}_2\mathbf{a}_2), \tag{10.63}$$

$$\overline{\overline{\zeta}}_g = \mathbf{z} \times \overline{\overline{\mathsf{I}}}_g + \frac{1}{2\eta}(\mathbf{a}_1\mathbf{b}_1 + \mathbf{b}_1\mathbf{a}_1), \tag{10.64}$$

$$\overline{\overline{\mu}}_g = \frac{1}{2\eta}(\overline{\overline{\mathsf{B}}}_g + \mathbf{a}_1\mathbf{b}_2 + \mathbf{b}_1\mathbf{a}_2), \tag{10.65}$$

where $\mathbf{x}, \mathbf{z}$ are arbitrary Gibbsian vectors, τ, η are arbitrary scalars, $\overline{\overline{\mathsf{I}}}_g$ is the Gibbsian unit dyadic and $\overline{\overline{\mathsf{B}}}_g$ is an arbitrary Gibbsian dyadic. Media defined by medium dyadics of the form (10.62)–(10.65) were called decomposable (DC) media.

As a continuation to the theory it was shown that yet another scalar parameter α can be added to the definition of the medium without changing the decomposition of the fields. In this case the definitions (10.62)–(10.65) are generalized (after some manipulations) to the form [21, 19]

$$\overline{\overline{\epsilon}}_g = \alpha(\mathbf{z} \times \overline{\overline{\mathsf{I}}}_g + \mathbf{a}_1\mathbf{b}_1 + \mathbf{b}_1\mathbf{a}_1) + \eta\left(-\overline{\overline{\mathsf{B}}}_g^T + \mathbf{a}_2\mathbf{b}_1 + \mathbf{b}_2\mathbf{a}_1\right), \tag{10.66}$$

$$\overline{\overline{\xi}}_g = \eta(\mathbf{x} \times \overline{\overline{\mathsf{I}}}_g + \mathbf{a}_2\mathbf{b}_2 + \mathbf{b}_2\mathbf{a}_2) + \alpha(\overline{\overline{\mathsf{B}}}_g + \mathbf{a}_1\mathbf{b}_2 + \mathbf{b}_1\mathbf{a}_2), \tag{10.67}$$

$$\overline{\overline{\zeta}}_g = \tau(\mathbf{z} \times \overline{\overline{\mathsf{I}}}_g + \mathbf{a}_1\mathbf{b}_1 + \mathbf{b}_1\mathbf{a}_1) - \alpha\left(-\overline{\overline{\mathsf{B}}}_g^T + \mathbf{a}_2\mathbf{b}_1 + \mathbf{b}_2\mathbf{a}_1\right), \tag{10.68}$$

$$\overline{\overline{\mu}}_g = -\alpha(\mathbf{x} \times \overline{\overline{\mathsf{I}}}_g + \mathbf{a}_2\mathbf{b}_2 + \mathbf{b}_2\mathbf{a}_2) + \tau(\overline{\overline{\mathsf{B}}}_g + \mathbf{a}_1\mathbf{b}_2 + \mathbf{b}_1\mathbf{a}_2). \tag{10.69}$$

Four-Dimensional Representation

It turns out that the rather complicated looking conditions (10.66)–(10.69) of the DC medium can be represented in a compact way in terms of the 4D formalism [48]. Actually, applying an approach similar to that of [54, 21] to two-form representation of fields, additional solutions to media satisfying the DC-medium definition can be obtained.

Let us recall that field two-forms $\mathbf{\Phi}, \mathbf{\Psi}$ with the plane-wave dependence $\exp(\boldsymbol{\nu}|\mathbf{x})$ satisfy the orthogonality conditions (5.232)

$$\mathbf{\Phi} \wedge \mathbf{\Phi} = 0, \quad \mathbf{\Phi} \wedge \mathbf{\Psi} = 0, \quad \mathbf{\Psi} \wedge \mathbf{\Psi} = 0 \tag{10.70}$$

in any medium. Applying the natural dot product between two two-forms the four-form conditions (10.70) can be expressed as the respective scalar conditions (5.233) as

$$\mathbf{\Phi}\cdot\mathbf{\Phi} = 0, \quad \mathbf{\Phi}\cdot\mathbf{\Psi} = \mathbf{\Phi}\cdot\overline{\overline{\mathsf{M}}}|\mathbf{\Phi} = 0, \quad \mathbf{\Psi}\cdot\mathbf{\Psi} = \mathbf{\Phi}|\overline{\overline{\mathsf{M}}}^T\cdot\overline{\overline{\mathsf{M}}}|\mathbf{\Phi} = 0. \tag{10.71}$$

Thus, for any medium bidyadic $\overline{\overline{\mathsf{M}}}$, the two-form $\mathbf{\Phi}$ of any plane wave satisfies a condition of the form

$$\mathbf{\Phi}|(\alpha\mathbf{e}_N\lfloor\overline{\overline{\mathsf{I}}}^{(2)T} + 2\beta\mathbf{e}_N\lfloor\overline{\overline{\mathsf{M}}} + \gamma\overline{\overline{\mathsf{M}}}^T\cdot\overline{\overline{\mathsf{M}}})|\mathbf{\Phi} = 0, \tag{10.72}$$

for all choices of the scalars α, β, γ. For the modified medium bidyadic this condition can be expressed as

$$\mathbf{\Phi}|(\alpha\overline{\overline{\mathsf{E}}} + \beta(\overline{\overline{\mathsf{M}}}_m + \overline{\overline{\mathsf{M}}}_m^T) + \gamma\overline{\overline{\mathsf{M}}}_m^T\cdot\overline{\overline{\mathsf{M}}}_m)|\mathbf{\Phi} = 0. \tag{10.73}$$

Let us now assume that the medium requires that the field two-form $\mathbf{\Phi}$ of a given plane wave satisfies either of the two polarization conditions

$$\mathbf{A}|\mathbf{\Phi} = 0, \tag{10.74}$$

$$\mathbf{B}|\mathbf{\Phi} = 0, \tag{10.75}$$

involving two bivectors **A** and **B**. Expanding the bivectors in their spatial and temporal parts as

$$\mathbf{A} = \mathbf{a}_1\wedge\mathbf{a}_2 + \mathbf{e}_4\wedge\mathbf{a}_3, \quad \mathbf{B} = \mathbf{b}_1\wedge\mathbf{b}_2 + \mathbf{e}_4\wedge\mathbf{b}_3, \tag{10.76}$$

the polarization conditions become

$$\mathbf{A}|\mathbf{\Phi} = (\mathbf{a}_1\wedge\mathbf{a}_2)|\mathbf{B} + \mathbf{a}_3|\mathbf{E} = 0, \tag{10.77}$$

$$\mathbf{B}|\mathbf{\Phi} = (\mathbf{b}_1\wedge\mathbf{b}_2)|\mathbf{B} + \mathbf{b}_3|\mathbf{E} = 0, \tag{10.78}$$

which have a certain similarity to the Gibbsian conditions (10.61).

Waves with fields satisfying (10.74) or (10.75) can be respectively called A-waves and B-waves and the medium, in analogy with the medium defined by (10.62)–(10.65), can be called a DC medium. Thus, any plane wave in such a medium is required to satisfy

$$(\mathbf{\Phi}|\mathbf{A})(\mathbf{B}|\mathbf{\Phi}) = \mathbf{\Phi}|(\mathbf{A}\mathbf{B})|\mathbf{\Phi} = 0, \tag{10.79}$$

for some bivectors **A**, **B**. Following the pattern of [54], let us now define the class of DC media by combining (10.73) and (10.79) and requiring that the condition

$$\mathbf{\Phi}|(\alpha\overline{\overline{\mathsf{E}}} + \beta(\overline{\overline{\mathsf{M}}}_m + \overline{\overline{\mathsf{M}}}_m^T) + \gamma\overline{\overline{\mathsf{M}}}_m^T \cdot \overline{\overline{\mathsf{M}}}_m)|\mathbf{\Phi} = \mathbf{\Phi}|(\mathbf{AB})|\mathbf{\Phi} \qquad (10.80)$$

be satisfied for all two-forms $\mathbf{\Phi}$, which implies that the symmetric parts of the dyadics in brackets on both sides of (10.80) must be the same. Thus, a condition defining possible DC media can be expressed as

$$\alpha\overline{\overline{\mathsf{E}}} + \beta(\overline{\overline{\mathsf{M}}}_m + \overline{\overline{\mathsf{M}}}_m^T) + \gamma\overline{\overline{\mathsf{M}}}_m^T \cdot \overline{\overline{\mathsf{M}}}_m = \mathbf{AB} + \mathbf{BA}, \qquad (10.81)$$

where a disturbing factor $1/2$ has been absorbed by the dyadic **AB**. To find solutions $\overline{\overline{\mathsf{M}}}_m$ for possible DC media, we must separate two cases: either $\gamma \neq 0$ or $\gamma = 0$.

- The case $\gamma \neq 0$ requires solving a symmetric quadratic bidyadic equation and the corresponding class of media can be called that of (proper) DC media.
- In the case $\gamma = 0$ the bidyadic equation (10.81) is reduced to an equation of the first order. This requires separate consideration and it defines the class of what will be called special decomposable (SDC) media.

10.2.2 DC-Medium Subclasses

Assuming $\gamma \neq 0$ in (10.81) we can set $\gamma = 1$ without losing generality. In this case the DC-medium condition (10.81) can be expressed in compact form as

$$\overline{\overline{\mathsf{M}}}{}'^T_m \cdot \overline{\overline{\mathsf{M}}}{}'_m = \alpha'\overline{\overline{\mathsf{E}}}, \qquad (10.82)$$

by defining

$$\overline{\overline{\mathsf{M}}}{}'_m = \overline{\overline{\mathsf{M}}}_m + \beta\overline{\overline{\mathsf{E}}} - \mathbf{DC}, \qquad (10.83)$$

$$\alpha' = \beta^2 - \alpha. \qquad (10.84)$$

Here, the bidyadic **C** equals

$$\mathbf{C} = \mathbf{A}, \qquad (10.85)$$

and the bidyadic **D** is related to **A** and **B** by

$$\mathbf{D} \cdot \overline{\overline{\mathsf{M}}}_m + \beta\mathbf{D} - \frac{1}{2}(\mathbf{D} \cdot \mathbf{D})\mathbf{C} = \mathbf{B}. \qquad (10.86)$$

The condition (10.82) has the form of modified closure relation (4.142), whose solutions can be split in P- and Q-solutions, as were shown by (4.144) and (4.145). Accordingly, the class of DC media can be split in two subclasses. The first subclass assumes that there exist a dyadic $\overline{\overline{\mathsf{P}}} \in \mathbb{E}_1\mathbb{F}_1$ so that

$$\overline{\overline{\mathsf{M}}}'_m = \mathbf{e}_N \lfloor \overline{\overline{\mathsf{P}}}^{(2)T}. \tag{10.87}$$

The second subclass assumes that there exists a dyadic $\overline{\overline{\mathsf{Q}}} \in \mathbb{E}_1\mathbb{E}_1$ so that

$$\overline{\overline{\mathsf{M}}}'_m = \overline{\overline{\mathsf{Q}}}^{(2)}. \tag{10.88}$$

Let us consider these two cases separately and respectively call them by the names PDC media and QDC media.

QDC Media

Redefining the coefficient α, the QDC medium solution of (10.82) as obtained from (10.83) to (10.84) must be of the form

$$\overline{\overline{\mathsf{M}}}_m = \alpha\overline{\overline{\mathsf{E}}} + \overline{\overline{\mathsf{Q}}}^{(2)} + \mathbf{DC}, \tag{10.89}$$

or

$$\overline{\overline{\mathsf{M}}} = \alpha\overline{\overline{\mathsf{I}}}^{(2)T} + \boldsymbol{\varepsilon}_N \lfloor \overline{\overline{\mathsf{Q}}}^{(2)} + \boldsymbol{\varepsilon}_N \lfloor \mathbf{DC}, \tag{10.90}$$

depending on a metric dyadic $\overline{\overline{\mathsf{Q}}} \in \mathbb{E}_1\mathbb{E}_1$, two bivectors $\mathbf{C}, \mathbf{D}$ and a scalar α. For $\alpha = 0$ the medium bidyadic has the form of (8.107), which corresponds to that of the extended Q medium. Thus, a QDC medium can also be called an extended Q-axion medium.

The definition (10.86) can in this case be expressed more explicitly as

$$\mathbf{B} = \mathbf{D}|\overline{\overline{\mathsf{Q}}}^{(2)} + \frac{1}{2}(\mathbf{D}\cdot\mathbf{D})\mathbf{C}. \tag{10.91}$$

Assuming that the bivectors $\mathbf{C}$ and $\mathbf{D}$ are given, the bidyadics in the polarization conditions (10.74) and (10.75) satisfied by the field two-form in the QDC medium can be written from (10.85) and (10.91) as

$$\mathbf{C}|\boldsymbol{\Phi} = 0, \qquad \left(\mathbf{D}|\overline{\overline{\mathsf{Q}}}^{(2)} + \frac{1}{2}(\mathbf{D}\cdot\mathbf{D})\mathbf{C}\right)|\boldsymbol{\Phi} = 0. \tag{10.92}$$

One can show that the 3D Gibbsian medium dyadics corresponding to the general QDC medium (10.90) can be expressed in the general form (10.66)–(10.69).

PDC Media

The second possibility in (10.82), that of the PDC medium, yields the following expansions for the medium bidyadics,

$$\overline{\overline{\mathsf{M}}}_m = \alpha\overline{\overline{\mathsf{E}}} + \mathbf{e}_N\lfloor\overline{\overline{\mathsf{P}}}^{(2)T} + \mathbf{DC}, \tag{10.93}$$

or

$$\overline{\overline{\mathsf{M}}} = \alpha\overline{\overline{\mathsf{I}}}^{(2)T} + \overline{\overline{\mathsf{P}}}^{(2)T} + \boldsymbol{\varepsilon}_N\lfloor\mathbf{DC}, \tag{10.94}$$

in terms of a dyadic $\overline{\overline{\mathsf{P}}} \in \mathbb{E}_1\mathbb{F}_1$, two bivectors $\mathbf{C}, \mathbf{D}$ and a scalar α. The definition (10.86) in this case is

$$\mathbf{B} = \mathbf{D}|\overline{\overline{\mathsf{P}}}^{(2)T} + \frac{1}{2}(\mathbf{D}\cdot\mathbf{D})\mathbf{C}. \tag{10.95}$$

Thus, for given bidyadics $\mathbf{C}, \mathbf{D}$, the polarization conditions (10.74) and (10.75) satisfied by the field two-form in the PDC medium can be written from (10.85) and (10.95) as

$$\mathbf{C}|\boldsymbol{\Phi} = 0, \quad \left(\mathbf{D}|\overline{\overline{\mathsf{P}}}^{(2)T} + \frac{1}{2}(\mathbf{D}\cdot\mathbf{D})\mathbf{C}\right)|\boldsymbol{\Phi} = 0. \tag{10.96}$$

For $\alpha = 0$ the medium bidyadic has the form of (8.171), which corresponds to that of the extended P medium. Thus, a PDC medium can also be called an extended P-axion medium. Expansion of (10.94) in terms of 3D medium dyadics is left as an exercise.

SDC Media

The SDC medium condition defined by $\gamma = 0$ and $\beta = 1$ in (10.81) yields a bidyadic equation of the first order,

$$\alpha\overline{\overline{\mathsf{E}}} + \overline{\overline{\mathsf{M}}}_m + \overline{\overline{\mathsf{M}}}_m^T = \mathbf{AB} + \mathbf{BA}. \tag{10.97}$$

This equation can be interpreted so that the symmetric part of the bidyadic $\overline{\overline{\mathsf{M}}}_m - \mathbf{AB}$ must be a multiple of $\overline{\overline{\mathsf{E}}}$. Thus, redefining α, the modified medium bidyadic must thus be of the form

$$\overline{\overline{\mathsf{M}}}_m = \alpha\overline{\overline{\mathsf{E}}} + \mathbf{AB} + \overline{\overline{\mathsf{A}}}, \tag{10.98}$$

where $\overline{\overline{\mathsf{A}}} \in \mathbb{E}_2\mathbb{E}_2$ is an arbitrary antisymmetric bidyadic. From (3.86) it is known that such an antisymmetric bidyadic can be expressed in terms of a trace-free dyadic $\overline{\mathsf{B}}_o \in \mathbb{E}_1\mathbb{F}_1$ as

$$\overline{\overline{\mathsf{A}}} = \mathbf{e}_N\lfloor(\overline{\mathsf{B}}_o{}^\wedge_\wedge\overline{\mathsf{I}})^T, \quad \mathrm{tr}\overline{\mathsf{B}}_o = 0. \tag{10.99}$$

It is easy to verify that a medium defined by (10.98) satisfies the decomposition condition (10.79). In fact, any plane wave in such a medium satisfies

$$\begin{aligned}\boldsymbol{\Phi}\cdot\boldsymbol{\Psi} - \alpha\boldsymbol{\Phi}\cdot\boldsymbol{\Phi} &= \boldsymbol{\Phi}|(\mathbf{e}_N\lfloor\overline{\overline{\mathsf{M}}} - \alpha\overline{\overline{\mathsf{E}}})|\boldsymbol{\Phi}\\ &= \boldsymbol{\Phi}|(\mathbf{AB} + \overline{\overline{\mathsf{A}}})|\boldsymbol{\Phi}\\ &= (\mathbf{A}|\boldsymbol{\Phi})(\mathbf{B}|\boldsymbol{\Phi}) = 0.\end{aligned} \tag{10.100}$$

Redefining again $\overline{\overline{\mathsf{A}}}$ and α, (10.98) can be expressed in the form

$$\overline{\overline{\mathsf{M}}}_m = \alpha\overline{\overline{\mathsf{E}}} + \overline{\overline{\mathsf{A}}} + \mathbf{AB} + \mathbf{BA}, \tag{10.101}$$

which corresponds to

$$\overline{\overline{\mathsf{M}}} = \alpha\overline{\overline{\mathsf{I}}}^{(2)T} + \varepsilon_N\lfloor\overline{\overline{\mathsf{A}}} + \varepsilon_N\lfloor(\mathbf{AB} + \mathbf{BA}). \tag{10.102}$$

Without sacrificing generality, the bivectors in (10.102) can be assumed to satisfy $\mathbf{A}\cdot\mathbf{B} = 0$, whence the third term is trace free. In this case the three terms correspond to the respective axion, skewon, and principal parts of the medium bidyadic $\overline{\overline{\mathsf{M}}}$. While the axion and skewon parts may be arbitrary, the principal part is restricted to be of the simple form as defined by two bivectors **A** and **B**. Following the pattern of the QDC and PDC media, an SDC medium defined by (10.98) can be called an extended skewon–axion medium.

In conclusion, the class of DC media as defined by the condition (10.81) consists of extended Q-axion (QDC) media, extended P-axion (PDC) media, and extended skewon–axion (SDC) media. This analysis has shown us that the present compact 4D formalism [48] has actually brought about two extra subclasses of solutions in addition to that of QDC media, first introduced through the more cumbersome 3D Gibbsian formalism in [54, 21].

10.2.3 Plane-Wave Properties

Let us finally consider properties of plane waves for the three medium subclasses defined above. Because the axion part plays no role in plane-wave propagation, it can again be ignored in all of the cases. Actually, the plane-wave problem has already been discussed for the extended-Q, extended-P, and extended-skewon media in the previous chapters. Thus, we may just adopt the results with some changes in symbols. In all of these cases the quartic dispersion equation $D(\nu) = 0$ can be split in two quadratic equations corresponding to the decomposed waves. The converse, starting from the splitting of the quartic equation and deriving the polarization decomposition, can also be done [89, 53].

QDC Media

In the case of QDC medium (extended Q-axion medium) the dispersion dyadic equals

$$\overline{\overline{\mathsf{D}}}(\nu) = \nu\nu\rfloor\rfloor\overline{\overline{\mathsf{M}}}_m = \nu\nu\rfloor\rfloor(\overline{\overline{\mathsf{Q}}}^{(2)} + \mathbf{DC}). \qquad (10.103)$$

Comparing with (8.147), the two dispersion equations become

$$\nu|\overline{\overline{\mathsf{Q}}}|\nu = 0, \qquad (10.104)$$

$$\nu|(\overline{\overline{\mathsf{Q}}} + \mathbf{CD}\lfloor\lfloor\overline{\overline{\mathsf{Q}}}^{-1})|\nu = 0, \qquad (10.105)$$

(10.104) corresponds to a Q medium without the generalizing term, that is, with $\mathbf{DC} = 0$. This is valid for fields satisfying the field condition

$$\mathbf{C}|\boldsymbol{\Phi} = \mathbf{A}|\boldsymbol{\Phi} = 0, \qquad (10.106)$$

whence (10.104) corresponds to the A-wave. To verify that (10.105) corresponds to the B-wave, that is, the condition (10.75), which from (10.91) equals

$$\left(\mathbf{D}\cdot\overline{\overline{\mathsf{Q}}}^{(2)} + \frac{1}{2}(\mathbf{D}\cdot\mathbf{D})\mathbf{C}\right)|\boldsymbol{\Phi} = 0, \qquad (10.107)$$

is left as an exercise. (10.105) presumes that $\overline{\overline{\mathsf{Q}}}$ be of full rank. In the opposite case it must be replaced by

$$(\varepsilon_N\varepsilon_N\lfloor\lfloor\overline{\overline{\mathsf{Q}}}^{(3)})||(\nu\nu\rfloor\rfloor\mathbf{AB}) = \nu\nu||((\varepsilon_N\varepsilon_N\lfloor\lfloor\overline{\overline{\mathsf{Q}}}^{(3)})\rfloor\rfloor\mathbf{AB}) = 0. \quad (10.108)$$

When $\overline{\overline{\mathsf{Q}}}^{(3)} = 0$, there is no dispersion equation for the B-wave in a QDC medium.

PDC Media

In the case of PDC medium (extended P-axion medium) the dispersion equation is decomposed as (8.184) and (8.185). In terms of the present symbols they have the form

$$\boldsymbol{\nu}|(\mathbf{D}\lfloor\overline{\overline{\mathsf{P}}}^T)|\boldsymbol{\nu} = 0, \tag{10.109}$$

$$\boldsymbol{\nu}|(\mathbf{C}\lfloor\overline{\overline{\mathsf{P}}}^{-1T})|\boldsymbol{\nu} = 0. \tag{10.110}$$

Omitting some details, in (8.191) it was shown that the field corresponding to the dispersion equation (10.110) satisfies the condition

$$\mathbf{C}|\boldsymbol{\Phi} = \mathbf{A}|\boldsymbol{\Phi} = 0, \tag{10.111}$$

that is, it represents the A-wave. Similarly, the solutions of the dispersion equation (10.109) correspond to the field condition (8.192), which in the present form reads

$$\left(\mathbf{D}|\overline{\overline{\mathsf{P}}}^{(2)T} + \frac{1}{2}(\mathbf{D}\cdot\mathbf{D})\mathbf{C}\right)|\boldsymbol{\Phi} = \mathbf{B}|\boldsymbol{\Phi} = 0, \tag{10.112}$$

due to the definition (10.95). Again, (10.110) assumes that $\overline{\overline{\mathsf{P}}}$ has full rank. In the converse case it must be replaced by

$$\boldsymbol{\nu}|(\mathbf{C}\lfloor(\overline{\overline{\mathsf{I}}}^{(4)}\lfloor\lfloor\overline{\overline{\mathsf{P}}}^{(3)}))|\boldsymbol{\nu} = 0, \tag{10.113}$$

and this is identically valid for $\overline{\overline{\mathsf{P}}}^{(3)} = 0$. For the P-axion medium without the extension, $\mathbf{DC} = 0$, the PDC medium is an NDE medium because both dispersion equations vanish.

SDC Media

Because the SDC medium equals the extended skewon–axion medium, properties of a plane wave in an SDC medium equal those discussed in Section 7.4. Expressing the dispersion dyadic as

$$\overline{\overline{\mathsf{D}}}(\boldsymbol{\nu}) = \boldsymbol{\nu}\boldsymbol{\nu}\rfloor\rfloor(\mathbf{e}_N\lfloor(\overline{\overline{\mathsf{B}}}_o{}^\wedge_\wedge\overline{\overline{\mathsf{I}}})^T + \mathbf{AB} + \mathbf{BA}), \tag{10.114}$$

where $\overline{\overline{\mathsf{B}}}_o \in \mathbb{E}_1\mathbb{F}_1$ is a trace-free dyadic, and denoting for simplicity

$$\boldsymbol{\nu}' = \overline{\overline{\mathsf{B}}}_o^T|\boldsymbol{\nu}, \quad \mathbf{a} = \boldsymbol{\nu}\rfloor\mathbf{A}, \quad \mathbf{b} = \boldsymbol{\nu}\rfloor\mathbf{B}, \tag{10.115}$$

after some steps left as an exercise we end up at the dispersion equation

$$\overline{\overline{\mathsf{D}}}^{(3)}(\boldsymbol{\nu}) = 2(\mathbf{e}_N\mathbf{e}_N\lfloor\lfloor\boldsymbol{\nu}\boldsymbol{\nu})(\boldsymbol{\nu}'|\mathbf{a})(\boldsymbol{\nu}'|\mathbf{b}) = 0. \tag{10.116}$$

Thus, $\boldsymbol{\nu}'$ must satisfy either $\boldsymbol{\nu}'|\mathbf{a} = 0$ or $\boldsymbol{\nu}'|\mathbf{b} = 0$.

In conclusion, in the SDC medium case, the dispersion equation is decomposed to two equations as

$$\nu'|\mathbf{a} = \nu|(\overline{\overline{\mathsf{B}}}_o\rfloor\mathbf{A})|\nu = 0, \tag{10.117}$$

$$\nu'|\mathbf{b} = \nu|(\overline{\overline{\mathsf{B}}}_o\rfloor\mathbf{B})|\nu = 0, \tag{10.118}$$

whence the plane-wave propagation depends on the two metric dyadics $\overline{\overline{\mathsf{B}}}_o\rfloor\mathbf{A}$ and $\overline{\overline{\mathsf{B}}}_o\rfloor\mathbf{B} \in \mathbb{E}_1\mathbb{E}_1$. The medium has no birefringence if the symmetric parts of these two dyadics are multiples of one another.

To find the field conditions, the equation for the potential one-form can be expanded as

$$\begin{aligned}\overline{\overline{\mathsf{D}}}(\nu)|\boldsymbol{\phi} &= (\nu\nu\rfloor\rfloor\overline{\overline{\mathsf{A}}} + \mathbf{ab} + \mathbf{ba})|\boldsymbol{\phi} \\ &= \mathbf{e}_N\lfloor(\nu \wedge \nu' \wedge \boldsymbol{\phi}) + \mathbf{a}(\mathbf{b}|\boldsymbol{\phi}) + \mathbf{b}(\mathbf{a}|\boldsymbol{\phi}) = 0.\end{aligned} \tag{10.119}$$

Multiplying by $\boldsymbol{\phi}|$ yields $(\mathbf{a}|\boldsymbol{\phi})(\mathbf{b}|\boldsymbol{\phi}) = 0$. Multiplying by $\nu'|$ leads to the condition

$$(\nu'|\mathbf{a})(\mathbf{b}|\boldsymbol{\phi}) + (\nu'|\mathbf{b})(\mathbf{a}|\boldsymbol{\phi}) = 0, \tag{10.120}$$

from which we obtain the two possibilities

$$\nu'|\mathbf{a} = \nu|(\overline{\overline{\mathsf{B}}}_o\rfloor\mathbf{A})|\nu = 0 \quad \Rightarrow \quad \mathbf{a}|\boldsymbol{\phi} = -\mathbf{A}|\boldsymbol{\Phi} = 0, \tag{10.121}$$

$$\nu'|\mathbf{b} = \nu|(\overline{\overline{\mathsf{B}}}_o\rfloor\mathbf{B})|\nu = 0 \quad \Rightarrow \quad \mathbf{b}|\boldsymbol{\phi} = -\mathbf{B}|\boldsymbol{\Phi} = 0. \tag{10.122}$$

This verifies that the plane waves corresponding to the two dispersion equations (10.117) and (10.118) are respectively A-waves and B-waves. For vanishing extension, $\mathbf{AB} \to 0$, the SDC medium becomes an NDE medium since the dispersion equation is satisfied by any ν.

PROBLEMS

10.1 Show that NDE media corresponding to Case 3 (**a** and **d** are quadratic functions while **b** and **c** do not depend on ν in (10.1)) are actually included in the Case 2 of NDE media.

10.2 Show that NDE media corresponding to Case 4 (**a** is a quadratic function while **b** and **d** are linear functions and **c** does not depend on ν in (10.1)) are actually included in the Case 1 of NDE media.

10.3 Assuming that the bidyadic **A** in (10.8) is simple, whence it can be expressed in terms of two vectors as $\mathbf{A} = \mathbf{a}_1 \wedge \mathbf{a}_2$, show that the dispersion dyadic $\overline{\overline{\mathsf{D}}}(\nu)$ is of variable rank for different ν.

10.4 Show that a medium bidyadic $\overline{\overline{\mathsf{M}}}$ restricted by a relation of the form (10.17) satisfies the dispersion equation for any $\boldsymbol{\nu}$.

10.5 Show by considering different possibilities for the coefficients A, B, and C that the condition (10.17) can be reduced to the form (10.21).

10.6 Show that for the Case 2 of NDE media the medium bidyadic can be represented in a form combining the skewon–axion medium and P-axion medium in one expression as

$$\overline{\overline{\mathsf{M}}} = A\overline{\overline{\mathsf{B}}}^{(2)T} + B(\overline{\overline{\mathsf{B}}}{}^{\wedge}_{\wedge}\overline{\overline{\mathsf{I}}})^T + C\overline{\overline{\mathsf{I}}}^{(2)T}$$

for some dyadic $\overline{\overline{\mathsf{B}}}$ and scalars A, B, C. Verify that the dispersion equation (5.240) is satisfied for this medium bidyadic identically for any $\boldsymbol{\nu}$.

10.7 Show that both Case 1 and Case 2 NDE media are closed in an affine transformation (6.16).

10.8 Show that when the medium is defined through the medium bidyadic $\overline{\overline{\mathsf{N}}}$ as (5.52) in the form

$$\overline{\overline{\mathsf{N}}} = \overline{\overline{\mathsf{P}}}_1^{(2)T} - \alpha_1\overline{\overline{\mathsf{I}}}^{(2)T}, \quad \mathrm{tr}\overline{\overline{\mathsf{P}}}_1^{(4)} = \Delta_{P_1} \neq -\alpha_1^2$$

the medium is P-axion medium, that is, $\overline{\overline{\mathsf{M}}} = \overline{\overline{\mathsf{N}}}^{-1}$ is of the similar form.

10.9 Show that the orthogonality conditions (10.36) follow from the form (10.39).

10.10 Derive the expressions (10.48) and (10.49) to construct the dispersion dyadic (10.50) of a Q-axion medium with less than full-rank dyadic $\overline{\overline{\mathsf{Q}}}$ by assuming the expression (10.45).

10.11 Verify that the Q-axion medium bidyadic with $\overline{\overline{\mathsf{Q}}}$ defined by the rank 3 dyadic (10.45) satisfies the dyadic dispersion equation (5.240).

10.12 Expanding the expression (10.50) for the dispersion dyadic of a Q-axion medium with dyadic $\overline{\overline{\mathsf{Q}}}$ of the form (10.45), show that it equals $\overline{\overline{\mathsf{D}}}(\boldsymbol{\nu}) = \overline{\overline{\mathsf{Q}}}^{(2)} \lfloor\lfloor \boldsymbol{\nu}\boldsymbol{\nu}$.

10.13 Define a condition for a Gibbsian anisotropic medium in which plane-wave fields are decomposed to TE_a and TM_b components satisfying the conditions $\mathbf{a} \cdot \mathbf{E}_a = 0$, $\mathbf{b} \cdot \mathbf{H}_b = 0$ for given Gibbsain vectors **a** and **b**, by starting from a condition of the form $\alpha \mathbf{E}_g \cdot \mathbf{B}_g + \beta \mathbf{H}_g \cdot \mathbf{D}_g = 0$.

10.14 Find the 3D representation for the SDC-medium bidyadic (10.94) by substituting the expansions

$$\overline{\overline{\mathsf{P}}} = \overline{\overline{\mathsf{P}}}_s + \boldsymbol{\pi}\mathbf{e}_4 + \boldsymbol{\varepsilon}_4\mathbf{p} + p\boldsymbol{\varepsilon}_4\mathbf{e}_4,$$

$$\mathbf{DC} = (\mathbf{d}_1 \wedge \mathbf{d}_2 + \mathbf{d}_3 \wedge \mathbf{e}_4)(\mathbf{c}_1 \wedge \mathbf{c}_2 + \mathbf{c}_3 \wedge \mathbf{e}_4),$$

in (5.65).

10.15 Find the 3D representation for the SDC-medium bidyadic (10.98) by substituting the expansions

$$\overline{\overline{\mathsf{B}}}_o = \overline{\overline{\mathsf{C}}}_s + \mathbf{e}_4\boldsymbol{\gamma}_s + \mathbf{c}_s\boldsymbol{\varepsilon}_4 - \mathbf{e}_4\boldsymbol{\varepsilon}_4(\text{tr}\overline{\overline{\mathsf{C}}}_s),$$

$$\mathbf{A} = \mathbf{e}_{123}\lfloor\boldsymbol{\alpha}_s + \mathbf{a}_s \wedge \mathbf{e}_4, \qquad \mathbf{B} = \mathbf{e}_{123}\lfloor\boldsymbol{\beta}_s + \mathbf{b}_s \wedge \mathbf{e}_4.$$

10.16 Verify that the QDC-medium dispersion equation (10.104) corresponds to the A-wave polarization condition.

10.17 Verify that the QDC-medium dispersion equation (10.105) corresponds to the B-wave polarization condition.

10.18 Derive the dyadic dispersion equation (10.116) for the SDC-medium case starting from (10.114).

APPENDIX A

Solutions to Problems

CHAPTER 1

1.1 Expand $(\alpha\mathbf{A} + \beta\mathbf{B}) \wedge (\alpha\mathbf{A} + \beta\mathbf{B}) = 2\alpha\beta\mathbf{A} \wedge \mathbf{B}$.

1.2 Starting from

$$\begin{aligned}\mathbf{B} \wedge \mathbf{B} &= (\mathbf{C} - \alpha\mathbf{A}) \wedge (\mathbf{C} - \alpha\mathbf{A}) \\ &= \mathbf{C} \wedge \mathbf{C} - 2\alpha\mathbf{C} \wedge \mathbf{A} + \alpha^2\mathbf{A} \wedge \mathbf{A} = 0,\end{aligned}$$

we have $\mathbf{C} \wedge \mathbf{C} = 2\alpha\mathbf{C} \wedge \mathbf{A} \quad \Rightarrow \quad \boldsymbol{\varepsilon}_N|(\mathbf{C} \wedge \mathbf{C}) = 2\alpha\boldsymbol{\varepsilon}_N|(\mathbf{C} \wedge \mathbf{A})$. From $\boldsymbol{\varepsilon}_N|(\mathbf{C} \wedge \mathbf{A}) \neq 0$, the coefficient α and bivector $\mathbf{B} = \mathbf{C} - \alpha\mathbf{A}$ can be solved as

$$\alpha = \frac{\boldsymbol{\varepsilon}_N|(\mathbf{C} \wedge \mathbf{C})}{\boldsymbol{\varepsilon}_N|(\mathbf{C} \wedge \mathbf{A})}, \quad \mathbf{B} = \mathbf{C} - \frac{\boldsymbol{\varepsilon}_N|(\mathbf{C} \wedge \mathbf{C})}{\boldsymbol{\varepsilon}_N|(\mathbf{C} \wedge \mathbf{A})}\mathbf{A}.$$

1.3 $\mathbf{B} \wedge \mathbf{B} = 0$ yields the condition

$$B_{12}B_{34} + B_{31}B_{24} + B_{14}B_{23} = 0.$$

1.4 Defining a vector basis so that $\mathbf{A} = \mathbf{e}_1 \wedge \mathbf{e}_2 = \mathbf{e}_{12}$ and expanding $\mathbf{B} = \sum_{i<j} B_{ij}\mathbf{e}_{ij}$, we have

$$\mathbf{A} \wedge \mathbf{B} = B_{34}\mathbf{e}_{12} \wedge \mathbf{e}_{34} = 0, \quad \Rightarrow \quad B_{34} = 0$$

From the expansion we see that $\mathbf{B}$ must be of the form $\mathbf{B} = \mathbf{e}_1 \wedge \mathbf{a} + \mathbf{e}_2 \wedge \mathbf{b}$ for some vectors $\mathbf{a}, \mathbf{b}$. From $\mathbf{B} \wedge \mathbf{B} = \mathbf{e}_1 \wedge \mathbf{a} \wedge \mathbf{e}_2 \wedge \mathbf{b} \neq 0$ it follows that the four vectors must be linearly independent.

Multiforms, Dyadics, and Electromagnetic Media, First Edition. Ismo V. Lindell.

1.5 Expressing $\mathbf{C} = \mathbf{a}_1 \wedge \mathbf{a}_2 + \mathbf{a}_3 \wedge \mathbf{a}_4$ and applying the bac-cab rule twice we have

$$\boldsymbol{\beta}\rfloor\mathbf{C} = \mathbf{a}_1(\boldsymbol{\beta}|\mathbf{a}_2) - \mathbf{a}_2(\boldsymbol{\beta}|\mathbf{a}_1) + \mathbf{a}_3(\boldsymbol{\beta}|\mathbf{a}_4) - \mathbf{a}_4(\boldsymbol{\beta}|\mathbf{a}_3) = 0.$$

(1) For nonsimple $\mathbf{C}$ we have $\mathbf{C} \wedge \mathbf{C} = 2\mathbf{a}_1 \wedge \mathbf{a}_2 \wedge \mathbf{a}_3 \wedge \mathbf{a}_4 \neq 0$, whence the vectors $\mathbf{a}_i$ make a basis with dual basis $\boldsymbol{\alpha}_i$. From $\boldsymbol{\alpha}_i|(\boldsymbol{\beta}\rfloor\mathbf{C}) = 0$, we obtain $\boldsymbol{\beta}|\mathbf{a}_i = 0$ for all i and $\boldsymbol{\beta} = \sum \boldsymbol{\alpha}_i \mathbf{a}_i|\boldsymbol{\beta} = 0$. (2) For $\boldsymbol{\beta} \neq 0$, the vectors $\mathbf{a}_i$ cannot form a basis, whence $\mathbf{C} \wedge \mathbf{C} = 0$.

1.6 Writing $\mathbf{A} = \mathbf{x} \wedge \mathbf{y}$, $\mathbf{B} = \mathbf{z} \wedge \mathbf{w}$ in terms of four vectors, from $\mathbf{A} \wedge \mathbf{B} = \mathbf{x} \wedge \mathbf{y} \wedge \mathbf{z} \wedge \mathbf{w} = 0$ it follows that they must be linearly dependent. Assuming $\mathbf{x} \wedge \mathbf{y} \wedge \mathbf{z} \neq 0$ we must have $\mathbf{w} = \alpha\mathbf{x} + \beta\mathbf{y} + \gamma\mathbf{z}$ for some scalars α, β, γ. We can write

$$\mathbf{A} = \mathbf{x} \wedge \mathbf{y} = \mathbf{a} \wedge \mathbf{c}, \qquad \mathbf{B} = \mathbf{z} \wedge (\alpha\mathbf{x} + \beta\mathbf{y}) = \mathbf{b} \wedge \mathbf{c}$$

by choosing $\mathbf{a} = \mathbf{x}/\beta$, $\mathbf{b} = \mathbf{z}$ and $\mathbf{c} = \alpha\mathbf{x} + \beta\mathbf{y}$. The case $\mathbf{x} \wedge \mathbf{y} \wedge \mathbf{z} = 0$ can be handled similarly.

1.7 Assuming $\mathbf{A} \wedge \mathbf{A} \neq 0$ and $\mathbf{B} \wedge \mathbf{B} \neq 0$ and requiring

$$(\mathbf{A} + \lambda\mathbf{B}) \wedge (\mathbf{A} + \lambda\mathbf{B}) = \mathbf{e}_N \varepsilon_N|((\mathbf{A} + \lambda\mathbf{B}) \wedge (\mathbf{A} + \lambda\mathbf{B})) = 0$$

for some scalar λ leads to the scalar equation

$$\varepsilon_N|(\mathbf{A} \wedge \mathbf{A}) + 2\lambda\varepsilon_N|(\mathbf{A} \wedge \mathbf{B}) + \lambda^2\varepsilon_N|(\mathbf{B} \wedge \mathbf{B}) = 0,$$

from which λ can be solved.

1.8 Because $\mathbf{A} \wedge \mathbf{A} = 0$ and $\varepsilon_N|(\mathbf{A} \wedge \mathbf{A}) = 0$ are equivalent, expanding

$$\begin{aligned}\varepsilon_N|(\mathbf{A} \wedge \mathbf{A}) &= \mathbf{A}|(\varepsilon_N\lfloor\mathbf{A}) = \mathbf{A}|\boldsymbol{\Phi} = (\mathbf{e}_N\lfloor(\varepsilon_N\lfloor\mathbf{A}))|\boldsymbol{\Phi} \\ &= \mathbf{e}_N|(\boldsymbol{\Phi} \wedge \boldsymbol{\Phi}) = 0\end{aligned}$$

we obtain $\boldsymbol{\Phi} \wedge \boldsymbol{\Phi} = 0$.

1.9 Applying the bac-cab formula we can write

$$\mathbf{A}\lfloor\boldsymbol{\alpha} = \alpha_1\mathbf{e}_2 - \alpha_2\mathbf{e}_1 + \alpha_3\mathbf{e}_4 - \alpha_4\mathbf{e}_3 = \sum a_i\mathbf{e}_i$$

Finding the coefficients α_i in terms of the coefficients a_i the solution becomes

$$\begin{aligned}\boldsymbol{\alpha} &= a_2\varepsilon_1 - a_1\varepsilon_2 + a_4\varepsilon_3 - a_3\varepsilon_4 = -(\varepsilon_1 \wedge \varepsilon_2 + \varepsilon_3 \wedge \varepsilon_4)\lfloor\mathbf{a} \\ &= -(\varepsilon_N\lfloor(\mathbf{e}_1 \wedge \mathbf{e}_2 + \mathbf{e}_3 \wedge \mathbf{e}_4))\lfloor\mathbf{a} = -(\varepsilon_N\lfloor\mathbf{A})\lfloor\mathbf{a}\end{aligned}$$

Because of $\mathbf{A} \wedge \mathbf{A} = 2\mathbf{e}_N$ the solution can be written as

$$\boldsymbol{\alpha} = -\frac{2\boldsymbol{\varepsilon}_N \lfloor \mathbf{A}}{\boldsymbol{\varepsilon}_N | \mathbf{A} \wedge \mathbf{A}} \lfloor \mathbf{a}.$$

1.10 From $\mathbf{A} \wedge \mathbf{A} = 2\mathbf{e}_{1234} = 2\mathbf{e}_N$ and expanding

$$\begin{aligned}(\mathbf{A} \wedge \mathbf{A})\lfloor \overline{\overline{\mathsf{I}}}^T &= 2\sum_i \mathbf{e}_N \lfloor \boldsymbol{\varepsilon}_i \mathbf{e}_i = 2(\mathbf{e}_{234}\mathbf{e}_1 + \mathbf{e}_{314}\mathbf{e}_2 + \mathbf{e}_{124}\mathbf{e}_3 - \mathbf{e}_{123}\mathbf{e}_4) \\ &= 2(\mathbf{e}_{34} \wedge (\mathbf{e}_2\mathbf{e}_1 - \mathbf{e}_1\mathbf{e}_2) + \mathbf{e}_{12} \wedge (\mathbf{e}_4\mathbf{e}_3 - \mathbf{e}_3\mathbf{e}_4)) \\ &= 2(\mathbf{e}_{34} \wedge (\mathbf{e}_{12} \lfloor \overline{\overline{\mathsf{I}}}^T) + \mathbf{e}_{12} \wedge (\mathbf{e}_{34} \lfloor \overline{\overline{\mathsf{I}}}^T)) \\ &= 2((\mathbf{e}_{12} + \mathbf{e}_{34}) \wedge ((\mathbf{e}_{12} + \mathbf{e}_{34}) \lfloor \overline{\overline{\mathsf{I}}}^T) \\ &= 2\mathbf{A} \wedge (\mathbf{A} \lfloor \overline{\overline{\mathsf{I}}}^T),\end{aligned}$$

we obtain

$$\begin{aligned}(\mathbf{A} \wedge \mathbf{A})\lfloor \boldsymbol{\alpha} = 2\mathbf{A} \wedge (\mathbf{A} \lfloor \boldsymbol{\alpha}) \quad \Rightarrow \quad \boldsymbol{\varepsilon}_N | \boldsymbol{\alpha} &= \boldsymbol{\varepsilon}_N \lfloor (\mathbf{A} \wedge \mathbf{A})\boldsymbol{\alpha} \\ &= 2(\boldsymbol{\varepsilon}_N \lfloor \mathbf{A}) \lfloor (\mathbf{A} \lfloor \boldsymbol{\alpha}),\end{aligned}$$

from which $\boldsymbol{\alpha}$ can be solved.

1.11 Choosing a basis of one-vectors we can assume $\boldsymbol{\alpha} = \boldsymbol{\varepsilon}_1, \boldsymbol{\beta} = \boldsymbol{\varepsilon}_2$, and expand $\mathbf{A} = \sum_{i<j} A_{ij}\mathbf{e}_{ij}$. From

$$\begin{aligned}\mathbf{A} \lfloor \boldsymbol{\varepsilon}_1 &= A_{12}\mathbf{e}_2 - A_{31}\mathbf{e}_3 + A_{14}\mathbf{e}_4 = 0 \\ \mathbf{A} \lfloor \boldsymbol{\varepsilon}_2 &= -A_{12}\mathbf{e}_1 2A_{23}\mathbf{e}_3 + A_{24}\mathbf{e}_4 = 0\end{aligned}$$

we obtain $A_{ij} = 0$ for all ij except $ij = 34$. Thus, we can write

$$\mathbf{A} = A_{34}\mathbf{e}_{34} = A_{34}\mathbf{e}_N \lfloor (\boldsymbol{\varepsilon}_1 \wedge \boldsymbol{\varepsilon}_2) = \mathbf{k}_N \lfloor (\boldsymbol{\alpha} \wedge \boldsymbol{\beta}), \quad \mathbf{k}_N = A_{34}\mathbf{e}_N.$$

1.12 Comparing the two expressions $\mathbf{A} = \sum_{i<j} A_{ij}\mathbf{e}_{ij}$ and $\mathbf{A} = \sum_k A_k\mathbf{E}_k$ we find that they are equal when the coefficients satisfy

$$\begin{aligned}2A_1 &= A_{12} + A_{34}, \quad 2A_2 = A_{23} + A_{14}, \quad 2A_3 = A_{31} + A_{24}, \\ 2A_4 &= A_{12} - A_{34}, \quad 2A_5 = A_{23} - A_{14}, \quad 2A_6 = A_{31} - A_{24}.\end{aligned}$$

Thus, the bivectors $\mathbf{E}_i$ form a basis. Considering different i and j we can find the orthogonality rules

$$\begin{aligned}\mathbf{E}_i \cdot \mathbf{E}_j = 0, \ \ i \neq j, \quad \mathbf{E}_i \cdot \mathbf{E}_i = 2, \ i = 1, 2, 3, \\ \mathbf{E}_i \cdot \mathbf{E}_i = -2, \ i = 4, 5, 6.\end{aligned}$$

1.13 To derive (1.106) we can apply (1.76), written in another form, and proceed as

$$\begin{aligned}\alpha\rfloor(\mathbf{a}\wedge\mathbf{b}\wedge\mathbf{c}) &= (\alpha|\mathbf{a})(\mathbf{b}\wedge\mathbf{c}) + (\alpha|\mathbf{b})(\mathbf{c}\wedge\mathbf{a}) + (\alpha|\mathbf{c})(\mathbf{a}\wedge\mathbf{b})\\ &= (\alpha|\mathbf{a})(\mathbf{b}\wedge\mathbf{c}) + \alpha|(\mathbf{bc}-\mathbf{cb})\wedge\mathbf{a}\\ &= (\alpha|\mathbf{a})(\mathbf{b}\wedge\mathbf{c}) - (\alpha\rfloor(\mathbf{b}\wedge\mathbf{c}))\wedge\mathbf{a},\end{aligned}$$

applying the bac-cab formula. Because this is a linear identity in the simple bivector $\mathbf{b}\wedge\mathbf{c}$ we can replace it by the general bivector $\mathbf{C}$ whence the rule becomes

$$\alpha\rfloor(\mathbf{a}\wedge\mathbf{C}) = (\alpha|\mathbf{a})\mathbf{C} - (\alpha\rfloor\mathbf{C})\wedge\mathbf{a}$$

which is equivalent to (1.106). To derive (1.107) we can start from (1.83) written as

$$\begin{aligned}\alpha\rfloor(\mathbf{a}\wedge\mathbf{b}\wedge\mathbf{c}\wedge\mathbf{d}) &= -(\alpha|\mathbf{a})(\mathbf{b}\wedge\mathbf{c}\wedge\mathbf{d}) + (\alpha|\mathbf{b})(\mathbf{c}\wedge\mathbf{d}\wedge\mathbf{a})\\ &\quad - (\alpha|\mathbf{c})(\mathbf{d}\wedge\mathbf{a}\wedge\mathbf{b}) + (\alpha|\mathbf{d})(\mathbf{a}\wedge\mathbf{b}\wedge\mathbf{c})\\ &= -\alpha|(\mathbf{ab}-\mathbf{ba})\wedge(\mathbf{c}\wedge\mathbf{d}) - \alpha|(\mathbf{cd}-\mathbf{dc})\wedge(\mathbf{a}\wedge\mathbf{b})\\ &= (\alpha\rfloor(\mathbf{a}\wedge\mathbf{b}))\wedge(\mathbf{c}\wedge\mathbf{d}) + (\alpha\rfloor(\mathbf{c}\wedge\mathbf{d}))\wedge(\mathbf{a}\wedge\mathbf{b}).\end{aligned}$$

Replacing $\mathbf{a}\wedge\mathbf{b}$ and $\mathbf{c}\wedge\mathbf{d}$ by the respective general bivectors $\mathbf{B}$ and $\mathbf{C}$ we arrive at the rule

$$\alpha\rfloor(\mathbf{B}\wedge\mathbf{C}) = (\alpha\rfloor\mathbf{B})\wedge\mathbf{C} + (\alpha\rfloor\mathbf{C})\wedge\mathbf{B},$$

which is equivalent to (1.107).

1.14 From (1.108) we see that if the bivector $\mathbf{B}$ is decomposed as $\mathbf{B} = \mathbf{B}_{\|} + \mathbf{B}_{\perp}$, we must have $\mathbf{B}_{\perp} = 0$, whence the solution can be expressed as

$$\mathbf{B} = \mathbf{B}_{\|} = -\frac{\mathbf{a}\wedge(\alpha\rfloor\mathbf{B})}{\mathbf{a}|\alpha},$$

for any one-form α satisfying $\mathbf{a}|\alpha \neq 0$. Thus any bivector of the form $\mathbf{B} = \mathbf{a}\wedge\mathbf{b}$ is a solution.

1.15 Choose a basis so that $\alpha = \varepsilon_1$ and expand $\mathbf{B} = \sum_{i<j} B_{ij}\mathbf{e}_{ij}$. From

$$\alpha\rfloor\mathbf{B} = \varepsilon_1\rfloor\sum B_{ij}\mathbf{e}_{ij} = -B_{12}\mathbf{e}_2 + B_{31}\mathbf{e}_3 - B_{14}\mathbf{e}_4 = 0$$

we conclude that $B_{12} = B_{31} = B_{14} = 0$, whence

$$\overline{\overline{\mathsf{B}}} = B_{23}\mathbf{e}_{23} + B_{24}\mathbf{e}_{24} + B_{34}\mathbf{e}_{34}$$

satisfying $\mathbf{B} \wedge \mathbf{B} = 0$. Writing

$$\mathbf{B} = \boldsymbol{\varepsilon}_1 \rfloor (B_{23}\mathbf{e}_{123} + B_{24}\mathbf{e}_{124} + B_{34}\mathbf{e}_{134})$$

the bivector has the form $\boldsymbol{\alpha} \rfloor \mathbf{k}$.

1.16 Starting from (1.83) and expanding

$$\begin{aligned}\boldsymbol{\alpha} \rfloor (\mathbf{a} \wedge \mathbf{b} \wedge \mathbf{c} \wedge \mathbf{d}) &= -(\boldsymbol{\alpha}|\mathbf{a})(\mathbf{b} \wedge \mathbf{c} \wedge \mathbf{d}) + (\boldsymbol{\alpha}|\mathbf{b})(\mathbf{c} \wedge \mathbf{d} \wedge \mathbf{a}) \\ &\quad - (\boldsymbol{\alpha}|\mathbf{c})(\mathbf{d} \wedge \mathbf{a} \wedge \mathbf{b}) + (\boldsymbol{\alpha}|\mathbf{d})(\mathbf{a} \wedge \mathbf{b} \wedge \mathbf{c}) \\ &= -((\boldsymbol{\alpha}|\mathbf{a})(\mathbf{b} \wedge \mathbf{c}) + (\boldsymbol{\alpha}|\mathbf{b})(\mathbf{c} \wedge \mathbf{a}) \\ &\quad + (\boldsymbol{\alpha}|\mathbf{c})(\mathbf{a} \wedge \mathbf{b})) \wedge \mathbf{d} + (\boldsymbol{\alpha}|\mathbf{d})(\mathbf{a} \wedge \mathbf{b} \wedge \mathbf{c}) \\ &= -(\boldsymbol{\alpha} \rfloor (\mathbf{a} \wedge \mathbf{b} \wedge \mathbf{c})) \wedge \mathbf{d} + (\boldsymbol{\alpha}|\mathbf{d})(\mathbf{a} \wedge \mathbf{b} \wedge \mathbf{c}),\end{aligned}$$

we may replace $\mathbf{a} \wedge \mathbf{b} \wedge \mathbf{c}$ by the general trivector $\mathbf{k}$ and arrive at the rule

$$\boldsymbol{\alpha} \rfloor (\mathbf{k} \wedge \mathbf{d}) = -(\boldsymbol{\alpha} \rfloor \mathbf{k}) \wedge \mathbf{d} + (\boldsymbol{\alpha}|\mathbf{d})\mathbf{k}.$$

Replacing $\mathbf{d}$ by $\mathbf{a}$ and assuming $\mathbf{a}|\boldsymbol{\alpha} \neq 0$ we can expand the trivector as

$$\mathbf{k} = \mathbf{k}_{\alpha} + \mathbf{k}_{a}, \quad \mathbf{k}_{\alpha} = \frac{\boldsymbol{\alpha} \rfloor (\mathbf{k} \wedge \mathbf{a})}{\mathbf{a}|\boldsymbol{\alpha}}, \quad \mathbf{k}_{a} = \frac{\mathbf{a} \wedge (\boldsymbol{\alpha} \rfloor \mathbf{k})}{\mathbf{a}|\boldsymbol{\alpha}}.$$

1.17 The identity can be more easily derived for the special case:

$$\mathbf{e}_N \lfloor (\boldsymbol{\alpha} \wedge \boldsymbol{\varepsilon}_{12}) = \mathbf{e}_{34} \lfloor \boldsymbol{\alpha} = -\boldsymbol{\alpha} \rfloor \mathbf{e}_{34} = -\boldsymbol{\alpha} \rfloor (\mathbf{e}_N \lfloor \boldsymbol{\varepsilon}_{12}).$$

Since choice of the basis is arbitrary, we can replace $\boldsymbol{\varepsilon}_{12}$ by any simple two-form $\boldsymbol{\Psi} = \boldsymbol{\beta} \wedge \boldsymbol{\gamma}$. Because of linearity, it remains valid for a sum of any two simple two-forms and, thus, for any two-form $\boldsymbol{\Psi}$.

1.18 Express $\mathbf{k}$ in terms of a one-form $\boldsymbol{\alpha}$ as $\mathbf{e}_N \lfloor \boldsymbol{\alpha}$ and expand

$$\begin{aligned}(\boldsymbol{\varepsilon}_N \lfloor \mathsf{A}) \rfloor \mathbf{k} &= \mathbf{k} \lfloor (\boldsymbol{\varepsilon}_N \lfloor \mathsf{A}) \\ &= (\mathbf{e}_N \lfloor \boldsymbol{\alpha}) \lfloor (\boldsymbol{\varepsilon}_N \lfloor \mathsf{A}) = \mathbf{e}_N \lfloor (\boldsymbol{\alpha} \wedge (\boldsymbol{\varepsilon}_N \lfloor \mathsf{A})) \\ &= \mathbf{e}_N \lfloor ((\boldsymbol{\varepsilon}_N \lfloor \mathsf{A}) \wedge \boldsymbol{\alpha}) = \mathbf{e}_N \lfloor (\boldsymbol{\varepsilon}_N \lfloor (\mathsf{A} \lfloor \boldsymbol{\alpha})) \\ &= -\mathsf{A} \lfloor \boldsymbol{\alpha} = \boldsymbol{\alpha} \rfloor \mathsf{A} = (\boldsymbol{\varepsilon}_N \lfloor \mathbf{k}) \rfloor \mathsf{A}.\end{aligned}$$

To check: $\mathbf{k} = \mathbf{e}_{234}$, $\mathsf{A} = \mathbf{e}_{12}$,

$$(\boldsymbol{\varepsilon}_N \lfloor \mathbf{e}_{12}) \rfloor \mathbf{e}_{234} = \boldsymbol{\varepsilon}_{34} \rfloor \mathbf{e}_{234} = \mathbf{e}_2$$

$$(\boldsymbol{\varepsilon}_N \lfloor \mathbf{e}_{234} \rfloor \mathbf{e}_{12} = -\boldsymbol{\varepsilon}_1 \rfloor \mathbf{e}_{12} = \mathbf{e}_2.$$

1.19 We can choose a vector basis $\{\mathbf{e}_i\}$ so that $\mathbf{A} = \mathbf{e}_{12} + \mathbf{e}_{34}$. Expanding $\mathbf{X} = \sum X_{ij}\mathbf{e}_{ij}$ we obtain $X_{12} + X_{34} = 0$, which leaves 5 free parameters X_{ij}, $ij \neq 34$ for the solution.

1.20 Expanding

$$\mathbf{a} = \mathbf{a}_s + a_4\mathbf{e}_4, \qquad \mathbf{b} = \mathbf{b}_s + b_4\mathbf{e}_4,$$

we can write

$$\begin{aligned}\mathbf{a} \wedge \mathbf{b} &= \mathbf{a}_s \wedge \mathbf{b}_s + (b_4\mathbf{a}_s - a_4\mathbf{b}_s) \wedge \mathbf{e}_4 \\ &= (b_4\mathbf{a}_s - a_4\mathbf{b}_s) \wedge \left(\frac{1}{b_4}\mathbf{b}_s + \mathbf{e}_4\right).\end{aligned}$$

Assuming $b_4\mathbf{a}_s - a_4\mathbf{b}_s \neq 0$ we can identify

$$\mathbf{c}_s = b_4\mathbf{a}_s - a_4\mathbf{b}_s, \qquad \mathbf{d} = \mathbf{b}/b_4.$$

In the case $b_4\mathbf{a}_s - a_4\mathbf{b}_s = 0$ we have $\mathbf{a} \wedge \mathbf{b} = \mathbf{a}_s \wedge \mathbf{b}_s$.

CHAPTER 2

2.1 (1) For $\overline{\overline{\mathsf{I}}} = \sum \mathbf{e}_i\boldsymbol{\varepsilon}_i$ and $\overline{\overline{\mathsf{A}}} = \sum \mathbf{e}_j\boldsymbol{\alpha}_j$ expand

$$\begin{aligned}\mathrm{tr}(\overline{\overline{\mathsf{I}}}{}^{\wedge}_{\wedge}\overline{\overline{\mathsf{A}}}) &= \sum_{i,j}(\mathbf{e}_i \wedge \mathbf{e}_j)(\boldsymbol{\varepsilon}_i \wedge \boldsymbol{\alpha}_j)||\overline{\overline{\mathsf{I}}}^{(2)T} = \sum_{i,j}(\mathbf{e}_i \wedge \mathbf{e}_j)|(\boldsymbol{\varepsilon}_i \wedge \boldsymbol{\alpha}_j) \\ &= \sum_i(\mathbf{e}_i|\boldsymbol{\varepsilon}_i)\sum_j(\mathbf{e}_j|\boldsymbol{\alpha}_j) - \sum_{i,j}(\mathbf{e}_i|\boldsymbol{\alpha}_j)(\mathbf{e}_j|\boldsymbol{\varepsilon}_i) \\ &= \mathrm{tr}\overline{\overline{\mathsf{I}}}\ \mathrm{tr}\overline{\overline{\mathsf{A}}} - \overline{\overline{\mathsf{I}}}^T||\overline{\overline{\mathsf{A}}} = 3\mathrm{tr}\overline{\overline{\mathsf{A}}}.\end{aligned}$$

(2) Applying the dyadic rule we can expand

$$\begin{aligned}\mathrm{tr}(\overline{\overline{\mathsf{I}}}{}^{\wedge}_{\wedge}\overline{\overline{\mathsf{A}}}) &= \overline{\overline{\mathsf{I}}}^{(2)}||(\overline{\overline{\mathsf{I}}}{}^{\wedge}_{\wedge}\overline{\overline{\mathsf{A}}})^T = (\overline{\overline{\mathsf{I}}}^{(2)}\lfloor\lfloor\overline{\overline{\mathsf{I}}}^T)||\overline{\overline{\mathsf{A}}}^T \\ &= ((\mathrm{tr}\overline{\overline{\mathsf{I}}})\overline{\overline{\mathsf{I}}} - \overline{\overline{\mathsf{I}}})||\overline{\overline{\mathsf{A}}}^T = (4-1)\overline{\overline{\mathsf{I}}}||\overline{\overline{\mathsf{A}}}^T = 3\mathrm{tr}\overline{\overline{\mathsf{A}}}.\end{aligned}$$

2.2 Applying the identity

$$\overline{\overline{\mathsf{I}}}^{(2)}\lfloor\lfloor\overline{\overline{\mathsf{B}}}^T = (\mathrm{tr}\overline{\overline{\mathsf{B}}})\overline{\overline{\mathsf{I}}} - \overline{\overline{\mathsf{I}}}|\overline{\overline{\mathsf{B}}}|\overline{\overline{\mathsf{I}}} = (\mathrm{tr}\overline{\overline{\mathsf{B}}})\overline{\overline{\mathsf{I}}} - \overline{\overline{\mathsf{B}}},$$

we obtain

$$\overline{\overline{\mathsf{B}}} = (\mathrm{tr}\overline{\overline{\mathsf{B}}})\overline{\overline{\mathsf{I}}} - \overline{\overline{\mathsf{A}}}.$$

Taking the trace, $\text{tr}\overline{\overline{\mathsf{B}}}$ can be solved as

$$\text{tr}\overline{\overline{\mathsf{B}}} = 4\text{tr}\overline{\overline{\mathsf{B}}} - \text{tr}\overline{\overline{\mathsf{A}}} \quad \Rightarrow \quad \text{tr}\overline{\overline{\mathsf{B}}} = \frac{1}{3}\text{tr}\overline{\overline{\mathsf{A}}},$$

whence the solution reads

$$\overline{\overline{\mathsf{B}}} = \frac{1}{3}\text{tr}\overline{\overline{\mathsf{A}}}\,\overline{\overline{\mathsf{I}}} - \overline{\overline{\mathsf{A}}}.$$

2.3 Applying the multivector expansion twice on the right side of

$$(\mathbf{a}_1\boldsymbol{\alpha}_1{}^\wedge_\wedge\mathbf{a}_2\boldsymbol{\alpha}_2{}^\wedge_\wedge\mathbf{a}_3\boldsymbol{\alpha}_3)\lfloor\lfloor\boldsymbol{\delta}\mathbf{d} = ((\mathbf{a}_1\wedge\mathbf{a}_2\wedge\mathbf{a}_3)\lfloor\boldsymbol{\delta})((\boldsymbol{\alpha}_1\wedge\boldsymbol{\alpha}_2\wedge\boldsymbol{\alpha}_3)\lfloor\mathbf{d})$$

produces $3\times 3 = 9$ terms invariant in cyclic permutations of the three dyadics $\mathbf{a}_i\boldsymbol{\alpha}_i$. The result can be expressed as

$$(\mathbf{a}_1\boldsymbol{\alpha}_1{}^\wedge_\wedge\mathbf{a}_2\boldsymbol{\alpha}_2{}^\wedge_\wedge\mathbf{a}_3\boldsymbol{\alpha}_3)\lfloor\lfloor\boldsymbol{\delta}\mathbf{d} = (\mathbf{a}_1\boldsymbol{\alpha}_1\wedge\mathbf{a}_2\boldsymbol{\alpha}_2)\mathbf{a}_3\boldsymbol{\alpha}_3||\boldsymbol{\delta}\mathbf{d}$$
$$-\,\mathbf{a}_1\boldsymbol{\alpha}_1{}^\wedge_\wedge(\mathbf{a}_2\boldsymbol{\alpha}_2|\mathbf{d}\boldsymbol{\delta}|\mathbf{a}_3\boldsymbol{\alpha}_3 + \mathbf{a}_3\boldsymbol{\alpha}_3|\mathbf{d}\boldsymbol{\delta}|\mathbf{a}_2\boldsymbol{\alpha}_2) + \cdots,$$

where $\cdots$ corresponds to two permutations of the previous three terms. Replacing $\mathbf{a}_i\boldsymbol{\alpha}_i$ by $\overline{\overline{\mathsf{A}}}_i$, the expansion rule becomes

$$(\overline{\overline{\mathsf{A}}}_1{}^\wedge_\wedge\overline{\overline{\mathsf{A}}}_2{}^\wedge_\wedge\overline{\overline{\mathsf{A}}}_3)\lfloor\lfloor\overline{\overline{\mathsf{D}}}^T = (\overline{\overline{\mathsf{A}}}_1{}^\wedge_\wedge\overline{\overline{\mathsf{A}}}_2)(\overline{\overline{\mathsf{A}}}_3||\overline{\overline{\mathsf{D}}}^T) + (\overline{\overline{\mathsf{A}}}_2{}^\wedge_\wedge\overline{\overline{\mathsf{A}}}_3)(\overline{\overline{\mathsf{A}}}_1||\overline{\overline{\mathsf{D}}}^T)$$
$$+\,(\overline{\overline{\mathsf{A}}}_3{}^\wedge_\wedge\overline{\overline{\mathsf{A}}}_1)(\overline{\overline{\mathsf{A}}}_2||\overline{\overline{\mathsf{D}}}^T) - \overline{\overline{\mathsf{A}}}_1{}^\wedge_\wedge(\overline{\overline{\mathsf{A}}}_2|\overline{\overline{\mathsf{D}}}|\overline{\overline{\mathsf{A}}}_3 + \overline{\overline{\mathsf{A}}}_3|\overline{\overline{\mathsf{D}}}|\overline{\overline{\mathsf{A}}}_2)$$
$$-\,\overline{\overline{\mathsf{A}}}_2{}^\wedge_\wedge(\overline{\overline{\mathsf{A}}}_3|\overline{\overline{\mathsf{D}}}|\overline{\overline{\mathsf{A}}}_1 + \overline{\overline{\mathsf{A}}}_1|\overline{\overline{\mathsf{D}}}|\overline{\overline{\mathsf{A}}}_3) - \overline{\overline{\mathsf{A}}}_3{}^\wedge_\wedge(\overline{\overline{\mathsf{A}}}_1|\overline{\overline{\mathsf{D}}}|\overline{\overline{\mathsf{A}}}_2 + \overline{\overline{\mathsf{A}}}_2|\overline{\overline{\mathsf{D}}}|\overline{\overline{\mathsf{A}}}_1).$$

For $\overline{\overline{\mathsf{A}}}_1 = \overline{\overline{\mathsf{A}}}_2 = \overline{\overline{\mathsf{A}}}_3 = \overline{\overline{\mathsf{A}}}$ this reduces to

$$\overline{\overline{\mathsf{A}}}^{(3)}\lfloor\lfloor\overline{\overline{\mathsf{D}}}^T = \overline{\overline{\mathsf{A}}}^{(2)}(\overline{\overline{\mathsf{A}}}||\overline{\overline{\mathsf{D}}}^T) - \overline{\overline{\mathsf{A}}}{}^\wedge_\wedge(\overline{\overline{\mathsf{A}}}|\overline{\overline{\mathsf{D}}}|\overline{\overline{\mathsf{A}}})$$

and for $\overline{\overline{\mathsf{D}}} = \overline{\overline{\mathsf{A}}}^{-1}$

$$\overline{\overline{\mathsf{A}}}^{(3)}\lfloor\lfloor\overline{\overline{\mathsf{A}}}^{-1T} = 2\overline{\overline{\mathsf{A}}}^{(2)}.$$

For $\overline{\overline{\mathsf{A}}} = \overline{\overline{\mathsf{I}}}$ we have

$$\text{tr}(\overline{\overline{\mathsf{I}}}^{(3)}\lfloor\lfloor\overline{\overline{\mathsf{I}}}^{-1T}) = (\overline{\overline{\mathsf{I}}}^{(3)}\lfloor\lfloor\overline{\overline{\mathsf{I}}}^{-1T})||\overline{\overline{\mathsf{I}}}^{(2)T} = 3\text{tr}\overline{\overline{\mathsf{I}}}^{(3)} = 3\cdot 4 = 12,$$
$$\text{tr}(2\overline{\overline{\mathsf{I}}}^{(2)}) = 2\cdot 6 = 12.$$

2.4 Inserting $\overline{\overline{\mathsf{C}}} = \overline{\overline{\mathsf{A}}}^T$ and $\overline{\overline{\mathsf{A}}}{}^\wedge_\wedge\overline{\overline{\mathsf{B}}} = \overline{\overline{\mathsf{D}}}$ in the dyadic rule and applying $\overline{\overline{\mathsf{A}}}||\overline{\overline{\mathsf{A}}}^{-1T} = \text{tr}(\overline{\overline{\mathsf{A}}}|\overline{\overline{\mathsf{A}}}^{-1}) = \text{tr}\overline{\overline{\mathsf{I}}} = 4$ yields

$$\overline{\overline{\mathsf{D}}}\lfloor\lfloor\overline{\overline{\mathsf{A}}}^{-1T} = 2\overline{\overline{\mathsf{B}}} + (\overline{\overline{\mathsf{B}}}||\overline{\overline{\mathsf{A}}}^{-1T})\overline{\overline{\mathsf{A}}}.$$

Operating this by $||\overline{\overline{\mathsf{A}}}^{-1T}$, the left side can be written as

$$(\overline{\overline{\mathsf{D}}}\lfloor\lfloor\overline{\overline{\mathsf{A}}}^{-1T})||\overline{\overline{\mathsf{A}}}^{-1T} = 2\overline{\overline{\mathsf{D}}}||(\overline{\overline{\mathsf{A}}}^{-1T})^{(2)}$$

while the right side becomes

$$2\overline{\overline{\mathsf{B}}}||\overline{\overline{\mathsf{A}}}^{-1T} + 4\overline{\overline{\mathsf{B}}}||\overline{\overline{\mathsf{A}}}^{-1T} = 6\overline{\overline{\mathsf{B}}}||\overline{\overline{\mathsf{A}}}^{-1T}.$$

Combining these we obtain $\overline{\overline{\mathsf{B}}}||\overline{\overline{\mathsf{A}}}^{-1T} = (1/3)\overline{\overline{\mathsf{D}}}||\overline{\overline{\mathsf{A}}}^{(-2)T}$ which inserted in the dyadic equation leads to the solution

$$\overline{\overline{\mathsf{B}}} = \frac{1}{2}\overline{\overline{\mathsf{D}}}\lfloor\lfloor\overline{\overline{\mathsf{A}}}^{-1T} - \frac{1}{6}(\overline{\overline{\mathsf{D}}}||(\overline{\overline{\mathsf{A}}}^{(-2)T})\overline{\overline{\mathsf{A}}}.$$

A condition between $\overline{\overline{\mathsf{A}}}$ and $\overline{\overline{\mathsf{D}}}$ is obtained by substituting the solution $\overline{\overline{\mathsf{B}}}$ in the original equation,

$$\overline{\overline{\mathsf{D}}} = \frac{1}{2}(\overline{\overline{\mathsf{D}}}\lfloor\lfloor\overline{\overline{\mathsf{A}}}^{-1T})^{\wedge}_{\wedge}\overline{\overline{\mathsf{A}}} - \frac{1}{3}(\overline{\overline{\mathsf{D}}}||(\overline{\overline{\mathsf{A}}}^{(-2)T})\overline{\overline{\mathsf{A}}}^{(2)}.$$

As a simple check, setting $\overline{\overline{\mathsf{D}}} = \overline{\overline{\mathsf{I}}}^{(2)}$ and $\overline{\overline{\mathsf{A}}} = \overline{\overline{\mathsf{I}}}$ leads to

$$\overline{\overline{\mathsf{B}}} = \frac{1}{2}\overline{\overline{\mathsf{I}}}^{(2)}\lfloor\lfloor\overline{\overline{\mathsf{I}}}^{-1T} - \frac{1}{6}(\mathrm{tr}(\overline{\overline{\mathsf{I}}}^{(2)})\overline{\overline{\mathsf{I}}} = \frac{1}{2}3\overline{\overline{\mathsf{I}}} - \frac{1}{6}6\overline{\overline{\mathsf{I}}} = \frac{1}{2}\overline{\overline{\mathsf{I}}},$$

which obviously is the correct solution.

2.5 (1) Expanding $\overline{\overline{\mathsf{A}}} = \sum \mathbf{a}_i\mathbf{b}_i$ we have

$$\begin{aligned}
\sum_j(\mathbf{b}_j\mathbf{a}_j)|\sum_i(\boldsymbol{\Phi}\lfloor\mathbf{a}_i\mathbf{b}_i) &= \sum_{i,j}\mathbf{b}_j(\boldsymbol{\Phi}|(\mathbf{a}_i\wedge\mathbf{a}_j))\mathbf{b}_i \\
&= \boldsymbol{\Phi}|\sum_{i,j}(\mathbf{a}_i\wedge\mathbf{a}_j)(\mathbf{b}_j\mathbf{b}_i) \\
&= \boldsymbol{\Phi}|\sum_{i<j}(\mathbf{a}_i\wedge\mathbf{a}_j)(\mathbf{b}_j\mathbf{b}_i - \mathbf{b}_i\mathbf{b}_j) \\
&= \boldsymbol{\Phi}|\sum_{i<j}(\mathbf{a}_i\wedge\mathbf{a}_j)(\mathbf{b}_i\wedge\mathbf{b}_j)\lfloor\overline{\overline{\mathsf{I}}}^T \\
&= \boldsymbol{\Phi}|\left(\sum_i\mathbf{a}_i\mathbf{b}_i\right)^{(2)}\lfloor\overline{\overline{\mathsf{I}}}^T.
\end{aligned}$$

(2) Because the identity is linear in $\boldsymbol{\Phi}$, it is valid if it can be proved for a simple one-form $\boldsymbol{\Phi} = \boldsymbol{\alpha}\wedge\boldsymbol{\beta}$:

$$\begin{aligned}
\overline{\overline{\mathsf{A}}}^T|((\boldsymbol{\alpha}\wedge\boldsymbol{\beta})\lfloor\overline{\overline{\mathsf{A}}}) &= \overline{\overline{\mathsf{A}}}^T|(\boldsymbol{\beta}(\boldsymbol{\alpha}|\overline{\overline{\mathsf{A}}}) - \boldsymbol{\alpha}(\boldsymbol{\beta}|\overline{\overline{\mathsf{A}}})) \\
&= (\boldsymbol{\beta}|\overline{\overline{\mathsf{A}}})(\boldsymbol{\alpha}|\overline{\overline{\mathsf{A}}}) - (\boldsymbol{\alpha}|\overline{\overline{\mathsf{A}}})(\boldsymbol{\beta}|\overline{\overline{\mathsf{A}}}) \\
&= ((\boldsymbol{\alpha}|\overline{\overline{\mathsf{A}}})\wedge(\boldsymbol{\beta}|\overline{\overline{\mathsf{A}}}))\lfloor\overline{\overline{\mathsf{I}}}^T = (\boldsymbol{\alpha}\wedge\boldsymbol{\beta})|\overline{\overline{\mathsf{A}}}^{(2)}\lfloor\overline{\overline{\mathsf{I}}}^T.
\end{aligned}$$

2.6 Applying the rule of Problem 2.1 we have

$$(\mathrm{tr}\overline{\overline{\mathsf{X}}})\overline{\overline{\mathsf{I}}} - \overline{\overline{\mathsf{X}}} = \lambda\overline{\overline{\mathsf{X}}}$$

taking the trace yields $4\mathrm{tr}\overline{\overline{\mathsf{X}}} - \mathrm{tr}\overline{\overline{\mathsf{X}}} - \lambda\mathrm{tr}\overline{\overline{\mathsf{X}}} = \mathrm{tr}\overline{\overline{\mathsf{X}}}(\lambda - 3) = 0$. One solution is $\lambda = -1$ and any $\overline{\overline{\mathsf{X}}}$ satisfying $\mathrm{tr}\overline{\overline{\mathsf{X}}} = 0$. The other solution is $\lambda = 3$ and $\overline{\overline{\mathsf{X}}} = \alpha\overline{\overline{\mathsf{I}}}$ for any scalar α.

2.7 Setting $\mathbf{B} = \mathbf{C} = \mathbf{A}$ in

$$\boldsymbol{\alpha}\rfloor(\mathbf{B} \wedge \mathbf{C}) = \mathbf{B} \wedge (\boldsymbol{\alpha}\rfloor\mathbf{C}) + \mathbf{C} \wedge (\boldsymbol{\alpha}\rfloor\mathbf{B}),$$

we obtain

$$A\boldsymbol{\alpha}\rfloor\mathbf{e}_N = \mathbf{A} \wedge (\boldsymbol{\alpha}\rfloor\mathbf{A}), \quad A = \frac{1}{2}\varepsilon_N|(\mathbf{A} \wedge \mathbf{A}).$$

Applying $\boldsymbol{\varepsilon}_N\lfloor(\boldsymbol{\alpha}\rfloor\mathbf{e}_N) = \boldsymbol{\alpha}$, this becomes

$$A\boldsymbol{\alpha} = \boldsymbol{\varepsilon}_N\lfloor(\mathbf{A} \wedge (\boldsymbol{\alpha}\rfloor\mathbf{A})) = -\boldsymbol{\varepsilon}_N\lfloor(\mathbf{A} \wedge (\mathbf{A}\lfloor\overline{\overline{\mathsf{I}}}^T))|\boldsymbol{\alpha}.$$

Since this is valid for any $\boldsymbol{\alpha}$, we obtain

$$A\overline{\overline{\mathsf{I}}}^T = -\boldsymbol{\varepsilon}_N\lfloor(\mathbf{A} \wedge (\mathbf{A}\lfloor\overline{\overline{\mathsf{I}}}^T)) = -((\boldsymbol{\varepsilon}_N\lfloor\mathbf{A})\lfloor\overline{\overline{\mathsf{I}}})|(\mathbf{A}\lfloor\overline{\overline{\mathsf{I}}}^T).$$

The condition for the existence of the inverse is now $A \neq 0$, or $\mathbf{A}$ must be a nonsimple bivector. In this case the inverse can be expressed in the form

$$(\mathbf{A}\lfloor\overline{\overline{\mathsf{I}}}^T)^{-1} = \boldsymbol{\Phi}\lfloor\overline{\overline{\mathsf{I}}}, \quad \boldsymbol{\Phi} = -\frac{\boldsymbol{\varepsilon}_N\lfloor\mathbf{A}}{\varepsilon_N|(\mathbf{A} \wedge \mathbf{A})/2}.$$

This result proves that the inverse of an antisymmetric dyadic is antisymmetric.

2.8 Applying $\mathbf{A} = \mathbf{a} \wedge \mathbf{b} + \mathbf{c} \wedge \mathbf{d}$ and $\mathbf{A} \wedge \mathbf{A} = 2\mathbf{a} \wedge \mathbf{b} \wedge \mathbf{c} \wedge \mathbf{d}$, we can make use of the bac-cab formula as

$$\overline{\overline{\mathsf{I}}}\rfloor\mathbf{A} = \mathbf{ba} - \mathbf{ab} + \mathbf{dc} - \mathbf{cd}.$$

Expanding

$$\begin{aligned}(\overline{\overline{\mathsf{I}}}\rfloor\mathbf{A})^{(2)} = {} & (\mathbf{a} \wedge \mathbf{b})(\mathbf{a} \wedge \mathbf{b}) + (\mathbf{c} \wedge \mathbf{d})(\mathbf{c} \wedge \mathbf{d}) - (\mathbf{b} \wedge \mathbf{d})(\mathbf{c} \wedge \mathbf{a}) \\ & -(\mathbf{b} \wedge \mathbf{c})(\mathbf{a} \wedge \mathbf{d}) - (\mathbf{a} \wedge \mathbf{d})(\mathbf{b} \wedge \mathbf{c}) - (\mathbf{c} \wedge \mathbf{a})(\mathbf{b} \wedge \mathbf{d})\end{aligned}$$

this can be compiled to the form

$$(\overline{\overline{\mathsf{I}}}\rfloor\mathbf{A})^{(2)} = \mathbf{AA} - \overline{\overline{\mathsf{D}}} - \overline{\overline{\mathsf{D}}}^T,$$

when denoting

$$\overline{\overline{\mathsf{D}}} = (\mathbf{a}\wedge\mathbf{b})(\mathbf{c}\wedge\mathbf{d}) + (\mathbf{b}\times\mathbf{c})(\mathbf{a}\wedge\mathbf{d}) + (\mathbf{c}\wedge\mathbf{a})(\mathbf{b}\wedge\mathbf{d}).$$

Actually we can write

$$\overline{\overline{\mathsf{D}}} + \overline{\overline{\mathsf{D}}}^T = (\mathbf{a}\wedge\mathbf{b}\wedge\mathbf{c}\wedge\mathbf{d})\lfloor\overline{\overline{\mathsf{I}}}^{(2)T} = \frac{1}{2}(\mathbf{A}\wedge\mathbf{A})\lfloor\overline{\overline{\mathsf{I}}}^{(2)T},$$

whence the rule finally emerges in the form

$$(\overline{\overline{\mathsf{I}}}\rfloor\mathbf{A})^{(2)} = \mathbf{A}\mathbf{A} - \frac{1}{2}(\mathbf{A}\wedge\mathbf{A})\lfloor\overline{\overline{\mathsf{I}}}^{(2)T}.$$

2.9 The condition $\overline{\overline{\mathsf{A}}}_1|\overline{\overline{\mathsf{A}}}_2 = 0$ can be decomposed in four 3D equations

$$\overline{\overline{\mathsf{B}}}_1|\overline{\overline{\mathsf{B}}}_2 + \mathbf{b}_1\boldsymbol{\beta}_2 = 0, \quad \boldsymbol{\beta}_1|\mathbf{b}_2 + B_1B_2 = 0,$$

$$\overline{\overline{\mathsf{B}}}_1|\mathbf{b}_2 + \mathbf{b}_1B_2 = 0, \quad \boldsymbol{\beta}_1|\overline{\overline{\mathsf{B}}}_2 + B_1\mathbf{b}_2 = 0,$$

which yield the condition $\overline{\overline{\mathsf{B}}}_2 = \mathbf{b}_2\boldsymbol{\beta}_2/B_2$ whence

$$\overline{\overline{\mathsf{A}}}_2 = \frac{1}{B_2}(\mathbf{b}_2 + B_2\mathbf{e}_4)(\boldsymbol{\beta}_2 + B_2\boldsymbol{\varepsilon}_4).$$

2.10 First we have

$$\mathrm{tr}(\overline{\overline{\mathsf{D}}}\lfloor\lfloor\overline{\overline{\mathsf{I}}}^T) = (\overline{\overline{\mathsf{D}}}\lfloor\lfloor\overline{\overline{\mathsf{I}}}^T)||\overline{\overline{\mathsf{I}}}^{(2)T} = \mathbf{D}||(\overline{\overline{\mathsf{I}}}{}^\wedge_\wedge\overline{\overline{\mathsf{I}}}^{(2)})^T = 3\mathrm{tr}\overline{\overline{\mathsf{D}}} = 0,$$

whence

$$\mathrm{tr}\overline{\overline{\mathsf{D}}} = \mathrm{tr}(\overline{\overline{\mathsf{I}}}^{(4)}\lfloor\lfloor\overline{\overline{\mathsf{A}}}^T) = (\overline{\overline{\mathsf{I}}}^{(4)}\lfloor\lfloor\overline{\overline{\mathsf{A}}}^T)||\overline{\overline{\mathsf{I}}}^{(3)T} = (\overline{\overline{\mathsf{I}}}^{(4)}\lfloor\lfloor\overline{\overline{\mathsf{I}}}^{(3)T})||\overline{\overline{\mathsf{A}}}^T)$$
$$= \mathrm{tr}\overline{\overline{\mathsf{A}}} = 0.$$

Expanding

$$\overline{\overline{\mathsf{D}}}\lfloor\lfloor\overline{\overline{\mathsf{I}}}^T = (\overline{\overline{\mathsf{I}}}^{(4)}\lfloor\lfloor\overline{\overline{\mathsf{A}}}^T)\lfloor\lfloor\overline{\overline{\mathsf{I}}}^T = \overline{\overline{\mathsf{I}}}^{(4)}\lfloor\lfloor(\overline{\overline{\mathsf{A}}}{}^\wedge_\wedge\overline{\overline{\mathsf{I}}})^T$$
$$= (\overline{\overline{\mathsf{I}}}^{(4)}\lfloor\lfloor\overline{\overline{\mathsf{I}}}^T)\lfloor\lfloor\overline{\overline{\mathsf{A}}}^T = \overline{\overline{\mathsf{I}}}^{(3)T}\lfloor\lfloor\overline{\overline{\mathsf{A}}}^T$$
$$= (\mathrm{tr}\overline{\overline{\mathsf{A}}})\overline{\overline{\mathsf{I}}}^{(2)} - \overline{\overline{\mathsf{I}}}{}^\wedge_\wedge\overline{\overline{\mathsf{A}}} = \overline{\overline{\mathsf{I}}}{}^\wedge_\wedge\overline{\overline{\mathsf{A}}},$$

and

$$(\overline{\overline{\mathsf{I}}}{}^\wedge_\wedge\overline{\overline{\mathsf{A}}})\lfloor\lfloor\overline{\overline{\mathsf{I}}}^T = (\mathrm{tr}\overline{\overline{\mathsf{A}}})\overline{\overline{\mathsf{I}}} + 2\overline{\overline{\mathsf{A}}} = 2\overline{\overline{\mathsf{A}}}$$

lead to the following chain of conclusions: $\overline{\overline{\mathsf{D}}}\lfloor\lfloor\overline{\overline{\mathsf{I}}}^T = 0$ requires $\overline{\overline{\mathsf{A}}}{}^\wedge_\wedge\overline{\overline{\mathsf{I}}} = 0$ and $\overline{\overline{\mathsf{A}}} = 0$, whence $\overline{\overline{\mathsf{D}}} = \overline{\overline{\mathsf{I}}}^{(4)}\lfloor\lfloor\overline{\overline{\mathsf{A}}}^T = 0$.

2.11 Applying the bac-cab rule (1.106), we can expand

$$\begin{aligned}\overline{\overline{\mathsf{I}}}^{(3)}\lfloor\lfloor\boldsymbol{\alpha}\mathbf{a} &= \boldsymbol{\alpha}\rfloor\overline{\overline{\mathsf{I}}}^{(3)}\lfloor\mathbf{a} = \boldsymbol{\alpha}\rfloor(\mathbf{a}\wedge\overline{\overline{\mathsf{I}}}^{(2)})\\ &= \mathbf{a}\wedge(\boldsymbol{\alpha}\rfloor\overline{\overline{\mathsf{I}}}^{(2)}) + (\mathbf{a}|\boldsymbol{\alpha})\overline{\overline{\mathsf{I}}}^{(2)}\\ &= \mathbf{a}\wedge(\overline{\overline{\mathsf{I}}}\wedge\boldsymbol{\alpha}) + (\mathbf{a}|\boldsymbol{\alpha})\overline{\overline{\mathsf{I}}}^{(2)}\\ &= -\overline{\overline{\mathsf{I}}}{}^\wedge_\wedge\mathbf{a}\boldsymbol{\alpha} + (\mathbf{a}|\boldsymbol{\alpha})\overline{\overline{\mathsf{I}}}^{(2)}.\end{aligned}$$

Replacing $\mathbf{a}\boldsymbol{\alpha}$ by $\overline{\overline{\mathsf{A}}}$ we arrive at the identity

$$\overline{\overline{\mathsf{I}}}^{(3)}\lfloor\lfloor\overline{\overline{\mathsf{A}}}^T = (\mathrm{tr}\overline{\overline{\mathsf{A}}})\overline{\overline{\mathsf{I}}}^{(2)} - \overline{\overline{\mathsf{I}}}{}^\wedge_\wedge\overline{\overline{\mathsf{A}}}.$$

The other identity can be derived along similar lines.

2.12 Setting $\overline{\overline{\mathsf{A}}} = \mathbf{a}\boldsymbol{\alpha}$ and $\overline{\overline{\mathsf{B}}} = \mathbf{b}\boldsymbol{\beta}$ and $\overline{\overline{\mathsf{I}}}^{(2)} = \mathbf{e}_{ij}\boldsymbol{\varepsilon}_{ij}$, we can write

$$(\overline{\overline{\mathsf{I}}}^{(2)}{}^\wedge_\wedge\overline{\overline{\mathsf{A}}})\lfloor\lfloor\overline{\overline{\mathsf{B}}}^T = (\overline{\overline{\mathsf{I}}}^{(2)}{}^\wedge_\wedge\mathbf{a}\boldsymbol{\alpha})\lfloor\lfloor\boldsymbol{\beta}\mathbf{b} = \boldsymbol{\beta}\rfloor(\mathbf{a}\wedge\mathbf{e}_{ij})(\boldsymbol{\varepsilon}_{ij}\wedge\boldsymbol{\alpha})\lfloor\mathbf{b}.$$

Applying the bac-cab rule the two terms become

$$\boldsymbol{\beta}\rfloor(\mathbf{a}\wedge\mathbf{e}_{ij}) = \mathbf{a}\wedge(\boldsymbol{\beta}\rfloor\mathbf{e}_{ij}) + (\mathbf{a}|\boldsymbol{\beta})\mathbf{e}_{ij},$$

$$(\boldsymbol{\varepsilon}_{ij}\wedge\boldsymbol{\alpha})\lfloor\mathbf{b} = (\boldsymbol{\varepsilon}_{ij}\lfloor\mathbf{b})\wedge\boldsymbol{\alpha} + (\mathbf{b}|\boldsymbol{\alpha})\boldsymbol{\varepsilon}_{ij}.$$

Combining these, we obtain a sum of four terms

$$\begin{aligned}(\overline{\overline{\mathsf{I}}}^{(2)}{}^\wedge_\wedge\overline{\overline{\mathsf{A}}})\lfloor\lfloor\overline{\overline{\mathsf{B}}}^T &= \mathbf{a}\wedge(\boldsymbol{\beta}\rfloor\overline{\overline{\mathsf{I}}}^{(2)}\lfloor\mathbf{b})\wedge\boldsymbol{\alpha} + (\mathbf{a}|\boldsymbol{\beta})(\overline{\overline{\mathsf{I}}}^{(2)}\lfloor\mathbf{b})\wedge\boldsymbol{\alpha}\\ &\quad + (\mathbf{b}|\boldsymbol{\alpha})\mathbf{a}\wedge(\boldsymbol{\beta}\rfloor\overline{\overline{\mathsf{I}}}^{(2)}) + (\mathbf{a}|\boldsymbol{\beta})(\mathbf{b}|\boldsymbol{\alpha})\overline{\overline{\mathsf{I}}}^{(2)},\end{aligned}$$

which can be expanded separately as

$$\begin{aligned}\mathbf{a}\wedge(\boldsymbol{\beta}\rfloor\overline{\overline{\mathsf{I}}}^{(2)}\lfloor\mathbf{b})\wedge\boldsymbol{\alpha} &= (\overline{\overline{\mathsf{I}}}^{(2)}\lfloor\lfloor\boldsymbol{\beta}\mathbf{b}){}^\wedge_\wedge\mathbf{a}\boldsymbol{\alpha} = (\overline{\overline{\mathsf{I}}}^{(2)}\lfloor\lfloor\overline{\overline{\mathsf{B}}}^T){}^\wedge_\wedge\overline{\overline{\mathsf{A}}}\\ &= (\mathrm{tr}\overline{\overline{\mathsf{B}}}\,\overline{\overline{\mathsf{I}}} - \overline{\overline{\mathsf{B}}}){}^\wedge_\wedge\overline{\overline{\mathsf{A}}}\\ (\mathbf{a}|\boldsymbol{\beta})(\overline{\overline{\mathsf{I}}}^{(2)}\lfloor\mathbf{b})\wedge\boldsymbol{\alpha} &= (\mathbf{a}|\boldsymbol{\beta})(\mathbf{b}\wedge\overline{\overline{\mathsf{I}}})\wedge\boldsymbol{\alpha} = -\overline{\overline{\mathsf{I}}}{}^\wedge_\wedge(\mathbf{b}\boldsymbol{\beta}|\mathbf{a}\boldsymbol{\alpha})\\ &= -\overline{\overline{\mathsf{I}}}{}^\wedge_\wedge(\overline{\overline{\mathsf{B}}}|\overline{\overline{\mathsf{A}}})\\ (\mathbf{b}|\boldsymbol{\alpha})\mathbf{a}\wedge(\boldsymbol{\beta}\rfloor\overline{\overline{\mathsf{I}}}^{(2)}) &= (\mathbf{b}|\boldsymbol{\alpha})\mathbf{a}\wedge(\overline{\overline{\mathsf{I}}}\wedge\boldsymbol{\beta}) = -\overline{\overline{\mathsf{I}}}{}^\wedge_\wedge(\mathbf{a}\boldsymbol{\alpha}|\mathbf{b}\boldsymbol{\beta})\\ &= -\overline{\overline{\mathsf{I}}}{}^\wedge_\wedge(\overline{\overline{\mathsf{A}}}|\overline{\overline{\mathsf{B}}})\\ (\mathbf{a}|\boldsymbol{\beta})(\mathbf{b}|\boldsymbol{\alpha})\overline{\overline{\mathsf{I}}}^{(2)} &= \mathrm{tr}(\mathbf{a}\boldsymbol{\alpha}|\mathbf{b}\boldsymbol{\beta})\overline{\overline{\mathsf{I}}}^{(2)}\\ &= \mathrm{tr}(\overline{\overline{\mathsf{A}}}|\overline{\overline{\mathsf{B}}})\overline{\overline{\mathsf{I}}}^{(2)}.\end{aligned}$$

Combining these, we obtain the required identity,

$$(\overline{\overline{\mathsf{I}}}^{(2)}{}^\wedge_\wedge\overline{\overline{\mathsf{A}}})\lfloor\lfloor\overline{\overline{\mathsf{B}}}^T = \mathrm{tr}(\overline{\overline{\mathsf{A}}}|\overline{\overline{\mathsf{B}}})\overline{\overline{\mathsf{I}}}^{(2)} + (\mathrm{tr}\overline{\overline{\mathsf{B}}})\overline{\overline{\mathsf{I}}}{}^\wedge_\wedge\overline{\overline{\mathsf{A}}} - \overline{\overline{\mathsf{I}}}{}^\wedge_\wedge(\overline{\overline{\mathsf{A}}}|\overline{\overline{\mathsf{B}}} + \overline{\overline{\mathsf{B}}}|\overline{\overline{\mathsf{A}}}) - \overline{\overline{\mathsf{A}}}{}^\wedge_\wedge\overline{\overline{\mathsf{B}}},$$

2.13 Applying the identities of the previous problems, we can expand

$$\begin{aligned}\overline{\overline{\mathsf{I}}}^{(4)}\lfloor\lfloor(\overline{\overline{\mathsf{A}}}{}^\wedge_\wedge\overline{\overline{\mathsf{B}}})^T &= (\overline{\overline{\mathsf{I}}}^{(4)}\lfloor\lfloor\overline{\overline{\mathsf{A}}}^T)\lfloor\lfloor\overline{\overline{\mathsf{B}}}^T = (\mathrm{tr}\overline{\overline{\mathsf{A}}})\overline{\overline{\mathsf{I}}}^{(3)}\lfloor\lfloor\overline{\overline{\mathsf{B}}}^T - (\overline{\overline{\mathsf{I}}}^{(2)}{}^\wedge_\wedge\overline{\overline{\mathsf{A}}}))\lfloor\lfloor\overline{\overline{\mathsf{B}}}^T\\ &= (\mathrm{tr}\overline{\overline{\mathsf{A}}}\ \mathrm{tr}\overline{\overline{\mathsf{B}}} - \mathrm{tr}(\overline{\overline{\mathsf{A}}}|\overline{\overline{\mathsf{B}}}))\overline{\overline{\mathsf{I}}}^{(2)}\\ &\quad + \overline{\overline{\mathsf{I}}}{}^\wedge_\wedge(\overline{\overline{\mathsf{A}}}|\overline{\overline{\mathsf{B}}} + \overline{\overline{\mathsf{B}}}|\overline{\overline{\mathsf{A}}} - (\mathrm{tr}\overline{\overline{\mathsf{A}}})\overline{\overline{\mathsf{B}}} - (\mathrm{tr}\overline{\overline{\mathsf{B}}})\overline{\overline{\mathsf{A}}}) + \overline{\overline{\mathsf{A}}}{}^\wedge_\wedge\overline{\overline{\mathsf{B}}}.\end{aligned}$$

This identity can be made more compact through the rules

$$\begin{aligned}(\overline{\overline{\mathsf{A}}}{}^\wedge_\wedge\overline{\overline{\mathsf{B}}})\lfloor\lfloor\overline{\overline{\mathsf{I}}}^T &= (\mathrm{tr}\overline{\overline{\mathsf{A}}})\overline{\overline{\mathsf{B}}} + (\mathrm{tr}\overline{\overline{\mathsf{B}}})\overline{\overline{\mathsf{A}}} - \overline{\overline{\mathsf{A}}}|\overline{\overline{\mathsf{B}}} - \overline{\overline{\mathsf{B}}}|\overline{\overline{\mathsf{A}}},\\ \mathrm{tr}(\overline{\overline{\mathsf{A}}}{}^\wedge_\wedge\overline{\overline{\mathsf{B}}}) &= \frac{1}{2}((\overline{\overline{\mathsf{A}}}{}^\wedge_\wedge\overline{\overline{\mathsf{B}}})\lfloor\lfloor\overline{\overline{\mathsf{I}}}^T)||\overline{\overline{\mathsf{I}}}^T = \mathrm{tr}\overline{\overline{\mathsf{A}}}\ \mathrm{tr}\overline{\overline{\mathsf{B}}} - \mathrm{tr}(\overline{\overline{\mathsf{A}}}|\overline{\overline{\mathsf{B}}}),\end{aligned}$$

which inserted gives us

$$\overline{\overline{\mathsf{I}}}^{(4)}\lfloor\lfloor(\overline{\overline{\mathsf{A}}}{}^\wedge_\wedge\overline{\overline{\mathsf{B}}})^T = \mathrm{tr}(\overline{\overline{\mathsf{A}}}{}^\wedge_\wedge\overline{\overline{\mathsf{B}}})\ \overline{\overline{\mathsf{I}}}^{(2)} - ((\overline{\overline{\mathsf{A}}}{}^\wedge_\wedge\overline{\overline{\mathsf{B}}})\lfloor\lfloor\overline{\overline{\mathsf{I}}}^T){}^\wedge_\wedge\overline{\overline{\mathsf{I}}} + \overline{\overline{\mathsf{A}}}{}^\wedge_\wedge\overline{\overline{\mathsf{B}}}.$$

Because of linear dependence, the bidyadic $\overline{\overline{\mathsf{A}}}{}^\wedge_\wedge\overline{\overline{\mathsf{B}}}$ can be replaced by an arbitrary bidyadic $\overline{\overline{\mathsf{C}}} \in \mathbb{E}_2\mathbb{F}_2$.

2.14 Operating the different terms of the identity as $\overline{\overline{\mathsf{I}}}^{(4)}\lfloor\lfloor()^T$ and applying the identities given of the previous problems we have

$$\begin{aligned}\overline{\overline{\mathsf{I}}}^{(4)}\lfloor\lfloor(\overline{\overline{\mathsf{I}}}^{(4)T}\lfloor\lfloor\overline{\overline{\mathsf{C}}}) &= \overline{\overline{\mathsf{C}}},\\ \overline{\overline{\mathsf{I}}}^{(4)}\lfloor\lfloor(\mathrm{tr}\overline{\overline{\mathsf{C}}}\ \overline{\overline{\mathsf{I}}}^{(2)T}) &= \mathrm{tr}\overline{\overline{\mathsf{C}}}\ \overline{\overline{\mathsf{I}}}^{(2)}\\ \overline{\overline{\mathsf{I}}}^{(4)}\lfloor\lfloor((\overline{\overline{\mathsf{C}}}^T\lfloor\lfloor\overline{\overline{\mathsf{I}}}){}^\wedge_\wedge\overline{\overline{\mathsf{I}}}^T) &= (\overline{\overline{\mathsf{I}}}^{(4)}\lfloor\lfloor\overline{\overline{\mathsf{I}}}^T)\lfloor\lfloor(\overline{\overline{\mathsf{C}}}^T\lfloor\lfloor\overline{\overline{\mathsf{I}}}) = \overline{\overline{\mathsf{I}}}^{(3)}\lfloor\lfloor(\overline{\overline{\mathsf{C}}}^T\lfloor\lfloor\overline{\overline{\mathsf{I}}})\\ &= \mathrm{tr}(\overline{\overline{\mathsf{C}}}^T\lfloor\lfloor\overline{\overline{\mathsf{I}}})\overline{\overline{\mathsf{I}}}^{(2)} - \overline{\overline{\mathsf{I}}}{}^\wedge_\wedge(\overline{\overline{\mathsf{C}}}\lfloor\lfloor\overline{\overline{\mathsf{I}}}^T)\\ &= 2\mathrm{tr}\overline{\overline{\mathsf{C}}}\ \overline{\overline{\mathsf{I}}}^{(2)} - (\overline{\overline{\mathsf{C}}}\lfloor\lfloor\overline{\overline{\mathsf{I}}}^T){}^\wedge_\wedge\overline{\overline{\mathsf{I}}}\\ \overline{\overline{\mathsf{I}}}^{(4)}\lfloor\lfloor\overline{\overline{\mathsf{C}}}^T &= \overline{\overline{\mathsf{I}}}^{(4)}\lfloor\lfloor\overline{\overline{\mathsf{C}}}^T.\end{aligned}$$

Combining these we arrive at the original identity. The second case follows similar lines.

2.15 Inserting $\overline{\overline{\mathsf{A}}} = \overline{\overline{\mathsf{B}}} + \overline{\overline{\mathsf{C}}}$ and applying linearity we obtain

$$f((\overline{\overline{\mathsf{B}}} + \overline{\overline{\mathsf{C}}})^{(2)}) = f(\overline{\overline{\mathsf{B}}}^{(2)}) + 2f(\overline{\overline{\mathsf{B}}}{}^\wedge_\wedge\overline{\overline{\mathsf{C}}}) + f(\overline{\overline{\mathsf{C}}}^{(2)}) = 2f(\overline{\overline{\mathsf{B}}}{}^\wedge_\wedge\overline{\overline{\mathsf{C}}}) = 0.$$

Assuming $\overline{\overline{\mathsf{B}}} = \mathbf{e}_i\boldsymbol{\varepsilon}_k$ and $\overline{\overline{\mathsf{C}}} = \mathbf{e}_j\boldsymbol{\varepsilon}_\ell$ we have $f(\mathbf{e}_{ij}\boldsymbol{\varepsilon}_{k\ell}) = 0$, whence expanding the general bidyadic $\overline{\overline{\mathsf{D}}} = \sum_{ij,k\ell} D_{ij,k\ell}\mathbf{e}_{ij}\boldsymbol{\varepsilon}_{k\ell}$ we obtain

$$f(\overline{\overline{\mathsf{D}}}) = f\left(\sum_{ij,k\ell} D_{ij,k\ell}\mathbf{e}_{ij}\boldsymbol{\varepsilon}_{k\ell}\right) = \sum_{ij,k\ell} D_{ij,k\ell}f(\mathbf{e}_{ij}\boldsymbol{\varepsilon}_{k\ell}) = 0.$$

2.16 To find the representation $\overline{\overline{\mathsf{B}}}_o{}^\wedge_\wedge\overline{\overline{\mathsf{I}}} = (\overline{\overline{\mathsf{S}}}{}^\wedge_\wedge(\mathbf{A}\lfloor\overline{\overline{\mathsf{I}}}^T))\rfloor\boldsymbol{\varepsilon}_N \in \mathbb{E}_2\mathbb{F}_2$, we replace $\overline{\overline{\mathsf{S}}}$ by $\mathbf{ss}$ and $\mathbf{A}$ by $\mathbf{e}_{12}$ and expand

$$
\begin{aligned}
(\mathbf{ss}{}^\wedge_\wedge(\mathbf{e}_{12}\lfloor\overline{\overline{\mathsf{I}}}^T))\rfloor\boldsymbol{\varepsilon}_N &= (\mathbf{ss}{}^\wedge_\wedge(\mathbf{e}_2\mathbf{e}_1 - \mathbf{e}_1\mathbf{e}_2))\rfloor\boldsymbol{\varepsilon}_N \\
&= (\mathbf{s}\wedge\mathbf{e}_2)(\mathbf{s}\wedge\mathbf{e}_1)\rfloor\boldsymbol{\varepsilon}_N - (\mathbf{s}\wedge\mathbf{e}_1)(\mathbf{s}\wedge\mathbf{e}_2)\rfloor\boldsymbol{\varepsilon}_N \\
&= -(\mathbf{s}\wedge\mathbf{e}_2)\mathbf{s}\rfloor\boldsymbol{\varepsilon}_{234} - (\mathbf{s}\wedge\mathbf{e}_1)\mathbf{s}\rfloor\boldsymbol{\varepsilon}_{134}
\end{aligned}
$$

Applying the multivector rule we can further expand

$$
\begin{aligned}
(\mathbf{ss}{}^\wedge_\wedge(\mathbf{e}_{12}\lfloor\overline{\overline{\mathsf{I}}}^T))\rfloor\boldsymbol{\varepsilon}_N &= -(\mathbf{s}\wedge\mathbf{e}_2)((\mathbf{s}|\boldsymbol{\varepsilon}_2)\boldsymbol{\varepsilon}_{34} + \boldsymbol{\varepsilon}_2\wedge(\mathbf{s}\rfloor\boldsymbol{\varepsilon}_{34})) \\
&\quad - (\mathbf{s}\wedge\mathbf{e}_1)((\mathbf{s}|\boldsymbol{\varepsilon}_1)\boldsymbol{\varepsilon}_{34} + \boldsymbol{\varepsilon}_1\wedge(\mathbf{s}\rfloor\boldsymbol{\varepsilon}_{34})) \\
&= -\mathbf{s}\wedge((\mathbf{e}_2\boldsymbol{\varepsilon}_2 + \mathbf{e}_1\boldsymbol{\varepsilon}_1)|\mathbf{s})\boldsymbol{\varepsilon}_{34} \\
&\quad - \mathbf{s}\wedge(\mathbf{e}_2\boldsymbol{\varepsilon}_2 + \mathbf{e}_1\boldsymbol{\varepsilon}_1)\wedge(\mathbf{s}\rfloor\boldsymbol{\varepsilon}_{34}).
\end{aligned}
$$

Working on the latter term as

$$
\begin{aligned}
\mathbf{s}\wedge(\mathbf{e}_2\boldsymbol{\varepsilon}_2 + \mathbf{e}_1\boldsymbol{\varepsilon}_1)\wedge(\mathbf{s}\rfloor\boldsymbol{\varepsilon}_{34}) &= \mathbf{s}\wedge\overline{\overline{\mathsf{I}}}\wedge(\mathbf{s}\rfloor\boldsymbol{\varepsilon}_{34}) \\
&\quad - \mathbf{s}\wedge(\mathbf{e}_3\boldsymbol{\varepsilon}_3 + \mathbf{e}_4\boldsymbol{\varepsilon}_4)\wedge(\mathbf{s}\rfloor\boldsymbol{\varepsilon}_{34}) \\
&= -(\mathbf{ss}\rfloor\boldsymbol{\varepsilon}_{34}){}^\wedge_\wedge\overline{\overline{\mathsf{I}}} + \mathbf{s}\wedge(\mathbf{e}_3\boldsymbol{\varepsilon}_3|\mathbf{s} \\
&\quad + \mathbf{e}_4\boldsymbol{\varepsilon}_4|\mathbf{s})\boldsymbol{\varepsilon}_{34},
\end{aligned}
$$

and inserting this in the previous expansion we obtain

$$
(\mathbf{ss}{}^\wedge_\wedge(\mathbf{e}_{12}\lfloor\overline{\overline{\mathsf{I}}}^T))\rfloor\boldsymbol{\varepsilon}_N = -\mathbf{s}\wedge(\overline{\overline{\mathsf{I}}}|\mathbf{s})\boldsymbol{\varepsilon}_{34} + (\mathbf{ss}\rfloor\boldsymbol{\varepsilon}_{34}){}^\wedge_\wedge\overline{\overline{\mathsf{I}}} = (\mathbf{ss}\rfloor(\mathbf{e}_{12}\rfloor\boldsymbol{\varepsilon}_N)){}^\wedge_\wedge\overline{\overline{\mathsf{I}}}.
$$

Replacing $\mathbf{ss}$ by $\overline{\overline{\mathsf{S}}}$ and $\mathbf{e}_{12}$ by $\mathbf{A}$ this can be written as the identity

$$
(\overline{\overline{\mathsf{S}}}{}^\wedge_\wedge(\mathbf{A}\lfloor\overline{\overline{\mathsf{I}}}^T))\rfloor\boldsymbol{\varepsilon}_N = (\overline{\overline{\mathsf{S}}}\rfloor(\mathbf{A}\rfloor\boldsymbol{\varepsilon}_N)){}^\wedge_\wedge\overline{\overline{\mathsf{I}}} = (((\overline{\overline{\mathsf{S}}}\wedge\mathbf{A})\rfloor\boldsymbol{\varepsilon}_N){}^\wedge_\wedge\overline{\overline{\mathsf{I}}}),
$$

valid for any symmetric dyadic $\overline{\overline{\mathsf{S}}}$ and a bivector $\mathbf{A}$. The dyadic $\overline{\overline{\mathsf{B}}}_o$ can now be identified as

$$
\overline{\overline{\mathsf{B}}}_o = \overline{\overline{\mathsf{S}}}\rfloor(\mathbf{A}\rfloor\boldsymbol{\varepsilon}_N) = (\overline{\overline{\mathsf{S}}}\wedge\mathbf{A})\rfloor\boldsymbol{\varepsilon}_N.
$$

2.17 Starting from the equation

$$
\overline{\overline{\mathsf{A}}}|(\mathbf{a}_s + \mathbf{e}_4 a_4) = \mathbf{b}_s + \mathbf{e}_4 b_4
$$

and splitting it in spatial and temporal parts as

$$
\begin{aligned}
\overline{\overline{\mathsf{A}}}_s|\mathbf{a}_s + \mathbf{a}_s a_4 &= \mathbf{b}_s \\
\boldsymbol{\alpha}_s|\mathbf{b}_s + a a_4 &= b_4
\end{aligned}
$$

we can solve for $\mathbf{a}_s, a_4$ as

$$\mathbf{a}_s = \overline{\overline{\mathsf{A}}}_s^{-1}|\mathbf{b}_s + \frac{1}{D}((\overline{\overline{\mathsf{A}}}_s^{-1}|\mathbf{a}_s)(\boldsymbol{\alpha}_s|\overline{\overline{\mathsf{A}}}_s^{-1})|\mathbf{b}_s - (\overline{\overline{\mathsf{A}}}_s^{-1}|\mathbf{a}_s)b_4)$$

$$a_4 = \frac{1}{D}(-(\boldsymbol{\alpha}_s|\overline{\overline{\mathsf{A}}}_s^{-1})|\mathbf{b}_s + b_4)$$

with $D = a - \boldsymbol{\alpha}_s|\overline{\overline{\mathsf{A}}}_s^{-1}|\mathbf{a}_s$. Combining these as

$$\mathbf{a}_s + \mathbf{e}_4 a_4 = \overline{\overline{\mathsf{A}}}^{-1}|(\mathbf{b}_s + \mathbf{e}_4 b_4)$$

we can identify the inverse dyadic,

$$\overline{\overline{\mathsf{A}}}^{-1} = \overline{\overline{\mathsf{A}}}_s^{-1} + \frac{1}{D}(\overline{\overline{\mathsf{A}}}_s^{-1}|\mathbf{a}_s\boldsymbol{\varepsilon}_4 - \mathbf{e}_4\boldsymbol{\alpha}_s|\overline{\overline{\mathsf{A}}}_s^{-1} - \mathbf{e}_4\boldsymbol{\varepsilon}_4),$$

which can be further written in the form of (2.134).

2.18 The inverse of the dyadic $\overline{\overline{\mathsf{B}}} = \overline{\overline{\mathsf{A}}} + \mathbf{ab} + \mathbf{ba}$ can be found by solving the equation $\overline{\overline{\mathsf{B}}}|\boldsymbol{\gamma} = \mathbf{c}$ for the one-form $\boldsymbol{\gamma}$ and expressing the solution in the form $\boldsymbol{\gamma} = \overline{\overline{\mathsf{B}}}^{-1}|\mathbf{c}$. Representing the solution as

$$\boldsymbol{\gamma} = \overline{\overline{\mathsf{A}}}^{-1}|\mathbf{c} - \overline{\overline{\mathsf{A}}}^{-1}|(\mathbf{ab} + \mathbf{ba})|\boldsymbol{\gamma},$$

let us multiply this by $\mathbf{a}|$ and $\mathbf{b}|$ and express the equations in the form

$$\begin{pmatrix} 1 + \mathbf{a}|\overline{\overline{\mathsf{A}}}^{-1}|\mathbf{b} & \mathbf{a}|\overline{\overline{\mathsf{A}}}^{-1}|\mathbf{a} \\ \mathbf{b}|\overline{\overline{\mathsf{A}}}^{-1}|\mathbf{b} & 1 + \mathbf{b}|\overline{\overline{\mathsf{A}}}^{-1}|\mathbf{a} \end{pmatrix} \begin{pmatrix} \mathbf{a}|\boldsymbol{\gamma} \\ \mathbf{b}|\boldsymbol{\gamma} \end{pmatrix} = \begin{pmatrix} \mathbf{a}|\overline{\overline{\mathsf{A}}}^{-1}|\mathbf{c} \\ \mathbf{b}|\overline{\overline{\mathsf{A}}}^{-1}|\mathbf{c} \end{pmatrix}.$$

Assuming that the determinant

$$\begin{aligned}\Delta &= (1 + \mathbf{a}|\overline{\overline{\mathsf{A}}}^{-1}|\mathbf{b})(1 + \mathbf{b}|\overline{\overline{\mathsf{A}}}^{-1}|\mathbf{a}) - (\mathbf{a}|\overline{\overline{\mathsf{A}}}^{-1}|\mathbf{a})(\mathbf{b}|\overline{\overline{\mathsf{A}}}^{-1}|\mathbf{b}) \\ &= 1 + \overline{\overline{\mathsf{A}}}^{-1}||(\mathbf{ab} + \mathbf{ba}) + \overline{\overline{\mathsf{A}}}^{(-2)}||(\mathbf{ab}{}^\wedge_\wedge\mathbf{ba})\end{aligned}$$

does not vanish, we can solve for $\mathbf{a}|\boldsymbol{\gamma}$ and $\mathbf{b}|\boldsymbol{\gamma}$ and substitute them in the expression

$$\boldsymbol{\gamma} = \overline{\overline{\mathsf{A}}}^{-1}|\mathbf{c} - (\overline{\overline{\mathsf{A}}}^{-1}|\mathbf{b}\ \ \overline{\overline{\mathsf{A}}}^{-1}|\mathbf{a}) \begin{pmatrix} \mathbf{a}|\boldsymbol{\gamma} \\ \mathbf{b}|\boldsymbol{\gamma} \end{pmatrix},$$

from which the inverse dyadic can be identified as

$$\begin{aligned}\overline{\overline{\mathsf{B}}}^{-1} &= \overline{\overline{\mathsf{A}}}^{-1} - \frac{1}{\Delta}\overline{\overline{\mathsf{A}}}^{-1}|((1 + \mathbf{b}|\overline{\overline{\mathsf{A}}}^{-1}|\mathbf{a})\mathbf{ba} - (\mathbf{a}|\overline{\overline{\mathsf{A}}}^{-1}|\mathbf{a})\mathbf{bb} \\ &\quad - (\mathbf{b}|\overline{\overline{\mathsf{A}}}^{-1}|\mathbf{b})\mathbf{aa} + (1 + \mathbf{a}|\overline{\overline{\mathsf{A}}}^{-1}|\mathbf{b})\mathbf{ab})|\overline{\overline{\mathsf{A}}}^{-1} \\ &= \overline{\overline{\mathsf{A}}}^{-1} - \frac{1}{\Delta}\overline{\overline{\mathsf{A}}}^{-1}|(\mathbf{ab} + \mathbf{ba} + \overline{\overline{\mathsf{A}}}^{-1T}\rfloor\rfloor(\mathbf{ab}{}^\wedge_\wedge\mathbf{ba}))|\overline{\overline{\mathsf{A}}}^{-1}.\end{aligned}$$

For the special case $\mathbf{b} = \mathbf{a}/2$ we obtain

$$(\overline{\overline{\mathsf{A}}} + \mathbf{aa})^{-1} = \overline{\overline{\mathsf{A}}}^{-1} - \frac{\overline{\overline{\mathsf{A}}}^{-1}|\mathbf{aa}|\overline{\overline{\mathsf{A}}}^{-1}}{1 + \overline{\overline{\mathsf{A}}}^{-1}||\mathbf{aa}},$$

which can be easily verified by direct multiplication.

2.19 Assuming $\Delta = \mathrm{tr}\overline{\overline{\mathsf{A}}}_s^{(3)} \neq 0$, from the Cayley–Hamilton equation

$$\overline{\overline{\mathsf{A}}}_s^3 - (\mathrm{tr}\overline{\overline{\mathsf{A}}}_s)\overline{\overline{\mathsf{A}}}_s^2 + (\mathrm{tr}\overline{\overline{\mathsf{A}}}_s^{(2)})\overline{\overline{\mathsf{A}}}_s - (\mathrm{tr}\overline{\overline{\mathsf{A}}}_s^{(3)})\overline{\overline{\mathsf{I}}}_s = 0$$

we obtain the inverse rule

$$\overline{\overline{\mathsf{A}}}_s^{-1} = \frac{1}{\Delta}(\overline{\overline{\mathsf{A}}}_s^2 - (\mathrm{tr}\overline{\overline{\mathsf{A}}}_s)\overline{\overline{\mathsf{A}}}_s + (\mathrm{tr}\overline{\overline{\mathsf{A}}}_s^{(2)})\overline{\overline{\mathsf{I}}}_s).$$

Applying the rule (2.101), written for $\overline{\overline{\mathsf{A}}}_s$ as

$$\overline{\overline{\mathsf{I}}}^{(3)}\lfloor\lfloor\overline{\overline{\mathsf{A}}}_s^{(2)T} = (\mathrm{tr}\overline{\overline{\mathsf{A}}}_s^{(2)})\overline{\overline{\mathsf{I}}} - (\mathrm{tr}\overline{\overline{\mathsf{A}}}_s)\overline{\overline{\mathsf{A}}}_s + \overline{\overline{\mathsf{A}}}_s^2$$

we can write

$$\begin{aligned}\overline{\overline{\mathsf{A}}}_s^{-1} &= \frac{1}{\Delta}(\overline{\overline{\mathsf{I}}}^{(3)}\lfloor\lfloor\overline{\overline{\mathsf{A}}}_s^{(2)T} - (\mathrm{tr}\overline{\overline{\mathsf{A}}}_s^{(2)})\overline{\overline{\mathsf{I}}} + (\mathrm{tr}\overline{\overline{\mathsf{A}}}_s^{(2)})\overline{\overline{\mathsf{I}}}_s) \\ &= \frac{1}{\Delta}(\overline{\overline{\mathsf{I}}}_s^{(3)}\lfloor\lfloor\overline{\overline{\mathsf{A}}}_s^{(2)T} + (\overline{\overline{\mathsf{I}}}_s^{(2)}{}^\wedge_\wedge\mathbf{e}_4\boldsymbol{\varepsilon}_4)\lfloor\lfloor\overline{\overline{\mathsf{A}}}_s^{(2)T} - (\mathrm{tr}\overline{\overline{\mathsf{A}}}_s^{(2)})\mathbf{e}_4\boldsymbol{\varepsilon}_4).\end{aligned}$$

Obviously, the term $(\overline{\overline{\mathsf{I}}}_s^{(2)}{}^\wedge_\wedge\mathbf{e}_4\boldsymbol{\varepsilon}_4)\lfloor\lfloor\overline{\overline{\mathsf{A}}}_s^{(2)T} \in \mathbb{E}_1\mathbb{F}_1$ must be of the form $\alpha\mathbf{e}_4\boldsymbol{\varepsilon}_4$. The coefficient α can be expanded as

$$\begin{aligned}\alpha &= ((\overline{\overline{\mathsf{I}}}_s^{(2)}{}^\wedge_\wedge\mathbf{e}_4\boldsymbol{\varepsilon}_4)\lfloor\lfloor\overline{\overline{\mathsf{A}}}_s^{(2)T})||\boldsymbol{\varepsilon}_4\mathbf{e}_4 \\ &= (\overline{\overline{\mathsf{I}}}_s^{(2)}{}^\wedge_\wedge\mathbf{e}_4\boldsymbol{\varepsilon}_4)\lfloor\lfloor(\boldsymbol{\varepsilon}_4\mathbf{e}_4)||\overline{\overline{\mathsf{A}}}_s^{(2)T} = \mathrm{tr}\overline{\overline{\mathsf{A}}}_s^{(2)}.\end{aligned}$$

Inserting this in the above expression, the inverse of the spatial dyadic is reduced to

$$\overline{\overline{\mathsf{A}}}_s^{-1} = \frac{1}{\Delta}\overline{\overline{\mathsf{I}}}_s^{(3)}\lfloor\lfloor\overline{\overline{\mathsf{A}}}_s^{(2)T}.$$

2.20 (1) Assuming $\mathbf{a} = \mathbf{a}_1$ is a vector in a basis $\{\mathbf{a}_i\}$, we have

$$\overline{\overline{\mathsf{A}}}^{(4)}|(\mathbf{a}_1 \wedge \mathbf{a}_2 \wedge \mathbf{a}_3 \wedge \mathbf{a}_4) = (\overline{\overline{\mathsf{A}}}|\mathbf{a}_1) \wedge (\overline{\overline{\mathsf{A}}}|\mathbf{a}_2)|(\overline{\overline{\mathsf{A}}}|\mathbf{a}_3)|(\overline{\overline{\mathsf{A}}}|\mathbf{a}_4) = 0$$

which implies $\overline{\overline{\mathsf{A}}}^{(4)}|\mathbf{e}_N = 0$. In other words, the dyadic $\overline{\overline{\mathsf{A}}}$ must be of rank 3 or lower.

(2) Assuming $\mathbf{a} = \mathbf{a}_1$ and $\mathbf{b} = \mathbf{a}_2$ are vectors in a basis $\{\mathbf{a}_i\}$, we have

$$\overline{\overline{A}}^{(3)}|(\mathbf{a}_i \wedge \mathbf{a}_j \wedge \mathbf{a}_k) = (\overline{\overline{A}}|\mathbf{a}_i) \wedge (\overline{\overline{A}}|\mathbf{a}_j)|(\overline{\overline{A}}|\mathbf{a}_k) = 0$$

for all four index combinations i, j, k since at least one of the indices must be either 1 or 2. Thus, $\overline{\overline{A}}^{(3)}|\mathbf{k} = 0$ for any trivector $\mathbf{k}$ and, hence, $\overline{\overline{A}}^{(3)} = 0$. In other words, the dyadic $\overline{\overline{A}}$ must be of rank 2 or lower.

2.21 The representation $\overline{\overline{C}}_s = \overline{\overline{I}}_s^{(3)} \lfloor\lfloor \overline{\overline{A}}_s^T = \mathbf{e}_{123}\boldsymbol{\varepsilon}_{123} \lfloor\lfloor \overline{\overline{A}}_s^T$ can be inverted as $\overline{\overline{A}}_s = \mathbf{e}_{123}\boldsymbol{\varepsilon}_{123} \lfloor\lfloor \overline{\overline{C}}_s^T = \overline{\overline{I}}_s^{(3)} \lfloor\lfloor \overline{\overline{C}}_s^T$. The corresponding relation between the inverses is obtained from

$$\overline{\overline{C}}_s|\overline{\overline{C}}_s^{-1} = (\mathbf{e}_{123} \lfloor \overline{\overline{A}}_s^T \rfloor \boldsymbol{\varepsilon}_{123})|\overline{\overline{C}}_s^{-1} = \overline{\overline{I}}_s^{(2)},$$

which implies

$$\boldsymbol{\varepsilon}_{123} \lfloor \overline{\overline{C}}_s^{-1} = \overline{\overline{A}}_s^{-1T} \rfloor \boldsymbol{\varepsilon}_{123} \quad \Rightarrow \quad \overline{\overline{C}}_s^{-1} = \overline{\overline{I}}_s^{(3)} \lfloor\lfloor \overline{\overline{A}}_s^{-1T}.$$

Invoking the inverse rule (2.136), we can write

$$\begin{aligned}
\mathrm{tr}\overline{\overline{A}}_s^{(3)} &= \mathrm{tr}(\overline{\overline{I}}_s^{(3)} \lfloor\lfloor \overline{\overline{C}}_s^T)^{(3)} = \Delta_{C_s}, \\
\overline{\overline{C}}_s^{-1} &= \overline{\overline{I}}_s^{(3)} \lfloor\lfloor \overline{\overline{A}}_s^{-1T} = \frac{1}{\Delta_{C_s}} \overline{\overline{I}}_s^{(3)} \lfloor\lfloor (\overline{\overline{I}}_s^{(3)T} \lfloor\lfloor \overline{\overline{A}}_s^{(2)}) = \frac{1}{\Delta_{C_s}} \overline{\overline{A}}_s^{(2)} \\
&= \frac{1}{\Delta_{C_s}} (\overline{\overline{I}}_s^{(3)} \lfloor\lfloor \overline{\overline{C}}_s^T)^{(2)},
\end{aligned}$$

which coincides with (2.137).

2.22 Expand

$$\begin{aligned}
\boldsymbol{\varepsilon}_N\boldsymbol{\varepsilon}_N \lfloor\lfloor (\overline{\overline{A}} {}^\wedge_\wedge \overline{\overline{S}}^{(2)}) &= (\boldsymbol{\varepsilon}_N\boldsymbol{\varepsilon}_N \lfloor\lfloor \overline{\overline{S}}^{(2)}) \lfloor\lfloor \overline{\overline{A}} = \Delta_S \overline{\overline{S}}^{(-2)} \lfloor\lfloor \overline{\overline{A}} \\
&= \Delta_S((\overline{\overline{S}}^{-1}||\overline{\overline{A}})\overline{\overline{S}}^{-1} - \overline{\overline{S}}^{-1}|\overline{\overline{A}}|\overline{\overline{S}}^{-1}) \\
&= -\Delta_S \overline{\overline{S}}^{-1}|(\mathbf{A} \lfloor \overline{\overline{I}}^T)|\overline{\overline{S}}^{-1} = -\Delta_S \overline{\overline{S}}^{-1}|(\mathbf{A} \lfloor \overline{\overline{S}}^{-1}).
\end{aligned}$$

Applying now the identity of Problem 2.5, we can proceed as

$$\begin{aligned}
\boldsymbol{\varepsilon}_N\boldsymbol{\varepsilon}_N \lfloor\lfloor (\overline{\overline{A}} {}^\wedge_\wedge \overline{\overline{S}}^{(2)}) &= -\Delta_S(\mathbf{A}|\overline{\overline{S}}^{-2}) \lfloor \overline{\overline{I}} \\
&= -(\mathbf{A} \lfloor (\mathbf{e}_N\mathbf{e}_N \lfloor\lfloor \overline{\overline{S}}^2)) \lfloor \overline{\overline{I}}.
\end{aligned}$$

One should note that the resulting identity does not require $\overline{\overline{S}}$ to have an inverse.

CHAPTER 3

3.1 Assume $\mathbf{a} = \mathbf{e}_i$, whence $(\overline{\overline{\mathsf{C}}}\lfloor\mathbf{e}_i)|\mathbf{e}_j = \overline{\overline{\mathsf{C}}}|\mathbf{e}_{ij} = 0$ for any basis bivector $\mathbf{e}_{ij}$. Thus,

$$\overline{\overline{\mathsf{C}}} = \overline{\overline{\mathsf{C}}}|\overline{\overline{\mathsf{I}}}^{(2)} = \sum_{j<j} \overline{\overline{\mathsf{C}}}|\mathbf{e}_{ij}\boldsymbol{\varepsilon}_{ij} = 0.$$

3.2 Applying the two identities

$$(\overline{\overline{\mathsf{I}}}{}^{\wedge}_{\wedge}\overline{\overline{\mathsf{X}}})\lfloor\lfloor\overline{\overline{\mathsf{I}}}^T = (\mathrm{tr}\overline{\overline{\mathsf{I}}})\overline{\overline{\mathsf{X}}} + (\mathrm{tr}\overline{\overline{\mathsf{X}}})\overline{\overline{\mathsf{I}}} - \overline{\overline{\mathsf{I}}}|\overline{\overline{\mathsf{I}}}|\overline{\overline{\mathsf{X}}} - \overline{\overline{\mathsf{X}}}|\overline{\overline{\mathsf{I}}}|\overline{\overline{\mathsf{I}}}$$
$$= 2\overline{\overline{\mathsf{X}}} + (\mathrm{tr}\overline{\overline{\mathsf{X}}})\overline{\overline{\mathsf{I}}},$$
$$\mathrm{tr}(\overline{\overline{\mathsf{I}}}{}^{\wedge}_{\wedge}\overline{\overline{\mathsf{X}}}) = 3\mathrm{tr}\overline{\overline{\mathsf{X}}},$$

valid for any dyadic $\overline{\overline{\mathsf{X}}}$, we have

$$(\overline{\overline{\mathsf{I}}}{}^{\wedge}_{\wedge}\overline{\overline{\mathsf{X}}})\lfloor\lfloor\overline{\overline{\mathsf{I}}}^T = \overline{\overline{\mathsf{Y}}}\lfloor\lfloor\overline{\overline{\mathsf{I}}}^T = 2\overline{\overline{\mathsf{X}}} + \frac{1}{3}\mathrm{tr}\overline{\overline{\mathsf{Y}}},$$

from which $\overline{\overline{\mathsf{X}}}$ can be solved as

$$\overline{\overline{\mathsf{X}}} = \frac{1}{2}\overline{\overline{\mathsf{Y}}}\lfloor\lfloor\overline{\overline{\mathsf{I}}}^T - \frac{\mathrm{tr}\overline{\overline{\mathsf{Y}}}}{6}\overline{\overline{\mathsf{I}}}.$$

Inserting this in the original equation an equation for $\overline{\overline{\mathsf{Y}}}$ is obtained,

$$\overline{\overline{\mathsf{Y}}} = \frac{1}{2}(\overline{\overline{\mathsf{Y}}}\lfloor\lfloor\overline{\overline{\mathsf{I}}}^T){}^{\wedge}_{\wedge}\overline{\overline{\mathsf{I}}} - \frac{\mathrm{tr}\overline{\overline{\mathsf{Y}}}}{3}\overline{\overline{\mathsf{I}}}^{(2)}.$$

If $\overline{\overline{\mathsf{Y}}}$ does not satisfy such an equation, a contradiction will arise and the solution is not valid. It is also a sufficient condition because it shows that $\overline{\overline{\mathsf{Y}}}$ is of the form $\overline{\overline{\mathsf{Z}}}{}^{\wedge}_{\wedge}\overline{\overline{\mathsf{I}}}$ for some dyadic $\overline{\overline{\mathsf{Z}}} \in \mathbb{E}_1\mathbb{F}_1$. Actually, the condition requires that the principal part of the bidyadic be zero, otherwise there being no solution $\overline{\overline{\mathsf{X}}}$.

3.3 Assuming linear independence, the six bivectors $\mathbf{A}_i$ make a basis. Denoting the fourth simple antisymmetric bidyadic by $\mathbf{BC} - \mathbf{CB}$ and expanding $\mathbf{B} = \sum_{i=1}^{6} \beta_i\mathbf{A}_i$, we obtain

$$\mathbf{BC} - \mathbf{CB} = \sum \beta_i(\mathbf{A}_i\mathbf{C} - \mathbf{CA}_i).$$

Because the sum of the four simple antisymmetric bivectors can be expressed in the form

$$\overline{\overline{\mathsf{C}}} = (\mathbf{A}_1 - \beta_2\mathbf{C})(\mathbf{A}_2 + \beta_1\mathbf{C}) - (\mathbf{A}_2 + \beta_1\mathbf{C})(\mathbf{A}_1 - \beta_2\mathbf{C}) + \cdots,$$

the resulting dyadic actually consists of three simple antisymmetric terms, only.

3.4 Assuming $\mathbf{A}_6 = \sum \alpha_i \mathbf{A}_i$ for $i = 1 \cdots 5$, the antisymmetric bidyadic can be written in the form

$$\overline{\overline{\mathsf{C}}} = (\mathbf{A}_1 - \alpha_1\mathbf{A}_5)(\mathbf{A}_2 + \alpha_2\mathbf{A}_5) - (\mathbf{A}_2 + \alpha_2\mathbf{A}_5)(\mathbf{A}_1 - \alpha_1\mathbf{A}_5)$$
$$+ (\mathbf{A}_3 - \alpha_3\mathbf{A}_5)(\mathbf{A}_4 + \alpha_4\mathbf{A}_5) - (\mathbf{A}_4 + \alpha_4\mathbf{A}_5)(\mathbf{A}_3 - \alpha_3\mathbf{A}_5),$$

as a sum of two simple antisymmetric dyadics.

3.5 Because the bivector unit dyadic can be expanded as $\overline{\overline{\mathsf{I}}}^{(2)} = \sum \mathbf{A}_i\mathbf{\Pi}_i$, the dyadic $\overline{\overline{\mathsf{D}}}$ satisfying $\overline{\overline{\mathsf{C}}}|\overline{\overline{\mathsf{D}}} = \overline{\overline{\mathsf{I}}}^{(2)} = \sum \mathbf{A}_i\mathbf{\Pi}_i$ can be written in the antisymmetric form

$$\overline{\overline{\mathsf{D}}} = (\mathbf{\Pi}_2\mathbf{\Pi}_1 - \mathbf{\Pi}_1\mathbf{\Pi}_2) + (\mathbf{\Pi}_4\mathbf{\Pi}_3 - \mathbf{\Pi}_3\mathbf{\Pi}_4) + (\mathbf{\Pi}_6\mathbf{\Pi}_5 - \mathbf{\Pi}_5\mathbf{\Pi}_6),$$

as can most easily be verified.

3.6 To find the inverse for the special skewon–axion bidyadic $\overline{\overline{\mathsf{D}}}$ we first derive a condition for the simple antisymmetric bidyadic $\overline{\overline{\mathsf{C}}}$ by expanding

$$\overline{\overline{\mathsf{C}}}^3 = (\mathbf{AB} - \mathbf{BA}) \cdot (\mathbf{AB} - \mathbf{BA}) \cdot (\mathbf{AB} - \mathbf{BA})$$
$$= ((\mathbf{A} \cdot \mathbf{B})(\mathbf{AB} + \mathbf{BA}) - (\mathbf{A} \cdot \mathbf{A})\mathbf{BB} - (\mathbf{B} \cdot \mathbf{B})\mathbf{AA}) \cdot (\mathbf{AB} - \mathbf{BA})$$
$$= \lambda^2(\mathbf{AB} - \mathbf{BA}) = \lambda^2\overline{\overline{\mathsf{C}}}$$
$$\lambda^2 = (\mathbf{A} \cdot \mathbf{B})^2 - (\mathbf{A} \cdot \mathbf{A})(\mathbf{B} \cdot \mathbf{B}).$$

Thus, a simple antisymmetric bidyadic satisfies the simple condition

$$\overline{\overline{\mathsf{C}}}^3 - \lambda^2\overline{\overline{\mathsf{C}}} = 0.$$

From this a condition for the bidyadic $\overline{\overline{\mathsf{D}}} = \overline{\overline{\mathsf{C}}} + \alpha\overline{\overline{\mathsf{E}}}$ can be formed as

$$\overline{\overline{\mathsf{C}}}^3 = (\overline{\overline{\mathsf{D}}} - \alpha\overline{\overline{\mathsf{E}}}) \cdot (\overline{\overline{\mathsf{D}}} - \alpha\overline{\overline{\mathsf{E}}}) \cdot (\overline{\overline{\mathsf{D}}} - \alpha\overline{\overline{\mathsf{E}}})$$
$$= \overline{\overline{\mathsf{D}}} \cdot \overline{\overline{\mathsf{D}}} \cdot \overline{\overline{\mathsf{D}}} - 3\alpha\overline{\overline{\mathsf{D}}} \cdot \overline{\overline{\mathsf{D}}} + 3\alpha^2\overline{\overline{\mathsf{D}}} - \alpha^3\overline{\overline{\mathsf{E}}}$$
$$= \lambda^2(\overline{\overline{\mathsf{D}}} - \alpha\overline{\overline{\mathsf{E}}}),$$

or

$$\overline{\overline{\mathsf{D}}} \cdot (\overline{\overline{\mathsf{D}}} \cdot \overline{\overline{\mathsf{D}}} - 3\alpha\overline{\overline{\mathsf{D}}} + (3\alpha^2 - \lambda^2)\overline{\overline{\mathsf{E}}}) + (\lambda^2\alpha - \alpha^3)\overline{\overline{\mathsf{E}}} = 0.$$

Written in the form

$$\overline{\overline{\mathsf{D}}}\cdot\overline{\overline{\mathsf{X}}} = \alpha(\alpha^2-\lambda^2)\overline{\overline{\mathsf{E}}}, \quad \overline{\overline{\mathsf{X}}} = \overline{\overline{\mathsf{D}}}\cdot\overline{\overline{\mathsf{D}}} - 3\alpha\overline{\overline{\mathsf{D}}} + (3\alpha^2-\lambda^2)\overline{\overline{\mathsf{E}}}$$

or

$$\overline{\overline{\mathsf{D}}}|(\boldsymbol{\varepsilon}_N\boldsymbol{\varepsilon}_N\lfloor\lfloor\overline{\overline{\mathsf{X}}}) = \alpha(\alpha^2-\lambda^2)\overline{\overline{\mathsf{I}}}^{(2)}.$$

the inverse of the bidyadic $\overline{\overline{\mathsf{D}}}$ can be identified as

$$\begin{aligned}\overline{\overline{\mathsf{D}}}^{-1} &= \frac{1}{\alpha(\alpha^2-\lambda^2)}\boldsymbol{\varepsilon}_N\boldsymbol{\varepsilon}_N\lfloor\lfloor\overline{\overline{\mathsf{X}}} \\ &= \frac{1}{\alpha(\alpha^2-\lambda^2)}\boldsymbol{\varepsilon}_N\boldsymbol{\varepsilon}_N\lfloor\lfloor(\overline{\overline{\mathsf{D}}}\cdot\overline{\overline{\mathsf{D}}} - 3\alpha\overline{\overline{\mathsf{D}}} + (3\alpha^2-\lambda^2)\overline{\overline{\mathsf{E}}}).\end{aligned}$$

3.7 Denoting

$$(\overline{\overline{\mathsf{I}}}^{(2)} + \mathbf{A\Phi} + \mathbf{B\Psi})^{-1} = \overline{\overline{\mathsf{I}}}^{(2)} - \overline{\overline{\mathsf{X}}},$$

and applying the inverse rule (3.137) twice we obtain

$$\overline{\overline{\mathsf{X}}} = \frac{\overline{\overline{\mathsf{N}}}}{D} = \frac{\mathbf{A\Phi}}{1+\mathbf{A}|\mathbf{\Phi}} + \frac{\left(\overline{\overline{\mathsf{I}}}^{(2)} - \frac{\mathbf{A\Phi}}{1+\mathbf{A}|\mathbf{\Phi}}\right)|\mathbf{B\Psi}|\left(\overline{\overline{\mathsf{I}}}^{(2)} - \frac{\mathbf{A\Phi}}{1+\mathbf{A}|\mathbf{\Phi}}\right)}{1+\mathbf{\Psi}|\left(\overline{\overline{\mathsf{I}}}^{(2)} - \frac{\mathbf{A\Phi}}{1+\mathbf{A}|\mathbf{\Phi}}\right)|\mathbf{B}}.$$

Combining the two terms, the numerator and the denominator can be written as

$$\overline{\overline{\mathsf{N}}} = \mathbf{A\Phi} + \mathbf{B\Psi} + (\mathbf{A}|\mathbf{\Phi})\mathbf{B\Psi} + (\mathbf{B}|\mathbf{\Psi})\mathbf{A\Phi} - \mathbf{A\Phi}|\mathbf{B\Psi} - \mathbf{B\Psi}|\mathbf{A\Phi}$$

$$D = 1 + \mathbf{A}|\mathbf{\Phi} + \mathbf{B}|\mathbf{\Psi} + (\mathbf{A}|\mathbf{\Phi})(\mathbf{B}|\mathbf{\Psi}) - (\mathbf{A}|\mathbf{\Psi})(\mathbf{B}|\mathbf{\Phi}).$$

The bidyadic $\overline{\overline{\mathsf{X}}} = \overline{\overline{\mathsf{N}}}/D$ can be shown to satisfy

$$(\mathbf{A\Phi} + \mathbf{B\Psi})|\overline{\overline{\mathsf{X}}} = \mathbf{A\Phi} + \mathbf{B\Psi} - \overline{\overline{\mathsf{X}}},$$

after which we can verify the result through multiplication,

$$\begin{aligned}(\overline{\overline{\mathsf{I}}}^{(2)} + \mathbf{A\Phi} + \mathbf{B\Psi})|(\overline{\overline{\mathsf{I}}}^{(2)} - \overline{\overline{\mathsf{X}}}) &= \overline{\overline{\mathsf{I}}}^{(2)} - \overline{\overline{\mathsf{X}}} + \mathbf{A\Phi} + \mathbf{B\Psi} \\ &\quad - (\mathbf{A\Phi} + \mathbf{B\Psi} - \overline{\overline{\mathsf{X}}}) = \overline{\overline{\mathsf{I}}}^{(2)}.\end{aligned}$$

3.8 The task is to substitute the dyadic (3.127)

$$\overline{\overline{\mathsf{A}}}_o = \overline{\overline{\mathsf{F}}}(\overline{\overline{\mathsf{B}}}_o) = \frac{1}{\mathrm{tr}\overline{\overline{\mathsf{B}}}_o^{(3)}}\left(\overline{\overline{\mathsf{B}}}_o^2 + \frac{1}{2}\mathrm{tr}\overline{\overline{\mathsf{B}}}_o^{(2)}\overline{\overline{\mathsf{I}}}\right)$$

in the expression (3.129)

$$\overline{\overline{\mathsf{B}}}_o = \overline{\overline{\mathsf{F}}}(\overline{\overline{\mathsf{A}}}_o) = \frac{1}{\mathrm{tr}\overline{\overline{\mathsf{A}}}_o^{(3)}}\left(\overline{\overline{\mathsf{A}}}_o^2 + \frac{1}{2}\mathrm{tr}\overline{\overline{\mathsf{A}}}_o^{(2)}\overline{\overline{\mathsf{I}}}\right),$$

as

$$\overline{\overline{\mathsf{B}}}_o = \overline{\overline{\mathsf{F}}}(\overline{\overline{\mathsf{A}}}_o) = \overline{\overline{\mathsf{F}}}(\overline{\overline{\mathsf{F}}}(\overline{\overline{\mathsf{B}}}_o))$$

$$= \mathrm{tr}\overline{\overline{\mathsf{B}}}_o^{(3)}\left(\frac{(\overline{\overline{\mathsf{B}}}_o^2 - \frac{1}{4}\mathrm{tr}\overline{\overline{\mathsf{B}}}_o^2\overline{\overline{\mathsf{I}}})^2 - \frac{1}{4}\mathrm{tr}(\overline{\overline{\mathsf{B}}}_o^2 - \frac{1}{4}\mathrm{tr}\overline{\overline{\mathsf{B}}}_o^2\overline{\overline{\mathsf{I}}})^2\overline{\overline{\mathsf{I}}}}{\mathrm{tr}(\overline{\overline{\mathsf{B}}}_o^2 - \frac{1}{4}\mathrm{tr}\overline{\overline{\mathsf{B}}}_o^2\overline{\overline{\mathsf{I}}})^{(3)}}\right),$$

to see whether this is satisfied for any trace-free dyadic $\overline{\overline{\mathsf{B}}}_o$. Invoking the Cayley–Hamilton equation (2.119) we obtain

$$(\overline{\overline{\mathsf{B}}}_o^2 - \frac{1}{4}\mathrm{tr}\overline{\overline{\mathsf{B}}}_o^2\overline{\overline{\mathsf{I}}})^2 = \overline{\overline{\mathsf{B}}}_o^4 - \frac{1}{2}\mathrm{tr}\overline{\overline{\mathsf{B}}}_o^2\,\overline{\overline{\mathsf{B}}}_o^2 + \frac{1}{16}(\mathrm{tr}\overline{\overline{\mathsf{B}}}_o^2)^2\overline{\overline{\mathsf{I}}}$$

$$= \mathrm{tr}\overline{\overline{\mathsf{B}}}^{(3)}\overline{\overline{\mathsf{B}}}_o - \left(\mathrm{tr}\overline{\overline{\mathsf{B}}}_o^{(4)} - \frac{1}{16}(\mathrm{tr}\overline{\overline{\mathsf{B}}}_o^2)^2\right)\overline{\overline{\mathsf{I}}}$$

$$\mathrm{tr}\left(\overline{\overline{\mathsf{B}}}_o^2 - \frac{1}{4}\mathrm{tr}\overline{\overline{\mathsf{B}}}_o^2\overline{\overline{\mathsf{I}}}\right)^2 = -4\mathrm{tr}\overline{\overline{\mathsf{B}}}_o^{(4)} + \frac{1}{4}(\mathrm{tr}\overline{\overline{\mathsf{B}}}_o^2)^2,$$

whence the numerator of the dyadic inside the brackets melts to $\mathrm{tr}\overline{\overline{\mathsf{B}}}^{(3)}\,\overline{\overline{\mathsf{B}}}_o$. Applying twice the Cayley–Hamilton equation, and skipping the algebraic steps, the denominator can be expanded as

$$\mathrm{tr}\left(\overline{\overline{\mathsf{B}}}_o^2 - \frac{1}{4}\mathrm{tr}\overline{\overline{\mathsf{B}}}_o^2\overline{\overline{\mathsf{I}}}\right)^{(3)} = \frac{1}{3}\mathrm{tr}\left(\overline{\overline{\mathsf{B}}}_o^2 - \frac{1}{4}\mathrm{tr}\overline{\overline{\mathsf{B}}}_o^2\overline{\overline{\mathsf{I}}}\right)^3 = (\mathrm{tr}\overline{\overline{\mathsf{B}}}_o^{(3)})^2,$$

which inserted reduces the expression to $\overline{\overline{\mathsf{B}}}_o = \overline{\overline{\mathsf{B}}}_o$.

3.9 Applying the identity

$$(\overline{\overline{\mathsf{A}}}{}^\wedge_\wedge\overline{\overline{\mathsf{B}}})|(\overline{\overline{\mathsf{C}}}{}^\wedge_\wedge\overline{\overline{\mathsf{D}}}) = (\overline{\overline{\mathsf{A}}}|\overline{\overline{\mathsf{C}}}){}^\wedge_\wedge(\overline{\overline{\mathsf{B}}}|\overline{\overline{\mathsf{D}}}) + (\overline{\overline{\mathsf{A}}}|\overline{\overline{\mathsf{D}}}){}^\wedge_\wedge(\overline{\overline{\mathsf{B}}}|\overline{\overline{\mathsf{C}}})$$

we can expand

$$(\overline{\overline{\mathsf{N}}}{}^\wedge_\wedge\overline{\overline{\mathsf{I}}})^2 = \overline{\overline{\mathsf{N}}}{}^2{}^\wedge_\wedge\overline{\overline{\mathsf{I}}} + \overline{\overline{\mathsf{N}}}{}^\wedge_\wedge\overline{\overline{\mathsf{N}}} = \overline{\overline{\mathsf{N}}}{}^\wedge_\wedge\overline{\overline{\mathsf{N}}},$$

$$(\overline{\overline{\mathsf{N}}}{}^\wedge_\wedge\overline{\overline{\mathsf{I}}})^3 = (\overline{\overline{\mathsf{N}}}{}^\wedge_\wedge\overline{\overline{\mathsf{N}}})|(\overline{\overline{\mathsf{N}}}{}^\wedge_\wedge\overline{\overline{\mathsf{I}}}) = 2\overline{\overline{\mathsf{N}}}{}^2{}^\wedge_\wedge\overline{\overline{\mathsf{N}}} = 0.$$

In terms of these we have

$$(\overline{\overline{\mathsf{D}}} - \alpha\overline{\overline{\mathsf{I}}}^{(2)})^3 = (\overline{\overline{\mathsf{N}}}{}^\wedge_\wedge\overline{\overline{\mathsf{I}}})^3 = 0,$$

which corresponds to the equation

$$\overline{\overline{\mathsf{D}}}^3 - 3\alpha\overline{\overline{\mathsf{D}}}^2 + 3\alpha^2\overline{\overline{\mathsf{D}}} - \alpha^3\overline{\overline{\mathsf{I}}}^{(2)} = 0.$$

For $\alpha \neq 0$ we obtain the inverse rule

$$\overline{\overline{\mathsf{D}}}^{-1} = \frac{1}{\alpha^3}(\overline{\overline{\mathsf{D}}}^2 - 3\alpha\overline{\overline{\mathsf{D}}} + 3\alpha^2\overline{\overline{\mathsf{I}}}^{(2)})$$
$$= \frac{1}{\alpha^3}(\alpha^2\overline{\overline{\mathsf{I}}}^{(2)} - \alpha\overline{\overline{\mathsf{N}}}{}^{\wedge}_{\wedge}\overline{\overline{\mathsf{I}}} + \overline{\overline{\mathsf{N}}}{}^{\wedge}_{\wedge}\overline{\overline{\mathsf{N}}}).$$

3.10 Assuming that the vector $\mathbf{a}$ and one-form $\boldsymbol{\alpha}$ satisfy $\mathbf{a}|\boldsymbol{\alpha} \neq 0$, we can temporarily denote $\mathbf{a}\boldsymbol{\alpha} = (\mathbf{a}|\boldsymbol{\alpha})\mathbf{e}_4\boldsymbol{\varepsilon}_4$, whence

$$\overline{\overline{\mathsf{C}}} = \alpha\overline{\overline{\mathsf{I}}}^{(2)} + (\mathbf{a}|\boldsymbol{\alpha})\overline{\overline{\mathsf{I}}}{}^{\wedge}_{\wedge}\mathbf{e}_4\boldsymbol{\varepsilon}_4$$
$$= \alpha\overline{\overline{\mathsf{I}}}{}^{(2)}_{s} + (\alpha + \mathbf{a}|\boldsymbol{\alpha})\overline{\overline{\mathsf{I}}}{}_{s}{}^{\wedge}_{\wedge}\mathbf{e}_4\boldsymbol{\varepsilon}_4.$$

Obviously, the inverse is

$$\overline{\overline{\mathsf{C}}}^{-1} = \frac{1}{\alpha}\overline{\overline{\mathsf{I}}}{}^{(2)}_{s} + \frac{1}{\alpha + \mathbf{a}|\boldsymbol{\alpha}}\overline{\overline{\mathsf{I}}}{}_{s}{}^{\wedge}_{\wedge}\mathbf{e}_4\boldsymbol{\varepsilon}_4$$
$$= \frac{1}{\alpha}\overline{\overline{\mathsf{I}}}^{(2)} - \frac{\mathbf{a}|\boldsymbol{\alpha}}{\alpha(\alpha + \mathbf{a}|\boldsymbol{\alpha})}\overline{\overline{\mathsf{I}}}{}^{\wedge}_{\wedge}\mathbf{e}_4\boldsymbol{\varepsilon}_4$$

which can be expressed with original quantities as

$$\overline{\overline{\mathsf{C}}}^{-1} = \frac{1}{\alpha}\overline{\overline{\mathsf{I}}}^{(2)} - \frac{1}{\alpha(\alpha + \mathbf{a}|\boldsymbol{\alpha})}\overline{\overline{\mathsf{I}}}{}^{\wedge}_{\wedge}\mathbf{a}\boldsymbol{\alpha}.$$

This result can be checked to yield $\overline{\overline{\mathsf{C}}}|\overline{\overline{\mathsf{C}}}^{-1} = \overline{\overline{\mathsf{I}}}^{(2)}$ also in the case $\mathbf{a}|\boldsymbol{\alpha} \neq 0$, whence the assumption can be released.

3.11 (1) Choose $\mathbf{a} = \mathbf{e}_1$ and expand $\overline{\overline{\mathsf{C}}} = \sum \mathbf{e}_{ij}\boldsymbol{\Phi}_{ij}$ in terms of six two-forms $\boldsymbol{\Phi}_{ij}$. From

$$\mathbf{e}_1 \wedge \overline{\overline{\mathsf{C}}} = \mathbf{e}_{123}\boldsymbol{\Phi}_{23} + \mathbf{e}_{124}\boldsymbol{\Phi}_{24} - \mathbf{e}_{314}\boldsymbol{\Phi}_{34} = 0$$

we obtain $\boldsymbol{\Phi}_{23} = \boldsymbol{\Phi}_{24} = \boldsymbol{\Phi}_{34} = 0$. Choosing $\mathbf{a} = \mathbf{e}_2, \mathbf{e}_3, \mathbf{e}_4$, we similarly obtain $\boldsymbol{\Phi}_{12} = \boldsymbol{\Phi}_{13} = \boldsymbol{\Phi}_{14} = 0$. Thus, $\overline{\overline{\mathsf{C}}} = 0$.

(2) Expand

$$\overline{\overline{\mathsf{C}}} = \overline{\overline{\mathsf{C}}}|\overline{\overline{\mathsf{I}}}^{(2)} = \sum_{i<j} \overline{\overline{\mathsf{C}}}|\mathbf{e}_{ij}\boldsymbol{\varepsilon}_{ij} = \sum_{i<j}(\overline{\overline{\mathsf{C}}}\rfloor\mathbf{e}_i)\rfloor\mathbf{e}_j\boldsymbol{\varepsilon}_{ij} = 0.$$

3.12 From $\boldsymbol{\alpha} \wedge \overline{\overline{\mathsf{M}}}\lfloor\boldsymbol{\alpha} = 0$ it follows that the dyadic $\overline{\overline{\mathsf{M}}}\lfloor\boldsymbol{\alpha}$ must be of the form $\overline{\overline{\mathsf{M}}}\lfloor\boldsymbol{\alpha} = \boldsymbol{\alpha} \wedge \overline{\overline{\mathsf{A}}}$ for some dyadic $\overline{\overline{\mathsf{A}}} \in \mathbb{F}_1\mathbb{E}_1$. Since the dyadic $\overline{\overline{\mathsf{A}}}$ satisfies $\boldsymbol{\alpha} \wedge \overline{\overline{\mathsf{A}}}|\boldsymbol{\alpha} = \overline{\overline{\mathsf{M}}}|(\boldsymbol{\alpha} \wedge \boldsymbol{\alpha}) = 0$, we must have

$$\overline{\overline{\mathsf{A}}}|\boldsymbol{\alpha} = A\boldsymbol{\alpha},$$

that is, $\boldsymbol{\alpha}$ is an eigen one-form of the dyadic $\overline{\overline{\mathsf{A}}}$ for some eigenvalue A. Since this must be valid for any one-form $\boldsymbol{\alpha}$, A must be independent of $\boldsymbol{\alpha}$ and we have

$$\overline{\overline{\mathsf{A}}} = \overline{\overline{\mathsf{A}}}|\overline{\overline{\mathsf{I}}}^T = \sum_i \overline{\overline{\mathsf{A}}}|\boldsymbol{\varepsilon}_i \mathbf{e}_i = \sum_i A\boldsymbol{\varepsilon}_i \mathbf{e}_i = A\overline{\overline{\mathsf{I}}}^T.$$

Thus, the bidyadic $\overline{\overline{\mathsf{M}}}$ must satisfy

$$\overline{\overline{\mathsf{M}}}\lfloor\boldsymbol{\alpha} = A\boldsymbol{\alpha}\wedge\overline{\overline{\mathsf{I}}}^T = A\overline{\overline{\mathsf{I}}}^{(2)T}\lfloor\boldsymbol{\alpha}$$

for any one-form $\boldsymbol{\alpha}$. Applying the result of the previous problem to the bidyadic $\overline{\overline{\mathsf{M}}} - A\overline{\overline{\mathsf{I}}}^{(2)T}$, we obtain $\overline{\overline{\mathsf{M}}} - A\overline{\overline{\mathsf{I}}}^{(2)T} = 0$, whence $\overline{\overline{\mathsf{M}}}$ must be a multiple of the unit bidyadic $\overline{\overline{\mathsf{I}}}^{(2)T}$. Operating as

$$\mathbf{e}_N\lfloor(\boldsymbol{\alpha}\wedge\overline{\overline{\mathsf{M}}}\lfloor\boldsymbol{\alpha}) = -\boldsymbol{\alpha}\rfloor(\mathbf{e}_N\lfloor\overline{\overline{\mathsf{M}}})\lfloor\boldsymbol{\alpha} = \overline{\overline{\mathsf{M}}}_m\lfloor\lfloor\boldsymbol{\alpha}\boldsymbol{\alpha},$$

where $\overline{\overline{\mathsf{M}}}_m = \mathbf{e}_N\lfloor\overline{\overline{\mathsf{M}}} \in \mathbb{E}_2\mathbb{E}_2$, we can prove the second property.

3.13 Because we have

$$(\mathbf{e}_N\lfloor\overline{\overline{\mathsf{I}}}^{(2)T})\lfloor\lfloor\boldsymbol{\alpha}\boldsymbol{\alpha} = \mathbf{e}_N\lfloor(\boldsymbol{\alpha}\wedge(\overline{\overline{\mathsf{I}}}^{(2)T}\lfloor\boldsymbol{\alpha})) = \mathbf{e}_N\lfloor(\boldsymbol{\alpha}\wedge\boldsymbol{\alpha}\wedge\overline{\overline{\mathsf{I}}}^T) = 0$$

identically for any one-form $\boldsymbol{\alpha}$, the solution $\overline{\overline{\mathsf{Q}}}$ must satisfy $\overline{\overline{\mathsf{Q}}}^{(2)}\lfloor\lfloor\boldsymbol{\alpha}\boldsymbol{\alpha} = 0$ for any $\boldsymbol{\alpha}$. From

$$\overline{\overline{\mathsf{Q}}}^{(2)}\lfloor\lfloor\boldsymbol{\alpha}\boldsymbol{\alpha} = (\overline{\overline{\mathsf{Q}}}||\boldsymbol{\alpha}\boldsymbol{\alpha})\overline{\overline{\mathsf{Q}}} - (\overline{\overline{\mathsf{Q}}}|\boldsymbol{\alpha})(\boldsymbol{\alpha}|\overline{\overline{\mathsf{Q}}}) = 0$$

we obtain two possibilities.

(1) Assuming $\overline{\overline{\mathsf{Q}}}||\boldsymbol{\alpha}\boldsymbol{\alpha} \neq 0$ is valid for some $\boldsymbol{\alpha}$, we can express

$$\overline{\overline{\mathsf{Q}}} = \frac{(\overline{\overline{\mathsf{Q}}}|\boldsymbol{\alpha})(\boldsymbol{\alpha}|\overline{\overline{\mathsf{Q}}})}{\overline{\overline{\mathsf{Q}}}||\boldsymbol{\alpha}\boldsymbol{\alpha}},$$

which corresponds to the solution $\overline{\overline{\mathsf{Q}}}^{(2)} = 0$, $Q = 0$, that is, $\overline{\overline{\mathsf{Q}}}$ is of the form $\overline{\overline{\mathsf{Q}}} = \mathbf{ab}$.

(2) For the converse case $\overline{\overline{\mathsf{Q}}}||\boldsymbol{\alpha}\boldsymbol{\alpha} = 0$ for all $\boldsymbol{\alpha}$ we must have $\overline{\overline{\mathsf{Q}}}$ antisymmetric, that is, it can be expressed in terms of a bivector $\mathbf{Q}$ as $\overline{\overline{\mathsf{Q}}} = \mathbf{Q}\lfloor\overline{\overline{\mathsf{I}}}^T$. Since we must have $\overline{\overline{\mathsf{Q}}}|\boldsymbol{\alpha} = 0$ for any $\boldsymbol{\alpha}$, this leads to $\mathbf{Q} = 0$, whence the second solution would be $\overline{\overline{\mathsf{Q}}} = 0$ and $Q = 0$.

Because $Q = 0$ in both cases, in particular, the equation $\overline{\overline{\mathsf{Q}}}^{(2)} = \mathbf{e}_N\lfloor\overline{\overline{\mathsf{I}}}^{(2)}$ does not have any solution.

3.14 (1) Multiplying (3.10) by $\overline{\overline{\mathsf{C}}}^{-6}|$ and dividing by a_6 leads to the equation

$$\overline{\overline{\mathsf{C}}}^{-6} + (a_5/a_6)\overline{\overline{\mathsf{C}}}^{-5} + (a_4/a_6)\overline{\overline{\mathsf{C}}}^{-4} + (a_3/a_6)\overline{\overline{\mathsf{C}}}^{-3} + (a_2/a_6)\overline{\overline{\mathsf{C}}}^{-2} + (a_1/a_6)\overline{\overline{\mathsf{C}}}^{-1} + (1/a_6)\overline{\overline{\mathsf{I}}}^{(2)} = 0.$$

(2) Because we have

$$\overline{\overline{\mathsf{D}}}^2 = (\mathbf{e}_N\lfloor\overline{\overline{\mathsf{C}}}^T\rfloor\boldsymbol{\varepsilon}_N)|(\mathbf{e}_N\lfloor\overline{\overline{\mathsf{C}}}^T\rfloor\boldsymbol{\varepsilon}_N) = \mathbf{e}_N\lfloor\overline{\overline{\mathsf{C}}}^{2T}\rfloor\boldsymbol{\varepsilon}_N = \overline{\overline{\mathsf{I}}}^{(4)}\lfloor\lfloor\overline{\overline{\mathsf{C}}}^{2T},$$

and similarly for other powers of $\overline{\overline{\mathsf{D}}}$, operating 3.10 by $\overline{\overline{\mathsf{I}}}^{(4)}\lfloor\lfloor()^T$ the Cayley–Hamilton equation for $\overline{\overline{\mathsf{D}}}$ is obtained when replacing $\overline{\overline{\mathsf{C}}}$ by $\overline{\overline{\mathsf{D}}}$.

3.15 Let us define $\mathbf{e}_1 = \mathbf{a}$ and $\boldsymbol{\varepsilon}_1 = \boldsymbol{\alpha}/4$ to satisfy $\mathbf{e}_1|\boldsymbol{\varepsilon}_1 = 1$. Thus, we have

$$\overline{\overline{\mathsf{B}}}_o = \frac{1}{2}(\overline{\overline{\mathsf{I}}} - \mathbf{a}\boldsymbol{\alpha}) = \frac{1}{2}(\overline{\overline{\mathsf{I}}} - 4\mathbf{e}_1\boldsymbol{\varepsilon}_1),$$
$$\overline{\overline{\mathsf{C}}} = \overline{\overline{\mathsf{B}}}_o{}^\wedge_\wedge\overline{\overline{\mathsf{I}}} = \overline{\overline{\mathsf{I}}}^{(2)} - 2\mathbf{e}_1\boldsymbol{\varepsilon}_1{}^\wedge_\wedge\overline{\overline{\mathsf{I}}}$$
$$= -\mathbf{e}_{12}\boldsymbol{\varepsilon}_{12} + \mathbf{e}_{23}\boldsymbol{\varepsilon}_{23} - \mathbf{e}_{31}\boldsymbol{\varepsilon}_{31} - \mathbf{e}_{14}\boldsymbol{\varepsilon}_{14} + \mathbf{e}_{24}\boldsymbol{\varepsilon}_{24} + \mathbf{e}_{34}\boldsymbol{\varepsilon}_{34},$$

whence

$$\overline{\overline{\mathsf{C}}}\rfloor\mathbf{e}_N = -\mathbf{e}_{12}\mathbf{e}_{34} + \mathbf{e}_{23}\mathbf{e}_{14} - \mathbf{e}_{31}\mathbf{e}_{24} - \mathbf{e}_{14}\mathbf{e}_{23} + \mathbf{e}_{24}\mathbf{e}_{31} + \mathbf{e}_{34}\mathbf{e}_{12}$$

is antisymmetric. Because of $(\mathbf{e}_1\boldsymbol{\varepsilon}_1{}^\wedge_\wedge\overline{\overline{\mathsf{I}}})^2 = \mathbf{e}_1\boldsymbol{\varepsilon}_1{}^\wedge_\wedge\overline{\overline{\mathsf{I}}}$, we have

$$\overline{\overline{\mathsf{D}}} = \overline{\overline{\mathsf{C}}}^2 = (\overline{\overline{\mathsf{B}}}_o{}^\wedge_\wedge\overline{\overline{\mathsf{I}}})^2 = (\overline{\overline{\mathsf{I}}}^{(2)} - 2\mathbf{e}_1\boldsymbol{\varepsilon}_1{}^\wedge_\wedge\overline{\overline{\mathsf{I}}})^2 = \overline{\overline{\mathsf{I}}}^{(2)}$$

and $\mathrm{tr}\overline{\overline{\mathsf{D}}}^p = 4$ for $p \geq 0$. The coefficients (3.43)–(3.45) become

$$a_0 = 1, \quad a_2 = -2, \quad a_4 = 1, \quad a_6 = 0,$$

and (3.42) is reduced to

$$\overline{\overline{\mathsf{D}}}^3 - 2\overline{\overline{\mathsf{D}}}^2 + \overline{\overline{\mathsf{D}}} = \overline{\overline{\mathsf{D}}}|(\overline{\overline{\mathsf{D}}} - \overline{\overline{\mathsf{I}}}^{(2)})^2 = 0,$$

which is satisfied for $\overline{\overline{\mathsf{D}}} = \overline{\overline{\mathsf{I}}}^{(2)}$.

3.16 Inserting the expressions of $\overline{\overline{\mathsf{M}}}$ and the assumed $\overline{\overline{\mathsf{M}}}^{-1}$ in $\overline{\overline{\mathsf{M}}}|\overline{\overline{\mathsf{M}}}^{-1} = \overline{\overline{\mathsf{I}}}^{(2)}$ and assuming that $\mathbf{\Pi}$ and $\mathbf{\Lambda}$ are linearly independent and $\alpha' = 1/\alpha$, we obtain a relation between the unknown and known bivectors as

$$\begin{pmatrix} \mathbf{\Pi}|\mathbf{C} + \alpha & \mathbf{\Lambda}|\mathbf{C} \\ \mathbf{\Pi}|\mathbf{D} & \mathbf{\Lambda}|\mathbf{D} + \alpha \end{pmatrix} \begin{pmatrix} \mathbf{C}' \\ \mathbf{D}' \end{pmatrix} = -\alpha' \begin{pmatrix} \mathbf{C} \\ \mathbf{D} \end{pmatrix}.$$

This can be solved as

$$\begin{pmatrix} \mathbf{C}' \\ \mathbf{D}' \end{pmatrix} = \frac{-1}{\alpha D} \begin{pmatrix} \mathbf{\Lambda}|\mathbf{D} + \alpha & -\mathbf{\Lambda}|\mathbf{C} \\ -\mathbf{\Pi}|\mathbf{D} & \mathbf{\Pi}|\mathbf{C} + \alpha \end{pmatrix} \begin{pmatrix} \mathbf{C} \\ \mathbf{D} \end{pmatrix},$$

$$D = (\mathbf{\Pi}|\mathbf{C} + \alpha)(\mathbf{\Lambda}|\mathbf{D} + \alpha) - (\mathbf{\Lambda}|\mathbf{C})(\mathbf{\Pi}|\mathbf{D}).$$

It is easy to verify the result by direct substitution. In the case $D = 0$, there is no inverse. For the special case $\mathbf{D} = 0$, the bivector $\mathbf{C}'$ is reduced to

$$\mathbf{C}' = \frac{-1}{\alpha(\mathbf{\Pi}|\mathbf{C} + \alpha)}\mathbf{C}.$$

3.17 Expanding from (3.43)–(3.45)

$$\begin{aligned} a_0 \mathrm{tr}\overline{\overline{\mathsf{D}}}^3 &= \mathrm{tr}\overline{\overline{\mathsf{D}}}^3 \\ a_2 \mathrm{tr}\overline{\overline{\mathsf{D}}}^2 &= -\frac{1}{2}\mathrm{tr}\overline{\overline{\mathsf{D}}}\,\mathrm{tr}\overline{\overline{\mathsf{D}}}^2 \\ a_4 \mathrm{tr}\overline{\overline{\mathsf{D}}} &= \frac{1}{8}(\mathrm{tr}\overline{\overline{\mathsf{D}}})^3 - \frac{1}{4}\mathrm{tr}\overline{\overline{\mathsf{D}}}^2\mathrm{tr}\overline{\overline{\mathsf{D}}} \\ a_6 \mathrm{tr}\overline{\overline{\mathsf{I}}}^{(2)} &= -\frac{1}{8}(\mathrm{tr}\overline{\overline{\mathsf{D}}})^3 + \frac{6}{8}\mathrm{tr}\overline{\overline{\mathsf{D}}}\,\mathrm{tr}\overline{\overline{\mathsf{D}}}^2 - \mathrm{tr}\overline{\overline{\mathsf{D}}}^3, \end{aligned}$$

we can show that

$$a_0 \mathrm{tr}\overline{\overline{\mathsf{D}}}^3 + a_2 \mathrm{tr}\overline{\overline{\mathsf{D}}}^2 + a_4 \mathrm{tr}\overline{\overline{\mathsf{D}}} + a_6 \mathrm{tr}\overline{\overline{\mathsf{I}}}^{(2)} = 0,$$

that is, the trace of (3.42) is an identity.

3.18 Denoting $x = \mathrm{tr}\overline{\overline{\mathsf{C}}}$ we can write

$$\begin{aligned} 6!a_6(x) = x^6 &- 15x^4\mathrm{tr}\overline{\overline{\mathsf{C}}}^2 + 40x^3\mathrm{tr}\overline{\overline{\mathsf{C}}}^3 + 45x^2(\mathrm{tr}\overline{\overline{\mathsf{C}}}^2)^2 \\ &- 120x\,\mathrm{tr}\overline{\overline{\mathsf{C}}}^2\,\mathrm{tr}\overline{\overline{\mathsf{C}}}^3 - 90x^2\mathrm{tr}\overline{\overline{\mathsf{C}}}^4 + 144x\,\mathrm{tr}\overline{\overline{\mathsf{C}}}^5 \\ &+ 40(\mathrm{tr}\overline{\overline{\mathsf{C}}}^3)^2 - 15(\mathrm{tr}\overline{\overline{\mathsf{C}}}^2)^3 + 90\mathrm{tr}\overline{\overline{\mathsf{C}}}^2\mathrm{tr}\overline{\overline{\mathsf{C}}}^4 - 120\mathrm{tr}\overline{\overline{\mathsf{C}}}^6, \end{aligned}$$

which when differentiated yields

$$\begin{aligned} 6!\partial_x a_6(x) &= 6x^5 - 60x^3\mathrm{tr}\overline{\overline{\mathsf{C}}}^2 + 120x^2\mathrm{tr}\overline{\overline{\mathsf{C}}}^3 + 90x(\mathrm{tr}\overline{\overline{\mathsf{C}}}^2)^2 \\ &\quad - 120\mathrm{tr}\overline{\overline{\mathsf{C}}}^2\,\mathrm{tr}\overline{\overline{\mathsf{C}}}^3 - 180x\mathrm{tr}\overline{\overline{\mathsf{C}}}^4 + 144\mathrm{tr}\overline{\overline{\mathsf{C}}}^5 \\ &= 6(x^5 - 10x^3\mathrm{tr}\overline{\overline{\mathsf{C}}}^2 + 20x^2\mathrm{tr}\overline{\overline{\mathsf{C}}}^3 + 15x(\mathrm{tr}\overline{\overline{\mathsf{C}}}^2)^2 \\ &\quad - 20\mathrm{tr}\overline{\overline{\mathsf{C}}}^2\,\mathrm{tr}\overline{\overline{\mathsf{C}}}^3 - 30x\mathrm{tr}\overline{\overline{\mathsf{C}}}^4 + 24\mathrm{tr}\overline{\overline{\mathsf{C}}}^5). \end{aligned}$$

The expression in brackets equals $-5!a_5(x)$. We can continue similarly for $i!\partial_x a_i$, $i < 6$.

3.19 Applying the rule (2.114) to $\overline{\overline{\mathsf{C}}} = \overline{\overline{\mathsf{B}}}_o^{(2)}$

$$\overline{\overline{\mathsf{I}}}^{(4)}\lfloor\lfloor\overline{\overline{\mathsf{B}}}_o^{(2)T} = \mathrm{tr}\overline{\overline{\mathsf{B}}}_o^{(2)}\,\overline{\overline{\mathsf{I}}}^{(2)} - (\overline{\overline{\mathsf{B}}}_o^{(2)}\lfloor\lfloor\overline{\overline{\mathsf{I}}}^T){}^\wedge_\wedge\overline{\overline{\mathsf{I}}} + \overline{\overline{\mathsf{B}}}_o^{(2)}$$
$$= \mathrm{tr}\overline{\overline{\mathsf{B}}}_o^{(2)}\,\overline{\overline{\mathsf{I}}}^{(2)} + \overline{\overline{\mathsf{B}}}_o^{2}{}^\wedge_\wedge\overline{\overline{\mathsf{I}}} + \overline{\overline{\mathsf{B}}}_o^{(2)}$$

and comparing with the rule (3.128) we have

$$\overline{\overline{\mathsf{I}}}^{(4)}\lfloor\lfloor\overline{\overline{\mathsf{B}}}_o^{(2)T} = \mathrm{tr}\overline{\overline{\mathsf{B}}}_o^{(3)}(\overline{\overline{\mathsf{B}}}_o{}^\wedge_\wedge\overline{\overline{\mathsf{I}}})^{-1} + \overline{\overline{\mathsf{B}}}_o^{(2)}.$$

Assuming $\mathrm{tr}\overline{\overline{\mathsf{B}}}_o^{(3)} \neq 0$, we obtain another form for the inverse bidyadic,

$$(\overline{\overline{\mathsf{B}}}_o{}^\wedge_\wedge\overline{\overline{\mathsf{I}}})^{-1} = \frac{1}{\mathrm{tr}\overline{\overline{\mathsf{B}}}_o^{(3)}}(\overline{\overline{\mathsf{I}}}^{(4)}\lfloor\lfloor\overline{\overline{\mathsf{B}}}_o^{(2)T} - \overline{\overline{\mathsf{B}}}_o^{(2)}).$$

3.20 Applying the bac-cab rule, valid for any one-form $\boldsymbol{\alpha}$, two-form $\boldsymbol{\Pi}$ and vector $\mathbf{a}$,

$$\mathbf{a}\rfloor(\boldsymbol{\alpha}\wedge\boldsymbol{\Pi}) = \boldsymbol{\alpha}\wedge(\mathbf{a}\rfloor\boldsymbol{\Pi}) + (\mathbf{a}|\boldsymbol{\alpha})\boldsymbol{\Pi},$$

and choosing $\mathbf{a}$ so that $\mathbf{a}|\boldsymbol{\alpha} = 1$, we can write from

$$\mathbf{a}\rfloor(\boldsymbol{\alpha}\wedge\overline{\overline{\mathsf{M}}}\lfloor\boldsymbol{\alpha}) = \boldsymbol{\alpha}\wedge(\mathbf{a}\rfloor\overline{\overline{\mathsf{M}}}\lfloor\boldsymbol{\alpha}) + \overline{\overline{\mathsf{M}}}\lfloor\boldsymbol{\alpha} = 0,$$

the representation $\overline{\overline{\mathsf{M}}}\lfloor\boldsymbol{\alpha} = \boldsymbol{\alpha}\wedge\overline{\overline{\mathsf{A}}}^T$, with $\overline{\overline{\mathsf{A}}}^T = -\mathbf{a}\rfloor\overline{\overline{\mathsf{M}}}\lfloor\boldsymbol{\alpha}$. Choosing $\mathbf{a} = \mathbf{e}_i$, $\boldsymbol{\alpha} = \boldsymbol{\varepsilon}_i$ and summing over i we obtain $\overline{\overline{\mathsf{A}}}^T = \overline{\overline{\mathsf{M}}}\lfloor\lfloor\overline{\overline{\mathsf{I}}}$. Since $\overline{\overline{\mathsf{M}}}$ is of full rank, so is $\overline{\overline{\mathsf{A}}}$. The dyadic $\overline{\overline{\mathsf{A}}}$ satisfies for any $\boldsymbol{\alpha}$

$$\boldsymbol{\alpha}\wedge\overline{\overline{\mathsf{A}}}^T|\boldsymbol{\alpha} = \overline{\overline{\mathsf{M}}}|(\boldsymbol{\alpha}\wedge\boldsymbol{\alpha}) = 0,$$

whence expanding $\overline{\overline{\mathsf{A}}} = \sum A_{i,j}\mathbf{e}_i\boldsymbol{\varepsilon}_j$ and taking $\boldsymbol{\alpha} = \boldsymbol{\varepsilon}_k$, we obtain

$$\boldsymbol{\varepsilon}_k\wedge\overline{\overline{\mathsf{A}}}|\boldsymbol{\varepsilon}_k = \sum A_{i,j}(\boldsymbol{\varepsilon}_k\wedge\boldsymbol{\varepsilon}_j)\delta_{ik} = \sum A_{i,j}\boldsymbol{\varepsilon}_{ij} = 0.$$

This implies $A_{i,j} = 0$ for $i \neq j$ and, thus, $\overline{\overline{\mathsf{A}}} = \sum A_{i,i}\mathbf{e}_i\boldsymbol{\varepsilon}_i$. Continuing with $\boldsymbol{\alpha} = \boldsymbol{\varepsilon}_k + \boldsymbol{\varepsilon}_\ell$, $k \neq \ell$, we have

$$\boldsymbol{\alpha}\wedge\overline{\overline{\mathsf{A}}}^T|\boldsymbol{\alpha} = \sum A_{i,i}(\boldsymbol{\varepsilon}_{ki}(\mathbf{e}_i|\boldsymbol{\varepsilon}_\ell) + \boldsymbol{\varepsilon}_{\ell i}(\mathbf{e}_i|\boldsymbol{\varepsilon}_k)) = A_{\ell,\ell}\boldsymbol{\varepsilon}_{k\ell} + A_{k,k}\boldsymbol{\varepsilon}_{\ell k}$$
$$= (A_{\ell,\ell} - A_{k,k})\boldsymbol{\varepsilon}_{k\ell} = 0, \quad \Rightarrow \quad A_{\ell,\ell} = A_{k,k}$$

from which we obtain $\overline{\overline{\mathsf{A}}} = A\overline{\overline{\mathsf{I}}}$. Thus, $\overline{\overline{\mathsf{M}}}\lfloor\lfloor\overline{\overline{\mathsf{I}}} = A\overline{\overline{\mathsf{I}}}^T$ and

$$\overline{\overline{\mathsf{M}}}\lfloor\boldsymbol{\alpha} = A\boldsymbol{\alpha}\wedge\overline{\overline{\mathsf{I}}}^T = A\overline{\overline{\mathsf{I}}}^{(2)T}\lfloor\boldsymbol{\alpha} \quad \Rightarrow \quad (\overline{\overline{\mathsf{M}}} - A\overline{\overline{\mathsf{I}}}^{(2)T})\lfloor\boldsymbol{\alpha} = 0.$$

Choosing basis one-forms $\boldsymbol{\alpha} = \boldsymbol{\varepsilon}_i$, we finally obtain

$$\begin{aligned}\overline{\overline{\mathsf{M}}} - A\overline{\overline{\mathsf{I}}}^{(2)T} &= (\overline{\overline{\mathsf{M}}} - A\overline{\overline{\mathsf{I}}}^{(2)T})|\sum \boldsymbol{\varepsilon}_{ij}\mathbf{e}_{ij} \\ &= \sum((\overline{\overline{\mathsf{M}}} - A\overline{\overline{\mathsf{I}}}^{(2)T})\lfloor\boldsymbol{\varepsilon}_i)|\boldsymbol{\varepsilon}_j\mathbf{e}_{ij} = 0,\end{aligned}$$

whence $\overline{\overline{\mathsf{M}}}$ must be a multiple of the unit bidyadic $\overline{\overline{\mathsf{I}}}^{(2)T}$. As a consequence we also obtain the following rule for the metric bidyadic $\overline{\overline{\mathsf{M}}}_m = \mathbf{e}_N\lfloor\overline{\overline{\mathsf{M}}}$: if $\overline{\overline{\mathsf{M}}}_m\lfloor\lfloor\boldsymbol{\nu}\boldsymbol{\nu} = \mathbf{e}_N\lfloor(\boldsymbol{\alpha}\wedge\overline{\overline{\mathsf{M}}}\lfloor\boldsymbol{\alpha}) = 0$ is satisfied for all $\boldsymbol{\alpha}$, the metric bidyadic must be of the form $\overline{\overline{\mathsf{M}}}_m = \mathbf{e}_N\lfloor\overline{\overline{\mathsf{M}}} = A\mathbf{e}_N\lfloor\overline{\overline{\mathsf{I}}}^{(2)T} = A\overline{\overline{\mathsf{E}}}$.

3.21 Applying the identity, the powers of $\overline{\overline{\mathsf{C}}} = \overline{\overline{\mathsf{B}}}_o{}^\wedge_\wedge\overline{\overline{\mathsf{I}}}$ can be expressed in the following symmetric pattern,

$$\begin{aligned}
2\overline{\overline{\mathsf{C}}} &= \overline{\overline{\mathsf{B}}}_o{}^\wedge_\wedge\overline{\overline{\mathsf{I}}} + \overline{\overline{\mathsf{I}}}{}^\wedge_\wedge\overline{\overline{\mathsf{B}}}_o \\
2\overline{\overline{\mathsf{C}}}^2 &= \overline{\overline{\mathsf{B}}}{}_o^2{}^\wedge_\wedge\overline{\overline{\mathsf{I}}} + 2\overline{\overline{\mathsf{B}}}_o{}^\wedge_\wedge\overline{\overline{\mathsf{B}}}_o + \overline{\overline{\mathsf{I}}}{}^\wedge_\wedge\overline{\overline{\mathsf{B}}}{}_o^2 \\
2\overline{\overline{\mathsf{C}}}^3 &= \overline{\overline{\mathsf{B}}}{}_o^3{}^\wedge_\wedge\overline{\overline{\mathsf{I}}} + 3\overline{\overline{\mathsf{B}}}{}_o^2{}^\wedge_\wedge\overline{\overline{\mathsf{B}}}_o + 3\overline{\overline{\mathsf{B}}}_o{}^\wedge_\wedge\overline{\overline{\mathsf{B}}}{}_o^2 + \overline{\overline{\mathsf{I}}}{}^\wedge_\wedge\overline{\overline{\mathsf{B}}}{}_o^3 \\
2\overline{\overline{\mathsf{C}}}^4 &= \overline{\overline{\mathsf{B}}}{}_o^4{}^\wedge_\wedge\overline{\overline{\mathsf{I}}} + 4\overline{\overline{\mathsf{B}}}{}_o^3{}^\wedge_\wedge\overline{\overline{\mathsf{B}}}_o + 6\overline{\overline{\mathsf{B}}}{}_o^2{}^\wedge_\wedge\overline{\overline{\mathsf{B}}}{}_o^2 + 4\overline{\overline{\mathsf{B}}}_o{}^\wedge_\wedge\overline{\overline{\mathsf{B}}}{}_o^3 + \overline{\overline{\mathsf{I}}}{}^\wedge_\wedge\overline{\overline{\mathsf{B}}}{}_o^4 \\
2\overline{\overline{\mathsf{C}}}^5 &= \overline{\overline{\mathsf{B}}}{}_o^5{}^\wedge_\wedge\overline{\overline{\mathsf{I}}} + 5\overline{\overline{\mathsf{B}}}{}_o^4{}^\wedge_\wedge\overline{\overline{\mathsf{B}}}_o + 10\overline{\overline{\mathsf{B}}}{}_o^3{}^\wedge_\wedge\overline{\overline{\mathsf{B}}}{}_o^2 + 10\overline{\overline{\mathsf{B}}}{}_o^2{}^\wedge_\wedge\overline{\overline{\mathsf{B}}}{}_o^3 + 5\overline{\overline{\mathsf{B}}}_o{}^\wedge_\wedge\overline{\overline{\mathsf{B}}}{}_o^4 + \overline{\overline{\mathsf{I}}}{}^\wedge_\wedge\overline{\overline{\mathsf{B}}}{}_o^5 \\
2\overline{\overline{\mathsf{C}}}^6 &= \overline{\overline{\mathsf{B}}}{}_o^6{}^\wedge_\wedge\overline{\overline{\mathsf{I}}} + 6\overline{\overline{\mathsf{B}}}{}_o^5{}^\wedge_\wedge\overline{\overline{\mathsf{B}}}_o + 15\overline{\overline{\mathsf{B}}}{}_o^4{}^\wedge_\wedge\overline{\overline{\mathsf{B}}}{}_o^2 + 20\overline{\overline{\mathsf{B}}}{}_o^3{}^\wedge_\wedge\overline{\overline{\mathsf{B}}}{}_o^3 + 15\overline{\overline{\mathsf{B}}}{}_o^2{}^\wedge_\wedge\overline{\overline{\mathsf{B}}}{}_o^4 \\
&\quad + 6\overline{\overline{\mathsf{B}}}_o{}^\wedge_\wedge\overline{\overline{\mathsf{B}}}{}_o^5 + \overline{\overline{\mathsf{I}}}{}^\wedge_\wedge\overline{\overline{\mathsf{B}}}{}_o^6.
\end{aligned}$$

The traces can be expanded as

$$\begin{aligned}
\mathrm{tr}\overline{\overline{\mathsf{C}}} &= 3\mathrm{tr}\overline{\overline{\mathsf{B}}}_o = 0 \\
\mathrm{tr}\overline{\overline{\mathsf{C}}}^2 &= \mathrm{tr}(\overline{\overline{\mathsf{B}}}{}_o^2{}^\wedge_\wedge\overline{\overline{\mathsf{I}}}) + 2\mathrm{tr}\overline{\overline{\mathsf{B}}}{}_o^{(2)} = 3\mathrm{tr}\overline{\overline{\mathsf{B}}}{}_o^2 + (\mathrm{tr}\overline{\overline{\mathsf{B}}}_o)^2 - \mathrm{tr}\overline{\overline{\mathsf{B}}}{}_o^2 \\
&= 2\mathrm{tr}\overline{\overline{\mathsf{B}}}{}_o^2, \\
\mathrm{tr}\overline{\overline{\mathsf{C}}}^3 &= 3\mathrm{tr}\overline{\overline{\mathsf{B}}}{}_o^3 + 3\mathrm{tr}(\overline{\overline{\mathsf{B}}}{}_o^2{}^\wedge_\wedge\overline{\overline{\mathsf{B}}}_o) = 3\mathrm{tr}\overline{\overline{\mathsf{B}}}{}_o^3 + 3(\mathrm{tr}\overline{\overline{\mathsf{B}}}{}_o^2)(\mathrm{tr}\overline{\overline{\mathsf{B}}}_o) - 3\mathrm{tr}\overline{\overline{\mathsf{B}}}{}_o^3 \\
&= 0, \\
\mathrm{tr}\overline{\overline{\mathsf{C}}}^4 &= 3\mathrm{tr}\overline{\overline{\mathsf{B}}}{}_o^4 + 4\mathrm{tr}(\overline{\overline{\mathsf{B}}}{}_o^3{}^\wedge_\wedge\overline{\overline{\mathsf{B}}}_o) + 6\mathrm{tr}(\overline{\overline{\mathsf{B}}}{}_o^2)^{(2)} = -\mathrm{tr}\overline{\overline{\mathsf{B}}}{}_o^4 + 3(\mathrm{tr}\overline{\overline{\mathsf{B}}}{}_o^2)^2 - 3\mathrm{tr}\overline{\overline{\mathsf{B}}}{}_o^4 \\
&= -4\mathrm{tr}\overline{\overline{\mathsf{B}}}{}_o^4 + 3(\mathrm{tr}\overline{\overline{\mathsf{B}}}{}_o^2)^2 \\
\mathrm{tr}\overline{\overline{\mathsf{C}}}^5 &= 3\mathrm{tr}\overline{\overline{\mathsf{B}}}{}_o^5 - 5\mathrm{tr}\overline{\overline{\mathsf{B}}}{}_o^5 + 10(\mathrm{tr}\overline{\overline{\mathsf{B}}}{}_o^3)(\mathrm{tr}\overline{\overline{\mathsf{B}}}{}_o^2) - 10\mathrm{tr}\overline{\overline{\mathsf{B}}}{}_o^5 \\
&= -12\mathrm{tr}\overline{\overline{\mathsf{B}}}{}_o^5 + 10(\mathrm{tr}\overline{\overline{\mathsf{B}}}{}_o^3)(\mathrm{tr}\overline{\overline{\mathsf{B}}}{}_o^2) \\
\mathrm{tr}\overline{\overline{\mathsf{C}}}^6 &= 3\mathrm{tr}\overline{\overline{\mathsf{B}}}{}_o^6 - 6\mathrm{tr}\overline{\overline{\mathsf{B}}}{}_o^6 + 15(\mathrm{tr}\overline{\overline{\mathsf{B}}}{}_o^4)(\mathrm{tr}\overline{\overline{\mathsf{B}}}{}_o^2) - 15\mathrm{tr}\overline{\overline{\mathsf{B}}}{}_o^6 + 20\mathrm{tr}((\overline{\overline{\mathsf{B}}}{}_o^3)^{(2)}) \\
&= -18\mathrm{tr}\overline{\overline{\mathsf{B}}}{}_o^6 + 15(\mathrm{tr}\overline{\overline{\mathsf{B}}}{}_o^4)(\mathrm{tr}\overline{\overline{\mathsf{B}}}{}_o^2) + 10(\mathrm{tr}\overline{\overline{\mathsf{B}}}{}_o^3)^2 - 10\mathrm{tr}\overline{\overline{\mathsf{B}}}{}_o^6 \\
&= -28\mathrm{tr}\overline{\overline{\mathsf{B}}}{}_o^6 + 15(\mathrm{tr}\overline{\overline{\mathsf{B}}}{}_o^4)(\mathrm{tr}\overline{\overline{\mathsf{B}}}{}_o^2) + 10(\mathrm{tr}\overline{\overline{\mathsf{B}}}{}_o^3)^2.
\end{aligned}$$

The trace-free dyadic $\overline{\overline{\mathsf{B}}}_o$ satisfies the Cayley–Hamilton equation

$$\overline{\overline{\mathsf{B}}}_o^4 + (\mathrm{tr}\overline{\overline{\mathsf{B}}}_o^{(2)})\overline{\overline{\mathsf{B}}}_o^2 - (\mathrm{tr}\overline{\overline{\mathsf{B}}}_o^{(3)})\overline{\overline{\mathsf{B}}}_o + (\mathrm{tr}\overline{\overline{\mathsf{B}}}_o^{(4)})\overline{\overline{\mathsf{I}}} = 0.$$

Inserting $\mathrm{tr}\overline{\overline{\mathsf{B}}}_o^{(2)} = -(1/2)\mathrm{tr}\overline{\overline{\mathsf{B}}}_o^2$ and taking the trace, we can solve

$$\mathrm{tr}\overline{\overline{\mathsf{B}}}_o^{(4)} = -\frac{1}{8}((\mathrm{tr}\overline{\overline{\mathsf{B}}}_o^2)^2 - 2\mathrm{tr}\overline{\overline{\mathsf{B}}}_o^4),$$

whence

$$\overline{\overline{\mathsf{B}}}_o^4 - \frac{1}{2}(\mathrm{tr}\overline{\overline{\mathsf{B}}}_o^2)\overline{\overline{\mathsf{B}}}_o^2 - (\mathrm{tr}\overline{\overline{\mathsf{B}}}_o^{(3)})\overline{\overline{\mathsf{B}}}_o + \frac{1}{8}((\mathrm{tr}\overline{\overline{\mathsf{B}}}_o^2)^2 - 2\mathrm{tr}\overline{\overline{\mathsf{B}}}_o^4)\overline{\overline{\mathsf{I}}} = 0.$$

Multiplying this by $\overline{\overline{\mathsf{B}}}_o|$ and taking the trace yields

$$\mathrm{tr}\overline{\overline{\mathsf{B}}}_o^5 - \frac{1}{2}(\mathrm{tr}\overline{\overline{\mathsf{B}}}_o^2)(\mathrm{tr}\overline{\overline{\mathsf{B}}}_o^3) - (\mathrm{tr}\overline{\overline{\mathsf{B}}}_o^{(3)})(\mathrm{tr}\overline{\overline{\mathsf{B}}}_o^2) = 0.$$

Applying $\mathrm{tr}\overline{\overline{\mathsf{B}}}_o^3 = 3\mathrm{tr}\overline{\overline{\mathsf{B}}}_o^{(3)}$ we obtain the expansion

$$\mathrm{tr}\overline{\overline{\mathsf{B}}}_o^5 = \frac{5}{6}(\mathrm{tr}\overline{\overline{\mathsf{B}}}_o^2)(\mathrm{tr}\overline{\overline{\mathsf{B}}}_o^3),$$

which yields the simple result

$$\mathrm{tr}\overline{\overline{\mathsf{C}}}^5 = -10(\mathrm{tr}\overline{\overline{\mathsf{B}}}_o^2)(\mathrm{tr}\overline{\overline{\mathsf{B}}}_o^3) + 10(\mathrm{tr}\overline{\overline{\mathsf{B}}}_o^3)(\mathrm{tr}\overline{\overline{\mathsf{B}}}_o^2) = 0.$$

Further, multiplying the Cayley–Hamilton equation by $\overline{\overline{\mathsf{B}}}_o^2|$ and taking the trace yields

$$\begin{aligned}\mathrm{tr}\overline{\overline{\mathsf{B}}}_o^6 &= \frac{1}{2}(\mathrm{tr}\overline{\overline{\mathsf{B}}}_o^2)(\mathrm{tr}\overline{\overline{\mathsf{B}}}_o^4) + \frac{1}{3}(\mathrm{tr}\overline{\overline{\mathsf{B}}}_o^3)^2 - \left(\frac{1}{8}(\mathrm{tr}\overline{\overline{\mathsf{B}}}_o^2)^2 - \frac{1}{4}(\mathrm{tr}\overline{\overline{\mathsf{B}}}_o^4)\mathrm{tr}\overline{\overline{\mathsf{B}}}_o^2\right)\\ &= \frac{3}{4}(\mathrm{tr}\overline{\overline{\mathsf{B}}}_o^2)(\mathrm{tr}\overline{\overline{\mathsf{B}}}_o^4) + \frac{1}{3}(\mathrm{tr}\overline{\overline{\mathsf{B}}}_o^3)^2 - \frac{1}{8}(\mathrm{tr}\overline{\overline{\mathsf{B}}}_o^2)^3,\end{aligned}$$

which allows us to expand

$$\begin{aligned}\mathrm{tr}\overline{\overline{\mathsf{C}}}^6 &= -21(\mathrm{tr}\overline{\overline{\mathsf{B}}}_o^2)(\mathrm{tr}\overline{\overline{\mathsf{B}}}_o^4) - \frac{28}{3}(\mathrm{tr}\overline{\overline{\mathsf{B}}}_o^3)^2 + \frac{7}{2}(\mathrm{tr}\overline{\overline{\mathsf{B}}}_o^2)^3\\ &\quad +15\mathrm{tr}(\overline{\overline{\mathsf{B}}}_o^4)(\mathrm{tr}\overline{\overline{\mathsf{B}}}_o^2) + 10\mathrm{tr}(\overline{\overline{\mathsf{B}}}_o^3)^2\\ &= -6(\mathrm{tr}\overline{\overline{\mathsf{B}}}_o^2)(\mathrm{tr}\overline{\overline{\mathsf{B}}}_o^4) + \frac{2}{3}(\mathrm{tr}\overline{\overline{\mathsf{B}}}_o^3)^2 + \frac{7}{2}(\mathrm{tr}\overline{\overline{\mathsf{B}}}_o^2)^3.\end{aligned}$$

In terms of these expressions for $\text{tr}\overline{\overline{\mathsf{C}}}^p$, the Cayley–Hamilton equation coefficients for the skewon bidyadic are simplified as

$$\begin{aligned}
a_0 &= 1\\
a_1 &= -a_0\text{tr}\overline{\overline{\mathsf{C}}} = 0\\
a_2 &= -\frac{1}{2}(a_1\text{tr}\overline{\overline{\mathsf{C}}} + a_0\text{tr}\overline{\overline{\mathsf{C}}}^2) = -\frac{1}{2}\text{tr}\overline{\overline{\mathsf{C}}}^2\\
a_3 &= -\frac{1}{3}(a_2\text{tr}\overline{\overline{\mathsf{C}}} + a_1\text{tr}\overline{\overline{\mathsf{C}}}^2 + a_0\text{tr}\overline{\overline{\mathsf{C}}}^3) = 0\\
a_4 &= -\frac{1}{4}(a_3\text{tr}\overline{\overline{\mathsf{C}}} + a_2\text{tr}\overline{\overline{\mathsf{C}}}^2 + a_1\text{tr}\overline{\overline{\mathsf{C}}}^3 + a_0\text{tr}\overline{\overline{\mathsf{C}}}^4) = \frac{1}{8}(\text{tr}\overline{\overline{\mathsf{C}}}^2)^2 - \frac{1}{4}\text{tr}\overline{\overline{\mathsf{C}}}^4\\
a_5 &= -\frac{1}{5}(a_4\text{tr}\overline{\overline{\mathsf{C}}} + a_3\text{tr}\overline{\overline{\mathsf{C}}}^2 + a_2\text{tr}\overline{\overline{\mathsf{C}}}^3 + a_1\text{tr}\overline{\overline{\mathsf{C}}}^4 + a_0\text{tr}\overline{\overline{\mathsf{C}}}^5) = 0\\
a_6 &= -\frac{1}{6}(a_5\text{tr}\overline{\overline{\mathsf{C}}} + a_4\text{tr}\overline{\overline{\mathsf{C}}}^2 + a_3\text{tr}\overline{\overline{\mathsf{C}}}^3 + a_2\text{tr}\overline{\overline{\mathsf{C}}}^4 + a_1\text{tr}\overline{\overline{\mathsf{C}}}^5 + a_0\text{tr}\overline{\overline{\mathsf{C}}}^6)\\
&= -\frac{1}{48}(\text{tr}\overline{\overline{\mathsf{C}}}^2)^3 + \frac{1}{8}\text{tr}\overline{\overline{\mathsf{C}}}^2\text{tr}\overline{\overline{\mathsf{C}}}^4 - \frac{1}{6}\text{tr}\overline{\overline{\mathsf{C}}}^6.
\end{aligned}$$

Thus, the Cayley–Hamilton equation (3.10) for any skewon bidyadic $\overline{\overline{\mathsf{C}}}$ can be expressed as

$$\begin{aligned}
&\overline{\overline{\mathsf{C}}}^6 - \frac{1}{2}(\text{tr}\overline{\overline{\mathsf{C}}}^2)\overline{\overline{\mathsf{C}}}^4 + \left(\frac{1}{8}\left(\text{tr}\overline{\overline{\mathsf{C}}}^2\right)^2 - \frac{1}{4}\text{tr}\overline{\overline{\mathsf{C}}}^4\right)\overline{\overline{\mathsf{C}}}^2\\
&\quad + \left(-\frac{1}{48}(\text{tr}\overline{\overline{\mathsf{C}}}^2)^3 + \frac{1}{8}\text{tr}\overline{\overline{\mathsf{C}}}^2\text{tr}\overline{\overline{\mathsf{C}}}^4 - \frac{1}{6}\text{tr}\overline{\overline{\mathsf{C}}}^6\right)\overline{\overline{\mathsf{I}}}^{(2)} = 0.
\end{aligned}$$

As a simple check, the trace of the left side can be shown to vanish.

3.22 For the skewon bidyadic $\overline{\overline{\mathsf{C}}} = \overline{\overline{\mathsf{B}}}_o{}^\wedge_\wedge\overline{\overline{\mathsf{I}}}$, the determinant has the form

$$\det\overline{\overline{\mathsf{C}}} = a_6 = -\frac{1}{48}(\text{tr}\overline{\overline{\mathsf{C}}}^2)^3 + \frac{1}{8}\text{tr}\overline{\overline{\mathsf{C}}}^2\text{tr}\overline{\overline{\mathsf{C}}}^4 - \frac{1}{6}\text{tr}\overline{\overline{\mathsf{C}}}^6,$$

in which we can insert

$$\begin{aligned}
\text{tr}\overline{\overline{\mathsf{C}}}^2 &= 2\text{tr}\overline{\overline{\mathsf{B}}}_o^2,\\
\text{tr}\overline{\overline{\mathsf{C}}}^4 &= -4\text{tr}\overline{\overline{\mathsf{B}}}_o^4 + 3(\text{tr}\overline{\overline{\mathsf{B}}}_o^2)^2,\\
\text{tr}\overline{\overline{\mathsf{C}}}^6 &= \frac{3}{4}\left(\text{tr}\overline{\overline{\mathsf{B}}}_o^2\right)\left(\text{tr}\overline{\overline{\mathsf{B}}}_o^4\right) + \frac{2}{3}\left(\text{tr}\overline{\overline{\mathsf{B}}}_o^3\right)^2 + \frac{7}{2}\left(\text{tr}\overline{\overline{\mathsf{B}}}_o^2\right)^3.
\end{aligned}$$

The expression can be expanded as

$$\begin{aligned}\det\overline{\overline{\mathsf{C}}} &= -\frac{1}{6}\left(\frac{1}{8}(\mathrm{tr}\overline{\overline{\mathsf{C}}}^2)^3 - \frac{3}{4}\mathrm{tr}\overline{\overline{\mathsf{C}}}^2\mathrm{tr}\overline{\overline{\mathsf{C}}}^4 + \mathrm{tr}\overline{\overline{\mathsf{C}}}^6\right).\\ &= -\frac{1}{6}\left(\frac{1}{8}\left(2\mathrm{tr}\overline{\overline{\mathsf{B}}}_o^2\right)^3 - \frac{3}{4}\left(2\mathrm{tr}\overline{\overline{\mathsf{B}}}_o^2\right)\left(-4\mathrm{tr}\overline{\overline{\mathsf{B}}}_o^4 + 3\left(\mathrm{tr}\overline{\overline{\mathsf{B}}}_o^2\right)^2\right)\right.\\ &\quad \left. -6\left(\mathrm{tr}\overline{\overline{\mathsf{B}}}_o^2\right)\left(\mathrm{tr}\overline{\overline{\mathsf{B}}}_o^4\right) + \frac{2}{3}\left(\mathrm{tr}\overline{\overline{\mathsf{B}}}_o^3\right)^2 + \frac{7}{2}\left(\mathrm{tr}\overline{\overline{\mathsf{B}}}_o^2\right)^3\right)\\ &= -\frac{1}{9}\left(\mathrm{tr}\overline{\overline{\mathsf{B}}}_o^3\right)^2 = -\left(\mathrm{tr}\overline{\overline{\mathsf{B}}}_o^{(3)}\right)^2.\end{aligned}$$

3.23 The identity can be extended as

$$\begin{aligned}\overline{\overline{\mathsf{I}}}^{(3)}\lfloor\lfloor(\alpha\mathbf{a}{}^\wedge_\wedge\mathbf{b}\boldsymbol{\beta}) &= ((\overline{\overline{\mathsf{I}}}^{(3)}\lfloor\lfloor(\boldsymbol{\alpha}\mathbf{a}))\lfloor\lfloor(\mathbf{b}\boldsymbol{\beta})\\ &= (\mathbf{a}|\boldsymbol{\alpha})\overline{\overline{\mathsf{I}}}^{(2)}\lfloor\lfloor\mathbf{b}\boldsymbol{\beta} - (\overline{\overline{\mathsf{I}}}{}^\wedge_\wedge\mathbf{a}\boldsymbol{\alpha})\lfloor\lfloor\boldsymbol{\beta}\mathbf{b}\\ &= ((\mathbf{a}|\boldsymbol{\alpha})(\mathbf{b}|\boldsymbol{\beta}) - \mathbf{a}\boldsymbol{\alpha}||\boldsymbol{\beta}\mathbf{b})\overline{\overline{\mathsf{I}}} - (\mathbf{a}|\boldsymbol{\alpha})\mathbf{b}\boldsymbol{\beta} - (\mathbf{b}|\boldsymbol{\beta})\mathbf{a}\boldsymbol{\alpha}\\ &\quad + \mathbf{a}\boldsymbol{\alpha}|\mathbf{b}\boldsymbol{\beta} + \mathbf{b}\boldsymbol{\beta}|\mathbf{a}\boldsymbol{\alpha}\\ &= \mathrm{tr}(\mathbf{a}\boldsymbol{\alpha}{}^\wedge_\wedge\mathbf{b}\boldsymbol{\beta})\overline{\overline{\mathsf{I}}} - (\mathbf{a}\boldsymbol{\alpha}{}^\wedge_\wedge\mathbf{b}\boldsymbol{\beta})\lfloor\lfloor\overline{\overline{\mathsf{I}}}^T.\end{aligned}$$

Since this is linear and valid for any bidyad $\mathbf{a}\boldsymbol{\alpha}{}^\wedge_\wedge\mathbf{b}\boldsymbol{\beta}$, it is valid more generally for any bidyadic $\overline{\overline{\mathsf{C}}}$ as

$$\overline{\overline{\mathsf{I}}}^{(3)}\lfloor\lfloor\overline{\overline{\mathsf{C}}}^T = \mathrm{tr}\overline{\overline{\mathsf{C}}}\,\overline{\overline{\mathsf{I}}} - \overline{\overline{\mathsf{C}}}\lfloor\lfloor\overline{\overline{\mathsf{I}}}^T.$$

Vanishing left side requires that trace of the right side vanish, or

$$4\mathrm{tr}\overline{\overline{\mathsf{C}}} = \mathrm{tr}(\overline{\overline{\mathsf{C}}}\lfloor\lfloor\overline{\overline{\mathsf{I}}}^T) = (\overline{\overline{\mathsf{C}}}\lfloor\lfloor\overline{\overline{\mathsf{I}}}^T)||\overline{\overline{\mathsf{I}}}^T = 2\mathrm{tr}\overline{\overline{\mathsf{C}}},$$

whence $\mathrm{tr}\overline{\overline{\mathsf{C}}} = 0$ and $\overline{\overline{\mathsf{C}}}\lfloor\lfloor\overline{\overline{\mathsf{I}}}^T = 0$. Thus, $\overline{\overline{\mathsf{I}}}^{(3)}\lfloor\lfloor\overline{\overline{\mathsf{C}}}^T = 0$ is equivalent to $\overline{\overline{\mathsf{C}}}\lfloor\lfloor\overline{\overline{\mathsf{I}}}^T = 0$ as a definition of the principal bidyadic.

3.24 The inverse of the metric bidyadic $\overline{\overline{\mathsf{D}}}$ involving only axion and skewon parts can be expressed as

$$\overline{\overline{\mathsf{D}}}^{-1} = \frac{1}{\alpha(\alpha^2-\lambda^2)}\boldsymbol{\varepsilon}_N\boldsymbol{\varepsilon}_N\lfloor\lfloor(\overline{\overline{\mathsf{D}}}\cdot\overline{\overline{\mathsf{D}}} - 3\alpha\overline{\overline{\mathsf{D}}} + (3\alpha^2-\lambda^2)\overline{\overline{\mathsf{E}}}).$$

Since $-3\alpha\overline{\overline{\mathsf{D}}} + (3\alpha^2-\lambda^2)\overline{\overline{\mathsf{E}}}$ contains no principal component, let us consider the term

$$\begin{aligned}\overline{\overline{\mathsf{D}}}\cdot\overline{\overline{\mathsf{D}}} &= \alpha^2\mathbf{e}_N\lfloor\overline{\overline{\mathsf{I}}}^{(2)T} + 2\alpha(\mathbf{AB}-\mathbf{BA})\\ &\quad + (\mathbf{A}\cdot\mathbf{B})(\mathbf{AB}+\mathbf{BA}) - (\mathbf{B}\cdot\mathbf{B})\mathbf{AA} - (\mathbf{A}\cdot\mathbf{A})\mathbf{BB},\end{aligned}$$

which is symmetric. Now it is obvious that, in the general case, it cannot be a multiple of $\mathbf{e}_N\lfloor\overline{\overline{\mathsf{I}}}^{(2)T}$ because it is not of full rank. In conclusion, the inverse of such a skewon–axion bidyadic contains a principal part. Similarly, the inverse of a principal bidyadic may contain an axion part.

3.25 For the skewon–axion bidyadic $\overline{\overline{\mathsf{D}}} = \overline{\overline{\mathsf{C}}}_o + \alpha\overline{\overline{\mathsf{I}}}$, $\overline{\overline{\mathsf{C}}}_o = \overline{\overline{\mathsf{B}}}_o {}^\wedge_\wedge \overline{\overline{\mathsf{I}}}$, $\mathrm{tr}\overline{\overline{\mathsf{B}}}_o = 0$ we multiply the rule (3.159) as

$$(\overline{\overline{\mathsf{I}}}^{(2)} + \overline{\overline{\mathsf{C}}}_o)^{-1}|(\overline{\overline{\mathsf{I}}}^{(2)} + \overline{\overline{\mathsf{C}}}_o) = \overline{\overline{\mathsf{I}}}^{(2)} + \overline{\overline{\mathsf{C}}}_o - \frac{\overline{\overline{\mathsf{X}}}}{A_6}$$

with $A_6 = 1 + a_2 + a_4 + a_6$ and

$$\begin{aligned}\overline{\overline{\mathsf{X}}} &= (\overline{\overline{\mathsf{C}}}_o - \overline{\overline{\mathsf{C}}}{}_o^2)|((1 + a_2 + a_4)\overline{\overline{\mathsf{I}}}^{(2)} + (1 + a_2)\overline{\overline{\mathsf{C}}}{}_o^2 + \overline{\overline{\mathsf{C}}}{}_o^4)|(\overline{\overline{\mathsf{I}}}^{(2)} + \overline{\overline{\mathsf{C}}}_o)\\
&= (1 + a_2 + a_4)(\overline{\overline{\mathsf{C}}}_o - \overline{\overline{\mathsf{C}}}{}_o^3) + (1 + a_2)(\overline{\overline{\mathsf{C}}}{}_o^3 - \overline{\overline{\mathsf{C}}}{}_o^5) + \overline{\overline{\mathsf{C}}}{}_o^5 - \overline{\overline{\mathsf{C}}}{}_o^7\\
&= (1 + a_2 + a_4)\overline{\overline{\mathsf{C}}}_o - a_4\overline{\overline{\mathsf{C}}}{}_o^3 - a_2\overline{\overline{\mathsf{C}}}{}_o^5 - \overline{\overline{\mathsf{C}}}{}_o^7\\
&= A_6\overline{\overline{\mathsf{C}}}_o - (a_6\overline{\overline{\mathsf{C}}} + a_4\overline{\overline{\mathsf{C}}}{}_o^3 + a_2\overline{\overline{\mathsf{C}}}{}_o^5 + \overline{\overline{\mathsf{C}}}{}_o^7).\end{aligned}$$

Applying the reduced Cayley–Hamilton equation

$$\overline{\overline{\mathsf{C}}}{}_o^6 + a_2\overline{\overline{\mathsf{C}}}{}_o^4 + a_4\overline{\overline{\mathsf{C}}}{}_o^2 + a_6\overline{\overline{\mathsf{I}}}^{(2)} = 0$$

we have

$$\overline{\overline{\mathsf{X}}} = (1 + a_2 + a_4 + a_6)\overline{\overline{\mathsf{C}}}_o = A_6\overline{\overline{\mathsf{C}}}_o,$$

whence we finally obtain

$$(\overline{\overline{\mathsf{I}}}^{(2)} + \overline{\overline{\mathsf{C}}}_o)^{-1}|(\overline{\overline{\mathsf{I}}}^{(2)} + \overline{\overline{\mathsf{C}}}_o) = \overline{\overline{\mathsf{I}}}^{(2)}.$$

CHAPTER 4

4.1 If $\det\overline{\overline{\mathsf{A}}}_s \neq 0$ or $\det\overline{\overline{\mathsf{B}}}_s \neq 0$, we respectively have $\overline{\overline{\mathsf{B}}}_s = 0$ or $\overline{\overline{\mathsf{A}}}_s = 0$, whence we must study the case $\det\overline{\overline{\mathsf{A}}}_s = \det\overline{\overline{\mathsf{B}}}_s = 0$. Expanding the dyadics in terms of four spatial vectors and one-forms as

$$\overline{\overline{\mathsf{A}}}_s = \mathbf{a}_1\boldsymbol{\varepsilon}_1 + \mathbf{a}_2\boldsymbol{\varepsilon}_2, \quad \overline{\overline{\mathsf{B}}}_s = \mathbf{b}_1\boldsymbol{\beta}_1 + \mathbf{b}_2\boldsymbol{\beta}_2,$$

where $\boldsymbol{\varepsilon}_1, \boldsymbol{\varepsilon}_2, \boldsymbol{\varepsilon}_3$ form a spatial basis. Assuming that $\mathbf{a}_1 \wedge \mathbf{a}_2$, $\mathbf{b}_1 \wedge \mathbf{b}_2$ and $\boldsymbol{\beta}_1 \wedge \boldsymbol{\beta}_2$ do not vanish, we would have both $\overline{\overline{\mathsf{A}}}{}_s^{(2)} \neq 0$ and $\overline{\overline{\mathsf{B}}}{}_s^{(2)} \neq 0$. However, the condition $\overline{\overline{\mathsf{A}}}_s|\overline{\overline{\mathsf{B}}}_s = 0$ requires $\boldsymbol{\varepsilon}_i|\mathbf{b}_1 = \boldsymbol{\varepsilon}_i|\mathbf{b}_2 = 0$ for $i = 1, 2$, whence we must have $\mathbf{b}_1 = b_1\boldsymbol{\varepsilon}_3$ and $\mathbf{b}_2 = b_2\boldsymbol{\varepsilon}_3$ which is contrary to the assumption $\mathbf{b}_1 \wedge \mathbf{b}_2 \neq 0$ and $\overline{\overline{\mathsf{B}}}{}_s^{(2)} \neq 0$. Thus, the case of both $\overline{\overline{\mathsf{A}}}{}_s^{(2)} \neq 0$ and $\overline{\overline{\mathsf{B}}}{}_s^{(2)} \neq 0$ is not possible.

4.2 Expressing $\overline{\overline{\mathsf{K}}} = \overline{\overline{\mathsf{I}}} - 2\overline{\overline{\Pi}}$, the dyadic $\overline{\overline{\Pi}}$ can be expanded as

$$\overline{\overline{\Pi}} = -\frac{1}{2}(\mathbf{e}_3\boldsymbol{\varepsilon}_4 + \mathbf{e}_4\boldsymbol{\varepsilon}_3 - \mathbf{e}_3\boldsymbol{\varepsilon}_3 - \mathbf{e}_4\boldsymbol{\varepsilon}_4) = \left(\frac{\mathbf{e}_3 - \mathbf{e}_4}{\sqrt{2}}\right)\left(\frac{\boldsymbol{\varepsilon}_3 - \boldsymbol{\varepsilon}_4}{\sqrt{2}}\right),$$

which satisfies $\overline{\overline{\Pi}}^2 = \overline{\overline{\Pi}}$ and, hence, $\overline{\overline{\mathsf{K}}}^2 = \overline{\overline{\mathsf{I}}}$. To express $\overline{\overline{\mathsf{K}}}$ in the form $\overline{\overline{\Pi}}' - \overline{\overline{\Pi}}$, we define

$$\begin{aligned}\overline{\overline{\Pi}}' &= \overline{\overline{\mathsf{I}}} - \overline{\overline{\Pi}}\\ &= \mathbf{e}_1\boldsymbol{\varepsilon}_1 + \mathbf{e}_2\boldsymbol{\varepsilon}_2 + \frac{1}{2}(\mathbf{e}_3\boldsymbol{\varepsilon}_4 + \mathbf{e}_4\boldsymbol{\varepsilon}_3 + \mathbf{e}_3\boldsymbol{\varepsilon}_3 + \mathbf{e}_4\boldsymbol{\varepsilon}_4)\\ &= \mathbf{e}_1\boldsymbol{\varepsilon}_1 + \mathbf{e}_2\boldsymbol{\varepsilon}_2 + \left(\frac{\mathbf{e}_3 + \mathbf{e}_4}{\sqrt{2}}\right)\left(\frac{\boldsymbol{\varepsilon}_3 + \boldsymbol{\varepsilon}_4}{\sqrt{2}}\right).\end{aligned}$$

The unipotent dyadic can now be given the standard form

$$\overline{\overline{\mathsf{K}}} = \mathbf{p}_1\boldsymbol{\pi}_1 + \mathbf{p}_2\boldsymbol{\pi}_2 + \mathbf{p}_3\boldsymbol{\pi}_3 - \mathbf{p}_4\boldsymbol{\pi}_4$$

when the vectors and one-forms are defined by

$$\mathbf{p}_1 = \mathbf{e}_1,\quad \mathbf{p}_2 = \mathbf{e}_2,\quad \mathbf{p}_3 = \frac{\mathbf{e}_3 + \mathbf{e}_4}{\sqrt{2}},\quad \mathbf{p}_4 = \frac{\mathbf{e}_3 - \mathbf{e}_4}{\sqrt{2}},$$

$$\boldsymbol{\pi}_1 = \boldsymbol{\varepsilon}_1,\quad \boldsymbol{\pi}_2 = \boldsymbol{\varepsilon}_2,\quad \boldsymbol{\pi}_3 = \frac{\boldsymbol{\varepsilon}_3 + \boldsymbol{\varepsilon}_4}{\sqrt{2}},\quad \boldsymbol{\pi}_4 = \frac{\boldsymbol{\varepsilon}_3 - \boldsymbol{\varepsilon}_4}{\sqrt{2}}.$$

4.3 From $\overline{\overline{\mathsf{A}}}^3 = 0$ we have $\overline{\overline{\mathsf{A}}}^4 = 0$, whence $\mathrm{tr}\overline{\overline{\mathsf{A}}}^3 = 0$ and $\mathrm{tr}\overline{\overline{\mathsf{A}}}^4 = 0$. From

$$(\overline{\overline{\mathsf{A}}}^3)^{(4)} = (\overline{\overline{\mathsf{A}}}^{(4)})^3 = \overline{\overline{\mathsf{I}}}^{(4)}(\mathrm{tr}\overline{\overline{\mathsf{A}}}^{(4)})^3 = 0$$

we have $\mathrm{tr}\overline{\overline{\mathsf{A}}}^{(4)} = 0$. The Cayley–Hamilton equation is then reduced to

$$(\mathrm{tr}\overline{\overline{\mathsf{A}}}^{(2)})\overline{\overline{\mathsf{A}}}^2 - (\mathrm{tr}\overline{\overline{\mathsf{A}}}^{(3)})\overline{\overline{\mathsf{A}}} = 0.$$

Multiplying this by $\overline{\overline{\mathsf{A}}}|$ leads to $\mathrm{tr}\overline{\overline{\mathsf{A}}}^{(3)} = 0$ and, further, to $\mathrm{tr}\overline{\overline{\mathsf{A}}}^{(2)} = 0$. From their expansions we obtain

$$\mathrm{tr}\overline{\overline{\mathsf{A}}}^{(2)} = \frac{1}{2}((\mathrm{tr}\overline{\overline{\mathsf{A}}})^2 - \mathrm{tr}\overline{\overline{\mathsf{A}}}^2) = 0 \quad \Rightarrow \quad \mathrm{tr}\overline{\overline{\mathsf{A}}}^2 = (\mathrm{tr}\overline{\overline{\mathsf{A}}})^2$$

$$\mathrm{tr}\overline{\overline{\mathsf{A}}}^{(3)} = \frac{1}{6}((\mathrm{tr}\overline{\overline{\mathsf{A}}})^3 - 3\mathrm{tr}\overline{\overline{\mathsf{A}}}\,\mathrm{tr}\overline{\overline{\mathsf{A}}}^2 + 2\mathrm{tr}\overline{\overline{\mathsf{A}}}^3) = -\frac{1}{3}(\mathrm{tr}\overline{\overline{\mathsf{A}}})^3 = 0,$$

whence we also have $\mathrm{tr}\overline{\overline{\mathsf{A}}} = 0$ and $\mathrm{tr}\overline{\overline{\mathsf{A}}}^2 = 0$.

4.4 The equivalence is seen by expanding

$$\overline{\overline{\mathsf{J}}}^{(2)}\lfloor\lfloor\overline{\overline{\mathsf{I}}}^T_{12} = \mathrm{tr}\overline{\overline{\mathsf{J}}}\,\overline{\overline{\mathsf{J}}} - \overline{\overline{\mathsf{J}}}^2,$$

$$\mathrm{tr}\overline{\overline{\mathsf{J}}}^{(2)} = \frac{1}{2}((\mathrm{tr}\overline{\overline{\mathsf{J}}})^2 - \mathrm{tr}\overline{\overline{\mathsf{J}}}^2),$$

and applying $\overline{\overline{\mathsf{J}}}^{(2)} = \overline{\overline{\mathsf{I}}}^{(2)}_{12}$ whence the two conditions become

$$\overline{\overline{\mathsf{I}}}^{(2)}_{12}\lfloor\lfloor\overline{\overline{\mathsf{I}}}^T_{12} = 2\overline{\overline{\mathsf{I}}}_{12} - \overline{\overline{\mathsf{I}}}_{12} = \overline{\overline{\mathsf{I}}}_{12} = \mathrm{tr}\overline{\overline{\mathsf{J}}}\,\overline{\overline{\mathsf{J}}} - \overline{\overline{\mathsf{J}}}^2 = \mathrm{tr}\overline{\overline{\mathsf{J}}}\,\overline{\overline{\mathsf{J}}} + \overline{\overline{\mathsf{I}}}_{12},$$

$$\mathrm{tr}\overline{\overline{\mathsf{J}}}^{(2)} = \frac{1}{2}((\mathrm{tr}\overline{\overline{\mathsf{J}}})^2 + 2) = 1 + \frac{1}{2}(\mathrm{tr}\overline{\overline{\mathsf{J}}})^2.$$

Thus, $\overline{\overline{\mathsf{J}}}^{(2)} = \overline{\overline{\mathsf{I}}}^{(2)}_{12}$ implies $\mathrm{tr}\overline{\overline{\mathsf{J}}} = 0$ and $\mathrm{tr}\overline{\overline{\mathsf{J}}}^{(2)} = 1$ and conversely.

4.5 From $\mathrm{tr}\overline{\overline{\mathsf{J}}} = 0$ and $\mathrm{tr}\overline{\overline{\mathsf{J}}}^{(2)} = 1$ the conditions for the coefficients become

$$J_{11} + J_{22} = 0, \quad J_{11}J_{22} - J_{12}J_{21} = 1.$$

Defining

$$J_{11} = -J_{22} = \sinh\theta, \quad J_{12} = A\cosh\theta, \quad J_{21} = -\frac{1}{A}\cosh\theta,$$

with $\theta = \sinh^{-1} J_{11}, A = J_{12}/\sqrt{1 + J_{11}^2}$, the above conditions are obviously satisfied for any choice of θ and A. Inserted in the expansion we obtain

$$\begin{aligned}\overline{\overline{\mathsf{J}}} &= \sinh\theta\,\mathbf{e}_1\boldsymbol{\varepsilon}_1 + A\cosh\theta\,\mathbf{e}_1\boldsymbol{\varepsilon}_2 - \frac{1}{A}\cosh\theta\,\mathbf{e}_2\boldsymbol{\varepsilon}_1 - \sinh\theta\,\mathbf{e}_2\boldsymbol{\varepsilon}_2 \\ &= \frac{1}{2A}\left(e^{\theta}(A\mathbf{e}_1 - \mathbf{e}_2)(A\boldsymbol{\varepsilon}_2 + \boldsymbol{\varepsilon}_1) + e^{-\theta}(A\mathbf{e}_1 + \mathbf{e}_2)(A\boldsymbol{\varepsilon}_2 - \boldsymbol{\varepsilon}_1)\right),\end{aligned}$$

when the vectors and one-forms are defined as

$$\mathbf{a}_1 = \frac{1}{A}(A\mathbf{e}_1 - \mathbf{e}_2), \quad \mathbf{a}_2 = \frac{e^{-\theta}}{A}(A\mathbf{e}_1 + \mathbf{e}_2),$$

$$\boldsymbol{\alpha}_1 = -\frac{1}{2}(A\boldsymbol{\varepsilon}_2 - \boldsymbol{\varepsilon}_1), \quad \boldsymbol{\alpha}_2 = \frac{e^{\theta}}{2}(A\boldsymbol{\varepsilon}_2 + \boldsymbol{\varepsilon}_1).$$

Because they also satisfy the orthogonality conditions

$$\mathbf{a}_i|\boldsymbol{\alpha}_j = \delta_{ij},$$

$\mathbf{a}_1, \mathbf{a}_2$ and $\boldsymbol{\alpha}_1, \boldsymbol{\alpha}_2$ form reciprocal bases. Thus, the most general 2D AC-dyadic can be written in the standard form of (4.100),

$$\overline{\overline{\mathsf{J}}} = \mathbf{a}_1\boldsymbol{\alpha}_2 - \mathbf{a}_2\boldsymbol{\alpha}_1,$$

satisfying

$$\overline{\overline{\mathsf{J}}}^2 = -(\mathbf{a}_1\boldsymbol{\alpha}_1 + \mathbf{a}_2\boldsymbol{\alpha}_2) = -\overline{\overline{\mathsf{I}}}_{12}.$$

4.6 Applying various dyadic rules we can expand

$$\begin{aligned}
\overline{\overline{\mathsf{J}}}^{(2)}\lfloor\lfloor\overline{\overline{\mathsf{I}}}^T &= (\mathrm{tr}\overline{\overline{\mathsf{J}}})\overline{\overline{\mathsf{J}}} - \overline{\overline{\mathsf{J}}}^2 = \overline{\overline{\mathsf{I}}},\\
\overline{\overline{\mathsf{J}}}^{(3)}\lfloor\lfloor\overline{\overline{\mathsf{I}}}^T &= (\mathrm{tr}\overline{\overline{\mathsf{J}}})\overline{\overline{\mathsf{J}}}^{(2)} - \overline{\overline{\mathsf{J}}}{}^\wedge_\wedge\overline{\overline{\mathsf{J}}}^2 = \overline{\overline{\mathsf{J}}}{}^\wedge_\wedge\overline{\overline{\mathsf{I}}},\\
\overline{\overline{\mathsf{J}}}^{(3)}\lfloor\lfloor\overline{\overline{\mathsf{I}}}^{(2)T} &= \frac{1}{2}(\overline{\overline{\mathsf{J}}}^{(3)}\lfloor\lfloor\overline{\overline{\mathsf{I}}}^T)\lfloor\lfloor\overline{\overline{\mathsf{I}}}^T = \frac{1}{2}(\overline{\overline{\mathsf{J}}}{}^\wedge_\wedge\overline{\overline{\mathsf{I}}})\lfloor\lfloor\overline{\overline{\mathsf{I}}}^T = \overline{\overline{\mathsf{J}}}\\
\overline{\overline{\mathsf{I}}}^{(3)}\lfloor\lfloor\overline{\overline{\mathsf{J}}}^T &= (\mathrm{tr}\overline{\overline{\mathsf{J}}})\overline{\overline{\mathsf{I}}}^{(2)} - \overline{\overline{\mathsf{I}}}{}^\wedge_\wedge\overline{\overline{\mathsf{J}}} = -\overline{\overline{\mathsf{I}}}{}^\wedge_\wedge\overline{\overline{\mathsf{J}}},\\
\overline{\overline{\mathsf{I}}}^{(4)}\lfloor\lfloor\overline{\overline{\mathsf{J}}}^{(3)T} &= \frac{1}{3}(\overline{\overline{\mathsf{I}}}^{(4)}\lfloor\lfloor\overline{\overline{\mathsf{J}}}^{(2)T})\lfloor\lfloor\overline{\overline{\mathsf{J}}}^T = \frac{1}{3}\overline{\overline{\mathsf{J}}}^{(2)}\lfloor\lfloor\overline{\overline{\mathsf{J}}}^T = -\overline{\overline{\mathsf{J}}},\\
\overline{\overline{\mathsf{I}}}^{(3)}\lfloor\lfloor\overline{\overline{\mathsf{J}}}^{(2)T} &= \frac{1}{2}(\overline{\overline{\mathsf{I}}}^{(3)}\lfloor\lfloor\overline{\overline{\mathsf{J}}}^T)\lfloor\lfloor\overline{\overline{\mathsf{J}}}^T = -\frac{1}{2}(\overline{\overline{\mathsf{I}}}{}^\wedge_\wedge\overline{\overline{\mathsf{J}}})\lfloor\lfloor\overline{\overline{\mathsf{J}}}^T = \overline{\overline{\mathsf{I}}}
\end{aligned}$$

4.7 Inserting $\overline{\overline{\overline{\overline{\mathsf{M}}}}} = M\overline{\overline{\mathsf{J}}}^{(2)T}$ in the rule (2.114)

$$\overline{\overline{\mathsf{I}}}^{(4)}\lfloor\lfloor\overline{\overline{\overline{\overline{\mathsf{M}}}}} = (\mathrm{tr}\overline{\overline{\overline{\overline{\mathsf{M}}}}})\overline{\overline{\mathsf{I}}}^{(2)} - (\overline{\overline{\overline{\overline{\mathsf{M}}}}}^T\lfloor\lfloor\overline{\overline{\mathsf{I}}}^T){}^\wedge_\wedge\overline{\overline{\mathsf{I}}} + \overline{\overline{\overline{\overline{\mathsf{M}}}}}^T,$$

and applying $\mathrm{tr}\overline{\overline{\mathsf{J}}} = 0$, whence

$$\mathrm{tr}\overline{\overline{\overline{\overline{\mathsf{M}}}}} = M\mathrm{tr}\overline{\overline{\mathsf{J}}}^{(2)} = \frac{M}{2}((\mathrm{tr}\overline{\overline{\mathsf{J}}})^2 - \mathrm{tr}\overline{\overline{\mathsf{J}}}^2) = \frac{M}{2}\mathrm{tr}\overline{\overline{\mathsf{I}}} = 2M,$$

$$(\overline{\overline{\overline{\overline{\mathsf{M}}}}}^T\lfloor\lfloor\overline{\overline{\mathsf{I}}}^T){}^\wedge_\wedge\overline{\overline{\mathsf{I}}} = M((\overline{\overline{\mathsf{J}}}^{(2)}\lfloor\lfloor\overline{\overline{\mathsf{I}}}^T){}^\wedge_\wedge\overline{\overline{\mathsf{I}}} = M((\mathrm{tr}\overline{\overline{\mathsf{J}}})\overline{\overline{\mathsf{J}}} - \overline{\overline{\mathsf{J}}}^2){}^\wedge_\wedge\overline{\overline{\mathsf{I}}} = M\overline{\overline{\mathsf{I}}}{}^\wedge_\wedge\overline{\overline{\mathsf{I}}} = 2\overline{\overline{\mathsf{I}}}^{(2)},$$

we obtain

$$\overline{\overline{\mathsf{I}}}^{(4)}\lfloor\lfloor\overline{\overline{\overline{\overline{\mathsf{M}}}}} = M(2\overline{\overline{\mathsf{I}}}^{(2)} - 2\overline{\overline{\mathsf{I}}}^{(2)} + \overline{\overline{\mathsf{J}}}^{(2)}) = M\overline{\overline{\mathsf{J}}}^{(2)} = \overline{\overline{\overline{\overline{\mathsf{M}}}}}^T.$$

Thus, $\overline{\overline{\overline{\overline{\mathsf{M}}}}}$ consists of its axion and principal parts, only:

$$\overline{\overline{\overline{\overline{\mathsf{M}}}}} = \frac{M}{3}\overline{\overline{\mathsf{I}}}^{(2)T} + \overline{\overline{\overline{\overline{\mathsf{M}}}}}_1, \quad \mathrm{tr}\overline{\overline{\overline{\overline{\mathsf{M}}}}}_1 = 0.$$

Assuming vanishing of the principal part, $\overline{\overline{\overline{\overline{\mathsf{M}}}}}_1 = 0$, leads to

$$\overline{\overline{\mathsf{J}}}^{(2)} = \frac{1}{3}\overline{\overline{\mathsf{I}}}^{(2)}, \quad \Rightarrow \quad (\overline{\overline{\mathsf{J}}}^{(2)})^2 = (\overline{\overline{\mathsf{J}}}^2)^{(2)} = \overline{\overline{\mathsf{I}}}^{(2)} = \frac{1}{9}\overline{\overline{\mathsf{I}}}^{(2)},$$

a contradiction. Thus, for real $\overline{\overline{\mathsf{J}}}$ such a bidyadic $\overline{\overline{\overline{\overline{\mathsf{M}}}}}$ contains a nonzero principal part.

4.8 We can expand

$$(\overline{\overline{\mathsf{I}}}^{(4)}\lfloor\lfloor\overline{\overline{\mathsf{J}}}^T)^2 = (\mathbf{e}_N\lfloor\overline{\overline{\mathsf{J}}}^T\rfloor\boldsymbol{\varepsilon}_N)|(\mathbf{e}_N\lfloor\overline{\overline{\mathsf{J}}}^T\rfloor\boldsymbol{\varepsilon}_N) = \mathbf{e}_N\lfloor\overline{\overline{\mathsf{J}}}^{2T}\rfloor\boldsymbol{\varepsilon}_N$$
$$= -\overline{\overline{\mathsf{I}}}^{(4)}\lfloor\lfloor\overline{\overline{\mathsf{I}}}^{(2)T} = -\overline{\overline{\mathsf{I}}}^{(2)}.$$

4.9 Expanding

$$\overline{\overline{\mathsf{J}}}^2 = -\overline{\overline{\mathsf{I}}}_s^{(2)} - \overline{\overline{\mathsf{I}}}_s{}^\wedge_\wedge\mathbf{e}_4\boldsymbol{\varepsilon}_4$$

we obtain four equations

$$\overline{\overline{\mathsf{A}}}_s|\overline{\overline{\mathsf{A}}}_s - \overline{\overline{\mathsf{C}}}_s|\overline{\overline{\mathsf{B}}}_s = -\overline{\overline{\mathsf{I}}}_s^{(2)}$$
$$\overline{\overline{\mathsf{B}}}_s|\overline{\overline{\mathsf{A}}}_s - \overline{\overline{\mathsf{D}}}_s|\overline{\overline{\mathsf{B}}}_s = 0$$
$$\overline{\overline{\mathsf{A}}}_s|\overline{\overline{\mathsf{C}}}_s - \overline{\overline{\mathsf{C}}}_s|\overline{\overline{\mathsf{D}}}_s = 0$$
$$\overline{\overline{\mathsf{B}}}_s|\overline{\overline{\mathsf{C}}}_s - \overline{\overline{\mathsf{D}}}_s|\overline{\overline{\mathsf{D}}}_s = \overline{\overline{\mathsf{I}}}_s.$$

Assuming that $\overline{\overline{\mathsf{B}}}_s^{-1}$ exists, the second equation yields $\overline{\overline{\mathsf{A}}}_s = \overline{\overline{\mathsf{B}}}_s^{-1}|\overline{\overline{\mathsf{D}}}_s|\overline{\overline{\mathsf{C}}}_s$, which substituted in the first equation yields the fourth equation. Similarly, from the second equation we have $\overline{\overline{\mathsf{D}}}_s = \overline{\overline{\mathsf{B}}}_s|\overline{\overline{\mathsf{A}}}_s|\overline{\overline{\mathsf{B}}}_s^{-1}$ which inserted in the third equation yields $\overline{\overline{\mathsf{A}}}_s|\overline{\overline{\mathsf{C}}}_s|\overline{\overline{\mathsf{B}}}_s - \overline{\overline{\mathsf{C}}}_s|\overline{\overline{\mathsf{B}}}_s|\overline{\overline{\mathsf{A}}}_s = 0$, which is valid because of the first equation. Thus, two of the dyadics can be chosen at will, say $\overline{\overline{\mathsf{A}}}_s$ and $\overline{\overline{\mathsf{B}}}_s$ whence $\overline{\overline{\mathsf{D}}}_s$ was obtained from the second equation and $\overline{\overline{\mathsf{C}}}_s$ from the first equation as

$$\overline{\overline{\mathsf{C}}}_s = \overline{\overline{\mathsf{B}}}_s^{-1} + \overline{\overline{\mathsf{A}}}_s^2|\overline{\overline{\mathsf{B}}}_s^{-1}.$$

The medium bidyadic can now be expressed in the form

$$\overline{\overline{\mathsf{M}}} = \overline{\overline{\mathsf{B}}}_s^{-1}\wedge\mathbf{e}_4 + (\overline{\overline{\mathsf{A}}}_s + \boldsymbol{\varepsilon}_4\wedge\overline{\overline{\mathsf{B}}}_s)|(\overline{\overline{\mathsf{I}}}_s^{(2)} + \overline{\overline{\mathsf{A}}}_s|\overline{\overline{\mathsf{B}}}_s^{-1}\wedge\mathbf{e}_4).$$

4.10 One can find the following representations:

$$\mathbf{x}_1 = \frac{1}{\sqrt{2}}(\mathbf{e}_1 + \mathbf{e}_2) \qquad \mathbf{x}_2 = \frac{j}{\sqrt{2}}(\mathbf{e}_1 - \mathbf{e}_2),$$
$$\mathbf{x}_3 = \frac{1}{\sqrt{2}}(\mathbf{e}_3 + \mathbf{e}_4) \qquad \mathbf{x}_4 = \frac{j}{\sqrt{2}}(\mathbf{e}_3 - \mathbf{e}_4),$$
$$\xi_1 = \frac{j}{\sqrt{2}}(\boldsymbol{\varepsilon}_1 - \boldsymbol{\varepsilon}_2) \qquad \xi_2 = \frac{1}{\sqrt{2}}(\boldsymbol{\varepsilon}_1 + \boldsymbol{\varepsilon}_2),$$
$$\xi_3 = \frac{j}{\sqrt{2}}(\boldsymbol{\varepsilon}_3 - \boldsymbol{\varepsilon}_4) \qquad \xi_4 = \frac{1}{\sqrt{2}}(\boldsymbol{\varepsilon}_3 + \boldsymbol{\varepsilon}_4).$$

4.11 The following dyadic can be shown to yield the connection

$$\overline{\overline{\mathsf{D}}} = \mathbf{e}_1\boldsymbol{\varepsilon}_1 + \mathbf{e}_2\boldsymbol{\varepsilon}_3 + \mathbf{e}_3\boldsymbol{\varepsilon}_2 + \mathbf{e}_4\boldsymbol{\varepsilon}_4 = \overline{\overline{\mathsf{D}}}^{-1}.$$

This can be verified as

$$\overline{\overline{\mathsf{D}}}|\overline{\overline{\mathsf{J}}}|\overline{\overline{\mathsf{D}}}^{-1} = (\mathbf{e}_1\boldsymbol{\varepsilon}_2 - \mathbf{e}_3\boldsymbol{\varepsilon}_1 + \mathbf{e}_2\boldsymbol{\varepsilon}_4 - \mathbf{e}_4\boldsymbol{\varepsilon}_3)|\overline{\overline{\mathsf{D}}}^{-1}$$
$$= \mathbf{e}_1\boldsymbol{\varepsilon}_3 - \mathbf{e}_3\boldsymbol{\varepsilon}_1 + \mathbf{e}_2\boldsymbol{\varepsilon}_4 - \mathbf{e}_4\boldsymbol{\varepsilon}_2 = \overline{\overline{\mathsf{D}}}'.$$

4.12 The unipotent spatial dyadic $\overline{\overline{\beta}}_s \in \mathbb{F}_1\mathbb{E}_1$ satisfies

$$\overline{\overline{\beta}}_s^2 = \overline{\overline{\mathsf{I}}}_s^T, \quad \Rightarrow \quad (\overline{\overline{\beta}}_s - \overline{\overline{\mathsf{I}}}_s^T)|(\overline{\overline{\beta}} + \overline{\overline{\mathsf{I}}}_s^T) = 0.$$

Excluding $\overline{\overline{\beta}}_s = \pm\overline{\overline{\mathsf{I}}}_s^T$, neither of the bracketed terms can have an inverse, whence they must be of rank lower than 3. A product of two such dyadics can be zero only if the rank of one of them is lower than 2. Let us assume

$$\overline{\overline{\beta}}_s \pm \overline{\overline{\mathsf{I}}}_s^T = B\boldsymbol{\beta}_1\mathbf{b}_1, \quad \boldsymbol{\beta}_1|\mathbf{b}_1 = 1$$

with either sign, whence

$$\overline{\overline{\beta}}_s^2 = \overline{\overline{\mathsf{I}}}_s^T + (\mp 2 + B\boldsymbol{\beta}_1|\mathbf{b}_1)\boldsymbol{\beta}_1\mathbf{b}_1 = \overline{\overline{\mathsf{I}}}_s^T$$

implies $B = \pm 2$. Thus, the dyadic has the form

$$\overline{\overline{\beta}}_s = \pm(\overline{\overline{\mathsf{I}}}_s^T - 2\boldsymbol{\beta}_1\mathbf{b}_1), \quad \boldsymbol{\beta}_1|\mathbf{b}_1 = 1,$$

is the general solution for the unipotent 3D dyadic when excluding multiples of the unit dyadic. Similarly we obtain for the unipotent spatial bidyadic $\overline{\overline{\alpha}}_s \in \mathbb{F}_2\mathbb{E}_2$ the representation

$$\overline{\overline{\alpha}}_s = \pm(\overline{\overline{\mathsf{I}}}_s^{(2)T} - 2\boldsymbol{\alpha}_{12}\mathbf{a}_{12}), \quad \boldsymbol{\alpha}_{12}|\mathbf{a}_{12} = 1.$$

4.13 Operating both sides of the equation by $\lfloor\lfloor\boldsymbol{\alpha}\boldsymbol{\alpha}$, the left side can be expanded as

$$\overline{\overline{\mathsf{Q}}}^{(2)}\lfloor\lfloor\boldsymbol{\alpha}\boldsymbol{\alpha} = (\overline{\overline{\mathsf{Q}}}||\boldsymbol{\alpha}\boldsymbol{\alpha})\overline{\overline{\mathsf{Q}}} - (\overline{\overline{\mathsf{Q}}}|\boldsymbol{\alpha})(\boldsymbol{\alpha}|\overline{\overline{\mathsf{Q}}}),$$

while the right side yields

$$(\mathbf{e}_N\lfloor\overline{\overline{\mathsf{I}}}^{(2)T})\lfloor\lfloor\boldsymbol{\alpha} = \boldsymbol{\alpha}\rfloor(\mathbf{e}_N\lfloor(\boldsymbol{\alpha}\wedge\overline{\overline{\mathsf{I}}}^T)) = \mathbf{e}_N\lfloor(\boldsymbol{\alpha}\wedge\boldsymbol{\alpha}\wedge\overline{\overline{\mathsf{I}}}^T) = 0.$$

From these we have

$$0 = ((\overline{\overline{\mathsf{Q}}}|\boldsymbol{\alpha})(\boldsymbol{\alpha}|\overline{\overline{\mathsf{Q}}}))^{(2)} = (\overline{\overline{\mathsf{Q}}}||\boldsymbol{\alpha}\boldsymbol{\alpha})^2\overline{\overline{\mathsf{Q}}}^{(2)} = (\overline{\overline{\mathsf{Q}}}||\boldsymbol{\alpha}\boldsymbol{\alpha})^2\mathbf{e}_N\lfloor\overline{\overline{\mathsf{I}}}^{(2)T} = 0,$$

which must be valid for any one-form $\boldsymbol{\alpha}$. This requires that $\overline{\overline{\mathsf{Q}}}$ be an antisymmetric dyadic, of the form $\overline{\overline{\mathsf{Q}}} = \mathbf{A}\lfloor\overline{\overline{\mathsf{I}}}^T$. Defining $\Delta_A = \boldsymbol{\varepsilon}_N|(\mathbf{A}\wedge\mathbf{A})/2$, the original equation for the bivector $\mathbf{A}$ then becomes

$$(\mathbf{A}\lfloor\overline{\overline{\mathsf{I}}}^T)^{(2)} = \mathbf{A}\mathbf{A} - \Delta_A\mathbf{e}_N\lfloor\overline{\overline{\mathsf{I}}}^{(2)T} = \mathbf{e}_N\lfloor\overline{\overline{\mathsf{I}}}^{(2)T}$$

or

$$\boldsymbol{\varepsilon}_N\lfloor\mathbf{A}\mathbf{A} - \Delta_A\overline{\overline{\mathsf{I}}}^{(2)T} = \overline{\overline{\mathsf{I}}}^{(2)T}.$$

Taking the trace leads to $2\Delta_A - 6\Delta_A = 6$, whence $\Delta_A = -3/2$ and $\boldsymbol{\varepsilon}_N\lfloor\mathbf{A}\mathbf{A} = (1/2)\overline{\overline{\mathsf{I}}}^{(2)T}$ which obviously cannot be satisfied by any bivector $\mathbf{A}$. Thus, the original equation has no solution $\overline{\overline{\mathsf{Q}}}$.

4.14 Assuming that $\overline{\overline{\mathsf{P}}}$ has full rank, multiplying the equation by $|\overline{\overline{\mathsf{P}}}^{(-2)T}$ we obtain

$$(\overline{\overline{\mathsf{Q}}}|\overline{\overline{\mathsf{P}}}^{-1})^{(2)} = \mathbf{e}_N\lfloor\overline{\overline{\mathsf{I}}}^{(2)T},$$

whence from the result of the previous problem we conclude that there is no solution $\overline{\overline{\mathsf{Q}}}$.

When $\overline{\overline{\mathsf{P}}}$ has rank 1 or lower, we have $\overline{\overline{\mathsf{P}}}^{(2)} = 0$, whence $\overline{\overline{\mathsf{Q}}}$ may be any metric dyadic of rank 1 or lower.

When $\overline{\overline{\mathsf{P}}}$ is of rank 2, we can write $\overline{\overline{\mathsf{P}}} = \mathbf{p}_1\boldsymbol{\varepsilon}_1 + \boldsymbol{p}_2\boldsymbol{\varepsilon}_2$, whence the equation becomes

$$\overline{\overline{\mathsf{Q}}}^{(2)} = \mathbf{e}_N\lfloor\boldsymbol{\varepsilon}_{12}\mathbf{p}_{12} = \mathbf{e}_{34}\mathbf{p}_{12}.$$

Here we assume that $\{\mathbf{e}_i\}$ and $\{\boldsymbol{\varepsilon}_j\}$ are reciprocal bases. This equation has an infinity of solutions, for example, $\overline{\overline{\mathsf{Q}}} = \mathbf{e}_3\mathbf{p}_1 + \mathbf{e}_4\mathbf{p}_2$ and $\overline{\overline{\mathsf{Q}}} = \mathbf{e}_3\mathbf{p}_2 - \mathbf{e}_4\mathbf{p}_1$ are two possibilities.

Finally, for $\overline{\overline{\mathsf{P}}}$ of rank 3, we may expand

$$\overline{\overline{\mathsf{P}}} = \mathbf{p}_1\boldsymbol{\varepsilon}_1 + \mathbf{p}_2\boldsymbol{\varepsilon}_2 + \mathbf{p}_3\boldsymbol{\varepsilon}_3, \quad \overline{\overline{\mathsf{Q}}} = \sum_{i=1}^{4}\mathbf{e}_i\mathbf{q}_i,$$

with $\mathbf{p}_1\wedge\mathbf{p}_2\wedge\mathbf{p}_3 \neq 0$. From the equation

$$\overline{\overline{\mathsf{Q}}}^{(2)} = \sum_{i<j}\mathbf{e}_{ij}\mathbf{q}_{ij} = \mathbf{e}_{34}\mathbf{p}_{12} + \mathbf{e}_{14}\mathbf{p}_{23} + \mathbf{e}_{24}\mathbf{p}_{31},$$

we obtain conditions for the unknown vectors $\mathbf{q}_i$,

$$\mathbf{q}_{12} = \mathbf{q}_{23} = \mathbf{q}_{31} = 0 \quad \mathbf{q}_{14} = \mathbf{p}_{23}, \; \mathbf{q}_{24} = \mathbf{p}_{31}, \; \mathbf{q}_{34} = \mathbf{p}_{12}.$$

From the first three conditions it follows that the three vectors $\mathbf{q}_1, \mathbf{q}_2, \mathbf{q}_3$ must be multiples of the same vector. But this is contradictory to the last three conditions because it implies that $\mathbf{q}_{14}, \mathbf{q}_{24,}$ and $\mathbf{q}_{34}$ would be multiples of the same bivector.

Thus, we may conclude that the equation has solutions only when the rank of $\overline{\overline{\mathsf{P}}}$ is at most 2, in which case the solution $\overline{\overline{\mathsf{Q}}}$ is not unique.

4.15 The inverse of the transformation dyadic $\overline{\overline{\mathsf{D}}}$ is

$$\overline{\overline{\mathsf{D}}}^{-1} = \overline{\overline{\mathsf{I}}}_s^{(2)} + (\overline{\overline{\mathsf{G}}}_s|\overline{\overline{\Gamma}}'_s)^{\wedge}_{\wedge}\mathbf{e}_4\boldsymbol{\varepsilon}_4,$$

as can be verified by direct multiplication because

$$(\overline{\overline{\mathsf{G}}}_s|\overline{\overline{\Gamma}}'_s)|(\overline{\overline{\mathsf{G}}}'_s|\overline{\overline{\Gamma}}_s) = \overline{\overline{\mathsf{G}}}_s|\overline{\overline{\Gamma}}_s = \overline{\overline{\mathsf{I}}}_s.$$

To prove the statement, we expand

$$\begin{aligned}
\overline{\overline{\mathsf{D}}}|\overline{\overline{\mathsf{J}}}|\overline{\overline{\mathsf{D}}}^{-1} &= \overline{\overline{\mathsf{D}}}|(\mathbf{e}_{123}\lfloor\overline{\overline{\Gamma}}_s \wedge \boldsymbol{\varepsilon}_4 + \mathbf{e}_4 \wedge \overline{\overline{\mathsf{G}}}_s\rfloor\boldsymbol{\varepsilon}_{123})|\overline{\overline{\mathsf{D}}}^{-1} \\
&= (\mathbf{e}_{123}\lfloor\overline{\overline{\Gamma}}_s \wedge \boldsymbol{\varepsilon}_4)|((\overline{\overline{\mathsf{G}}}_s|\overline{\overline{\Gamma}}'_s)^{\wedge}_{\wedge}\mathbf{e}_4\boldsymbol{\varepsilon}_4) \\
&\quad + ((\overline{\overline{\mathsf{G}}}'_s|\overline{\overline{\Gamma}}_s)^{\wedge}_{\wedge}\mathbf{e}_4\boldsymbol{\varepsilon}_4)|(\mathbf{e}_4 \wedge \overline{\overline{\mathsf{G}}}_s\rfloor\boldsymbol{\varepsilon}_{123}) \\
&= (\mathbf{e}_{123}\lfloor\overline{\overline{\Gamma}}_s)|(\overline{\overline{\mathsf{G}}}_s|\overline{\overline{\Gamma}}'_s \wedge \boldsymbol{\varepsilon}_4) + (\mathbf{e}_4 \wedge \overline{\overline{\mathsf{G}}}'_s|\overline{\overline{\Gamma}}_s)|(\overline{\overline{\mathsf{G}}}_s\rfloor\boldsymbol{\varepsilon}_{123}) \\
&= \mathbf{e}_{123}\lfloor\overline{\overline{\Gamma}}'_s \wedge \boldsymbol{\varepsilon}_4 + \mathbf{e}_4 \wedge \overline{\overline{\mathsf{G}}}'_s\rfloor\boldsymbol{\varepsilon}_{123}.
\end{aligned}$$

Thus, we have arrived at the AC bidyadic

$$\overline{\overline{\mathsf{J}}}' = \mathbf{e}_{123}\lfloor\overline{\overline{\Gamma}}'_s \wedge \boldsymbol{\varepsilon}_4 + \mathbf{e}_4 \wedge \overline{\overline{\mathsf{G}}}'_s\rfloor\boldsymbol{\varepsilon}_{123}$$

with symmetric spatial metric dyadics $\overline{\overline{\mathsf{G}}}'_s$ and $\overline{\overline{\Gamma}}'_s$.

4.16 The Cayley–Hamilton equation (2.119) for the unipotent dyadic $\overline{\overline{\mathsf{A}}} = \overline{\overline{\mathsf{K}}}$ is simplified to

$$\overline{\overline{\mathsf{I}}} - (\mathrm{tr}\overline{\overline{\mathsf{K}}})\overline{\overline{\mathsf{K}}} + (\mathrm{tr}\overline{\overline{\mathsf{K}}}^{(2)})\overline{\overline{\mathsf{I}}} - (\mathrm{tr}\overline{\overline{\mathsf{K}}}^{(3)})\overline{\overline{\mathsf{K}}} + (\mathrm{tr}\overline{\overline{\mathsf{K}}}^{(4)})\overline{\overline{\mathsf{I}}} = 0.$$

Obvious solutions are $\overline{\overline{\mathsf{K}}} = \pm\overline{\overline{\mathsf{I}}}$ with $\mathrm{tr}\overline{\overline{\mathsf{K}}} = \pm 4$. For other solutions the equations are split to

$$1 + \mathrm{tr}\overline{\overline{\mathsf{K}}}^{(2)} + \mathrm{tr}\overline{\overline{\mathsf{K}}}^{(4)} = 0, \qquad \mathrm{tr}\overline{\overline{\mathsf{K}}} + \mathrm{tr}\overline{\overline{\mathsf{K}}}^{(3)} = 0.$$

Applying the expansions (2.33)–(2.35) for $\overline{\overline{\mathsf{A}}} = \overline{\overline{\mathsf{K}}}$, these equations can respectively be reduced to

$$(\mathrm{tr}\overline{\overline{\mathsf{K}}})^2((\mathrm{tr}\overline{\overline{\mathsf{K}}})^2 - 4) = 0, \qquad \mathrm{tr}\overline{\overline{\mathsf{K}}}((\mathrm{tr}\overline{\overline{\mathsf{K}}})^2 - 4) = 0,$$

which can be compared with those of the AC dyadic, (4.88) and (4.89). Thus, the possible values of $\mathrm{tr}\overline{\overline{\mathsf{K}}}$ for the unipotent dyadic are $\mathrm{tr}\overline{\overline{\mathsf{K}}} = 0$, $\mathrm{tr}\overline{\overline{\mathsf{K}}} = \pm 2$ and $\mathrm{tr}\overline{\overline{\mathsf{K}}} = \pm 4$.

4.17 Since it was shown that (4.122) and (4.125) imply

$$\mathrm{tr}\overline{\overline{\mathsf{J}}}^5 = \mathrm{tr}\overline{\overline{\mathsf{J}}}^3 = \mathrm{tr}\overline{\overline{\mathsf{J}}} = 0,$$

applying (4.123) we can expand the coefficients a_i from (3.13) to (3.19) for $\overline{\overline{\mathsf{C}}} = \overline{\overline{\mathsf{J}}}$ as

$$\begin{aligned} a_0 &= 1 \\ 2a_2 &= -\mathrm{tr}\overline{\overline{\mathsf{C}}}^2 = 6 \\ 4a_4 &= -(a_2\mathrm{tr}\overline{\overline{\mathsf{C}}}^2 + \mathrm{tr}\overline{\overline{\mathsf{C}}}^4) = 12 \\ 6a_6 &= -(a_4\mathrm{tr}\overline{\overline{\mathsf{C}}}^2 + a_2\mathrm{tr}\overline{\overline{\mathsf{C}}}^4 + \mathrm{tr}\overline{\overline{\mathsf{C}}}^6) = 6. \end{aligned}$$

while $a_1 = a_3 = a_5 = 0$. Inserting these in (3.10) yields

$$\overline{\overline{\mathsf{J}}}^6 + 3\overline{\overline{\mathsf{J}}}^4 + 3\overline{\overline{\mathsf{J}}}^2 + \overline{\overline{\mathsf{I}}}^{(2)} = (\overline{\overline{\mathsf{J}}}^2 + \overline{\overline{\mathsf{I}}}^{(2)})^3 = 0.$$

CHAPTER 5

5.1 Since (5.29) and (5.31) are three-form equations, applying the bar-product rule from Appendix B,

$$\boldsymbol{\alpha}_s|\mathbf{a}_s = \boldsymbol{\alpha}_g|\overline{\overline{\Gamma}}_s|\mathbf{a}_s = \boldsymbol{\alpha}_g \cdot \mathbf{a}_s$$

the corresponding Gibbsian scalar equations can be written as

$$\mathbf{e}_{123}|(\mathbf{d}_s \wedge \mathbf{B}) = \mathbf{d}_s|(\mathbf{e}_{123}\lfloor\mathbf{B}) = \nabla\cdot\mathbf{B}_g = 0,$$

$$\mathbf{e}_{123}|(\mathbf{d}_s \wedge \mathbf{D} - \varrho) = \mathbf{d}_s|(\mathbf{e}_{123}\lfloor\mathbf{D}) - \mathbf{e}_{123}|\varrho = \nabla\cdot\mathbf{D}_g - \varrho_g = 0.$$

Applying the wedge-product rule

$$\mathbf{e}_{123}\lfloor(\boldsymbol{\alpha}_s \wedge \boldsymbol{\beta}_s) = \boldsymbol{\alpha}_g \times \boldsymbol{\beta}_g$$

the two-form equations (5.30) and (5.32) become vector equations as

$$\mathbf{e}_{123}\lfloor(\mathbf{d}_s \wedge \mathbf{E} + \partial_\tau\mathbf{B}) = \nabla\times\mathbf{E}_g + \partial_\tau\mathbf{B}_g = 0,$$

$$\mathbf{e}_{123}\lfloor(\mathbf{d}_s \wedge \mathbf{H} - \partial_\tau\mathbf{D} - \mathbf{J}) = \nabla\times\mathbf{H}_g - \partial_\tau\mathbf{D}_g - \mathbf{J}_g = 0.$$

5.2 From $\mathbf{d}\wedge\boldsymbol{\Psi}=0$ we have the potential representation $\boldsymbol{\Psi}=\mathbf{d}\wedge\boldsymbol{\psi}$, whence the second Maxwell equation yields

$$\mathbf{d}\wedge\boldsymbol{\Phi}=\mathbf{d}\wedge\overline{\overline{\mathsf{N}}}|\boldsymbol{\Psi}=(\mathbf{d}\wedge\overline{\overline{\mathsf{N}}}\lfloor\mathbf{d})|\boldsymbol{\psi}=0.$$

The equation corresponding to (5.62), (5.63) can be written as

$$(\overline{\overline{\mathsf{N}}}_m\lfloor\lfloor\mathbf{dd})|\boldsymbol{\psi}=0.$$

5.3 Inserting (5.83)–(5.86) in (5.65) we can expand

$$\begin{aligned}
\overline{\overline{\mathsf{M}}} &= \overline{\overline{\alpha}}+\overline{\overline{\epsilon}}'\wedge\mathbf{e}_4+\boldsymbol{\varepsilon}_4\wedge\overline{\overline{\mu}}^{-1}+\boldsymbol{\varepsilon}_4\wedge\overline{\overline{\beta}}\wedge\mathbf{e}_e\\
&= \boldsymbol{\varepsilon}_{123}\lfloor\overline{\overline{\xi}}_g|\overline{\overline{\mu}}_g^{-1}\rfloor\mathbf{e}_{123}+\boldsymbol{\varepsilon}_{123}\lfloor(\overline{\overline{\epsilon}}_g-\overline{\overline{\xi}}_g|\overline{\overline{\mu}}_g^{-1}|\overline{\overline{\zeta}}_g)\wedge\mathbf{e}_4\\
&\quad+\boldsymbol{\varepsilon}_4\wedge(\boldsymbol{\varepsilon}_{123}\lfloor\overline{\overline{\mu}}_g)^{-1}-\boldsymbol{\varepsilon}_4\wedge\overline{\overline{\mu}}_g^{-1}|\overline{\overline{\zeta}}_g\wedge\mathbf{e}_4\\
&= \boldsymbol{\varepsilon}_{123}\lfloor\boldsymbol{\varepsilon}_{123}\lfloor\overline{\overline{\epsilon}}_g\wedge\mathbf{e}_4+\boldsymbol{\varepsilon}_{123}\lfloor\overline{\overline{\xi}}_g|\overline{\overline{\mu}}_g^{-1}|(\overline{\overline{\mathsf{I}}}\rfloor\mathbf{e}_{123}-\overline{\overline{\zeta}}_g\wedge\mathbf{e}_4)\\
&\quad+\boldsymbol{\varepsilon}_4\wedge\overline{\overline{\mu}}_g^{-1}|(\overline{\overline{\mathsf{I}}}\rfloor\mathbf{e}_{123}-\overline{\overline{\zeta}}_g\wedge\mathbf{e}_4)\\
&= \boldsymbol{\varepsilon}_{123}\lfloor\overline{\overline{\epsilon}}_g\wedge\mathbf{e}_4+(\boldsymbol{\varepsilon}_{123}\lfloor\overline{\overline{\xi}}_g+\boldsymbol{\varepsilon}_4\wedge\overline{\overline{\mathsf{I}}}^T)|\overline{\overline{\mu}}_g^{-1}|(\overline{\overline{\mathsf{I}}}\rfloor\mathbf{e}_{123}-\overline{\overline{\zeta}}_g\wedge\mathbf{e}_4).
\end{aligned}$$

Here the inverse $\overline{\overline{\mu}}_g^{-1}=(\overline{\overline{\mu}}_g)^{-1}$ is assumed to be in space $\mathbb{F}_1\mathbb{F}_1$.

5.4 The skewon bidyadic $\overline{\overline{\mathsf{M}}}_2=(\overline{\overline{\mathsf{B}}}_o{}^{\wedge}_{\wedge}\overline{\overline{\mathsf{I}}})^T$ obeys the rule

$$\overline{\overline{\mathsf{I}}}^{(4)T}\lfloor\lfloor\overline{\overline{\mathsf{M}}}_2=(\mathrm{tr}\overline{\overline{\mathsf{M}}}_2)\overline{\overline{\mathsf{I}}}^{(2)T}-(\overline{\overline{\mathsf{M}}}_2\lfloor\lfloor\overline{\overline{\mathsf{I}}}){}^{\wedge}_{\wedge}\overline{\overline{\mathsf{I}}}^T+\overline{\overline{\mathsf{M}}}_2=-\overline{\overline{\mathsf{M}}}_2.$$

Equating

$$\mathrm{tr}(\overline{\overline{\mathsf{I}}}^{(4)}\lfloor\lfloor\overline{\overline{\mathsf{M}}}_2)=(\overline{\overline{\mathsf{I}}}^{(4)}\lfloor\lfloor\overline{\overline{\mathsf{I}}}^{(2)T})||\overline{\overline{\mathsf{M}}}_2=\overline{\overline{\mathsf{I}}}^{(2)}||\overline{\overline{\mathsf{M}}}_2=\mathrm{tr}\overline{\overline{\mathsf{M}}}_2$$

to $-\mathrm{tr}\overline{\overline{\mathsf{M}}}_2$ yields $\mathrm{tr}\overline{\overline{\mathsf{M}}}_2=0$, whence we obtain

$$\overline{\overline{\mathsf{M}}}_2=\frac{1}{2}(\overline{\overline{\mathsf{M}}}_2\lfloor\lfloor\overline{\overline{\mathsf{I}}}){}^{\wedge}_{\wedge}\overline{\overline{\mathsf{I}}}^T,\quad\Rightarrow\quad\overline{\overline{\mathsf{B}}}_o=\frac{1}{2}(\overline{\overline{\mathsf{M}}}_2\lfloor\lfloor\overline{\overline{\mathsf{I}}})^T.$$

5.5 Applying the orthogonality of the dyadic expressions

$$\begin{aligned}
(\overline{\overline{\mathsf{I}}}_s^{(2)T}+\overline{\overline{\xi}}\wedge\mathbf{e}_4)|(\boldsymbol{\varepsilon}_4\wedge\overline{\overline{\mathsf{I}}}_s^T+\overline{\overline{\xi}}) &= \overline{\overline{\xi}}+(\overline{\overline{\xi}}\wedge\mathbf{e}_4)|(\boldsymbol{\varepsilon}_4\wedge\overline{\overline{\mathsf{I}}}_s^T)\\
&= \overline{\overline{\xi}}-\overline{\overline{\xi}}=0,\\
-(\overline{\overline{\mathsf{I}}}_s^{(2)T}-\overline{\overline{\zeta}}\wedge\mathbf{e}_4)|(\boldsymbol{\varepsilon}_4\wedge\overline{\overline{\mathsf{I}}}^T-\overline{\overline{\zeta}}) &= \overline{\overline{\zeta}}+(\overline{\overline{\zeta}}\wedge\mathbf{e}_4)|(\boldsymbol{\varepsilon}_4\wedge\overline{\overline{\mathsf{I}}}_s^T)\\
&= \overline{\overline{\zeta}}-\overline{\overline{\zeta}}=0,
\end{aligned}$$

one can directly verify that the bidyadic $\overline{\overline{\mathsf{N}}}$ defined by the 3D expansion (5.77) is the inverse of the bidyadic $\overline{\overline{\mathsf{M}}}$ defined by

(5.75). Similarly, we can apply the orthogonality properties

$$(\overline{\overline{\mathsf{I}}}\rfloor \mathbf{e}_{123} - \overline{\overline{\zeta}}_g \wedge \mathbf{e}_4)|(\boldsymbol{\varepsilon}_4 \wedge \overline{\overline{\mathsf{I}}}^T - \boldsymbol{\varepsilon}_{123}\lfloor \overline{\overline{\zeta}}_g) = -\overline{\overline{\zeta}}_g + \overline{\overline{\zeta}}_g = 0,$$
$$(\overline{\overline{\mathsf{I}}} \wedge \boldsymbol{\varepsilon}_4 + \overline{\overline{\xi}}_g\rfloor \boldsymbol{\varepsilon}_{123})|(\mathbf{e}_{123}\lfloor \overline{\overline{\mathsf{I}}}^T + \mathbf{e}_4 \wedge \overline{\overline{\xi}}_g) = -\overline{\overline{\xi}}_g + \overline{\overline{\xi}}_g = 0,$$

to verify (5.90).

5.6 Since $\overline{\overline{\mathsf{M}}}_m$ is symmetric, it defines a principal–axion medium. To show that the axion part vanishes, let us write $\overline{\overline{\mathsf{Q}}} = \sum \mathbf{q}_i\mathbf{q}_i$. Any symmetric dyadic $\overline{\overline{\mathsf{S}}} \in \mathbb{E}_1\mathbb{E}_1$ can be written as a sum of symmetric dyadic products of vectors because of

$$\overline{\overline{\mathsf{S}}} = \sum \mathbf{a}_i\mathbf{b}_i = \frac{1}{2}\sum(\mathbf{a}_i\mathbf{b}_i + \mathbf{b}_i\mathbf{a}_i)$$
$$= \frac{1}{4}\sum((\mathbf{a}_i + \mathbf{b}_i)(\mathbf{a}_i + \mathbf{b}_i) - (\mathbf{a}_i - \mathbf{b}_i)(\mathbf{a}_i - \mathbf{b}_i))$$

(the minus sign makes some of the vectors $\mathbf{q}_i$ imaginary). Forming the trace of the medium bidyadic we obtain

$$\mathrm{tr}\overline{\overline{\mathsf{M}}} = \mathrm{tr}(\boldsymbol{\varepsilon}_N\lfloor \overline{\overline{\mathsf{Q}}}^{(2)}) = \sum_{i<j}(\boldsymbol{\varepsilon}_N\lfloor \mathbf{q}_{ij})|\mathbf{q}_{ij} = \boldsymbol{\varepsilon}_N|\sum_{i<j}(\mathbf{q}_{ij} \wedge \mathbf{q}_{ij}) = 0.$$

5.7 When the bivector $\mathbf{Q}$ is not simple, we can write $\mathbf{Q} = \mathbf{e}_1 \wedge \mathbf{e}_2 + \mathbf{e}_3 \wedge \mathbf{e}_4$, with $\mathbf{e}_1 \wedge \mathbf{e}_2 \wedge \mathbf{e}_3 \wedge \mathbf{e}_4 = \mathbf{e}_N \neq 0$, whence

$$\overline{\overline{\mathsf{Q}}} = (\mathbf{e}_{12} + \mathbf{e}_{34})\lfloor \overline{\overline{\mathsf{I}}}^T = \mathbf{e}_2\mathbf{e}_1 - \mathbf{e}_1\mathbf{e}_2 + \mathbf{e}_4\mathbf{e}_3 - \mathbf{e}_3\mathbf{e}_4$$

and

$$\overline{\overline{\mathsf{M}}}_m = \overline{\overline{\mathsf{Q}}}^{(2)} = \mathbf{e}_{12}\mathbf{e}_{12} + \mathbf{e}_{34}\mathbf{e}_{34} - \mathbf{e}_{24}\mathbf{e}_{31} - \mathbf{e}_{23}\mathbf{e}_{14} - \mathbf{e}_{14}\mathbf{e}_{23} - \mathbf{e}_{31}\mathbf{e}_{24}.$$

Since this is symmetric, we have a principal–axion medium, $\overline{\overline{\mathsf{M}}} = \overline{\overline{\mathsf{M}}}_1 + \overline{\overline{\mathsf{M}}}_3$. The axion component can be found as $\overline{\overline{\mathsf{M}}}_3 = \overline{\overline{\mathsf{I}}}^{(2)T}\mathrm{tr}\overline{\overline{\mathsf{M}}}/6$. Let us form

$$\overline{\overline{\mathsf{M}}} = \boldsymbol{\varepsilon}_N\lfloor \overline{\overline{\mathsf{M}}}_m = \boldsymbol{\varepsilon}_{34}\mathbf{e}_{12} + \boldsymbol{\varepsilon}_{12}\mathbf{e}_{34} - \boldsymbol{\varepsilon}_{31}\mathbf{e}_{31} - \boldsymbol{\varepsilon}_{14}\mathbf{e}_{14}$$
$$- \mathbf{e}_{23}\mathbf{e}_{23} - \mathbf{e}_{24}\mathbf{e}_{24},$$

the trace of which equals $\mathrm{tr}\overline{\overline{\mathsf{M}}} = -4$. For a simple bivector $\mathbf{Q}$ we can drop the term $\mathbf{e}_3 \wedge \mathbf{e}_4$. From $\overline{\overline{\mathsf{M}}}_m = \mathbf{e}_{12}\mathbf{e}_{12}$ we have $\mathrm{tr}\overline{\overline{\mathsf{M}}} = \mathrm{tr}(\boldsymbol{\varepsilon}_{34}\mathbf{e}_{12}) = 0$, whence in this case there is no axion component.

5.8 Inserting the expansions (5.75) and (5.77) in (5.179) we obtain

$$\boldsymbol{\varepsilon}_2 \wedge \boldsymbol{\varepsilon}_3 \wedge \boldsymbol{\varepsilon}_4 \wedge \overline{\overline{\mu}}_b^{-1}|(\overline{\overline{\mathsf{I}}}_s^{(2)T} - \overline{\overline{\zeta}}_b \wedge \boldsymbol{\varepsilon}_4) = 0,$$
$$-\boldsymbol{\varepsilon}_2 \wedge \boldsymbol{\varepsilon}_3 \wedge \boldsymbol{\varepsilon}_4 \wedge \overline{\overline{\epsilon}}_b^{-1}|(\overline{\overline{\mathsf{I}}}_s^{(2)T} + \overline{\overline{\xi}}_b \wedge \boldsymbol{\varepsilon}_4) = 0,$$

which reduce to

$$\boldsymbol{\varepsilon}_2 \wedge \boldsymbol{\varepsilon}_3 \wedge \overline{\overline{\mu}}_b^{-1} = 0, \quad \boldsymbol{\varepsilon}_2 \wedge \boldsymbol{\varepsilon}_3 \wedge \overline{\overline{\epsilon}}_b^{-1} = 0.$$

These correspond to the Gibbsian conditions

$$\mathbf{e}_1 \cdot \overline{\overline{\mu}}_{gb}^{-1} = 0, \quad \mathbf{e}_1 \cdot \overline{\overline{\epsilon}}_{gb}^{-1} = 0$$

which require the field conditions (5.183) at the interface.

5.9 Expanding

$$\begin{aligned}-\mu\overline{\overline{\mathsf{M}}} &= \boldsymbol{\varepsilon}_N \lfloor (\overline{\overline{\mathsf{G}}}_s - \mu\epsilon \mathbf{e}_4\mathbf{e}_4)^{(2)} = \boldsymbol{\varepsilon}_N \lfloor (\overline{\overline{\mathsf{G}}}_s^{(2)} - \mu\epsilon \overline{\overline{\mathsf{G}}}_s {}^{\wedge}_{\wedge} \mathbf{e}_4\mathbf{e}_4) \\ &= \boldsymbol{\varepsilon}_N \lfloor (\mathbf{e}_{12}\mathbf{e}_{12} + \mathbf{e}_{23}\mathbf{e}_{23} + \mathbf{e}_{31}\mathbf{e}_{31}) \\ &\quad - \mu\epsilon\boldsymbol{\varepsilon}_N \lfloor (\mathbf{e}_{14}\mathbf{e}_{14} + \mathbf{e}_{24}\mathbf{e}_{24} + \mathbf{e}_{34}\mathbf{e}_{34}) \\ &= \boldsymbol{\varepsilon}_{34}\mathbf{e}_{12} + \boldsymbol{\varepsilon}_{14}\mathbf{e}_{23} + \boldsymbol{\varepsilon}_{24}\mathbf{e}_{31}) - \mu\epsilon(\boldsymbol{\varepsilon}_{23}\mathbf{e}_{14} + \boldsymbol{\varepsilon}_{31}\mathbf{e}_{24} + \boldsymbol{\varepsilon}_{12}\mathbf{e}_{34})\end{aligned}$$

The trace of each term vanishes, whence $\mathrm{tr}\overline{\overline{\mathsf{M}}} = 0$.

5.10 The boundary conditions are

$$\boldsymbol{\varepsilon}_2 \wedge \boldsymbol{\varepsilon}_3 \wedge \mathbf{E}_a = 0, \quad \boldsymbol{\varepsilon}_1 \wedge \boldsymbol{\varepsilon}_3 \wedge \mathbf{H}_a = 0,$$

which are more general than the SH boundary conditions. For Gibbsian field vectors they take the form

$$\mathbf{e}_1 \cdot \mathbf{E}_{ga} = 0, \quad \mathbf{e}_2 \cdot \mathbf{H}_{ga} = 0.$$

5.11 Denoting $\mathbf{A} = \mathbf{e}_N \lfloor \boldsymbol{\Phi}$ and $\mathbf{B} = \mathbf{e}_N \lfloor \boldsymbol{\Psi}$, omitting the summation signs $\sum_i$ and $\sum_j$, we can expand

$$\begin{aligned}((\mathbf{e}_N \lfloor \overline{\overline{\mathsf{U}}}) \lfloor \lfloor \overline{\overline{\mathsf{I}}}^T)^2 |\mathbf{a} &= ((\mathbf{A}\boldsymbol{\Psi} - \mathbf{B}\boldsymbol{\Phi}) \lfloor \lfloor \boldsymbol{\varepsilon}_i \mathbf{e}_i) | ((\mathbf{A}\boldsymbol{\Psi} - \mathbf{B}\boldsymbol{\Phi}) \lfloor \lfloor \boldsymbol{\varepsilon}_j \mathbf{e}_j) | \mathbf{a} \\ &= \mathbf{f}(\boldsymbol{\Phi}, \boldsymbol{\Psi}) - \mathbf{g}(\boldsymbol{\Phi}, \boldsymbol{\Psi}) - \mathbf{g}(\boldsymbol{\Psi}, \boldsymbol{\Phi}) + \mathbf{f}(\boldsymbol{\Psi}, \boldsymbol{\Phi}),\end{aligned}$$

with

$$\begin{aligned}\mathbf{f}(\boldsymbol{\Phi}, \boldsymbol{\Psi}) &= (\mathbf{A} \lfloor \boldsymbol{\varepsilon}_i)(\boldsymbol{\Psi} \lfloor \mathbf{e}_i) | (\mathbf{A} \lfloor \boldsymbol{\varepsilon}_j)(\boldsymbol{\Psi} \lfloor \mathbf{e}_j) | \mathbf{a} \\ &= \mathbf{e}_N \lfloor (\boldsymbol{\Phi} \wedge \mathbf{e}_i)(\boldsymbol{\Psi} \lfloor \mathbf{e}_i) | ((\mathbf{e}_N \lfloor (\boldsymbol{\Phi} \wedge \boldsymbol{\varepsilon}_j))(\boldsymbol{\Psi} | (\mathbf{e}_j \wedge \mathbf{a})), \\ \mathbf{g}(\boldsymbol{\Phi}, \boldsymbol{\Psi}) &= (\mathbf{A} \lfloor \boldsymbol{\varepsilon}_i)(\boldsymbol{\Psi} \lfloor \mathbf{e}_i) | (\mathbf{B} \lfloor \boldsymbol{\varepsilon}_i)(\boldsymbol{\Phi} \lfloor \mathbf{e}_i) | \mathbf{a} \\ &= \mathbf{e}_N \lfloor (\boldsymbol{\Phi} \wedge \boldsymbol{\varepsilon}_i)(\boldsymbol{\Psi} \lfloor \mathbf{e}_i) | (\mathbf{e}_N \lfloor (\boldsymbol{\Psi} \wedge \boldsymbol{\varepsilon}_i))(\boldsymbol{\Phi} | (\mathbf{e}_i \wedge \mathbf{a})).\end{aligned}$$

We can do the summation with respect to j as

$$(\mathbf{e}_N \lfloor (\boldsymbol{\Phi} \wedge \varepsilon_j)(\boldsymbol{\Psi} | (\mathbf{e}_j \wedge \mathbf{a})) = (\mathbf{e}_N \lfloor (\boldsymbol{\Phi} \wedge \varepsilon_j)(\mathbf{e}_j | (\mathbf{a} \rfloor \boldsymbol{\Psi})) \\ = \mathbf{e}_N \lfloor (\boldsymbol{\Phi} \wedge (\mathbf{a} \rfloor \boldsymbol{\Psi})),$$

whence the functions become

$$\begin{aligned} \mathbf{f}(\boldsymbol{\Phi}, \boldsymbol{\Psi}) &= \mathbf{e}_N \lfloor (\boldsymbol{\Phi} \wedge \mathbf{e}_i)(\boldsymbol{\Psi} \lfloor \mathbf{e}_i) | (\mathbf{e}_N \lfloor (\boldsymbol{\Phi} \wedge (\mathbf{a} \rfloor \boldsymbol{\Psi}))) \\ &= -\mathbf{e}_N \lfloor (\boldsymbol{\Phi} \wedge \mathbf{e}_i)\mathbf{e}_N | (\boldsymbol{\Phi} \wedge (\boldsymbol{\Psi} \lfloor \mathbf{a}) \wedge (\boldsymbol{\Psi} \lfloor \mathbf{e}_i)), \\ \mathbf{g}(\boldsymbol{\Phi}, \boldsymbol{\Psi}) &= \mathbf{e}_N \lfloor (\boldsymbol{\Phi} \wedge \varepsilon_i)(\boldsymbol{\Psi} \lfloor \mathbf{e}_i) | (\mathbf{e}_N \lfloor (\boldsymbol{\Psi} \wedge (\mathbf{a} \rfloor \boldsymbol{\Phi}))) \\ &= -\mathbf{e}_N \lfloor (\boldsymbol{\Phi} \wedge \varepsilon_i)\mathbf{e}_N | (\boldsymbol{\Psi} \wedge (\boldsymbol{\Phi} \lfloor \mathbf{a}) \wedge (\boldsymbol{\Psi} \lfloor \mathbf{e}_i)). \end{aligned}$$

At this point we can apply identities of the form

$$\begin{aligned} \boldsymbol{\Psi} \wedge (\boldsymbol{\Psi} \lfloor \mathbf{b}) &= \frac{1}{2}(\boldsymbol{\Psi} \cdot \boldsymbol{\Psi})\varepsilon_N \lfloor \mathbf{b} \\ (\boldsymbol{\Psi} \lfloor \mathbf{b}) \wedge (\boldsymbol{\Psi} \lfloor \mathbf{c}) &= (\boldsymbol{\Psi} \lfloor \bar{\bar{\mathrm{I}}})^{(2)} \lfloor (\mathbf{b} \wedge \mathbf{c}) = \boldsymbol{\Psi}\boldsymbol{\Psi} | (\mathbf{b} \wedge \mathbf{c}) \\ &\quad - \frac{1}{2}(\boldsymbol{\Psi} \cdot \boldsymbol{\Psi})\varepsilon_N \lfloor (\mathbf{b} \wedge \mathbf{c}), \end{aligned}$$

to the two functions and do the second summation as

$$\begin{aligned} \mathbf{f}(\boldsymbol{\Phi}, \boldsymbol{\Psi}) &= -\mathbf{e}_N \lfloor (\boldsymbol{\Phi} \wedge \mathbf{e}_i)\mathbf{e}_N | (\boldsymbol{\Phi} \wedge (\boldsymbol{\Psi} \lfloor \mathbf{a}) \wedge (\boldsymbol{\Psi} \lfloor \mathbf{e}_i)) \\ &= -\mathbf{e}_N \lfloor (\boldsymbol{\Phi} \wedge \mathbf{e}_i)\mathbf{e}_N | (\boldsymbol{\Phi} \wedge \boldsymbol{\Psi})(\boldsymbol{\Psi} | (\mathbf{a} \wedge \mathbf{e}_i)) \\ &\quad + \frac{1}{2}(\boldsymbol{\Psi} \cdot \boldsymbol{\Psi})\mathbf{e}_N \lfloor (\boldsymbol{\Phi} \wedge \mathbf{e}_i)\mathbf{e}_N | (\boldsymbol{\Phi} \wedge \varepsilon_N \lfloor (\mathbf{a} \wedge \mathbf{e}_i)), \\ &= -(\boldsymbol{\Phi} \cdot \boldsymbol{\Psi})\mathbf{e}_N \lfloor (\boldsymbol{\Phi} \wedge (\boldsymbol{\Psi} \lfloor \mathbf{a})) + \frac{1}{2}(\boldsymbol{\Psi} \cdot \boldsymbol{\Psi})\mathbf{e}_N \lfloor (\boldsymbol{\Phi} \wedge (\boldsymbol{\Phi} \lfloor \mathbf{a})), \\ &= -(\boldsymbol{\Phi} \cdot \boldsymbol{\Psi})\mathbf{e}_N \lfloor (\boldsymbol{\Phi} \wedge (\boldsymbol{\Psi} \lfloor \mathbf{a})) + \frac{1}{4}(\boldsymbol{\Psi} \cdot \boldsymbol{\Psi})(\boldsymbol{\Phi} \cdot \boldsymbol{\Phi})\mathbf{a}, \\ \mathbf{g}(\boldsymbol{\Phi}, \boldsymbol{\Psi}) &= -\frac{1}{2}(\boldsymbol{\Psi} \cdot \boldsymbol{\Psi})\mathbf{e}_N | (\boldsymbol{\Phi} \wedge \varepsilon_i)\mathbf{e}_N \lfloor ((\boldsymbol{\Phi} \lfloor \mathbf{a}) \wedge (\varepsilon_N \lfloor \mathbf{e}_i)). \\ &= \frac{1}{2}(\boldsymbol{\Psi} \cdot \boldsymbol{\Psi})\mathbf{e}_N \lfloor (\boldsymbol{\Phi} \wedge \varepsilon_i)(\boldsymbol{\Phi} \lfloor \mathbf{a}) | \mathbf{e}_i \\ &= \frac{1}{2}(\boldsymbol{\Psi} \cdot \boldsymbol{\Psi})\mathbf{e}_N \lfloor (\boldsymbol{\Phi} \wedge (\boldsymbol{\Phi} \lfloor \mathbf{a})) \\ &= \frac{1}{4}(\boldsymbol{\Psi} \cdot \boldsymbol{\Psi})(\boldsymbol{\Phi} \cdot \boldsymbol{\Phi})\mathbf{a}. \end{aligned}$$

Summing the four terms and applying the identity

$$\boldsymbol{\Psi} \wedge (\boldsymbol{\Phi} \lfloor \mathbf{a}) + \boldsymbol{\Phi} \wedge (\boldsymbol{\Psi} \lfloor \mathbf{a}) = (\boldsymbol{\Phi} \cdot \boldsymbol{\Psi})\varepsilon_N \lfloor \mathbf{a}$$

we finally obtain

$$((\mathbf{e}_N\lfloor\overline{\overline{\mathsf{U}}})\lfloor\lfloor\overline{\overline{\mathsf{I}}}^T)^2|\mathbf{a} = (\boldsymbol{\Psi}\cdot\boldsymbol{\Psi})(\boldsymbol{\Phi}\cdot\boldsymbol{\Phi})\mathbf{a}$$
$$-(\boldsymbol{\Phi}\cdot\boldsymbol{\Psi})\mathbf{e}_N\lfloor(\boldsymbol{\Phi}\wedge(\boldsymbol{\Psi}\lfloor\mathbf{a})) - (\boldsymbol{\Psi}\cdot\boldsymbol{\Phi})\mathbf{e}_N\lfloor(\boldsymbol{\Psi}\wedge(\boldsymbol{\Phi}\lfloor\mathbf{a}))$$
$$= ((\boldsymbol{\Psi}\cdot\boldsymbol{\Psi})(\boldsymbol{\Phi}\cdot\boldsymbol{\Phi}) - (\boldsymbol{\Phi}\cdot\boldsymbol{\Psi})^2)\mathbf{a},$$

which is valid for any vector **a**. Thus, we can write

$$((\mathbf{e}_N\lfloor\overline{\overline{\mathsf{U}}})\lfloor\lfloor\overline{\overline{\mathsf{I}}}^T)^2 = ((\boldsymbol{\Psi}\cdot\boldsymbol{\Psi})(\boldsymbol{\Phi}\cdot\boldsymbol{\Phi}) - (\boldsymbol{\Phi}\cdot\boldsymbol{\Psi})^2)\overline{\overline{\mathsf{I}}}.$$

5.12 A skewon–axion medium is defined by $\overline{\overline{\mathsf{M}}} = (\overline{\overline{\mathsf{B}}}{}^{\wedge}_{\wedge}\overline{\overline{\mathsf{I}}})^T$ where $\overline{\overline{\mathsf{B}}} \in \mathbb{E}_1\mathbb{F}_1$ may be any dyadic. The dispersion dyadic can be expanded as

$$\overline{\overline{\mathsf{D}}}(\boldsymbol{\nu}) = \overline{\overline{\mathsf{M}}}_m\lfloor\lfloor\boldsymbol{\nu}\boldsymbol{\nu} = -\boldsymbol{\nu}\rfloor(\mathbf{e}_N\lfloor(\overline{\overline{\mathsf{B}}}{}^{\wedge}_{\wedge}\overline{\overline{\mathsf{I}}})^T)\lfloor\boldsymbol{\nu} = \mathbf{e}_N\lfloor(\boldsymbol{\nu}\wedge(\overline{\overline{\mathsf{B}}}{}^{\wedge}_{\wedge}\overline{\overline{\mathsf{I}}})\lfloor\boldsymbol{\nu})$$
$$= \mathbf{e}_N\lfloor(\boldsymbol{\nu}\wedge(\overline{\overline{\mathsf{B}}}^T|\boldsymbol{\nu})\wedge\overline{\overline{\mathsf{I}}}^T)$$

Denoting $\boldsymbol{\nu}' = \overline{\overline{\mathsf{B}}}^T|\boldsymbol{\nu}$, we have

$$\overline{\overline{\mathsf{D}}}(\boldsymbol{\nu}) = \mathbf{e}_N\lfloor(\boldsymbol{\nu}\wedge\boldsymbol{\nu}')\lfloor\overline{\overline{\mathsf{I}}}^T = \mathbf{A}\lfloor\overline{\overline{\mathsf{I}}}^T,$$

which is an antisymmetric dyadic with simple bivector **A** because

$$\mathbf{A}\cdot\mathbf{A} = (\mathbf{e}_N\lfloor(\boldsymbol{\nu}\wedge\boldsymbol{\nu}'))|\boldsymbol{\varepsilon}_N\lfloor(\mathbf{e}_N\lfloor(\boldsymbol{\nu}\wedge\boldsymbol{\nu}')) = (\mathbf{e}_N\lfloor(\boldsymbol{\nu}\wedge\boldsymbol{\nu}'))|(\boldsymbol{\nu}\wedge\boldsymbol{\nu}')$$
$$= \mathbf{e}_N\lfloor(\boldsymbol{\nu}\wedge\boldsymbol{\nu}'\wedge\boldsymbol{\nu}\wedge\boldsymbol{\nu}') = 0.$$

Expressing it in the form $\mathbf{A} = \mathbf{a}\wedge\mathbf{b}$, we can write

$$\overline{\overline{\mathsf{D}}}(\boldsymbol{\nu}) = (\mathbf{a}\wedge\mathbf{b})\lfloor\overline{\overline{\mathsf{I}}}^T = \mathbf{ba} - \mathbf{ab}.$$

Because we have

$$\overline{\overline{\mathsf{D}}}^{(2)}(\boldsymbol{\nu}) = (\mathbf{a}\wedge\mathbf{b})(\mathbf{a}\wedge\mathbf{b}), \quad \overline{\overline{\mathsf{D}}}^{(3)}(\boldsymbol{\nu}) = 0,$$

the dyadic dispersion equation is satisfied identically for any $\boldsymbol{\nu}$.

5.13 To expand the dyadic $\overline{\overline{\mathsf{X}}} = \boldsymbol{\gamma}_e\wedge(\overline{\overline{\mathsf{I}}}^T\rfloor\boldsymbol{\Phi}) \in \mathbb{F}_4\mathbb{F}_1$ in 3D source and field quantities we insert

$$\boldsymbol{\gamma}_e = \varrho_e - \mathbf{J}_e\wedge\boldsymbol{\varepsilon}_4, \quad \boldsymbol{\Phi} = \mathbf{B} + \mathbf{E}\wedge\boldsymbol{\varepsilon}_4$$

and proceed

$$\overline{\overline{\mathsf{X}}} = (\varrho_e - \mathbf{J}_e\wedge\boldsymbol{\varepsilon}_4)\wedge(\overline{\overline{\mathsf{I}}}^T\rfloor(\mathbf{B} + \mathbf{E}\wedge\boldsymbol{\varepsilon}_4)$$
$$= \varrho_e\wedge(\boldsymbol{\varepsilon}_4\mathbf{e}_4)\rfloor(\mathbf{E}\wedge\boldsymbol{\varepsilon}_4) - (\mathbf{J}_e\wedge\boldsymbol{\varepsilon}_4)\wedge(\overline{\overline{\mathsf{I}}}^T\rfloor(\mathbf{B} - \mathbf{E}\wedge\boldsymbol{\varepsilon}_4))$$
$$= \varrho_e\wedge\boldsymbol{\varepsilon}_4\mathbf{E} - \boldsymbol{\varepsilon}_4\wedge(\mathbf{J}_e\wedge\overline{\overline{\mathsf{I}}}^T\rfloor\mathbf{B}) - (\boldsymbol{\varepsilon}_4\wedge\mathbf{J}_e)\wedge(\boldsymbol{\varepsilon}_4\mathbf{E} - \mathbf{E}\boldsymbol{\varepsilon}_4)$$
$$= -\boldsymbol{\varepsilon}_4\wedge(\varrho_e\mathbf{E} + \mathbf{J}_e\wedge\overline{\overline{\mathsf{I}}}_s^T\rfloor\mathbf{B} - (\mathbf{J}_e\wedge\mathbf{E})\boldsymbol{\varepsilon}_4).$$

5.14 The stress–energy dyadic

$$\overline{\overline{\mathsf{T}}} = \frac{1}{2}(\boldsymbol{\Psi} \wedge \overline{\overline{\mathsf{I}}}^T \rfloor \boldsymbol{\Phi} - \boldsymbol{\Phi} \wedge \overline{\overline{\mathsf{I}}}^T \rfloor \boldsymbol{\Psi})$$

vanishes for any fields satisfying $\boldsymbol{\Psi} = M\boldsymbol{\Phi}$. To prove the converse, we expand

$$\begin{aligned} 2\overline{\overline{\mathsf{T}}}|\mathbf{e}_4 &= -(\boldsymbol{\Psi} \wedge \overline{\overline{\mathsf{I}}}^T \wedge \mathbf{e}_4)|\boldsymbol{\Phi} + (\boldsymbol{\Phi} \wedge \overline{\overline{\mathsf{I}}}^T \wedge \mathbf{e}_4)|\boldsymbol{\Psi} \\ &= -\boldsymbol{\Psi} \wedge \mathbf{E} - \boldsymbol{\Phi} \wedge \mathbf{H}. \end{aligned}$$

Requiring this to be zero leads to

$$\begin{aligned} -\mathbf{H} \wedge \mathbf{E} + \mathbf{E} \wedge \mathbf{H} = 2\mathbf{E} \wedge \mathbf{H} = 0, \\ \mathbf{D} \wedge \mathbf{E} + \mathbf{B} \wedge \mathbf{H} = 0. \end{aligned}$$

Assuming these to be valid for any nonzero $\mathbf{E}, \mathbf{B}$, there must exist a scalar M such that $\mathbf{H} = M\mathbf{E}$, whence from $(\mathbf{D} - M\mathbf{B}) \wedge \mathbf{E} = 0$ for any $\mathbf{E}$ we conclude that $\mathbf{D} - M\mathbf{B} = 0$.

5.15 Expanding $\boldsymbol{\Phi} \wedge \boldsymbol{\Phi} = 2\mathbf{B} \wedge \mathbf{E} \wedge \varepsilon_4 = 0$ yields $\mathbf{B} \wedge \mathbf{E} = 0$. Applying the bac-cab rule

$$\mathbf{a}\rfloor(\mathbf{E} \wedge \mathbf{B}) = \mathbf{E} \wedge (\mathbf{a}\rfloor\mathbf{B}) + (\mathbf{a}|\mathbf{E})\mathbf{B} = 0,$$

for a spatial vector $\mathbf{a}$ satisfying $\mathbf{a}|\mathbf{E} \neq 0$ leads to the representation

$$\mathbf{B} = \boldsymbol{\alpha} \wedge \mathbf{E}, \quad \boldsymbol{\alpha} = \frac{\mathbf{a}\rfloor\mathbf{B}}{\mathbf{a}|\mathbf{E}}.$$

5.16 Expanded in 3D medium dyadics the condition takes the form

$$\begin{aligned} &-A\mathbf{e}_{123}\mathbf{e}_{123}\lfloor\lfloor\overline{\overline{\mu}}_g^{-1} - A\mathbf{e}_4 \wedge \overline{\overline{\xi}}_g|\overline{\overline{\mu}}_g^{-1}\rfloor\mathbf{e}_{123} + A\mathbf{e}_{123}\lfloor\overline{\overline{\mu}}_g^{-1}|\overline{\overline{\zeta}}_g \wedge \mathbf{e}_4 \\ &+ A(\overline{\overline{\epsilon}}_g - \overline{\overline{\xi}}_g|\overline{\overline{\mu}}_g^{-1}|\overline{\overline{\zeta}}_g)^\wedge_\wedge\mathbf{e}_4\mathbf{e}_4) + B\mathbf{e}_{123}\mathbf{e}_{123}\lfloor\lfloor\overline{\overline{\epsilon}}_g^{-1T} \\ &- B(\mathbf{e}_4 \wedge \overline{\overline{\zeta}}_g|\overline{\overline{\epsilon}}_g^{-1}\rfloor\mathbf{e}_{123})^T + B(\mathbf{e}_{123}\lfloor\overline{\overline{\epsilon}}_g^{-1}|\overline{\overline{\xi}}_g \wedge \mathbf{e}_4)^T \\ &- B(\overline{\overline{\mu}}_g^T - \overline{\overline{\xi}}_g^T|\overline{\overline{\epsilon}}_g^{-1T}|\overline{\overline{\zeta}}_g^T)^\wedge_\wedge\mathbf{e}_4\mathbf{e}_4 = 0. \end{aligned}$$

Equating the four spatiotemporal terms separately leads to

$$\begin{aligned} -A\overline{\overline{\mu}}_g^{-1} + B\overline{\overline{\epsilon}}_g^{-1T} = 0, &\quad \Rightarrow \quad A\overline{\overline{\epsilon}}_g = B\overline{\overline{\mu}}_g^T \\ -A\overline{\overline{\xi}}_g|\overline{\overline{\mu}}_g^{-1} - B\overline{\overline{\xi}}_g^T|\overline{\overline{\epsilon}}_g^{-1T} = 0, &\quad \Rightarrow \quad \overline{\overline{\xi}}_g^T = -\overline{\overline{\xi}}_g \\ A\overline{\overline{\mu}}_g^{-1}|\overline{\overline{\zeta}}_g + B\overline{\overline{\epsilon}}_g^{-1T}|\overline{\overline{\zeta}}_g^T = 0, &\quad \Rightarrow \quad \overline{\overline{\zeta}}_g^T = -\overline{\overline{\zeta}}_g. \end{aligned}$$

The fourth condition

$$A(\overline{\overline{\epsilon}}_g - \overline{\overline{\xi}}_g|\overline{\overline{\mu}}_g^{-1}|\overline{\overline{\zeta}}_g) - B(\overline{\overline{\mu}}_g^T - \overline{\overline{\xi}}_g^T|\overline{\overline{\epsilon}}_g^{-1T}|\overline{\overline{\zeta}}_g^T) = 0$$

is valid whenever the three other conditions are valid.

5.17 Starting from the Maxwell equation we can proceed as

$$\begin{aligned} 0 = \mathbf{e}_N\lfloor(\boldsymbol{\nu}\wedge\boldsymbol{\Psi}) &= -\boldsymbol{\nu}\rfloor(\overline{\overline{\mathsf{Q}}}^{(2)} + \mathbf{AB})|(\boldsymbol{\nu}\wedge\boldsymbol{\phi}) \\ &= -\boldsymbol{\nu}\rfloor((\overline{\overline{\mathsf{Q}}}|\boldsymbol{\nu})\wedge(\overline{\overline{\mathsf{Q}}}|\boldsymbol{\phi}) + \mathbf{AB}|(\boldsymbol{\nu}\wedge\boldsymbol{\phi})) \\ &= -(\boldsymbol{\nu}|\overline{\overline{\mathsf{Q}}}|\boldsymbol{\phi})\overline{\overline{\mathsf{Q}}}|\boldsymbol{\nu} + (\boldsymbol{\nu}|\overline{\overline{\mathsf{Q}}}|\boldsymbol{\nu})\overline{\overline{\mathsf{Q}}}|\boldsymbol{\phi} - (\boldsymbol{\nu}\rfloor\mathbf{A})(\mathbf{B}\lfloor\boldsymbol{\nu})|\boldsymbol{\phi}, \end{aligned}$$

which yields the equation for $\boldsymbol{\phi}$.

5.18 The dispersion dyadic has the form

$$\overline{\overline{\mathsf{D}}}(\boldsymbol{\nu}) = M\overline{\overline{\mathsf{E}}}\lfloor\lfloor\boldsymbol{\nu}\boldsymbol{\nu} + \mathbf{a}_1\mathbf{c}_1 + \mathbf{a}_2\mathbf{c}_2$$

with $\mathbf{a}_i = \mathbf{A}_i\lfloor\boldsymbol{\nu},\ \mathbf{c}_i = \mathbf{C}_i\lfloor\boldsymbol{\nu}$. Because

$$\begin{aligned} \overline{\overline{\mathsf{E}}}\lfloor\lfloor\boldsymbol{\nu}\boldsymbol{\nu} &= -\boldsymbol{\nu}\rfloor(\mathbf{e}_N\lfloor\overline{\overline{\mathsf{I}}}^{(2)T})\lfloor\boldsymbol{\nu} = -\boldsymbol{\nu}\rfloor(\mathbf{e}_N\lfloor(\boldsymbol{\nu}\wedge\overline{\overline{\mathsf{I}}}^T)) \\ &= \mathbf{e}_N\lfloor(\boldsymbol{\nu}\wedge\boldsymbol{\nu}\wedge\overline{\overline{\mathsf{I}}}^T) = 0, \end{aligned}$$

the axion term falls out. Because we have

$$\begin{aligned} \overline{\overline{\mathsf{D}}}^{(2)}(\boldsymbol{\nu}) &= (\mathbf{a}_1\wedge\mathbf{a}_2)(\mathbf{c}_1\wedge\mathbf{c}_2) \\ \overline{\overline{\mathsf{D}}}^{(3)}(\boldsymbol{\nu}) &= 0, \end{aligned}$$

such a medium does not have a dispersion equation and the wave one-form $\boldsymbol{\nu}$ of a plane wave can be freely chosen.

5.19 Since the first term does not affect the dispersion dyadic, we have

$$\overline{\overline{\mathsf{D}}}(\boldsymbol{\nu}) = \mathbf{a}_1\mathbf{c}_1 + \mathbf{a}_2\mathbf{c}_2 + \mathbf{a}_3\mathbf{c}_3,$$

where we denote $\mathbf{a}_i = \mathbf{A}_i\lfloor\boldsymbol{\nu}, \mathbf{c}_i = \mathbf{C}_i\lfloor\boldsymbol{\nu}$. These vectors satisfy $\mathbf{a}_i|\boldsymbol{\nu} = 0$, $\mathbf{c}_i|\boldsymbol{\nu} = 0$. The double-wedge powers are

$$\begin{aligned} \overline{\overline{\mathsf{D}}}^{(2)}(\boldsymbol{\nu}) &= \mathbf{a}_{12}\mathbf{c}_{12} + \mathbf{a}_{23}\mathbf{c}_{23} + \mathbf{a}_{31}\mathbf{c}_{31}, \\ \overline{\overline{\mathsf{D}}}^{(3)}(\boldsymbol{\nu}) &= \mathbf{a}_{123}\mathbf{c}_{123}. \end{aligned}$$

The dyadic dispersion equation

$$\overline{\overline{\mathsf{D}}}^{(3)}(\boldsymbol{\nu}) = \mathbf{a}_{123}\mathbf{c}_{123} = 0$$

is split in two equations as

$$(\mathbf{A}_1\lfloor\nu)\wedge(\mathbf{A}_2\lfloor\nu)\wedge(\mathbf{A}_3\lfloor\nu)=0,$$
$$(\mathbf{C}_1\lfloor\nu)\wedge(\mathbf{C}_2\lfloor\nu)\wedge(\mathbf{C}_3\lfloor\nu)=0.$$

which appear cubic in ν. Applying the bac-cab identity

$$(\mathbf{B}\lfloor\alpha)\wedge\mathbf{c}=\alpha\rfloor(\mathbf{B}\wedge\mathbf{c})-(\mathbf{c}|\alpha)\mathbf{B}$$

we can expand

$$\begin{aligned}\mathbf{a}_2\wedge\mathbf{a}_3&=(\mathbf{A}_2\lfloor\nu)\wedge\mathbf{a}_3=\nu\rfloor(\mathbf{A}_2\wedge\mathbf{a}_3)-\mathbf{A}_2(\mathbf{a}_3|\nu)\\&=\nu\rfloor(\mathbf{a}_3\wedge\mathbf{A}_2).\end{aligned}$$

In particular we have $(\mathbf{a}_2\wedge\mathbf{a}_3)\lfloor\nu=0$. Applying another bac-cab rule

$$\mathbf{B}\wedge(\mathbf{C}\lfloor\alpha)+\mathbf{C}\wedge(\mathbf{B}\lfloor\alpha)=(\mathbf{B}\wedge\mathbf{C})\lfloor\alpha=(\mathbf{B}\cdot\mathbf{C})\mathbf{e}_N\lfloor\alpha,$$

we can write

$$\begin{aligned}\mathbf{a}_1\wedge\mathbf{a}_2\wedge\mathbf{a}_3&=(\mathbf{A}_1\wedge\nu)\wedge\mathbf{a}_2\wedge\mathbf{a}_3\\&=-\mathbf{A}_1\wedge((\mathbf{a}_2\wedge\mathbf{a}_3)\lfloor\nu)+\mathbf{A}_1\cdot(\mathbf{a}_2\wedge\mathbf{a}_3)\mathbf{e}_N\lfloor\nu.\\&=\mathbf{A}_1\cdot(\mathbf{a}_2\wedge\mathbf{a}_3)\mathbf{e}_N\lfloor\nu.\end{aligned}$$

The dispersion equations are actually only of the second order in ν, and they have the scalar form

$$\mathbf{A}_1\cdot(\mathbf{a}_2\wedge\mathbf{a}_3)=\mathbf{A}_1\cdot((\mathbf{A}_2\lfloor\nu)\wedge(\mathbf{A}_3\lfloor\nu))=0,$$
$$\mathbf{C}_1\cdot(\mathbf{c}_2\wedge\mathbf{c}_3)=\mathbf{C}_1\cdot((\mathbf{C}_2\lfloor\nu)\wedge(\mathbf{C}_3\lfloor\nu))=0.$$

5.20 Inserting $\overline{\overline{\mathsf{D}}}(\nu)=(\overline{\overline{\mathsf{G}}}|\nu)(\overline{\overline{\mathsf{G}}}|\nu)/\mu$ in (5.254) and denoting $\mathbf{a}=\overline{\overline{\mathsf{G}}}|\nu$ yields

$$\boldsymbol{\Phi}=\lambda\boldsymbol{\varepsilon}_N\lfloor(\mathbf{a}\wedge\mathbf{d}),\qquad\lambda=\mathbf{a}|\alpha/\mu.$$

Because $\mathbf{d}$ satisfies $\mathbf{d}|\nu=0$, we can express it as $\mathbf{d}=\nu\rfloor\mathbf{A}$ in terms of some bivector $\mathbf{A}$. Applying a bac-cab rule we can expand

$$\mathbf{a}\wedge\mathbf{d}=\mathbf{a}\wedge(\nu\rfloor\mathbf{A})=\nu\rfloor(\mathbf{a}\wedge\mathbf{A})-(\mathbf{a}|\nu)\mathbf{A}.$$

Taking $\mathbf{a}|\nu=\overline{\overline{\mathsf{G}}}||\nu\nu=0$ into account we can further expand

$$\begin{aligned}\boldsymbol{\Phi}&=\lambda\boldsymbol{\varepsilon}_N\lfloor(\nu\rfloor(\mathbf{a}\wedge\mathbf{A}))=\lambda\nu\wedge(\boldsymbol{\varepsilon}_N\lfloor(\mathbf{A}\wedge\mathbf{a}))\\&=\lambda\nu\wedge((\boldsymbol{\varepsilon}_N\lfloor\mathbf{A})\lfloor\mathbf{a})=\nu\wedge(\mathbf{a}\rfloor\boldsymbol{\Xi}),\end{aligned}$$

with $\boldsymbol{\Xi}=-\lambda\boldsymbol{\varepsilon}_N\lfloor\mathbf{A}$. This coincides with the expression (5.268).

5.21 Writing $\overline{\overline{\mathsf{C}}} = \mathbf{ac} + \mathbf{bd}$ in terms of four vectors we can expand

$$\begin{aligned}\overline{\overline{\mathsf{C}}}\rfloor(\varepsilon_N\lfloor(\overline{\overline{\mathsf{C}}}^{(2)T})) &= (\mathbf{ac}+\mathbf{bd})\rfloor(\varepsilon_N\lfloor(\mathbf{c}\wedge\mathbf{d})(\mathbf{a}\wedge\mathbf{b}))\\ &= -(\mathbf{a}(\varepsilon_N\lfloor(\mathbf{c}\wedge\mathbf{d}\wedge\mathbf{c})+\mathbf{b}(\varepsilon_N\lfloor(\mathbf{c}\wedge\mathbf{d}\wedge\mathbf{d}))(\mathbf{a}\wedge\mathbf{b})\\ &= 0.\end{aligned}$$

CHAPTER 6

6.1 Because $\mathrm{tr}(\overline{\overline{\mathsf{M}}}_1|\overline{\overline{\mathsf{M}}}_2) = \mathrm{tr}(\overline{\overline{\mathsf{M}}}_2|\overline{\overline{\mathsf{M}}}_1)$, we have

$$\begin{aligned}\mathrm{tr}\overline{\overline{\mathsf{M}}}_a &= \mathrm{tr}(\overline{\overline{\mathsf{A}}}^{(-2)T}|\overline{\overline{\mathsf{M}}}|\overline{\overline{\mathsf{A}}}^{(2)T)})\\ &= (\overline{\overline{\mathsf{M}}}|\overline{\overline{\mathsf{A}}}^{(2)T}|\overline{\overline{\mathsf{A}}}^{(-2)T}) = \mathrm{tr}\overline{\overline{\mathsf{M}}}.\\ \overline{\overline{\mathsf{I}}}^{(4)}\lfloor\lfloor\overline{\overline{\mathsf{M}}}_a &= \mathbf{e}_N\lfloor(\overline{\overline{\mathsf{A}}}^{(-2)T}|\overline{\overline{\mathsf{M}}}|\overline{\overline{\mathsf{A}}}^{(2)T}\rfloor\varepsilon_N\\ &= (\mathbf{e}_N\lfloor(\overline{\overline{\mathsf{A}}}^{(-2)T}\rfloor\varepsilon_N)|(\mathbf{e}_N\lfloor\overline{\overline{\mathsf{M}}}\rfloor\varepsilon_N)|(\mathbf{e}_N\lfloor|\overline{\overline{\mathsf{A}}}^{(2)T}\rfloor\varepsilon_N)\\ &= \frac{\overline{\overline{\mathsf{A}}}^{(2)}}{\mathrm{tr}\overline{\overline{\mathsf{A}}}^{(4)}}|(\overline{\overline{\mathsf{I}}}^{(4)}\lfloor\lfloor\overline{\overline{\mathsf{M}}})|(\overline{\overline{\mathsf{A}}}^{(-2)}\mathrm{tr}\overline{\overline{\mathsf{A}}}^{(4)})\\ &= \overline{\overline{\mathsf{A}}}^{(2)}|(\overline{\overline{\mathsf{I}}}^{(4)}\lfloor\lfloor\overline{\overline{\mathsf{M}}})|\overline{\overline{\mathsf{A}}}^{(-2)}.\end{aligned}$$

6.2 The transformed medium bidyadic has the expansion

$$\begin{aligned}\overline{\overline{\mathsf{M}}}_a &= \overline{\overline{\mathsf{T}}}^{(-2)T}|\overline{\overline{\mathsf{M}}}|\overline{\overline{\mathsf{T}}}^{(2)T} = \overline{\overline{\mathsf{T}}}^{(2)T}|\overline{\overline{\mathsf{M}}}|\overline{\overline{\mathsf{T}}}^{(2)T}\\ &= (\overline{\overline{\mathsf{I}}}_s^{(2)T} - \varepsilon_4\mathbf{e}_4{}^\wedge_\wedge\overline{\overline{\mathsf{I}}}_s^T)|\overline{\overline{\mathsf{M}}}|(\overline{\overline{\mathsf{I}}}_s^{(2)T} - \varepsilon_4\mathbf{e}_4{}^\wedge_\wedge\overline{\overline{\mathsf{I}}}_s^T)\\ &= \overline{\overline{\alpha}} - \overline{\overline{\epsilon}}'\wedge\mathbf{e}_4 - \varepsilon_4\wedge\overline{\overline{\mu}}^{-1} + \varepsilon_4\wedge\overline{\overline{\beta}}\wedge\mathbf{e}_4,\end{aligned}$$

which yields the transformation rules

$$\overline{\overline{\alpha}}_a = \overline{\overline{\alpha}},\quad \overline{\overline{\epsilon}}'_a = -\overline{\overline{\epsilon}}',\quad \overline{\overline{\mu}}_a^{-1} = -\overline{\overline{\mu}}^{-1},\quad \overline{\overline{\beta}}_a = \overline{\overline{\beta}}.$$

Thus, a medium dyadic of the form

$$\overline{\overline{\mathsf{M}}} = \overline{\overline{\alpha}} + \varepsilon_4\wedge\overline{\overline{\beta}}\wedge\mathbf{e}_4$$

is invariant in time inversion. Since $\overline{\overline{\mu}}^{-1} = 0$, there is no Gibbsian representation of such a medium.

6.3

$$\begin{aligned}\overline{\overline{\mathsf{Q}}}(\overline{\overline{\mathsf{Q}}}(\overline{\overline{\mathsf{X}}})) &= (\overline{\overline{\mathsf{I}}} - (\overline{\overline{\mathsf{I}}} - \overline{\overline{\mathsf{X}}})|(\overline{\overline{\mathsf{I}}} + \overline{\overline{\mathsf{X}}})^{-1})|(\overline{\overline{\mathsf{I}}} + (\overline{\overline{\mathsf{I}}} - \overline{\overline{\mathsf{X}}})|(\overline{\overline{\mathsf{I}}} + \overline{\overline{\mathsf{X}}})^{-1})^{-1}\\ &= ((\overline{\overline{\mathsf{I}}} + \overline{\overline{\mathsf{X}}}) - (\overline{\overline{\mathsf{I}}} - \overline{\overline{\mathsf{X}}}))|((\overline{\overline{\mathsf{I}}} + \overline{\overline{\mathsf{X}}}) + (\overline{\overline{\mathsf{I}}} - \overline{\overline{\mathsf{X}}}))^{-1}\\ &= 2\overline{\overline{\mathsf{X}}}|(2\overline{\overline{\mathsf{I}}})^{-1} = \overline{\overline{\mathsf{X}}}\\ \overline{\overline{\mathsf{Q}}}(-\overline{\overline{\mathsf{X}}}) &= (\overline{\overline{\mathsf{I}}} + \overline{\overline{\mathsf{X}}})|(\overline{\overline{\mathsf{I}}} - \overline{\overline{\mathsf{X}}})^{-1}) = \overline{\overline{\mathsf{Q}}}^{-1}(\overline{\overline{\mathsf{X}}})\end{aligned}$$

$$\begin{aligned}\overline{\overline{\mathsf{Q}}}(\overline{\overline{\mathsf{G}}}|\overline{\overline{\mathsf{X}}}^T|\overline{\overline{\Gamma}}) &= (\overline{\overline{\mathsf{I}}} - \overline{\overline{\mathsf{G}}}|\overline{\overline{\mathsf{X}}}^T|\overline{\overline{\Gamma}})|(\overline{\overline{\mathsf{I}}} + \overline{\overline{\mathsf{G}}}|\overline{\overline{\mathsf{X}}}^T|\overline{\overline{\Gamma}})^{-1} \\ &= (\overline{\overline{\mathsf{G}}}|(\overline{\overline{\mathsf{I}}} - \overline{\overline{\mathsf{X}}})^T|\overline{\overline{\Gamma}})|(\overline{\overline{\mathsf{G}}}|(\overline{\overline{\mathsf{I}}} + \overline{\overline{\mathsf{X}}})^T|\overline{\overline{\Gamma}})^{-1} \\ &= \overline{\overline{\mathsf{G}}}|(\overline{\overline{\mathsf{I}}} - \overline{\overline{\mathsf{X}}})^T|\overline{\overline{\Gamma}}|\overline{\overline{\mathsf{G}}}|(\overline{\overline{\mathsf{I}}} + \overline{\overline{\mathsf{X}}})^{-1T}|\overline{\overline{\Gamma}} \\ &= \overline{\overline{\mathsf{G}}}|\overline{\overline{\mathsf{Q}}}^T(\overline{\overline{\mathsf{X}}})|\overline{\overline{\Gamma}}.\end{aligned}$$

6.4 Replacing **a** by $\mathbf{a}+\mathbf{b}$ in the definition, the rotation dyadic $\overline{\overline{\mathsf{R}}}$ satisfies

$$(\overline{\overline{\mathsf{R}}}|\mathbf{a})\cdot(\overline{\overline{\mathsf{R}}}|\mathbf{b}) = \mathbf{a}\cdot\mathbf{b} \quad \Rightarrow \quad \mathbf{a}|(\overline{\overline{\mathsf{R}}}^T|\overline{\overline{\Gamma}}|\overline{\overline{\mathsf{R}}} - \overline{\overline{\Gamma}})|\mathbf{b} = 0$$

for any vectors **a, b**, whence the rotation dyadic satisfies

$$\overline{\overline{\mathsf{R}}}^T|\overline{\overline{\Gamma}}|\overline{\overline{\mathsf{R}}} = \overline{\overline{\Gamma}} \quad \Rightarrow \quad \overline{\overline{\mathsf{G}}}|\overline{\overline{\mathsf{R}}}^T|\overline{\overline{\Gamma}} = \overline{\overline{\mathsf{R}}}^{-1}.$$

Operating both sides of the latter equation by $\overline{\overline{\mathsf{Q}}}()$

$$\overline{\overline{\mathsf{Q}}}(\overline{\overline{\mathsf{G}}}|\overline{\overline{\mathsf{R}}}^T|\overline{\overline{\Gamma}}) = \overline{\overline{\mathsf{Q}}}(\overline{\overline{\mathsf{R}}}^{-1}),$$

and applying properties of the quotient function in Problem 6.3 we obtain

$$\overline{\overline{\mathsf{G}}}|\overline{\overline{\mathsf{Q}}}^T(\overline{\overline{\mathsf{R}}})|\overline{\overline{\Gamma}} = -\overline{\overline{\mathsf{Q}}}(\overline{\overline{\mathsf{R}}}),$$

which equals

$$\overline{\overline{\mathsf{Q}}}^T(\overline{\overline{\mathsf{R}}})|\overline{\overline{\Gamma}} = (\overline{\overline{\Gamma}}|\overline{\overline{\mathsf{Q}}}(\overline{\overline{\mathsf{R}}}))^T = -\overline{\overline{\Gamma}}|\overline{\overline{\mathsf{Q}}}(\overline{\overline{\mathsf{R}}}).$$

Thus, $\overline{\overline{\Gamma}}|\overline{\overline{\mathsf{Q}}}(\overline{\overline{\mathsf{R}}})$ is an antisymmetric dyadic and it can be expressed in terms of some two-form $\mathbf{\Pi}$ as $\overline{\overline{\Gamma}}|\overline{\overline{\mathsf{Q}}}(\overline{\overline{\mathsf{R}}}) = \overline{\overline{\mathsf{I}}}^T\rfloor\mathbf{\Pi}$. This leads to the representation

$$\overline{\overline{\mathsf{Q}}}(\overline{\overline{\mathsf{R}}}) = \overline{\overline{\mathsf{G}}}\rfloor\mathbf{\Pi}, \quad \Rightarrow \quad \overline{\overline{\mathsf{R}}} = \overline{\overline{\mathsf{Q}}}(\overline{\overline{\mathsf{G}}}\rfloor\mathbf{\Pi})$$

or

$$\overline{\overline{\mathsf{R}}} = (\overline{\overline{\mathsf{I}}} - \overline{\overline{\mathsf{G}}}\rfloor\mathbf{\Pi})|(\overline{\overline{\mathsf{I}}} + \overline{\overline{\mathsf{G}}}\rfloor\mathbf{\Pi})^{-1}.$$

6.5 Denoting $s = \sin\theta$ and $c = \cos\theta$ for brevity, we can expand both sides of (6.45) as

$$\begin{aligned}\begin{pmatrix} Z_d\mathbf{D}_d \\ \mathbf{B}_d \end{pmatrix} &= \begin{pmatrix} Z_d\overline{\overline{\epsilon}}_d & Z_d\overline{\overline{\xi}}_d \\ \overline{\overline{\zeta}}_d & \overline{\overline{\mu}}_d \end{pmatrix} \Big| \begin{pmatrix} \mathbf{E}_d \\ \mathbf{H}_d \end{pmatrix} = \begin{pmatrix} Z_d\overline{\overline{\epsilon}}_d & \overline{\overline{\xi}}_d \\ \overline{\overline{\zeta}}_d & \overline{\overline{\mu}}_d/Z_d \end{pmatrix} \Big| \begin{pmatrix} \mathbf{E}_d \\ Z_d\mathbf{H}_d \end{pmatrix} \\ &= \begin{pmatrix} Zc & s \\ -Zs & c \end{pmatrix} \Big| \begin{pmatrix} \mathbf{D} \\ \mathbf{B} \end{pmatrix} = \begin{pmatrix} Zc & s \\ -Zs & c \end{pmatrix} \Big| \begin{pmatrix} \overline{\overline{\epsilon}} & \overline{\overline{\xi}} \\ \overline{\overline{\zeta}} & \overline{\overline{\mu}} \end{pmatrix} \Big| \begin{pmatrix} \mathbf{E} \\ \mathbf{H} \end{pmatrix}.\end{aligned}$$

On the second line we can insert from (6.46)

$$\begin{pmatrix} \mathbf{E}_d \\ Z_d\mathbf{H}_d \end{pmatrix} = \begin{pmatrix} c & s \\ -s & c \end{pmatrix} \begin{pmatrix} \mathbf{E} \\ Z\mathbf{H} \end{pmatrix} = \begin{pmatrix} c & Zs \\ -s & Zc \end{pmatrix} \begin{pmatrix} \mathbf{E} \\ \mathbf{H} \end{pmatrix}.$$

Because $(\mathbf{E} \;\; \mathbf{H})$ is arbitrary, we can eliminate it and the result becomes

$$\begin{pmatrix} Z_d\overline{\overline{\epsilon}}_d & \overline{\overline{\xi}}_d \\ \overline{\overline{\zeta}}_d & \overline{\overline{\mu}}_d/Z_d \end{pmatrix} \begin{pmatrix} c & Zs \\ -s & Zc \end{pmatrix} = \begin{pmatrix} Zc & s \\ -Zs & c \end{pmatrix} | \begin{pmatrix} \overline{\overline{\epsilon}} & \overline{\overline{\xi}} \\ \overline{\overline{\zeta}} & \overline{\overline{\mu}} \end{pmatrix},$$

or

$$\begin{pmatrix} Z_d\overline{\overline{\epsilon}}_d & \overline{\overline{\xi}}_d \\ \overline{\overline{\zeta}}_d & \overline{\overline{\mu}}_d/Z_d \end{pmatrix} = \frac{1}{Z} \begin{pmatrix} Zc & s \\ -Zs & c \end{pmatrix} | \begin{pmatrix} \overline{\overline{\epsilon}} & \overline{\overline{\xi}} \\ \overline{\overline{\zeta}} & \overline{\overline{\mu}} \end{pmatrix} \begin{pmatrix} Zc & -Zs \\ s & c \end{pmatrix}.$$

This can be written componentwise as

$$\begin{aligned} \overline{\overline{\epsilon}}_d Z_d &= \overline{\overline{\epsilon}} c^2 Z + (\overline{\overline{\xi}} + \overline{\overline{\zeta}}) sc + \overline{\overline{\mu}} s^2/Z, \\ \overline{\overline{\xi}}_d &= -\overline{\overline{\epsilon}} scZ + \overline{\overline{\xi}} c^2 - \overline{\overline{\zeta}} s^2 + \overline{\overline{\mu}} sc/Z, \\ \overline{\overline{\zeta}}_d &= -\overline{\overline{\epsilon}} scZ - \overline{\overline{\xi}} s^2 + \overline{\overline{\zeta}} c^2 + \overline{\overline{\mu}} sc/Z, \\ \overline{\overline{\mu}}_d/Z_d &= \overline{\overline{\epsilon}} s^2 Z - (\overline{\overline{\xi}} + \overline{\overline{\zeta}}) sc + \overline{\overline{\mu}} c^2/Z, \end{aligned}$$

and, finally, in the form

$$\begin{aligned} \overline{\overline{\epsilon}}_d Z_d + \overline{\overline{\mu}}_d/Z_d &= \overline{\overline{\epsilon}} Z + \overline{\overline{\mu}}/Z, \\ \overline{\overline{\xi}}_d - \overline{\overline{\zeta}}_d &= \overline{\overline{\xi}} - \overline{\overline{\zeta}}, \\ \overline{\overline{\epsilon}}_d Z_d - \overline{\overline{\mu}}_d/Z_d &= (\overline{\overline{\epsilon}} Z - \overline{\overline{\mu}}/Z)(c^2 - s^2) + (\overline{\overline{\xi}} + \overline{\overline{\zeta}}) 2sc, \\ \overline{\overline{\xi}}_d + \overline{\overline{\zeta}}_d &= -(\overline{\overline{\epsilon}} Z - \overline{\overline{\mu}}/Z) 2sc + (\overline{\overline{\xi}} + \overline{\overline{\zeta}})(c^2 - s^2), \end{aligned}$$

which coincide with (6.61)–(6.64).

6.6 Inserting the Hehl–Obukhov decomposition in the reciprocity-transformation rule for the modified medium bidyadic (6.103) we obtain

$$\overline{\overline{\mathsf{M}}}_{mr} = \overline{\overline{\mathsf{M}}}{}_m^{*T} = \overline{\overline{\mathsf{M}}}{}_{m1}^{*T} + \overline{\overline{\mathsf{M}}}{}_{m2}^{*T} + \overline{\overline{\mathsf{M}}}{}_{m3}^{*T}.$$

The axion part $\overline{\overline{\mathsf{M}}}_3$ is transformed to

$$\begin{aligned} \overline{\overline{\mathsf{M}}}{}_{m3}^{*T} &= M_3(\mathbf{e}_N \lfloor \overline{\overline{\mathsf{I}}}{}^{(2)})^* = M_3 \mathbf{e}_N^* \lfloor \overline{\overline{\mathsf{I}}}{}^{(2)*} = -M_3 \mathbf{e}_N \lfloor \overline{\overline{\mathsf{I}}}{}^{(2)} \\ &= -\overline{\overline{\mathsf{M}}}_{m3}, \end{aligned}$$

whence the axion part is transformed to an axion bidyadic which equals the original bidyadic with a change in sign. The skewon part $\overline{\overline{\mathsf{M}}}_2 = (\overline{\overline{\mathsf{B}}}_o{}^\wedge_\wedge\overline{\overline{\mathsf{I}}})^T$ is transformed to

$$\overline{\overline{\mathsf{M}}}{}^{*T}_{m2} = (\mathbf{e}_N\lfloor(\overline{\overline{\mathsf{B}}}_o{}^\wedge_\wedge\overline{\overline{\mathsf{I}}})^T)^{*T} = -(\mathbf{e}_N\lfloor(\overline{\overline{\mathsf{B}}}_o{}^\wedge_\wedge\overline{\overline{\mathsf{I}}})^T)^* = -\mathbf{e}^*_N\lfloor(\overline{\overline{\mathsf{B}}}{}^*_o{}^\wedge_\wedge\overline{\overline{\mathsf{I}}}{}^*)^T$$
$$= \mathbf{e}_N\lfloor(\overline{\overline{\mathsf{B}}}{}^*_o{}^\wedge_\wedge\overline{\overline{\mathsf{I}}})^T,$$

which yields an antisymmetric bidyadic because of $\mathrm{tr}\overline{\overline{\mathsf{B}}}_{or} = \mathrm{tr}\overline{\overline{\mathsf{B}}}{}^*_o = (\mathrm{tr}\overline{\overline{\mathsf{B}}}_o)^* = 0$. The principal part $\overline{\overline{\mathsf{M}}}_1$ is transformed to

$$\overline{\overline{\mathsf{M}}}{}^{*T}_{m1} = \overline{\overline{\mathsf{M}}}{}^*_{m1},$$

whence the transformed principal bidyadic is symmetric. To test whether this is devoid of an axion component, let us consider the trace:

$$\mathrm{tr}(\boldsymbol{\varepsilon}_N\lfloor\overline{\overline{\mathsf{M}}}_{m1r}) = \mathrm{tr}(\boldsymbol{\varepsilon}_N\lfloor\overline{\overline{\mathsf{M}}}{}^{*T}_{m1}) = \overline{\overline{\mathsf{I}}}{}^{(2)}||(\boldsymbol{\varepsilon}_N\lfloor\overline{\overline{\mathsf{M}}}{}^{*T}_{m1}) = (\overline{\overline{\mathsf{I}}}{}^{(2)}||(\boldsymbol{\varepsilon}^*_N\lfloor\overline{\overline{\mathsf{M}}}{}^T_{m1})^*$$
$$= -(\overline{\overline{\mathsf{I}}}{}^{(2)}||(\boldsymbol{\varepsilon}_N\lfloor\overline{\overline{\mathsf{M}}}{}^T_{m1})^* = -\mathrm{tr}(\boldsymbol{\varepsilon}_N\lfloor\overline{\overline{\mathsf{M}}}_{m1})^*.$$

Since the last term is zero, so is the first one, whence there is no axion component in the transformed principal bidyadic.

In conclusion, axion, skewon, and principal parts of a medium bidyadic are transformed individually to axion, skewon, and principal parts of the transformed medium bidyadic.

6.7 Because of the identity

$$(\overline{\overline{\mathsf{C}}}{}^{(2)}|(\boldsymbol{\alpha}\wedge\boldsymbol{\beta}))\wedge(\overline{\overline{\mathsf{C}}}{}^{(2)}|(\boldsymbol{\gamma}\wedge\boldsymbol{\delta})) = (\overline{\overline{\mathsf{C}}}|\boldsymbol{\alpha})\wedge(\overline{\overline{\mathsf{C}}}|\boldsymbol{\beta})\wedge(\overline{\overline{\mathsf{C}}}|\boldsymbol{\gamma})\wedge(\overline{\overline{\mathsf{C}}}|\boldsymbol{\delta})$$
$$= \overline{\overline{\mathsf{C}}}{}^{(4)}|(\boldsymbol{\alpha}\wedge\boldsymbol{\beta}\wedge\boldsymbol{\gamma}\wedge\boldsymbol{\delta}),$$

the electromagnetic four-form $\boldsymbol{\Psi}\wedge\boldsymbol{\Phi}$ is transformed as

$$\boldsymbol{\Psi}'\wedge\boldsymbol{\Phi}' = (\overline{\overline{\mathsf{C}}}{}^{(2)}|\boldsymbol{\Psi})\wedge(\overline{\overline{\mathsf{C}}}{}^{(2)}|\boldsymbol{\Phi}) = \overline{\overline{\mathsf{C}}}{}^{(4)}|(\boldsymbol{\Psi}\wedge\boldsymbol{\Phi}) = \sigma^4\boldsymbol{\Psi}\wedge\boldsymbol{\Phi}.$$

To find the conformal transformation of the stress–energy dyadic we consider the transformed dyadic

$$\boldsymbol{\Psi}'\wedge\overline{\overline{\mathsf{I}}}{}^T\rfloor\rfloor\boldsymbol{\Phi}' = (\overline{\overline{\mathsf{C}}}{}^{(2)}|\boldsymbol{\Psi})\wedge(\overline{\overline{\mathsf{C}}}|\overline{\overline{\mathsf{C}}}{}^{-1})\rfloor\rfloor(\overline{\overline{\mathsf{C}}}{}^{(2)}|\boldsymbol{\Phi})$$
$$= \overline{\overline{\mathsf{C}}}{}^{(3)}|(\boldsymbol{\Psi}\wedge(\overline{\overline{\mathsf{C}}}{}^{-1}\rfloor\rfloor\overline{\overline{\mathsf{C}}}{}^{(2)})|\boldsymbol{\Phi}).$$

To proceed we can expand

$$\mathbf{a}|(\overline{\overline{\mathsf{C}}}{}^{-1}\rfloor\overline{\overline{\mathsf{C}}}{}^{(2)})|\boldsymbol{\Phi} = \overline{\overline{\mathsf{C}}}\wedge(\mathbf{a}|\overline{\overline{\mathsf{C}}}{}^{-1}|\overline{\overline{\mathsf{C}}})|\boldsymbol{\Phi} = (\overline{\overline{\mathsf{C}}}\wedge\mathbf{a})|\boldsymbol{\Phi}$$
$$= -\boldsymbol{\Phi}|(\mathbf{a}\wedge\lfloor\lfloor\overline{\overline{\mathsf{C}}}{}^T)$$
$$= \mathbf{a}|(\boldsymbol{\Phi}\lfloor\overline{\overline{\mathsf{C}}}{}^T) = \mathbf{a}|(\boldsymbol{\Phi}\lfloor\overline{\overline{\mathsf{I}}})|\overline{\overline{\mathsf{C}}}{}^T = \mathbf{a}|(\overline{\overline{\mathsf{I}}}{}^T\rfloor\boldsymbol{\Phi})|\overline{\overline{\mathsf{C}}}{}^T,$$

which is valid for any vector **a** and, thus, yields the rule

$$(\overline{\overline{\mathsf{C}}}^{-1}\rfloor\overline{\overline{\mathsf{C}}}^{(2)})|\mathbf{\Phi} = (\overline{\overline{\mathsf{I}}}^T\rfloor\mathbf{\Phi})|\overline{\overline{\mathsf{C}}}^T.$$

In terms of this, we obtain

$$\mathbf{\Psi}'\wedge\overline{\overline{\mathsf{I}}}^T\rfloor\mathbf{\Phi}' = \overline{\overline{\mathsf{C}}}^{(3)}|(\mathbf{\Psi}\wedge\overline{\overline{\mathsf{I}}}^T\rfloor\mathbf{\Phi})|\overline{\overline{\mathsf{C}}}^T,$$

which finally leads to the following transformation rule for the Maxwell stress–energy dyadic,

$$\overline{\overline{\mathsf{T}}}' = \frac{1}{2}(\mathbf{\Psi}'\wedge\overline{\overline{\mathsf{I}}}^T\rfloor\mathbf{\Phi}' - \mathbf{\Phi}'\wedge\overline{\overline{\mathsf{I}}}^T\rfloor\mathbf{\Psi}') = \overline{\overline{\mathsf{C}}}^{(3)}|\overline{\overline{\mathsf{T}}}|\overline{\overline{\mathsf{C}}}^T.$$

6.8 Considering the Hehl–Obukhov decomposition of a medium bidyadic as $\overline{\overline{\mathsf{M}}} = \overline{\overline{\mathsf{M}}}_1 + \overline{\overline{\mathsf{M}}}_2 + \overline{\overline{\mathsf{M}}}_3$, from (6.181) it follows that the axion part $\overline{\overline{\mathsf{M}}}_3 = M\overline{\overline{\mathsf{I}}}^{(2)T}$ remains invariant in the conformal transformation:

$$\overline{\overline{\mathsf{M}}}_3' = \overline{\overline{\mathsf{C}}}^{(2)}|\overline{\overline{\mathsf{M}}}_3|\overline{\overline{\mathsf{C}}}^{(-2)} = M\overline{\overline{\mathsf{C}}}^{(2)}|\overline{\overline{\mathsf{I}}}^{(2)T}|\overline{\overline{\mathsf{C}}}^{(-2)} = \overline{\overline{\mathsf{M}}}_3.$$

Because of the symmetry involved in the transformation (6.184), each of the three parts is transformed individually in the conformal transformation. Thus a medium with only the principal part $\overline{\overline{\mathsf{M}}} = \overline{\overline{\mathsf{M}}}_1$ is transformed to a medium whose medium bidyadic has only the principal part. The same applies to the skewon part.

To check the last statement, we consider the transformation of a medium bidyadic of the form $\overline{\overline{\mathsf{M}}}_2 = (\overline{\overline{\mathsf{B}}}_o{}^\wedge_\wedge\overline{\overline{\mathsf{I}}})^T$ with a trace-free dyadic $\overline{\overline{\mathsf{B}}}_o$. Applying (6.181) we can expand

$$\overline{\overline{\mathsf{M}}}_2' = \overline{\overline{\mathsf{C}}}^{(2)}|\overline{\overline{\mathsf{M}}}_2|\overline{\overline{\mathsf{C}}}^{(-2)} = \overline{\overline{\mathsf{C}}}^{(2)}|(\overline{\overline{\mathsf{B}}}_o^T{}^\wedge_\wedge\overline{\overline{\mathsf{I}}}^T)|\overline{\overline{\mathsf{C}}}^{(-2)}$$
$$= (\overline{\overline{\mathsf{C}}}|\overline{\overline{\mathsf{B}}}_o^T|\overline{\overline{\mathsf{C}}}^{-1}){}^\wedge_\wedge(\overline{\overline{\mathsf{C}}}|\overline{\overline{\mathsf{I}}}^T|\overline{\overline{\mathsf{C}}}^{-1}) = (\overline{\overline{\mathsf{B}}}_o'{}^\wedge_\wedge\overline{\overline{\mathsf{I}}})^T,$$

with the transformed dyadic defined by

$$\overline{\overline{\mathsf{B}}}_o' = (\overline{\overline{\mathsf{C}}}|\overline{\overline{\mathsf{B}}}_o^T|\overline{\overline{\mathsf{C}}}^{-1})^T = \overline{\overline{\mathsf{C}}}^{-1T}|\overline{\overline{\mathsf{B}}}_o|\overline{\overline{\mathsf{C}}}^T.$$

Because $\mathrm{tr}\overline{\overline{\mathsf{B}}}_o = 0$ implies $\mathrm{tr}\overline{\overline{\mathsf{B}}}_o' = 0$, the transformed bidyadic $\overline{\overline{\mathsf{M}}}_2'$ corresponds to a skewon medium.

6.9 From the relations between the medium dyadics and their duals

$$\overline{\overline{\epsilon}}_d Z_d = Z\overline{\overline{\epsilon}}\cos^2\theta + (\overline{\overline{\mu}}/Z)\sin^2\theta + (\overline{\overline{\xi}} + \overline{\overline{\zeta}})\sin\theta\cos\theta$$
$$\overline{\overline{\mu}}_d/Z_d = Z\overline{\overline{\epsilon}}\sin^2\theta + (\overline{\overline{\mu}}/Z)\cos^2\theta - (\overline{\overline{\xi}} + \overline{\overline{\zeta}})\sin\theta\cos\theta$$
$$\overline{\overline{\xi}}_d = -(\overline{\overline{\epsilon}}Z - \overline{\overline{\mu}}/Z)\sin\theta\cos\theta + \overline{\overline{\xi}}\cos^2\theta - \overline{\overline{\zeta}}\sin^2\theta$$
$$\overline{\overline{\zeta}}_d = -(\overline{\overline{\epsilon}}Z - \overline{\overline{\mu}}/Z)\sin\theta\cos\theta - \overline{\overline{\xi}}\sin^2\theta + \overline{\overline{\zeta}}\cos^2\theta,$$

the 4×4 matrix can be identified as

$$\mathcal{R}(\theta) = \begin{pmatrix} A & B \\ C & D \end{pmatrix}$$

in terms of four 2×2 matrices

$$A = \begin{pmatrix} c^2 & s^2 \\ s^2 & c^2 \end{pmatrix}, \; B = sc\begin{pmatrix} 1 & 1 \\ -1 & -1 \end{pmatrix} = -C^T, \; D = \begin{pmatrix} c^2 & -s^2 \\ -s^2 & c^2 \end{pmatrix},$$

where $c = \cos\theta$ and $s = \sin\theta$. Forming the matrix product

$$\mathcal{R}(\theta)\mathcal{R}^T(\theta) = \begin{pmatrix} AA^T + BB^T & AC^T + BD^T \\ CA^T + DB^T & CC^T + DD^T \end{pmatrix}$$

we have

$$AA^T + BB^T = \begin{pmatrix} c^4 + s^4 + 2s^2c^2 & 2s^2c^2 - 2s^2c^2 \\ 2s^2c^2 - 2s^2c^2 & c^4 + s^4 + 2s^2c^2 \end{pmatrix} = \begin{pmatrix} 1 & 0 \\ 0 & 1 \end{pmatrix}$$

$$AC^T + BD^T = CA^T + DB^T = \begin{pmatrix} 0 & 0 \\ 0 & 0 \end{pmatrix}$$

$$DD^T + CC^T = \begin{pmatrix} c^4 + s^4 + 2s^2c^2 & -2s^2c^2 + 2s^2c^2 \\ -2s^2c^2 + 2s^2c^2 & c^4 + s^4 + 2s^2c^2 \end{pmatrix} = \begin{pmatrix} 1 & 0 \\ 0 & 1 \end{pmatrix}.$$

Thus, $\mathcal{R}(\theta)\mathcal{R}^T(\theta)$ becomes a 4×4 unit matrix.

6.10 The transformation rule for the corresponding modified medium bidyadics becomes

$$\begin{aligned} \overline{\overline{\mathsf{M}}}'_m &= \mathbf{e}_N\lfloor\lfloor\overline{\overline{\mathsf{M}}}' = Y^2\overline{\overline{\mathsf{M}}}^{-1T}\rfloor\mathbf{e}_N = Y^2(\varepsilon_N\lfloor\overline{\overline{\mathsf{M}}}_m)^{-1T}\rfloor\mathbf{e}_N \\ &= Y^2\mathbf{e}_N\mathbf{e}_N\lfloor\lfloor\overline{\overline{\mathsf{M}}}_m^{-1T}. \end{aligned}$$

Making the transformation twice returns the original medium dyadic

$$\begin{aligned} \overline{\overline{\mathsf{M}}}''_m &= Y^2\mathbf{e}_N\mathbf{e}_N\lfloor\lfloor\overline{\overline{\mathsf{M}}}'^{-1T}_m = \mathbf{e}_N\mathbf{e}_N\lfloor\lfloor(\mathbf{e}_N\mathbf{e}_N\lfloor\lfloor\overline{\overline{\mathsf{M}}}_m^{-1T})^{-1T} = (\overline{\overline{\mathsf{M}}}_m^{-1T})^{-1T} \\ &= \overline{\overline{\mathsf{M}}}_m. \end{aligned}$$

From (5.88) we have the expansion

$$\overline{\overline{\mathsf{M}}}'_m = \overline{\overline{\epsilon}}'_{g}{}^{\wedge}_{\wedge}\mathbf{e}_4\mathbf{e}_4 - (\mathbf{e}_{123}\lfloor\overline{\overline{\mathsf{I}}}^T + \mathbf{e}_4 \wedge \overline{\overline{\xi}}'_g)|\overline{\overline{\mu}}'^{-1}_g|(\overline{\overline{\mathsf{I}}}\rfloor\mathbf{e}_{123} - \overline{\overline{\zeta}}'_g \wedge \mathbf{e}_4),$$

and, from (5.89),

$$\overline{\overline{\mathsf{M}}}_m^{-1} = -\boldsymbol{\varepsilon}_{123}\boldsymbol{\varepsilon}_{123}\lfloor\lfloor\overline{\overline{\mu}}_g - (\boldsymbol{\varepsilon}_4 \wedge \overline{\overline{\mathsf{I}}}^T - \boldsymbol{\varepsilon}_{123}\lfloor\overline{\overline{\zeta}}_g)|\overline{\overline{\epsilon}}_g^{-1}|(\overline{\overline{\mathsf{I}}} \wedge \boldsymbol{\varepsilon}_4 + \overline{\overline{\xi}}_g\rfloor\boldsymbol{\varepsilon}_{123}).$$

Operating the latter as

$$\mathbf{e}_N\mathbf{e}_N\lfloor\lfloor\overline{\overline{\mathsf{M}}}_m^{-1T} = \mathbf{e}_{123}\mathbf{e}_{123}\lfloor\lfloor\overline{\overline{\epsilon}}_g^{-1T} - \mathbf{e}_4 \wedge \overline{\overline{\xi}}_g^T|\overline{\overline{\epsilon}}_g^{-1T}\rfloor\mathbf{e}_{123} + \mathbf{e}_{123}\lfloor\overline{\overline{\epsilon}}_g^{-1T}|\overline{\overline{\zeta}}_g^T \wedge \mathbf{e}_4 - (\overline{\overline{\mu}}_g^T - \overline{\overline{\xi}}_g^T|\overline{\overline{\epsilon}}_g^{-1T}|\overline{\overline{\zeta}}_g^T)^\wedge_\wedge\mathbf{e}_4\mathbf{e}_4,$$

we obtain the relations

$$\overline{\overline{\mu}}'^{-1}_g = -Y^2\overline{\overline{\epsilon}}_g^{-1T},$$

$$\overline{\overline{\xi}}'_g|\overline{\overline{\mu}}'^{-1}_g = Y^2\overline{\overline{\xi}}_g^T|\overline{\overline{\epsilon}}_g^{-1T},$$

$$\overline{\overline{\mu}}'^{-1}_g|\overline{\overline{\zeta}}'_g = Y^2\overline{\overline{\epsilon}}_g^{-1T}|\overline{\overline{\zeta}}_g^T,$$

$$\overline{\overline{\epsilon}}'_g - \overline{\overline{\xi}}'_g|\overline{\overline{\mu}}'^{-1}_g|\overline{\overline{\zeta}}'_g = -Y^2(\overline{\overline{\mu}}_g^T - \overline{\overline{\xi}}_g^T|\overline{\overline{\epsilon}}_g^{-1T}|\overline{\overline{\zeta}}_g^T),$$

which lead to the following transformation rules:

$$\begin{pmatrix} \overline{\overline{\epsilon}}'_g & \overline{\overline{\xi}}'_g \\ \overline{\overline{\zeta}}'_g & \overline{\overline{\mu}}'_g \end{pmatrix} = \begin{pmatrix} -Y^2\overline{\overline{\mu}}_g^T & -\overline{\overline{\xi}}_g^T \\ -\overline{\overline{\zeta}}_g^T & -\overline{\overline{\epsilon}}_g^T/Y^2 \end{pmatrix}.$$

6.11 The transformation rule (6.59) for the modified medium bidyadic is

$$Z_d\overline{\overline{\mathsf{M}}}_{md} = (\overline{\overline{\mathsf{E}}} + \cot\theta\, Z\overline{\overline{\mathsf{M}}}_m)|(\cot\theta\,\overline{\overline{\mathsf{E}}} - Z\overline{\overline{\mathsf{M}}}_m)^{-1}\rfloor\mathbf{e}_N.$$

To be able to find an analytic expression for the inverse bidyadic, let us choose the transformation parameters so that

$$\cot\theta\,\overline{\overline{\mathsf{E}}} - Z(\alpha\overline{\overline{\mathsf{E}}} + \overline{\overline{\mathsf{Q}}}^{(2)}) = -Z\overline{\overline{\mathsf{Q}}}^{(2)},$$

that is, $\cot\theta = Z\alpha$, whence we obtain

$$\begin{aligned}
\overline{\overline{\mathsf{M}}}_{md} &= \frac{-1}{Z_dZ}(\overline{\overline{\mathsf{E}}} + Z^2\alpha(\alpha\overline{\overline{\mathsf{E}}} + \overline{\overline{\mathsf{Q}}}^{(2)}))|\overline{\overline{\mathsf{Q}}}^{(-2)} \\
&= \frac{-1}{Z_dZ}((1 + Z^2\alpha^2)\overline{\overline{\mathsf{E}}} + Z^2\alpha\overline{\overline{\mathsf{Q}}}^{(2)})|\overline{\overline{\mathsf{Q}}}^{(-2)}\rfloor\mathbf{e}_N \\
&= \frac{-1}{Z_dZ}((1 + Z^2\alpha^2)\mathbf{e}_N\mathbf{e}_N\lfloor\lfloor\overline{\overline{\mathsf{Q}}}^{(-2)} + Z^2\alpha\overline{\overline{\mathsf{E}}}) \\
&= \frac{-1}{Z_dZ}(Z^2\alpha\overline{\overline{\mathsf{E}}} + (1 + Z^2\alpha^2)\overline{\overline{\mathsf{Q}}}^{(2T)}/\Delta_Q).
\end{aligned}$$

From this we can identify

$$\alpha_d = -\frac{Z}{Z_d}\alpha, \quad \overline{\overline{\mathsf{Q}}}_d = \sqrt{\frac{-(1+Z^2\alpha^2)}{Z_d Z \Delta_Q}}\overline{\overline{\mathsf{Q}}}^T.$$

For the special choice $Z_d = -Z = \pm 1/\sqrt{\Delta_Q - \alpha^2}$ we obtain

$$\overline{\overline{\mathsf{M}}}_d = \alpha\overline{\overline{\mathsf{E}}} + \overline{\overline{\mathsf{Q}}}^{(2)T}.$$

6.12 Equation (6.60) is of the general form

$$A\overline{\overline{\mathsf{M}}}_d|\overline{\overline{\mathsf{M}}} + B\overline{\overline{\mathsf{M}}}_d + C\overline{\overline{\mathsf{M}}} + D\overline{\overline{\mathsf{I}}}^{(2)T} = 0.$$

Multiplying this equation as $\overline{\overline{\mathsf{M}}}|()|\overline{\overline{\mathsf{M}}}^{-1}$ and subtracting the original equation leaves us with a condition which can be written as

$$(\overline{\overline{\mathsf{M}}}|\overline{\overline{\mathsf{M}}}_d - \overline{\overline{\mathsf{M}}}_d\overline{\overline{\mathsf{M}}})|(A\overline{\overline{\mathsf{M}}} + B\overline{\overline{\mathsf{I}}}^{(2)T}) = 0.$$

The bidyadics $\overline{\overline{\mathsf{M}}}, \overline{\overline{\mathsf{M}}}_d$ commute in the general case, that is, when the bidyadic

$$A\overline{\overline{\mathsf{M}}} + B\overline{\overline{\mathsf{I}}}^{(2)T} = Z_d \sin\theta(-Z\sin\theta\overline{\overline{\mathsf{M}}} + \cos\theta\overline{\overline{\mathsf{I}}}^{(2)T})$$

has an inverse.

6.13 Starting from the PEMC condition (6.75) and applying (6.39) yields a condition of the form

$$\boldsymbol{\varepsilon}_3 \wedge (A\boldsymbol{\Psi}_d - B\boldsymbol{\Phi}_d) = 0,$$

with

$$A = \frac{Z_d}{Z}(\cos\theta - MZ\sin\theta), \quad B = \frac{1}{Z}(\sin\theta + MZ\cos\theta).$$

To be self-dual, the duality transformation must depend on the PEMC parameter M as $B = MA$. This gives us a lot of freedom to choose. For the case $\sin\theta = 0$, $Z_d = -Z$ we obtain $A = MZ$ and $B = 1/Z$, whence we must have $Z = 1/M$.

CHAPTER 7

7.1 Starting from $\overline{\overline{\mathsf{M}}} = \epsilon\overline{\overline{\mathsf{C}}}_s \wedge \mathbf{e}_4 + \frac{1}{\mu}\boldsymbol{\varepsilon}_4 \wedge \overline{\overline{\mathsf{C}}}_s^{-1}$ and inserting $\overline{\overline{\mathsf{C}}}_s = \mathbf{e}_{123}\lfloor\overline{\overline{\mathsf{B}}}_s$, $\overline{\overline{\mathsf{C}}}_s^{-1} = \overline{\overline{\mathsf{B}}}_s^{-1}\rfloor\mathbf{e}_{123}$, we obtain

$$\overline{\overline{\mathsf{M}}}\lfloor\lfloor\overline{\overline{\mathsf{I}}} = -\sum_i \mathbf{e}_i\rfloor\overline{\overline{\mathsf{M}}}\lfloor\boldsymbol{\varepsilon}_i$$
$$= -\epsilon\mathbf{e}_i\rfloor(\boldsymbol{\varepsilon}_{123}\lfloor\overline{\overline{\mathsf{B}}}_s \wedge \mathbf{e}_4)\lfloor\boldsymbol{\varepsilon}_i - \frac{1}{\mu}\mathbf{e}_i\rfloor(\boldsymbol{\varepsilon}_4 \wedge \overline{\overline{\mathsf{B}}}_s^{-1}\rfloor\mathbf{e}_{123})\lfloor\boldsymbol{\varepsilon}_i$$
$$= \epsilon\boldsymbol{\varepsilon}_{123}\lfloor(\mathbf{e}_i \wedge \overline{\overline{\mathsf{B}}}_s \wedge \mathbf{e}_4)\lfloor\boldsymbol{\varepsilon}_i + \frac{1}{\mu}\mathbf{e}_i\rfloor(\boldsymbol{\varepsilon}_4 \wedge \overline{\overline{\mathsf{B}}}_s^{-1} \wedge \boldsymbol{\varepsilon}_i)\rfloor\mathbf{e}_{123},$$

when omitting the summing sign. Applying the bac-cab rules

$$(\overline{\overline{\mathsf{B}}}_s \wedge \mathbf{e}_4)\lfloor\boldsymbol{\varepsilon}_i = (\overline{\overline{\mathsf{B}}}_s|\boldsymbol{\varepsilon}_i)\mathbf{e}_4 - \overline{\overline{\mathsf{B}}}_s(\mathbf{e}_4|\boldsymbol{\varepsilon}_i)$$
$$\mathbf{e}_i\rfloor(\boldsymbol{\varepsilon}_4 \wedge \overline{\overline{\mathsf{B}}}_s^{-1}) = \boldsymbol{\varepsilon}_4(\mathbf{e}_i|\overline{\overline{\mathsf{B}}}_s^{-1}) - (\mathbf{e}_i|\boldsymbol{\varepsilon}_4)\overline{\overline{\mathsf{B}}}_s^{-1}$$

we can proceed as

$$\overline{\overline{\mathsf{M}}}\lfloor\lfloor\overline{\overline{\mathsf{I}}} = \epsilon\boldsymbol{\varepsilon}_{123}\lfloor((\mathbf{e}_i \wedge \overline{\overline{\mathsf{B}}}_s|\boldsymbol{\varepsilon}_i)\mathbf{e}_4 - \mathbf{e}_4 \wedge \overline{\overline{\mathsf{B}}}_s)$$
$$+ \frac{1}{\mu}(\boldsymbol{\varepsilon}_4(\mathbf{e}_i|\overline{\overline{\mathsf{B}}}_s^{-1} \wedge \boldsymbol{\varepsilon}_i) - \overline{\overline{\mathsf{B}}}_s^{-1} \wedge \boldsymbol{\varepsilon}_4)\rfloor\mathbf{e}_{123}$$
$$= \epsilon\boldsymbol{\varepsilon}_{123}\lfloor((\mathbf{e}_i \wedge \overline{\overline{\mathsf{B}}}_s|\boldsymbol{\varepsilon}_i)\mathbf{e}_4) + \frac{1}{\mu}(\boldsymbol{\varepsilon}_4(\mathbf{e}_i|\overline{\overline{\mathsf{B}}}_s^{-1} \wedge \boldsymbol{\varepsilon}_i))\rfloor\mathbf{e}_{123}.$$

Requiring $\overline{\overline{\mathsf{M}}}\lfloor\lfloor\overline{\overline{\mathsf{I}}} = 0$, the two terms must vanish separately. Expanding $\overline{\overline{\mathsf{B}}}_s = \sum_{j,k} B_{j,k}\mathbf{e}_j\mathbf{e}_k$, $\overline{\overline{\mathsf{B}}}_s^{-1} = \sum_{j,k} \beta_{j,k}\boldsymbol{\varepsilon}_j\boldsymbol{\varepsilon}_k$ we have

$$\sum_i \mathbf{e}_i \wedge \overline{\overline{\mathsf{B}}}_s|\boldsymbol{\varepsilon}_i = \sum_{i,j} B_{j,i}\mathbf{e}_{ij} = 0, \quad \sum_i \mathbf{e}_i|\overline{\overline{\mathsf{B}}}_s^{-1} \wedge \boldsymbol{\varepsilon}_i = \sum_{i,k} \beta_{i,k}\boldsymbol{\varepsilon}_{ki} = 0,$$

from which it follows that $B_{j,i} = 0$ for $i \neq j$ and $\beta_{i,k} = 0$ for $i \neq k$. Thus, both $\overline{\overline{\mathsf{B}}}_s$ and $\overline{\overline{\mathsf{B}}}_s^{-1}$ must be symmetric spatial dyadics.

7.2 Inserting the expression for the inverse of the symmetric spatial metric dyadic $\overline{\overline{\mathsf{G}}}_s = \mathbf{e}_1\mathbf{e}_1 + \mathbf{e}_2\mathbf{e}_2 + \mathbf{e}_3\mathbf{e}_3$,

$$\overline{\overline{\mathsf{G}}}_s^{-1} = \frac{\boldsymbol{\varepsilon}_{123}\boldsymbol{\varepsilon}_{123}\lfloor\lfloor\overline{\overline{\mathsf{G}}}^{(2)T}}{\boldsymbol{\varepsilon}_{123}\boldsymbol{\varepsilon}_{123}||\overline{\overline{\mathsf{G}}}^{(3)}} = \boldsymbol{\varepsilon}_{123}\boldsymbol{\varepsilon}_{123}\lfloor\lfloor\overline{\overline{\mathsf{G}}}^{(2)}$$

the last term of (7.17) becomes

$$\boldsymbol{\varepsilon}_4 \wedge (\mu\overline{\overline{\mathsf{G}}}_s)^{-1}\rfloor\mathbf{e}_{123} = \frac{1}{\mu}\boldsymbol{\varepsilon}_4 \wedge (\boldsymbol{\varepsilon}_{123}\boldsymbol{\varepsilon}_{123}\lfloor\lfloor\overline{\overline{\mathsf{G}}}^{(2)})\rfloor\mathbf{e}_{123} = -\frac{1}{\mu}\boldsymbol{\varepsilon}_N\lfloor\overline{\overline{\mathsf{G}}}_s^{(2)}.$$

The last two terms of (7.17) can be combined as

$$\begin{aligned}
&\epsilon'\boldsymbol{\varepsilon}_{123}\lfloor\overline{\overline{\mathsf{G}}}_s\wedge\mathbf{e}_4+\boldsymbol{\varepsilon}_4\wedge(\mu\overline{\overline{\mathsf{G}}}_s)^{-1}\rfloor\mathbf{e}_{123}\\
&=\epsilon'\boldsymbol{\varepsilon}_{123}\lfloor\overline{\overline{\mathsf{G}}}_s\wedge\mathbf{e}_4-\frac{1}{\mu}\boldsymbol{\varepsilon}_N\lfloor\overline{\overline{\mathsf{G}}}_s^{(2)}\\
&=-\frac{1}{\mu}\boldsymbol{\varepsilon}_N\lfloor(\mu\epsilon'\mathbf{e}_4\wedge\overline{\overline{\mathsf{G}}}_s\wedge\mathbf{e}_4+\overline{\overline{\mathsf{G}}}_s^{(2)})\\
&=-\frac{1}{\mu}\boldsymbol{\varepsilon}_N\lfloor(-\mu\epsilon'\overline{\overline{\mathsf{G}}}_s{}^\wedge_\wedge\mathbf{e}_4\mathbf{e}_4+\overline{\overline{\mathsf{G}}}_s^{(2)})\\
&=-\frac{1}{\mu}\boldsymbol{\varepsilon}_N\lfloor(-\mu\epsilon'\mathbf{e}_4\mathbf{e}_4+\overline{\overline{\mathsf{G}}}_s)^{(2)}.
\end{aligned}$$

Defining $\overline{\overline{\mathsf{G}}}=\overline{\overline{\mathsf{G}}}_s-\mu\epsilon'\mathbf{e}_4\mathbf{e}_4$ and $\overline{\overline{\mathsf{T}}}=\overline{\overline{\mathsf{I}}}_s-\mathbf{e}_4\boldsymbol{\varepsilon}_4$, (7.17) reduces to

$$\overline{\overline{\mathsf{M}}}=A\overline{\overline{\mathsf{I}}}^{(2)T}+B\overline{\overline{\mathsf{T}}}^{(2)T}-\frac{1}{\mu}\boldsymbol{\varepsilon}_N\lfloor\overline{\overline{\mathsf{G}}}^{(2)}.$$

7.3 The condition (7.33) is clear from Problem 7.2. To derive (7.34), let us expand

$$\begin{aligned}
\boldsymbol{\varepsilon}_N\lfloor\overline{\overline{\mathsf{G}}}^{(2)}&=\boldsymbol{\varepsilon}_N\lfloor(\overline{\overline{\mathsf{G}}}_s^{(2)}+\mu\epsilon'\mathbf{e}_4\wedge\overline{\overline{\mathsf{G}}}_s\wedge\mathbf{e}_4)\\
&=-\boldsymbol{\varepsilon}_4\wedge(\boldsymbol{\varepsilon}_{123}\lfloor\overline{\overline{\mathsf{G}}}_s^{(2)})-\mu\epsilon'\boldsymbol{\varepsilon}_{123}\lfloor\overline{\overline{\mathsf{G}}}_s\wedge\mathbf{e}_4,
\end{aligned}$$

and substitute it in

$$\begin{aligned}
\overline{\overline{\mathsf{T}}}^{(2)T}|(\boldsymbol{\varepsilon}_N\lfloor\overline{\overline{\mathsf{G}}}^{(2)})&=(\overline{\overline{\mathsf{I}}}_s^{(2)T}+\boldsymbol{\varepsilon}_4\wedge\overline{\overline{\mathsf{I}}}_s^T\wedge\mathbf{e}_4)|(\boldsymbol{\varepsilon}_N\lfloor\overline{\overline{\mathsf{G}}}^{(2)})\\
&=-\mu\epsilon'\boldsymbol{\varepsilon}_{123}\lfloor\overline{\overline{\mathsf{G}}}_s\wedge\mathbf{e}_4+\boldsymbol{\varepsilon}_4\wedge(\boldsymbol{\varepsilon}_{123}\lfloor\overline{\overline{\mathsf{G}}}_s^{(2)}),\\
(\boldsymbol{\varepsilon}_N\lfloor\overline{\overline{\mathsf{G}}}^{(2)})|\overline{\overline{\mathsf{T}}}^{(2)T}&=(\boldsymbol{\varepsilon}_N\lfloor\overline{\overline{\mathsf{G}}}^{(2)})|(\overline{\overline{\mathsf{I}}}_s^{(2)T}+\boldsymbol{\varepsilon}_4\wedge\overline{\overline{\mathsf{I}}}_s^T\wedge\mathbf{e}_4)\\
&=-\boldsymbol{\varepsilon}_4\wedge(\boldsymbol{\varepsilon}_{123}\lfloor\overline{\overline{\mathsf{G}}}_s^{(2)})+\mu\epsilon'\boldsymbol{\varepsilon}_{123}\lfloor\overline{\overline{\mathsf{G}}}_s\wedge\mathbf{e}_4.
\end{aligned}$$

Comparing these two expansions we obtain

$$\overline{\overline{\mathsf{T}}}^{(2)T}|(\boldsymbol{\varepsilon}_N\lfloor\overline{\overline{\mathsf{G}}}^{(2)})=-(\boldsymbol{\varepsilon}_N\lfloor\overline{\overline{\mathsf{G}}}^{(2)})|\overline{\overline{\mathsf{T}}}^{(2)T},$$

whence the condition (7.31) is verified.

7.4

$$\begin{aligned}
\overline{\overline{\mathsf{T}}}^{(2)}&=(\overline{\overline{\mathsf{I}}}_s-\mathbf{e}_4\boldsymbol{\varepsilon}_4)^{(2)}=(\overline{\overline{\mathsf{I}}}-2\mathbf{e}_4\boldsymbol{\varepsilon}_4)^{(2)}\\
&=\overline{\overline{\mathsf{I}}}^{(2)}-2\mathbf{e}_4\boldsymbol{\varepsilon}_4{}^\wedge_\wedge\overline{\overline{\mathsf{I}}}=\frac{1}{2}(\overline{\overline{\mathsf{I}}}-4\mathbf{e}_4\boldsymbol{\varepsilon}_4){}^\wedge_\wedge\overline{\overline{\mathsf{I}}}\\
&=\overline{\overline{\mathsf{B}}}_o{}^\wedge_\wedge\overline{\overline{\mathsf{I}}}.
\end{aligned}$$

The dyadic $\overline{\overline{\mathsf{B}}}_o=\frac{1}{2}(\overline{\overline{\mathsf{I}}}-4\mathbf{e}_4\boldsymbol{\varepsilon}_4)$ satisfies $\mathrm{tr}\overline{\overline{\mathsf{B}}}_o=0$.

7.5 The eigenfields satisfy

$$\overline{\overline{\mathsf{M}}}|\boldsymbol{\Phi}_\pm = \left(A\overline{\overline{\mathsf{I}}}^{(2)T} + B\overline{\overline{\mathsf{T}}}^{(2)T} + \epsilon'\overline{\overline{\mathsf{C}}}_s \wedge \mathbf{e}_4 + \frac{1}{\mu}\boldsymbol{\varepsilon}_4 \wedge \overline{\overline{\mathsf{C}}}_s^{-1}\right)$$
$$|(\mathbf{B}_\pm + \mathbf{E}_\pm \wedge \boldsymbol{\varepsilon}_4)$$
$$= (A+B)\mathbf{B}_\pm + (A-T)\mathbf{E}_\pm \wedge \boldsymbol{\varepsilon}_4 + \epsilon'\overline{\overline{\mathsf{C}}}_s|\mathbf{E}_\pm$$
$$+ \frac{1}{\mu}\boldsymbol{\varepsilon}_4 \wedge \overline{\overline{\mathsf{C}}}_s^{-1}|\mathbf{B}_\pm$$
$$= M_\pm(\mathbf{B}_\pm + \mathbf{E}_\pm \wedge \boldsymbol{\varepsilon}_4).$$

Separating the spatial and temporal parts we obtain the relations

$$(A+B)\mathbf{B}_\pm + \epsilon'\overline{\overline{\mathsf{C}}}_s|\mathbf{E}_\pm = M_\pm\mathbf{B}_\pm,$$
$$(A-B)\mathbf{E}_\pm - \frac{1}{\mu}\overline{\overline{\mathsf{C}}}_s^{-1})|\mathbf{B}_\pm = M_\pm\mathbf{E}_\pm,$$

or

$$\mathbf{B}_\pm = \frac{\epsilon'}{M_\pm - A - B}\overline{\overline{\mathsf{C}}}_s|\mathbf{E}_\pm,$$
$$\mathbf{E}_\pm = \frac{1}{\mu(M_\pm - A + B)}\overline{\overline{\mathsf{C}}}_s^{-1}|\mathbf{B}_\pm.$$

These are the same equations if $M_\pm = A \pm \sqrt{B^2 - \epsilon'/\mu}$. Doing the same for the medium bidyadics $\overline{\overline{\mathsf{M}}}_\pm$ we obtain,

$$\mathbf{B}_\pm = \frac{\epsilon_\pm}{M_\pm}\overline{\overline{\mathsf{C}}}_s|\mathbf{E}_\pm,$$
$$\mathbf{E}_\pm = \frac{1}{\mu_\pm M_\pm}\overline{\overline{\mathsf{C}}}_s^{-1}|\mathbf{B}_\pm,$$

which imply $M_\pm^2 = \epsilon_\pm/\mu_\pm$. Comparing the eigenfield relations of the two cases we can solve

$$\epsilon_\pm = \epsilon'\frac{M_\pm}{M_\pm - A - B} = \epsilon'\frac{A \pm \sqrt{B^2 - \epsilon'/\mu}}{-B \pm \sqrt{B^2 - \epsilon'/\mu}},$$
$$\mu_\pm = \mu\frac{M_\pm - A + B}{M_\pm} = \mu\frac{B \pm \sqrt{B^2 - \epsilon'/\mu}}{A \pm \sqrt{B^2 - \epsilon'/\mu}}.$$

As a simple check, for $A = B = 0$ we have $\epsilon_\pm = \epsilon'$ and $\mu_\pm = \mu$.

7.6 Since $\mathbf{A}_s\lfloor\lfloor\overline{\overline{\mathsf{I}}}^T \in \mathbb{E}_1\mathbb{E}_1$ is antisymmetric and $\overline{\overline{\mathsf{G}}}^{(2)}\lfloor\lfloor\boldsymbol{\nu}\boldsymbol{\nu} \in \mathbb{E}_1\mathbb{E}_1$ is symmetric, the antisymmetric part of the dispersion equation (7.61) is

$$(\mathbf{A}_s\lfloor\overline{\overline{\mathsf{I}}}^T)^{\wedge}_{\wedge}(\overline{\overline{\mathsf{G}}}^{(2)}\lfloor\lfloor\boldsymbol{\nu}\boldsymbol{\nu})^{(2)} = 0.$$

Denoting the symmetric dyadic $\overline{\overline{\mathsf{S}}} = \overline{\overline{\mathsf{G}}}^{(2)}\lfloor\lfloor\boldsymbol{\nu}\boldsymbol{\nu} \in \mathbb{E}_1\mathbb{E}_1$, we can apply the identity of the Problem 2.22,

$$\boldsymbol{\varepsilon}_N\boldsymbol{\varepsilon}_N\lfloor\lfloor((\mathbf{A}\lfloor\overline{\overline{\mathsf{I}}}^T)^{\wedge}_{\wedge}\overline{\overline{\mathsf{S}}}^{(2)}) = \boldsymbol{\Phi}\lfloor\overline{\overline{\mathsf{I}}},$$

with $\boldsymbol{\Phi} = -\mathbf{A}|(\boldsymbol{\varepsilon}_N\boldsymbol{\varepsilon}_N\lfloor\lfloor\overline{\overline{\mathsf{S}}}^{(2)}) \in \mathbb{F}_2$. Thus, equivalent to the above antisymmetric equation is vanishing of the two-form

$$\begin{aligned}\boldsymbol{\Phi} &= -\mathbf{A}|(\boldsymbol{\varepsilon}_N\boldsymbol{\varepsilon}_N\lfloor\lfloor\overline{\overline{\mathsf{S}}}^{(2)}) = -\mathbf{A}|(\boldsymbol{\varepsilon}_N\lfloor\overline{\overline{\mathsf{S}}}^{(2)})\rfloor\boldsymbol{\varepsilon}_N\\ &= -(\boldsymbol{\varepsilon}_N\lfloor\mathbf{A})|\overline{\overline{\mathsf{S}}}^{(2)}\rfloor\boldsymbol{\varepsilon}_N = -\alpha(\boldsymbol{\nu}\wedge\boldsymbol{\varepsilon}_4)|\overline{\overline{\mathsf{S}}}^{(2)}\rfloor\boldsymbol{\varepsilon}_N.\end{aligned}$$

Here we can apply the rule

$$(\boldsymbol{\alpha}\wedge\boldsymbol{\beta})|\overline{\overline{\mathsf{S}}}^{(2)} = (\boldsymbol{\alpha}|\overline{\overline{\mathsf{S}}})\wedge(\boldsymbol{\beta}|\overline{\overline{\mathsf{S}}})$$

and continue as

$$\begin{aligned}\boldsymbol{\Phi} &= -\alpha(\boldsymbol{\nu}|\overline{\overline{\mathsf{S}}})\wedge(\boldsymbol{\varepsilon}_4|\overline{\overline{\mathsf{S}}})\rfloor\boldsymbol{\varepsilon}_N\\ &= -\alpha(\boldsymbol{\nu}|(\overline{\overline{\mathsf{G}}}^{(2)}\lfloor\lfloor\boldsymbol{\nu}\boldsymbol{\nu}))\wedge(\boldsymbol{\varepsilon}_4|(\overline{\overline{\mathsf{G}}}^{(2)}\lfloor\lfloor\boldsymbol{\nu}\boldsymbol{\nu}))\rfloor\boldsymbol{\varepsilon}_N.\end{aligned}$$

But this vanishes identically, since $\boldsymbol{\nu}|(\overline{\overline{\mathsf{G}}}^{(2)}\lfloor\lfloor\boldsymbol{\nu}\boldsymbol{\nu}) = -(\boldsymbol{\nu}\wedge\boldsymbol{\nu})|\overline{\overline{\mathsf{G}}}^{(2)}\lfloor\boldsymbol{\nu} = 0$ for all $\boldsymbol{\nu}$. Thus, the antisymmetric part of the dispersion equation vanishes identically.

7.7 The symmetric part of the dispersion equation (7.61) is

$$\mathbf{A}_s\mathbf{A}_s{}^{\wedge}_{\wedge}(\overline{\overline{\mathsf{G}}}^{(2)}\lfloor\lfloor\boldsymbol{\nu}\boldsymbol{\nu}) + (\overline{\overline{\mathsf{G}}}^{(2)}\lfloor\lfloor\boldsymbol{\nu}\boldsymbol{\nu})^{(3)} = 0.$$

Making use of (5.262) for $p = 4$ and $\overline{\overline{\mathsf{G}}}^{(4)} = -\mu\epsilon'\mathbf{e}_N\mathbf{e}_N$, we can expand

$$\begin{aligned}(\overline{\overline{\mathsf{G}}}^{(2)}\lfloor\lfloor\boldsymbol{\nu}\boldsymbol{\nu})^{(3)} &= ((\overline{\overline{\mathsf{G}}}||\boldsymbol{\nu}\boldsymbol{\nu})\overline{\overline{\mathsf{G}}} - (\overline{\overline{\mathsf{G}}}|\boldsymbol{\nu})(\boldsymbol{\nu}|\overline{\overline{\mathsf{G}}}))^{(3)}\\ &= (\overline{\overline{\mathsf{G}}}||\boldsymbol{\nu}\boldsymbol{\nu})^3\overline{\overline{\mathsf{G}}}^{(3)} - (\overline{\overline{\mathsf{G}}}||\boldsymbol{\nu}\boldsymbol{\nu})^2\overline{\overline{\mathsf{G}}}^{(2)}{}^{\wedge}_{\wedge}(\overline{\overline{\mathsf{G}}}|\boldsymbol{\nu})(\boldsymbol{\nu}|\overline{\overline{\mathsf{G}}})\\ &= (\overline{\overline{\mathsf{G}}}||\boldsymbol{\nu}\boldsymbol{\nu})^2(\overline{\overline{\mathsf{G}}}^{(4)}\lfloor\lfloor\boldsymbol{\nu}\boldsymbol{\nu}) = -\mu\epsilon'(\overline{\overline{\mathsf{G}}}||\boldsymbol{\nu}\boldsymbol{\nu})^2\mathbf{e}_N\mathbf{e}_N\lfloor\lfloor\boldsymbol{\nu}\boldsymbol{\nu}.\end{aligned}$$

Writing $\overline{\overline{\mathsf{G}}}^{(2)} = \sum_i \mathbf{B}_i\mathbf{B}_i$, we can expand

$$\mathbf{A}_s\mathbf{A}_s{}^{\wedge}_{\wedge}(\overline{\overline{\mathsf{G}}}^{(2)}\lfloor\lfloor\boldsymbol{\nu}\boldsymbol{\nu}) = \sum_i(\mathbf{A}_s\wedge(\mathbf{B}_i\lfloor\boldsymbol{\nu}))(\mathbf{A}_s\wedge(\mathbf{B}_i\lfloor\boldsymbol{\nu})).$$

Applying the identity

$$\mathbf{A}\wedge(\mathbf{B}\rfloor\alpha)+\mathbf{B}\wedge(\mathbf{A}\rfloor\alpha)=(\mathbf{A}\wedge\mathbf{B})\rfloor\alpha=(\mathbf{A}\cdot\mathbf{B})\mathbf{e}_N\rfloor\alpha$$

valid for any bivectors $\mathbf{A},\mathbf{B}$ and one-form $\boldsymbol{\alpha}$, we have

$$\mathbf{A}_s\wedge(\mathbf{B}_i\rfloor\boldsymbol{\nu})=(\mathbf{A}_s\cdot\mathbf{B}_i)\boldsymbol{\varepsilon}_N\rfloor\boldsymbol{\nu},$$

because of $\mathbf{A}_s\rfloor\boldsymbol{\nu}=0$. Thus, we can write

$$\mathbf{A}_s\mathbf{A}_s{}^{\wedge}_{\wedge}(\overline{\overline{\mathsf{G}}}^{(2)}\rfloor\rfloor\boldsymbol{\nu}\boldsymbol{\nu})=\mathbf{A}_s\cdot\overline{\overline{\mathsf{G}}}^{(2)}\cdot\mathbf{A}_s(\mathbf{e}_N\mathbf{e}_N\rfloor\rfloor\boldsymbol{\nu}\boldsymbol{\nu})$$

and the dyadic dispersion equation is equivalent to the scalar equation

$$\mu^3D(\boldsymbol{\nu})=\mathbf{A}_s\cdot\overline{\overline{\mathsf{G}}}^{(2)}\cdot\mathbf{A}_s-\mu\epsilon'(\overline{\overline{\mathsf{G}}}||\boldsymbol{\nu}\boldsymbol{\nu})^2=0.$$

Expanding the first term as

$$\begin{aligned}
\mathbf{A}_s\cdot\overline{\overline{\mathsf{G}}}^{(2)}\cdot\mathbf{A}_s &= \boldsymbol{\varepsilon}_N\boldsymbol{\varepsilon}_N||(\mathbf{A}_s\mathbf{A}_s)^{\wedge}_{\wedge}\overline{\overline{\mathsf{G}}}^{(2)})=(\boldsymbol{\varepsilon}_N\rfloor\mathbf{A}_s)(\boldsymbol{\varepsilon}_N\rfloor\mathbf{A}_s)||\overline{\overline{\mathsf{G}}}^{(2)}\\
&= \alpha^2(\boldsymbol{\nu}\boldsymbol{\nu}^{\wedge}_{\wedge}\boldsymbol{\varepsilon}_4\boldsymbol{\varepsilon}_4)||\overline{\overline{\mathsf{G}}}^{(2)}\\
&= \alpha^2(\boldsymbol{\nu}\boldsymbol{\nu}^{\wedge}_{\wedge}\boldsymbol{\varepsilon}_4\boldsymbol{\varepsilon}_4)||(\overline{\overline{\mathsf{G}}}_s^{(2)}-\mu\epsilon'\overline{\overline{\mathsf{G}}}_s{}^{\wedge}_{\wedge}\mathbf{e}_4\mathbf{e}_4)\\
&= -\mu\epsilon'(2\mu B\nu_4)^2(\overline{\overline{\mathsf{G}}}_s||\boldsymbol{\nu}\boldsymbol{\nu}),
\end{aligned}$$

where we denote $\nu_4=\mathbf{e}_4|\boldsymbol{\nu}$. Inserting $\overline{\overline{\mathsf{G}}}||\boldsymbol{\nu}\boldsymbol{\nu}=\overline{\overline{\mathsf{G}}}_s||\boldsymbol{\nu}\boldsymbol{\nu}-\mu\epsilon'\nu_4^2$, the dispersion equation becomes

$$\begin{aligned}
\mu^3D(\boldsymbol{\nu}) &= -\mu\epsilon'(2\mu B\nu_4)^2(\overline{\overline{\mathsf{G}}}_s||\boldsymbol{\nu}\boldsymbol{\nu})-\mu\epsilon'(\overline{\overline{\mathsf{G}}}||\boldsymbol{\nu}\boldsymbol{\nu})^2\\
&= -\mu\epsilon'((2\mu B\nu_4)^2(\overline{\overline{\mathsf{G}}}_s||\boldsymbol{\nu}\boldsymbol{\nu})+(\overline{\overline{\mathsf{G}}}_s||\boldsymbol{\nu}\boldsymbol{\nu})^2)\\
&\quad -2\mu\epsilon'(\overline{\overline{\mathsf{G}}}_s||\boldsymbol{\nu}\boldsymbol{\nu})\nu_4^2+(\mu\epsilon')^2\nu_4^4))\\
&= -\mu\epsilon'((\overline{\overline{\mathsf{G}}}_s||\boldsymbol{\nu}\boldsymbol{\nu})^2+((2\mu B)^2-2\mu\epsilon')(\overline{\overline{\mathsf{G}}}_s||\boldsymbol{\nu}\boldsymbol{\nu})\nu_4^2\\
&\quad +(\mu\epsilon')^2\nu_4^4)\\
&= -\mu\epsilon'((\overline{\overline{\mathsf{G}}}_s||\boldsymbol{\nu}\boldsymbol{\nu})^2-2(\mu\epsilon'-2(\mu B)^2)(\overline{\overline{\mathsf{G}}}_s||\boldsymbol{\nu}\boldsymbol{\nu})\nu_4^2\\
&\quad +(\mu\epsilon')^2\nu_4^4)\\
&= 0.
\end{aligned}$$

The last expression can be given the factorized form

$$\mu^3D(\boldsymbol{\nu})=\mu\epsilon'D_+(\boldsymbol{\nu})D_-(\boldsymbol{\nu}),\quad D_\pm(\boldsymbol{\nu})=\overline{\overline{\mathsf{G}}}_s||\boldsymbol{\nu}\boldsymbol{\nu}-\mu_\pm\epsilon'_\pm\nu_4^2$$

by solving the second-order equation for $x = (\overline{\overline{\mathsf{G}}}_s||\mathbf{v}\mathbf{v})/v_4^2 = \mu_\pm \epsilon'_\pm$,

$$x^2 - 2(\mu\epsilon' - 2(\mu B)^2)x + (\mu\epsilon')^2 = 0.$$

The solutions can be expressed as

$$\begin{aligned} x &= \mu\epsilon' - 2(\mu B)^2 \pm 2\mu B\sqrt{(\mu B)^2 - \mu\epsilon'} \\ &= -(\mu B \pm \sqrt{(\mu B)^2 - \mu\epsilon'})^2. \end{aligned}$$

7.8 From the expression (7.20) we obtain $(\overline{\overline{\mathsf{T}}}^{(2)T})^2 = \overline{\overline{\mathsf{I}}}^{(2)T}$, whence $\overline{\overline{\mathsf{T}}}^{(2)T} = \overline{\overline{\mathsf{U}}}$. Also, from

$$(\boldsymbol{\varepsilon}_N \lfloor \overline{\overline{\mathsf{G}}}^{(2)})^2 = (\boldsymbol{\varepsilon}_N \boldsymbol{\varepsilon}_N \lfloor\lfloor \overline{\overline{\mathsf{G}}}^{(2)})|\overline{\overline{\mathsf{G}}}^{(2)} = \Delta_G \overline{\overline{\mathsf{G}}}^{(-2)T}|\overline{\overline{\mathsf{G}}}^{(2)} = -\mu\epsilon' \overline{\overline{\mathsf{I}}}^{(2)T}$$

we have $-\boldsymbol{\varepsilon}_N \lfloor \overline{\overline{\mathsf{G}}}^{(2)}/\mu = C\overline{\overline{\mathsf{J}}}$ with $C = \sqrt{\epsilon'/\mu}$. Studying the transformed medium bidyadic $\overline{\overline{\mathsf{M}}}_D = \overline{\overline{\mathsf{D}}}|\overline{\overline{\mathsf{M}}}|\overline{\overline{\mathsf{D}}}^{-1}$ termwise, we have

$$\begin{aligned} (\overline{\overline{\mathsf{D}}}|\overline{\overline{\mathsf{U}}}|\overline{\overline{\mathsf{D}}}^{-1})^2 &= \overline{\overline{\mathsf{D}}}|\overline{\overline{\mathsf{U}}}^2|\overline{\overline{\mathsf{D}}}^{-1} = \overline{\overline{\mathsf{I}}}^{(2)T}, \\ (\overline{\overline{\mathsf{D}}}|\overline{\overline{\mathsf{J}}}|\overline{\overline{\mathsf{D}}}^{-1})^2 &= \overline{\overline{\mathsf{D}}}|\overline{\overline{\mathsf{J}}}^2|\overline{\overline{\mathsf{D}}}^{-1} = -\overline{\overline{\mathsf{I}}}^{(2)T}, \end{aligned}$$

whence

$$\overline{\overline{\mathsf{M}}}_D = A\overline{\overline{\mathsf{I}}}^{(2)T} + B\overline{\overline{\mathsf{U}}}_D + C\overline{\overline{\mathsf{J}}}_D,$$

where $\overline{\overline{\mathsf{U}}}_D$ is a unipotent bidyadic and $\overline{\overline{\mathsf{J}}}_D$ is an AC bidyadic. However, the medium so defined is not necessarily a GBI medium. In the special case when $\overline{\overline{\mathsf{D}}}$ is of the form $\overline{\overline{\mathsf{A}}}^{(2)}$ the similarity transformation is an affine transformation and a GDI medium is transformed to a GBI medium.

7.9 Substituting $\overline{\overline{\mathsf{M}}}_d = \overline{\overline{\mathsf{M}}}$ in the duality transformation rule (6.59) requires that the self-dual medium bidyadic must satisfy the second-order equation

$$\overline{\overline{\mathsf{M}}}^2 + \cot\theta \frac{Z - Z_d}{ZZ_d}\overline{\overline{\mathsf{M}}} + \frac{1}{ZZ_d}\overline{\overline{\mathsf{I}}}^{(2)} = 0.$$

If parameters $Z - Z_d$ and θ can be defined so that this equation equals (7.31) satisfied by the general GBI medium bidyadic, the medium is self-dual. Thus, the following conditions for the transformation parameters are obtained:

$$\cot\theta\frac{Z - Z_d}{ZZ_d} = -2A, \quad \frac{1}{ZZ_d} = A^2 - B^2 + \frac{\epsilon'}{\mu}.$$

Since there are two equations for three parameters, it is obvious that a solution can be found by adding a third condition. Considering the axion medium, $\overline{\overline{\mathsf{M}}} = A\overline{\overline{\mathsf{I}}}^{(2)T}$, we can assume the involutory transformation $\theta = \pi/2, Z_d = Z$ in which case the condition becomes $\overline{\overline{\mathsf{M}}}^2 = \overline{\overline{\mathsf{I}}}^{(2)T}/Z^2$, which is satisfied by choosing $Z = 1/A$ or $Z = -1/A$.

7.10 For the simple skewon–axion medium the conditions are

$$\mathbf{D} = (A+B)\mathbf{B} = \alpha\mathbf{B}, \quad \mathbf{H} = (-A+B)\mathbf{E} = -\beta\mathbf{E}.$$

The Maxwell equations outside sources have the form

$$\mathbf{d}_s \wedge \mathbf{B} = 0, \quad \mathbf{d}_s \wedge \mathbf{E} + \partial_\tau \mathbf{B} = 0,$$

$$\mathbf{d}_s \wedge \mathbf{D} = 0, \quad \mathbf{d}_s \wedge \mathbf{H} - \partial_\tau \mathbf{D} = 0,$$

the latter of which in the skewon–axion medium become

$$\alpha\mathbf{d}_s \wedge \mathbf{B} = 0, \quad -\beta\mathbf{d}_s \wedge \mathbf{E} - \alpha\partial_\tau \mathbf{B} = 0.$$

For $B \neq 0$ we have $\alpha \neq \beta$ whence we must have $\mathbf{d} \wedge \mathbf{E} = 0$ and $\partial_\tau \mathbf{B} = 0$. Assuming $\mathbf{B} = 0$ an initial value, we have $\mathbf{B} = 0$ and $\mathbf{D} = 0$ for all τ while $\mathbf{H} = -\beta\mathbf{E} = \mathbf{d}_s\varphi$ where φ may be any scalar potential function. Because at an interface $\boldsymbol{\varepsilon}_3|\mathbf{x} = 0$ we have $\boldsymbol{\varepsilon}_3 \wedge \mathbf{D} = 0$ and $\boldsymbol{\varepsilon}_3 \wedge \mathbf{B} = 0$, the DB conditions are satisfied at the interface of an SS medium.

7.11 The dyadic rule (7.122) can be obtained by writing $\overline{\overline{\mathsf{B}}}_g = \sum \mathbf{r}_i\mathbf{s}_i$ and expanding

$$\begin{aligned}
\mathbf{p} \times \mathbf{s}_i\mathbf{r}_i + \mathbf{r}_i\mathbf{s}_i \times \mathbf{p} &= (\mathbf{p} \times \mathbf{s}_i)\mathbf{r}_i - \mathbf{r}_i(\mathbf{p} \times \mathbf{s}_i) \\
&= (\mathbf{r}_i \times (\mathbf{p} \times \mathbf{s}_i)) \times \overline{\overline{\mathsf{I}}}_g \\
&= (\mathbf{p}(\mathbf{r}_i \cdot \mathbf{s}_i) - (\mathbf{p} \cdot \mathbf{r}_i)\mathbf{s}_i) \times \overline{\overline{\mathsf{I}}}_g, \\
&= (\mathbf{p}\mathrm{tr}\overline{\overline{\mathsf{B}}}_g - \mathbf{p} \cdot \overline{\overline{\mathsf{B}}}_g) \times \overline{\overline{\mathsf{I}}}_g.
\end{aligned}$$

7.12 Inserting the expressions (7.110)–(7.113), the medium equations can be written as

$$\begin{aligned}
\mathbf{D} &= \overline{\overline{\alpha}}|\mathbf{B} + \overline{\overline{\epsilon}}'|\mathbf{E} \\
&= \mathrm{tr}\overline{\overline{\mathsf{B}}}_s\mathbf{B} - (\boldsymbol{\varepsilon}_{123}\lfloor\overline{\overline{\mathsf{B}}}_s\rfloor\mathbf{e}_{123})|\mathbf{B} - \boldsymbol{\beta}_s \wedge \mathbf{E}, \\
\mathbf{H} &= \overline{\overline{\mu}}^{-1}|\mathbf{B} + \overline{\overline{\beta}}|\mathbf{E} \\
&= -\mathbf{b}_s\rfloor\mathbf{B} - (\overline{\overline{\mathsf{B}}}_s + b\overline{\overline{\mathsf{I}}}_s^T)|\mathbf{E}.
\end{aligned}$$

Applying the rules of Appendix B, we obtain

$$\begin{aligned}\mathbf{D}_g &= \mathbf{e}_{123}\lfloor\mathbf{D} = \mathrm{tr}\overline{\overline{\mathsf{B}}}_s\mathbf{B}_g - \overline{\overline{\mathsf{B}}}_s|\mathbf{B}_g - \mathbf{e}_{123}\lfloor(\boldsymbol{\beta}_s\wedge\mathbf{E}),\\ &= \mathrm{tr}\overline{\overline{\mathsf{B}}}_s\mathbf{B}_g - \overline{\overline{\mathsf{B}}}_g\cdot\mathbf{B}_g - \boldsymbol{\beta}_g\times\mathbf{E}_g,\\ \mathbf{H}_g &= \overline{\overline{\mathsf{G}}}_s|\mathbf{H} = -\overline{\overline{\mathsf{G}}}_s|(\mathbf{b}_s\rfloor\mathbf{B}) - (\overline{\overline{\mathsf{B}}}{}_g^T + b\overline{\overline{\mathsf{G}}}_s)\cdot\mathbf{E}_g.\\ &= -\mathbf{b}_s\times\mathbf{B}_g - \overline{\overline{\mathsf{B}}}{}_g^T\cdot\mathbf{E}_g - b\mathbf{E}_g\end{aligned}$$

with $\overline{\overline{\mathsf{B}}}_g = \overline{\overline{\mathsf{B}}}_s|\overline{\overline{\mathsf{G}}}_s$. Here we have applied

$$\begin{aligned}\overline{\overline{\mathsf{G}}}_s|(\mathbf{b}_s\rfloor\mathbf{B}) &= \overline{\overline{\mathsf{G}}}_s|(\mathbf{b}_s\rfloor(\boldsymbol{\varepsilon}_{123}\lfloor\mathbf{B}_g)) = -\overline{\overline{\mathsf{G}}}_s|(\boldsymbol{\varepsilon}_{123}\lfloor(\mathbf{B}_g\wedge\mathbf{b}_s))\\ &= -\boldsymbol{\beta}_g\times\mathbf{b}_s.\end{aligned}$$

7.13 The dispersion dyadic can be written as

$$\overline{\overline{\mathsf{D}}}(\boldsymbol{\nu}) = \mathbf{F}(\boldsymbol{\nu})\lfloor\overline{\overline{\mathsf{I}}}{}^T, \quad \mathbf{F} = \mathbf{e}_N\lfloor(\boldsymbol{\alpha}\wedge\boldsymbol{\beta}).$$

$\mathbf{F}$ is a simple bivector, which can be expressed as $\mathbf{F} = \mathbf{a}\wedge\mathbf{b}$, and $\overline{\overline{\mathsf{D}}}(\boldsymbol{\nu})$ is an antisymmetric dyadic. Thus, the dyadic dispersion equation

$$\overline{\overline{\mathsf{D}}}{}^{(3)}(\boldsymbol{\nu}) = ((\mathbf{a}\wedge\mathbf{b})\lfloor\overline{\overline{\mathsf{I}}}{}^T)^{(3)} = (\mathbf{ba}-\mathbf{ab})^{(3)} = 0$$

is valid for any $\boldsymbol{\nu}$. From the antisymmetry of $\overline{\overline{\mathsf{D}}}$ we have

$$\overline{\overline{\mathsf{D}}}{}^T(\boldsymbol{\nu}) = (\overline{\overline{\mathsf{M}}}_m\lfloor\lfloor\boldsymbol{\nu}\boldsymbol{\nu})^T = \overline{\overline{\mathsf{M}}}{}_m^T\lfloor\lfloor\boldsymbol{\nu}\boldsymbol{\nu} = -\overline{\overline{\mathsf{M}}}_m\lfloor\lfloor\boldsymbol{\nu}\boldsymbol{\nu},$$

or

$$(\overline{\overline{\mathsf{M}}}{}_m^T + \overline{\overline{\mathsf{M}}}_m)\lfloor\lfloor\boldsymbol{\nu}\boldsymbol{\nu} = 0,$$

which must be valid for any $\boldsymbol{\nu}$. From this it follows that the bidyadic in brackets must be of the form $\alpha\overline{\overline{\mathsf{E}}}$ and, thus, the symmetric part of $\overline{\overline{\mathsf{M}}}_m$ must be a multiple of $\overline{\overline{\mathsf{E}}}$, which rules out the principal part of $\overline{\overline{\mathsf{M}}}$.

7.14 Defining $\boldsymbol{\alpha} = A\boldsymbol{\varepsilon}_4 + \boldsymbol{\varepsilon}_1$ in (7.126) and (7.127), the conditions (7.130) for the fields in the skewon–axion medium become

$$\begin{aligned}A\boldsymbol{\varepsilon}_4\wedge\mathbf{B} + \boldsymbol{\varepsilon}_1\wedge\mathbf{E}\wedge\boldsymbol{\varepsilon}_4 &= 0\\ A\boldsymbol{\varepsilon}_4\wedge\mathbf{D} - \boldsymbol{\varepsilon}_1\wedge\mathbf{H}\wedge\boldsymbol{\varepsilon}_4 &= 0,\end{aligned}$$

which equal

$$\begin{aligned}A\mathbf{B} + \boldsymbol{\varepsilon}_1\wedge\mathbf{E} &= 0,\\ A\mathbf{D} - \boldsymbol{\varepsilon}_1\wedge\mathbf{H} &= 0.\end{aligned}$$

Multiplying these by $\boldsymbol{\varepsilon}_3\wedge$ we arrive at the SHDB conditions (7.153) and (7.154) and, in Gibbsian form, at (7.155) and (7.156).

7.15 Inserting $\overline{\overline{\mathsf{M}}}_o = \overline{\overline{\mathsf{M}}} - \frac{1}{6}\mathrm{tr}\overline{\overline{\mathsf{M}}}\,\overline{\overline{\mathsf{I}}}^{(2)T}$ in the eigenrule we obtain

$$-\left((\overline{\overline{\mathsf{M}}}\lfloor\lfloor\overline{\overline{\mathsf{I}}})^{\wedge}_{\wedge}\overline{\overline{\mathsf{I}}}^T - \frac{1}{6}\mathrm{tr}\overline{\overline{\mathsf{M}}}(\overline{\overline{\mathsf{I}}}^{(2)T}\lfloor\lfloor\overline{\overline{\mathsf{I}}})^{\wedge}_{\wedge}\overline{\overline{\mathsf{I}}}^T\right) + \overline{\overline{\mathsf{M}}} - \frac{1}{6}\mathrm{tr}\overline{\overline{\mathsf{M}}}\,\overline{\overline{\mathsf{I}}}^{(2)T}$$
$$= -\left(\overline{\overline{\mathsf{M}}} - \frac{1}{6}\mathrm{tr}\overline{\overline{\mathsf{M}}}\,\overline{\overline{\mathsf{I}}}^{(2)T}\right)$$

Combining terms we arrive at the condition (7.95).

7.16 The inverse bidyadic has the form

$$\overline{\overline{\mathsf{M}}}^{-1} = (\overline{\overline{\mathsf{A}}}_o{}^{\wedge}_{\wedge}\overline{\overline{\mathsf{I}}})^T - \frac{(\overline{\overline{\mathsf{A}}}_o{}^{\wedge}_{\wedge}\overline{\overline{\mathsf{I}}})^T|\boldsymbol{\Delta}\mathbf{B}|(\overline{\overline{\mathsf{A}}}_o{}^{\wedge}_{\wedge}\overline{\overline{\mathsf{I}}})^T}{1 + \mathbf{B}|(\overline{\overline{\mathsf{A}}}_o{}^{\wedge}_{\wedge}\overline{\overline{\mathsf{I}}})|\boldsymbol{\Delta}},$$

where $\overline{\overline{\mathsf{A}}}_o$ is the dyadic defined by (3.127). The result can be verified by multiplication.

7.17 From (5.102)–(5.105) the conditions for the skewon-medium dyadics are obtained,

$$\overline{\overline{\alpha}} = \boldsymbol{\varepsilon}_{123}\mathbf{e}_{123}\lfloor\lfloor\overline{\overline{\beta}}^T,$$
$$\overline{\overline{\epsilon}}' = -\boldsymbol{\varepsilon}_{123}\mathbf{e}_{123}\lfloor\lfloor\overline{\overline{\epsilon}}'^T,$$
$$\overline{\overline{\mu}}^{-1} = -\boldsymbol{\varepsilon}_{123}\mathbf{e}_{123}\lfloor\lfloor\overline{\overline{\mu}}^{-1T},$$
$$\overline{\overline{\beta}} = \boldsymbol{\varepsilon}_{123}\mathbf{e}_{123}\lfloor\lfloor\overline{\overline{\alpha}}^T.$$

Applying the relations to the Gibbsian dyadics (5.83)–(5.86) as

$$\overline{\overline{\alpha}} = \boldsymbol{\varepsilon}_{123}\lfloor(\overline{\overline{\xi}}_g|\overline{\overline{\mu}}_g^{-1})\rfloor\mathbf{e}_{123},$$
$$\overline{\overline{\epsilon}}' = \boldsymbol{\varepsilon}_{123}\lfloor(\overline{\overline{\epsilon}}_g - \overline{\overline{\xi}}_g|\overline{\overline{\mu}}_g^{-1}|\overline{\overline{\zeta}}_g),$$
$$\overline{\overline{\mu}}^{-1} = \boldsymbol{\varepsilon}_{123}\lfloor\overline{\overline{\mu}}_g^{-1},$$
$$\overline{\overline{\beta}} = -\overline{\overline{\mu}}_g^{-1}|\overline{\overline{\zeta}}_g,$$

the skewon-medium conditions become

$$\overline{\overline{\xi}}_g|\overline{\overline{\mu}}_g^{-1} = -(\overline{\overline{\mu}}_g^{-1}|\overline{\overline{\zeta}}_g)^T,$$
$$\overline{\overline{\epsilon}}_g - \overline{\overline{\xi}}_g|\overline{\overline{\mu}}_g^{-1}|\overline{\overline{\zeta}}_g = -(\overline{\overline{\epsilon}}_g - \overline{\overline{\xi}}_g|\overline{\overline{\mu}}_g^{-1}|\overline{\overline{\zeta}}_g)^T,$$
$$\overline{\overline{\mu}}_g^{-1} = -\overline{\overline{\mu}}_g^{-1T},$$
$$\overline{\overline{\mu}}_g^{-1}|\overline{\overline{\zeta}}_g = -(\overline{\overline{\xi}}_g|\overline{\overline{\mu}}_g^{-1})^T.$$

From these we finally obtain the simple conditions

$$\overline{\overline{\mu}}_g^{-1T} = -\overline{\overline{\mu}}_g^{-1}, \quad \overline{\overline{\xi}}_g^T = \overline{\overline{\zeta}}_g, \quad \overline{\overline{\epsilon}}_g^T = -\overline{\overline{\epsilon}}_g^T.$$

Since both $\overline{\overline{\mu}}_g^{-1}$ and $\overline{\overline{\epsilon}}_g$ of a skewon medium are antisymmetric Gibbsian dyadics, they do not have inverses.

7.18 Applying the relations (5.102)–(5.105), the medium bidyadic of the special principal–axion medium can be written as

$$\overline{\overline{\mathsf{M}}} = \overline{\overline{\alpha}} + \boldsymbol{\varepsilon}_4 \wedge \overline{\overline{\beta}} \wedge \mathbf{e}_4 = -\boldsymbol{\varepsilon}_{123}\lfloor\overline{\overline{\beta}}^T\rfloor\mathbf{e}_{123} + \boldsymbol{\varepsilon}_4 \wedge \overline{\overline{\beta}} \wedge \mathbf{e}_4,$$

and the corresponding modified medium bidyadic as

$$\begin{aligned}\overline{\overline{\mathsf{M}}}_m &= -\mathbf{e}_N\lfloor(\boldsymbol{\varepsilon}_{123}\lfloor\overline{\overline{\beta}}^T\rfloor\mathbf{e}_{123}) + \mathbf{e}_N\lfloor(\boldsymbol{\varepsilon}_4 \wedge \overline{\overline{\beta}} \wedge \mathbf{e}_4)\\ &= \mathbf{e}_4 \wedge \overline{\overline{\beta}}^T\rfloor\mathbf{e}_{123} - \mathbf{e}_{123}\lfloor\overline{\overline{\beta}} \wedge \mathbf{e}_4\\ &= -(\mathbf{e}_{123}\lfloor\overline{\overline{\beta}} \wedge \mathbf{e}_4)^T - \mathbf{e}_{123}\lfloor\overline{\overline{\beta}} \wedge \mathbf{e}_4,\end{aligned}$$

which is symmetric. The dispersion dyadic becomes

$$\begin{aligned}\overline{\overline{\mathsf{D}}}(\boldsymbol{\nu}) &= \overline{\overline{\mathsf{M}}}_m\lfloor\lfloor\boldsymbol{\nu}\boldsymbol{\nu} = (\mathbf{e}_4 \wedge \overline{\overline{\beta}}^T\rfloor\mathbf{e}_{123})\lfloor\lfloor\boldsymbol{\nu}\boldsymbol{\nu} - (\mathbf{e}_{123}\lfloor\overline{\overline{\beta}} \wedge \mathbf{e}_4)\lfloor\lfloor\boldsymbol{\nu}\boldsymbol{\nu}\\ &= -\boldsymbol{\nu}\rfloor(\mathbf{e}_4 \wedge \overline{\overline{\beta}}^T\rfloor\mathbf{e}_{123})\lfloor\boldsymbol{\nu} + \boldsymbol{\nu}\rfloor(\mathbf{e}_{123}\lfloor\overline{\overline{\beta}} \wedge \mathbf{e}_4)\lfloor\boldsymbol{\nu}\\ &= -((\mathbf{e}_4(\boldsymbol{\nu}|\overline{\overline{\beta}}^T) - (\boldsymbol{\nu}|\mathbf{e}_4)\overline{\overline{\beta}}^T)\rfloor\mathbf{e}_{123})\lfloor\boldsymbol{\nu}\\ &\quad + \boldsymbol{\nu}\rfloor(\mathbf{e}_{123}\lfloor((\overline{\overline{\beta}}|\boldsymbol{\nu})\mathbf{e}_4 - (\boldsymbol{\nu}|\mathbf{e}_4)\overline{\overline{\beta}})),\\ &= \overline{\overline{\mathsf{A}}}_s + \overline{\overline{\mathsf{A}}}_s^T + \mathbf{a}_s\mathbf{e}_4 + \mathbf{e}_4\mathbf{a}_s,\end{aligned}$$

where the spatial dyadic $\overline{\overline{\mathsf{A}}}_s \in \mathbb{E}_1\mathbb{E}_1$ and the spatial vector $\mathbf{a}_s$ are defined by $\overline{\overline{\mathsf{A}}}_s = -(\mathbf{e}_4|\boldsymbol{\nu})(\overline{\overline{\beta}}^T \wedge \boldsymbol{\nu})\rfloor\mathbf{e}_{123}$ and $\mathbf{a}_s = \mathbf{e}_{123}\lfloor(\boldsymbol{\nu} \wedge \overline{\overline{\beta}}|\boldsymbol{\nu})$. The dyadic dispersion equation becomes

$$\begin{aligned}\overline{\overline{\mathsf{D}}}^{(3)}(\boldsymbol{\nu}) &= (\overline{\overline{\mathsf{A}}}_a + \overline{\overline{\mathsf{A}}}_s^T)^{(3)} + (\overline{\overline{\mathsf{A}}}_a + \overline{\overline{\mathsf{A}}}_s^T)^{(2)}{}^\wedge_\wedge(\mathbf{a}_s\mathbf{e}_4 + \mathbf{e}_4\mathbf{a}_s)\\ &\quad -(\overline{\overline{\mathsf{A}}}_a + \overline{\overline{\mathsf{A}}}_s^T){}^\wedge_\wedge(\mathbf{a}_s\mathbf{a}_s{}^\wedge_\wedge\mathbf{e}_4\mathbf{e}_4) = 0.\end{aligned}$$

This is equivalent to four spatial equations as

$$\begin{aligned}(\overline{\overline{\mathsf{A}}}_s + \overline{\overline{\mathsf{A}}}_s^T)^{(3)} &= 0,\\ (\overline{\overline{\mathsf{A}}}_a + \overline{\overline{\mathsf{A}}}_s^T)^{(2)} \wedge \mathbf{a}_s &= 0,\\ \mathbf{a}_s \wedge (\overline{\overline{\mathsf{A}}}_s + \overline{\overline{\mathsf{A}}}_s^T)^{(2)} &= 0,\\ (\overline{\overline{\mathsf{A}}}_s + \overline{\overline{\mathsf{A}}}_s^T){}^\wedge_\wedge\mathbf{a}_s\mathbf{a}_s &= 0.\end{aligned}$$

From the last one we conclude that the dyadic $\overline{\overline{\mathsf{A}}}_s$ must satisfy

$$\overline{\overline{\mathsf{A}}}_s + \overline{\overline{\mathsf{A}}}_s^T = \mathbf{a}_s\mathbf{b}_s + \mathbf{b}_s\mathbf{a}_s.$$

for some vector $\mathbf{b}_s$. It turns out that the other three equations are then satisfied automatically. Thus, the last equation carries all the information of the dispersion equation. For $\mathbf{e}_4|\boldsymbol{\nu} = \nu_4 \neq 0$ it can be expressed as

$$\left[\mathbf{e}_{123}\lfloor(\boldsymbol{\nu}_s \wedge \overline{\overline{\beta}}) - (\overline{\overline{\beta}}^T \wedge \boldsymbol{\nu}_s)\rfloor\mathbf{e}_{123}\right]^{\wedge}_{\wedge}(\mathbf{e}_{123}\lfloor(\boldsymbol{\nu}_s \wedge \overline{\overline{\beta}}|\boldsymbol{\nu}_s))$$
$$(\mathbf{e}_{123}\lfloor(\boldsymbol{\nu}_s \wedge \overline{\overline{\beta}}|\boldsymbol{\nu}_s) = 0,$$

where $\boldsymbol{\nu}_s$ is the spatial part of $\boldsymbol{\nu}$. After a few applications of the bac-cab rule the dyadic dispersion equation can be reduced to

$$(\mathbf{e}_{123}\mathbf{e}_{123}\lfloor\lfloor\boldsymbol{\nu}_s\boldsymbol{\nu}_s)\mathbf{e}_{123}\lfloor(\boldsymbol{\nu}_s \wedge (\overline{\overline{\beta}}|\boldsymbol{\nu}_s) \wedge (\overline{\overline{\beta}}^2|\boldsymbol{\nu}_s)) = 0,$$

or, for $\boldsymbol{\nu}_s \neq 0$ to

$$\boldsymbol{\nu}_s \wedge (\overline{\overline{\beta}}|\boldsymbol{\nu}_s) \wedge (\overline{\overline{\beta}}^2|\boldsymbol{\nu}_s) = 0.$$

It appears that, for a given spatial one-form $\boldsymbol{\varepsilon}_s$, there is no solution $\boldsymbol{\nu}_s = \nu\boldsymbol{\varepsilon}_s$ because $\boldsymbol{\varepsilon}_s$, $\overline{\overline{\beta}}|\boldsymbol{\varepsilon}_s$, and $\overline{\overline{\beta}}^2|\boldsymbol{\varepsilon}_s$ are not linearly dependent in general. Because, for such a $\boldsymbol{\nu}$, $\overline{\overline{\mathsf{D}}}^{-1}(\boldsymbol{\nu})$ exists, the fields of the plane wave must vanish. However, when $\boldsymbol{\nu}$ is an eigen one-form of $\overline{\overline{\beta}}$, or a linear combination of two eigen one-forms, the dispersion equation is satisfied.

7.19 The medium conditions for the Gibbsian fields become

$$\mathbf{D}_g = \mathbf{e}_{123}\lfloor\mathbf{D} = \mathbf{e}_{123}\lfloor\overline{\overline{\alpha}}|(\boldsymbol{\varepsilon}_{123}\lfloor\mathbf{B}_g)$$
$$= -\overline{\overline{\beta}}^T|\mathbf{B}_g = -\overline{\overline{\beta}}_g^T \cdot \mathbf{B}_g,$$
$$\mathbf{H}_g = \overline{\overline{\mathsf{G}}}_s|\mathbf{H} = \overline{\overline{\mathsf{G}}}_s|\overline{\overline{\beta}}|\mathbf{E} = \overline{\overline{\beta}}_g \cdot \mathbf{E}_g.$$

Here we define

$$\overline{\overline{\beta}}_g = \overline{\overline{\mathsf{G}}}_s|\overline{\overline{\beta}}, \quad \overline{\overline{\mathsf{G}}}_s = \mathbf{e}_1\mathbf{e}_1 + \mathbf{e}_2\mathbf{e}_2 + \mathbf{e}_3\mathbf{e}_3.$$

The equations for a time-harmonic plane wave are now

$$\mathbf{p} \times \mathbf{E}_g = \mathbf{B}_g, \quad \mathbf{p} \times \mathbf{H}_g = -\mathbf{D}_g, \quad \mathbf{p} = \mathbf{k}/\omega.$$

Applying the medium conditions we can expand

$$\mathbf{p} \times \mathbf{H}_g = \mathbf{p} \times \overline{\overline{\beta}}_g \cdot \mathbf{E}_g = -\mathbf{D}_g = \overline{\overline{\beta}}_g^T \cdot \mathbf{B}_g = (\overline{\overline{\beta}}_g^T \times \mathbf{p}) \cdot \mathbf{E}_g,$$

whence the field satisfies

$$\overline{\overline{\mathsf{B}}}_g(\mathbf{p})\cdot\mathbf{E}_g = 0, \quad \overline{\overline{\mathsf{B}}}_g = \mathbf{p}\times\overline{\overline{\beta}}_g - \overline{\overline{\beta}}{}_g^T\times\mathbf{p} = \mathbf{p}\times\overline{\overline{\beta}}_g + (\mathbf{p}\times\overline{\overline{\beta}}_g)^T.$$

Applying

$$(\mathbf{p}\times\overline{\overline{\beta}}_g)^{(2)} = (\mathbf{p}\times\overline{\overline{\mathsf{I}}})^{(2)}\cdot\overline{\overline{\beta}}{}_g^{(2)} = \mathbf{pp}\cdot\overline{\overline{\beta}}{}_g^{(2)} = \det\overline{\overline{\beta}}_g\,\overline{\overline{\beta}}{}_g^{-1T},$$

where we assume that $\overline{\overline{\beta}}_g$ is of full rank, we can expand

$$\begin{aligned}
\det\overline{\overline{\mathsf{B}}}_g &= \frac{1}{6}\overline{\overline{\beta}}_g{}^\times_\times\overline{\overline{\beta}}_g : \overline{\overline{\beta}}_g \\
&= -(\mathbf{p}\times\overline{\overline{\beta}}_g)^{(2)} : (\overline{\overline{\beta}}{}_g^T\times\mathbf{p}) + (\mathbf{p}\times\overline{\overline{\beta}}_g) : (\overline{\overline{\beta}}{}_g^T\times\mathbf{p})^{(2)} \\
&= -2(\mathbf{p}\times\overline{\overline{\beta}}_g)^{(2)} : (\overline{\overline{\beta}}{}_g^T\times\mathbf{p}) = -2(\mathbf{pp}\cdot\overline{\overline{\beta}}{}_g^{(2)}) : (\overline{\overline{\beta}}{}_g^T\times\mathbf{p}) \\
&= = 2\det\overline{\overline{\beta}}_g\,\mathbf{p}\cdot\overline{\overline{\beta}}{}_g^{-1T}\cdot(\mathbf{p}\times\overline{\overline{\beta}}_g\cdot\mathbf{p}) \\
&= 2\det\overline{\overline{\beta}}_g\,(\overline{\overline{\beta}}{}_g^{-1}\cdot\mathbf{p})\times\mathbf{p}\cdot(\overline{\overline{\beta}}_g\cdot\mathbf{p}).
\end{aligned}$$

The determinant vanishes only when the three vectors $\mathbf{p}$, $\overline{\overline{\beta}}_g\cdot\mathbf{p}$, and $\overline{\overline{\beta}}{}_g^{-1}\cdot\mathbf{p}$ are linearly dependent. Since for $\mathbf{p} = p\mathbf{u}$ where $\mathbf{u}$ is a unit vector, the determinant of the dyadic $\overline{\overline{\mathsf{B}}}_g$ does not vanish in general, we must have $\mathbf{E}_g = 0$ which implies vanishing of all field components.

7.20 Applying $\mathbf{b}|\boldsymbol{\phi} = 0$ in the dispersion equation (7.178) we obtain $\boldsymbol{\nu}\wedge(\boldsymbol{\nu}|\overline{\overline{\mathsf{B}}})\wedge\boldsymbol{\phi} = 0$, whence the potential must be of the form

$$\boldsymbol{\phi} = \alpha\boldsymbol{\nu} + \beta\boldsymbol{\nu}|\overline{\overline{\mathsf{B}}}.$$

Thus, we must have

$$\mathbf{b}|\boldsymbol{\phi} = \alpha\mathbf{b}|\boldsymbol{\nu} + \beta\boldsymbol{\nu}|\overline{\overline{\mathsf{B}}}|\mathbf{b} = \beta\boldsymbol{\nu}|(\overline{\overline{\mathsf{B}}}\lfloor\mathbf{B})|\boldsymbol{\nu} = 0$$

because of $\mathbf{b}|\boldsymbol{\nu} = 0$.

CHAPTER 8

8.1 Applying the 3D inverse rule (2.137) for a spatial metric bidyadic $\overline{\overline{\mathsf{C}}}_s\in\mathbb{F}_2\mathbb{F}_2$,

$$\overline{\overline{\mathsf{C}}}{}_s^{-1} = \frac{(\mathbf{e}_{123}\mathbf{e}_{123}\lfloor\lfloor\overline{\overline{\mathsf{C}}}{}_s^T)^{(2)}}{\boldsymbol{\varepsilon}_{123}\boldsymbol{\varepsilon}_{123}||(\mathbf{e}_{123}\mathbf{e}_{123}\lfloor\lfloor\overline{\overline{\mathsf{C}}}_s)^{(3)}},$$

and substituting $\overline{\overline{\mathsf{B}}}_s = \overline{\overline{\mathsf{C}}}_s^{-1} \in \mathbb{E}_2\mathbb{E}_2$ and $\overline{\overline{\mathsf{C}}}_s = \overline{\overline{\mathsf{B}}}_s^{-1}$, yields the rule for a full-rank spatial metric bidyadic $\overline{\overline{\mathsf{B}}}_s$ expressed as a double-wedge square of some spatial dyadic. This is not the case for the general bidyadic in the 4D space.

8.2 Starting from the expressions (8.31)–(8.34), we have

$$\overline{\overline{\mu}}^{-1} = -\boldsymbol{\varepsilon}_{123}\lfloor\overline{\overline{\mathsf{Q}}}_s^{(2)} = -\Delta_{Q_s}\overline{\overline{\mathsf{Q}}}^{-1T}\rfloor\mathbf{e}_{123},$$

whence the Gibbsian dyadic becomes

$$\overline{\overline{\mu}}_g = \mathbf{e}_{123}\lfloor\overline{\overline{\mu}} = -\frac{1}{\Delta_{Q_s}}\mathbf{e}_{123}\lfloor(\boldsymbol{\varepsilon}_{123}\lfloor\overline{\overline{\mathsf{Q}}}_s^T) = \mu\overline{\overline{\mathsf{Q}}}_s^T.$$

The other medium dyadics are obtained as

$$\begin{aligned}
\overline{\overline{\xi}}_g &= \mathbf{e}_{123}\lfloor\overline{\overline{\xi}} = \mathbf{e}_{123}\lfloor\overline{\overline{\alpha}}|\overline{\overline{\mu}} = -\frac{1}{\Delta_{Q_s}}(\overline{\overline{\mathsf{Q}}}_s \wedge \mathbf{a}_s)|(\boldsymbol{\varepsilon}_{123}\lfloor\overline{\overline{\mathsf{Q}}}_s^T)\\
&= -(\overline{\overline{\mathsf{Q}}}_s \wedge \mathbf{a}_s)|\overline{\overline{\mathsf{Q}}}_s^{(-2)}\rfloor\mathbf{e}_{123} = -(\overline{\overline{\mathsf{I}}}_s \wedge (\mathbf{a}_s|\overline{\overline{\mathsf{Q}}}_s^{-1}))\rfloor\mathbf{e}_{123}\\
&= -\overline{\overline{\mathsf{I}}}_s\rfloor(\mathbf{a}_s|\overline{\overline{\mathsf{Q}}}_s^{-1})\rfloor\mathbf{e}_{123} = \overline{\overline{\mathsf{I}}}_s\rfloor\mathbf{X}_s\\
\overline{\overline{\zeta}}_g &= \mathbf{e}_{123}\lfloor\overline{\overline{\zeta}} = -\mathbf{e}_{123}\lfloor\overline{\overline{\mu}}|\overline{\overline{\beta}} = \frac{1}{\Delta_{Q_s}}\overline{\overline{\mathsf{Q}}}_s^T|(\boldsymbol{\varepsilon}_{123}\lfloor(\mathbf{b}_s \wedge \overline{\overline{\mathsf{Q}}}_s)\\
&= \mathbf{e}_{123}\lfloor\overline{\overline{\mathsf{Q}}}^{(-2)}|(\mathbf{b}_s \wedge \overline{\overline{\mathsf{Q}}}_s) = \mathbf{e}_{123}\lfloor((\overline{\overline{\mathsf{Q}}}^{-1}|\mathbf{b}_s) \wedge \overline{\overline{\mathsf{I}}}_s^T)\\
&= \mathbf{e}_{123}\lfloor(\overline{\overline{\mathsf{Q}}}^{-1}|\mathbf{b}_s)\lfloor\overline{\overline{\mathsf{I}}}_s^T = \mathbf{Z}_s\lfloor\overline{\overline{\mathsf{I}}}_s^T = \overline{\overline{\mathsf{I}}}_s\rfloor\mathbf{Z}_s\\
\overline{\overline{\epsilon}}_g &= \mathbf{e}_{123}\lfloor(\overline{\overline{\epsilon}}' - \overline{\overline{\alpha}}|\overline{\overline{\mu}}|\overline{\overline{\beta}}) = c\overline{\overline{\mathsf{Q}}}_s - \mathbf{b}_s\mathbf{a}_s - \overline{\overline{\xi}}_g|\overline{\overline{\beta}}\\
&= c\overline{\overline{\mathsf{Q}}}_s - \mathbf{b}_s\mathbf{a}_s + \frac{1}{\Delta_{Q_s}}(\overline{\overline{\mathsf{Q}}}_s \wedge \mathbf{a}_s)|(\boldsymbol{\varepsilon}_{123}\lfloor\overline{\overline{\mathsf{Q}}}_s^T)|(\mathbf{e}_{123}\lfloor(\mathbf{b}_s \wedge \overline{\overline{\mathsf{Q}}}_s)).
\end{aligned}$$

The last expression can be expanded as

$$\begin{aligned}
&\frac{1}{\Delta_{Q_s}}(\overline{\overline{\mathsf{Q}}}_s \wedge \mathbf{a}_s)|(\boldsymbol{\varepsilon}_{123}\lfloor\overline{\overline{\mathsf{Q}}}_s^T)|(\mathbf{e}_{123}\lfloor(\mathbf{b}_s \wedge \overline{\overline{\mathsf{Q}}}_s))\\
&\qquad = (\overline{\overline{\mathsf{Q}}}_s \wedge \mathbf{a}_s)|\overline{\overline{\mathsf{Q}}}_s^{(-2)}|(\mathbf{b}_s \wedge \overline{\overline{\mathsf{Q}}}_s)\\
&= (\overline{\overline{\mathsf{Q}}}_s \wedge \mathbf{a}_s)|((\overline{\overline{\mathsf{Q}}}_s^{-1}|\mathbf{b}_s) \wedge \overline{\overline{\mathsf{I}}}_s^T) = \mathbf{b}_s\mathbf{a}_s - (\mathbf{a}_s|\overline{\overline{\mathsf{Q}}}_s^{-1}|\mathbf{b}_s)\overline{\overline{\mathsf{Q}}}_s,
\end{aligned}$$

whence

$$\overline{\overline{\epsilon}}_g = c\overline{\overline{\mathsf{Q}}}_s - (\mathbf{a}_s|\overline{\overline{\mathsf{Q}}}_s^{-1}|\mathbf{b}_s)\overline{\overline{\mathsf{Q}}}_s = \epsilon\overline{\overline{\mathsf{Q}}}_s.$$

8.3 Let us assume Gibbsian medium dyadics of the form

$$\overline{\overline{\epsilon}}_g = \epsilon\overline{\overline{\mathsf{W}}}, \quad \overline{\overline{\mu}}_g = \mu\overline{\overline{\mathsf{W}}}^T, \quad \overline{\overline{\xi}}_g = \overline{\overline{\mathsf{I}}}_s\rfloor\mathbf{X} = \mathbf{X}\lfloor\overline{\overline{\mathsf{I}}}_s^T, \quad \overline{\overline{\zeta}}_g = \overline{\overline{\mathsf{I}}}_s\rfloor\mathbf{Z} = \mathbf{Z}\lfloor\overline{\overline{\mathsf{I}}}_s^T,$$

where $\overline{\overline{\mathsf{W}}} \in \mathbb{E}_1\mathbb{E}_1$ is a spatial metric dyadic and $\mathbf{X}, \mathbf{Z}$ are two spatial bivectors. The task is to build the corresponding modified bidyadic $\overline{\overline{\mathsf{M}}}_m \in \mathbb{F}_2\mathbb{E}_2$ by applying the expansion (5.88)

$$\overline{\overline{\mathsf{M}}}_m = \overline{\overline{\epsilon}}_g{}^\wedge_\wedge\mathbf{e}_4\mathbf{e}_4 - (\mathbf{e}_{123}\lfloor\overline{\overline{\mathsf{I}}}_s^T + \mathbf{e}_4\wedge\overline{\overline{\xi}}_g)|\overline{\overline{\mu}}_g^{-1}|(\overline{\overline{\mathsf{I}}}_s\rfloor\mathbf{e}_{123} - \overline{\overline{\zeta}}_g\wedge\mathbf{e}_4).$$

Expanding

$$\overline{\overline{\epsilon}}_g{}^\wedge_\wedge\mathbf{e}_4\mathbf{e}_4 = \epsilon\overline{\overline{\mathsf{W}}}{}^\wedge_\wedge\mathbf{e}_4\mathbf{e}_4$$

$$\overline{\overline{\mu}}_g^{-1} = \frac{1}{\mu}\overline{\overline{\mathsf{W}}}^{-1T} = \frac{1}{\mu\Delta_W}\boldsymbol{\varepsilon}_{123}\boldsymbol{\varepsilon}_{123}\lfloor\lfloor\overline{\overline{\mathsf{W}}}^{(2)}$$

$$\mathbf{e}_{123}\lfloor\overline{\overline{\mathsf{I}}}_s^T + \mathbf{e}_4\wedge\overline{\overline{\xi}}_g = \mathbf{e}_{123}\lfloor\overline{\overline{\mathsf{I}}}_s^T + \mathbf{e}_4\wedge(\mathbf{X}\lfloor\overline{\overline{\mathsf{I}}}_s^T)$$

$$\overline{\overline{\mathsf{I}}}_s\rfloor\mathbf{e}_{123} - \overline{\overline{\zeta}}_g\wedge\mathbf{e}_4 = \overline{\overline{\mathsf{I}}}_s\rfloor\mathbf{e}_{123} - (\overline{\overline{\mathsf{I}}}_s\rfloor\mathbf{Z})\wedge\mathbf{e}_4$$

and, inserting the in the above expression, when defining $\mu = -1/\Delta_W$ for simplicity, we have

$$\overline{\overline{\mathsf{M}}}_m = \epsilon\overline{\overline{\mathsf{W}}}{}^\wedge_\wedge\mathbf{e}_4\mathbf{e}_4$$

$$+(\mathbf{e}_{123}\lfloor\overline{\overline{\mathsf{I}}}_s^T + \mathbf{e}_4\wedge(\mathbf{X}\lfloor\overline{\overline{\mathsf{I}}}_s^T))|\boldsymbol{\varepsilon}_{123}\boldsymbol{\varepsilon}_{123}\lfloor\lfloor\overline{\overline{\mathsf{W}}}^{(2)}|(\overline{\overline{\mathsf{I}}}_s\rfloor\mathbf{e}_{123} - (\overline{\overline{\mathsf{I}}}_s\rfloor\mathbf{Z})\wedge\mathbf{e}_4)$$

$$= \epsilon\overline{\overline{\mathsf{W}}}{}^\wedge_\wedge\mathbf{e}_4\mathbf{e}_4 + \overline{\overline{\mathsf{W}}}^{(2)} - ((\overline{\overline{\mathsf{W}}}^{(2)}\rfloor\boldsymbol{\varepsilon}_{123})\rfloor\mathbf{Z})\wedge\mathbf{e}_4$$

$$+\mathbf{e}_4\wedge(\mathbf{X}\lfloor(\boldsymbol{\varepsilon}_{123}\lfloor\overline{\overline{\mathsf{W}}}^{(2)})) - \mathbf{e}_4\wedge(\mathbf{X}\lfloor(\boldsymbol{\varepsilon}_{123}\lfloor\overline{\overline{\mathsf{W}}}^{(2)}\rfloor\boldsymbol{\varepsilon}_{123})\rfloor\mathbf{Z})\wedge\mathbf{e}_4.$$

Defining $\boldsymbol{\alpha} = \boldsymbol{\varepsilon}_{123}\lfloor\mathbf{X} = \mathbf{X}\rfloor\boldsymbol{\varepsilon}_{123}$ and $\boldsymbol{\beta} = \boldsymbol{\varepsilon}_{123}\lfloor\mathbf{Z} = \mathbf{Z}\rfloor\boldsymbol{\varepsilon}_{123}$, we can expand further

$$\mathbf{X}\lfloor(\boldsymbol{\varepsilon}_{123}\lfloor\overline{\overline{\mathsf{W}}}^{(2)}\rfloor\boldsymbol{\varepsilon}_{123})\rfloor\mathbf{Z} = (-\mathbf{X}\rfloor\boldsymbol{\varepsilon}_{123})\rfloor\overline{\overline{\mathsf{W}}}^{(2)}\lfloor(-\boldsymbol{\varepsilon}_{123}\lfloor\mathbf{Z})$$

$$= \boldsymbol{\alpha}\rfloor\overline{\overline{\mathsf{W}}}^{(2)}\lfloor\boldsymbol{\beta} = -\overline{\overline{\mathsf{W}}}^{(2)}\lfloor\lfloor\boldsymbol{\alpha}\boldsymbol{\beta} = -(\overline{\overline{\mathsf{W}}}||\boldsymbol{\alpha}\boldsymbol{\beta})\overline{\overline{\mathsf{W}}} + (\overline{\overline{\mathsf{W}}}|\boldsymbol{\beta})(\boldsymbol{\alpha}|\overline{\overline{\mathsf{W}}})$$

$$\mathbf{X}\lfloor(\boldsymbol{\varepsilon}_{123}\lfloor\overline{\overline{\mathsf{W}}}^{(2)}) = -\boldsymbol{\alpha}\rfloor\overline{\overline{\mathsf{W}}}^{(2)} = -\overline{\overline{\mathsf{W}}}\wedge(\boldsymbol{\alpha}|\overline{\overline{\mathsf{W}}})$$

$$(\overline{\overline{\mathsf{W}}}^{(2)}\rfloor\boldsymbol{\varepsilon}_{123})\rfloor\mathbf{Z} = -\overline{\overline{\mathsf{W}}}^{(2)}\lfloor\boldsymbol{\beta} = -(\overline{\overline{\mathsf{W}}}|\boldsymbol{\beta})\wedge\overline{\overline{\mathsf{W}}}.$$

Denoting $\mathbf{a} = \boldsymbol{\alpha}|\overline{\overline{\mathsf{W}}}$, $\mathbf{b} = \overline{\overline{\mathsf{W}}}|\boldsymbol{\beta}$ and combining these we have

$$\begin{aligned}\overline{\overline{\mathsf{M}}}_m &= \epsilon\overline{\overline{\mathsf{W}}}{}^{\wedge}_{\wedge}\mathbf{e}_4\mathbf{e}_4 + \overline{\overline{\mathsf{W}}}{}^{(2)} - \mathbf{b}\wedge\overline{\overline{\mathsf{W}}}\wedge\mathbf{e}_4 + \mathbf{e}_4\wedge\overline{\overline{\mathsf{W}}}\wedge\mathbf{a}\\ &\quad +\mathbf{e}_4\wedge((\overline{\overline{\mathsf{W}}}||\boldsymbol{\alpha\beta})\overline{\overline{\mathsf{W}}} - \mathbf{ba})\wedge\mathbf{e}_4\\ &= \lambda\overline{\overline{\mathsf{W}}}{}^{\wedge}_{\wedge}\mathbf{e}_4\mathbf{e}_4 + \overline{\overline{\mathsf{W}}}{}^{(2)} + \overline{\overline{\mathsf{W}}}{}^{\wedge}_{\wedge}\mathbf{be}_4 - \overline{\overline{\mathsf{W}}}{}^{\wedge}_{\wedge}\mathbf{e}_4\mathbf{a} - \mathbf{e}_4\mathbf{a}{}^{\wedge}_{\wedge}\mathbf{be}_4\\ &= (\overline{\overline{\mathsf{W}}} + \mathbf{be}_4 - \mathbf{e}_4\mathbf{a} + \lambda\mathbf{e}_4\mathbf{e}_4)^{(2)}, \qquad \lambda = \epsilon - \overline{\overline{\mathsf{W}}}||\boldsymbol{\alpha\beta}.\end{aligned}$$

Thus, $\overline{\overline{\mathsf{M}}}_m$ has the form of a Q-medium bidyadic.

8.4

$$\begin{aligned}3D(\boldsymbol{\nu}) &= \boldsymbol{\varepsilon}_N\boldsymbol{\varepsilon}_N||(\overline{\overline{\mathsf{Q}}}{}^{(2)}{}^{\wedge}_{\wedge}(\overline{\overline{\mathsf{Q}}}{}^{(2)}\lfloor\lfloor\boldsymbol{\nu\nu})^{(2)})\\ &= \boldsymbol{\varepsilon}_N\boldsymbol{\varepsilon}_N||(\overline{\overline{\mathsf{Q}}}{}^{(2)}{}^{\wedge}_{\wedge}((\overline{\overline{\mathsf{Q}}}||\boldsymbol{\nu\nu})\overline{\overline{\mathsf{Q}}} - \overline{\overline{\mathsf{Q}}}|\boldsymbol{\nu\nu}|\overline{\overline{\mathsf{Q}}})^{(2)})\\ &= (\overline{\overline{\mathsf{Q}}}||\boldsymbol{\nu\nu})\boldsymbol{\varepsilon}_N\boldsymbol{\varepsilon}_N||(\overline{\overline{\mathsf{Q}}}{}^{(2)}{}^{\wedge}_{\wedge}((\overline{\overline{\mathsf{Q}}}||\boldsymbol{\nu\nu})\overline{\overline{\mathsf{Q}}}{}^{(2)} - \overline{\overline{\mathsf{Q}}}{}^{\wedge}_{\wedge}(\overline{\overline{\mathsf{Q}}}|\boldsymbol{\nu\nu}|\overline{\overline{\mathsf{Q}}})))\\ &= (\overline{\overline{\mathsf{Q}}}||\boldsymbol{\nu\nu})((\overline{\overline{\mathsf{Q}}}||\boldsymbol{\nu\nu})6\boldsymbol{\varepsilon}_N\boldsymbol{\varepsilon}_N||\overline{\overline{\mathsf{Q}}}{}^{(4)} - 3(\boldsymbol{\varepsilon}_N\boldsymbol{\varepsilon}_N\lfloor\lfloor\overline{\overline{\mathsf{Q}}}{}^{(3)})\\ &\qquad ||(\overline{\overline{\mathsf{Q}}}|\boldsymbol{\nu\nu}|\overline{\overline{\mathsf{Q}}}))\\ &= (\overline{\overline{\mathsf{Q}}}||\boldsymbol{\nu\nu})(\boldsymbol{\varepsilon}_N\boldsymbol{\varepsilon}_N||\overline{\overline{\mathsf{Q}}}{}^{(4)})(6(\overline{\overline{\mathsf{Q}}}||\boldsymbol{\nu\nu}) - 3\overline{\overline{\mathsf{Q}}}{}^{-1T}||(\overline{\overline{\mathsf{Q}}}|\boldsymbol{\nu\nu}|\overline{\overline{\mathsf{Q}}}))\\ &= 3(\overline{\overline{\mathsf{Q}}}||\boldsymbol{\nu\nu})^2(\boldsymbol{\varepsilon}_N\boldsymbol{\varepsilon}_N||\overline{\overline{\mathsf{Q}}}{}^{(4)}) = 0.\end{aligned}$$

8.5 Let us expand

$$\begin{aligned}\overline{\overline{\mathsf{D}}}(\boldsymbol{\nu})|\boldsymbol{\phi} &= -(\boldsymbol{\nu}\rfloor\overline{\overline{\mathsf{Q}}}{}^{(2)}\lfloor\boldsymbol{\nu})|(\boldsymbol{\nu}|\overline{\overline{\mathsf{Q}}}\rfloor\boldsymbol{\Gamma}) = \boldsymbol{\nu}\rfloor(\overline{\overline{\mathsf{Q}}}{}^{(2)}\lfloor(\boldsymbol{\nu}|\overline{\overline{\mathsf{Q}}}\rfloor\boldsymbol{\Gamma}))|\boldsymbol{\nu}\\ &= -\boldsymbol{\nu}\rfloor(\overline{\overline{\mathsf{Q}}}{}^{(2)}\lfloor(\boldsymbol{\Gamma}\lfloor(\boldsymbol{\nu}|\overline{\overline{\mathsf{Q}}}))|\boldsymbol{\nu},\end{aligned}$$

and invoke the bac-cab rule

$$\mathbf{A}\lfloor(\boldsymbol{\Gamma}\lfloor\mathbf{a}) = (\mathbf{A}\wedge\mathbf{a})\lfloor\boldsymbol{\Gamma} - (\mathbf{A}|\boldsymbol{\Gamma})\mathbf{a}$$

in the form

$$\overline{\overline{\mathsf{Q}}}{}^{(2)}\lfloor(\boldsymbol{\Gamma}\lfloor(\boldsymbol{\nu}|\overline{\overline{\mathsf{Q}}})) = (\overline{\overline{\mathsf{Q}}}{}^{(2)}\wedge(\boldsymbol{\nu}|\overline{\overline{\mathsf{Q}}}))\lfloor\boldsymbol{\Gamma} - (\overline{\overline{\mathsf{Q}}}{}^{(2)}|\boldsymbol{\Gamma})\boldsymbol{\nu}|\overline{\overline{\mathsf{Q}}}.$$

Inserting this, we can proceed as

$$\begin{aligned}\overline{\overline{\mathsf{D}}}(\boldsymbol{\nu})|\boldsymbol{\phi} &= -\boldsymbol{\nu}\rfloor((\overline{\overline{\mathsf{Q}}}{}^{(2)}\wedge(\boldsymbol{\nu}|\overline{\overline{\mathsf{Q}}}))\lfloor\boldsymbol{\Gamma} - (\overline{\overline{\mathsf{Q}}}{}^{(2)}|\boldsymbol{\Gamma})\boldsymbol{\nu}|\overline{\overline{\mathsf{Q}}})|\boldsymbol{\nu}\\ &= -(\boldsymbol{\nu}\wedge\boldsymbol{\nu})\rfloor\overline{\overline{\mathsf{Q}}}{}^{(3)}\lfloor\boldsymbol{\Gamma} + (\boldsymbol{\nu}\rfloor\overline{\overline{\mathsf{Q}}}{}^{(2)}|\boldsymbol{\Gamma})(\boldsymbol{\nu}|\overline{\overline{\mathsf{Q}}}|\boldsymbol{\nu}).\end{aligned}$$

Both terms of the last expression can be seen to vanish.

8.6 Let us assume that a bidyadic $\overline{\overline{\mathsf{M}}}_m = \overline{\overline{\mathsf{Q}}}^{(2)} + \sqrt{\Delta_Q}\,\overline{\overline{\mathsf{E}}}$ is symmetric. Expanding

$$(\overline{\overline{\mathsf{Q}}}^{(2)} + \sqrt{\Delta_Q}\,\overline{\overline{\mathsf{E}}})^T \cdot (\overline{\overline{\mathsf{Q}}}^{(2)} - \sqrt{\Delta_Q}\,\overline{\overline{\mathsf{E}}})$$
$$= \overline{\overline{\mathsf{Q}}}^{(2)T} \cdot \overline{\overline{\mathsf{Q}}}^{(2)} + \sqrt{\Delta_Q}(\overline{\overline{\mathsf{Q}}}^{(2)} - \overline{\overline{\mathsf{Q}}}^{(2)T}) - \Delta_Q \overline{\overline{\mathsf{E}}}$$
$$= \Delta_Q \mathbf{e}_N \lfloor \overline{\overline{\mathsf{Q}}}^{(-2)} | \overline{\overline{\mathsf{Q}}}^{(2)} - \Delta_Q \mathbf{e}_N \lfloor \overline{\overline{\mathsf{I}}}^{(2)T} = 0,$$

from this we can conclude that the bidyadic $\overline{\overline{\mathsf{Q}}}^{(2)} \pm \sqrt{\Delta_Q}\,\overline{\overline{\mathsf{E}}}$ cannot have an inverse for either sign. In fact, assuming that one of these has an inverse, by multiplying this expression by the inverse, vanishing of the other one follows which would require that $\overline{\overline{\mathsf{Q}}}^{(2)}$ be a multiple of $\mathbf{e}_N \lfloor \overline{\overline{\mathsf{I}}}^{(2)T}$ which is not possible (see Problem 4.14).

8.7 The Gibbsian relations (8.122)–(8.125) can be expressed more compactly in dyadic-matrix form as

$$\begin{pmatrix} \overline{\overline{\epsilon}}_g & \overline{\overline{\xi}}_g \\ \overline{\overline{\zeta}}_g & \overline{\overline{\mu}}_g \end{pmatrix} = \begin{pmatrix} 0 & \epsilon \\ -\mu & 0 \end{pmatrix} \left[\mathcal{A} + \begin{pmatrix} -\mathbf{p}_1/\mu \\ \mathbf{p}_2/\epsilon \end{pmatrix} (\mathbf{q}_1 \ \mathbf{q}_2) \right],$$

where $\mathcal{A}$ is the antisymmetric dyadic matrix

$$\mathcal{A} = \begin{pmatrix} -\overline{\overline{\mathsf{I}}} \rfloor \mathbf{Z}_s/\mu & -\overline{\overline{\mathsf{Q}}}_s^T \\ \overline{\overline{\mathsf{Q}}}_s & \overline{\overline{\mathsf{I}}} \rfloor \mathbf{X}_s/\epsilon \end{pmatrix}.$$

Since fields of a plane wave in any medium satisfy the orthogonality conditions $\mathbf{E}_g \cdot \mathbf{B}_g = 0$ and $\mathbf{H}_g \cdot \mathbf{D}_g = 0$, we can form the following chain of relations in Gibbsian symbols:

$$0 = -\frac{1}{\mu}\mathbf{E}_g \cdot \mathbf{B}_g + \frac{1}{\epsilon}\mathbf{H}_g \cdot \mathbf{D}_g$$
$$= -\frac{1}{\mu}\mathbf{E}_g \cdot (\overline{\overline{\zeta}}_g \cdot \mathbf{E}_g + \overline{\overline{\mu}}_g \cdot \mathbf{H}_g) + \frac{1}{\epsilon}\mathbf{H}_g \cdot (\overline{\overline{\epsilon}}_g \cdot \mathbf{E}_g + \overline{\overline{\xi}}_g \cdot \mathbf{H}_g)$$
$$= (\mathbf{E}_g \ \mathbf{H}_g) \cdot \begin{pmatrix} 0 & -1/\mu \\ 1/\epsilon & 0 \end{pmatrix} \begin{pmatrix} \overline{\overline{\epsilon}}_g & \overline{\overline{\xi}}_g \\ \overline{\overline{\zeta}}_g & \overline{\overline{\mu}}_g \end{pmatrix} \cdot \begin{pmatrix} \mathbf{E}_g \\ \mathbf{H}_g \end{pmatrix}$$
$$= (\mathbf{E}_g \ \mathbf{H}_g) \cdot \left[\mathcal{A} + \begin{pmatrix} -\mathbf{p}_1/\mu \\ \mathbf{p}_2/\epsilon \end{pmatrix} (\mathbf{q}_1 \ \mathbf{q}_2) \right] \cdot \begin{pmatrix} \mathbf{E}_g \\ \mathbf{H}_g \end{pmatrix}$$
$$= (\mathbf{E}_g \ \mathbf{H}_g) \cdot \begin{pmatrix} -\mathbf{p}_1/\mu \\ \mathbf{p}_2/\epsilon \end{pmatrix} (\mathbf{q}_1 \ \mathbf{q}_2) \cdot \begin{pmatrix} \mathbf{E}_g \\ \mathbf{H}_g \end{pmatrix}$$
$$= \left(-\frac{1}{\mu}\mathbf{p}_1 \cdot \mathbf{E}_g + \frac{1}{\epsilon}\mathbf{p}_2 \cdot \mathbf{H}_g \right) (\mathbf{q}_1 \cdot \mathbf{E}_g + \mathbf{q}_2 \cdot \mathbf{H}_g).$$

Thus, the Gibbsian field vectors must satisfy either of the two conditions

$$\epsilon \mathbf{p}_1 \cdot \mathbf{E}_g - \mu \mathbf{p}_2 \cdot \mathbf{H}_g = 0, \qquad \mathbf{q}_1 \cdot \mathbf{E}_g + \mathbf{q}_2 \cdot \mathbf{H}_g = 0,$$

which for the field one-forms take the corresponding form of

$$\epsilon \mathbf{p}_1 | \mathbf{E} - \mu \mathbf{p}_2 | \mathbf{H} = 0, \qquad \mathbf{q}_1 | \mathbf{E} + \mathbf{q}_2 | \mathbf{H} = 0.$$

8.8 Expanding

$$\overline{\overline{\mathsf{M}}} \lfloor\lfloor \overline{\overline{\mathsf{I}}} = \overline{\overline{\mathsf{P}}}^{(2)} \lfloor\lfloor \overline{\overline{\mathsf{I}}} = (\mathrm{tr}\overline{\overline{\mathsf{P}}})\overline{\overline{\mathsf{P}}} - \overline{\overline{\mathsf{P}}}^2 = 0,$$

we may distinguish the two possibilities, $\mathrm{tr}\overline{\overline{\mathsf{P}}} = 0$ and $\mathrm{tr}\overline{\overline{\mathsf{P}}} \neq 0$. Assuming first $\mathrm{tr}\overline{\overline{\mathsf{P}}} \neq 0$, we can define $\overline{\overline{\Pi}} = \overline{\overline{\mathsf{P}}}/\mathrm{tr}\overline{\overline{\mathsf{P}}}$ satisfying $\overline{\overline{\Pi}}^2 = \overline{\overline{\Pi}}$ and $\mathrm{tr}\overline{\overline{\Pi}} = 1$. Thus, $\overline{\overline{\Pi}}$ is an example of a projection dyadic. From (4.42) and (4.45) it follows that this one must be of the form $\overline{\overline{\Pi}} = \mathbf{p}\boldsymbol{\pi}$ with $\mathbf{p}|\boldsymbol{\pi} = 1$. Thus, we have $\overline{\overline{\mathsf{P}}} = \mathrm{tr}\overline{\overline{\mathsf{P}}}(\mathbf{p}\boldsymbol{\pi})$ which implies $\overline{\overline{\mathsf{M}}} = \overline{\overline{\mathsf{P}}}^{(2)T} = 0$.

This leaves us the second choice, $\mathrm{tr}\overline{\overline{\mathsf{P}}} = 0$. In this case $\overline{\overline{\mathsf{P}}}$ satisfies $\overline{\overline{\mathsf{P}}}^2 = 0$, whence it must be nilpotent dyadic. From (4.19) we know that such a dyadic can be expressed as

$$\overline{\overline{\mathsf{P}}} = \mathbf{p}_1\boldsymbol{\pi}_1 + \mathbf{p}_2\boldsymbol{\pi}_2, \quad \boldsymbol{\pi}_i|\mathbf{p}_j = 0, \quad i,j = 1,2$$

Thus, any principal P-medium dyadic must be of the form

$$\overline{\overline{\mathsf{M}}} = \boldsymbol{\Delta}\mathbf{A}, \quad \boldsymbol{\Delta} = \boldsymbol{\pi}_1 \wedge \boldsymbol{\pi}_2, \quad \mathbf{A} = \mathbf{p}_1 \wedge \mathbf{p}_2.$$

This is a valid solution, since it satisfies

$$\overline{\overline{\mathsf{M}}} \lfloor\lfloor \overline{\overline{\mathsf{I}}} = (\boldsymbol{\pi}_1|\mathbf{p}_1)\boldsymbol{\pi}_2\mathbf{p}_2 + (\boldsymbol{\pi}_2|\mathbf{p}_2)\boldsymbol{\pi}_1\mathbf{p}_1 - \boldsymbol{\pi}_1\mathbf{p}_1|\boldsymbol{\pi}_2\mathbf{p}_2 - \boldsymbol{\pi}_2\mathbf{p}_2|\boldsymbol{\pi}_1\mathbf{p}_1$$
$$= 0.$$

8.9 For symmetric $\overline{\overline{\mathsf{Q}}}^{(2)}$ we can expand

$$\overline{\overline{\mathsf{J}}}^{T2} = (\boldsymbol{\varepsilon}_N \lfloor \overline{\overline{\mathsf{Q}}}^{(2)})|(\boldsymbol{\varepsilon}_N \lfloor \overline{\overline{\mathsf{Q}}}^{(2)}) = (\boldsymbol{\varepsilon}_N \boldsymbol{\varepsilon}_N \lfloor\lfloor \overline{\overline{\mathsf{Q}}}^{(2)})|\overline{\overline{\mathsf{Q}}}^{(2)}$$
$$= \Delta_Q \overline{\overline{\mathsf{Q}}}^{(-2)T}|\overline{\overline{\mathsf{Q}}}^{(2)} = \Delta_Q \overline{\overline{\mathsf{I}}}^{(2)T}$$

whence for $\Delta_Q = -1$ we have $\overline{\overline{\mathsf{J}}}^{T2} = -\overline{\overline{\mathsf{I}}}^{(2)T}$. Defining $M = \sqrt{A^2 + B^2}$ and $\tan\psi = B/A$, we can write

$$\overline{\overline{\mathsf{M}}} = M(\cos\psi\, \overline{\overline{\mathsf{I}}}^{(2)T} + \sin\psi\, \overline{\overline{\mathsf{J}}}^T) = Me^{\psi \overline{\overline{\mathsf{J}}}^T}$$

8.10 Expanding

$$\begin{aligned}\overline{\overline{\mathsf{M}}}^2 &= M^2 e^{2\psi\overline{\overline{\mathsf{J}}}^T} = M^2(\cos 2\psi\ \overline{\overline{\mathsf{I}}}^{(2)T} + \sin 2\psi\ \overline{\overline{\mathsf{J}}}^T)\\ &= M^2(2\cos^2\psi\ \overline{\overline{\mathsf{I}}}^{(2)T} + 2\cos\psi\sin\psi\ \overline{\overline{\mathsf{J}}}^T - \overline{\overline{\mathsf{I}}}^{(2)T})\\ &= 2M\cos\psi\ \overline{\overline{\mathsf{M}}} - M^2\overline{\overline{\mathsf{I}}}^{(2)T},\end{aligned}$$

the equation satisfied by the bidyadic $\overline{\overline{\mathsf{M}}}$ can be written as

$$(\overline{\overline{\mathsf{M}}} - Me^{j\psi}\ \overline{\overline{\mathsf{I}}}^{(2)})|(\overline{\overline{\mathsf{M}}} - Me^{-j\psi}\ \overline{\overline{\mathsf{I}}}^{(2)}) = 0.$$

From this, the two eigenvalues can be identified as $M_\pm = Me^{\pm j\psi}$ and the eigen two-forms can be constructed as

$$\begin{aligned}\mathbf{\Phi}_\pm &= (\overline{\overline{\mathsf{M}}} - Me^{\mp j\psi}\overline{\overline{\mathsf{I}}}^{(2)T})|\mathbf{\Theta}_\pm,\\ &= M(e^{\psi\overline{\overline{\mathsf{J}}}^T} - e^{\mp j\psi}\overline{\overline{\mathsf{I}}}^{(2)T})|\mathbf{\Theta}_\pm\\ &= M(\sin\psi\ \overline{\overline{\mathsf{J}}}^T \pm j\sin\psi\overline{\overline{\mathsf{I}}}^{(2)T})|\mathbf{\Theta}_\pm\\ &= \pm jM\sin\psi(\overline{\overline{\mathsf{I}}}^{(2)T} \pm j\overline{\overline{\mathsf{J}}}^T)|\mathbf{\Theta}_\pm.\end{aligned}$$

Choosing $\mathbf{\Theta}_\pm = \pm\mathbf{\Phi}/j2M\sin\psi = \pm\mathbf{\Phi}/(M_+ - M_-)$, we can write

$$\mathbf{\Phi}_\pm = \frac{1}{2}(\overline{\overline{\mathsf{I}}}^{(2)T} \pm j\overline{\overline{\mathsf{J}}}^T)|\mathbf{\Phi}, \quad \Rightarrow \quad \mathbf{\Phi} = \mathbf{\Phi}_+ + \mathbf{\Phi}_-$$

which yields the decomposition rule valid for $M_+ \neq M_-$. The orthogonality of the two eigen two-forms follows from

$$\begin{aligned}4\mathbf{\Phi}_+ \cdot \mathbf{\Phi}_- &= \mathbf{\Phi}|(\overline{\overline{\mathsf{I}}}^{(2)T} + j\overline{\overline{\mathsf{J}}}^T)^T|(\mathbf{e}_N\lfloor(\overline{\overline{\mathsf{I}}}^{(2)T} - j\overline{\overline{\mathsf{J}}}^T))|\mathbf{\Phi}\\ &= \mathbf{\Phi}|(\mathbf{e}_N\lfloor(\overline{\overline{\mathsf{I}}}^{(2)T} + j\overline{\overline{\mathsf{J}}}^T))^T|(\overline{\overline{\mathsf{I}}}^{(2)T} - j\overline{\overline{\mathsf{J}}}^T)|\mathbf{\Phi}\\ &= \mathbf{\Phi}|(\mathbf{e}_N\lfloor(\overline{\overline{\mathsf{I}}}^{(2)T} + j\overline{\overline{\mathsf{J}}}^T))|(\overline{\overline{\mathsf{I}}}^{(2)T} - j\overline{\overline{\mathsf{J}}}^T)|\mathbf{\Phi}\\ &= \mathbf{\Phi}|(\mathbf{e}_N\lfloor(\overline{\overline{\mathsf{I}}}^{(2)T} + \overline{\overline{\mathsf{J}}}^{T2})|\mathbf{\Phi}\\ &= 0.\end{aligned}$$

Here we have applied the symmetry of the bidyadic $\mathbf{e}_N\lfloor(\overline{\overline{\mathsf{I}}}^{(2)T} + j\overline{\overline{\mathsf{J}}}^T) = \mathbf{e}_N\lfloor\overline{\overline{\mathsf{I}}}^{(2)T} + j\overline{\overline{\mathsf{Q}}}^{(2)}$.

8.11 Substituting the decomposition of the previous problem in terms of the eigenfields $\mathbf{\Phi}_\pm$ in

$$\mathbf{d}\wedge(\overline{\overline{\mathsf{M}}}|\mathbf{\Phi}) = \gamma_e, \quad \mathbf{d}\wedge\mathbf{\Phi} = \gamma_m,$$

the resulting equations

$$M_+\mathbf{d}\wedge\mathbf{\Phi}_+ + M_-\mathbf{d}\wedge\mathbf{\Phi}_- = \gamma_e,$$
$$\mathbf{d}\wedge\mathbf{\Phi}_+ + \mathbf{d}\wedge\mathbf{\Phi}_- = \gamma_m,$$

can be split in two uncoupled equations as

$$\mathbf{d}\wedge\mathbf{\Phi}_+ = \gamma_+, \qquad \mathbf{d}\wedge\mathbf{\Phi}_- = \gamma_-$$

with source three-forms

$$\gamma_{\mp} = \pm(M_{\pm}\gamma_m - \gamma_e)/(M_+ - M_-).$$

The bidyadic $\overline{\mathsf{U}}(\mathbf{\Psi},\mathbf{\Phi}) = -\overline{\mathsf{U}}(\mathbf{\Phi},\mathbf{\Psi})$ vanishes for each eigenfield, $\overline{\mathsf{U}}(\mathbf{\Psi}_{\pm},\mathbf{\Phi}_{\pm}) = M_{\pm}\overline{\mathsf{U}}(\mathbf{\Phi}_{\pm},\mathbf{\Phi}_{\pm}) = 0$, whence the eigenfields alone do not carry any energy. For total fields, energy transportation is possible through the interaction of both eigenfields:

$$\begin{aligned}\overline{\mathsf{U}}(\mathbf{\Psi},\mathbf{\Phi}) &= \overline{\mathsf{U}}(\mathbf{\Psi}_+,\mathbf{\Phi}_-) + \overline{\mathsf{U}}(\mathbf{\Psi}_-,\mathbf{\Phi}_+)\\ &= M_+\overline{\mathsf{U}}(\mathbf{\Phi}_+,\mathbf{\Phi}_-) + M_-\overline{\mathsf{U}}(\mathbf{\Phi}_-,\mathbf{\Phi}_+)\\ &= (M_+ - M_-)\overline{\mathsf{U}}(\mathbf{\Phi}_+,\mathbf{\Phi}_-).\end{aligned}$$

8.12 The dispersion function corresponding to the extended Q-medium has the form

$$D(\mathbf{v}) = \varepsilon_N\varepsilon_N\lfloor\lfloor((\overline{\overline{\mathsf{Q}}}^{(2)} + \mathbf{AB})^{\wedge}_{\wedge}(\mathbf{vv}\rfloor\rfloor(\overline{\overline{\mathsf{Q}}}^{(2)} + \mathbf{AB}))^{(2)}).$$

Applying the general rule

$$\mathbf{vv}\rfloor\rfloor\overline{\overline{\mathsf{Q}}}^{(p)} = (\mathbf{vv}\lfloor\lfloor\overline{\overline{\mathsf{Q}}})\overline{\overline{\mathsf{Q}}}^{(p-1)} - \overline{\overline{\mathsf{Q}}}^{(p-2)}{}^{\wedge}_{\wedge}(\overline{\overline{\mathsf{Q}}}\rfloor\mathbf{vv}\lfloor\overline{\overline{\mathsf{Q}}}),$$

where $\overline{\overline{\mathsf{Q}}}^{(p-2)}{}^{\wedge}_{\wedge}$ can be omitted for $p = 2$, and expanding

$$(\mathbf{vv}\rfloor\rfloor(\overline{\overline{\mathsf{Q}}}^{(2)} + \mathbf{AB}))^{(2)} = (\mathbf{vv}\rfloor\rfloor\overline{\overline{\mathsf{Q}}}^{(2)})^{(2)} + (\mathbf{vv}\rfloor\rfloor\overline{\overline{\mathsf{Q}}}^{(2)})^{\wedge}_{\wedge}(\mathbf{vv}\rfloor\rfloor\mathbf{AB}),$$

we obtain the four-term expression

$$D(\mathbf{v}) = \sum D_i = \varepsilon_N\varepsilon_N\lfloor\lfloor\Big((\overline{\overline{\mathsf{Q}}}^{(2)}{}^{\wedge}_{\wedge}(\mathbf{vv}\rfloor\rfloor\overline{\overline{\mathsf{Q}}}^{(2)})^{(2)} + (\mathbf{AB})^{\wedge}_{\wedge}(\mathbf{vv}\rfloor\rfloor\overline{\overline{\mathsf{Q}}}^{(2)})^{(2)}$$
$$+ (\overline{\overline{\mathsf{Q}}}^{(2)}{}^{\wedge}_{\wedge}(\mathbf{vv}\rfloor\rfloor\overline{\overline{\mathsf{Q}}}^{(2)})^{\wedge}_{\wedge}(\mathbf{vv}\rfloor\rfloor\mathbf{AB}) + (\mathbf{AB})^{\wedge}_{\wedge}(\mathbf{vv}\rfloor\rfloor\overline{\overline{\mathsf{Q}}}^{(2)})^{\wedge}_{\wedge}(\mathbf{vv}\rfloor\rfloor\mathbf{AB})\Big).$$

The fourth term vanishes identically:

$$\begin{aligned}
D_4 &= \boldsymbol{\varepsilon}_N\boldsymbol{\varepsilon}_N||((\mathbf{AB})^{\wedge}_{\wedge}(\boldsymbol{v}\boldsymbol{v}\rfloor\rfloor\overline{\overline{\mathsf{Q}}}^{(2)})^{\wedge}_{\wedge}(\boldsymbol{v}\boldsymbol{v}\rfloor\rfloor\mathbf{AB}))\\
&= (\boldsymbol{v}\boldsymbol{v}\rfloor\rfloor\overline{\overline{\mathsf{Q}}}^{(2)})^{\wedge}_{\wedge}(\mathbf{A}\wedge(\boldsymbol{v}\rfloor\mathbf{A})(\mathbf{B}\wedge(\boldsymbol{v}\rfloor\mathbf{B}))\\
&= \frac{1}{4}\boldsymbol{\varepsilon}_N\boldsymbol{\varepsilon}_N||((\boldsymbol{v}\boldsymbol{v}\rfloor\rfloor\overline{\overline{\mathsf{Q}}}^{(2)})^{\wedge}_{\wedge}(\boldsymbol{v}\boldsymbol{v}\rfloor\rfloor(\mathbf{AB}^{\wedge}_{\wedge}\mathbf{AB})))\\
&= \frac{1}{4}(\boldsymbol{\varepsilon}_N\boldsymbol{\varepsilon}_N||(\mathbf{AB}^{\wedge}_{\wedge}\mathbf{AB}))(\boldsymbol{v}\boldsymbol{v}\rfloor\rfloor\overline{\overline{\mathsf{Q}}}^{(2)})^{\wedge}_{\wedge}(\boldsymbol{v}\boldsymbol{v}\rfloor\rfloor\mathbf{e}_N\mathbf{e}_N)\\
&= \frac{1}{4}(\boldsymbol{\varepsilon}_N\boldsymbol{\varepsilon}_N||(\mathbf{AB}^{\wedge}_{\wedge}\mathbf{AB}))(\boldsymbol{v}\boldsymbol{v}\rfloor\rfloor\overline{\overline{\mathsf{Q}}}^{(2)})^{\wedge}_{\wedge}(\boldsymbol{v}\boldsymbol{v}\rfloor\rfloor\mathbf{e}_N\mathbf{e}_N) = 0,
\end{aligned}$$

because $(\boldsymbol{v}\rfloor\mathbf{C})\wedge(\boldsymbol{v}\rfloor\mathbf{e}_N) = 0$ for any bivector $\mathbf{C}$ (check by setting $\boldsymbol{v} = \boldsymbol{\varepsilon}_1$). Applying the expansion

$$\begin{aligned}
(\boldsymbol{v}\boldsymbol{v}\rfloor\rfloor\overline{\overline{\mathsf{Q}}}^{(2)})^{(2)} &= (\boldsymbol{v}\boldsymbol{v}||\overline{\overline{\mathsf{Q}}})^2\overline{\overline{\mathsf{Q}}}^{(2)} - (\boldsymbol{v}\boldsymbol{v}||\overline{\overline{\mathsf{Q}}})\overline{\overline{\mathsf{Q}}}^{\wedge}_{\wedge}(\overline{\overline{\mathsf{Q}}}|\boldsymbol{v}\boldsymbol{v}|\overline{\overline{\mathsf{Q}}})\\
&= (\boldsymbol{v}\boldsymbol{v}||\overline{\overline{\mathsf{Q}}})(\boldsymbol{v}\boldsymbol{v}\rfloor\rfloor\overline{\overline{\mathsf{Q}}}^{(3)})\\
&= (\boldsymbol{v}\boldsymbol{v}||\overline{\overline{\mathsf{Q}}})(\boldsymbol{v}\boldsymbol{v}\rfloor\rfloor(\mathbf{e}_N\mathbf{e}_N\lfloor\lfloor\overline{\overline{\mathsf{Q}}}^{-1T}))(\boldsymbol{\varepsilon}_N\boldsymbol{\varepsilon}_N||\overline{\overline{\mathsf{Q}}}^{(4)}),
\end{aligned}$$

the first term becomes

$$\begin{aligned}
D_1 &= \boldsymbol{\varepsilon}_N\boldsymbol{\varepsilon}_N||(\overline{\overline{\mathsf{Q}}}^{(2)\wedge}_{\ \ \wedge}(\boldsymbol{v}\boldsymbol{v}\rfloor\rfloor\overline{\overline{\mathsf{Q}}}^{(2)})^{(2)})\\
&= (\boldsymbol{v}\boldsymbol{v}||\overline{\overline{\mathsf{Q}}})(\boldsymbol{\varepsilon}_N\boldsymbol{\varepsilon}_N||\overline{\overline{\mathsf{Q}}}^{(4)})\boldsymbol{\varepsilon}_N\boldsymbol{\varepsilon}_N||(\overline{\overline{\mathsf{Q}}}^{(2)\wedge}_{\ \ \wedge}(\boldsymbol{v}\boldsymbol{v}\rfloor\rfloor(\mathbf{e}_N\mathbf{e}_N\lfloor\lfloor\overline{\overline{\mathsf{Q}}}^{-1T})))\\
&= (\boldsymbol{v}\boldsymbol{v}||\overline{\overline{\mathsf{Q}}})(\boldsymbol{\varepsilon}_N\boldsymbol{\varepsilon}_N||\overline{\overline{\mathsf{Q}}}^{(4)})(\overline{\overline{\mathsf{Q}}}^{(2)}||(\boldsymbol{\varepsilon}_N\boldsymbol{\varepsilon}_N\lfloor\lfloor(\mathbf{e}_N\mathbf{e}_N\lfloor\lfloor(\overline{\overline{\mathsf{Q}}}^{-1T\wedge}_{\ \ \ \ \wedge}\boldsymbol{v}\boldsymbol{v})))\\
&= (\boldsymbol{v}\boldsymbol{v}||\overline{\overline{\mathsf{Q}}})(\boldsymbol{\varepsilon}_N\boldsymbol{\varepsilon}_N||\overline{\overline{\mathsf{Q}}}^{(4)})(\overline{\overline{\mathsf{Q}}}^{(2)}\lfloor\lfloor(\overline{\overline{\mathsf{Q}}}^{-1T})||\boldsymbol{v}\boldsymbol{v}\\
&= (\boldsymbol{v}\boldsymbol{v}||\overline{\overline{\mathsf{Q}}})(\boldsymbol{\varepsilon}_N\boldsymbol{\varepsilon}_N||\overline{\overline{\mathsf{Q}}}^{(4)})(4\overline{\overline{\mathsf{Q}}} - \overline{\overline{\mathsf{Q}}}|\overline{\overline{\mathsf{Q}}}^{-1}|\overline{\overline{\mathsf{Q}}})||\boldsymbol{v}\boldsymbol{v}\\
&= (\boldsymbol{v}\boldsymbol{v}||\overline{\overline{\mathsf{Q}}})^2(\boldsymbol{\varepsilon}_N\boldsymbol{\varepsilon}_N||\overline{\overline{\mathsf{Q}}}^{(4)}).
\end{aligned}$$

This yields $D(\boldsymbol{v})$ for the Q-medium case, that is, for $\mathbf{AB} = 0$. For the generalized Q-medium the second and third terms must be added. The second term can be expanded as

$$\begin{aligned}
D_2 &= \boldsymbol{\varepsilon}_N\boldsymbol{\varepsilon}_N||((\mathbf{AB})^{\wedge}_{\wedge}(\boldsymbol{v}\boldsymbol{v}\rfloor\rfloor\overline{\overline{\mathsf{Q}}}^{(2)})^{(2)})\\
&= (\boldsymbol{v}\boldsymbol{v}||\overline{\overline{\mathsf{Q}}})(\boldsymbol{\varepsilon}_N\boldsymbol{\varepsilon}_N||\overline{\overline{\mathsf{Q}}}^{(4)})\boldsymbol{\varepsilon}_N\boldsymbol{\varepsilon}_N||((\mathbf{AB})^{\wedge}_{\wedge}(\boldsymbol{v}\boldsymbol{v}\rfloor\rfloor(\mathbf{e}_N\mathbf{e}_N\lfloor\lfloor\overline{\overline{\mathsf{Q}}}^{-1T})))\\
&= (\boldsymbol{v}\boldsymbol{v}||\overline{\overline{\mathsf{Q}}})(\boldsymbol{\varepsilon}_N\boldsymbol{\varepsilon}_N||\overline{\overline{\mathsf{Q}}}^{(4)})((\boldsymbol{\varepsilon}_N\boldsymbol{\varepsilon}_N\lfloor\lfloor\mathbf{AB})||(\mathbf{e}_N\mathbf{e}_N\lfloor\lfloor(\overline{\overline{\mathsf{Q}}}^{-1T\wedge}_{\ \ \ \ \wedge}\boldsymbol{v}\boldsymbol{v}))\\
&= (\boldsymbol{v}\boldsymbol{v}||\overline{\overline{\mathsf{Q}}})(\boldsymbol{\varepsilon}_N\boldsymbol{\varepsilon}_N||\overline{\overline{\mathsf{Q}}}^{(4)})(\mathbf{AB}\lfloor\lfloor\overline{\overline{\mathsf{Q}}}^{-1T})||\boldsymbol{v}\boldsymbol{v},
\end{aligned}$$

and the third term as

$$\begin{aligned}
D_3 &= \varepsilon_N\varepsilon_N||(\overline{\overline{\mathsf{Q}}}^{(2)}{}^{\wedge}_{\wedge}(\mathbf{vv}\rfloor\rfloor\overline{\overline{\mathsf{Q}}}^{(2)}){}^{\wedge}_{\wedge}(\mathbf{vv}\rfloor\rfloor\mathbf{AB}))\\
&= \varepsilon_N\varepsilon_N||(3(\overline{\overline{\mathsf{Q}}}||\mathbf{vv})\overline{\overline{\mathsf{Q}}}^{(3)} - \overline{\overline{\mathsf{Q}}}^{(2)}{}^{\wedge}_{\wedge}(\overline{\overline{\mathsf{Q}}}|\mathbf{vv}|\overline{\overline{\mathsf{Q}}})){}^{\wedge}_{\wedge}(\mathbf{vv}\rfloor\rfloor\mathbf{AB}))\\
&= (\varepsilon_N\varepsilon_N\lfloor\lfloor(2(\overline{\overline{\mathsf{Q}}}||\mathbf{vv})\overline{\overline{\mathsf{Q}}}^{(3)} + \mathbf{vv}\rfloor\rfloor\overline{\overline{\mathsf{Q}}}^{(4)})||(\mathbf{vv}\rfloor\rfloor\mathbf{AB})\\
&= 2(\overline{\overline{\mathsf{Q}}}||\mathbf{vv})(\varepsilon_N\varepsilon_N\lfloor\lfloor\overline{\overline{\mathsf{Q}}}^{(3)})||(\mathbf{vv}\rfloor\rfloor\mathbf{AB})\\
&= 2(\overline{\overline{\mathsf{Q}}}||\mathbf{vv})(\varepsilon_N\varepsilon_N||\overline{\overline{\mathsf{Q}}}^{(4)})(\overline{\overline{\mathsf{Q}}}^{-1T}\rfloor\rfloor\mathbf{AB})||\mathbf{vv}.
\end{aligned}$$

Summing up, we finally obtain

$$D(\mathbf{v}) = (\varepsilon_N\varepsilon_N||\overline{\overline{\mathsf{Q}}}^{(4)})(\mathbf{vv}||\overline{\overline{\mathsf{Q}}})(\overline{\overline{\mathsf{Q}}} + \mathbf{AB}\lfloor\lfloor\overline{\overline{\mathsf{Q}}}^{-1T})||\mathbf{vv},$$

as the dispersion equation for the generalized Q-medium.

8.13 Expanding the first one of the two terms in (8.167), with (8.166) inserted, yields

$$\begin{aligned}
&\varepsilon_N|(\mathbf{A}\wedge\overline{\overline{\mathsf{Q}}}^{(2)})|(\mathbf{v}_2\mathbf{v}_2{}^{\wedge}_{\wedge}\overline{\overline{\mathsf{Q}}}^{-1})|\mathbf{A} = \varepsilon_N|(\mathbf{A}\wedge((\overline{\overline{\mathsf{Q}}}|\mathbf{v}_2\mathbf{v}_2){}^{\wedge}_{\wedge}(\overline{\overline{\mathsf{Q}}}|\overline{\overline{\mathsf{Q}}}^{-1}))|\mathbf{A}\\
&= -\varepsilon_N|((\mathbf{A}\wedge(\overline{\overline{\mathsf{Q}}}|\mathbf{v}_2)\wedge\overline{\overline{\mathsf{I}}})\wedge\mathbf{v}_2)|\mathbf{A} = \varepsilon_N|(\mathbf{A}\wedge(\mathbf{v}_2\rfloor\mathbf{A})\wedge(\overline{\overline{\mathsf{Q}}}|\mathbf{v}_2))\\
&= \frac{1}{2}(\mathbf{A}\cdot\mathbf{A})\varepsilon_N|((\mathbf{v}_2\rfloor\mathbf{e}_N)\wedge(\overline{\overline{\mathsf{Q}}}|\mathbf{v}_2)) = \frac{1}{2}(\mathbf{A}\cdot\mathbf{A})(\mathbf{v}_2|\overline{\overline{\mathsf{Q}}}|\mathbf{v}_2)
\end{aligned}$$

here we have applied the rule

$$\mathbf{A}\wedge(\mathbf{v}_2\rfloor\mathbf{A}) = \frac{1}{2}\mathbf{v}_2\rfloor(\mathbf{A}\wedge\mathbf{A}) = \frac{1}{2}\mathbf{v}_2\rfloor\mathbf{e}_N(\mathbf{A}\cdot\mathbf{A}).$$

The second term can be expanded as

$$\begin{aligned}
&\frac{1}{2}(\mathbf{A}\cdot\mathbf{A})\mathbf{B}|(\mathbf{v}_2\mathbf{v}_2{}^{\wedge}_{\wedge}\overline{\overline{\mathsf{Q}}}^{-1})|\mathbf{A} = \frac{1}{2}(\mathbf{A}\cdot\mathbf{A})(\mathbf{v}_2\mathbf{v}_2{}^{\wedge}_{\wedge}\overline{\overline{\mathsf{Q}}}^{-1})||\mathbf{BA}\\
&= \frac{1}{2}(\mathbf{A}\cdot\mathbf{A})\mathbf{v}_2\mathbf{v}_2||(\overline{\overline{\mathsf{Q}}}^{-1}\rfloor\rfloor\mathbf{BA}) = -\frac{1}{2}(\mathbf{A}\cdot\mathbf{A})\overline{\overline{\mathsf{Q}}}||\mathbf{v}_2\mathbf{v}_2,
\end{aligned}$$

the last step of which is due to the dispersion equation (8.156). Summing the two results yields zero, whence the condition (8.167) is verified.

8.14 Assuming that $\overline{\overline{\alpha}}$ has a spatial inverse $\overline{\overline{\alpha}}^{-1}$, the medium conditions for the P-medium can be written in the form

$$\begin{pmatrix}\mathbf{B}\\ \mathbf{H}\end{pmatrix} = \begin{pmatrix}\overline{\overline{\mathsf{A}}} & \overline{\overline{\mathsf{B}}}\\ \overline{\overline{\mathsf{C}}} & \overline{\overline{\mathsf{D}}}\end{pmatrix}|\begin{pmatrix}\mathbf{D}\\ \mathbf{E}\end{pmatrix} = \begin{pmatrix}\overline{\overline{\alpha}}^{-1} & -\overline{\overline{\alpha}}^{-1}|\overline{\overline{\epsilon}}'\\ \overline{\overline{\mu}}^{-1}|\overline{\overline{\alpha}}^{-1} & \overline{\overline{\beta}} - \overline{\overline{\mu}}^{-1}|\overline{\overline{\alpha}}^{-1}|\overline{\overline{\epsilon}}'\end{pmatrix}|\begin{pmatrix}\mathbf{D}\\ \mathbf{E}\end{pmatrix}$$

whence, inserting (8.44)–(8.47), the medium dyadics can be expanded as

$$\overline{\overline{\mathsf{A}}} = \overline{\overline{\mathsf{P}}}^{(-2)T}$$
$$\overline{\overline{\mathsf{B}}} = -\overline{\overline{\mathsf{P}}}^{(-2)T}|(\boldsymbol{\pi}_s \wedge \overline{\overline{\mathsf{P}}}_s^T) = (\overline{\overline{\mathsf{P}}}_s^{-1T}|\boldsymbol{\pi}_s) \wedge \overline{\overline{\mathsf{I}}}_s^T$$
$$\overline{\overline{\mathsf{C}}} = -(\overline{\overline{\mathsf{P}}}_s^T \wedge \mathbf{p}_s)|\overline{\overline{\mathsf{P}}}_s^{(-2)T} = -\overline{\overline{\mathsf{I}}}_s^T \wedge (\mathbf{p}_s|\overline{\overline{\mathsf{P}}}_s^{-1T})$$
$$\overline{\overline{\mathsf{D}}} = \boldsymbol{\pi}_s\mathbf{p}_s - p\overline{\overline{\mathsf{P}}}_s^T - (\overline{\overline{\mathsf{P}}}_s^T \wedge \mathbf{p}_s)|\overline{\overline{\mathsf{P}}}_s^{(-2)T}|(\boldsymbol{\pi}_s \wedge \overline{\overline{\mathsf{P}}}_s^T)$$
$$= (\boldsymbol{\pi}_s|\overline{\overline{\mathsf{P}}}_s^{-1}|\mathbf{p}_s - p)\overline{\overline{\mathsf{P}}}_s^T.$$

The corresponding expressions for the Gibbsian dyadics $\overline{\overline{\mathsf{A}}}_g - \overline{\overline{\mathsf{D}}}_g$ are obtained (see Appendix B) through

$$\begin{pmatrix} \overline{\overline{\mathsf{A}}}_g & \overline{\overline{\mathsf{B}}}_g \\ \overline{\overline{\mathsf{C}}}_g & \overline{\overline{\mathsf{D}}}_g \end{pmatrix} = \begin{pmatrix} (\mathbf{e}_{123}\lfloor\overline{\overline{\mathsf{A}}}\rfloor\boldsymbol{\varepsilon}_{123})|\overline{\overline{\mathsf{G}}}_s & \mathbf{e}_{123}\lfloor\overline{\overline{\mathsf{B}}} \\ (\overline{\overline{\mathsf{G}}}_s|\overline{\overline{\mathsf{C}}}\rfloor\boldsymbol{\varepsilon}_{123})|\overline{\overline{\mathsf{G}}}_s & \overline{\overline{\mathsf{G}}}_s|\overline{\overline{\mathsf{D}}} \end{pmatrix},$$

after which (8.50)–(8.53) are straightforwardly obtained.

8.15 In terms of the spatial expansions (8.172) for the bivectors **C** and **D** the orthogonality condition (8.191) can be expressed as

$$\mathbf{e}_{123}|(\boldsymbol{\gamma}_s \wedge \mathbf{B}) + \mathbf{c}_s|\mathbf{E} = 0,$$

while the one corresponding to (8.192) has a more complicated form. Let us first expand

$$\mathbf{D}\cdot\mathbf{D} = (\mathbf{e}_{123}\lfloor\boldsymbol{\delta}_s + \mathbf{d}_s \wedge \mathbf{e}_s)|\boldsymbol{\varepsilon}_N\lfloor(\mathbf{e}_{123}\lfloor\boldsymbol{\delta}_s + \mathbf{d}_s \wedge \mathbf{e}_s)$$
$$= (\mathbf{e}_{123}\lfloor\boldsymbol{\delta}_s + \mathbf{d}_s \wedge \mathbf{e}_s)|(-\boldsymbol{\varepsilon}_4 \wedge \boldsymbol{\delta}_s + \boldsymbol{\varepsilon}_{123}\lfloor\mathbf{d}_s) = 2\boldsymbol{\delta}_s|\mathbf{d}_s,$$

and insert it, together with the spatial expansion (8.42) in (8.192). The result can be written as

$$(\mathbf{e}_{123}|(\boldsymbol{\delta}_s \wedge \overline{\overline{\mathsf{P}}}_s^{(2)T} + \mathbf{d}_s|\overline{\overline{\mathsf{P}}}_s^T \wedge \mathbf{p}_s - (\boldsymbol{\delta}_s|\mathbf{d}_s)\mathbf{e}_{123}\lfloor\boldsymbol{\gamma}_s)|\mathbf{B}$$
$$+(-(\mathbf{e}_{123}|(\boldsymbol{\delta}_s \wedge \boldsymbol{\pi}_s\overline{\overline{\mathsf{P}}}_s^T) + p\mathbf{d}_s|\overline{\overline{\mathsf{P}}}_s^T - (\mathbf{d}_s|\boldsymbol{\pi}_s)\mathbf{p}_s - (\boldsymbol{\delta}_s|\mathbf{d}_s)\mathbf{c}_s)|\mathbf{E} = 0.$$

8.16 From (8.143) we obtain

$$\overline{\overline{\mathsf{M}}}_d = \frac{1}{Z^2}(\overline{\overline{\mathsf{Q}}}^{(2)} + \mathbf{AB})^{-1}\rfloor\mathbf{e}_N$$
$$= \frac{1}{Z^2}\overline{\overline{\mathsf{Q}}}^{(-2)}\rfloor\mathbf{e}_N - \frac{1}{Z^2}\frac{(\overline{\overline{\mathsf{Q}}}^{(-2)}|\mathbf{A})(\mathbf{B}|\overline{\overline{\mathsf{Q}}}^{(-2)})}{1 - \mathbf{B}|\overline{\overline{\mathsf{Q}}}^{(-2)}|\mathbf{A}}\rfloor\mathbf{e}_N$$

and

$$\overline{\overline{\mathsf{M}}}_{md} = \frac{1}{Z^2\Delta_Q}\overline{\overline{\mathsf{Q}}}^{(2)T} - \frac{1}{Z^2\Delta_Q}\frac{(\overline{\overline{\mathsf{Q}}}^{(2)T}\rfloor\varepsilon_N|\mathbf{A})(\mathbf{B}|\varepsilon_N\lfloor\overline{\overline{\mathsf{Q}}}^{(2)T})}{\Delta_Q - \mathbf{B}|\varepsilon_N\lfloor\overline{\overline{\mathsf{Q}}}^{(2)T}||\rfloor\varepsilon_N\mathbf{A}}$$

$$= \frac{1}{Z^2\Delta_Q}\overline{\overline{\mathsf{Q}}}^{(2)T} - \frac{1}{Z^2\Delta_Q}\frac{(\overline{\overline{\mathsf{Q}}}^{(2)T}\rfloor\varepsilon_N|\mathbf{A})(\mathbf{B}|\varepsilon_N\lfloor\overline{\overline{\mathsf{Q}}}^{(2)T})}{\Delta_Q - \mathbf{B}|\varepsilon_N\lfloor\overline{\overline{\mathsf{Q}}}^{(2)T}||\rfloor\varepsilon_N\mathbf{A}}$$

$$= \frac{1}{Z^2\Delta_Q}\overline{\overline{\mathsf{Q}}}^{(2)T} - \frac{1}{Z^2\Delta_Q}\frac{(\mathbf{A}\cdot\overline{\overline{\mathsf{Q}}}^{(2)})(\overline{\overline{\mathsf{Q}}}^{(2)}\cdot\mathbf{B})}{\Delta_Q - \mathbf{A}\cdot\overline{\overline{\mathsf{Q}}}^{(2)}\cdot\mathbf{B}}.$$

The self-dual property requires

$$\overline{\overline{\mathsf{Q}}}^{(2)} = \overline{\overline{\mathsf{Q}}}^{(2)T}/Z^2\Delta_Q \quad \Rightarrow \quad \overline{\overline{\mathsf{Q}}}^{(2)T} = \pm\overline{\overline{\mathsf{Q}}}^{(2)}, \quad Z^2 = \pm 1/\Delta_Q$$

whence $\overline{\overline{\mathsf{Q}}}^{(2)}$ must be a symmetric or an antisymmetric bidyadic. Further, from

$$\mathbf{AB} = \mathbf{A}_d\mathbf{B}_d = \mp\frac{(\mathbf{A}\cdot\overline{\overline{\mathsf{Q}}}^{(2)})(\overline{\overline{\mathsf{Q}}}^{(2)}\cdot\mathbf{B})}{\Delta_Q - \mathbf{A}\cdot\overline{\overline{\mathsf{Q}}}^{(2)}\cdot\mathbf{B}}.$$

It follows that $\mathbf{A}$ and $\mathbf{B}$ must be eigenbivectors of the bidyadic $\overline{\overline{\mathsf{Q}}}^{(2)}$

$$\mathbf{A}\cdot\overline{\overline{\mathsf{Q}}}^{(2)} = \lambda_a\mathbf{A}, \quad \overline{\overline{\mathsf{Q}}}^{(2)}\cdot\mathbf{B} = \lambda_b\mathbf{B}.$$

Because of

$$\mathbf{A}\cdot\overline{\overline{\mathsf{Q}}}^{(2)}\cdot\mathbf{B} = \lambda_a\mathbf{A}\cdot\mathbf{B} = \lambda_b\mathbf{A}\cdot\mathbf{B},$$

either $\mathbf{A}\cdot\mathbf{B} = 0$ or $\lambda_a = \lambda_b = \lambda$. In the latter case the eigenvalue must satisfy the equation

$$\lambda^2 = \mp(\Delta_Q - \mathbf{A}\cdot\overline{\overline{\mathsf{Q}}}^{(2)}\cdot\mathbf{B}) = \mp\Delta_Q \pm \lambda(\mathbf{A}\cdot\mathbf{B}).$$

8.17 Plane-wave fields in a P-medium satisfy

$$\boldsymbol{\nu}\wedge\boldsymbol{\Psi} = \boldsymbol{\nu}\wedge(\overline{\overline{\mathsf{P}}}^T|\boldsymbol{\nu})\wedge(\overline{\overline{\mathsf{P}}}^T|\boldsymbol{\phi}) = 0,$$

whence there is no restriction to $\boldsymbol{\nu}$ and, for some coefficients A, B, the potential and fields satisfy

$$\boldsymbol{\phi} = \overline{\overline{\mathsf{P}}}^{-1T}|(A\boldsymbol{\nu} + B(\overline{\overline{\mathsf{P}}}^T|\boldsymbol{\nu}))$$

$$\boldsymbol{\Phi} = \boldsymbol{\nu}\wedge\overline{\overline{\mathsf{P}}}^{-1T}|(A\boldsymbol{\nu} + B(\overline{\overline{\mathsf{P}}}^T|\boldsymbol{\nu})) = A\boldsymbol{\nu}\wedge(\overline{\overline{\mathsf{P}}}^{-1T}|\boldsymbol{\nu})$$

$$\boldsymbol{\Psi} = \overline{\overline{\mathsf{P}}}^{(2)T}|\boldsymbol{\Phi} = A(\overline{\overline{\mathsf{P}}}^T|\boldsymbol{\nu})\wedge\boldsymbol{\nu}.$$

Inserting the special dyadic $\overline{\overline{\mathsf{P}}}$ whose inverse can be expressed as

$$\overline{\overline{\mathsf{P}}}^{-1} = \frac{1}{P}(\mathbf{e}_1\boldsymbol{\varepsilon}_1 + \mathbf{e}_2\boldsymbol{\varepsilon}_2) + \overline{\overline{\mathsf{P}}}'_{34}$$

$$\overline{\overline{\mathsf{P}}}'_{34} = (P\mathbf{e}_3\boldsymbol{\varepsilon}_3 + P\mathbf{e}_4\boldsymbol{\varepsilon}_4 + \overline{\overline{\mathsf{P}}}_{34})^{-1}$$

$$= \frac{1}{\Delta}((P + P_{4,4})\mathbf{e}_3\boldsymbol{\varepsilon}_3 - P_{3,4}\mathbf{e}_3\boldsymbol{\varepsilon}_4 - P_{4,3}\mathbf{e}_4\boldsymbol{\varepsilon}_3 + (P + P_{3,3})\mathbf{e}_4\boldsymbol{\varepsilon}_4)$$

$$\Delta = (P + P_{3,3})(P + P_{4,4}) - P_{3,4}P_{4.3},$$

we obtain

$$\boldsymbol{\Phi} = \frac{A}{P}(\boldsymbol{\nu} \wedge (-\boldsymbol{\varepsilon}_3\mathbf{e}_3 - \boldsymbol{\varepsilon}_4\mathbf{e}_4))|\boldsymbol{\nu} + A\boldsymbol{\nu} \wedge (\overline{\overline{\mathsf{P}}}'^{\,T}_{34}|\boldsymbol{\nu})$$

$$\boldsymbol{\Psi} = A(\overline{\overline{\mathsf{P}}}_{34}^{\,T}|\boldsymbol{\nu}) \wedge \boldsymbol{\nu}.$$

It follows that the field two-forms satisfy in the medium

$$\boldsymbol{\varepsilon}_{34} \wedge \boldsymbol{\Phi} = \boldsymbol{\varepsilon}_{34} \wedge \mathbf{B} = 0, \quad \Rightarrow \quad \boldsymbol{\varepsilon}_3 \wedge \mathbf{B} = 0,$$

$$\boldsymbol{\varepsilon}_{34} \wedge \boldsymbol{\Psi} = \boldsymbol{\varepsilon}_{34} \wedge \mathbf{D} = 0, \quad \Rightarrow \quad \boldsymbol{\varepsilon}_3 \wedge \mathbf{D} = 0,$$

producing DB-boundary conditions at an interface $\boldsymbol{\varepsilon}_3|\mathbf{x} = 0$.

CHAPTER 9

9.1 Expanding

$$\overline{\overline{\mathsf{M}}}^2 = M^2\overline{\overline{\mathsf{I}}}^{(2)T} + (2M + \mathbf{A}\cdot\mathbf{B})\boldsymbol{\varepsilon}_N\lfloor\mathbf{AB}$$

$$= (2M + \mathbf{A}\cdot\mathbf{B})\overline{\overline{\mathsf{M}}} - M(M + \mathbf{A}\cdot\mathbf{B})\overline{\overline{\mathsf{I}}}^{(2)T}$$

the bidyadic $\overline{\overline{\mathsf{M}}}$ is seen to satisfy the second-order equation.

9.2 Inserting the bidyadic $\overline{\overline{\mathsf{D}}}$ and its inverse in (9.45) we can proceed as

$$\overline{\overline{\mathsf{M}}}_1 = \overline{\overline{\mathsf{D}}}^{-1}|\overline{\overline{\mathsf{M}}}|\overline{\overline{\mathsf{D}}}$$

$$= (\overline{\overline{\alpha}} + \overline{\overline{\epsilon}}' \wedge \mathbf{e}_4 + \boldsymbol{\varepsilon}_4 \wedge \overline{\overline{\mu}}^{-1} + \boldsymbol{\varepsilon}_4 \wedge \overline{\overline{\beta}} \wedge \mathbf{e}_4$$

$$-\overline{\overline{\alpha}}|\overline{\overline{\mu}}|(\overline{\overline{\mu}}^{-1} + \overline{\overline{\beta}} \wedge \mathbf{e}_4))|\overline{\overline{\mathsf{D}}}$$

$$= (\overline{\overline{\epsilon}}' \wedge \mathbf{e}_4 + \boldsymbol{\varepsilon}_4 \wedge \overline{\overline{\mu}}^{-1} + \boldsymbol{\varepsilon}_4 \wedge \overline{\overline{\beta}} \wedge \mathbf{e}_4 - \overline{\overline{\alpha}}|\overline{\overline{\mu}}|\overline{\overline{\beta}} \wedge \mathbf{e}_4)|\overline{\overline{\mathsf{D}}}$$

$$= (\overline{\overline{\epsilon}}' - \overline{\overline{\alpha}}|\overline{\overline{\mu}}|\overline{\overline{\beta}}) \wedge \mathbf{e}_4 + \boldsymbol{\varepsilon}_4 \wedge \overline{\overline{\mu}}^{-1} + \boldsymbol{\varepsilon}_4 \wedge (\overline{\overline{\beta}} - \overline{\overline{\mu}}^{-1}|\overline{\overline{\alpha}}|\overline{\overline{\mu}}) \wedge \mathbf{e}_4.$$

This is valid for any $\overline{\overline{\mathsf{M}}}$. Applying $\overline{\overline{\mu}}^{-1}|\overline{\overline{\alpha}} = \overline{\overline{\beta}}|\overline{\overline{\mu}}^{-1}$ from (9.40) we have

$$\overline{\overline{\mathsf{D}}}^{-1}|\overline{\overline{\mathsf{M}}}|\overline{\overline{\mathsf{D}}} = (\overline{\overline{\epsilon}}' \wedge \mathbf{e}_4 - \overline{\overline{\alpha}}|\overline{\overline{\mu}}|\overline{\overline{\beta}}) \wedge \mathbf{e}_4 + \boldsymbol{\varepsilon}_4 \wedge \overline{\overline{\mu}}^{-1}.$$

From (9.40) and (9.41) we obtain

$$\overline{\overline{\alpha}}|\overline{\overline{\mu}}|\overline{\overline{\beta}} = \overline{\overline{\mu}}|\overline{\overline{\beta}}^2 = \overline{\overline{\mu}}|(\overline{\overline{\mu}}^{-1}|\overline{\overline{\epsilon}}' - M^2\overline{\overline{\mathsf{I}}}_s^T) = \overline{\overline{\epsilon}}' - M^2\overline{\overline{\mu}},$$

whence, finally, we arrive at the result

$$\overline{\overline{\mathsf{M}}}_1 = M^2\overline{\overline{\mu}} \wedge \mathbf{e}_4 + \boldsymbol{\varepsilon}_4 \wedge \overline{\overline{\mu}}^{-1}.$$

9.3 In analogy to the previous problem, we can define the transformation bidyadics

$$\overline{\overline{\mathsf{D}}} = \overline{\overline{\mathsf{I}}}^{(2)T} + \boldsymbol{\varepsilon}_4 \wedge \overline{\overline{\beta}}|\overline{\overline{\epsilon}}'^{-1}, \quad \overline{\overline{\mathsf{D}}}^{-1} = \overline{\overline{\mathsf{I}}}^{(2)T} - \boldsymbol{\varepsilon}_4 \wedge \overline{\overline{\beta}}|\overline{\overline{\epsilon}}'^{-1}$$

and proceed similarly,

$$\begin{aligned}
\overline{\overline{\mathsf{M}}}_1 &= \overline{\overline{\mathsf{D}}}^{-1}|\overline{\overline{\mathsf{M}}}|\overline{\overline{\mathsf{D}}} \\
&= (\overline{\overline{\alpha}} + \overline{\overline{\epsilon}}' \wedge \mathbf{e}_4 + \boldsymbol{\varepsilon}_4 \wedge \overline{\overline{\mu}}^{-1} + \boldsymbol{\varepsilon}_4 \wedge \overline{\overline{\beta}} \wedge \mathbf{e}_4 \\
&\quad - \boldsymbol{\varepsilon}_4 \wedge \overline{\overline{\beta}}|\overline{\overline{\epsilon}}'^{-1}|(\overline{\overline{\alpha}} + \overline{\overline{\epsilon}}' \wedge \mathbf{e}_4))|\overline{\overline{\mathsf{D}}} \\
&= \overline{\overline{\alpha}} + \overline{\overline{\epsilon}}' \wedge \mathbf{e}_4 + \boldsymbol{\varepsilon}_4 \wedge (\overline{\overline{\mu}}^{-1} - \overline{\overline{\beta}}|\overline{\overline{\epsilon}}'^{-1}|\overline{\overline{\alpha}}) - \epsilon'|\overline{\overline{\beta}}|\overline{\overline{\epsilon}}'^{-1}. \\
&= \overline{\overline{\epsilon}}' \wedge \mathbf{e}_4 + \boldsymbol{\varepsilon}_4 \wedge (\overline{\overline{\mu}}^{-1} - \overline{\overline{\beta}}|\overline{\overline{\epsilon}}'^{-1}|\overline{\overline{\alpha}}),
\end{aligned}$$

where we have applied $\epsilon'|\overline{\overline{\beta}}|\overline{\overline{\epsilon}}'^{-1} = \overline{\overline{\alpha}}$ due to (9.39). Invoking (9.38) we can expand

$$\overline{\overline{\beta}}|\overline{\overline{\epsilon}}'^{-1}|\overline{\overline{\alpha}} = \overline{\overline{\epsilon}}'^{-1}|\overline{\overline{\alpha}}^2 = \overline{\overline{\mu}}^{-1} - M^2\overline{\overline{\epsilon}}'^{-1},$$

whence the result becomes

$$\overline{\overline{\mathsf{M}}}_1 = \overline{\overline{\epsilon}}' \wedge \mathbf{e}_4 + M^2\boldsymbol{\varepsilon}_4 \wedge \overline{\overline{\epsilon}}'^{-1}.$$

9.4 Applying the trigonometric rules

$$\tan(\theta/2) = \sin\theta/(1+\cos\theta) = (1-\cos\theta)/\sin\theta = 1/\cot(\theta/2)$$

(9.33) can be written as

$$M_\pm = \mp\frac{1}{Z}\begin{pmatrix}\tan(\theta/2)\\ \cot(\theta/2)\end{pmatrix} = \mp\frac{\sin\theta}{Z(1\pm\cos\theta)}$$

whence (9.34) can be verified by

$$\begin{pmatrix} -\cos\theta & -(1/Z)\sin\theta \\ -Z\sin\theta & \cos\theta \end{pmatrix}\begin{pmatrix} M_\pm \\ 1 \end{pmatrix} = \begin{pmatrix} \pm\frac{\sin\theta\cos\theta}{Z(1\pm\cos\theta)} - \frac{1}{Z}\sin\theta \\ \pm\frac{\sin^2\theta}{1\pm\cos\theta} + \cos\theta \end{pmatrix}$$

$$= \begin{pmatrix} -\frac{\sin\theta}{Z(1\pm\cos\theta)} \\ \pm 1 \end{pmatrix} = \pm\begin{pmatrix} M_\pm \\ 1 \end{pmatrix}.$$

9.5 Assuming $\overline{\overline{\epsilon}}' = 0$ and $\overline{\overline{\mu}}^{-1} = 0$, (9.38) and (9.41) yield

$$\overline{\overline{\alpha}}^2 + M^2\overline{\overline{\mathrm{I}}}_s^{(2)T} = (\overline{\overline{\alpha}} - jM\overline{\overline{\mathrm{I}}}^{(2)T})|(\overline{\overline{\alpha}} + jM\overline{\overline{\mathrm{I}}}^{(2)T}) = 0,$$

$$\overline{\overline{\beta}}^2 + M^2\overline{\overline{\mathrm{I}}}_s^T = (\overline{\overline{\beta}} - jM\overline{\overline{\mathrm{I}}}^T)|(\overline{\overline{\beta}} + jM\overline{\overline{\mathrm{I}}}^T) = 0.$$

Since we know (see Problem 4.12) that one of the dyadics in each product must be of rank 1 or 0, without losing generality we can set

$$\overline{\overline{\alpha}} = jM(\overline{\overline{\mathrm{I}}}_s^{(2)T} - 2\boldsymbol{\Theta}\mathbf{A}),$$

$$\overline{\overline{\beta}} = \pm jM(\overline{\overline{\mathrm{I}}}_s^T - 2\boldsymbol{\beta}\mathbf{b}),$$

for some spatial quantities: two-form $\boldsymbol{\Theta}$, one-form $\boldsymbol{\beta}$, bivector $\mathbf{A}$, and vector $\mathbf{b}$. The above conditions require

$$\boldsymbol{\Theta}\mathbf{A} = 0 \quad \text{or} \quad \boldsymbol{\Theta}|\mathbf{A} = 1,$$

$$\boldsymbol{\beta}\mathbf{b} = 0 \quad \text{or} \quad \boldsymbol{\beta}|\mathbf{b} = 1.$$

The four basic cases can be treated separately as

(1) $\boldsymbol{\Theta}\mathbf{A} = 0$ and $\boldsymbol{\beta}\mathbf{b} = 0$, whence

$$\overline{\overline{\mathsf{M}}} = jM(\overline{\overline{\mathrm{I}}}_s^{(2)T} \pm \overline{\overline{\mathrm{I}}}_s{}^\wedge_\wedge\boldsymbol{\varepsilon}_4\mathbf{e}_4) = jM(\overline{\overline{\mathrm{I}}}_s \pm \mathbf{e}_4\boldsymbol{\varepsilon}_4)^{(2)T}.$$

In this case $\overline{\overline{\mathsf{M}}}$ is either a multiple of $\overline{\overline{\mathrm{I}}}^{(2)T}$ or $\overline{\overline{\mathrm{T}}}^{(2)T}$ where $\overline{\overline{\mathrm{T}}} = \overline{\overline{\mathrm{I}}}_s - \mathbf{e}_4\boldsymbol{\varepsilon}_4$ is the time reversion dyadic.

(2) $\boldsymbol{\Theta}\mathbf{A} = 0$, $\boldsymbol{\beta}|\mathbf{b} = 1$, whence

$$\overline{\overline{\mathsf{M}}} = jM((\overline{\overline{\mathrm{I}}}_s \pm \mathbf{e}_4\boldsymbol{\varepsilon}_4)^{(2)T} \mp 2\boldsymbol{\beta}\mathbf{b}{}^\wedge_\wedge\boldsymbol{\varepsilon}_4\mathbf{e}_4).$$

(3) $\boldsymbol{\Theta}|\mathbf{A} = 1$, $\boldsymbol{\beta}\mathbf{b} = 0$, whence

$$\overline{\overline{\mathsf{M}}} = jM((\overline{\overline{\mathrm{I}}}_s \pm \mathbf{e}_4\boldsymbol{\varepsilon}_4)^{(2)T} - 2\boldsymbol{\Theta}\mathbf{A}).$$

(4) $\boldsymbol{\Theta}|\mathbf{A} = \boldsymbol{\beta}|\mathbf{b} = 1$, whence

$$\overline{\overline{\mathsf{M}}} = jM((\overline{\overline{\mathrm{I}}}_s \pm \mathbf{e}_4\boldsymbol{\varepsilon}_4)^{(2)T} - 2(\boldsymbol{\Theta}\mathbf{A} \pm \boldsymbol{\beta}\mathbf{b}{}^{\wedge}_{\wedge}\boldsymbol{\varepsilon}_4\mathbf{e}_4)).$$

The results can be easily verified to satisfy (9.36).

9.6

$$\begin{aligned}\overline{\overline{\mathrm{I}}}^{(4)}\lfloor\lfloor\overline{\overline{\mathsf{M}}}' &= \mathbf{e}_N\lfloor(\overline{\overline{\mathsf{D}}}|\overline{\overline{\mathsf{M}}}|\overline{\overline{\mathsf{D}}}^{-1})\rfloor\boldsymbol{\varepsilon}_N \\ &= (\mathbf{e}_N\lfloor\overline{\overline{\mathsf{D}}}\rfloor\boldsymbol{\varepsilon}_N)|(\mathbf{e}_N\lfloor\overline{\overline{\mathsf{M}}}\rfloor\boldsymbol{\varepsilon}_N)|(\mathbf{e}_N\lfloor\overline{\overline{\mathsf{D}}}^{-1}\rfloor\boldsymbol{\varepsilon}_N) \\ &= (\overline{\overline{\mathrm{I}}}^{(4)}\lfloor\lfloor\overline{\overline{\mathsf{D}}})|(\overline{\overline{\mathrm{I}}}^{(4)}\lfloor\lfloor\overline{\overline{\mathsf{M}}})|(\overline{\overline{\mathrm{I}}}^{(4)}\lfloor\lfloor\overline{\overline{\mathsf{D}}})^{-1}.\end{aligned}$$

Here we have applied the rule

$$\begin{aligned}\mathbf{e}_N\lfloor\overline{\overline{\mathsf{D}}}^{-1}\rfloor\boldsymbol{\varepsilon}_N &= (\mathbf{e}_N\lfloor\overline{\overline{\mathrm{I}}}^{(2)T})|\overline{\overline{\mathsf{D}}}^{-1}|(\overline{\overline{\mathrm{I}}}^{(2)T}\rfloor\boldsymbol{\varepsilon}_N) \\ &= (\boldsymbol{\varepsilon}_N\lfloor\overline{\overline{\mathrm{I}}}^{(2)})^{-1}|\overline{\overline{\mathsf{D}}}^{-1}|(\overline{\overline{\mathrm{I}}}^{(2)}\rfloor\mathbf{e}_N)^{-1} \\ &= (\mathbf{e}_N\lfloor\overline{\overline{\mathsf{D}}}\rfloor\boldsymbol{\varepsilon}_N)^{-1} = (\overline{\overline{\mathrm{I}}}^{(4)}\lfloor\lfloor\overline{\overline{\mathsf{D}}})^{-1}.\end{aligned}$$

9.7 The two bidyadics satisfy the properties

$$\overline{\overline{\mathsf{M}}}^2 - M^2\overline{\overline{\mathrm{I}}}^{(2)T}, \quad \overline{\overline{\mathsf{M}}}_1^2 - M^2\overline{\overline{\mathrm{I}}}^{(2)T}, \quad \overline{\overline{\mathsf{M}}}|\overline{\overline{\mathsf{M}}}_1 = -\overline{\overline{\mathsf{M}}}_1|\overline{\overline{\mathsf{M}}}.$$

Defining $\overline{\overline{\mathsf{D}}} = (\overline{\overline{\mathsf{M}}} + \overline{\overline{\mathsf{M}}}_1)/M$ we have

$$\overline{\overline{\mathsf{D}}}^2 = \frac{1}{M^2}(\overline{\overline{\mathsf{M}}} + \overline{\overline{\mathsf{M}}}_1)^2 = \frac{1}{M^2}(\overline{\overline{\mathsf{M}}}^2 + \overline{\overline{\mathsf{M}}}_1^2) = -2\overline{\overline{\mathrm{I}}}^{(2)T},$$

whence $\overline{\overline{\mathsf{D}}}^{-1} = -\overline{\overline{\mathsf{D}}}/2$. For this choice of $\overline{\overline{\mathsf{D}}}$ we obtain

$$\begin{aligned}\overline{\overline{\mathsf{D}}}|\overline{\overline{\mathsf{M}}}_1|\overline{\overline{\mathsf{D}}}^{-1} &= -\frac{1}{2M^2}(\overline{\overline{\mathsf{M}}} + \overline{\overline{\mathsf{M}}}_1)|\overline{\overline{\mathsf{M}}}_1|(\overline{\overline{\mathsf{M}}} + \overline{\overline{\mathsf{M}}}_1) \\ &= -\frac{1}{2M^2}(\overline{\overline{\mathsf{M}}}|\overline{\overline{\mathsf{M}}}_1|\overline{\overline{\mathsf{M}}} + \overline{\overline{\mathsf{M}}}|\overline{\overline{\mathsf{M}}}_1^2 + \overline{\overline{\mathsf{M}}}_1^2|\overline{\overline{\mathsf{M}}} + \overline{\overline{\mathsf{M}}}_1^3) \\ &= -\frac{1}{2M^2}(-\overline{\overline{\mathsf{M}}}^2|\overline{\overline{\mathsf{M}}}_1 - M^2\overline{\overline{\mathsf{M}}} - M^2\overline{\overline{\mathsf{M}}} - M^2\overline{\overline{\mathsf{M}}}_1) \\ &= -\frac{1}{2M^2}(-2M^2\overline{\overline{\mathsf{M}}}) = \overline{\overline{\mathsf{M}}}.\end{aligned}$$

9.8 Let us expand

$$\begin{aligned}\overline{\overline{\mathsf{M}}} &= \overline{\overline{\epsilon}} \wedge \mathbf{e}_4 + \boldsymbol{\varepsilon}_4 \wedge \overline{\overline{\mu}}^{-1}, \\ \overline{\overline{\mathsf{M}}}^2 &= -\overline{\overline{\epsilon}}|\overline{\overline{\mu}}^{-1} + \boldsymbol{\varepsilon}_4 \wedge \overline{\overline{\mu}}^{-1}|\overline{\overline{\epsilon}} \wedge \mathbf{e}_4 \\ \overline{\overline{\mathsf{M}}}^3 &= -\overline{\overline{\epsilon}}|\overline{\overline{\mu}}^{-1}|\overline{\overline{\epsilon}} \wedge \mathbf{e}_4 - \boldsymbol{\varepsilon}_4 \wedge \overline{\overline{\mu}}^{-1}|\overline{\overline{\epsilon}}|\overline{\overline{\mu}}^{-1}.\end{aligned}$$

Requiring that (9.78) be satisfied leads to four conditions for the spatial medium dyadics $\overline{\overline{\epsilon}}$ and $\overline{\overline{\mu}}^{-1}$,

$$A\overline{\overline{\epsilon}}|\overline{\overline{\mu}}^{-1} = C\overline{\overline{\mathsf{I}}}_s^{(2)T},$$
$$\overline{\overline{\epsilon}}|\overline{\overline{\mu}}^{-1}|\overline{\overline{\epsilon}} = B\overline{\overline{\epsilon}},$$
$$\overline{\overline{\mu}}^{-1}|\overline{\overline{\epsilon}}|\overline{\overline{\mu}}^{-1} = B\overline{\overline{\mu}}^{-1},$$
$$A\overline{\overline{\mu}}^{-1}|\overline{\overline{\epsilon}} = C\overline{\overline{\mathsf{I}}}_s^{T}.$$

Because $\overline{\overline{\mu}}^{-1}$ is of full rank, the dyadic $\overline{\overline{\mu}}$ exists, whence the four conditions reduce to two as

$$\overline{\overline{\epsilon}} = B\overline{\overline{\mu}}, \quad A\overline{\overline{\epsilon}} = C\overline{\overline{\mu}}.$$

The coefficients must satisfy $C = AB$. Because we now have

$$\overline{\overline{\mathsf{M}}}^2 = -B\overline{\overline{\mathsf{I}}}^{(2)T},$$

the medium bidyadic actually satisfies the quadratic equation.

9.9 The eigenproblem

$$(A\overline{\overline{\mathsf{I}}}^{(2)T} + \varepsilon_N\lfloor\mathbf{AB} - \varepsilon_N\lfloor\mathbf{BA})|\mathbf{\Phi}_i = M_i\mathbf{\Phi}_i$$

has a solution of the form

$$\mu_c = A, \quad \mathbf{A}|\mathbf{\Phi}_c = \mathbf{B}|\mathbf{\Phi}_c = 0$$

for any $\mathbf{\Phi}_c$ satisfying the two conditions. The other solutions must have the form

$$\mathbf{\Phi}_i = \alpha_i\varepsilon_N\lfloor\mathbf{A} + \beta_i\varepsilon_N\lfloor\mathbf{B}$$

and M_i, α_i, β_i satisfy the equation

$$\begin{pmatrix} A - M_i + \mathbf{A}\cdot\mathbf{B} & \mathbf{B}\cdot\mathbf{B} \\ -\mathbf{A}\cdot\mathbf{A} & A - M_i - \mathbf{A}\cdot\mathbf{B} \end{pmatrix}\begin{pmatrix}\alpha_i \\ \beta_i\end{pmatrix} = \begin{pmatrix}0 \\ 0\end{pmatrix}.$$

Requiring that the determinant of the matrix vanish, we obtain for $i = \pm$

$$M_\pm = A \pm \lambda, \quad \lambda = \sqrt{(\mathbf{A}\cdot\mathbf{B}^2 - (\mathbf{A}\cdot\mathbf{A})(\mathbf{B}\cdot\mathbf{B}},$$

while the corresponding field two-forms are defined by

$$\mathbf{\Phi}_\pm = \varepsilon_N\lfloor\mathbf{A}(\mathbf{B}\cdot\mathbf{B}) - \varepsilon_N\lfloor\mathbf{B}(A - M_\pm + \mathbf{A}\cdot\mathbf{B})$$
$$= \varepsilon_N\lfloor(\mathbf{AB} - \mathbf{BA})\cdot\mathbf{B} \pm \lambda\varepsilon_N\lfloor\mathbf{B}.$$

9.10 Let us expand powers of the skewon bidyadic $\overline{\overline{\mathsf{C}}}_o$

$$\begin{aligned}\overline{\overline{\mathsf{C}}}_o^2 &= \boldsymbol{\varepsilon}_N\lfloor(\mathbf{AB}-\mathbf{BA})|\boldsymbol{\varepsilon}_N\lfloor(\mathbf{AB}-\mathbf{BA})\\ &= \boldsymbol{\varepsilon}_N\lfloor(\mathbf{A}(\mathbf{B}\cdot\mathbf{AB}-\mathbf{B}\cdot\mathbf{BA})-\mathbf{B}(\mathbf{A}\cdot\mathbf{AB}-\mathbf{A}\cdot\mathbf{BA}))\\ &= (\mathbf{A}\cdot\mathbf{B})\boldsymbol{\varepsilon}_N\lfloor(\mathbf{AB}+\mathbf{BA})-(\mathbf{A}\cdot\mathbf{A})\boldsymbol{\varepsilon}_N\lfloor\mathbf{BB}-(\mathbf{B}\cdot\mathbf{B})\boldsymbol{\varepsilon}_N\lfloor\mathbf{AA}\\ \overline{\overline{\mathsf{C}}}_o^3 &= \boldsymbol{\varepsilon}_N\lfloor(\mathbf{AB}-\mathbf{BA})\cdot\overline{\overline{\mathsf{C}}}_o^2\\ &= (\mathbf{A}\cdot\mathbf{B})\boldsymbol{\varepsilon}_N\lfloor((\mathbf{B}\cdot\mathbf{A})(\mathbf{AB}-\mathbf{BA})+(\mathbf{B}\cdot\mathbf{B})\mathbf{AA}-(\mathbf{A}\cdot\mathbf{A})\mathbf{BB})\\ &\quad-(\mathbf{A}\cdot\mathbf{A})\boldsymbol{\varepsilon}_N\lfloor((\mathbf{B}\cdot\mathbf{B})\mathbf{AB}-(\mathbf{A}\cdot\mathbf{B})\mathbf{BB})\\ &\quad-(\mathbf{B}\cdot\mathbf{B})\boldsymbol{\varepsilon}_N\lfloor((\mathbf{B}\cdot\mathbf{A})\mathbf{AA}-(\mathbf{A}\cdot\mathbf{A})\mathbf{BA})\\ &= ((\mathbf{A}\cdot\mathbf{B})^2-(\mathbf{A}\cdot\mathbf{A})(\mathbf{B}\cdot\mathbf{B}))\boldsymbol{\varepsilon}_N\lfloor(\mathbf{AB}-\mathbf{BA}).\\ &= \frac{1}{2}\mathrm{tr}\overline{\overline{\mathsf{C}}}_o^2\,\overline{\overline{\mathsf{C}}}_o.\end{aligned}$$

The last step is obtained from

$$\begin{aligned}\mathrm{tr}\overline{\overline{\mathsf{C}}}_o^2 &= (\mathbf{A}\cdot\mathbf{B})(\mathbf{A}\cdot\mathbf{B}+\mathbf{B}\cdot\mathbf{A})-(\mathbf{A}\cdot\mathbf{A})(\mathbf{B}\cdot\mathbf{B})-(\mathbf{B}\cdot\mathbf{B})(\mathbf{A}\cdot\mathbf{A})\\ &= 2((\mathbf{A}\cdot\mathbf{B})^2-(\mathbf{A}\cdot\mathbf{A})(\mathbf{B}\cdot\mathbf{B})).\end{aligned}$$

Thus, the medium bidyadic $\overline{\overline{\mathsf{M}}} = A\overline{\overline{\mathsf{I}}}^{(2)T}+\overline{\overline{\mathsf{C}}}_o$ satisfies the cubic equation

$$(\overline{\overline{\mathsf{M}}}-A\overline{\overline{\mathsf{I}}}^{(2)T})^3-\frac{1}{2}\mathrm{tr}(\overline{\overline{\mathsf{M}}}-A\overline{\overline{\mathsf{I}}}^{(2)})^2\,(\overline{\overline{\mathsf{M}}}-A\overline{\overline{\mathsf{I}}}^{(2)T})=0,$$

where $A = \mathrm{tr}\overline{\overline{\mathsf{M}}}/6$.

9.11 Expanding

$$\begin{aligned}&(M_a-M_b)(M_b-M_c)(M_c-M_a)(\overline{\overline{\Pi}}_a+\overline{\overline{\Pi}}_b+\overline{\overline{\Pi}}_c)\\ &= -(M_b-M_c)(\overline{\overline{\mathsf{M}}}-M_b\overline{\overline{\mathsf{I}}}^{(2)T})|(\overline{\overline{\mathsf{M}}}-M_c\overline{\overline{\mathsf{I}}}^{(2)T})\\ &\quad-(M_c-M_a)(\overline{\overline{\mathsf{M}}}-M_c\overline{\overline{\mathsf{I}}}^{(2)T})|(\overline{\overline{\mathsf{M}}}-M_a\overline{\overline{\mathsf{I}}}^{(2)T})\\ &\quad-(M_a-M_b)(\overline{\overline{\mathsf{M}}}-M_a\overline{\overline{\mathsf{I}}}^{(2)T})|(\overline{\overline{\mathsf{M}}}-M_b\overline{\overline{\mathsf{I}}}^{(2)T})\\ &= -((M_b-M_c)+(M_c-M_a)+(M_a-M_b))\overline{\overline{\mathsf{M}}}^2\\ &\quad+((M_b^2-M_c^2)+(M_c^2-M_a^2)+(M_a^2-M_b^2))\overline{\overline{\mathsf{M}}}\\ &\quad-(((M_b-M_c)M_bM_c+(M_c-M_a)M_cM_a\\ &\quad+(M_a-M_b)M_aM_b)\overline{\overline{\mathsf{I}}}^{(2)T}\\ &= (M_a-M_b)(M_b-M_c)(M_c-M_a)\overline{\overline{\mathsf{I}}}^{(2)T}.\end{aligned}$$

The last step can be verified backwards by expanding the last expression.

9.12 Generalizing the procedure of Problem 9.10, we can expand different quantities as

$$\begin{aligned}
\overline{\overline{\mathsf{C}}} &= \boldsymbol{\varepsilon}_N \lfloor (\mathbf{AC} + \mathbf{BD}) \\
\mathrm{tr}\overline{\overline{\mathsf{C}}} &= \mathbf{A}\cdot\mathbf{C} + \mathbf{B}\cdot\mathbf{D}, \\
\mathbf{e}_N \lfloor \overline{\overline{\mathsf{C}}}^2 &= \mathbf{AC}(\mathbf{A}\cdot\mathbf{C}) + \mathbf{AD}(\mathbf{B}\cdot\mathbf{C}) + \mathbf{BC}(\mathbf{A}\cdot\mathbf{D}) + \mathbf{BD}(\mathbf{B}\cdot\mathbf{D}), \\
\mathrm{tr}\overline{\overline{\mathsf{C}}}^2 &= (\mathbf{A}\cdot\mathbf{C})^2 + 2(\mathbf{A}\cdot\mathbf{D})(\mathbf{B}\cdot\mathbf{C}) + (\mathbf{B}\cdot\mathbf{D})^2, \\
\mathbf{e}_N \lfloor \overline{\overline{\mathsf{C}}}^3 &= (\mathbf{AD}(\mathbf{B}\cdot\mathbf{C}) + \mathbf{BC}(\mathbf{A}\cdot\mathbf{D}))(\mathbf{A}\cdot\mathbf{C} + \mathbf{B}\cdot\mathbf{D}) \\
&\quad + (\mathbf{AC} + \mathbf{BD})(\mathbf{A}\cdot\mathbf{D})(\mathbf{B}\cdot\mathbf{C}) + \mathbf{AC}(\mathbf{A}\cdot\mathbf{C})^2 + \mathbf{BD}(\mathbf{B}\cdot\mathbf{D})^2 \\
&= \mathbf{e}_N \lfloor \overline{\overline{\mathsf{C}}}^2 \mathrm{tr}\overline{\overline{\mathsf{C}}} - (\mathbf{AC}(\mathbf{A}\cdot\mathbf{C}) + \mathbf{BD}(\mathbf{B}\cdot\mathbf{D}))(\mathbf{A}\cdot\mathbf{C} + \mathbf{B}\cdot\mathbf{D}) \\
&\quad + (\mathbf{AC} + \mathbf{BD})(\mathbf{A}\cdot\mathbf{D})(\mathbf{B}\cdot\mathbf{C}) + \mathbf{AC}(\mathbf{A}\cdot\mathbf{C})^2 + \mathbf{BD}(\mathbf{B}\cdot\mathbf{D})^2, \\
&= \mathbf{e}_N \lfloor \overline{\overline{\mathsf{C}}}^2 \mathrm{tr}\overline{\overline{\mathsf{C}}} + \mathbf{e}_N \lfloor \overline{\overline{\mathsf{C}}}((\mathbf{A}\cdot\mathbf{D})(\mathbf{B}\cdot\mathbf{C}) - (\mathbf{A}\cdot\mathbf{C})(\mathbf{B}\cdot\mathbf{D}).
\end{aligned}$$

Interpreting the last factor of $\overline{\overline{\mathsf{C}}}$ from

$$(\mathrm{tr}\overline{\overline{\mathsf{C}}})^2 - \mathrm{tr}\overline{\overline{\mathsf{C}}}^2 = 2(\mathbf{A}\cdot\mathbf{C})(\mathbf{B}\cdot\mathbf{D}) - 2(\mathbf{A}\cdot\mathbf{D})(\mathbf{B}\cdot\mathbf{C}),$$

the bidyadic $\overline{\overline{\mathsf{C}}}$ can finally be shown to satisfy the cubic equation (9.110)

$$\overline{\overline{\mathsf{C}}}^3 - (\mathrm{tr}\overline{\overline{\mathsf{C}}})\overline{\overline{\mathsf{C}}}^2 + \frac{1}{2}((\mathrm{tr}\overline{\overline{\mathsf{C}}})^2 - \mathrm{tr}\overline{\overline{\mathsf{C}}}^2)\overline{\overline{\mathsf{C}}} = 0.$$

9.13 Inserting (9.119) with (9.127) in (9.120) and (9.121) we obtain

$$\begin{aligned}
\overline{\overline{\mathsf{M}}}^2 &= -(c\overline{\overline{\mathrm{I}}}_s^{(2)T} + \boldsymbol{\Delta}_s(\mathbf{a}_s|\overline{\overline{\mu}}^{-1})) + \boldsymbol{\varepsilon}_4 \wedge (c\overline{\overline{\mathrm{I}}}_s^T + (\overline{\overline{\mu}}^{-1}|\boldsymbol{\Delta}_s)\mathbf{a}_s) \wedge \mathbf{e}_4 \\
\overline{\overline{\mathsf{M}}}^4 &= (c\overline{\overline{\mathrm{I}}}_s^{(2)T} + \boldsymbol{\Delta}_s(\mathbf{a}_s|\overline{\overline{\mu}}^{-1}))^2 - \boldsymbol{\varepsilon}_4 \wedge (c\overline{\overline{\mathrm{I}}}_s^T + (\overline{\overline{\mu}}^{-1}|\boldsymbol{\Delta}_s)\mathbf{a}_s)^2 \wedge \mathbf{e}_4 \\
&= c^2\overline{\overline{\mathrm{I}}}_s^{(2)T} + \lambda\Delta_s(\mathbf{a}_s|\overline{\overline{\mu}}^{-1}) - \boldsymbol{\varepsilon}_4 \wedge (c^2\overline{\overline{\mathrm{I}}}_s^T + \lambda(\overline{\overline{\mu}}^{-1}|\boldsymbol{\Delta}_s)\mathbf{a}_s) \wedge \mathbf{e}_4,
\end{aligned}$$

with $\lambda = 2c + \mathbf{a}_s|\overline{\overline{\mu}}^{-1}|\boldsymbol{\Delta}_s$. Equation (9.118) can be split in two components as

$$\begin{aligned}
(c^2 + 2Ac + B)\overline{\overline{\mathrm{I}}}^{(2)T} + (\lambda + 2A)\boldsymbol{\Delta}_s(\mathbf{a}_s|\overline{\overline{\mu}}^{-1}) &= 0 \\
\boldsymbol{\varepsilon}_4 \wedge ((c^2 + 2Ac + B)\overline{\overline{\mathrm{I}}}_s^T + (\lambda + 2A)(\overline{\overline{\mu}}^{-1}|\boldsymbol{\Delta}_s)\mathbf{a}_s) \wedge \mathbf{e}_4 &= 0
\end{aligned}$$

These are valid for

$$\begin{aligned}
\lambda + 2A = 0 \quad &\Rightarrow \quad A = -(c + \frac{1}{2}\mathbf{a}_s|\overline{\overline{\mu}}^{-1}|\boldsymbol{\Delta}_s) \\
c^2 + 2Ac + B = 0 \quad &\Rightarrow \quad B = c(c + \mathbf{a}_s|\overline{\overline{\mu}}^{-1}|\boldsymbol{\Delta}_s),
\end{aligned}$$

which yield (9.128) and (9.129).

9.14 In the case of the principal BQ medium the modified medium bidyadic $\overline{\overline{\mathsf{M}}}_m = \mathbf{e}_N\lfloor\overline{\overline{\mathsf{M}}}$ is symmetric. From this it follows that also the bidyadics $\mathbf{e}_N\lfloor\overline{\overline{\mathsf{M}}}^n$ must be symmetric. In fact, we can expand

$$\mathbf{e}_N\lfloor\overline{\overline{\mathsf{M}}}^2 = (\mathbf{e}_N\lfloor\overline{\overline{\mathsf{M}}})|(\varepsilon_N\lfloor(\mathbf{e}_N\lfloor\overline{\overline{\mathsf{M}}})) = (\mathbf{e}_N\lfloor\overline{\overline{\mathsf{M}}})|(\varepsilon_N\lfloor\overline{\overline{\mathsf{I}}}^{(2)})|(\mathbf{e}_N\lfloor\overline{\overline{\mathsf{M}}}),$$

which is obviously symmetric. We can proceed similarly for higher powers of $\overline{\overline{\mathsf{M}}}$. As a conclusion, for principal BQ media, the bidyadics $\mathbf{e}_N\lfloor\overline{\overline{\Pi}}_i$ are symmetric. This implies that, the "natural" dot product of two two-forms $\mathbf{\Phi}_1\cdot\mathbf{\Phi}_2 = \mathbf{\Phi}_1|(\mathbf{e}_N\lfloor\mathbf{\Phi}_2) = \mathbf{\Phi}_2\cdot\mathbf{\Phi}_1$ of two eigen two-forms associated with different eigenvalues is zero. This is seen from the expansion

$$\mathbf{\Phi}_i\cdot\mathbf{\Phi}_j = \mathbf{\Phi}_i|(\mathbf{e}_N\lfloor\overline{\overline{\Pi}}_j)|\mathbf{\Phi} = (\mathbf{e}_N\lfloor\mathbf{\Phi}_i)|\overline{\overline{\Pi}}_j|\mathbf{\Phi} = (\mathbf{e}_N\lfloor\overline{\overline{\Pi}}_i|\mathbf{\Phi})|\overline{\overline{\Pi}}_j|\mathbf{\Phi}$$

$$= \mathbf{\Phi}|(\mathbf{e}_N\lfloor\overline{\overline{\Pi}}_i)^T|\overline{\overline{\Pi}}_j|\mathbf{\Phi} = \mathbf{\Phi}|(\mathbf{e}_N\lfloor(\overline{\overline{\Pi}}_i|\overline{\overline{\Pi}}_j))|\mathbf{\Phi} = 0, \quad i \neq j,$$

and orthogonality of the projection dyadics. Thus, the eigenfields are orthogonal in the principal BQ medium.

9.15 Starting from the representation (9.148)–(9.151) find the 3D conditions for the BQ medium corresponding to (9.162) and (9.163).

$$\overline{\overline{\mathsf{A}}}_s = \overline{\overline{\alpha}}^2 - \overline{\overline{\epsilon}}'|\overline{\overline{\mu}}^{-1},$$
$$\overline{\overline{\mathsf{B}}}_s = \overline{\overline{\alpha}}|\overline{\overline{\epsilon}}' - \overline{\overline{\epsilon}}'|\overline{\overline{\beta}},$$
$$\overline{\overline{\mathsf{C}}}_s = \overline{\overline{\mu}}^{-1}|\overline{\overline{\alpha}} - \overline{\overline{\beta}}|\overline{\overline{\mu}}^{-1},$$
$$\overline{\overline{\mathsf{D}}}_s = \overline{\overline{\mu}}^{-1}|\overline{\overline{\epsilon}}' - \overline{\overline{\beta}}^2,$$

$$\overline{\overline{\mu}}^{-1}|\overline{\overline{\epsilon}}|\overline{\overline{\mu}}^{-1}|(\overline{\overline{\xi}} + \overline{\overline{\zeta}})|\overline{\overline{\mu}}^{-1}|\overline{\overline{\epsilon}} = M_a^2 M_b^2 \overline{\overline{\mu}}^{-1}|(\overline{\overline{\xi}} + \overline{\overline{\zeta}}),$$

$$\overline{\overline{\mu}}^{-1}|\overline{\overline{\epsilon}}|(\overline{\overline{\mu}}^{-1}|(\overline{\overline{\xi}} + \overline{\overline{\zeta}}))^2 = (\overline{\overline{\mu}}^{-1}|\overline{\overline{\epsilon}} + M_a^2\overline{\overline{\mathsf{I}}}_s^T)|(\overline{\overline{\mu}}^{-1}|\overline{\overline{\epsilon}} + M_b^2\overline{\overline{\mathsf{I}}}_s^T)$$

9.16 Inserting (9.167)–(9.170) in (9.147) we obtain

$$\overline{\overline{\mathsf{M}}}^2 = -\overline{\overline{\epsilon}}|\overline{\overline{\mu}}^{-1} + \overline{\overline{\xi}}|\overline{\overline{\mu}}^{-1}|\overline{\overline{\epsilon}}\wedge\mathbf{e}_4 - \overline{\overline{\epsilon}}|\overline{\overline{\mu}}^{-1}|\overline{\overline{\xi}}\wedge\mathbf{e}_4 + \varepsilon_4\wedge\overline{\overline{\mu}}^{-1}|\overline{\overline{\epsilon}}\wedge\mathbf{e}_4$$
$$= -(\overline{\overline{\epsilon}}|\overline{\overline{\mu}}^{-1})|(\overline{\overline{\mathsf{I}}}_s^{(2)T} + \overline{\overline{\xi}}\wedge\mathbf{e}_4) + (\overline{\overline{\xi}} + \varepsilon_4\wedge\overline{\overline{\mathsf{I}}}_s^T)|(\overline{\overline{\mu}}^{-1}|\overline{\overline{\epsilon}})\wedge\mathbf{e}_4.$$

Because $(\overline{\overline{\mathsf{I}}}_s^{(2)T} + \overline{\overline{\xi}}\wedge\mathbf{e}_4)|(\overline{\overline{\xi}} + \varepsilon_4\wedge\overline{\overline{\mathsf{I}}}_s^T) = \overline{\overline{\xi}} - \overline{\overline{\xi}} = 0$, we obtain

$$\overline{\overline{\mathsf{M}}}^4 = (-\overline{\overline{\epsilon}}|\overline{\overline{\mu}}^{-1} + \overline{\overline{\xi}}|\overline{\overline{\mu}}^{-1}|\overline{\overline{\epsilon}}\wedge\mathbf{e}_4 - \overline{\overline{\epsilon}}|\overline{\overline{\mu}}^{-1}|\overline{\overline{\xi}}\wedge\mathbf{e}_4 + \varepsilon_4\wedge\overline{\overline{\mu}}^{-1}|\overline{\overline{\epsilon}}\wedge\mathbf{e}_4)^2$$
$$= (\overline{\overline{\epsilon}}|\overline{\overline{\mu}}^{-1})^2|(\overline{\overline{\mathsf{I}}}_s^{(2)T} + \overline{\overline{\xi}}\wedge\mathbf{e}_4) - (\overline{\overline{\xi}} + \varepsilon_4\wedge\overline{\overline{\mathsf{I}}}_s^T)|(\overline{\overline{\mu}}^{-1}|\overline{\overline{\epsilon}})^2\wedge\mathbf{e}_4.$$

Inserting these in (9.118) the equation can be written as

$$0 = (\overline{\overline{\epsilon}}|\overline{\overline{\mu}}^{-1})^2 + 2A(\overline{\overline{\epsilon}}|\overline{\overline{\mu}}^{-1}) + B\overline{\overline{\mathsf{I}}}_s^{(2)T}$$
$$-\boldsymbol{\varepsilon}_4 \wedge ((\overline{\overline{\mu}}^{-1}|\overline{\overline{\epsilon}})^2 + 2A(\overline{\overline{\mu}}^{-1}|\overline{\overline{\epsilon}}) + B\overline{\overline{\mathsf{I}}}_s^T) \wedge \mathbf{e}_4$$
$$+((\overline{\overline{\epsilon}}|\overline{\overline{\mu}}^{-1})^2 + 2A(\overline{\overline{\epsilon}}|\overline{\overline{\mu}}^{-1}))|\overline{\overline{\xi}} \wedge \mathbf{e}_4 - \overline{\overline{\xi}}|((\overline{\overline{\mu}}^{-1}|\overline{\overline{\epsilon}})^2 + 2A(\overline{\overline{\mu}}^{-1}|\overline{\overline{\epsilon}})) \wedge \mathbf{e}_4.$$

The three lines must equal zero separately. The second line yields

$$(\overline{\overline{\mu}}^{-1}|\overline{\overline{\epsilon}} + A\overline{\overline{\mathsf{I}}}_s^T)^2 - (A^2 - B)\overline{\overline{\mathsf{I}}}_s^T = 0,$$

whence

$$\overline{\overline{\epsilon}} = -A\overline{\overline{\mu}} \pm \sqrt{A^2 - B}\,(\overline{\overline{\mu}}|\overline{\overline{\mathsf{U}}}_s^T),$$

where $\overline{\overline{\mathsf{U}}}_s^T \in \mathbb{F}_1\mathbb{E}_1$ is a spatial unipotent dyadic. Applying this we have

$$\overline{\overline{\epsilon}}|\overline{\overline{\mu}}^{-1} = -A\overline{\overline{\mathsf{I}}}^{(2)T} \pm \sqrt{A^2 - B}\,(\overline{\overline{\mu}}|\overline{\overline{\mathsf{U}}}_s^T|\overline{\overline{\mu}}^{-1}).$$

The first line also vanishes because $\overline{\overline{\mu}}|\overline{\overline{\mathsf{U}}}_s^T|\overline{\overline{\mu}}^{-1} \in \mathbb{F}_2\mathbb{E}_2$ is another spatial unipotent dyadic. Applying these to the third line, it becomes

$$\overline{\overline{\mathsf{I}}}_s^{(2)T}|\overline{\overline{\xi}} \wedge \mathbf{e}_4 - \overline{\overline{\xi}}|\overline{\overline{\mathsf{I}}}_s^T \wedge \mathbf{e}_4 = \overline{\overline{\xi}} \wedge \mathbf{e}_4 - \overline{\overline{\xi}} \wedge \mathbf{e}_4 = 0.$$

Since this is valid for any dyadic $\overline{\overline{\xi}}$, the medium bidyadic (5.75) which in the present case reads

$$\overline{\overline{\mathsf{M}}} = \overline{\overline{\epsilon}} \wedge \mathbf{e}_4 + (\boldsymbol{\varepsilon}_4 \wedge \overline{\overline{\mathsf{I}}}_s^T + \overline{\overline{\xi}})|\overline{\overline{\mu}}^{-1}|(\overline{\overline{\mathsf{I}}}_s^{(2)T} + \overline{\overline{\xi}} \wedge \mathbf{e}_4),$$

with $\overline{\overline{\epsilon}}$ and $\overline{\overline{\mu}}$ related in the above manner, represents a BQ medium for any $\overline{\overline{\xi}}$.

9.17 Starting from $\overline{\overline{\mathsf{D}}} = \overline{\overline{\mathsf{I}}}^{(2)T} - \overline{\overline{\mathsf{F}}}_s \wedge \mathbf{e}_4$ and $\overline{\overline{\mathsf{D}}}^{-1} = \overline{\overline{\mathsf{I}}}^{(2)T} + \overline{\overline{\mathsf{F}}}_s \wedge \mathbf{e}_4$, and applying the expansion (5.75), we can expand

$$\begin{aligned}\overline{\overline{\mathsf{D}}}|\overline{\overline{\mathsf{M}}}|\overline{\overline{\mathsf{D}}}^{-1} &= (\overline{\overline{\mathsf{I}}}^{(2)T} - \overline{\overline{\mathsf{F}}}_s \wedge \mathbf{e}_4)|\overline{\overline{\mathsf{M}}}|(\overline{\overline{\mathsf{I}}}^{(2)T} + \overline{\overline{\mathsf{F}}}_s \wedge \mathbf{e}_4)\\
&= \overline{\overline{\mathsf{M}}} + \overline{\overline{\mathsf{F}}}_s|\overline{\overline{\mu}}^{-1} - \overline{\overline{\mathsf{F}}}_s|\overline{\overline{\mu}}^{-1}|\overline{\overline{\zeta}} \wedge \mathbf{e}_4 + \overline{\overline{\xi}}|\overline{\overline{\mu}}^{-1}|\overline{\overline{\mathsf{F}}}_s \wedge \mathbf{e}_4\\
&\quad +\boldsymbol{\varepsilon}_4 \wedge \overline{\overline{\mu}}^{-1}|\overline{\overline{\mathsf{F}}}_s \wedge \mathbf{e}_4 + \overline{\overline{\mathsf{F}}}_s|\overline{\overline{\mu}}^{-1}|\overline{\overline{\mathsf{F}}}_s \wedge \mathbf{e}_4)\\
&= (\overline{\overline{\xi}} + \overline{\overline{\mathsf{F}}}_s)|\overline{\overline{\mu}}^{-1} + (\overline{\overline{\epsilon}} - (\overline{\overline{\xi}} + \overline{\overline{\mathsf{F}}}_s)|\overline{\overline{\mu}}^{-1}|(\overline{\overline{\zeta}} - \overline{\overline{\mathsf{F}}}_s)) \wedge \mathbf{e}_4\\
&\quad +\boldsymbol{\varepsilon}_4 \wedge \overline{\overline{\mu}}^{-1} - \boldsymbol{\varepsilon}_4 \wedge \overline{\overline{\mu}}^{-1}|(\overline{\overline{\zeta}} - \overline{\overline{\mathsf{F}}}_s) \wedge \mathbf{e}_4\\
&= \overline{\overline{\mathsf{M}}}_1.\end{aligned}$$

The parameter dyadics are transformed as

$$\overline{\overline{\epsilon}}_1 = \overline{\overline{\epsilon}}, \quad \overline{\overline{\xi}}_1 = \overline{\overline{\xi}} + \overline{\overline{\mathsf{F}}}_s, \quad \overline{\overline{\zeta}}_1 = \overline{\overline{\zeta}} - \overline{\overline{\mathsf{F}}}_s, \quad \overline{\overline{\mu}}_1 = \overline{\overline{\mu}}.$$

Choosing now $\overline{\overline{\mathsf{F}}}_s = (\overline{\overline{\zeta}} - \overline{\overline{\xi}})/2$, we obtain $\overline{\overline{\xi}}_1 = \overline{\overline{\zeta}}_1 = (\overline{\overline{\xi}} + \overline{\overline{\zeta}})/2$. Thus, we can represent the general medium dyadic $\overline{\overline{\mathsf{M}}}$ as

$$\overline{\overline{\mathsf{M}}} = \overline{\overline{\mathsf{D}}}^{-1}|\overline{\overline{\mathsf{M}}}_1|\overline{\overline{\mathsf{D}}}$$

in terms of a medium dyadic $\overline{\overline{\mathsf{M}}}_1$ satisfying $\overline{\overline{\xi}}_1 = \overline{\overline{\zeta}}_1$.

9.18 The potential one-form $\boldsymbol{\phi}$ of a plane wave in a CU-medium defined by the medium bidyadic (9.114) satisfies $(\boldsymbol{\nu} \wedge \overline{\overline{\mathsf{M}}}\lfloor \boldsymbol{\nu})|\boldsymbol{\phi} = 0$. Let us expand

$$\begin{aligned}(\boldsymbol{\nu} \wedge \overline{\overline{\mathsf{M}}}\lfloor \boldsymbol{\nu})|\boldsymbol{\phi} &= (\boldsymbol{\nu} \wedge (\overline{\overline{\mathsf{B}}}{}_{o\wedge}^{2\wedge}\overline{\overline{\mathsf{I}}})^T \lfloor \boldsymbol{\nu} + 2\boldsymbol{\nu} \wedge \overline{\overline{\mathsf{B}}}{}_o^{(2)T}\lfloor \boldsymbol{\nu})|\boldsymbol{\phi} \\ &= (\boldsymbol{\nu} \wedge (\overline{\overline{\mathsf{B}}}{}_o^{2T}|\boldsymbol{\nu}) \wedge \overline{\overline{\mathsf{I}}}^T + 2\boldsymbol{\nu} \wedge (\overline{\overline{\mathsf{B}}}{}_o^T|\boldsymbol{\nu}) \wedge \overline{\overline{\mathsf{B}}}{}_o^T)\boldsymbol{\phi} \\ &= (\boldsymbol{\nu} \wedge \boldsymbol{\nu}_2 \wedge \boldsymbol{\phi} + 2\boldsymbol{\nu} \wedge \boldsymbol{\nu}_1 \wedge \overline{\overline{\mathsf{B}}}{}_o^T)|\boldsymbol{\phi} = 0,\end{aligned}$$

denoting $\boldsymbol{\nu}_i = \overline{\overline{\mathsf{B}}}{}_o^{iT}|\boldsymbol{\nu}$. From

$$-\boldsymbol{\nu}_1 \wedge (\boldsymbol{\nu} \wedge \overline{\overline{\mathsf{M}}}\lfloor \boldsymbol{\nu})|\boldsymbol{\phi} = \boldsymbol{\nu} \wedge \boldsymbol{\nu}_1 \wedge \boldsymbol{\nu}_2 \wedge \boldsymbol{\phi} = 0,$$

the potential one-form can be represented in the form

$$\boldsymbol{\phi} = A\boldsymbol{\nu} + B\boldsymbol{\nu}_1 + C\boldsymbol{\nu}_2.$$

Substituting this in the potential equation we obtain

$$\boldsymbol{\nu} \wedge \boldsymbol{\nu}_1 \wedge (B\boldsymbol{\nu}_2 + 2C\boldsymbol{\nu}_3) = 0.$$

Multiplying by $\boldsymbol{\nu}_3\wedge$ and $\boldsymbol{\nu}_2\wedge$ we obtain the respective conditions

$$B\boldsymbol{\nu} \wedge \boldsymbol{\nu}_1 \wedge \boldsymbol{\nu}_2 \wedge \boldsymbol{\nu}_3 = 0, \qquad C\boldsymbol{\nu} \wedge \boldsymbol{\nu}_1 \wedge \boldsymbol{\nu}_2 \wedge \boldsymbol{\nu}_3 = 0.$$

For $\boldsymbol{\Phi} = B\boldsymbol{\nu} \wedge \boldsymbol{\nu}_1 + C\boldsymbol{\nu} \wedge \boldsymbol{\nu}_2 \neq 0$ at least one of the coefficients B and C must be nonzero, whence we must have

$$\boldsymbol{\nu} \wedge \boldsymbol{\nu}_1 \wedge \boldsymbol{\nu}_2 \wedge \boldsymbol{\nu}_3 = 0,$$

which, when multiplied by $\mathbf{e}_N|$, can be considered as the scalar dispersion equation for $\boldsymbol{\nu}$. This requires that for any solution $\boldsymbol{\nu}$ the one-forms $\boldsymbol{\nu}$, $\overline{\overline{\mathsf{B}}}{}_o^T|\boldsymbol{\nu}$, $\overline{\overline{\mathsf{B}}}{}_o^{2T}|\boldsymbol{\nu}$, and $\overline{\overline{\mathsf{B}}}{}_o^{3T}|\boldsymbol{\nu}$ must be linearly dependent. Obviously, this cannot be true for the general one-form $\boldsymbol{\nu}$ when $\overline{\overline{\mathsf{B}}}_o$ is of full rank. Thus, the dispersion equation has no solution except, for example, when $\boldsymbol{\nu}$ is an eigen one-form of the dyadic

$\overline{\overline{\mathsf{B}}}_o^T$, or, more generally, a linear combination of three eigen one-forms.

CHAPTER 10

10.1 For Case 3 the dispersion dyadic representation (10.1) is of the form

$$\overline{\overline{\mathsf{M}}}_m\lfloor\lfloor\nu\nu = \overline{\overline{\mathsf{D}}}(\nu) = \mathbf{a}(\nu)\mathbf{c} + \mathbf{b}\mathbf{d}(\nu)$$

where the vectors $\mathbf{c}$ and $\mathbf{b}$ are independent of ν. From

$$\mathbf{b}\wedge\overline{\overline{\mathsf{D}}}(\nu)|\nu = (\mathbf{b}\wedge\mathbf{a})(\mathbf{c}|\nu) = 0,$$

valid for all ν, we conclude that either $\mathbf{c} = 0$ or $\mathbf{b}\wedge\mathbf{a} = 0$. In the latter case, we must have $\mathbf{a}(\nu) = a(\nu)\mathbf{b}$ for some scalar $a(\nu)$. Thus, the dispersion dyadic must be of the form

$$\overline{\overline{\mathsf{D}}}(\nu) = \mathbf{b}\mathbf{d}'(\nu)$$

with either $\mathbf{d}'(\nu) = \mathbf{d}(\nu)$ or $\mathbf{d}'(\nu) = a(\nu)\mathbf{c} + \mathbf{d}(\nu)$. From $\nu|\overline{\overline{\mathsf{D}}}(\nu) = 0$ we obtain $(\nu|\mathbf{b})\mathbf{d}'(\nu) = 0$ for all ν, whence either $\mathbf{b} = 0$ or $\mathbf{d}'(\nu) = 0$. Since both lead to $\overline{\overline{\mathsf{D}}}(\nu) = 0$ for all ν, Case 3 yields the axion medium which is already included in Case 1 or Case 2.

10.2 For Case 4, the dispersion dyadic representation (10.1) is of the form

$$\overline{\overline{\mathsf{M}}}_m\lfloor\lfloor\nu\nu = \overline{\overline{\mathsf{D}}}(\nu) = \mathbf{a}(\nu)\mathbf{c} + \mathbf{b}(\nu)\mathbf{d}(\nu),$$

where $\mathbf{c}$ is independent of ν. Requiring $\overline{\overline{\mathsf{D}}}(\nu)|\nu = 0$ and assuming $\mathbf{a}(\nu)\wedge\mathbf{b}(\nu) \neq 0$ we obtain $\mathbf{c}|\nu = 0$ and $\mathbf{d}|\nu = 0$ for all ν. This requires $\mathbf{c} = 0$ and $\mathbf{d} = \mathbf{D}\lfloor\nu$, where $\mathbf{D}$ is a bidyadic, whence the dispersion dyadic must be of the form

$$\overline{\overline{\mathsf{D}}}(\nu) = \mathbf{b}(\nu)\mathbf{D}\lfloor\nu.$$

From $\nu|\overline{\overline{\mathsf{D}}}(\nu) = 0$ we obtain either $\mathbf{D} = 0$ or $\mathbf{b}(\nu) = \mathbf{B}\lfloor\nu$. Since we now have

$$\overline{\overline{\mathsf{D}}}(\nu) = \mathbf{B}\mathbf{D}\lfloor\lfloor\nu\nu,$$

this belongs to Case 1 (10.8). Alternatively, from $\mathbf{a}(\boldsymbol{\nu}) \wedge \mathbf{b}(\boldsymbol{\nu}) = 0$ we obtain the relation $\mathbf{a}(\boldsymbol{\nu}) = a(\boldsymbol{\nu})\mathbf{b}(\boldsymbol{\nu})$, where $a(\boldsymbol{\nu})$ is a linear function of $\boldsymbol{\nu}$. In this case the dispersion dyadic becomes

$$\overline{\overline{\mathsf{D}}}(\boldsymbol{\nu}) = \mathbf{b}(\boldsymbol{\nu})(a(\boldsymbol{\nu})\mathbf{c} + \mathbf{d}(\boldsymbol{\nu})).$$

From $\boldsymbol{\nu}|\overline{\overline{\mathsf{D}}}(\boldsymbol{\nu}) = 0$ we have $\boldsymbol{\nu}|\mathbf{b}(\boldsymbol{\nu}) = 0$ which leads to $\mathbf{b}(\boldsymbol{\nu}) = \mathbf{B}\lfloor\boldsymbol{\nu}$ and the previous case, or to $\overline{\overline{\mathsf{D}}}(\boldsymbol{\nu}) = 0$ which corresponds to the axion medium case which is also included in Case 1 solution (10.10).

10.3 Considering a 2D subspace of one-forms $\boldsymbol{\nu}$ defined by $\mathbf{a}_1|\boldsymbol{\nu} = \mathbf{a}_2|\boldsymbol{\nu} = 0$, for $\boldsymbol{\nu}$ taken from that subspace we have

$$\mathbf{a} = \mathbf{A}\lfloor\boldsymbol{\nu} = (\mathbf{a}_2|\boldsymbol{\nu})\mathbf{a}_1 - (\mathbf{a}_1|\boldsymbol{\nu})\mathbf{a}_2 = 0.$$

Thus, the dispersion dyadic $\overline{\overline{\mathsf{D}}}(\boldsymbol{\nu})$ is of rank 1 for such a subspace of $\boldsymbol{\nu}$ and of rank 2 for $\boldsymbol{\nu}$ taken from the complementary subspace if the other three bivectors are not simple.

10.4 Defining a bivector $\mathbf{D} = \mathbf{e}_N\lfloor(\boldsymbol{\alpha} \wedge \boldsymbol{\beta})$, the dispersion dyadic can be expressed as

$$\overline{\overline{\mathsf{D}}}(\boldsymbol{\nu}) = \mathbf{e}_N\lfloor(\boldsymbol{\nu} \wedge \overline{\overline{\mathsf{M}}}\lfloor\boldsymbol{\nu}) = \mathbf{D}\lfloor\overline{\overline{\mathsf{P}}}^T = (\mathbf{D}\lfloor\overline{\overline{\mathsf{I}}}^T)|\overline{\overline{\mathsf{P}}}^T,$$

where $\mathbf{D}\lfloor\overline{\overline{\mathsf{I}}}^T \in \mathbb{E}_1\mathbb{E}_1$ is an antisymmetric dyadic. The bivector $\mathbf{D}$ is simple because it satisfies

$$\begin{aligned}\boldsymbol{\varepsilon}_N|(\mathbf{D} \wedge \mathbf{D}) &= \boldsymbol{\varepsilon}_N|((\mathbf{e}_N\lfloor(\boldsymbol{\alpha} \wedge \boldsymbol{\beta})) \wedge (\mathbf{e}_N\lfloor(\boldsymbol{\alpha} \wedge \boldsymbol{\beta}))) \\ &= \mathbf{e}_N|(\boldsymbol{\alpha} \wedge \boldsymbol{\beta} \wedge \boldsymbol{\alpha} \wedge \boldsymbol{\beta}) = 0.\end{aligned}$$

Since a simple bivector can be expressed in terms of two vectors as $\mathbf{D} = \mathbf{d}_1 \wedge \mathbf{d}_2$, the antisymmetric dyadic satisfies

$$\begin{aligned}(\mathbf{D}\lfloor\overline{\overline{\mathsf{I}}}^T)^{(3)} &= ((\mathbf{d}_1 \wedge \mathbf{d}_2)\lfloor\overline{\overline{\mathsf{I}}}^T)^{(3)} = (\mathbf{d}_2\mathbf{d}_1 - \mathbf{d}_1\mathbf{d}_2)^{(3)} \\ &= \frac{1}{3}(\mathbf{d}_2\mathbf{d}_1 - \mathbf{d}_1\mathbf{d}_2)^{\wedge}_{\wedge}(\mathbf{d}_1\mathbf{d}_2{}^{\wedge}_{\wedge}\mathbf{d}_2\mathbf{d}_1) = 0.\end{aligned}$$

From this and the rule $(\overline{\overline{\mathsf{A}}}|\overline{\overline{\mathsf{B}}})^{(p)} = \overline{\overline{\mathsf{A}}}^{(p)}|\overline{\overline{\mathsf{B}}}^{(p)}$ it follows that the dispersion equation

$$\overline{\overline{\mathsf{D}}}^{(3)}(\boldsymbol{\nu}) = (\mathbf{D}\lfloor\overline{\overline{\mathsf{I}}}^T)^{(3)}|\overline{\overline{\mathsf{P}}}^{(3)T} = 0$$

is satisfied for any $\boldsymbol{\nu}$ and any medium bidyadic $\overline{\overline{\mathsf{M}}}$ of the form (10.17).

10.5 Considering the relation (10.20)

$$A\boldsymbol{\alpha} + B\boldsymbol{\beta} + C(\overline{\overline{\mathsf{P}}}^T|\boldsymbol{v}) = 0.$$

For $C = 0$ the one-forms $\boldsymbol{\alpha}$ and $\boldsymbol{\beta}$ satisfy $\boldsymbol{\alpha} \wedge \boldsymbol{\beta} = 0$, whence from (10.17) we see that $\boldsymbol{v} \wedge \overline{\overline{\mathsf{M}}}\lfloor \boldsymbol{v} = 0$, and, hence, $\overline{\overline{\mathsf{D}}}(\boldsymbol{v}) = 0$ which corresponds to the axion medium. For $C \neq 0$ we can choose $C = -1$, whence the relation can be rewritten as

$$\overline{\overline{\mathsf{P}}}^T|\boldsymbol{v} = A\boldsymbol{\alpha} + B\boldsymbol{\beta}.$$

Without sacrifying generality ($\boldsymbol{\alpha}$ and $\boldsymbol{\beta}$ appear interchangeable at this stage), we may additionally set $B = 1$, whence for non-axion solutions we must have

$$\boldsymbol{\beta} = \overline{\overline{\mathsf{P}}}^T|\boldsymbol{v} - A\boldsymbol{\alpha}.$$

Inserting this in (10.17), the condition (10.21) is obtained.

10.6 Combined Case 2 medium bidyadic of the form

$$\overline{\overline{\mathsf{M}}} = A\overline{\overline{\mathsf{B}}}^{(2)T} + B(\overline{\overline{\mathsf{B}}}{}^\wedge_\wedge\overline{\overline{\mathsf{I}}})^T + C\overline{\overline{\mathsf{I}}}^{(2)T}$$

corresponds to the skewon–axion medium only when $A = 0$, because for $A \neq 0$ even if $B \neq 0$ it yields the P-axion medium dyadic (10.31). This is seen by defining new symbols as

$$\overline{\overline{\mathsf{P}}} = \overline{\overline{\mathsf{B}}} + \frac{B}{A}\overline{\overline{\mathsf{I}}}, \quad M = A, \quad \alpha = C - \frac{B^2}{A}.$$

The dispersion dyadic for the combined bidyadic can be expanded as

$$\begin{aligned}
\overline{\overline{\mathsf{D}}}(\boldsymbol{v}) &= -\mathbf{e}_N\lfloor(\boldsymbol{v} \wedge \overline{\overline{\mathsf{M}}}\lfloor \boldsymbol{v}) \\
&= -\mathbf{e}_N\lfloor(A\boldsymbol{v} \wedge \overline{\overline{\mathsf{B}}}^{(2)T}\lfloor \boldsymbol{v} + B\boldsymbol{v} \wedge (\overline{\overline{\mathsf{B}}}{}^\wedge_\wedge\overline{\overline{\mathsf{I}}})^T\lfloor \boldsymbol{v}) \\
&= -\mathbf{e}_N\lfloor(A\boldsymbol{v} \wedge (\overline{\overline{\mathsf{B}}}^T|\boldsymbol{v}) \wedge \overline{\overline{\mathsf{B}}}^T + B\boldsymbol{v} \wedge (\overline{\overline{\mathsf{B}}}^T|\boldsymbol{v}) \wedge \overline{\overline{\mathsf{I}}}^T) \\
&= (-\mathbf{e}_N\lfloor(\boldsymbol{v} \wedge (\overline{\overline{\mathsf{B}}}^T|\boldsymbol{v})))\lfloor(A\overline{\overline{\mathsf{B}}} + B\overline{\overline{\mathsf{I}}})^T \\
&= (\mathbf{A}\lfloor\overline{\overline{\mathsf{I}}}^T)|(A\overline{\overline{\mathsf{B}}} + B\overline{\overline{\mathsf{I}}})^T.
\end{aligned}$$

Because the bivector $\mathbf{A} = -\mathbf{e}_N\lfloor(\boldsymbol{v} \wedge (\overline{\overline{\mathsf{B}}}^T|\boldsymbol{v}))$ is simple and thus satisfies $(\mathbf{A}\lfloor\overline{\overline{\mathsf{I}}}^T)^{(3)} = 0$, the dispersion equation (5.240)

$$\overline{\overline{\mathsf{D}}}^{(3)}(\boldsymbol{v}) = (\mathbf{A}\lfloor\overline{\overline{\mathsf{I}}}^T)^{(3)}|(A\overline{\overline{\mathsf{B}}} + B\overline{\overline{\mathsf{I}}})^{(3)T} = 0$$

is satisfied for any $\boldsymbol{v}$.

10.7 Applying (6.16), Case 1 medium bidyadics are mapped as

$$\overline{\overline{\mathsf{M}}}_a = \overline{\overline{\mathsf{A}}}^{(-2)T}|(\mathbf{\Pi C} + \mathbf{\Delta D} + \alpha\overline{\overline{\mathrm{I}}}^{(2)T})|\overline{\overline{\mathsf{A}}}^{(2)T}$$
$$= \mathbf{\Pi}_a\mathbf{C}_a + \mathbf{\Delta}_a\mathbf{D}_a + \alpha\overline{\overline{\mathrm{I}}}^{(2)},$$

with

$$\mathbf{\Pi}_a = \overline{\overline{\mathsf{A}}}^{(-2)T}|\mathbf{\Pi}, \quad \mathbf{\Delta}_a = \overline{\overline{\mathsf{A}}}^{(-2)T}|\mathbf{\Delta},$$
$$\mathbf{C}_a = \overline{\overline{\mathsf{A}}}^{(2)}|\mathbf{C}, \quad \mathbf{D}_a = \overline{\overline{\mathsf{A}}}^{(2)}|\mathbf{D},$$

and Case 2 medium bidyadics (see the previous problem) are mapped as

$$\overline{\overline{\mathsf{M}}}_a = \overline{\overline{\mathsf{A}}}^{(-2)T}|(A\overline{\overline{\mathsf{B}}}^{(2)T} + B(\overline{\overline{\mathsf{B}}}{}^\wedge_\wedge\overline{\overline{\mathrm{I}}})^T + C\overline{\overline{\mathrm{I}}}^{(2)T})|\overline{\overline{\mathsf{A}}}^{(2)T}$$
$$= A\overline{\overline{\mathsf{B}}}_a^{(2)T} + B(\overline{\overline{\mathsf{B}}}_a{}^\wedge_\wedge\overline{\overline{\mathrm{I}}})^T + C\overline{\overline{\mathrm{I}}}^{(2)T},$$

with

$$\overline{\overline{\mathsf{B}}}_a = \overline{\overline{\mathsf{A}}}|\overline{\overline{\mathsf{B}}}|\overline{\overline{\mathsf{A}}}^{-1}.$$

Thus, in both cases, the form of the medium bidyadic is retained in the transformation.

10.8 Starting from the property

$$(\overline{\overline{\mathsf{N}}} - \alpha_1\overline{\overline{\mathrm{I}}}^{(2)T})^T \cdot (\overline{\overline{\mathsf{N}}} - \alpha_1\overline{\overline{\mathrm{I}}}^{(2)T}) = \overline{\overline{\mathsf{P}}}_1^{(2)}|(\mathbf{e}_N\lfloor\overline{\overline{\mathsf{P}}}_1^{(2)T}) = \Delta_{P_1}\overline{\overline{\mathsf{E}}}.$$

Multiplying by $\overline{\overline{\mathsf{M}}}^T|$ from the left and by $|\overline{\overline{\mathsf{M}}}$ from the right we obtain

$$(\overline{\overline{\mathrm{I}}}^{(2)T} - \alpha_1\overline{\overline{\mathsf{M}}})^T \cdot (\overline{\overline{\mathrm{I}}}^{(2)T} - \alpha_1\overline{\overline{\mathsf{M}}}) = \Delta_{P_1}\overline{\overline{\mathsf{M}}}^T \cdot \overline{\overline{\mathsf{M}}}$$

which can be rewritten in the form

$$(\overline{\overline{\mathsf{M}}} - \alpha\overline{\overline{\mathrm{I}}}^{(2)T})^T \cdot (\overline{\overline{\mathsf{M}}} - \alpha\overline{\overline{\mathrm{I}}}^{(2)T}) = \beta\overline{\overline{\mathsf{E}}}$$

with

$$\alpha = \frac{\alpha_1}{\alpha_1^2 - \Delta_{P_1}}, \quad \beta = \frac{\Delta_{P_1}}{(\alpha_1^2 - \Delta_{P_1})^2}$$

Thus, the bidyadic $\overline{\overline{\mathsf{M}}}$ must correspond to either a P-axion or a Q-axion medium. Since there is no dispersion equation for the medium defined through the bidyadic $\overline{\overline{\mathsf{N}}}$, there is no dispersion equation when defined through the corresponding bidyadic $\overline{\overline{\mathsf{M}}} = \overline{\overline{\mathsf{N}}}^{-1}$. Now it is known that a full-rank Q-axion medium has a dispersion equation, whence $\overline{\overline{\mathsf{M}}}$ must represent a P-axion medium.

10.9 Applying the bac-cab rule

$$\mathbf{c} \wedge (\mathbf{D}\lfloor \boldsymbol{\nu}) = (\mathbf{c}|\boldsymbol{\nu})\mathbf{D} - (\mathbf{c} \wedge \mathbf{D})\lfloor \boldsymbol{\nu}$$

to $\mathbf{c} = \mathbf{C}\lfloor \boldsymbol{\nu}$ we obtain

$$(\mathbf{C}\lfloor \boldsymbol{\nu}) \wedge (\mathbf{D}\lfloor \boldsymbol{\nu}) = -((\mathbf{C}\lfloor \boldsymbol{\nu}) \wedge \mathbf{D})\lfloor \boldsymbol{\nu}.$$

Thus, from (10.39) we have

$$\begin{aligned}\boldsymbol{\Phi} &= (\boldsymbol{\nu}|\mathbf{p})\varepsilon_N\lfloor((\mathbf{C}\lfloor \boldsymbol{\nu}) \wedge (\mathbf{D}\lfloor \boldsymbol{\nu})) = -(\boldsymbol{\nu}|\mathbf{p})\varepsilon_N\lfloor(((\mathbf{C}\lfloor \boldsymbol{\nu}) \wedge \mathbf{D})\lfloor \boldsymbol{\nu}) \\ &= (\boldsymbol{\nu}|\mathbf{p})(\varepsilon_N\lfloor((\mathbf{C}\lfloor \boldsymbol{\nu}) \wedge \mathbf{D})) \wedge \boldsymbol{\nu},\end{aligned}$$

and

$$\begin{aligned}\mathbf{C}|\boldsymbol{\Phi} &= (\boldsymbol{\nu}|\mathbf{p})(\varepsilon_N\lfloor((\mathbf{C}\lfloor \boldsymbol{\nu}) \wedge \mathbf{D})) \wedge \boldsymbol{\nu}|\mathbf{C} \\ &= -(\boldsymbol{\nu}|\mathbf{p})(\varepsilon_N\lfloor((\mathbf{C}\lfloor \boldsymbol{\nu}) \wedge \mathbf{D}))|(\mathbf{C}\lfloor \boldsymbol{\nu}) \\ &= -(\boldsymbol{\nu}|\mathbf{p})\varepsilon_N|((\mathbf{C}\lfloor \boldsymbol{\nu}) \wedge \mathbf{D} \wedge (\mathbf{C}\lfloor \boldsymbol{\nu})) = 0.\end{aligned}$$

Since $\boldsymbol{\Phi}$ is a symmetric function of $\mathbf{C}$ and $\mathbf{D}$ we also have $\mathbf{D}|\boldsymbol{\Phi} = 0$.

10.10 The dispersion dyadic can be expanded as

$$\begin{aligned}\overline{\overline{\mathsf{D}}}(\boldsymbol{\nu}) &= \overline{\overline{\mathsf{M}}}_m\lfloor\lfloor \boldsymbol{\nu}\boldsymbol{\nu} = \overline{\overline{\mathsf{Q}}}^{(2)}\lfloor\lfloor \boldsymbol{\nu}\boldsymbol{\nu} \\ &= (\mathbf{e}_{12}\lfloor \boldsymbol{\nu})(\mathbf{q}_{12}\lfloor \boldsymbol{\nu}) + (\mathbf{e}_{23}\lfloor \boldsymbol{\nu})(\mathbf{q}_{23}\lfloor \boldsymbol{\nu}) + (\mathbf{e}_{31}\lfloor \boldsymbol{\nu})(\mathbf{q}_{31}\lfloor \boldsymbol{\nu}) \\ &= \mathbf{e}_1\mathbf{d}_1 + \mathbf{e}_2\mathbf{d}_2 + \mathbf{e}_3\mathbf{d}_3,\end{aligned}$$

where we denote

$$\begin{aligned}\mathbf{d}_1 &= -\nu_2(\mathbf{q}_{12}\lfloor \boldsymbol{\nu}) + \nu_3(\mathbf{q}_{31}\lfloor \boldsymbol{\nu}), \\ \mathbf{d}_2 &= -\nu_3(\mathbf{q}_{23}\lfloor \boldsymbol{\nu}) + \nu_1(\mathbf{q}_{12}\lfloor \boldsymbol{\nu}), \\ \mathbf{d}_3 &= -\nu_1(\mathbf{q}_{31}\lfloor \boldsymbol{\nu}) + \nu_2(\mathbf{q}_{23}\lfloor \boldsymbol{\nu}),\end{aligned}$$

and $\nu_i = \mathbf{e}_i|\boldsymbol{\nu}$. Because of $\boldsymbol{\nu}|\overline{\overline{\mathsf{D}}}(\boldsymbol{\nu}) = 0$, we have

$$\nu_1\mathbf{d}_1 + \nu_2\mathbf{d}_2 + \nu_3\mathbf{d}_3 = 0.$$

Assuming $\nu_3 \neq 0$ we can write

$$\mathbf{d}_3 = -(\nu_1\mathbf{d}_1 + \nu_2\mathbf{d}_2)/\nu_3,$$

insert it in the expression of $\overline{\overline{\mathsf{D}}}(\boldsymbol{\nu})$ above, and expand

$$\begin{aligned}
\nu_3\overline{\overline{\mathsf{D}}}(\boldsymbol{\nu}) &= (\nu_3\mathbf{e}_1 - \nu_1\mathbf{e}_3)\mathbf{d}_1 + (\nu_3\mathbf{e}_2 - \nu_2\mathbf{e}_3)\mathbf{d}_2 \\
&= -\boldsymbol{\nu}\rfloor(\mathbf{e}_3 \wedge \mathbf{e}_1\mathbf{d}_1 + \mathbf{e}_3 \wedge \mathbf{e}_2\mathbf{d}_2) \\
&= -\boldsymbol{\nu}\rfloor(\mathbf{e}_3 \wedge \mathbf{e}_1(\nu_3\mathbf{q}_{31} - \nu_2\mathbf{q}_{12}) + \mathbf{e}_3 \wedge \mathbf{e}_2(\nu_1\mathbf{q}_{12} - \nu_3\mathbf{q}_{23}))\lfloor\boldsymbol{\nu} \\
&= -\boldsymbol{\nu}\rfloor(\mathbf{e}_3 \wedge \mathbf{e}_1(\boldsymbol{\nu}|(\mathbf{e}_3\mathbf{q}_3 + \mathbf{e}_2\mathbf{q}_2) \wedge \mathbf{q}_1) \\
&\quad +\mathbf{e}_3 \wedge \mathbf{e}_2(\boldsymbol{\nu}|(\mathbf{e}_1\mathbf{q}_1 + \mathbf{e}_3\mathbf{q}_3) \wedge \mathbf{q}_2))\lfloor\boldsymbol{\nu} \\
&= -\boldsymbol{\nu}\rfloor(\mathbf{e}_3 \wedge \mathbf{e}_1(\boldsymbol{\nu}|\overline{\overline{\mathsf{Q}}} \wedge \mathbf{q}_1) + \mathbf{e}_3 \wedge \mathbf{e}_2(\boldsymbol{\nu}|\overline{\overline{\mathsf{Q}}} \wedge \mathbf{q}_2))\lfloor\boldsymbol{\nu}.
\end{aligned}$$

This is of the form $\overline{\overline{\mathsf{D}}} = \mathbf{ac} + \mathbf{bd}$ where the two dyads equal those of (10.48) and (10.49). Combining the two terms as

$$\begin{aligned}
\nu_3\overline{\overline{\mathsf{D}}}(\boldsymbol{\nu}) &= -\boldsymbol{\nu}\rfloor(\mathbf{e}_3(\boldsymbol{\nu}|\overline{\overline{\mathsf{Q}}})\,{}^{\wedge}_{\wedge}(\mathbf{e}_1\mathbf{q}_1 + \mathbf{e}_2\mathbf{q}_2))\lfloor\boldsymbol{\nu} \\
&= (\mathbf{e}_3(\boldsymbol{\nu}|\overline{\overline{\mathsf{Q}}})\,{}^{\wedge}_{\wedge}\overline{\overline{\mathsf{Q}}})\lfloor\lfloor\boldsymbol{\nu\nu},
\end{aligned}$$

the expression (10.50) is obtained.

10.11 The double-wedge cube of the dispersion dyadic can be written as

$$\begin{aligned}
\overline{\overline{\mathsf{D}}}^{(3)}(\boldsymbol{\nu}) &= \left(\sum \mathbf{e}_i\mathbf{d}_i\right)^{(3)} = \mathbf{e}_{123}, \\
\mathbf{d}_{123}(\boldsymbol{\nu}) &= -\nu_2\nu_3\nu_1(\mathbf{q}_{12}\lfloor\boldsymbol{\nu}) \wedge (\mathbf{q}_{23}\lfloor\boldsymbol{\nu}) \wedge (\mathbf{q}_{31}\lfloor\boldsymbol{\nu}) \\
&\quad +\nu_3\nu_1\nu_2(\mathbf{q}_{31}\lfloor\boldsymbol{\nu}) \wedge (\mathbf{q}_{12}\lfloor\boldsymbol{\nu}) \wedge (\mathbf{q}_{23}\lfloor\boldsymbol{\nu}),
\end{aligned}$$

whence $\mathbf{d}_{123}(\boldsymbol{\nu}) = 0$ identically, as expected. This implies that the three vector functions $\mathbf{d}_i(\boldsymbol{\nu})$ must be linearly dependent.

10.12 The expression (10.50) can be verified by applying the rule

$$(\overline{\overline{\mathsf{A}}}\,{}^{\wedge}_{\wedge}\overline{\overline{\mathsf{B}}})\lfloor\lfloor\overline{\overline{\mathsf{C}}} = (\overline{\overline{\mathsf{A}}}||\overline{\overline{\mathsf{C}}}^T)\overline{\overline{\mathsf{B}}} + (\overline{\overline{\mathsf{B}}}||\overline{\overline{\mathsf{C}}}^T)\overline{\overline{\mathsf{A}}} - \overline{\overline{\mathsf{A}}}|\overline{\overline{\mathsf{C}}}^T|\overline{\overline{\mathsf{B}}} - \overline{\overline{\mathsf{B}}}|\overline{\overline{\mathsf{C}}}^T|\overline{\overline{\mathsf{A}}},$$

valid for dyadics $\overline{\overline{\mathsf{A}}}, \overline{\overline{\mathsf{B}}} \in \mathbb{E}_1\mathbb{E}_1$ and $\overline{\overline{\mathsf{C}}} \in \mathbb{F}_1\mathbb{F}_1$, by expanding

$$\begin{aligned}
((\mathbf{e}_3(\boldsymbol{\nu}|\overline{\overline{\mathsf{Q}}}))\,{}^{\wedge}_{\wedge}\overline{\overline{\mathsf{Q}}})\lfloor\lfloor\boldsymbol{\nu\nu} &= (\mathbf{e}_3|\boldsymbol{\nu})(\boldsymbol{\nu}|\overline{\overline{\mathsf{Q}}}|\boldsymbol{\nu})\overline{\overline{\mathsf{Q}}} + (\overline{\overline{\mathsf{Q}}}||\boldsymbol{\nu\nu})\mathbf{e}_3(\boldsymbol{\nu}|\overline{\overline{\mathsf{Q}}}) \\
&\quad -\mathbf{e}_3(\boldsymbol{\nu}|\overline{\overline{\mathsf{Q}}}|\boldsymbol{\nu})(\boldsymbol{\nu}|\overline{\overline{\mathsf{Q}}}) - (\overline{\overline{\mathsf{Q}}}|\boldsymbol{\nu})(\boldsymbol{\nu}|\mathbf{e}_3)(\boldsymbol{\nu}|\overline{\overline{\mathsf{Q}}}) \\
&= (\mathbf{e}_3|\boldsymbol{\nu})((\boldsymbol{\nu}|\overline{\overline{\mathsf{Q}}}|\boldsymbol{\nu})\overline{\overline{\mathsf{Q}}} - (\overline{\overline{\mathsf{Q}}}|\boldsymbol{\nu})(\boldsymbol{\nu}|\overline{\overline{\mathsf{Q}}})) \\
&= (\mathbf{e}_3|\boldsymbol{\nu})\overline{\overline{\mathsf{Q}}}^{(2)}\lfloor\lfloor\boldsymbol{\nu\nu} \\
&= (\mathbf{e}_3|\boldsymbol{\nu})\overline{\overline{\mathsf{D}}}(\boldsymbol{\nu}).
\end{aligned}$$

10.13 Requiring that an orthogonality condition of the form

$$\alpha \mathbf{E}_g \cdot \mathbf{B}_g + \beta \mathbf{H}_g \cdot \mathbf{D}_g = 0,$$

satisfies by the fields of a plane wave in any medium, implies the polarization condition $(\mathbf{a} \cdot \mathbf{E}_g)(\mathbf{b} \cdot \mathbf{H}_g) = 0$ in an anisotropic medium, we set

$$\begin{aligned} &\alpha \mathbf{E}_g \cdot \mathbf{B}_g + \beta \mathbf{H}_g \cdot \mathbf{D}_g - \mathbf{E}_g \cdot \mathbf{ab} \cdot \mathbf{H}_g \\ &= \mathbf{E}_g \cdot (\alpha \overline{\overline{\mu}}_g + \beta \overline{\overline{\epsilon}}_g^T - \mathbf{ab}) \cdot \mathbf{H}_g = 0. \end{aligned}$$

To be satisfied for any field vectors $\mathbf{E}_g, \mathbf{H}_g$ requires that the medium dyadics satisfy the condition

$$\alpha \overline{\overline{\mu}}_g + \beta \overline{\overline{\epsilon}}_g^T - \mathbf{ab} = 0.$$

10.14 Substituting the 3D expansions in (5.65) for the PDC medium yields

$$\begin{aligned} \overline{\overline{\alpha}} &= M\overline{\overline{\mathsf{P}}}_s^{(2)} + \alpha \overline{\overline{\mathsf{I}}}_s^{(2)T} + (\boldsymbol{\varepsilon}_{123} \lfloor \mathbf{d}_3)(\mathbf{c}_1 \wedge \mathbf{c}_2), \\ \overline{\overline{\epsilon}}' &= -\boldsymbol{\pi} \wedge \overline{\overline{\mathsf{P}}}_s + (\boldsymbol{\varepsilon}_{123} \lfloor \mathbf{d}_3)\mathbf{c}_3, \\ \overline{\overline{\mu}}^{-1} &= -\overline{\overline{\mathsf{P}}}_s \wedge \mathbf{p} - \boldsymbol{\varepsilon}_{123} \lfloor (\mathbf{d}_1 \wedge \mathbf{d}_2)(\mathbf{c}_1 \wedge \mathbf{c}_2), \\ \overline{\overline{\beta}} &= \boldsymbol{\pi}\mathbf{p} - p\overline{\overline{\mathsf{P}}}_s - \alpha \overline{\overline{\mathsf{I}}}_s^T - \boldsymbol{\varepsilon}_{123} \lfloor (\mathbf{d}_1 \wedge \mathbf{d}_2)\mathbf{c}_3. \end{aligned}$$

The dyadics $\overline{\overline{\epsilon}}'$ and $\overline{\overline{\mu}}^{-1}$ have a restricted form. In the case $\mathbf{DC} = 0$ they do not have an inverse, which is in contrast to the QDC medium.

10.15 Applying $\overline{\overline{\mathsf{I}}} = \overline{\overline{\mathsf{I}}}_s + \mathbf{e}_4 \boldsymbol{\varepsilon}_4$ and $\overline{\overline{\mathsf{I}}}^{(2)} = \overline{\overline{\mathsf{I}}}_s^{(2)} + \overline{\overline{\mathsf{I}}}_s {}^\wedge_\wedge \mathbf{e}_4 \boldsymbol{\varepsilon}_4$, we have

$$\begin{aligned} \boldsymbol{\varepsilon}_N \lfloor \overline{\overline{\mathsf{A}}} &= (\overline{\overline{\mathsf{B}}}_o {}^\wedge_\wedge \overline{\overline{\mathsf{I}}})^T \\ &= (\overline{\overline{\mathsf{C}}}_s {}^\wedge_\wedge \overline{\overline{\mathsf{I}}}_s)^T - \boldsymbol{\varepsilon}_4 \wedge \overline{\overline{\mathsf{I}}}_s^T \wedge \mathbf{c}_s - \boldsymbol{\gamma}_s \wedge \overline{\overline{\mathsf{I}}}_s^T \wedge \mathbf{e}_4 \\ &\quad - \boldsymbol{\varepsilon}_4 \wedge (\overline{\overline{\mathsf{C}}}_s - \mathrm{tr}\overline{\overline{\mathsf{C}}}_s \overline{\overline{\mathsf{I}}}_s^T) \wedge \mathbf{e}_4. \end{aligned}$$

Further, we can expand

$$\begin{aligned} \boldsymbol{\varepsilon}_N \lfloor (\mathbf{AB} + \mathbf{BA}) = {} & (-\boldsymbol{\varepsilon}_4 \wedge \boldsymbol{\alpha}_s + \boldsymbol{\varepsilon}_{123} \lfloor \mathbf{a}_3)(\mathbf{e}_{123} \lfloor \boldsymbol{\beta}_s + \mathbf{b}_s \wedge \mathbf{e}_4) \\ & + (-\boldsymbol{\varepsilon}_4 \wedge \boldsymbol{\beta}_s + \boldsymbol{\varepsilon}_{123} \lfloor \mathbf{b}_s)(\mathbf{e}_{123} \lfloor \boldsymbol{\alpha}_s + \mathbf{a}_s \wedge \mathbf{e}_4). \end{aligned}$$

Inserting the expansions in (10.102) we can identify one set of 3D medium dyadics as

$$\begin{aligned} \overline{\overline{\alpha}} &= \alpha \overline{\overline{\mathsf{I}}}_s^{(2)T} + (\overline{\overline{\mathsf{C}}}_s {}^\wedge_\wedge \overline{\overline{\mathsf{I}}}_s)^T + \boldsymbol{\varepsilon}_{123} \mathbf{e}_{123} \lfloor \lfloor (\mathbf{a}_s \boldsymbol{\beta}_s + \mathbf{b}_s \boldsymbol{\alpha}_s), \\ \overline{\overline{\epsilon}}' &= -\boldsymbol{\gamma}_s \wedge \overline{\overline{\mathsf{I}}}_s^T + \boldsymbol{\varepsilon}_{123} \lfloor (\mathbf{a}_s \mathbf{b}_s + \mathbf{b}_s \mathbf{a}_s), \end{aligned}$$

$$\overline{\overline{\mu}}^{-1} = -\overline{\overline{\mathsf{I}}}_s^T \wedge \mathbf{c}_s - (\boldsymbol{\alpha}_s\boldsymbol{\beta}_s + \boldsymbol{\beta}_s\boldsymbol{\alpha}_s)\rfloor\mathbf{e}_{123},$$
$$\overline{\overline{\beta}} = -\alpha\overline{\overline{\mathsf{I}}}_s^T - (\overline{\overline{\mathsf{C}}}_s - \mathrm{tr}\overline{\overline{\mathsf{C}}}_s\,\overline{\overline{\mathsf{I}}}_s)^T - (\boldsymbol{\alpha}_s\mathbf{b}_s + \boldsymbol{\beta}_s\mathbf{a}_s).$$

It can be seen that the dyadics $\mathbf{e}_{123}\lfloor\overline{\overline{\epsilon}}'$ and $\overline{\overline{\mu}}^{-1}\rfloor\boldsymbol{\varepsilon}_{123}$ may have arbitrary antisymmetric parts while their symmetric parts are incomplete consisting of only two dyads. The decomposition can be written for the 3D fields $\mathbf{E}, \mathbf{B}$ as

$$(\mathbf{B}|(\mathbf{e}_{123}\lfloor\boldsymbol{\alpha}_s) + \mathbf{E}|\mathbf{a}_s)(\mathbf{B}|(\mathbf{e}_{123}\lfloor\boldsymbol{\beta}_s) + \mathbf{E}|\mathbf{b}_s) = 0.$$

10.16 To verify that (10.104) corresponds to the A-wave, applying (10.89) we can write for the QDC medium

$$\overline{\overline{\mathsf{D}}}(\boldsymbol{\nu})|\boldsymbol{\phi} = \boldsymbol{\nu}\boldsymbol{\nu}\rfloor\rfloor(M\overline{\overline{\mathsf{Q}}}^{(2)} + \mathbf{DC} - \beta\mathbf{e}_N\lfloor\overline{\overline{\mathsf{I}}}^{(2)T})|\boldsymbol{\phi}$$
$$= M(\boldsymbol{\nu}\boldsymbol{\nu}||\overline{\overline{\mathsf{Q}}})\overline{\overline{\mathsf{Q}}}|\boldsymbol{\phi} - M(\overline{\overline{\mathsf{Q}}}|\boldsymbol{\nu})\boldsymbol{\nu}|\overline{\overline{\mathsf{Q}}}|\boldsymbol{\phi} + \mathbf{dc}|\boldsymbol{\phi} = 0,$$

where we denote $\mathbf{d} = \boldsymbol{\nu}\rfloor\mathbf{D}$ and $\mathbf{c} = \boldsymbol{\nu}\rfloor\mathbf{C}$. Choosing the Lorenz condition $\boldsymbol{\nu}|\overline{\overline{\mathsf{Q}}}|\boldsymbol{\phi} = 0$, the potential satisfies

$$M(\boldsymbol{\nu}\boldsymbol{\nu}||\overline{\overline{\mathsf{Q}}})\overline{\overline{\mathsf{Q}}}|\boldsymbol{\phi} + \mathbf{d}(\mathbf{c}|\boldsymbol{\phi}) = 0.$$

The solution corresponding to the dispersion condition (10.104) leads to

$$\mathbf{c}|\boldsymbol{\phi} = -\mathbf{C}|\boldsymbol{\Phi} = -\mathbf{A}|\boldsymbol{\Phi} = 0,$$

because of the definition (10.58). Thus, (10.104) corresponds to the A-wave.

10.17 To verify that the field corresponding to the dispersion equation (10.105) satisfies the B-wave condition $\mathbf{B}|\boldsymbol{\Phi} = 0$ we can start from $\mathbf{c}|\boldsymbol{\phi} \neq 0$ and $\boldsymbol{\nu}\boldsymbol{\nu}||\overline{\overline{\mathsf{Q}}} \neq 0$ with $\mathbf{d}$ and $\mathbf{c}$ as in the previous solution. Assuming the Lorenz condition $\boldsymbol{\nu}|\overline{\overline{\mathsf{Q}}}|\boldsymbol{\phi} = 0$ in

$$\overline{\overline{\mathsf{D}}}(\boldsymbol{\nu})|\boldsymbol{\phi} = M(\boldsymbol{\nu}\boldsymbol{\nu}||\overline{\overline{\mathsf{Q}}})\overline{\overline{\mathsf{Q}}}|\boldsymbol{\phi} - M(\overline{\overline{\mathsf{Q}}}|\boldsymbol{\nu})(\boldsymbol{\nu}|\overline{\overline{\mathsf{Q}}})|\boldsymbol{\phi} + \mathbf{dc}|\boldsymbol{\phi}$$
$$= M(\boldsymbol{\nu}\boldsymbol{\nu}||\overline{\overline{\mathsf{Q}}})\overline{\overline{\mathsf{Q}}}|\boldsymbol{\phi} + \mathbf{dc}|\boldsymbol{\phi} = 0$$

we see that the potential must be of the form

$$\boldsymbol{\phi} = \lambda\overline{\overline{\mathsf{Q}}}^{-1}|\mathbf{d},$$

where λ is some scalar coefficient. The Lorenz condition then requires that $\boldsymbol{\nu}$ must satisfy $\boldsymbol{\nu}|\mathbf{d} = 0$. Invoking (10.91) and the rule

$$2\mathbf{D} \wedge (\boldsymbol{\nu}\rfloor\mathbf{D}) = \boldsymbol{\nu}\rfloor(\mathbf{D} \wedge \mathbf{D}) = (\boldsymbol{\nu}\rfloor\mathbf{e}_N)(\mathbf{D} \cdot \mathbf{D}),$$

we finally obtain

$$\mathbf{B}|\boldsymbol{\Phi} = \lambda\mathbf{B}|(\boldsymbol{\nu} \wedge \overline{\overline{\mathsf{Q}}}^{-1}|\mathbf{d}) = \frac{\lambda}{2}(\mathbf{D}\cdot\mathbf{D})(M(\boldsymbol{\nu}|\overline{\overline{\mathsf{Q}}}|\boldsymbol{\nu}) + \mathbf{c}|\overline{\overline{\mathsf{Q}}}^{-1}|\mathbf{d}) = 0,$$

because of (10.105). Thus, the corresponding solution can be identified as the B-wave.

10.18 Let us expand

$$\boldsymbol{\nu}\boldsymbol{\nu}\rfloor\rfloor\overline{\overline{\mathsf{A}}} = -\boldsymbol{\nu}\rfloor(\mathbf{e}_N\lfloor(\overline{\overline{\mathsf{B}}}_o{}^\wedge_\wedge\overline{\overline{\mathsf{I}}})^T\lfloor\boldsymbol{\nu}) = \mathbf{e}_N\lfloor(\boldsymbol{\nu}\wedge(\overline{\overline{\mathsf{B}}}_o|\boldsymbol{\nu}))\lfloor\overline{\overline{\mathsf{I}}})^T = \mathbf{F}\lfloor\overline{\overline{\mathsf{I}}}^T,$$

where the bivector defined by $\mathbf{F} = \mathbf{e}_N\lfloor(\boldsymbol{\nu}\wedge\boldsymbol{\nu}')$ is simple since it satisfies $\mathbf{F}\cdot\mathbf{F} = 0$. Because of this, the antisymmetric dyadic $\mathbf{F}\lfloor\overline{\overline{\mathsf{I}}}^T \in \mathbb{E}_1\mathbb{E}_1$ satisfies

$$(\mathbf{F}\lfloor\overline{\overline{\mathsf{I}}}^T)^{(2)} = \mathbf{F}\mathbf{F}, \quad (\mathbf{F}\lfloor\overline{\overline{\mathsf{I}}}^T)^{(3)} = 0.$$

Applying (10.115) we can further expand

$$\begin{aligned}\overline{\overline{\mathsf{D}}}^{(3)}(\boldsymbol{\nu}) &= (\mathbf{F}\lfloor\overline{\overline{\mathsf{I}}}^T + \mathbf{ab} + \mathbf{ba}))^{(3)} \\ &= (\mathbf{F}\mathbf{F})^\wedge_\wedge(\mathbf{ab}+\mathbf{ba}) + (\mathbf{F}\lfloor\overline{\overline{\mathsf{I}}}^T)^\wedge_\wedge(\mathbf{ab}+\mathbf{ba})^{(2)}.\end{aligned}$$

Applying the orthogonality $\boldsymbol{\nu}|\mathbf{a} = \boldsymbol{\nu}|\mathbf{b} = 0$, we obtain

$$\mathbf{F}\wedge\mathbf{a} = (\mathbf{e}_N\lfloor(\boldsymbol{\nu}\wedge\boldsymbol{\nu}'))\wedge\mathbf{a} = -(\mathbf{e}_N\lfloor\boldsymbol{\nu})(\boldsymbol{\nu}'|\mathbf{a}),$$

and similarly for $\mathbf{F}\wedge\mathbf{b}$, whence we have

$$(\mathbf{F}\mathbf{F})^\wedge_\wedge(\mathbf{ab}+\mathbf{ba}) = 2(\mathbf{e}_N\mathbf{e}_N\lfloor\lfloor\boldsymbol{\nu}\boldsymbol{\nu})(\boldsymbol{\nu}'|\mathbf{a})(\boldsymbol{\nu}'|\mathbf{b}).$$

The last term of the above expression of $\overline{\overline{\mathsf{D}}}^{(3)}(\boldsymbol{\nu})$ vanishes due to

$$(\mathbf{F}\lfloor\overline{\overline{\mathsf{I}}}^T)^\wedge_\wedge(\mathbf{ab}{}^\wedge_\wedge\mathbf{ba}) = (\mathbf{e}_N\lfloor\boldsymbol{\nu})\boldsymbol{\nu}'|(\mathbf{ab}-\mathbf{ba})\wedge(\mathbf{a}\wedge\mathbf{b}) = 0,$$

which leads to the dispersion equation (10.116).

APPENDIX B

Transformation to Gibbsian Formalism

In this appendix we consider the connection between spatial multivectors and multiforms on one hand and Gibbsian vectors on the other hand. Explicit formulas allow one to interpret results obtained through the 4D analysis in terms of the Gibbsian formalism. The converse, interpreting a given Gibbsian vector in terms of the spatial part of a 4D quantity, is also possible if the nature of the quantity is given.

In the following the subscript $()_g$ denotes Gibbsian vectors and dyadics while the spatial part of a 4D vector or one-form is marked by the subscript $()_s$. The connection depends on a chosen vector basis $\mathbf{e}_i$ which defines the spatial metric dyadic

$$\overline{\overline{\mathsf{G}}}_s = \mathbf{e}_1\mathbf{e}_1 + \mathbf{e}_2\mathbf{e}_2 + \mathbf{e}_3\mathbf{e}_3, \tag{B.1}$$

with its spatial inverse

$$\overline{\overline{\Gamma}}_s = \overline{\overline{\mathsf{G}}}_s^{-1} = \boldsymbol{\varepsilon}_1\boldsymbol{\varepsilon}_1 + \boldsymbol{\varepsilon}_2\boldsymbol{\varepsilon}_2 + \boldsymbol{\varepsilon}_3\boldsymbol{\varepsilon}_3 = \boldsymbol{\varepsilon}_{123}\boldsymbol{\varepsilon}_{123}\lfloor\lfloor\overline{\overline{\mathsf{G}}}^{(2)}. \tag{B.2}$$

The dyadic satisfies

$$\overline{\overline{\mathsf{G}}}_s^{(2)} = \mathbf{e}_{12}\mathbf{e}_{12} + \mathbf{e}_{23}\mathbf{e}_{23} + \mathbf{e}_{31}\mathbf{e}_{31} = \mathbf{e}_{123}\mathbf{e}_{123}\lfloor\lfloor\overline{\overline{\Gamma}}_s. \tag{B.3}$$

$\overline{\overline{\mathsf{G}}}_s$ serves as the Gibbsian unit dyadic $\overline{\overline{\mathsf{I}}}_g = \overline{\overline{\mathsf{G}}}_s \in \mathbb{E}_1\mathbb{E}_1$ satisfying

$$\overline{\overline{\mathsf{I}}}_g \cdot \mathbf{x}_g = \mathbf{x}_g \cdot \overline{\overline{\mathsf{I}}}_g = \mathbf{x}_g \tag{B.4}$$

Multiforms, Dyadics, and Electromagnetic Media, First Edition. Ismo V. Lindell.

for any Gibbsian vector $\mathbf{x}_g$. The following set of transformation rules can be easily verified applying the basis expansions

$$\boldsymbol{\alpha}_s = \alpha_1 \boldsymbol{\varepsilon}_1 + \alpha_2 \boldsymbol{\varepsilon}_2 + \alpha_3 \boldsymbol{\varepsilon}_3, \tag{B.5}$$

$$\mathbf{a}_s = a_1 \mathbf{e}_1 + a_2 \mathbf{e}_2 + a_3 \mathbf{e}_3, \tag{B.6}$$

$$\boldsymbol{\theta}_s = \theta_{12} \boldsymbol{\varepsilon}_{12} + \theta_{23} \boldsymbol{\varepsilon}_{23} + \theta_{31} \boldsymbol{\varepsilon}_{31}, \tag{B.7}$$

$$\mathbf{A}_s = A_{12} \mathbf{e}_{12} + A_{23} \mathbf{e}_{23} + A_{31} \mathbf{e}_{31}. \tag{B.8}$$

VECTOR AND ONE-FORM

Any spatial vector can be immediately understood in Gibbsian sense, $\mathbf{a}_g = \mathbf{a}_s$. Transformation of a spatial one-form to a Gibbsian vector $\boldsymbol{\alpha}_g$ is given through

$$\boldsymbol{\alpha}_g = \overline{\overline{\mathsf{G}}}_s | \boldsymbol{\alpha}_s = \alpha_1 \mathbf{e}_1 + \alpha_2 \mathbf{e}_2 + \alpha_3 \mathbf{e}_3. \tag{B.9}$$

It can be inverted as

$$\boldsymbol{\alpha}_s = \overline{\overline{\Gamma}}_s | \boldsymbol{\alpha}_g. \tag{B.10}$$

BIVECTOR AND TWO-FORM

A spatial two-form $\boldsymbol{\theta}_s$ can be transformed to a Gibbsian vector as

$$\boldsymbol{\theta}_g = \mathbf{e}_{123} \lfloor \boldsymbol{\theta}_s = \theta_{12} \mathbf{e}_3 + \theta_{23} \mathbf{e}_1 + \theta_{31} \mathbf{e}_2, \tag{B.11}$$

and the inverse transformation is

$$\boldsymbol{\theta}_s = \boldsymbol{\varepsilon}_{123} \lfloor \boldsymbol{\theta}_g. \tag{B.12}$$

The Gibbsian vector corresponding to a spatial bivector $\mathbf{A}_s$ can be obtained by first transforming the bivector to the one-form $\boldsymbol{\varepsilon}_{123} \lfloor \mathbf{A}_s$, after which it can be transformed to the vector,

$$\mathbf{A}_g = \overline{\overline{\mathsf{G}}}_s | (\boldsymbol{\varepsilon}_{123} \lfloor \mathbf{A}_s) = \mathbf{e}_1 A_{23} + \mathbf{e}_2 A_{31} + \mathbf{e}_3 A_{12}. \tag{B.13}$$

The same result is obtained if the bivector is first transformed to a two-form as $\overline{\overline{\Gamma}}{}_s^{(2)} | \mathbf{A}_s$ and

$$\mathbf{A}_g = \mathbf{e}_{123} \lfloor (\overline{\overline{\Gamma}}{}_s^{(2)} | \mathbf{A}_s) = \overline{\overline{\mathsf{G}}}_s | (\boldsymbol{\varepsilon}_{123} \lfloor \mathbf{A}_s). \tag{B.14}$$

The inverse transformation is

$$\mathbf{A}_s = \mathbf{e}_{123}\lfloor(\overline{\overline{\Gamma}}_s|\mathbf{A}_g). \tag{B.15}$$

TRIVECTOR AND THREE-FORM

Spatial trivector $\mathbf{k}_s = \mathbf{e}_{123}k$ and three-form $\boldsymbol{\kappa}_s = \boldsymbol{\varepsilon}_{123}\kappa$ correspond to scalars in Gibbsian formalism:

$$k = \boldsymbol{\varepsilon}_{123}|\mathbf{k}_s, \tag{B.16}$$

$$\kappa = \mathbf{e}_{123}|\boldsymbol{\kappa}_s. \tag{B.17}$$

Conversely, a Gibbsian scalar α can be interpreted as a spatial trivector $\alpha\mathbf{e}_{123}$ or a spatial three-form $\alpha\boldsymbol{\varepsilon}_{123}$.

BAR PRODUCT

By definition, the bar product of a spatial vector $\mathbf{a}_s$ and a spatial one-form $\boldsymbol{\alpha}_s$ equals the dot product of the corresponding Gibbsian vectors as

$$\mathbf{a}_s|\boldsymbol{\alpha}_s = \mathbf{a}_s|\overline{\overline{\Gamma}}_s|\boldsymbol{\alpha}_g = \mathbf{a}_s \cdot \boldsymbol{\alpha}_g = \sum a_i\alpha_i. \tag{B.18}$$

The bar product of a spatial bivector $\mathbf{A}_s$ and a spatial two-form $\boldsymbol{\theta}_s$ can be shown to equal the dot product of the corresponding Gibbsian vectors as

$$\begin{aligned}\mathbf{A}_s|\boldsymbol{\theta}_s &= \mathbf{A}_s|(\boldsymbol{\varepsilon}_{123}\lfloor(\mathbf{e}_{123}\lfloor\boldsymbol{\theta}_s)) = (\boldsymbol{\varepsilon}_{123}\lfloor\mathbf{A}_s)|(\mathbf{e}_{123}\lfloor\boldsymbol{\theta}_s)\\ &= (\overline{\overline{\mathsf{G}}}_s|(\boldsymbol{\varepsilon}_{123}\lfloor\mathbf{A}_s))|\overline{\overline{\Gamma}}_s|(\mathbf{e}_{123}\lfloor\boldsymbol{\theta}_s) = \mathbf{A}_g \cdot \boldsymbol{\theta}_g.\end{aligned} \tag{B.19}$$

WEDGE PRODUCT

The wedge product of two spatial one-forms $\boldsymbol{\alpha}_s, \boldsymbol{\beta}_s$ has the following relation to the cross product of the corresponding Gibbsian vectors,

$$\begin{aligned}\boldsymbol{\alpha}_g \times \boldsymbol{\beta}_g &= \mathbf{e}_{123}\lfloor(\boldsymbol{\alpha}_s \wedge \boldsymbol{\beta}_s)\\ &= \mathbf{e}_1(\alpha_2\beta_3 - \alpha_3\beta_2) + \mathbf{e}_2(\alpha_3\beta_1 - \alpha_3\beta_1) + \mathbf{e}_3(\alpha_1\beta_2 - \alpha_2\beta_3).\end{aligned} \tag{B.20}$$

The wedge product of three one-forms transforms to a scalar as

$$\mathbf{e}_{123}|(\boldsymbol{\alpha}_s \wedge \boldsymbol{\beta}_s \wedge \boldsymbol{\gamma}_s) = (\mathbf{e}_{123}\lfloor(\boldsymbol{\alpha}_s \wedge \boldsymbol{\beta}_s))|\boldsymbol{\gamma}_s = (\boldsymbol{\alpha}_g \times \boldsymbol{\beta}_g)|\boldsymbol{\gamma}_s$$
$$= \alpha_g \times \boldsymbol{\beta}_g \cdot \boldsymbol{\gamma}_g. \tag{B.21}$$

The cross product of two spatial vectors can be interpreted in terms of their wedge product as

$$\mathbf{a}_s \times \mathbf{b}_s = \mathbf{e}_{123}\lfloor((\overline{\overline{\Gamma}}_s|\mathbf{a}_s) \wedge (\overline{\overline{\Gamma}}_s|\mathbf{b}_s)) = \mathbf{e}_{123}\lfloor(\overline{\overline{\Gamma}}{}_s^{(2)}|(\mathbf{a}_s \wedge \mathbf{b}_s)))$$
$$= \overline{\overline{\mathsf{G}}}_s|(\boldsymbol{\varepsilon}_{123}\lfloor(\mathbf{a}_s \wedge \mathbf{b}_s)). \tag{B.22}$$

CONTRACTION PRODUCTS

Contraction product of a spatial vector and a spatial two-form is related to the cross product of their Gibbsian counterparts as

$$\mathbf{a}_s\rfloor\boldsymbol{\theta}_s = \boldsymbol{\varepsilon}_1(a_2\theta_{12} - a_3\theta_{31}) + \boldsymbol{\varepsilon}_2(a_3\theta_{23} - a_1\theta_{12}) + \boldsymbol{\varepsilon}_3(a_1\theta_{31} - a_2\theta_{23}),$$
$$= \mathbf{a}_s\rfloor(\mathbf{e}_{123}\lfloor\boldsymbol{\theta}_g) = -(\boldsymbol{\varepsilon}_{123}\lfloor\boldsymbol{\theta}_g)\lfloor\mathbf{a}_s = \boldsymbol{\varepsilon}_{123}\lfloor(\mathbf{a}_s \wedge \boldsymbol{\theta}_g)$$
$$= \overline{\overline{\Gamma}}_s|(\mathbf{a}_s \times \boldsymbol{\theta}_g). \tag{B.23}$$

For the spatial one-form $\boldsymbol{\alpha}_s$ and bivector $\mathbf{A}_s$ we obtain

$$\boldsymbol{\alpha}_s\rfloor\mathbf{A}_s = \boldsymbol{\alpha}_g \times \mathbf{A}_g. \tag{B.24}$$

Contraction with a spatial three-form $\boldsymbol{\kappa}_s = \kappa\boldsymbol{\varepsilon}_{123}$ can be expressed as

$$\mathbf{a}_s\rfloor\boldsymbol{\kappa}_s = \boldsymbol{\varepsilon}_{123}\lfloor(\kappa\mathbf{a}_s), \tag{B.25}$$

$$\mathbf{A}_s\rfloor\boldsymbol{\kappa}_s = \kappa\boldsymbol{\varepsilon}_{123}\lfloor(\mathbf{e}_{123}\lfloor(\overline{\overline{\Gamma}}_s|\mathbf{A}_g)) = \overline{\overline{\Gamma}}_s|(\kappa\mathbf{A}_g). \tag{B.26}$$

BAC-CAB RULE

As an example let us consider the bac-cab rule which for spatial vectors and one-forms reads

$$\mathbf{a}_s\rfloor(\boldsymbol{\beta}_s \wedge \boldsymbol{\gamma}_s) = \boldsymbol{\beta}_s(\mathbf{a}_s|\boldsymbol{\gamma}_s) - \boldsymbol{\gamma}_s(\mathbf{a}_s|\boldsymbol{\beta}_s). \tag{B.27}$$

Expanding first

$$\boldsymbol{\beta}_s \wedge \boldsymbol{\gamma}_s = \mathbf{x}_g\rfloor\boldsymbol{\varepsilon}_{123}, \quad \mathbf{x}_g = \boldsymbol{\beta}_g \times \boldsymbol{\gamma}_g, \tag{B.28}$$

we obtain

$$\mathbf{a}_s\rfloor(\boldsymbol{\beta}_s \wedge \boldsymbol{\gamma}_s) = \mathbf{a}_s\rfloor(\mathbf{x}_g\rfloor\boldsymbol{\varepsilon}_{123}) = (\mathbf{a}_s \wedge \mathbf{x}_g)\rfloor\boldsymbol{\varepsilon}_{123} = \overline{\overline{\Gamma}}_s|(\mathbf{a}_g \times \mathbf{x}_g). \quad \text{(B.29)}$$

Thus, the bac-cab rule becomes the corresponding Gibbsian rule,

$$\begin{aligned}\mathbf{a}_g \times (\boldsymbol{\beta}_g \times \boldsymbol{\gamma}_g) = \mathbf{a}_g \times \mathbf{x}_g &= \boldsymbol{\beta}_g(\mathbf{a}_s|\boldsymbol{\gamma}_s) - \boldsymbol{\gamma}_g(\mathbf{a}_s|\boldsymbol{\beta}_s) \\ &= \boldsymbol{\beta}_g(\mathbf{a}_g \cdot \boldsymbol{\gamma}_g) - \boldsymbol{\gamma}_g(\mathbf{a}_g \cdot \boldsymbol{\beta}_g). \end{aligned} \quad \text{(B.30)}$$

CROSS PRODUCT OF TWO GIBBSIAN TWO-FORMS

The cross product of two Gibbsian vectors has many counterparts depending how we interpret the vector quantities. Assuming that the Gibbsian vectors correspond to two spatial two-forms, their cross product has the form

$$\begin{aligned}\boldsymbol{\theta}_g \times \boldsymbol{\Pi}_g &= (\mathbf{e}_{123}\lfloor\boldsymbol{\theta}_s) \times (\mathbf{e}_{123}\lfloor\boldsymbol{\Pi}_s) \\ &= \mathbf{e}_{123}\lfloor(((\mathbf{e}_{123}\lfloor\boldsymbol{\theta}_s)|\overline{\overline{\Gamma}}_s) \wedge ((\mathbf{e}_{123}\lfloor\boldsymbol{\Pi}_s)|\overline{\overline{\Gamma}}_s)). \end{aligned} \quad \text{(B.31)}$$

Now one can show through basis expansions that the following relation is valid

$$\boldsymbol{\theta}_g \times \boldsymbol{\Pi}_g = \overline{\overline{\mathsf{G}}}_s|(\boldsymbol{\theta}_s\lfloor(\mathbf{e}_{123}\lfloor\boldsymbol{\Pi}_s)). \quad \text{(B.32)}$$

As an example, the magnetic force law in Gibbsian form,

$$d\mathbf{F}_g = \mathbf{J}_g \times \mathbf{B}_g dV \quad \text{(B.33)}$$

where $d\mathbf{F}_g$ is the force in the differential volume dV corresponds to

$$\overline{\overline{\Gamma}}_s|(\mathbf{J}_g \times \mathbf{B}_g dV) = \mathbf{J}\lfloor(\mathbf{e}_{123}\lfloor\mathbf{B}) = -(\mathbf{e}_{123}\lfloor\mathbf{B})\rfloor\mathbf{J}dV = (dV\mathbf{e}_{123}\lfloor\mathbf{J})\rfloor\mathbf{B}. \quad \text{(B.34)}$$

Here $dV\mathbf{e}_{123}$ can be interpreted as a trivector volume element which makes $dV\mathbf{e}_{123}\lfloor\mathbf{J}$ the differential current moment vector.

APPENDIX C

Multivector and Dyadic Identities

In this appendix some useful identities are collected for convenience. Where not otherwise stated, the dimension n of the vector space equals 4. p and q are numbers in the range $0, \ldots, n$. Subscript $()_s$ denotes spatial part, corresponding to coordinates $1, 2, 3$, while the temporal coordinate is denoted by the index 4. Subindex $()_g$ refers to Gibbsian vector or dyadic quantity. Where not indicated otherwise, dyadics are elements of the space $\mathbb{E}_1\mathbb{F}_1$. Elements of $\mathbb{E}_2\mathbb{F}_2, \mathbb{F}_2\mathbb{E}_2, \mathbb{E}_2\mathbb{E}_2$, and $\mathbb{F}_2\mathbb{F}_2$ are called bidyadics.

NOTATION

$$\text{vectors } \mathbf{a}, \mathbf{b}, \mathbf{c}, \ldots \in \mathbb{E}_1, \quad \text{one-forms } \boldsymbol{\alpha}, \boldsymbol{\beta}, \boldsymbol{\gamma}, \ldots \in \mathbb{F}_1$$

$$\text{bivectors } \mathbf{A}, \mathbf{B}, \mathbf{C}, \ldots \in \mathbb{E}_2, \quad \text{one-forms } \boldsymbol{\Phi}, \boldsymbol{\Psi}, \boldsymbol{\Gamma}, \ldots \in \mathbb{F}_2$$

$$\text{trivectors } \mathbf{k}, \mathbf{p}, \ldots \in \mathbb{E}_3, \quad \text{three-forms } \boldsymbol{\kappa}, \boldsymbol{\pi}, \ldots \in \mathbb{F}_3$$

$$\text{quadrivectors } \mathbf{a}_N, \mathbf{b}_N, \ldots \in \mathbb{E}_4, \quad \text{four-forms } \boldsymbol{\kappa}_N, \boldsymbol{\pi}_N, \ldots \in \mathbb{F}_4$$

$$p-\text{vectors } \mathbf{a}^p \in \mathbb{E}_p, \quad q-\text{forms } \boldsymbol{\alpha}^q \in \mathbb{F}_q$$

$$p\text{-index } J = i_1 i_2 \cdots i_p, \quad N = 1234, \quad \mathbf{a}_N = \mathbf{a}_{1234} = \mathbf{a}_1 \wedge \mathbf{a}_2 \wedge \mathbf{a}_3 \wedge \mathbf{a}_4 \in \mathbb{E}_4$$

$$\text{vector basis } \mathbf{e}_1, \mathbf{e}_2, \mathbf{e}_3, \mathbf{e}_4, \quad \text{reciprocal one-form basis } \boldsymbol{\varepsilon}_1, \boldsymbol{\varepsilon}_2, \boldsymbol{\varepsilon}_3, \boldsymbol{\varepsilon}_4$$

Multiforms, Dyadics, and Electromagnetic Media, First Edition. Ismo V. Lindell.

$$\text{bi-index ordering } i<j: \quad ij = 12,\ 23,\ 31,\ 14,\ 24,\ 34$$

$$\text{dyadics} \quad \overline{\overline{\mathsf{A}}}, \overline{\overline{\mathsf{B}}}, \ldots \in \mathbb{E}_1\mathbb{F}_1, \quad \overline{\overline{\mathsf{A}}}^T, \overline{\overline{\mathsf{B}}}^T, \ldots \in \mathbb{F}_1\mathbb{E}_1$$

$$\text{metric dyadics} \quad \overline{\overline{\mathsf{A}}}, \overline{\overline{\mathsf{B}}}, \ldots \in \mathbb{E}_1\mathbb{E}_1, \quad \overline{\overline{\Gamma}}, \overline{\overline{\Pi}}, \ldots \in \mathbb{F}_1\mathbb{F}_1$$

$$\mathbf{d} = \boldsymbol{\varepsilon}_1\partial_{x_1} + \boldsymbol{\varepsilon}_2\partial_{x_2} + \boldsymbol{\varepsilon}_3\partial_{x_3} + \boldsymbol{\varepsilon}_4\partial_{x_4} = \mathbf{d}_s + \boldsymbol{\varepsilon}_4\partial_{x_4}$$

MULTIVECTORS AND MULTIFORMS

Multiplication

$$\mathbf{a}^p|\boldsymbol{\alpha}^p = \boldsymbol{\alpha}^p|\mathbf{a}^p \in \mathbb{E}_0$$

$$\mathbf{a}^p \wedge \mathbf{b}^q = (-1)^{pq}\mathbf{b}^q \wedge \mathbf{a}^p \in \mathbb{E}_{p+q} \quad (= 0 \text{ for } p+q > 4)$$

$$p > q, \quad \mathbf{a}^p\lfloor\boldsymbol{\alpha}^q = (-1)^{q(p-q)}\boldsymbol{\alpha}^q\rfloor\mathbf{a}^p \in \mathbb{E}_{p-q}$$

$$p = q + r, \quad (\mathbf{a}^p\lfloor\boldsymbol{\alpha}^q)|\boldsymbol{\beta}^r = \mathbf{a}^p|(\boldsymbol{\alpha}^q \wedge \boldsymbol{\beta}^r)$$

$$p > q + r, \quad (\mathbf{a}^p\lfloor\boldsymbol{\alpha}^q)\lfloor\boldsymbol{\beta}^r = \mathbf{a}^p\lfloor(\boldsymbol{\alpha}^q \wedge \boldsymbol{\beta}^r)$$

Reciprocal bases

$$\mathbf{e}_i|\boldsymbol{\varepsilon}_j = \boldsymbol{\varepsilon}_j|\mathbf{e}_i = \delta_{i,j}$$

$$\mathbf{e}_{K(i)} = \mathbf{e}_1 \wedge \mathbf{e}_2 \wedge \cdots \wedge \mathbf{e}_{i-1} \wedge \mathbf{e}_{i+1} \wedge \cdots \wedge \mathbf{e}_n \in \mathbb{E}_{n-1}$$

$$\mathbf{e}_{K(ij)} = \mathbf{e}_1 \wedge \mathbf{e}_2 \wedge \cdots \wedge \mathbf{e}_{i-1} \wedge \mathbf{e}_{i+1} \wedge \cdots \wedge \mathbf{e}_{j-1} \wedge \mathbf{e}_{j+1} \wedge \cdots \wedge \mathbf{e}_n \in \mathbb{E}_{n-2}$$

$$\mathbf{e}_N = \mathbf{e}_{1234} = \mathbf{e}_1 \wedge \mathbf{e}_2 \wedge \mathbf{e}_3 \wedge \mathbf{e}_4 = (-1)^{i-1}\mathbf{e}_i \wedge \mathbf{e}_{K(i)} = (-1)^{4-i}\mathbf{e}_{K(i)} \wedge \mathbf{e}_i$$

$$\mathbf{e}_N\lfloor\boldsymbol{\varepsilon}_i = (-1)^{i-1}\mathbf{e}_{K(i)}, \quad \mathbf{e}_N\lfloor\boldsymbol{\varepsilon}_{K(i)} = (-1)^{4-i}\mathbf{e}_i$$

$$\mathbf{e}_N\lfloor\boldsymbol{\varepsilon}_{ij} = (-1)^{i+j-1}\mathbf{e}_{K(ij)}, \quad \mathbf{e}_N\lfloor\boldsymbol{\varepsilon}_{K(ij)} = (-1)^{i+j-1}\mathbf{e}_{ij}$$

$$(\boldsymbol{\alpha}\rfloor\mathbf{e}_N) \wedge \mathbf{a} = (\boldsymbol{\alpha}|\mathbf{a})\mathbf{e}_N, \quad (\boldsymbol{\Phi}\rfloor\mathbf{e}_N) \wedge \mathbf{A} = (\boldsymbol{\Phi}|\mathbf{A})\mathbf{e}_N$$

$$\mathbf{e}_N\lfloor(\mathbf{a}^p\rfloor\boldsymbol{\varepsilon}_N) = \mathbf{a}^p, \quad \mathbf{e}_N\lfloor(\boldsymbol{\varepsilon}_N\lfloor\mathbf{a}^p) = (-1)^p\mathbf{a}^p$$

$$\mathbf{e}_N\lfloor(\boldsymbol{\varepsilon}_{123}\lfloor\mathbf{a}^p) = \mathbf{a}^p \wedge \mathbf{e}_4, \quad p < 3$$

$$\mathbf{e}_N\lfloor(\mathbf{a}\rfloor\boldsymbol{\Phi}) = -\mathbf{a} \wedge (\mathbf{e}_N\lfloor\boldsymbol{\Phi}), \quad \boldsymbol{\Phi} \in \mathbb{F}_2$$

$$\mathbf{e}_N\lfloor(\boldsymbol{\varepsilon}_i \wedge \boldsymbol{\alpha}) = (-1)^{i-1}\mathbf{e}_{K(i)}\lfloor\boldsymbol{\alpha}.$$

Bivector products

$$(\mathbf{a}\wedge\mathbf{b})|(\boldsymbol{\alpha}\wedge\boldsymbol{\beta}) = (\mathbf{a}|\boldsymbol{\alpha})(\mathbf{b}|\boldsymbol{\beta}) - (\mathbf{a}|\boldsymbol{\beta})(\mathbf{b}|\boldsymbol{\alpha})$$

$$(\mathbf{a}\wedge\mathbf{b})|(\boldsymbol{\alpha}\wedge\boldsymbol{\beta}) = \mathbf{a}|\{\mathbf{b}\rfloor(\boldsymbol{\alpha}\wedge\boldsymbol{\beta})\} = \{(\mathbf{a}\wedge\mathbf{b})\lfloor\boldsymbol{\alpha}\}|\boldsymbol{\beta}$$

$$(\mathbf{a}\wedge\mathbf{b})\lfloor\boldsymbol{\alpha} = \mathbf{b}(\mathbf{a}|\boldsymbol{\alpha}) - \mathbf{a}(\mathbf{b}|\boldsymbol{\alpha}) = (\mathbf{ba}-\mathbf{ab})|\boldsymbol{\alpha} = -\boldsymbol{\alpha}\rfloor(\mathbf{a}\wedge\mathbf{b})$$

$$\boldsymbol{\alpha}\wedge(\mathbf{a}\rfloor(\boldsymbol{\beta}\wedge\boldsymbol{\gamma})) = -(\mathbf{a}\rfloor(\boldsymbol{\beta}\wedge\boldsymbol{\gamma}))\wedge\boldsymbol{\alpha} = \mathbf{a}\rfloor(\boldsymbol{\alpha}\wedge\boldsymbol{\beta}\wedge\boldsymbol{\gamma}) - (\mathbf{a}|\boldsymbol{\alpha})(\boldsymbol{\beta}\wedge\boldsymbol{\gamma}).$$

Bac–cab rules

$$\boldsymbol{\alpha}\rfloor(\mathbf{b}\wedge\mathbf{c}) = \mathbf{b}(\boldsymbol{\alpha}|\mathbf{c}) - \mathbf{c}(\boldsymbol{\alpha}|\mathbf{b}) = (\mathbf{c}\wedge\mathbf{b})\lfloor\boldsymbol{\alpha}$$

$$\boldsymbol{\alpha}\rfloor(\mathbf{b}\wedge\mathbf{C}) = \mathbf{b}\wedge(\boldsymbol{\alpha}\rfloor\mathbf{C}) + \mathbf{C}(\boldsymbol{\alpha}|\mathbf{b}) = (\mathbf{C}\wedge\mathbf{b})\lfloor\boldsymbol{\alpha}, \quad \mathbf{C}\in\mathbb{E}_2$$

$$\boldsymbol{\alpha}\rfloor(\mathbf{B}\wedge\mathbf{C}) = \mathbf{B}\wedge(\boldsymbol{\alpha}\rfloor\mathbf{C}) + \mathbf{C}\wedge(\boldsymbol{\alpha}\rfloor\mathbf{B}) = -(\mathbf{B}\wedge\mathbf{C})\lfloor\boldsymbol{\alpha}, \quad \mathbf{B},\mathbf{C}\in\mathbb{E}_2$$

$$\mathbf{A}\rfloor(\boldsymbol{\beta}\wedge\boldsymbol{\Gamma}) = \boldsymbol{\beta}(\mathbf{A}|\boldsymbol{\Gamma}) + \boldsymbol{\Gamma}\lfloor(\mathbf{A}\lfloor\boldsymbol{\beta}), \quad \mathbf{A}\in\mathbb{E}_2, \boldsymbol{\Gamma}\in\mathbb{F}_2$$

$$\boldsymbol{\alpha}\rfloor(\mathbf{b}^p\wedge\mathbf{c}^q) = \mathbf{b}^p\wedge(\boldsymbol{\alpha}\rfloor\mathbf{c}^q) + (-1)^{pq}\mathbf{c}^q\wedge(\boldsymbol{\alpha}\rfloor\mathbf{b}^p)$$

$$(\boldsymbol{\alpha}\wedge\boldsymbol{\beta})\rfloor(\mathbf{B}\wedge\mathbf{C}) = \mathbf{B}(\boldsymbol{\alpha}\wedge\boldsymbol{\beta})|\mathbf{C} + \mathbf{C}(\boldsymbol{\alpha}\wedge\boldsymbol{\beta})|\mathbf{B} + (\boldsymbol{\beta}\rfloor\mathbf{B})\wedge(\boldsymbol{\alpha}\rfloor\mathbf{C}) \\ +(\boldsymbol{\beta}\rfloor\mathbf{C})\wedge(\boldsymbol{\alpha}\rfloor\mathbf{B})$$

$$(\boldsymbol{\alpha}\wedge\boldsymbol{\beta})\rfloor(\mathbf{b}^p\wedge\mathbf{c}^q)$$

$$= (\mathbf{b}^p\lfloor(\boldsymbol{\alpha}\wedge\boldsymbol{\beta}))\wedge\mathbf{c}^q + \mathbf{b}^p\wedge((\boldsymbol{\alpha}\wedge\boldsymbol{\beta})\rfloor\mathbf{c}^q) + (\boldsymbol{\alpha}\rfloor\mathbf{b}^p)\wedge(\mathbf{c}^q\lfloor\boldsymbol{\beta}) \\ -(\boldsymbol{\beta}\rfloor\mathbf{b}^p)\wedge(\mathbf{c}^q\lfloor\boldsymbol{\alpha}).$$

Trivector products

$$(\mathbf{a}\wedge\mathbf{b}\wedge\mathbf{c})|(\boldsymbol{\alpha}\wedge\boldsymbol{\beta}\wedge\boldsymbol{\gamma}) = ((\mathbf{a}\wedge\mathbf{b}\wedge\mathbf{c})\lfloor\boldsymbol{\alpha})|(\boldsymbol{\beta}\wedge\boldsymbol{\gamma}) \\ = \boldsymbol{\alpha}|((\mathbf{a}\wedge\mathbf{b}\wedge\mathbf{c})\lfloor(\boldsymbol{\beta}\wedge\boldsymbol{\gamma}))$$

$$= \boldsymbol{\alpha}|(\mathbf{a}(\mathbf{b}\wedge\mathbf{c}) + \mathbf{b}(\mathbf{c}\wedge\mathbf{a}) + \mathbf{c}(\mathbf{a}\wedge\mathbf{b}))|(\boldsymbol{\beta}\wedge\boldsymbol{\gamma})$$

$$= (\mathbf{a}|\boldsymbol{\alpha})(\mathbf{b}\wedge\mathbf{c})|(\boldsymbol{\beta}\wedge\boldsymbol{\gamma}) - (\mathbf{a}\rfloor(\boldsymbol{\beta}\wedge\boldsymbol{\gamma}))|(\boldsymbol{\alpha}\rfloor(\mathbf{b}\wedge\mathbf{c}))$$

$$(\mathbf{a}\wedge\mathbf{C})|(\boldsymbol{\alpha}\wedge\boldsymbol{\Gamma}) = (\mathbf{a}|\boldsymbol{\alpha})(\mathbf{C}|\boldsymbol{\Gamma}) - (\mathbf{a}\rfloor\boldsymbol{\Gamma})|(\boldsymbol{\alpha}\rfloor\mathbf{C})$$

$$(\mathbf{a}\wedge\mathbf{b}\wedge\mathbf{c})\lfloor\boldsymbol{\alpha} = \{(\mathbf{a}\wedge\mathbf{b})\mathbf{c} + (\mathbf{b}\wedge\mathbf{c})\mathbf{a} + (\mathbf{c}\wedge\mathbf{a})\mathbf{b}\}|\boldsymbol{\alpha}$$

$$(\mathbf{a}\wedge\mathbf{b}\wedge\mathbf{c})\lfloor\boldsymbol{\Gamma} = \{\mathbf{a}(\mathbf{b}\wedge\mathbf{c}) + \mathbf{b}(\mathbf{c}\wedge\mathbf{a}) + \mathbf{c}(\mathbf{a}\wedge\mathbf{b})\}|\boldsymbol{\Gamma}$$

$$\boldsymbol{\alpha}\wedge(\boldsymbol{\gamma}\lfloor\mathbf{A}) = (\boldsymbol{\alpha}\rfloor\mathbf{A})\rfloor\boldsymbol{\gamma} + (\boldsymbol{\alpha}\wedge\boldsymbol{\gamma})\lfloor\mathbf{A}, \quad \boldsymbol{\gamma}\in\mathbb{F}_3, \ \mathbf{A}\in\mathbb{E}_2$$

$$\boldsymbol{\alpha}\wedge(\boldsymbol{\varepsilon}_N\lfloor\mathbf{k}) = \boldsymbol{\varepsilon}_N\lfloor(\boldsymbol{\alpha}\rfloor\mathbf{k}), \quad \mathbf{k}\in\mathbb{E}_3.$$

Quadrivector products

$$(\mathbf{a}\wedge\mathbf{b}\wedge\mathbf{c}\wedge\mathbf{d})|(\alpha\wedge\beta\wedge\gamma\wedge\delta) = ((\mathbf{a}\wedge\mathbf{b}\wedge\mathbf{c}\wedge\mathbf{d})\lfloor\alpha)|(\beta\wedge\gamma\wedge\delta)$$

$$= \alpha|(\mathbf{a}(\mathbf{b}\wedge\mathbf{c}\wedge\mathbf{d}) - \mathbf{b}(\mathbf{c}\wedge\mathbf{d}\wedge\mathbf{a}) + \mathbf{c}(\mathbf{d}\wedge\mathbf{a}\wedge\mathbf{b}) \\ -\mathbf{d}(\mathbf{a}\wedge\mathbf{b}\wedge\mathbf{c}))|(\beta\wedge\gamma\wedge\delta)$$

$$(\mathbf{a}\wedge\mathbf{b}\wedge\mathbf{c}\wedge\mathbf{d})\lfloor\alpha = \mathbf{d}\wedge((\mathbf{b}\wedge\mathbf{c})\mathbf{a} + (\mathbf{c}\wedge\mathbf{a})\mathbf{b} + (\mathbf{a}\wedge\mathbf{b})\mathbf{c})|\alpha \\ -(\mathbf{a}\wedge\mathbf{b}\wedge\mathbf{c})(\mathbf{d}|\alpha)$$

$$(\mathbf{a}\wedge\mathbf{b}\wedge\mathbf{c}\wedge\mathbf{d})\lfloor(\beta\wedge\gamma\wedge\delta)$$

$$= (((\mathbf{a}(\mathbf{b}\wedge\mathbf{c} + \mathbf{b}(\mathbf{c}\wedge\mathbf{a}) + \mathbf{c}(\mathbf{a}\wedge\mathbf{b}))\wedge\mathbf{d} - \mathbf{d}(\mathbf{a}\wedge\mathbf{b}\wedge\mathbf{c}))|(\beta\wedge\gamma\wedge\delta)$$

$$(\mathbf{a}\wedge\mathbf{b}\wedge\mathbf{c}\wedge\mathbf{d})|(\alpha\wedge\beta\wedge\gamma\wedge\delta) = (\gamma\wedge\delta)|((\mathbf{a}\wedge\mathbf{b}\wedge\mathbf{c}\wedge\mathbf{d})\lfloor(\alpha\wedge\beta))$$

$$= (\gamma\wedge\delta)|((\mathbf{a}\wedge\mathbf{b})(\mathbf{c}\wedge\mathbf{d}) - (\mathbf{a}\wedge\mathbf{c})(\mathbf{b}\wedge\mathbf{d}) + (\mathbf{a}\wedge\mathbf{d})(\mathbf{b}\wedge\mathbf{c})$$

$$+(\mathbf{b}\wedge\mathbf{c})(\mathbf{a}\wedge\mathbf{d}) - (\mathbf{b}\wedge\mathbf{d})(\mathbf{a}\wedge\mathbf{c}) + (\mathbf{c}\wedge\mathbf{d})(\mathbf{a}\wedge\mathbf{b}))|(\alpha\wedge\beta)$$

$$(\mathbf{a}\wedge\mathbf{b}\wedge\mathbf{c}\wedge\mathbf{d})\lfloor(\alpha\wedge\beta) = (((\mathbf{b}\wedge\mathbf{c})\mathbf{a} + (\mathbf{c}\wedge\mathbf{a})\mathbf{b} + (\mathbf{a}\wedge\mathbf{b})\mathbf{c})\wedge\mathbf{d})|(\alpha\wedge\beta)$$

$$-(\mathbf{d}\wedge(\mathbf{a}(\mathbf{b}\wedge\mathbf{c}) + \mathbf{b}(\mathbf{c}\wedge\mathbf{a}) + \mathbf{c}(\mathbf{a}\wedge\mathbf{b})))|(\alpha\wedge\beta).$$

Special cases for $n = 3$

$$\mathbf{a}(\mathbf{b}\wedge\mathbf{c}\wedge\mathbf{d}) - \mathbf{b}(\mathbf{c}\wedge\mathbf{d}\wedge\mathbf{a}) + \mathbf{c}(\mathbf{d}\wedge\mathbf{a}\wedge\mathbf{b}) - \mathbf{d}(\mathbf{a}\wedge\mathbf{b}\wedge\mathbf{c}) = 0$$

$$(\mathbf{a}(\mathbf{b}\wedge\mathbf{c}) + \mathbf{b}(\mathbf{c}\wedge\mathbf{a}) + \mathbf{c}(\mathbf{a}\wedge\mathbf{b}))\wedge\mathbf{d} = \mathbf{d}(\mathbf{a}\wedge\mathbf{b}\wedge\mathbf{c})$$

$$((\mathbf{a}\wedge\mathbf{b})\mathbf{c} + (\mathbf{c}\wedge\mathbf{a})\mathbf{b} + (\mathbf{b}\wedge\mathbf{c})\mathbf{a})\wedge\mathbf{d} = \mathbf{d}\wedge(\mathbf{a}(\mathbf{b}\wedge\mathbf{c}) + \mathbf{b}(\mathbf{c}\wedge\mathbf{a}) + \mathbf{c}(\mathbf{a}\wedge\mathbf{b}))$$

$$((\mathbf{a}\wedge\mathbf{b})\rfloor\kappa)\wedge((\mathbf{b}\wedge\mathbf{c})\rfloor\kappa) = (\mathbf{b}\rfloor\kappa)((\mathbf{a}\wedge\mathbf{b}\wedge\mathbf{c})|\kappa), \quad \kappa\in\mathbb{F}_3.$$

Quintivector products

$$\alpha\rfloor(\mathbf{a}\wedge\mathbf{b}\wedge\mathbf{c}\wedge\mathbf{d}\wedge\mathbf{e}) = \alpha|(\mathbf{a}(\mathbf{b}\wedge\mathbf{c}\wedge\mathbf{d}\wedge\mathbf{e}) - \mathbf{b}(\mathbf{a}\wedge\mathbf{c}\wedge\mathbf{d}\wedge\mathbf{e})$$

$$+\mathbf{c}(\mathbf{a}\wedge\mathbf{b}\wedge\mathbf{d}\wedge\mathbf{e}) - \mathbf{d}(\mathbf{a}\wedge\mathbf{b}\wedge\mathbf{c}\wedge\mathbf{e}) + \mathbf{e}(\mathbf{a}\wedge\mathbf{b}\wedge\mathbf{c}\wedge\mathbf{d}))$$

$$= (\alpha\rfloor(\mathbf{a}\wedge\mathbf{b}\wedge\mathbf{c}))\wedge(\mathbf{d}\wedge\mathbf{e}) + (\mathbf{a}\wedge\mathbf{b}\wedge\mathbf{c})\wedge(\alpha\rfloor(\mathbf{d}\wedge\mathbf{e})) = 0$$

$$\mathbf{a}(\mathbf{b}\wedge\mathbf{c}\wedge\mathbf{d}\wedge\mathbf{e}) - \mathbf{b}(\mathbf{a}\wedge\mathbf{c}\wedge\mathbf{d}\wedge\mathbf{e}) + \mathbf{c}(\mathbf{a}\wedge\mathbf{b}\wedge\mathbf{d}\wedge\mathbf{e})$$

$$-\mathbf{d}(\mathbf{a}\wedge\mathbf{b}\wedge\mathbf{c}\wedge\mathbf{e}) + \mathbf{e}(\mathbf{a}\wedge\mathbf{b}\wedge\mathbf{c}\wedge\mathbf{d}) = 0$$

$$(((\mathbf{a}(\mathbf{b}\wedge\mathbf{c}) + \mathbf{b}(\mathbf{c}\wedge\mathbf{a}) + \mathbf{c}(\mathbf{a}\wedge\mathbf{b}))\wedge\mathbf{d}) - \mathbf{d}(\mathbf{a}\wedge\mathbf{b}\wedge\mathbf{c}))\wedge\mathbf{e} \\ +\mathbf{e}(\mathbf{a}\wedge\mathbf{b}\wedge\mathbf{c}\wedge\mathbf{d}) = 0$$

$$(\alpha\rfloor(\mathbf{a}\wedge\mathbf{b}\wedge\mathbf{c}))\wedge(\mathbf{d}\wedge\mathbf{e}) = -(\mathbf{a}\wedge\mathbf{b}\wedge\mathbf{c})\wedge(\alpha\rfloor(\mathbf{d}\wedge\mathbf{e})).$$

p-vector rules

$$
\begin{aligned}
(\mathbf{a}_1 \wedge \mathbf{a}_2 \wedge \cdots \wedge \mathbf{a}_p)\lfloor\boldsymbol{\alpha} &= (-1)^{p-1}\boldsymbol{\alpha}\rfloor(\mathbf{a}_1 \wedge \mathbf{a}_2 \wedge \cdots \wedge \mathbf{a}_p)\\
&= \boldsymbol{\alpha}|\sum_{i=1}^{p}(-1)^{i-1}\mathbf{a}_i\mathbf{a}_{K_p(i)}
\end{aligned}
$$

$$= \boldsymbol{\alpha}|(\mathbf{a}_1\mathbf{a}_{K_p(1)} - \mathbf{a}_2\mathbf{a}_{K_p(2)} + \cdots + (-1)^{p-1}\mathbf{a}_p\mathbf{a}_{K_p(p)}), \quad 2 \le p \le n$$

$$\mathbf{a}_{K_p(i)} = \mathbf{a}_1 \wedge \cdots \wedge \mathbf{a}_{i-1} \wedge \mathbf{a}_{i+1} \wedge \cdots \wedge \mathbf{a}_p$$

$$(\mathbf{a}^p\rfloor\boldsymbol{\varepsilon}_N)|(\boldsymbol{\alpha}^p\rfloor\mathbf{e}_N) = \mathbf{a}^p|\boldsymbol{\alpha}^p.$$

DYADICS

Basic rules

$$\mathbf{a}\boldsymbol{\alpha} \in \mathbb{E}_1\mathbb{F}_1, \quad \mathbf{a}\mathbf{b} \in \mathbb{E}_1\mathbb{E}_1, \quad (\mathbf{a}\boldsymbol{\alpha})^T = \boldsymbol{\alpha}\mathbf{a} \in \mathbb{F}_1\mathbb{E}_1, \quad \boldsymbol{\alpha}\boldsymbol{\beta} \in \mathbb{F}_1\mathbb{F}_1$$

$$\overline{\overline{\mathsf{A}}} = \sum \mathbf{a}_i\boldsymbol{\alpha}_i = \mathbf{a}_1\boldsymbol{\alpha}_1 + \mathbf{a}_2\boldsymbol{\alpha}_2 + \mathbf{a}_3\boldsymbol{\alpha}_3 + \mathbf{a}_4\boldsymbol{\alpha}_4$$

$$\overline{\overline{\mathsf{A}}}|\mathbf{a} = \mathbf{a}|\overline{\overline{\mathsf{A}}}^T, \quad \overline{\overline{\mathsf{A}}}|\overline{\overline{\mathsf{B}}} = \sum \mathbf{a}_i\boldsymbol{\alpha}_i|\sum \mathbf{b}_j\boldsymbol{\beta}_j = \sum_{i,j}(\boldsymbol{\alpha}_i|\mathbf{b}_j)\mathbf{a}_i\boldsymbol{\beta}_j$$

$$\overline{\overline{\mathsf{I}}} = \sum \mathbf{e}_i\boldsymbol{\varepsilon}_i, \quad \overline{\overline{\mathsf{I}}}|\overline{\overline{\mathsf{A}}} = \overline{\overline{\mathsf{A}}}|\overline{\overline{\mathsf{I}}} = \overline{\overline{\mathsf{A}}}$$

$$(\overline{\overline{\mathsf{A}}}|\overline{\overline{\mathsf{B}}})|\overline{\overline{\mathsf{C}}} = \overline{\overline{\mathsf{A}}}|(\overline{\overline{\mathsf{B}}}|\overline{\overline{\mathsf{C}}}), \quad \overline{\overline{\mathsf{A}}}^p = \overline{\overline{\mathsf{A}}}|\overline{\overline{\mathsf{A}}}^{p-1}, \quad \overline{\overline{\mathsf{A}}}^0 = \overline{\overline{\mathsf{I}}}.$$

Double-bar products

$$(\mathbf{a}\boldsymbol{\alpha})||(\mathbf{b}\boldsymbol{\beta})^T = (\mathbf{a}\boldsymbol{\alpha})||(\boldsymbol{\beta}\mathbf{b}) = (\mathbf{a}|\boldsymbol{\beta})(\mathbf{b}|\boldsymbol{\alpha})$$

$$\mathrm{tr}\,\overline{\overline{\mathsf{A}}} = \mathrm{tr}\sum \mathbf{a}_i\boldsymbol{\alpha}_i = \sum \mathbf{a}_i|\boldsymbol{\alpha}_i = \overline{\overline{\mathsf{A}}}||\overline{\overline{\mathsf{I}}}^T$$

$$\overline{\overline{\mathsf{A}}}||\overline{\overline{\mathsf{B}}}^T = \overline{\overline{\mathsf{B}}}||\overline{\overline{\mathsf{A}}}^T = \mathrm{tr}(\overline{\overline{\mathsf{A}}}|\overline{\overline{\mathsf{B}}}) = (\overline{\overline{\mathsf{A}}}|\overline{\overline{\mathsf{B}}})||\overline{\overline{\mathsf{I}}}^T.$$

Double-wedge products

$$(\mathbf{a}\boldsymbol{\alpha}){}^\wedge_\wedge(\mathbf{b}\boldsymbol{\beta}) = (\mathbf{a} \wedge \mathbf{b})(\boldsymbol{\alpha} \wedge \boldsymbol{\beta}), \quad (\mathbf{a}\mathbf{b}){}^\wedge_\wedge(\mathbf{c}\mathbf{d}) = (\mathbf{a} \wedge \mathbf{c})(\mathbf{b} \wedge \mathbf{d})$$

$$\overline{\overline{\mathsf{A}}}{}^\wedge_\wedge\overline{\overline{\mathsf{B}}} = \overline{\overline{\mathsf{B}}}{}^\wedge_\wedge\overline{\overline{\mathsf{A}}}, \quad (\overline{\overline{\mathsf{A}}}{}^\wedge_\wedge\overline{\overline{\mathsf{B}}})^T = \overline{\overline{\mathsf{A}}}^T{}^\wedge_\wedge\overline{\overline{\mathsf{B}}}^T$$

$$(\overline{\overline{\mathsf{A}}}{}^\wedge_\wedge\overline{\overline{\mathsf{B}}}){}^\wedge_\wedge\overline{\overline{\mathsf{C}}} = \overline{\overline{\mathsf{A}}}{}^\wedge_\wedge(\overline{\overline{\mathsf{B}}}{}^\wedge_\wedge\overline{\overline{\mathsf{C}}}) = \overline{\overline{\mathsf{A}}}{}^\wedge_\wedge\overline{\overline{\mathsf{B}}}{}^\wedge_\wedge\overline{\overline{\mathsf{C}}} = \overline{\overline{\mathsf{B}}}{}^\wedge_\wedge\overline{\overline{\mathsf{A}}}{}^\wedge_\wedge\overline{\overline{\mathsf{C}}} = \cdots$$

$$(\overline{\overline{\mathsf{A}}}{}^\wedge_\wedge\overline{\overline{\mathsf{B}}})|(\mathbf{a} \wedge \mathbf{b}) = (\overline{\overline{\mathsf{A}}}|\mathbf{a}) \wedge (\overline{\overline{\mathsf{B}}}|\mathbf{b}) + (\overline{\overline{\mathsf{B}}}|\mathbf{a}) \wedge (\overline{\overline{\mathsf{A}}}|\mathbf{b})$$

$$(\overline{\overline{\mathsf{A}}}{}^{\wedge}_{\wedge}\overline{\overline{\mathsf{B}}})\lfloor\mathbf{a} = (\overline{\overline{\mathsf{A}}}|\mathbf{a})\wedge\overline{\overline{\mathsf{B}}} + (\overline{\overline{\mathsf{B}}}|\mathbf{a})\wedge\overline{\overline{\mathsf{A}}}$$

$$\mathbf{a}\rfloor(\overline{\overline{\mathsf{A}}}{}^{\wedge}_{\wedge}\overline{\overline{\mathsf{B}}}) = \overline{\overline{\mathsf{A}}}\wedge(\mathbf{a}|\overline{\overline{\mathsf{B}}}) + \overline{\overline{\mathsf{B}}}\wedge(\mathbf{a}|\overline{\overline{\mathsf{A}}}).$$

Double-wedge powers

$$\overline{\overline{\mathsf{A}}}{}^{(2)} = \frac{1}{2}\overline{\overline{\mathsf{A}}}{}^{\wedge}_{\wedge}\overline{\overline{\mathsf{A}}} = \frac{1}{2}\sum \mathbf{a}_i\boldsymbol{\alpha}_i{}^{\wedge}_{\wedge}\sum \mathbf{a}_j\boldsymbol{\alpha}_j = \sum_{i<j}(\mathbf{a}_i\wedge\mathbf{a}_j)(\boldsymbol{\alpha}_i\wedge\boldsymbol{\alpha}_j) = \sum_{i<j}\mathbf{a}_{ij}\boldsymbol{\alpha}_{ij}$$

$$\overline{\overline{\mathsf{A}}}{}^{(p)} = \frac{1}{p!}\overline{\overline{\mathsf{A}}}{}^{\wedge}_{\wedge}\overline{\overline{\mathsf{A}}}{}^{\wedge}_{\wedge}\cdots{}^{\wedge}_{\wedge}\overline{\overline{\mathsf{A}}} = \frac{r!s!}{p!}\overline{\overline{\mathsf{A}}}{}^{(r)}{}^{\wedge}_{\wedge}\overline{\overline{\mathsf{A}}}{}^{(s)}, \quad r+s=p$$

$$\overline{\overline{\mathsf{A}}}{}^{(4)} = \mathbf{a}_N\boldsymbol{\alpha}_N = \mathbf{a}_{1234}\boldsymbol{\alpha}_{1234} = (\mathbf{a}_N|\boldsymbol{\alpha}_N)\mathbf{e}_N\boldsymbol{\varepsilon}_N = (\det\overline{\overline{\mathsf{A}}})\overline{\overline{\mathsf{I}}}{}^{(4)}$$

$$(\overline{\overline{\mathsf{A}}}|\mathbf{a})\wedge(\overline{\overline{\mathsf{A}}}|\mathbf{b}) = \overline{\overline{\mathsf{A}}}{}^{(2)}|(\mathbf{a}\wedge\mathbf{b})$$

$$(\overline{\overline{\mathsf{A}}}|\mathbf{a}_1)\wedge(\overline{\overline{\mathsf{A}}}|\mathbf{a}_2)\cdots\wedge(\overline{\overline{\mathsf{A}}}|\mathbf{a}_p) = \overline{\overline{\mathsf{A}}}{}^{(p)}|(\mathbf{a}_1\wedge\mathbf{a}_2\wedge\cdots\wedge\mathbf{a}_p)$$

$$\overline{\overline{\mathsf{A}}}{}^{(2)}\lfloor\mathbf{a} = (\overline{\overline{\mathsf{A}}}|\mathbf{a})\wedge\overline{\overline{\mathsf{A}}}, \qquad \boldsymbol{\alpha}\rfloor\overline{\overline{\mathsf{A}}}{}^{(2)} = \overline{\overline{\mathsf{A}}}\wedge(\boldsymbol{\alpha}|\overline{\overline{\mathsf{A}}})$$

$$\overline{\overline{\mathsf{A}}}{}^{(3)}\lfloor(\mathbf{a}\wedge\mathbf{b}) = (\overline{\overline{\mathsf{A}}}|\mathbf{a})\wedge(\overline{\overline{\mathsf{A}}}|\mathbf{b})\wedge\overline{\overline{\mathsf{A}}} = (\overline{\overline{\mathsf{A}}}{}^{(2)}|(\mathbf{a}\wedge\mathbf{b}))\wedge\overline{\overline{\mathsf{A}}}$$

$$(\overline{\overline{\mathsf{A}}}|\overline{\overline{\mathsf{B}}})^{(p)} = \overline{\overline{\mathsf{A}}}{}^{(p)}|\overline{\overline{\mathsf{B}}}{}^{(p)}, \qquad (\overline{\overline{\mathsf{A}}}{}^{q})^{(p)} = (\overline{\overline{\mathsf{A}}}{}^{(p)})^{q}$$

$$\left(\frac{1}{2}\overline{\overline{\mathsf{A}}}{}^{\wedge}_{\wedge}\overline{\overline{\mathsf{B}}}\right)^{(2)} = \frac{1}{2}\overline{\overline{\mathsf{A}}}{}^{(2)}{}^{\wedge}_{\wedge}\overline{\overline{\mathsf{B}}}{}^{(2)}$$

$$(\overline{\overline{\mathsf{A}}}{}^{(2)})^{(2)} = \frac{1}{8}\overline{\overline{\mathsf{A}}}{}^{\wedge}_{\wedge}\overline{\overline{\mathsf{A}}}{}^{\wedge}_{\wedge}\overline{\overline{\mathsf{A}}}{}^{\wedge}_{\wedge}\overline{\overline{\mathsf{A}}} = 3(\det\overline{\overline{\mathsf{A}}})\overline{\overline{\mathsf{I}}}{}^{(4)}$$

$$(\overline{\overline{\mathsf{I}}}\rfloor\mathbf{A})^{(2)} = (\mathbf{A}\lfloor\overline{\overline{\mathsf{I}}}{}^{T})^{(2)} = \mathbf{A}\mathbf{A} - \frac{1}{2}(\mathbf{A}\wedge\mathbf{A})\lfloor\overline{\overline{\mathsf{I}}}{}^{(2)T}, \quad \mathbf{A}\in\mathbb{E}_2.$$

Unit dyadics

$$\overline{\overline{\mathsf{I}}} = \overline{\overline{\mathsf{I}}}_s + \mathbf{e}_4\boldsymbol{\varepsilon}_4, \quad \overline{\overline{\mathsf{I}}}_s = \mathbf{e}_1\boldsymbol{\varepsilon}_1 + \mathbf{e}_2\boldsymbol{\varepsilon}_2 + \mathbf{e}_2\boldsymbol{\varepsilon}_3$$

$$\overline{\overline{\mathsf{I}}}{}^{(2)} = \left(\sum \mathbf{e}_i\boldsymbol{\varepsilon}_i\right)^{(2)} = \sum_{i<j}\mathbf{e}_{ij}\boldsymbol{\varepsilon}_{ij}$$

$$\overline{\overline{\mathsf{I}}}{}^{(p)} = \left(\sum \mathbf{e}_i\boldsymbol{\varepsilon}_i\right)^{(p)} = \sum \mathbf{e}_J\boldsymbol{\varepsilon}_J, \quad J = \{i_1 i_2\cdots i_p\}, \ i_1 < i_2\cdots < i_p$$

$$\overline{\overline{\mathsf{I}}}{}^{(4)} = \mathbf{e}_N\boldsymbol{\varepsilon}_N = \frac{\mathbf{k}_N\boldsymbol{\kappa}_N}{\mathbf{k}_N|\boldsymbol{\kappa}_N}$$

$$\overline{\overline{\mathsf{I}}}{}^{(2)}|(\mathbf{a}\wedge\mathbf{b}) = (\overline{\overline{\mathsf{I}}}|\mathbf{a})\wedge(\overline{\overline{\mathsf{I}}}|\mathbf{b}) = \mathbf{a}\wedge\mathbf{b}$$

$$\overline{\overline{\mathsf{I}}}^{(p)}|(\mathbf{a}_1 \wedge \mathbf{a}_2 \wedge \cdots \wedge \mathbf{a}_p) = \mathbf{a}_1 \wedge \mathbf{a}_2 \wedge \cdots \mathbf{a}_p$$

$$\mathrm{tr}\overline{\overline{\mathsf{I}}} = 4, \quad \mathrm{tr}(\overline{\overline{\mathsf{I}}}^{(2)}) = 6, \quad \mathrm{tr}\overline{\overline{\mathsf{I}}}^{(3)} = 4, \quad \mathrm{tr}\overline{\overline{\mathsf{I}}}^{(4)} = 1$$

$$\overline{\overline{\mathsf{I}}}{}^{\wedge}_{\wedge}\overline{\overline{\mathsf{A}}} = (\mathrm{tr}\overline{\overline{\mathsf{A}}})\overline{\overline{\mathsf{I}}}^{(2)} - \overline{\overline{\mathsf{I}}}^{(3)}\lfloor\lfloor\overline{\overline{\mathsf{A}}}^T$$

$$\mathrm{tr}\overline{\overline{\mathsf{A}}}^{(2)} = \frac{1}{2}((\mathrm{tr}\overline{\overline{\mathsf{A}}})^2 - \mathrm{tr}\overline{\overline{\mathsf{A}}}^2)$$

$$\mathrm{tr}\overline{\overline{\mathsf{A}}}^{(3)} = \frac{1}{6}((\mathrm{tr}\overline{\overline{\mathsf{A}}})^3 - 3\mathrm{tr}\overline{\overline{\mathsf{A}}}\ \mathrm{tr}\overline{\overline{\mathsf{A}}}^2 + 2\mathrm{tr}\overline{\overline{\mathsf{A}}}^3)$$

$$\mathrm{tr}\overline{\overline{\mathsf{A}}}^{(4)} = \frac{1}{24}((\mathrm{tr}\overline{\overline{\mathsf{A}}})^4 - 6(\mathrm{tr}\overline{\overline{\mathsf{A}}})^2\ \mathrm{tr}\overline{\overline{\mathsf{A}}}^2 + 8\mathrm{tr}\overline{\overline{\mathsf{A}}}\ \mathrm{tr}\overline{\overline{\mathsf{A}}}^3 + 3(\mathrm{tr}\overline{\overline{\mathsf{A}}}^2)^2 - 6\mathrm{tr}\overline{\overline{\mathsf{A}}}^4)$$

$$\overline{\overline{\mathsf{I}}}_s^{(2)} = \mathbf{e}_{12}\boldsymbol{\varepsilon}_{12} + \mathbf{e}_{23}\boldsymbol{\varepsilon}_{23} + \mathbf{e}_{31}\boldsymbol{\varepsilon}_{31}, \quad \overline{\overline{\mathsf{I}}}_s^{(3)} = \mathbf{e}_{123}\boldsymbol{\varepsilon}_{123}$$

$$\overline{\overline{\mathsf{I}}}^{(2)} = \overline{\overline{\mathsf{I}}}_s^{(2)} + \overline{\overline{\mathsf{I}}}_s{}^{\wedge}_{\wedge}\mathbf{e}_4\boldsymbol{\varepsilon}_4$$

$$\overline{\overline{\mathsf{I}}}^{(3)} = \overline{\overline{\mathsf{I}}}_s^{(3)} + \overline{\overline{\mathsf{I}}}_s^{(2)}{}^{\wedge}_{\wedge}\mathbf{e}_4\boldsymbol{\varepsilon}_4$$

$$\overline{\overline{\mathsf{E}}} = \mathbf{e}_N\lfloor\overline{\overline{\mathsf{I}}}^{(2)T} = \overline{\overline{\mathsf{I}}}^{(2)}\rfloor\mathbf{e}_N = (\mathbf{e}_N\lfloor\overline{\overline{\mathsf{I}}}^{(2)T})^T = (\overline{\overline{\mathsf{I}}}^{(2)}\rfloor\mathbf{e}_N)^T$$

$$= \mathbf{e}_{12}\mathbf{e}_{34} + \mathbf{e}_{23}\mathbf{e}_{14} + \mathbf{e}_{31}\mathbf{e}_{24} + \mathbf{e}_{14}\mathbf{e}_{23} + \mathbf{e}_{24}\mathbf{e}_{31} + \mathbf{e}_{34}\mathbf{e}_{12}$$

$$\overline{\overline{\mathsf{I}}}^{(2)}\lfloor\lfloor\overline{\overline{\mathsf{I}}}^T = 3\overline{\overline{\mathsf{I}}}, \quad \overline{\overline{\mathsf{I}}}^{(3)}\lfloor\lfloor\overline{\overline{\mathsf{I}}}^T = 2\overline{\overline{\mathsf{I}}}^{(2)}, \quad \overline{\overline{\mathsf{I}}}^{(3)}\lfloor\lfloor\overline{\overline{\mathsf{I}}}^{(2)T} = 3\overline{\overline{\mathsf{I}}}$$

$$\overline{\overline{\mathsf{I}}}^{(4)}\lfloor\lfloor\overline{\overline{\mathsf{I}}}^T = \overline{\overline{\mathsf{I}}}^{(3)}, \quad \overline{\overline{\mathsf{I}}}^{(4)}\lfloor\lfloor\overline{\overline{\mathsf{I}}}^{(2)T} = \overline{\overline{\mathsf{I}}}^{(2)}, \quad \overline{\overline{\mathsf{I}}}^{(4)}\lfloor\lfloor\overline{\overline{\mathsf{I}}}^{(3)T} = \overline{\overline{\mathsf{I}}}.$$

Multivectors and unit dyadics

$$(\mathbf{a} \wedge \mathbf{b})\lfloor\overline{\overline{\mathsf{I}}}^T = \mathbf{ba} - \mathbf{ab} = -\overline{\overline{\mathsf{I}}}\rfloor(\mathbf{a} \wedge \mathbf{b}) = (\overline{\overline{\mathsf{I}}}\rfloor(\mathbf{a} \wedge \mathbf{b}))^T$$

$$(\mathbf{a} \wedge \mathbf{b} \wedge \mathbf{c})\lfloor\overline{\overline{\mathsf{I}}}^T = (\mathbf{a} \wedge \mathbf{b})\mathbf{c} + (\mathbf{b} \wedge \mathbf{c})\mathbf{a} + (\mathbf{c} \wedge \mathbf{a})\mathbf{b} = \overline{\overline{\mathsf{I}}}^{(2)}\rfloor(\mathbf{a} \wedge \mathbf{b} \wedge \mathbf{c})$$

$$(\mathbf{a} \wedge \mathbf{b} \wedge \mathbf{c})\lfloor\overline{\overline{\mathsf{I}}}^{(2)T} = \mathbf{a}(\mathbf{b} \wedge \mathbf{c}) + \mathbf{b}(\mathbf{c} \wedge \mathbf{a}) + \mathbf{c}(\mathbf{a} \wedge \mathbf{b}) = \overline{\overline{\mathsf{I}}}\rfloor(\mathbf{a} \wedge \mathbf{b} \wedge \mathbf{c})$$

$$(\mathbf{a} \wedge \mathbf{b} \wedge \mathbf{c} \wedge \mathbf{d})\lfloor\overline{\overline{\mathsf{I}}}^T = -\overline{\overline{\mathsf{I}}}^{(3)}\rfloor(\mathbf{a} \wedge \mathbf{b} \wedge \mathbf{c} \wedge \mathbf{d})$$

$$= \mathbf{d} \wedge ((\mathbf{a} \wedge \mathbf{b})\mathbf{c} + (\mathbf{b} \wedge \mathbf{c})\mathbf{a} + (\mathbf{c} \wedge \mathbf{a})\mathbf{b}) - (\mathbf{a} \wedge \mathbf{b} \wedge \mathbf{c})\mathbf{d}$$

$$(\mathbf{a} \wedge \mathbf{b} \wedge \mathbf{c} \wedge \mathbf{d})\lfloor\overline{\overline{\mathsf{I}}}^{(2)T} = (\mathbf{a} \wedge \mathbf{b})(\mathbf{c} \wedge \mathbf{d}) + (\mathbf{c} \wedge \mathbf{a})(\mathbf{b} \wedge \mathbf{d}) + (\mathbf{b} \wedge \mathbf{c})(\mathbf{a} \wedge \mathbf{d})$$

$$+(\mathbf{a} \wedge \mathbf{d})(\mathbf{b} \wedge \mathbf{c}) + (\mathbf{b} \wedge \mathbf{d})(\mathbf{c} \wedge \mathbf{a}) + (\mathbf{c} \wedge \mathbf{d})(\mathbf{a} \wedge \mathbf{b}) = \overline{\overline{\mathsf{I}}}^{(2)}\rfloor(\mathbf{a} \wedge \mathbf{b} \wedge \mathbf{c} \wedge \mathbf{d})$$

$$= \{(\mathbf{b} \wedge \mathbf{c})\mathbf{a} + (\mathbf{c} \wedge \mathbf{a})\mathbf{b} + (\mathbf{a} \wedge \mathbf{b})\mathbf{c}\} \wedge \mathbf{d} - \mathbf{d} \wedge \{\mathbf{a}(\mathbf{b} \wedge \mathbf{c}) + \mathbf{b}(\mathbf{c} \wedge \mathbf{a}) + \mathbf{c}(\mathbf{a} \wedge \mathbf{b})\}$$

$$(\mathbf{a} \wedge \mathbf{b} \wedge \mathbf{c} \wedge \mathbf{d})\lfloor\overline{\overline{\mathsf{I}}}^{(3)T} = -\overline{\overline{\mathsf{I}}}\rfloor(\mathbf{a} \wedge \mathbf{b} \wedge \mathbf{c} \wedge \mathbf{d})$$

$$= (\mathbf{a}(\mathbf{b}\wedge\mathbf{c}) + \mathbf{b}(\mathbf{c}\wedge\mathbf{a}) + \mathbf{c}(\mathbf{a}\wedge\mathbf{b}))\wedge\mathbf{d} - \mathbf{d}(\mathbf{a}\wedge\mathbf{b}\wedge\mathbf{c})$$

$$\overline{\overline{\mathsf{I}}}^{(2)}\rfloor\mathbf{e}_N = \sum \mathbf{e}_{ij}\boldsymbol{\varepsilon}_{ij}\rfloor\mathbf{e}_N = \sum(-1)^{i+j-1}\mathbf{e}_{ij}\mathbf{e}_{K(ij)}, \quad i<j$$

$$(\mathbf{A}\wedge\mathbf{B})\lfloor\overline{\overline{\mathsf{I}}}^T = \mathbf{A}\wedge\overline{\overline{\mathsf{I}}}\rfloor\mathbf{B} + \mathbf{B}\wedge\overline{\overline{\mathsf{I}}}\rfloor\mathbf{A} = -\overline{\overline{\mathsf{I}}}^{(3)}\rfloor(\mathbf{A}\wedge\mathbf{B}), \quad \mathbf{A},\mathbf{B}\in\mathbb{E}_2$$

$$\overline{\overline{\mathsf{I}}}\rfloor(\mathbf{A}\wedge\mathbf{B}) = \mathbf{A}\lfloor\overline{\overline{\mathsf{I}}}^T\wedge\mathbf{B} + \mathbf{B}\lfloor\overline{\overline{\mathsf{I}}}^T\wedge\mathbf{A} = -(\mathbf{A}\wedge\mathbf{B})\rfloor\overline{\overline{\mathsf{I}}}^{(3)T}$$

$$(\mathbf{A}\wedge\mathbf{B})\lfloor\overline{\overline{\mathsf{I}}}^{(2)T} = \mathbf{A}\mathbf{B} + \mathbf{B}\mathbf{A} - (\mathbf{A}\lfloor\overline{\overline{\mathsf{I}}}^T)^{\wedge}_{\wedge}(\mathbf{B}\lfloor\overline{\overline{\mathsf{I}}}^T) = \overline{\overline{\mathsf{I}}}^{(2)}\rfloor(\mathbf{A}\wedge\mathbf{B})$$

$$\frac{1}{2}(\mathbf{A}\wedge\mathbf{A})\lfloor\overline{\overline{\mathsf{I}}}^{(2)T} = \mathbf{A}\mathbf{A} - (\mathbf{A}\lfloor\overline{\overline{\mathsf{I}}}^T)^{(2)} = \frac{1}{2}\overline{\overline{\mathsf{I}}}^{(2)}\rfloor(\mathbf{A}\wedge\mathbf{A})$$

$$\overline{\overline{\mathsf{A}}}^T|(\boldsymbol{\Phi}\lfloor\overline{\overline{\mathsf{A}}}) = \boldsymbol{\Phi}|\overline{\overline{\mathsf{A}}}^{(2)}\lfloor\overline{\overline{\mathsf{I}}}^T, \quad \overline{\overline{\mathsf{A}}}\in\mathbb{E}_1\mathbb{F}_1,\ \boldsymbol{\Phi}\in\mathbb{F}_2$$

$$\mathbf{a}\wedge(\mathbf{k}_N\lfloor\overline{\overline{\mathsf{I}}}^T) = \mathbf{k}_N\mathbf{a}, \qquad (\overline{\overline{\mathsf{I}}}\rfloor\mathbf{k}_N)\wedge\mathbf{a} = \mathbf{a}\mathbf{k}_N \quad \mathbf{a}\in\mathbb{E}_1$$

$$\mathbf{A}\wedge(\mathbf{k}_N\lfloor\overline{\overline{\mathsf{I}}}^{(2)T}) = \mathbf{k}_N\mathbf{A}, \qquad (\overline{\overline{\mathsf{I}}}^{(2)}\rfloor\mathbf{k}_N)\wedge\mathbf{A} = \mathbf{A}\mathbf{k}_N, \quad \mathbf{A}\in\mathbb{E}_2$$

$$(\mathbf{A}\lfloor\overline{\overline{\mathsf{I}}}^T\wedge\mathbf{B} + \mathbf{B}\lfloor\overline{\overline{\mathsf{I}}}^T\wedge\mathbf{A})\wedge\mathbf{a} = \mathbf{a}(\mathbf{A}\wedge\mathbf{B})$$

$$(\boldsymbol{\varepsilon}_N\lfloor\mathbf{A})\lfloor(\mathbf{A}\lfloor\overline{\overline{\mathsf{I}}}^T) = -\boldsymbol{\varepsilon}_N|(\mathbf{A}\wedge\mathbf{A})\overline{\overline{\mathsf{I}}}^T$$

$$(\mathbf{A}\lfloor\overline{\overline{\mathsf{I}}}^T)^{\wedge}_{\wedge}(\mathbf{B}\lfloor\overline{\overline{\mathsf{I}}}^T) = \mathbf{A}\mathbf{B} + \mathbf{B}\mathbf{A} - (\mathbf{A}\wedge\mathbf{B})\lfloor\overline{\overline{\mathsf{I}}}^{(2)T}$$

$$\mathbf{A}_s\wedge\overline{\overline{\mathsf{I}}}_s\rfloor\mathbf{B}_s + \mathbf{B}_s\wedge\overline{\overline{\mathsf{I}}}_s\rfloor\mathbf{A}_s = 0, \quad \mathbf{A}_s\lfloor\overline{\overline{\mathsf{I}}}_s^T\wedge\mathbf{A} = 0$$

$$(\mathbf{A}_s\lfloor\overline{\overline{\mathsf{I}}}_s^T)^{\wedge}_{\wedge}(\mathbf{B}_s\lfloor\overline{\overline{\mathsf{I}}}_s^T) = \mathbf{A}_s\mathbf{B}_s + \mathbf{B}_s\mathbf{A}_s, \quad (\mathbf{A}_s\lfloor\overline{\overline{\mathsf{I}}}_s^T)^{(2)} = \mathbf{A}_s\mathbf{A}_s.$$

$$\overline{\overline{\mathsf{S}}}^{\wedge}_{\wedge}(\mathbf{A}\lfloor\overline{\overline{\mathsf{I}}}^T) = (((\overline{\overline{\mathsf{S}}}\wedge\mathbf{A})\rfloor\boldsymbol{\varepsilon}_N)^{\wedge}_{\wedge}\overline{\overline{\mathsf{I}}})\rfloor\mathbf{e}_N, \quad \overline{\overline{\mathsf{S}}} = \overline{\overline{\mathsf{S}}}^T\in\mathbb{E}_1\mathbb{E}_1$$

Double multiplications

$$\mathrm{tr}(\overline{\overline{\mathsf{A}}}^{\wedge}_{\wedge}\overline{\overline{\mathsf{B}}}) = (\mathrm{tr}\overline{\overline{\mathsf{A}}})(\mathrm{tr}\overline{\overline{\mathsf{B}}}) - \mathrm{tr}(\overline{\overline{\mathsf{A}}}|\overline{\overline{\mathsf{B}}})$$

$$(\overline{\overline{\mathsf{A}}}^{\wedge}_{\wedge}\overline{\overline{\mathsf{B}}})|(\overline{\overline{\mathsf{C}}}^{\wedge}_{\wedge}\overline{\overline{\mathsf{D}}}) = (\overline{\overline{\mathsf{A}}}|\overline{\overline{\mathsf{C}}})^{\wedge}_{\wedge}(\overline{\overline{\mathsf{B}}}|\overline{\overline{\mathsf{D}}}) + (\overline{\overline{\mathsf{A}}}|\overline{\overline{\mathsf{D}}})^{\wedge}_{\wedge}(\overline{\overline{\mathsf{B}}}|\overline{\overline{\mathsf{C}}})$$

$$\mathrm{tr}((\overline{\overline{\mathsf{A}}}^{\wedge}_{\wedge}\overline{\overline{\mathsf{B}}})|(\overline{\overline{\mathsf{C}}}^{\wedge}_{\wedge}\overline{\overline{\mathsf{D}}})) = (\overline{\overline{\mathsf{A}}}^{\wedge}_{\wedge}\overline{\overline{\mathsf{B}}})||(\overline{\overline{\mathsf{C}}}^{\wedge}_{\wedge}\overline{\overline{\mathsf{D}}})^T$$

$$= (\overline{\overline{\mathsf{A}}}||\overline{\overline{\mathsf{C}}}^T)(\overline{\overline{\mathsf{B}}}||\overline{\overline{\mathsf{D}}}^T) + (\overline{\overline{\mathsf{A}}}||\overline{\overline{\mathsf{D}}}^T)(\overline{\overline{\mathsf{B}}}||\overline{\overline{\mathsf{C}}}^T) - (\overline{\overline{\mathsf{A}}}|\overline{\overline{\mathsf{D}}})||(\overline{\overline{\mathsf{B}}}|\overline{\overline{\mathsf{C}}})^T - (\overline{\overline{\mathsf{A}}}|\overline{\overline{\mathsf{C}}})||(\overline{\overline{\mathsf{B}}}|\overline{\overline{\mathsf{D}}})^T$$

$$(\overline{\overline{\mathsf{A}}}^{\wedge}_{\wedge}\overline{\overline{\mathsf{B}}})\lfloor\lfloor\overline{\overline{\mathsf{C}}}^T = (\overline{\overline{\mathsf{A}}}||\overline{\overline{\mathsf{C}}}^T)\overline{\overline{\mathsf{B}}} + (\overline{\overline{\mathsf{B}}}||\overline{\overline{\mathsf{C}}}^T)\overline{\overline{\mathsf{A}}} - \overline{\overline{\mathsf{A}}}|\overline{\overline{\mathsf{C}}}|\overline{\overline{\mathsf{B}}} - \overline{\overline{\mathsf{B}}}|\overline{\overline{\mathsf{C}}}|\overline{\overline{\mathsf{A}}}$$

$$\overline{\overline{\mathsf{A}}}^{(2)}\lfloor\lfloor\overline{\overline{\mathsf{B}}}^T = (\overline{\overline{\mathsf{A}}}||\overline{\overline{\mathsf{B}}}^T)\overline{\overline{\mathsf{A}}} - \overline{\overline{\mathsf{A}}}|\overline{\overline{\mathsf{B}}}|\overline{\overline{\mathsf{A}}}$$

$$\overline{\overline{\mathsf{A}}}^{(2)}\lfloor\lfloor\overline{\overline{\mathsf{I}}}^T = (\mathrm{tr}\overline{\overline{\mathsf{A}}})\overline{\overline{\mathsf{A}}} - \overline{\overline{\mathsf{A}}}^2, \qquad \overline{\overline{\mathsf{I}}}^{(2)}\lfloor\lfloor\overline{\overline{\mathsf{A}}}^T = (\mathrm{tr}\overline{\overline{\mathsf{A}}})\overline{\overline{\mathsf{I}}} - \overline{\overline{\mathsf{A}}}$$

$$(\overline{\overline{\mathsf{A}}}^{\wedge}_{\wedge}\overline{\overline{\mathsf{B}}}^{\wedge}_{\wedge}\overline{\overline{\mathsf{C}}})\lfloor\lfloor\overline{\overline{\mathsf{D}}}^T = (\overline{\overline{\mathsf{A}}}^{\wedge}_{\wedge}\overline{\overline{\mathsf{B}}})(\overline{\overline{\mathsf{C}}}||\overline{\overline{\mathsf{D}}}^T) + (\overline{\overline{\mathsf{B}}}^{\wedge}_{\wedge}\overline{\overline{\mathsf{C}}})(\overline{\overline{\mathsf{A}}}||\overline{\overline{\mathsf{D}}}^T) + (\overline{\overline{\mathsf{C}}}^{\wedge}_{\wedge}\overline{\overline{\mathsf{A}}})(\overline{\overline{\mathsf{B}}}||\overline{\overline{\mathsf{D}}}^T)$$

$$-\overline{\overline{\mathsf{A}}}^{\wedge}_{\wedge}(\overline{\overline{\mathsf{B}}}|\overline{\overline{\mathsf{D}}}|\overline{\overline{\mathsf{C}}} + \overline{\overline{\mathsf{C}}}|\overline{\overline{\mathsf{D}}}|\overline{\overline{\mathsf{B}}}) - \overline{\overline{\mathsf{B}}}^{\wedge}_{\wedge}(\overline{\overline{\mathsf{C}}}|\overline{\overline{\mathsf{D}}}|\overline{\overline{\mathsf{A}}} + \overline{\overline{\mathsf{A}}}|\overline{\overline{\mathsf{D}}}|\overline{\overline{\mathsf{C}}}) - \overline{\overline{\mathsf{C}}}^{\wedge}_{\wedge}(\overline{\overline{\mathsf{A}}}|\overline{\overline{\mathsf{D}}}|\overline{\overline{\mathsf{B}}} + \overline{\overline{\mathsf{B}}}|\overline{\overline{\mathsf{D}}}|\overline{\overline{\mathsf{A}}})$$

$$\overline{\overline{\mathsf{A}}}^{(3)}\lfloor\lfloor\overline{\overline{\mathsf{B}}}^T = (\overline{\overline{\mathsf{A}}}||\overline{\overline{\mathsf{B}}}^T)\overline{\overline{\mathsf{A}}}^{(2)} - \overline{\overline{\mathsf{A}}}{}^\wedge_\wedge(\overline{\overline{\mathsf{A}}}|\overline{\overline{\mathsf{B}}}|\overline{\overline{\mathsf{A}}})$$

$$\overline{\overline{\mathsf{A}}}^{(3)}\lfloor\lfloor\overline{\overline{\mathsf{I}}}^T = (\mathrm{tr}\overline{\overline{\mathsf{A}}})\overline{\overline{\mathsf{A}}}^{(2)} - \overline{\overline{\mathsf{A}}}{}^\wedge_\wedge\overline{\overline{\mathsf{A}}}^2, \qquad \overline{\overline{\mathsf{I}}}^{(3)}\lfloor\lfloor\overline{\overline{\mathsf{A}}}^T = (\mathrm{tr}\overline{\overline{\mathsf{A}}})\overline{\overline{\mathsf{I}}}^{(2)} - \overline{\overline{\mathsf{I}}}{}^\wedge_\wedge\overline{\overline{\mathsf{A}}}$$

$$\overline{\overline{\mathsf{I}}}^{(3)}\lfloor\lfloor(\overline{\overline{\mathsf{A}}}{}^\wedge_\wedge\overline{\overline{\mathsf{B}}})^T = \mathrm{tr}(\overline{\overline{\mathsf{A}}}{}^\wedge_\wedge\overline{\overline{\mathsf{B}}})\overline{\overline{\mathsf{I}}} - (\overline{\overline{\mathsf{A}}}{}^\wedge_\wedge\overline{\overline{\mathsf{B}}})\lfloor\lfloor\overline{\overline{\mathsf{I}}}^T$$

$$= (\mathrm{tr}\overline{\overline{\mathsf{A}}})(\mathrm{tr}\overline{\overline{\mathsf{B}}})\overline{\overline{\mathsf{I}}} - \mathrm{tr}(\overline{\overline{\mathsf{A}}}|\overline{\overline{\mathsf{B}}})\overline{\overline{\mathsf{I}}} - (\mathrm{tr}\overline{\overline{\mathsf{A}}})\overline{\overline{\mathsf{B}}} - (\mathrm{tr}\overline{\overline{\mathsf{B}}})\overline{\overline{\mathsf{A}}} + \overline{\overline{\mathsf{A}}}|\overline{\overline{\mathsf{B}}} + \overline{\overline{\mathsf{B}}}|\overline{\overline{\mathsf{A}}}$$

$$\overline{\overline{\mathsf{A}}}^{(3)}\lfloor\lfloor\overline{\overline{\mathsf{C}}}^T = \mathrm{tr}(\overline{\overline{\mathsf{A}}}^{(2)}|\overline{\overline{\mathsf{C}}})\overline{\overline{\mathsf{A}}} - \overline{\overline{\mathsf{A}}}|((\overline{\overline{\mathsf{C}}}\lfloor\lfloor\overline{\overline{\mathsf{A}}}^T)|\overline{\overline{\mathsf{A}}}, \qquad \overline{\overline{\mathsf{C}}} \in \mathbb{E}_2\mathbb{F}_2$$

$$\overline{\overline{\mathsf{I}}}^{(3)}\lfloor\lfloor\overline{\overline{\mathsf{C}}}^T = (\mathrm{tr}\overline{\overline{\mathsf{C}}})\overline{\overline{\mathsf{I}}} - \overline{\overline{\mathsf{C}}}\lfloor\lfloor\overline{\overline{\mathsf{I}}}^T, \qquad \overline{\overline{\mathsf{C}}} \in \mathbb{E}_2\mathbb{F}_2$$

$$(\overline{\overline{\mathsf{A}}}{}^\wedge_\wedge\overline{\overline{\mathsf{B}}}{}^\wedge_\wedge\overline{\overline{\mathsf{C}}}{}^\wedge_\wedge\overline{\overline{\mathsf{D}}})\lfloor\lfloor\overline{\overline{\mathsf{E}}}^T = (\overline{\overline{\mathsf{A}}}{}^\wedge_\wedge\overline{\overline{\mathsf{B}}}{}^\wedge_\wedge\overline{\overline{\mathsf{C}}})(\overline{\overline{\mathsf{D}}}||\overline{\overline{\mathsf{E}}}^T) + (\overline{\overline{\mathsf{A}}}{}^\wedge_\wedge\overline{\overline{\mathsf{B}}}{}^\wedge_\wedge\overline{\overline{\mathsf{D}}})(\overline{\overline{\mathsf{C}}}||\overline{\overline{\mathsf{E}}}^T)$$

$$+(\overline{\overline{\mathsf{A}}}{}^\wedge_\wedge\overline{\overline{\mathsf{C}}}{}^\wedge_\wedge\overline{\overline{\mathsf{D}}})(\overline{\overline{\mathsf{B}}}||\overline{\overline{\mathsf{E}}}^T) + (\overline{\overline{\mathsf{B}}}{}^\wedge_\wedge\overline{\overline{\mathsf{C}}}{}^\wedge_\wedge\overline{\overline{\mathsf{D}}})(\overline{\overline{\mathsf{A}}}||\overline{\overline{\mathsf{E}}}^T) - (\overline{\overline{\mathsf{A}}}{}^\wedge_\wedge\overline{\overline{\mathsf{B}}}){}^\wedge_\wedge(\overline{\overline{\mathsf{C}}}|\overline{\overline{\mathsf{E}}}|\overline{\overline{\mathsf{D}}} + \overline{\overline{\mathsf{D}}}|\overline{\overline{\mathsf{E}}}|\overline{\overline{\mathsf{C}}})$$

$$-(\overline{\overline{\mathsf{A}}}{}^\wedge_\wedge\overline{\overline{\mathsf{C}}}){}^\wedge_\wedge(\overline{\overline{\mathsf{B}}}|\overline{\overline{\mathsf{E}}}|\overline{\overline{\mathsf{D}}} + \overline{\overline{\mathsf{D}}}|\overline{\overline{\mathsf{E}}}|\overline{\overline{\mathsf{B}}}) - (\overline{\overline{\mathsf{A}}}{}^\wedge_\wedge\overline{\overline{\mathsf{D}}}){}^\wedge_\wedge(\overline{\overline{\mathsf{B}}}|\overline{\overline{\mathsf{E}}}|\overline{\overline{\mathsf{C}}} + \overline{\overline{\mathsf{C}}}|\overline{\overline{\mathsf{E}}}|\overline{\overline{\mathsf{B}}}) - (\overline{\overline{\mathsf{B}}}{}^\wedge_\wedge\overline{\overline{\mathsf{C}}}){}^\wedge_\wedge(\overline{\overline{\mathsf{A}}}|\overline{\overline{\mathsf{E}}}|\overline{\overline{\mathsf{D}}}$$

$$+\overline{\overline{\mathsf{D}}}|\overline{\overline{\mathsf{E}}}|\overline{\overline{\mathsf{A}}}) - (\overline{\overline{\mathsf{B}}}{}^\wedge_\wedge\overline{\overline{\mathsf{D}}}){}^\wedge_\wedge(\overline{\overline{\mathsf{A}}}|\overline{\overline{\mathsf{E}}}|\overline{\overline{\mathsf{C}}} + \overline{\overline{\mathsf{C}}}|\overline{\overline{\mathsf{E}}}|\overline{\overline{\mathsf{A}}}) - (\overline{\overline{\mathsf{C}}}{}^\wedge_\wedge\overline{\overline{\mathsf{D}}}){}^\wedge_\wedge(\overline{\overline{\mathsf{A}}}|\overline{\overline{\mathsf{E}}}|\overline{\overline{\mathsf{B}}} + \overline{\overline{\mathsf{B}}}|\overline{\overline{\mathsf{E}}}|\overline{\overline{\mathsf{A}}})$$

$$\overline{\overline{\mathsf{A}}}^{(4)}\lfloor\lfloor\overline{\overline{\mathsf{B}}}^T = (\overline{\overline{\mathsf{A}}}||\overline{\overline{\mathsf{B}}}^T)\overline{\overline{\mathsf{A}}}^{(3)} - \overline{\overline{\mathsf{A}}}^{(2)}{}^\wedge_\wedge(\overline{\overline{\mathsf{A}}}|\overline{\overline{\mathsf{B}}}|\overline{\overline{\mathsf{A}}})$$

$$\overline{\overline{\mathsf{A}}}^{(4)}\lfloor\lfloor\overline{\overline{\mathsf{I}}}^T = (\mathrm{tr}\,\overline{\overline{\mathsf{A}}})\overline{\overline{\mathsf{A}}}^{(3)} - \overline{\overline{\mathsf{A}}}^{(2)}{}^\wedge_\wedge\overline{\overline{\mathsf{A}}}^2, \qquad \overline{\overline{\mathsf{I}}}^{(4)}\lfloor\lfloor\overline{\overline{\mathsf{A}}}^T = (\mathrm{tr}\,\overline{\overline{\mathsf{A}}})\overline{\overline{\mathsf{I}}}^{(3)} - \overline{\overline{\mathsf{I}}}^{(2)}{}^\wedge_\wedge\overline{\overline{\mathsf{A}}}$$

$$\overline{\overline{\mathsf{I}}}^{(4)}\lfloor\lfloor(\overline{\overline{\mathsf{A}}}{}^\wedge_\wedge\overline{\overline{\mathsf{B}}})^T = \mathrm{tr}(\overline{\overline{\mathsf{A}}}{}^\wedge_\wedge\overline{\overline{\mathsf{B}}})\overline{\overline{\mathsf{I}}}^{(2)} - ((\overline{\overline{\mathsf{A}}}{}^\wedge_\wedge\overline{\overline{\mathsf{B}}})\lfloor\lfloor\overline{\overline{\mathsf{I}}}^T){}^\wedge_\wedge\overline{\overline{\mathsf{I}}} + \overline{\overline{\mathsf{A}}}{}^\wedge_\wedge\overline{\overline{\mathsf{B}}}$$

$$\overline{\overline{\mathsf{I}}}^{(4)}\lfloor\lfloor\overline{\overline{\mathsf{C}}}^T = (\mathrm{tr}\overline{\overline{\mathsf{C}}})\overline{\overline{\mathsf{I}}}^{(2)} - (\overline{\overline{\mathsf{C}}}\lfloor\lfloor\overline{\overline{\mathsf{I}}}^T){}^\wedge_\wedge\overline{\overline{\mathsf{I}}} + \overline{\overline{\mathsf{C}}}, \qquad \overline{\overline{\mathsf{C}}} \in \mathbb{E}_2\mathbb{F}_2$$

$$\overline{\overline{\mathsf{I}}}^{(4)}\lfloor\lfloor(\overline{\overline{\mathsf{A}}}{}^\wedge_\wedge\overline{\overline{\mathsf{B}}}{}^\wedge_\wedge\overline{\overline{\mathsf{C}}})^T = \mathrm{tr}(\overline{\overline{\mathsf{A}}}{}^\wedge_\wedge\overline{\overline{\mathsf{B}}}{}^\wedge_\wedge\overline{\overline{\mathsf{C}}})\overline{\overline{\mathsf{I}}} - (\overline{\overline{\mathsf{A}}}{}^\wedge_\wedge\overline{\overline{\mathsf{B}}}{}^\wedge_\wedge\overline{\overline{\mathsf{C}}})\lfloor\lfloor\overline{\overline{\mathsf{I}}}^{(2)T}$$

$$\overline{\overline{\mathsf{I}}}^{(4)}\lfloor\lfloor\overline{\overline{\mathsf{D}}}^T = (\mathrm{tr}\overline{\overline{\mathsf{D}}})\overline{\overline{\mathsf{I}}} - \overline{\overline{\mathsf{D}}}\lfloor\lfloor\overline{\overline{\mathsf{I}}}^{(2)T}, \qquad \overline{\overline{\mathsf{D}}} \in \mathbb{E}_3\mathbb{F}_3$$

$$\overline{\overline{\mathsf{A}}}^{(p+1)}\lfloor\lfloor\overline{\overline{\mathsf{A}}}^{-1T} = (n-p)\overline{\overline{\mathsf{A}}}^{(p)},$$

$$\overline{\overline{\mathsf{A}}}^{(p)}{}^\wedge_\wedge\overline{\overline{\mathsf{I}}}^{(4-p)} = (\mathrm{tr}\,\overline{\overline{\mathsf{A}}}^{(p)})\,\overline{\overline{\mathsf{I}}}^{(4)}$$

$$\overline{\overline{\mathsf{A}}}^{(p)}\lfloor\lfloor\overline{\overline{\mathsf{I}}}^{(p-1)T} = (\overline{\overline{\mathsf{I}}}^{(p)}\lfloor\lfloor\overline{\overline{\mathsf{A}}}^{(p-1)T})|\overline{\overline{\mathsf{A}}}, \qquad p > 1$$

$$\overline{\overline{\mathsf{I}}}^{(p+1)}\lfloor\lfloor\overline{\overline{\mathsf{A}}}^{(p)T} = (\mathrm{tr}\overline{\overline{\mathsf{A}}}^{(p)})\overline{\overline{\mathsf{I}}} - (\overline{\overline{\mathsf{I}}}^{(p)}\lfloor\lfloor\overline{\overline{\mathsf{A}}}^{(p-1)T})|\overline{\overline{\mathsf{A}}}, \quad p > 1$$

$$= (\mathrm{tr}\overline{\overline{\mathsf{A}}}^{(p)})\overline{\overline{\mathsf{I}}} + (\mathrm{tr}\overline{\overline{\mathsf{A}}}^{(p-1)})(-\overline{\overline{\mathsf{A}}}) + (\mathrm{tr}\overline{\overline{\mathsf{A}}}^{(p-2)})\overline{\overline{\mathsf{A}}}^2 + \cdots + (-\overline{\overline{\mathsf{A}}})^p$$

$$0 = \mathrm{tr}\overline{\overline{\mathsf{A}}}^{(4)}\,\overline{\overline{\mathsf{I}}} - \mathrm{tr}\overline{\overline{\mathsf{A}}}^{(3)}\,\overline{\overline{\mathsf{A}}} + \mathrm{tr}\overline{\overline{\mathsf{A}}}^{(2)}\,\overline{\overline{\mathsf{A}}}^2 - \mathrm{tr}\overline{\overline{\mathsf{A}}}\,\overline{\overline{\mathsf{A}}}^3 + \overline{\overline{\mathsf{A}}}^4.$$

Inverse dyadics

$$(\overline{\overline{\mathsf{A}}}|\overline{\overline{\mathsf{B}}})^{-1} = \overline{\overline{\mathsf{B}}}^{-1}|\overline{\overline{\mathsf{A}}}^{-1}, \qquad \det\overline{\overline{\mathsf{A}}} = \mathrm{tr}\overline{\overline{\mathsf{A}}}^{(4)}$$

$$(\overline{\overline{\mathsf{A}}}^{(p)})^{-1} = (\overline{\overline{\mathsf{A}}}^{-1})^{(p)} = \text{(def)}\ \overline{\overline{\mathsf{A}}}^{(-p)}, \quad 1 < p < 4,$$

$$\overline{\overline{\mathsf{A}}}^{-1} = \frac{\overline{\overline{\mathsf{I}}}^{(4)} \lfloor\lfloor \overline{\overline{\mathsf{A}}}^{(3)T}}{\det\overline{\overline{\mathsf{A}}}}, \quad \overline{\overline{\mathsf{A}}} \in \mathbb{E}_1\mathbb{F}_1$$

$$\overline{\overline{\mathsf{A}}}^{-1} = \frac{\boldsymbol{\varepsilon}_N\boldsymbol{\varepsilon}_N \lfloor\lfloor \overline{\overline{\mathsf{A}}}^{(3)T}}{\boldsymbol{\varepsilon}_N\boldsymbol{\varepsilon}_N || \overline{\overline{\mathsf{A}}}^{(4)}}, \quad \overline{\overline{\mathsf{A}}} \in \mathbb{E}_1\mathbb{E}_1$$

$$\overline{\overline{\mathsf{A}}}^{(-p)} = \frac{\overline{\overline{\mathsf{I}}}^{(4)} \lfloor\lfloor \overline{\overline{\mathsf{A}}}^{(3)T}}{\det \overline{\overline{\mathsf{A}}}}, \quad \overline{\overline{\mathsf{A}}} \in \mathbb{E}_1\mathbb{F}_1$$

$$\overline{\overline{\mathsf{A}}}^{(-p)} = \frac{\boldsymbol{\varepsilon}_N\boldsymbol{\varepsilon}_N \lfloor\lfloor \overline{\overline{\mathsf{A}}}^{(3)T}}{\boldsymbol{\varepsilon}_N\boldsymbol{\varepsilon}_N || \overline{\overline{\mathsf{A}}}^{(4)}}, \quad \overline{\overline{\mathsf{A}}} \in \mathbb{E}_1\mathbb{E}_1$$

$$(\mathbf{e}_N \lfloor \overline{\overline{\mathsf{I}}}^{(p)T})^{-1} = (-1)^{p(4-p)} \boldsymbol{\varepsilon}_N \lfloor \overline{\overline{\mathsf{I}}}^{(4-p)}$$

$$(\overline{\overline{\mathsf{A}}} + \mathbf{a}\boldsymbol{\alpha})^{-1} = \overline{\overline{\mathsf{A}}}^{-1} - \frac{\overline{\overline{\mathsf{A}}}^{-1}|\mathbf{a}\boldsymbol{\alpha}|\overline{\overline{\mathsf{A}}}^{-1}}{1 + \boldsymbol{\alpha}|\overline{\overline{\mathsf{A}}}^{-1}|\mathbf{a}}, \quad \overline{\overline{\mathsf{A}}} \in \mathbb{E}_1\mathbb{F}_1$$

$$(\overline{\overline{\mathsf{C}}} + \mathbf{AB})^{-1} = \overline{\overline{\mathsf{C}}}^{-1} - \frac{\overline{\overline{\mathsf{C}}}^{-1}|\mathbf{AB}|\overline{\overline{\mathsf{C}}}^{-1}}{1 + \mathbf{B}|\overline{\overline{\mathsf{C}}}^{-1}|\mathbf{A}}, \quad \overline{\overline{\mathsf{C}}} \in \mathbb{E}_2\mathbb{E}_2$$

$$(\mathbf{A} \lfloor \overline{\overline{\mathsf{I}}}^T)^{-1} = -2(\boldsymbol{\varepsilon}_N \lfloor \mathbf{A}) \lfloor \overline{\overline{\mathsf{I}}} \,/ (\boldsymbol{\varepsilon}_N | (\mathbf{A} \wedge \mathbf{A}))$$

$$(\overline{\overline{\mathsf{B}}}_o {}^\wedge_\wedge \overline{\overline{\mathsf{I}}})^{-1} = \frac{1}{\mathrm{tr}\overline{\overline{\mathsf{B}}}_o^{(3)}} (\overline{\overline{\mathsf{B}}}_o^2 {}^\wedge_\wedge \overline{\overline{\mathsf{I}}} + \mathrm{tr}\overline{\overline{\mathsf{B}}}_o^{(2)}\, \overline{\overline{\mathsf{I}}}^{(2)}), \quad \mathrm{tr}\overline{\overline{\mathsf{B}}}_o = 0$$

$$(\overline{\overline{\mathsf{A}}}_s + \mathbf{e}_4\mathbf{a}_s + \mathbf{b}_s\mathbf{e}_4 + c\mathbf{e}_4\mathbf{e}_4)^{-1} = \overline{\overline{\mathsf{A}}}_s^{-1} + \frac{(\overline{\overline{\mathsf{A}}}_s^{-1}|\mathbf{b}_s - \boldsymbol{\varepsilon}_4)(\mathbf{a}_s|\overline{\overline{\mathsf{A}}}_s^{-1} - \boldsymbol{\varepsilon}_4)}{c - \mathbf{a}_s|\overline{\overline{\mathsf{A}}}_s^{-1}|\mathbf{b}_s}$$

$$(\overline{\overline{\mathsf{A}}}_s + \mathbf{e}_4\boldsymbol{\alpha}_s + \mathbf{b}_s\boldsymbol{\varepsilon}_4 + c\mathbf{e}_4\boldsymbol{\varepsilon}_4)^{-1} = \overline{\overline{\mathsf{A}}}_s^{-1} + \frac{(\overline{\overline{\mathsf{A}}}_s^{-1}|\mathbf{b}_s - \mathbf{e}_4)(\boldsymbol{\alpha}_s|\overline{\overline{\mathsf{A}}}_s^{-1} - \boldsymbol{\varepsilon}_4)}{c - \boldsymbol{\alpha}_s|\overline{\overline{\mathsf{A}}}_s^{-1}|\mathbf{b}_s}$$

$$\det(\overline{\overline{\mathsf{A}}}_s + \mathbf{e}_4\boldsymbol{\alpha}_s + \mathbf{b}_s\boldsymbol{\varepsilon}_4 + c\mathbf{e}_4\boldsymbol{\varepsilon}_4) = \mathbf{e}_{123}|(c\overline{\overline{\mathsf{A}}}_s^{(3)T} - \overline{\overline{\mathsf{A}}}_s^{(2)T} {}^\wedge_\wedge \boldsymbol{\alpha}_s\mathbf{b}_s)|\boldsymbol{\varepsilon}_{123}.$$

Inverse of spatial dyadics

$$\overline{\overline{\mathsf{A}}}_s^{-1} = \frac{1}{\mathrm{tr}\overline{\overline{\mathsf{A}}}_s^{(3)}} \overline{\overline{\mathsf{I}}}_s^{(3)} \lfloor\lfloor \overline{\overline{\mathsf{A}}}_s^{(2)T}, \quad \overline{\overline{\mathsf{A}}}_s \in \mathbb{E}_1\mathbb{F}_1$$

$$\overline{\overline{\mathsf{C}}}_s^{-1} = \frac{1}{\mathrm{tr}(\overline{\overline{\mathsf{I}}}_s^{(3)} \lfloor\lfloor \overline{\overline{\mathsf{C}}}_s^T)^{(3)}} (\overline{\overline{\mathsf{C}}}_s^T \rfloor\rfloor \overline{\overline{\mathsf{I}}}_s^{(3)})^{(2)}, \quad \overline{\overline{\mathsf{C}}}_s \in \mathbb{E}_2\mathbb{F}_2.$$

METRIC DYADICS

Spatial metric dyadics

$$\overline{\overline{\mathsf{G}}}_s = \mathbf{e}_1\mathbf{e}_1 + \mathbf{e}_2\mathbf{e}_2 + \mathbf{e}_3\mathbf{e}_3, \quad \overline{\overline{\Gamma}}_s = \boldsymbol{\varepsilon}_1\boldsymbol{\varepsilon}_1 + \boldsymbol{\varepsilon}_2\boldsymbol{\varepsilon}_2 + \boldsymbol{\varepsilon}_3\boldsymbol{\varepsilon}_3$$

$$\overline{\overline{\mathsf{G}}}_s^{(2)} = \mathbf{e}_{12}\mathbf{e}_{12} + \mathbf{e}_{23}\mathbf{e}_{23} + \mathbf{e}_{31}\mathbf{e}_{31}, \quad \overline{\overline{\Gamma}}_s^{(2)} = \boldsymbol{\varepsilon}_{12}\boldsymbol{\varepsilon}_{12} + \boldsymbol{\varepsilon}_{23}\boldsymbol{\varepsilon}_{23} + \boldsymbol{\varepsilon}_{31}\boldsymbol{\varepsilon}_{31}$$

$$\overline{\overline{\mathsf{G}}}_s^{(3)} = \mathbf{e}_{123}\mathbf{e}_{123}, \quad \overline{\overline{\Gamma}}_s^{(3)} = \boldsymbol{\varepsilon}_{123}\boldsymbol{\varepsilon}_{123}.$$

4D Minkowski space

$$\overline{\overline{\mathsf{G}}} = \mathbf{e}_1\mathbf{e}_1 + \mathbf{e}_2\mathbf{e}_2 + \mathbf{e}_3\mathbf{e}_3 - \mathbf{e}_4\mathbf{e}_4 = \overline{\overline{\Gamma}}^{-1}$$

$$\overline{\overline{\Gamma}} = \boldsymbol{\varepsilon}_1\boldsymbol{\varepsilon}_1 + \boldsymbol{\varepsilon}_2\boldsymbol{\varepsilon}_2 + \boldsymbol{\varepsilon}_3\boldsymbol{\varepsilon}_3 - \boldsymbol{\varepsilon}_4\boldsymbol{\varepsilon}_4 = \overline{\overline{\mathsf{G}}}^{-1}$$

$$\overline{\overline{\mathsf{G}}}^{(2)} = \mathbf{e}_{12}\mathbf{e}_{12} + \mathbf{e}_{23}\mathbf{e}_{23} + \mathbf{e}_{31}\mathbf{e}_{31} - \overline{\overline{\mathsf{G}}}_s{}^\wedge_\wedge\mathbf{e}_4\mathbf{e}_4 = \overline{\overline{\Gamma}}^{(-2)}$$

$$\overline{\overline{\Gamma}}^{(2)} = \boldsymbol{\varepsilon}_{12}\boldsymbol{\varepsilon}_{12} + \boldsymbol{\varepsilon}_{23}\boldsymbol{\varepsilon}_{23} + \boldsymbol{\varepsilon}_{31}\boldsymbol{\varepsilon}_{31} - \overline{\overline{\Gamma}}_s{}^\wedge_\wedge\boldsymbol{\varepsilon}_4\boldsymbol{\varepsilon}_4 = \overline{\overline{\mathsf{G}}}^{(-2)}$$

$$\overline{\overline{\mathsf{G}}}^{(3)} = \mathbf{e}_{123}\mathbf{e}_{123} - \overline{\overline{\mathsf{G}}}_s^{(2)}{}^\wedge_\wedge\mathbf{e}_4\mathbf{e}_4 = \overline{\overline{\Gamma}}^{(-3)}$$

$$\overline{\overline{\Gamma}}^{(3)} = \boldsymbol{\varepsilon}_{123}\boldsymbol{\varepsilon}_{123} - \overline{\overline{\Gamma}}_s^{(2)}{}^\wedge_\wedge\boldsymbol{\varepsilon}_4\boldsymbol{\varepsilon}_4$$

$$\overline{\overline{\mathsf{G}}}^{(4)} = -\overline{\overline{\mathsf{G}}}_s^{(3)}{}^\wedge_\wedge\mathbf{e}_4\mathbf{e}_4 = -\mathbf{e}_N\mathbf{e}_N, \quad \overline{\overline{\Gamma}}^{(4)} = -\overline{\overline{\Gamma}}_s^{(3)}{}^\wedge_\wedge\boldsymbol{\varepsilon}_4\boldsymbol{\varepsilon}_4 = -\boldsymbol{\varepsilon}_N\boldsymbol{\varepsilon}_N.$$

ELECTROMAGNETICS

Maxwell equations

$$\mathbf{d} \wedge \boldsymbol{\Phi} = \gamma_m, \quad \mathbf{d} \wedge \boldsymbol{\Psi} = \gamma_e$$

$$\boldsymbol{\Phi} = \mathbf{B} + \mathbf{E} \wedge \boldsymbol{\varepsilon}_4, \quad \boldsymbol{\Psi} = \mathbf{D} - \mathbf{H} \wedge \boldsymbol{\varepsilon}_4$$

$$\gamma_m = \varrho_m - \mathbf{J}_m \wedge \boldsymbol{\varepsilon}_4, \quad \gamma_e = \varrho_e - \mathbf{J}_e \wedge \boldsymbol{\varepsilon}_4$$

$$\mathbf{d}_s \wedge \mathbf{E} + \partial_\tau \mathbf{B} = -\mathbf{J}_m, \quad -\mathbf{d}_s \wedge \mathbf{H} + \partial_\tau \mathbf{D} = -\mathbf{J}_e$$

$$\mathbf{d}_s \wedge \mathbf{B} = \varrho_m, \quad \mathbf{d}_s \wedge \mathbf{D} = \varrho_e$$

$$\mathbf{d}_s \wedge \mathbf{J}_m + \partial_\tau \varrho_m = 0, \quad \mathbf{d}_s \wedge \mathbf{J}_e + \partial_\tau \varrho_e = 0$$

$$\gamma_m = 0, \ \Rightarrow \ \boldsymbol{\Phi} = \mathbf{d} \wedge \boldsymbol{\alpha}, \quad \boldsymbol{\alpha} = \mathbf{A} - \phi\boldsymbol{\varepsilon}_4.$$

Medium equations

$$\mathbf{\Psi} = \overline{\overline{\mathsf{M}}}|\mathbf{\Phi}, \quad \mathbf{\Phi} = \overline{\overline{\mathsf{N}}}|\mathbf{\Psi}, \quad \overline{\overline{\mathsf{M}}}, \overline{\overline{\mathsf{N}}} \in \mathbb{F}_2\mathbb{E}_2,$$

$$\mathbf{e}_N\lfloor\mathbf{\Psi} = \overline{\overline{\mathsf{M}}}_m|\mathbf{\Phi}, \quad \mathbf{e}_N\lfloor\mathbf{\Phi} = \overline{\overline{\mathsf{N}}}_m|\mathbf{\Psi} = (\mathbf{e}_N\mathbf{e}_N\lfloor\lfloor\overline{\overline{\mathsf{M}}}_m^{-1})|\mathbf{\Psi}.$$

Modified medium dyadic $\overline{\overline{\mathsf{M}}}_m,\ (\mathbf{e}_N\mathbf{e}_N\lfloor\lfloor\overline{\overline{\mathsf{M}}}_m^{-1}) \in \mathbb{E}_2\mathbb{E}_2,\ \overline{\overline{\mathsf{M}}}_m^{-1} \in \mathbb{F}_2\mathbb{F}_2$

$$\overline{\overline{\mathsf{M}}}_m = \mathbf{e}_N\lfloor\overline{\overline{\mathsf{M}}}, \quad \overline{\overline{\mathsf{M}}} = \boldsymbol{\varepsilon}_N\lfloor\overline{\overline{\mathsf{M}}}_m$$

$$\overline{\overline{\mathsf{M}}}^{-1} = \overline{\overline{\mathsf{M}}}_m^{-1}\rfloor\mathbf{e}_N, \quad \overline{\overline{\mathsf{M}}}_m^{-1} = \overline{\overline{\mathsf{M}}}^{-1}\rfloor\boldsymbol{\varepsilon}_N.$$

Expansions

$$\begin{pmatrix}\mathbf{D}\\ \mathbf{H}\end{pmatrix} = \begin{pmatrix}\overline{\overline{\alpha}} & \overline{\overline{\epsilon}}'\\ \overline{\overline{\mu}}^{-1} & \overline{\overline{\beta}}\end{pmatrix}|\begin{pmatrix}\mathbf{B}\\ \mathbf{E}\end{pmatrix},$$

$$\overline{\overline{\alpha}} \in \mathbb{F}_2\mathbb{E}_2, \quad \overline{\overline{\epsilon}}' \in \mathbb{F}_2\mathbb{E}_1, \quad \overline{\overline{\mu}}^{-1} \in \mathbb{F}_1\mathbb{E}_2, \quad \overline{\overline{\beta}} \in \mathbb{F}_1\mathbb{E}_1$$

$$\overline{\overline{\mathsf{M}}} = \overline{\overline{\alpha}} + \overline{\overline{\epsilon}}' \wedge \mathbf{e}_4 + \boldsymbol{\varepsilon}_4 \wedge \overline{\overline{\mu}}^{-1} + \boldsymbol{\varepsilon}_4 \wedge \overline{\overline{\beta}} \wedge \mathbf{e}_4$$

$$\begin{pmatrix}\mathbf{D}\\ \mathbf{B}\end{pmatrix} = \begin{pmatrix}\overline{\overline{\epsilon}} & \overline{\overline{\xi}}\\ \overline{\overline{\zeta}} & \overline{\overline{\mu}}\end{pmatrix}|\begin{pmatrix}\mathbf{E}\\ \mathbf{H}\end{pmatrix}$$

$$\overline{\overline{\epsilon}},\ \overline{\overline{\mu}},\ \overline{\overline{\xi}}, \overline{\overline{\zeta}}\ \in \mathbb{F}_2\mathbb{E}_1$$

$$\overline{\overline{\mathsf{M}}} = \overline{\overline{\epsilon}} \wedge \mathbf{e}_4 + (\boldsymbol{\varepsilon}_4 \wedge \overline{\overline{\mathsf{I}}}^T + \overline{\overline{\xi}})|\overline{\overline{\mu}}^{-1}|(\overline{\overline{\mathsf{I}}}^{(2)T} - \overline{\overline{\zeta}} \wedge \mathbf{e}_4)$$

$$\overline{\overline{\mathsf{M}}}^{-1} = -\overline{\overline{\mu}} \wedge \mathbf{e}_4 - (\boldsymbol{\varepsilon}_4 \wedge \overline{\overline{\mathsf{I}}}^T - \overline{\overline{\zeta}})|\overline{\overline{\epsilon}}^{-1}|(\overline{\overline{\mathsf{I}}}^{(2)T} + \overline{\overline{\xi}} \wedge \mathbf{e}_4)$$

$$(\overline{\overline{\epsilon}} \wedge \mathbf{e}_4 + \boldsymbol{\varepsilon}_4 \wedge \overline{\overline{\mu}}^{-1})^{-1} = -(\overline{\overline{\mu}} \wedge \mathbf{e}_4 + \boldsymbol{\varepsilon}_4 \wedge \overline{\overline{\epsilon}}^{-1})$$

$$\begin{pmatrix}\mathbf{e}_{123}\lfloor\mathbf{D}\\ \mathbf{e}_{123}\lfloor\mathbf{B}\end{pmatrix} = \begin{pmatrix}\overline{\overline{\epsilon}}_g & \overline{\overline{\xi}}_g\\ \overline{\overline{\zeta}}_g & \overline{\overline{\mu}}_g\end{pmatrix}|\begin{pmatrix}\mathbf{E}\\ \mathbf{H}\end{pmatrix}.$$

Gibbsian dyadics $\overline{\overline{\epsilon}}_g,\ \overline{\overline{\mu}}_g,\ \overline{\overline{\xi}}_g,\ \overline{\overline{\zeta}}_g\ \in \mathbb{E}_1\mathbb{E}_1$

$$\overline{\overline{\mathsf{M}}} = \boldsymbol{\varepsilon}_{123}\lfloor\overline{\overline{\epsilon}}_g \wedge \mathbf{e}_4 + (\boldsymbol{\varepsilon}_{123}\lfloor\overline{\overline{\xi}}_g + \boldsymbol{\varepsilon}_4 \wedge \overline{\overline{\mathsf{I}}}^T)|\overline{\overline{\mu}}_g^{-1}|(\,\overline{\overline{\mathsf{I}}}\rfloor\mathbf{e}_{123} - \overline{\overline{\zeta}}_g \wedge \mathbf{e}_4)$$

$$\overline{\overline{\mathsf{M}}}_m = \overline{\overline{\epsilon}}_g{}^\wedge_\wedge\mathbf{e}_4\mathbf{e}_4 - (\mathbf{e}_{123}\lfloor\overline{\overline{\mathsf{I}}}^T + \mathbf{e}_4 \wedge \overline{\overline{\xi}}_g)|\overline{\overline{\mu}}_g^{-1}|(\,\overline{\overline{\mathsf{I}}}\rfloor\mathbf{e}_{123} - \overline{\overline{\zeta}}_g \wedge \mathbf{e}_4)$$

$$\overline{\overline{\mathsf{M}}}_m^{-1} = -\boldsymbol{\varepsilon}_{123}\boldsymbol{\varepsilon}_{123}\lfloor\lfloor\overline{\overline{\mu}}_g - (\boldsymbol{\varepsilon}_4 \wedge \overline{\overline{\mathsf{I}}}^T - \boldsymbol{\varepsilon}_{123}\lfloor\overline{\overline{\zeta}}_g)|\overline{\overline{\epsilon}}_g^{-1}|(\overline{\overline{\mathsf{I}}} \wedge \boldsymbol{\varepsilon}_4 + \overline{\overline{\xi}}_g\rfloor\boldsymbol{\varepsilon}_{123})$$

Relations

$$\overline{\overline{\epsilon}}' = \overline{\overline{\epsilon}} - \overline{\overline{\overline{\xi}}}|\overline{\overline{\mu}}^{-1}|\overline{\overline{\overline{\zeta}}}, \quad \overline{\overline{\alpha}} = \overline{\overline{\overline{\xi}}}|\overline{\overline{\mu}}^{-1}, \quad \overline{\overline{\overline{\beta}}} = -\overline{\overline{\mu}}^{-1}|\overline{\overline{\overline{\zeta}}}$$

$$\overline{\overline{\epsilon}} = \overline{\overline{\epsilon}}' - \overline{\overline{\alpha}}|\overline{\overline{\mu}}|\overline{\overline{\overline{\beta}}}, \quad \overline{\overline{\overline{\xi}}} = \overline{\overline{\alpha}}|\overline{\overline{\mu}}, \quad \overline{\overline{\overline{\zeta}}} = -\overline{\overline{\mu}}|\overline{\overline{\overline{\beta}}}$$

$$\overline{\overline{\epsilon}}' = \boldsymbol{\varepsilon}_{123}\lfloor(\overline{\overline{\epsilon}}_g - \overline{\overline{\overline{\xi}}}_g|\overline{\overline{\mu}}_g^{-1}|\overline{\overline{\overline{\zeta}}}_g), \quad \overline{\overline{\alpha}} = \boldsymbol{\varepsilon}_{123}\lfloor(\overline{\overline{\overline{\xi}}}_g|\overline{\overline{\mu}}_g^{-1})\rfloor\mathbf{e}_{123},$$

$$\overline{\overline{\overline{\beta}}} = -\overline{\overline{\mu}}_g^{-1}|\overline{\overline{\overline{\zeta}}}_g, \quad \overline{\overline{\mu}} = \boldsymbol{\varepsilon}_{123}\lfloor\overline{\overline{\mu}}_g$$

$$\overline{\overline{\epsilon}}_g = \mathbf{e}_{123}\lfloor\overline{\overline{\overline{\epsilon}}}, \quad \overline{\overline{\epsilon}} = \boldsymbol{\varepsilon}_{123}\lfloor\overline{\overline{\overline{\epsilon}}}_g, \quad \text{etc}$$

$$\overline{\overline{\epsilon}}^{-1} = \overline{\overline{\epsilon}}_g^{-1}\rfloor\mathbf{e}_{123}, \quad \overline{\overline{\epsilon}}_g^{-1} = \overline{\overline{\epsilon}}^{-1}\rfloor\boldsymbol{\varepsilon}_{123}, \quad \text{etc.}$$

References

[1] I. V. Lindell, *Differential Forms in Electromagnetics*, New York: John Wiley & Sons, 2004.

[2] F. Capolino (ed.), *Theory and Phenomena of Metamaterials*, Boca Raton: CRC Press, 2009; Chapter 4 "Differential forms and electromagnetic materials" by I. V. Lindell, pp. 4-1–4-16.

[3] B. Jancewicz, "A variable metric electrodynamics. The Coulomb and Biot-Savart laws in anisotropic media," *Ann. Phys.*, vol. 245, pp. 227–274, 1996.

[4] K. F. Warnick and P. Russer, "Differential forms and electromagnetic field theory", *Prog. Electromagn. Res.*, vol. 148, pp. 83–112, 2014.

[5] J. W. Gibbs and E. B. Wilson, *Vector Analysis*, New York: Scribner, 1909; reprint, New York: Dover, 1960.

[6] J. W. Gibbs, "On multiple algebra," *Proc. Am. Assoc. Adv. Sci.*, vol. 35, pp. 37–66, 1886; reprint in *The Scientific Papers of J. Willard Gibbs*, New York: Dover, 1961.

[7] H. Reichardt, *Vorlesungen über Vektor- und Tensorrechnung*, Berlin: VEB Deutscher Verlag der Wissenschaften, 1957.

[8] P. Szekeres, *A Course in Modern Mathematical Physics*, Cambridge University Press, 2004.

[9] F. W. Hehl, Yu. N. Obukhov, *Foundations of Classical Electrodynamics*, Boston: Birkhäuser, 2003.

[10] I. V. Lindell, "Inverse for the skewon medium dyadic," *Prog. Electromagn. Res.*, vol. 63, pp. 21–32, 2006.

[11] A. Favaro, Recent Advances in Electromagnetic Theory, PhD Thesis, Imperial College, London, 2012.

[12] I. V. Lindell, L. Bergamin, and A. Favaro, "Decomposable medium conditions in four-dimensional representation," *IEEE Trans. Antennas Propag.*, vol. 60, no. 1, pp. 367–376, 2012.

[13] G. A. Deschamps, "Electromagnetics and differential forms," *Proc. IEEE*, vol. 69, no. 6, pp. 676–696, June 1981.

[14] H. Flanders, *Differential Forms with Applications to the Physical Sciences*, New York: Academic Press, 1963.

[15] H. Cartan, *Differential Forms*, New York: Dover, 1970.

Multiforms, Dyadics, and Electromagnetic Media, First Edition. Ismo V. Lindell.

[16] D. Baldomir and P. Hammond, *Geometry of Electromagnetic Systems*, Oxford: Clarendon Press, 1996.

[17] G. A. Deschamps, "Exterior differential forms," in *Mathematics Applied to Sciences*, E. Roubine (ed.), Berlin: Springer, 1970, pp. 111–161.

[18] D. G. B. Edelen, *Applied Exterior Calculus*, Mineola, New York: Dover, 2005.

[19] C. W. Misner, K. S. Thorne, and J. A. Wheeler, *Gravitation*, San Francisco, CA: Freeman, 1973.

[20] S. Sternberg, *Curvature in Mathematics and Physics*, Mineola, NY: Dover, 2012.

[21] J. C. Maxwell, *A Treatise on Electricity and Magnetism*, vol. 2, 3rd edn, Oxford: Clarendon Press, 1891, pp. 278–283.

[22] O. Heaviside, *Electrical Papers*, New York: Chelsea, 1970; reprint of the 1st edition, London 1892 (vol. 1, p. 447; vol. 2, pp. 172–175). The original articles were respectively published in *The Electrician*, 1885, p. 306, and *Philos. Mag.*, August 1886, p. 118.

[23] H. Hertz, "Die Kräfte elektrischer Schwingungen, behandelt nach der Maxwellschen Theorie," *Wiedemanns Annalen der Physik*, vol. 36, pp. 1–22, 1889; Translation "The forces of electric oscillations, treated according to Maxwell's theory," *Electric Waves*, pp. 137–159; New York: Dover, 1962.

[24] J. A. Kong, *Electromagnetic Wave Theory*, 2nd edn, New York: John Wiley & Sons, 1990.

[25] I. V. Lindell, *Methods for Electromagnetic Field Analysis*, Oxford: Clarendon Press, 1992, Chapter 2.

[26] F. W. Hehl, Yu. N. Obukhov, G. Rubilar, and M. Blagojevic, "On the theory of the skewon field: From electrodynamics to gravity," *Phys. Lett. A*, vol. 347, pp. 14–24, 2005.

[27] F. W. Hehl, Yu. N. Obukhov, J.-P. Rivera, and H. Schmid, "Relativistic nature of a magnetoelectric modulus of Cr_2O_3 crystals: a four-dimensional pseudoscalar and its measurement," *Phys. Rev.*, A77 022106, 2008 (15 pp.).

[28] I. V. Lindell, A. H. Sihvola, S. A. Tretyakov, and A. J. Viitanen, *Electromagnetic Waves in Chiral and Bi-isotropic Media*, Boston: Artech House, 1994.

[29] D. Sievenpiper, L. Zhang, R. F. J. Broas, N. G. Alexopolous, and E. Yablonovitch, "High-impedance electromagnetic surfaces with a forbidden frequency band," *IEEE Trans. Microwave Theory Tech.*, vol. 47, no. 11, pp. 2059–2074, 1999.

[30] I. V. Lindell and A. Sihvola, "Perfect electromagnetic conductor," *J. Electromagn. Waves Appl.*, vol. 19, no. 7, pp. 861–869, 2005.

[31] I. V. Lindell and A. Sihvola, "Realization of the PEMC boundary," *IEEE Trans. Antennas Propag.*, vol. 53, no. 9, pp. 3012–3018, September 2005.

[32] A. Shahvarpour, T. Kodera, A. Parsa, and C. Caloz, "Arbitrary electromagnetic conductor boundaries using Faraday rotation in a grounded ferrite slab," *IEEE Trans. Microwave Theory Tech.*, vol. 58, no. 11, pp. 2781–2793, 2010.

[33] I. V. Lindell and A. Sihvola, "Electromagnetic boundary conditions and its realization with anisotropic metamaterial," *Phys. Rev. E*, vol. 57, no. 2, p. 026604 (7 p.), 2009.

[34] I. V. Lindell and A. Sihvola, "Electromagnetic boundary conditions defined in terms of normal field components," *IEEE Trans. Antennas Propag.*, vol. 58, no. 4, pp. 1128–1135, April 2010.

[35] V. H. Rumsey, "Some new forms of Huygens' principle," *IRE Trans. Antennas Propag.*, vol. 7, special supplement, pp. S102–S116, 1959.

[36] B. Zhang, H. Chen, B.-I. Wu, and J. A. Kong, "Extraordinary surface voltage effect in the invisibility cloak with an active device inside," *Phys. Rev. Lett.*, vol. 100, 063904 (4 pp.), February 2008.

[37] E. Martini, S. Maci, and A. Yaghjian, "DB boundary conditions at the inner surface of an arbitrarily shaped cloak," *Proc. EuCAP*, Rome, Italy, April 2011; pp. 3609–3611.

[38] D. Zaluski, S. Hrabar, and D. Muha, "Practical realization of DB metasurface," *Appl. Phys. Lett.*, vol. 104, 234106 (5 pp.), 2014.

[39] P.-S. Kildal, "Definition of artificially soft and hard surfaces for electromagnetic waves," *Electron. Lett.*, vol. 24, pp. 168–170, 1988.

[40] I. V. Lindell and B. Jancewicz, "Maxwell stress dyadic in differential-form formalism," *IEE Proc. Sci. Meas. Technol.*, vol. 147, no. 1, pp. 19–26, January 2000.

[41] V. G. Veselago, "The electrodynamics of substances with simultaneously negative values of ϵ and μ," *Soviet Phys. Uspekhi*, vol. 10, pp. 509–514, 1968.

[42] W. H. Greub, *Linear Algebra*, 3rd edn, Berlin: Springer, 1967.

[43] I. V. Lindell, A. Sihvola, P. Ylä-Oijala, and H. Wallén, "Zero backscattering from self-dual objects of finite size," *IEEE Trans. Antennas Propag.*, vol. 57, no. 9, pp. 2725–2731, September 2009.

[44] H. A. Lorentz, "The theorem of Poynting concerning the energy in the electromagnetic field and two general propositions concerning the propagation of light," *Amsterdammer Akademie der Wetenschappen*, vol. 4, p. 176, 1896.

[45] C. Altman and K. Suchy, *Reciprocity, Spatial Mapping and Time Reversal Electromagnetics*, Dordrecht: Kluwer, 1991.

[46] V. H. Rumsey, "The reaction concept in electromagnetic theory," *Phys. Rev.*, ser. 2, vol. 94, no. 6, pp. 1483–1491, June 1954.

[47] H. Bateman, "The transformation of the electrodynamical equations," *Proc. London Math. Soc.*, vol. 8, pp. 223–264, 1910.

[48] E. Cunningham, "The principle of relativity in electrodynamics and an extension thereof," *Proc. London Math. Soc.*, vol. 8, pp. 77–98, 1910.

[49] H. A. Kastrup, "On the advancements of conformal transformations and their associated symmetries in geometry and theoretical physics," *Ann. Phys.* (Berlin), vol. 17, no. 9–10, pp. 631–690, 2008.

[50] I. M. Besieris, A. M. Shaarawi, and A. M. Attiya, "Bateman conformal transformations within the framework of the bidirectional spectral representation," *Prog. Electromagn. Res.*, vol. 48, pp. 201–231, 2004.

[51] P. R. Girard, *Quaternions, Clifford Algebras and Relativistic Physics*, Basel: Birkhäuser, 2007, pp. 52–54, 82–85.

[52] I. V. Lindell and A. H. Sihvola, "The quotient function and its applications," *Am. J. Phys.*, vol. 66, no. 3, pp. 197–202, March 1998.

[53] Yu. N. Obukhov and F. W. Hehl, "Measuring a piecewise constant axion field in classical electrodynamics," *Phys. Lett.* vol. A341 pp. 357–365, 2005.

[54] I. V. Lindell, A. Sihvola, "Simple skewon medium realization of DB boundary condition," *PIER Lett.*, vol. 30, pp. 29–39, 2012.

[55] I. V. Lindell and A. Sihvola, "Transformation method for problems involving perfect electromagnetic conductor (PEMC) surfaces," *IEEE Trans. Antennas Propag.*, vol. 53, no. 9, pp. 3005–3011, September 2005.

[56] A. Sihvola and I. V. Lindell, "Perfect electromagnetic conductor medium," *Ann. Phys. (Berlin)*, vol. 17, no.9–10, pp. 787–802, 2008.

[57] C. Paiva and S. Matos, "Minkowskian isotropic media and the perfect electromagnetic conductor," *IEEE Trans. Antennas Propag.*, vol. 60, no. 7, pp. 3231–3245, 2012.

[58] I. V. Lindell, "The class of bi-anisotropic IB media," *Prog. Electromagn. Res.*, vol. 57, pp. 1–18, 2006.

[59] I. V. Lindell, "On electromagnetic fields in skewon-axion media," *International Conference on Electromagnetics in Advanced Applications, ICEAA'12*, Cape Town, South Africa, September 2012, pp. 58–61.

[60] I. V. Lindell and A. Sihvola, "Soft-and-hard/DB boundary conditions realized by a skewon-axion medium," *IEEE Trans. Antennas Propag.*, vol. 61, no. 2, pp. 768–774, 2013.

[61] I. V. Lindell, L. Bergamin, and A. Favaro, "The class of electromagnetic P-media and its generalization," *Prog. Electromagn. Res.B*, vol. 28, pp. 143–162, 2011.

[62] I. V. Lindell and H. Wallén, "Differential-form electromagnetics and bi-anisotropic Q-media," *J. Elecromagn. Waves Appl.*, vol. 18, no. 7, pp. 957–968, 2004.

[63] I. V. Lindell and H. Wallén, "Generalized Q-media and field decomposition in differential-form approach," *J. Elecromagn. Waves Appl.*, vol. 18, no. 8, pp. 1045–1056, 2004.

[64] I. V. Lindell and F. Olyslager, "Generalized decomposition of electromagnetic fields in bi-anisotropic media," *IEEE Trans. Antennas Propag.*, vol. 46, no. 10, pp. 1584–1585, October 1998.

[65] I. V. Lindell, "Class of electromagnetic SD media," *Metamaterials*, vol. 2, no. 2–3, pp. 54–70, 2008.

[66] D. H. Delphenich, "On linear electromagnetic constitutive laws that define almost-complex structures," *Annalen Phys.*, vol. 16, pp. 207–217, 2007.

[67] I. V. Lindell and F. Olyslager, "Green dyadics for self-dual bi-anisotropic media," *J. Electromagn. Waves Appl.*, vol. 14, pp. 153–163, 2000.

[68] I. V. Lindell, "Class of bi-quadratic (BQ) electromagnetic media," *PIER B*, vol. 7, pp. 281–297, 2008.

[69] P. C. Clemmow, "The resolution of a dipole field into transverse electric and magnetic waves," *IEE Proc.*, vol. 110, no. 1 pp. 107–111, 1963.

[70] P. C. Clemmow, *The Plane-Wave Spectrum Representation of Electromagnetic Fields*, New York: Pergamon Press, 1966.

[71] F. Olyslager and I. V. Lindell, "Field decomposition and factorization of the Helmholtz determinant operator for bianisotropic media," *IEEE Trans. Antennas Propag.*, vol. 49, no. 4, pp. 660–665, April 2001.

[72] F. Olyslager and I. V. Lindell, "Electromagnetics and exotic media: a quest for the Holy Grail," *IEEE Ant. Propag. Mag.*, vol. 44, no. 2, pp. 48–58, April 2002.

[73] M. Dahl, "Determination of the electromagnetic medium from the Fresnel surface," *J. Phys. A: Math. Theor.*, vol. 45, p. 405203, 2012.

[74] M. Dahl, "Characterisation and representation of non-dissipative electromagnetic medium with two Lorentz null cones," *J. Math. Phys.*, vol. 54, p. 011501, 2013.

[75] Ph. Boulanger and M. Hayes, *Bivectors and Waves in Mechanics and Optics*, London: Chapman and Hall, 1993.

[76] S. Brehmer and H. Haar, *Differentialformen und Vektoranalysis*, Berlin: VEB Deutscher Verlag der Wissenschaften, 1973.

[77] É. Cartan, *Les systèmes differentiels extérieurs et leur applications géométriques*, Paris: Hermann & Cie., 1945.

[78] N. V. Efimov and E. R. Rozendorn, *Linear Algebra and Multidimensional Geometry*, Moscow: Mir Publishers, 1975; Translation of the original Russian 1974 edition.

[79] A. Favaro and L. Bergamin, "The non-birefringent limit of all linear, skewonless media and its unique light-cone structure," *Annalen der Physik (Berlin)*, vol. 523, no. 5, pp. 383–401, 2011.

[80] J. W. Gibbs, *Element of Vector Analysis*, privately printed in two parts 1881, 1884; reprint in *The Scientific Papers of J. Willard Gibbs*, New York: Dover, 1961.

[81] H. Grassmann, *Extension Theory*, Providence, R I: American Mathematical Society, 2000; Translation of *Die Ausdehnungslehre*, Berlin: Verlag von Th. Chr. Fr. Ensin, 1862.

[82] B. Jancewicz, "Plane electromagnetic wave in PEMC," *J. Electro. Waves Appl.*, vol. 20, no. 12, pp. 647–659, 2006.

[83] P.-S. Kildal, "Fundamental properties of canonical soft and hard surfaces, perfect magnetic conductors and the newly introduced DB surface", *Proceedings of the International Conference on Electromagnetics in Advanced Applications, ICEAA'09*, Torino, Italy, August 2009, pp. 607–610.

[84] A. Kurosh, *Higher Algebra*, Moscow: Mir Publishers, 1980.

[85] I. V. Lindell, "Differential forms and bi-anisotropic media," *Electromagnetics*, vol. 26, pp. 191–201, 2006.

[86] I. V. Lindell, "Electromagnetic fields and self-dual media in differential-form representation," *Prog. Electromagn. Res.*, vol. 58, pp. 319–333, 2006.

[87] I. V. Lindell and A. Sihvola, "Negative-definite media, a class of bi-anisotropic metamaterials," *Microwave Opt. Technol. Lett.*, vol. 48, no. 3, pp. 602–606, 2006.

[88] P. Lounesto, *Clifford Algebras and Spinors*, Cambridge: Cambridge University Press, 1997.

[89] Yu. N. Obukhov and F.W. Hehl, "Possible skewon effects on light propagation," *Phys. Rev. D*, vol. 70, 125015, 2004.

[90] Yu. N. Obukhov, "Electromagnetic energy and momentum in moving media", *Ann. Phys. (Berlin)*, vol. 17 , pp. 830–851, 2008.

[91] F. Olyslager and I. V. Lindell, "Green's dyadics and factorization of the Helmholtz determinant operator for a class of bianisotropic media," *Microwave Opt. Technol. Lett.*, vol. 21, no. 4, pp. 304–309, May 1999.

[92] E. J. Post, *Formal Structure of Electromagnetics*, Amsterdam: North Holland, 1962; reprint, Mineola, NY: Dover, 1997.

[93] E. J. Post, "The constitutive map and some of its ramifications," *Ann. Phys.*, vol. 71, pp. 497–518, 1972.

[94] A. Serdukov, I. Semchenko, S. Tratyakov, and A. Sihvola, *Electromagnetics of Bi-Anisotropic Materials: Theory and Applications*. Amsterdam: Gordon and Breach, 2001.

[95] J. Van Bladel, *Electromagnetic Fields*, 2nd edn, Piscataway, NJ: IEEE Press, 2007.

[96] K. F. Warnick, R. Sefridge, and D. V. Arnold, "Electromagnetic boundary conditions and differential forms," *IEEE Proc. Microwaves Ant. Propag.*, vol. 142, no. 4, pp. 326–332, 1995.

[97] K. F. Warnick and D. V. Arnold, "Electromagnetic Green functions using differential forms," *J. Electromagn. Waves Appl.*, vol. 10, no. 3, pp. 427–438, 1996.

Index

Multiforms, Dyadics, and Electromagnetic Media, First Edition. Ismo V. Lindell.

IEEE PRESS SERIES ON ELECTROMAGNETIC WAVE THEORY

Andreas C. Cangellaris, *Series Editor*
University of Illinois, Urbana-Champaign, Illinois

Field Theory of Guided Waves, Second Edition
Robert E. Collin

Field Computation by Moment Methods
Roger F. Harrington

Radiation and Scattering of Waves
Leopold B. Felsen, Nathan Marcuvitz

Methods in Electromagnetic Wave Propagation, Second Edition
D. S. J. Jones

Mathematical Foundations for Electromagnetic Theory
Donald G. Dudley

The Transmission-Line Modeling Method: TLM
Christos Christopoulos

The Plane Wave Spectrum Representation of Electromagnetic Fields, Revised Edition
P. C. Clemmow

General Vector and Dyadic Analysis: Applied Mathematics in Field Theory
Chen-To Tai

Computational Methods for Electromagnetics
Andrew F. Peterson

Plane-Wave Theory of Time-Domain Fields: Near-Field Scanning Applications
Thorkild B. Hansen

Foundations for Microwave Engineering, Second Edition
Robert Collin

Time-Harmonic Electromagnetic Fields
Robert F. Harrington

Antenna Theory & Design, Revised Edition
Robert S. Elliott

Differential Forms in Electromagnetics
Ismo V. Lindell

Conformal Array Antenna Theory and Design
Lars Josefsson, Patrik Persson

Multigrid Finite Element Methods for Electromagnetic Field Modeling
Yu Zhu, Andreas C. Cangellaris

Electromagnetic Theory
Julius Adams Stratton

Electromagnetic Fields, Second Edition
Jean G. Van Bladel

Electromagnetic Fields in Cavities: Deterministic and Statistical Theories
David A. Hill

Discontinuities in the Electromagnetic Field
M. Mithat Idemen

Understanding Geometric Algebra for Electromagnetic Theory
John W. Arthur

The Power and Beauty of Electromagnetic Theory
Frederic R. Morgenthaler

Electromagnetic Modeling and Simulation
Levent Sevgi

Multiforms, Dyadics, and Electromagnetic Media
Ismo V. Lindell